MILLIMETER WAVE OPTICAL DIELECTRIC INTEGRATED GUIDES AND CIRCUITS • *Shiban K. Koul*

MICROWAVE DEVICES, CIRCUITS AND THEIR INTERACTION • *Charles A. Lee and G. Conrad Dalman*

ADVANCES IN MICROSTRIP AND PRINTED ANTENNAS • *Kai-Fong Lee and Wei Chen (eds.)*

SPHEROIDAL WAVE FUNCTIONS IN ELECTROMAGNETIC THEORY • *Le-Wei Li, Xiao-Kang Kang, and Mook-Seng Leong*

OPTICAL FILTER DESIGN AND ANALYSIS: A SIGNAL PROCESSING APPROACH • *Christi K. Madsen and Jian H. Zhao*

THEORY AND PRACTICE OF INFRARED TECHNOLOGY FOR NONDESTRUCTIVE TESTING • *Xavier P. V. Maldague*

OPTOELECTRONIC PACKAGING • *A. R. Mickelson, N. R. Basavanhally, and Y. C. Lee (eds.)*

OPTICAL CHARACTER RECOGNITION • *Shunji Mori, Hirobumi Nishida, and Hiromitsu Yamada*

ANTENNAS FOR RADAR AND COMMUNICATIONS: A POLARIMETRIC APPROACH • *Harold Mott*

INTEGRATED ACTIVE ANTENNAS AND SPATIAL POWER COMBINING • *Julio A. Navarro and Kai Chang*

ANALYSIS METHODS FOR RF, MICROWAVE, AND MILLIMETER-WAVE PLANAR TRANSMISSION LINE STRUCTURES • *Cam Nguyen*

FREQUENCY CONTROL OF SEMICONDUCTOR LASERS • *Motoichi Ohtsu (ed.)*

SOLAR CELLS AND THEIR APPLICATIONS • *Larry D. Partain (ed.)*

ANALYSIS OF MULTICONDUCTOR TRANSMISSION LINES • *Clayton R. Paul*

INTRODUCTION TO ELECTROMAGNETIC COMPATIBILITY • *Clayton R. Paul*

ELECTROMAGNETIC OPTIMIZATION BY GENETIC ALGORITHMS • *Yahya Rahmat-Samii and Eric Michielssen (eds.)*

INTRODUCTION TO HIGH-SPEED ELECTRONICS AND OPTOELECTRONICS • *Leonard M. Riaziat*

NEW FRONTIERS IN MEDICAL DEVICE TECHNOLOGY • *Arye Rosen and Harel Rosen (eds.)*

ELECTROMAGNETIC PROPAGATION IN MULTI-MODE RANDOM MEDIA • *Harrison E. Rowe*

ELECTROMAGNETIC PROPAGATION IN ONE-DIMENSIONAL RANDOM MEDIA • *Harrison E. Rowe*

NONLINEAR OPTICS • *E. G. Sauter*

COPLANAR WAVEGUIDE CIRCUITS, COMPONENTS, AND SYSTEMS • *Rainee N. Simons*

ELECTROMAGNETIC FIELDS IN UNCONVENTIONAL MATERIALS AND STRUCTURES • *Onkar N. Singh and Akhlesh Lakhtakia (eds.)*

FUNDAMENTALS OF GLOBAL POSITIONING SYSTEM RECEIVERS: A SOFTWARE APPROACH • *James Bao-yen Tsui*

InP-BASED MATERIALS AND DEVICES: PHYSICS AND TECHNOLOGY • *Osamu Wada and Hideki Hasegawa (eds.)*

COMPACT AND BROADBAND MICROSTRIP ANTENNAS • *Kin-Lu Wong*

DESIGN OF NONPLANAR MICROSTRIP ANTENNAS AND TRANSMISSION LINES • *Kin-Lu Wong*

FREQUENCY SELECTIVE SURFACE AND GRID ARRAY • *T. K. Wu (ed.)*

ACTIVE AND QUASI-OPTICAL ARRAYS FOR SOLID-STATE POWER COMBINING • *Robert A. York and Zoya B. Popović (eds.)*

OPTICAL SIGNAL PROCESSING, COMPUTING AND NEURAL NETWORKS • *Francis T. S. Yu and Suganda Jutamulia*

SiGe, GaAs, AND InP HETEROJUNCTION BIPOLAR TRANSISTORS • *Jiann Yuan*

ELECTRODYNAMICS OF SOLIDS AND MICROWAVE SUPERCONDUCTIVITY • *Shu-Ang Zhou*

Fiber-Optic
Communication Systems

Third Edition

Fiber-Optic Communication Systems

Third Edition

GOVIND P. AGRAWAL

The Institute of Optics
University of Rochester
Rochester, NY

WILEY-INTERSCIENCE

A JOHN WILEY & SONS, INC., PUBLICATION

Library of Congress Cataloging-in-Publication Data is available.

ISBN 0-471-21571-6

Printed in the United States of America

10 9 8 7 6 5 4 3 2

For My Parents

Contents

Preface

Since the publication of the first edition of this book in 1992, the state of the art of fiber-optic communication systems has advanced dramatically despite the relatively short period of only 10 years between the first and third editions. For example, the highest capacity of commercial fiber-optic links available in 1992 was only 2.5 Gb/s. A mere 4 years later, the wavelength-division-multiplexed (WDM) systems with the total capacity of 40 Gb/s became available commercially. By 2001, the capacity of commercial WDM systems exceeded 1.6 Tb/s, and the prospect of lightwave systems operating at 3.2 Tb/s or more were in sight. During the last 2 years, the capacity of transoceanic lightwave systems installed worldwide has exploded. Moreover, several other undersea networks were in the construction phase in December 2001. A global network covering 250,000 km with a capacity of 2.56 Tb/s (64 WDM channels at 10 Gb/s over 4 fiber pairs) is scheduled to be operational in 2002. Several conference papers presented in 2001 have demonstrated that lightwave systems operating at a bit rate of more than 10 Tb/s are within reach. Just a few years ago it was unimaginable that lightwave systems would approach the capacity of even 1 Tb/s by 2001.

The second edition of this book appeared in 1997. It has been well received by the scientific community involved with lightwave technology. Because of the rapid advances that have occurred over the last 5 years, the publisher and I deemed it necessary to bring out the third edition if the book were to continue to provide a comprehensive and up-to-date account of fiber-optic communication systems. The result is in your hands. The primary objective of the book remains the same. Specifically, it should be able to serve both as a textbook and a reference monograph. For this reason, the emphasis is on the physical understanding, but the engineering aspects are also discussed throughout the text.

Because of the large amount of material that needed to be added to provide comprehensive coverage, the book size has increased considerably compared with the first edition. Although all chapters have been updated, the major changes have occurred in Chapters 6–9. I have taken this opportunity to rearrange the material such that it is better suited for a two-semester course on optical communications. Chapters 1–5 provide the basic foundation while Chapters 6–10 cover the issues related to the design of advanced lightwave systems. More specifically, after the introduction of the elementary concepts in Chapter 1, Chapters 2–4 are devoted to the three primary components of a fiber-optic communications—optical fibers, optical transmitters, and optical receivers. Chapter 5 then focuses on the system design issues. Chapters 6 and 7 are devoted to the advanced techniques used for the management of fiber losses and chromatic dispersion, respectively. Chapter 8 focuses on the use of wavelength- and time-division multiplexing techniques for optical networks. Code-division multiplexing is also a part of this chapter. The use of optical solitons for fiber-optic systems is discussed in Chapter 9. Coherent lightwave systems are now covered in the last chapter. More than 30% of the material in Chapter 6–9 is new because of the rapid development of the WDM technology over the last 5 years. The contents of the book reflect the state of the art of lightwave transmission systems in 2001.

The primary role of this book is as a graduate-level textbook in the field of *optical communications*. An attempt is made to include as much recent material as possible so that students are exposed to the recent advances in this exciting field. The book can also serve as a reference text for researchers already engaged in or wishing to enter the field of optical fiber communications. The reference list at the end of each chapter is more elaborate than what is common for a typical textbook. The listing of recent research papers should be useful for researchers using this book as a reference. At the same time, students can benefit from it if they are assigned problems requiring reading of the original research papers. A set of problems is included at the end of each chapter to help both the teacher and the student. Although written primarily for graduate students, the book can also be used for an undergraduate course at the senior level with an appropriate selection of topics. Parts of the book can be used for several other related courses. For example, Chapter 2 can be used for a course on optical waveguides, and Chapter 3 can be useful for a course on optoelectronics.

Many universities in the United States and elsewhere offer a course on optical communications as a part of their curriculum in electrical engineering, physics, or optics. I have taught such a course since 1989 to the graduate students of the Institute of Optics, and this book indeed grew out of my lecture notes. I am aware that it is used as a textbook by many instructors worldwide—a fact that gives me immense satisfaction. I am acutely aware of a problem that is a side effect of an enlarged revised edition. How can a teacher fit all this material in a one-semester course on *optical communications*? I have to struggle with the same question. In fact, it is impossible to cover the entire book in one semester. The best solution is to offer a two-semester course covering Chapters 1 through 5 during the first semester, leaving the remainder for the second semester. However, not many universities may have the luxury of offering a two-semester course on optical communications. The book can be used for a one-semester course provided that the instructor makes a selection of topics. For example, Chapter 3 can be skipped if the students have taken a laser course previously. If only parts of Chapters 6 through 10 are covered to provide students a glimpse of the recent advances, the material can fit in a single one-semester course offered either at the senior level for undergraduates or to graduate students.

This edition of the book features a compact disk (CD) on the back cover provided by the Optiwave Corporation. The CD contains a state-of-the art software package suitable for designing modern lightwave systems. It also contains additional problems for each chapter that can be solved by using the software package. Appendix E provides more details about the software and the problems. It is my hope that the CD will help to train the students and will prepare them better for an industrial job.

A large number of persons have contributed to this book either directly or indirectly. It is impossible to mention all of them by name. I thank my graduate students and the students who took my course on optical communication systems and helped improve my class notes through their questions and comments. Thanks are due to many instructors who not only have adopted this book as a textbook for their courses but have also pointed out the misprints in previous editions, and thus have helped me in improving the book. I am grateful to my colleagues at the Institute of Optics for numerous discussions and for providing a cordial and productive atmosphere. I appreciated the help of Karen Rolfe, who typed the first edition of this book and made numerous revisions with a smile. Last, but not least, I thank my wife, Anne, and my daughters, Sipra, Caroline, and Claire, for understanding why I needed to spend many weekends on the book instead of spending time with them.

Govind P. Agrawal

Rochester, NY

December 2001

Chapter 1

Introduction

A communication system transmits information from one place to another, whether separated by a few kilometers or by transoceanic distances. Information is often carried by an electromagnetic carrier wave whose frequency can vary from a few megahertz to several hundred terahertz. Optical communication systems use high carrier frequencies (~ 100 THz) in the visible or near-infrared region of the electromagnetic spectrum. They are sometimes called lightwave systems to distinguish them from microwave systems, whose carrier frequency is typically smaller by five orders of magnitude (~ 1 GHz). Fiber-optic communication systems are lightwave systems that employ optical fibers for information transmission. Such systems have been deployed worldwide since 1980 and have indeed revolutionized the technology behind telecommunications. Indeed, the lightwave technology, together with microelectronics, is believed to be a major factor in the advent of the "information age." The objective of this book is to describe fiber-optic communication systems in a comprehensive manner. The emphasis is on the fundamental aspects, but the engineering issues are also discussed. The purpose of this introductory chapter is to present the basic concepts and to provide the background material. Section 1.1 gives a historical perspective on the development of optical communication systems. In Section 1.2 we cover concepts such as analog and digital signals, channel multiplexing, and modulation formats. Relative merits of guided and unguided optical communication systems are discussed in Section 1.3. The last section focuses on the building blocks of a fiber-optic communication system.

1.1 Historical Perspective

The use of light for communication purposes dates back to antiquity if we interpret optical communications in a broad sense [1]. Most civilizations have used mirrors, fire beacons, or smoke signals to convey a single piece of information (such as victory in a war). Essentially the same idea was used up to the end of the eighteenth century through signaling lamps, flags, and other semaphore devices. The idea was extended further, following a suggestion of Claude Chappe in 1792, to transmit mechanically

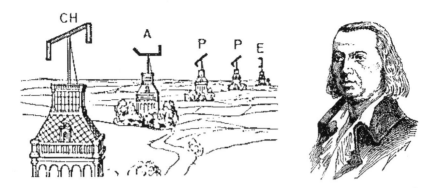

Figure 1.1: Schematic illustration of the optical telegraph and its inventor Claude Chappe. (After Ref. [2]; ©1944 American Association for the Advancement of Science; reprinted with permission.)

coded messages over long distances (~ 100 km) by the use of intermediate relay stations [2], acting as *regenerators* or *repeaters* in the modern-day language. Figure 1.1 shows the basic idea schematically. The first such "optical telegraph" was put in service between Paris and Lille (two French cities about 200 km apart) in July 1794. By 1830, the network had expanded throughout Europe [1]. The role of light in such systems was simply to make the coded signals visible so that they could be intercepted by the relay stations. The opto-mechanical communication systems of the nineteenth century were inherently slow. In modern-day terminology, the effective bit rate of such systems was less than 1 bit per second ($B < 1$ b/s).

1.1.1   Need for Fiber-Optic Communications

The advent of telegraphy in the 1830s replaced the use of light by electricity and began the era of electrical communications [3]. The bit rate B could be increased to ~ 10 b/s by the use of new coding techniques, such as the *Morse code*. The use of intermediate relay stations allowed communication over long distances (~ 1000 km). Indeed, the first successful transatlantic telegraph cable went into operation in 1866. Telegraphy used essentially a digital scheme through two electrical pulses of different durations (dots and dashes of the Morse code). The invention of the telephone in 1876 brought a major change inasmuch as electric signals were transmitted in analog form through a continuously varying electric current [4]. Analog electrical techniques were to dominate communication systems for a century or so.

The development of worldwide telephone networks during the twentieth century led to many advances in the design of electrical communication systems. The use of coaxial cables in place of wire pairs increased system capacity considerably. The first coaxial-cable system, put into service in 1940, was a 3-MHz system capable of transmitting 300 voice channels or a single television channel. The bandwidth of such systems is limited by the frequency-dependent cable losses, which increase rapidly for frequencies beyond 10 MHz. This limitation led to the development of microwave communication systems in which an electromagnetic carrier wave with frequencies in

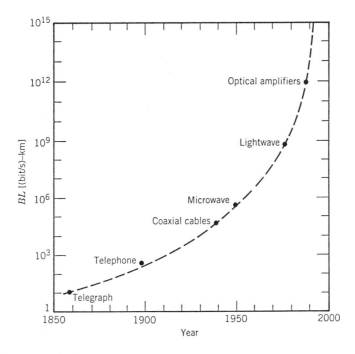

Figure 1.2: Increase in bit rate–distance product *BL* during the period 1850–2000. The emergence of a new technology is marked by a solid circle.

the range of 1–10 GHz is used to transmit the signal by using suitable modulation techniques.

The first microwave system operating at the carrier frequency of 4 GHz was put into service in 1948. Since then, both coaxial and microwave systems have evolved considerably and are able to operate at bit rates ~ 100 Mb/s. The most advanced coaxial system was put into service in 1975 and operated at a bit rate of 274 Mb/s. A severe drawback of such high-speed coaxial systems is their small *repeater spacing* (~ 1 km), which makes the system relatively expensive to operate. Microwave communication systems generally allow for a larger repeater spacing, but their bit rate is also limited by the carrier frequency of such waves. A commonly used figure of merit for communication systems is the *bit rate–distance product*, *BL*, where *B* is the bit rate and *L* is the repeater spacing. Figure 1.2 shows how the *BL* product has increased through technological advances during the last century and a half. Communication systems with $BL \sim 100$ (Mb/s)-km were available by 1970 and were limited to such values because of fundamental limitations.

It was realized during the second half of the twentieth century that an increase of several orders of magnitude in the *BL* product would be possible if optical waves were used as the carrier. However, neither a coherent optical source nor a suitable transmission medium was available during the 1950s. The invention of the laser and its demonstration in 1960 solved the first problem [5]. Attention was then focused on finding ways for using laser light for optical communications. Many ideas were

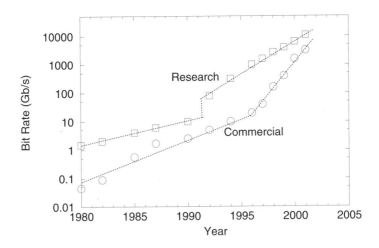

Figure 1.3: Increase in the capacity of lightwave systems realized after 1980. Commercial systems (circles) follow research demonstrations (squares) with a few-year lag. The change in the slope after 1992 is due to the advent of WDM technology.

advanced during the 1960s [6], the most noteworthy being the idea of light confinement using a sequence of gas lenses [7].

It was suggested in 1966 that optical fibers might be the best choice [8], as they are capable of guiding the light in a manner similar to the guiding of electrons in copper wires. The main problem was the high losses of optical fibers—fibers available during the 1960s had losses in excess of 1000 dB/km. A breakthrough occurred in 1970 when fiber losses could be reduced to below 20 dB/km in the wavelength region near 1 μm [9]. At about the same time, GaAs semiconductor lasers, operating continuously at room temperature, were demonstrated [10]. The simultaneous availability of *compact* optical sources and a *low-loss* optical fibers led to a worldwide effort for developing fiber-optic communication systems [11]. Figure 1.3 shows the increase in the capacity of lightwave systems realized after 1980 through several generations of development. As seen there, the commercial deployment of lightwave systems followed the research and development phase closely. The progress has indeed been rapid as evident from an increase in the bit rate by a factor of 100,000 over a period of less than 25 years. Transmission distances have also increased from 10 to 10,000 km over the same time period. As a result, the bit rate–distance product of modern lightwave systems can exceed by a factor of 10^7 compared with the first-generation lightwave systems.

1.1.2 Evolution of Lightwave Systems

The research phase of fiber-optic communication systems started around 1975. The enormous progress realized over the 25-year period extending from 1975 to 2000 can be grouped into several distinct generations. Figure 1.4 shows the increase in the *BL* product over this time period as quantified through various laboratory experiments [12]. The straight line corresponds to a doubling of the *BL* product every year. In every

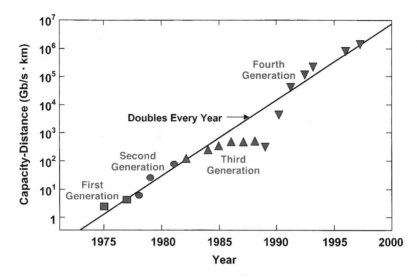

Figure 1.4: Increase in the *BL* product over the period 1975 to 1980 through several generations of lightwave systems. Different symbols are used for successive generations. (After Ref. [12]; ©2000 IEEE; reprinted with permission.)

generation, *BL* increases initially but then begins to saturate as the technology matures. Each new generation brings a fundamental change that helps to improve the system performance further.

The first generation of lightwave systems operated near 0.8 μm and used GaAs semiconductor lasers. After several field trials during the period 1977–79, such systems became available commercially in 1980 [13]. They operated at a bit rate of 45 Mb/s and allowed repeater spacings of up to 10 km. The larger repeater spacing compared with 1-km spacing of coaxial systems was an important motivation for system designers because it decreased the installation and maintenance costs associated with each repeater.

It was clear during the 1970s that the repeater spacing could be increased considerably by operating the lightwave system in the wavelength region near 1.3 μm, where fiber loss is below 1 dB/km. Furthermore, optical fibers exhibit minimum dispersion in this wavelength region. This realization led to a worldwide effort for the development of InGaAsP semiconductor lasers and detectors operating near 1.3 μm. The second generation of fiber-optic communication systems became available in the early 1980s, but the bit rate of early systems was limited to below 100 Mb/s because of dispersion in multimode fibers [14]. This limitation was overcome by the use of *single-mode* fibers. A laboratory experiment in 1981 demonstrated transmission at 2 Gb/s over 44 km of single-mode fiber [15]. The introduction of commercial systems soon followed. By 1987, second-generation lightwave systems, operating at bit rates of up to 1.7 Gb/s with a repeater spacing of about 50 km, were commercially available.

The repeater spacing of the second-generation lightwave systems was limited by the fiber losses at the operating wavelength of 1.3 μm (typically 0.5 dB/km). Losses of silica fibers become minimum near 1.55 μm. Indeed, a 0.2-dB/km loss was realized in 1979 in this spectral region [16]. However, the introduction of third-generation lightwave systems operating at 1.55 μm was considerably delayed by a large fiber dispersion near 1.55 μm. Conventional InGaAsP semiconductor lasers could not be used because of pulse spreading occurring as a result of simultaneous oscillation of several longitudinal modes. The dispersion problem can be overcome either by using dispersion-shifted fibers designed to have minimum dispersion near 1.55 μm or by limiting the laser spectrum to a single longitudinal mode. Both approaches were followed during the 1980s. By 1985, laboratory experiments indicated the possibility of transmitting information at bit rates of up to 4 Gb/s over distances in excess of 100 km [17]. Third-generation lightwave systems operating at 2.5 Gb/s became available commercially in 1990. Such systems are capable of operating at a bit rate of up to 10 Gb/s [18]. The best performance is achieved using dispersion-shifted fibers in combination with lasers oscillating in a single longitudinal mode.

A drawback of third-generation 1.55-μm systems is that the signal is regenerated periodically by using electronic repeaters spaced apart typically by 60–70 km. The repeater spacing can be increased by making use of a homodyne or heterodyne detection scheme because its use improves receiver sensitivity. Such systems are referred to as coherent lightwave systems. Coherent systems were under development worldwide during the 1980s, and their potential benefits were demonstrated in many system experiments [19]. However, commercial introduction of such systems was postponed with the advent of fiber amplifiers in 1989.

The fourth generation of lightwave systems makes use of *optical amplification* for increasing the repeater spacing and of *wavelength-division multiplexing* (WDM) for increasing the bit rate. As evident from different slopes in Fig. 1.3 before and after 1992, the advent of the WDM technique started a revolution that resulted in doubling of the system capacity every 6 months or so and led to lightwave systems operating at a bit rate of 10 Tb/s by 2001. In most WDM systems, fiber losses are compensated periodically using erbium-doped fiber amplifiers spaced 60–80 km apart. Such amplifiers were developed after 1985 and became available commercially by 1990. A 1991 experiment showed the possibility of data transmission over 21,000 km at 2.5 Gb/s, and over 14,300 km at 5 Gb/s, using a recirculating-loop configuration [20]. This performance indicated that an amplifier-based, all-optical, submarine transmission system was feasible for intercontinental communication. By 1996, not only transmission over 11,300 km at a bit rate of 5 Gb/s had been demonstrated by using actual submarine cables [21], but commercial transatlantic and transpacific cable systems also became available. Since then, a large number of submarine lightwave systems have been deployed worldwide.

Figure 1.5 shows the international network of submarine systems around 2000 [22]. The 27,000-km fiber-optic link around the globe (known as FLAG) became operational in 1998, linking many Asian and European countries [23]. Another major lightwave system, known as *Africa One* was operating by 2000; it circles the African continent and covers a total transmission distance of about 35,000 km [24]. Several WDM systems were deployed across the Atlantic and Pacific oceans during 1998–2001 in response to the Internet-induced increase in the data traffic; they have increased the total capacity by orders of magnitudes. A truly global network covering 250,000 km with a

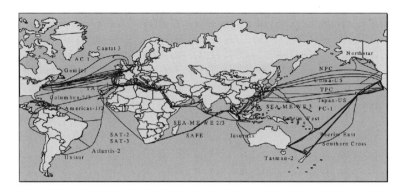

Figure 1.5: International undersea network of fiber-optic communication systems around 2000. (After Ref. [22]; ©2000 Academic; reprinted with permission.)

capacity of 2.56 Tb/s (64 WDM channels at 10 Gb/s over 4 fiber pairs) is scheduled to be operational in 2002 [25]. Clearly, the fourth-generation systems have revolutionized the whole field of fiber-optic communications.

The current emphasis of WDM lightwave systems is on increasing the system capacity by transmitting more and more channels through the WDM technique. With increasing WDM signal bandwidth, it is often not possible to amplify all channels using a single amplifier. As a result, new kinds of amplification schemes are being explored for covering the spectral region extending from 1.45 to 1.62 μm. This approach led in 2000 to a 3.28-Tb/s experiment in which 82 channels, each operating at 40 Gb/s, were transmitted over 3000 km, resulting in a *BL* product of almost 10,000 (Tb/s)-km. Within a year, the system capacity could be increased to nearly 11 Tb/s (273 WDM channels, each operating at 40 Gb/s) but the transmission distance was limited to 117 km [26]. In another record experiment, 300 channels, each operating at 11.6 Gb/s, were transmitted over 7380 km, resulting in a *BL* product of more than 25,000 (Tb/s)-km [27]. Commercial terrestrial systems with the capacity of 1.6 Tb/s were available by the end of 2000, and the plans were underway to extend the capacity toward 6.4 Tb/s. Given that the first-generation systems had a capacity of 45 Mb/s in 1980, it is remarkable that the capacity has jumped by a factor of more than 10,000 over a period of 20 years.

The fifth generation of fiber-optic communication systems is concerned with extending the wavelength range over which a WDM system can operate simultaneously. The conventional wavelength window, known as the C band, covers the wavelength range 1.53–1.57 μm. It is being extended on both the long- and short-wavelength sides, resulting in the L and S bands, respectively. The Raman amplification technique can be used for signals in all three wavelength bands. Moreover, a new kind of fiber, known as the *dry fiber* has been developed with the property that fiber losses are small over the entire wavelength region extending from 1.30 to 1.65 μm [28]. Availability of such fibers and new amplification schemes may lead to lightwave systems with thousands of WDM channels.

The fifth-generation systems also attempt to increase the bit rate of each channel within the WDM signal. Starting in 2000, many experiments used channels operating at 40 Gb/s; migration toward 160 Gb/s is also likely in the future. Such systems require an extremely careful management of fiber dispersion. An interesting approach is based on the concept of *optical solitons*—pulses that preserve their shape during propagation in a lossless fiber by counteracting the effect of dispersion through the fiber nonlinearity. Although the basic idea was proposed [29] as early as 1973, it was only in 1988 that a laboratory experiment demonstrated the feasibility of data transmission over 4000 km by compensating the fiber loss through Raman amplification [30]. Erbium-doped fiber amplifiers were used for soliton amplification starting in 1989. Since then, many system experiments have demonstrated the eventual potential of soliton communication systems. By 1994, solitons were transmitted over 35,000 km at 10 Gb/s and over 24,000 km at 15 Gb/s [31]. Starting in 1996, the WDM technique was also used for solitons in combination with dispersion management. In a 2000 experiment, up to 27 WDM channels, each operating at 20 Gb/s, were transmitted over 9000 km using a hybrid amplification scheme [32].

Even though the fiber-optic communication technology is barely 25 years old, it has progressed rapidly and has reached a certain stage of maturity. This is also apparent from the publication of a large number of books on optical communications and WDM networks since 1995 [33]–[55]. This third edition of a book, first published in 1992, is intended to present an up-to-date account of fiber-optic communications systems with emphasis on recent developments.

1.2 Basic Concepts

This section introduces a few basic concepts common to all communication systems. We begin with a description of analog and digital signals and describe how an analog signal can be converted into digital form. We then consider time- and frequency-division multiplexing of input signals, and conclude with a discussion of various modulation formats.

1.2.1 Analog and Digital Signals

In any communication system, information to be transmitted is generally available as an electrical signal that may take *analog* or *digital* form [56]. In the analog case, the signal (e. g., electric current) varies continuously with time, as shown schematically in Fig. 1.6(a). Familiar examples include audio and video signals resulting when a microphone converts voice or a video camera converts an image into an electrical signal. By contrast, the digital signal takes only a few discrete values. In the *binary representation* of a digital signal only two values are possible. The simplest case of a binary digital signal is one in which the electric current is either on or off, as shown in Fig. 1.6(b). These two possibilities are called "bit 1" and "bit 0" (*bit* is a contracted form of *binary digit*). Each bit lasts for a certain period of time T_B, known as the bit period or *bit slot*. Since one bit of information is conveyed in a time interval T_B, the bit rate B, defined as the number of bits per second, is simply $B = T_B^{-1}$. A well-known example of digital signals is provided by computer data. Each letter of the alphabet together with

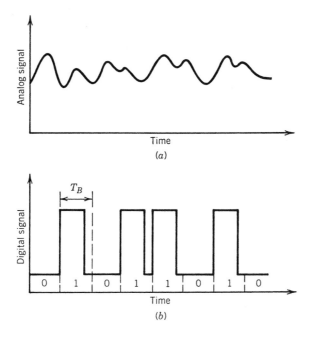

Figure 1.6: Representation of (a) an analog signal and (b) a digital signal.

other common symbols (decimal numerals, punctuation marks, etc.) is assigned a code number (ASCII code) in the range 0–127 whose binary representation corresponds to a 7-bit digital signal. The original ASCII code has been extended to represent 256 characters transmitted through 8-bit bytes. Both analog and digital signals are characterized by their bandwidth, which is a measure of the spectral contents of the signal. The *signal bandwidth* represents the range of frequencies contained within the signal and is determined mathematically through its Fourier transform.

An analog signal can be converted into digital form by sampling it at regular intervals of time [56]. Figure 1.7 shows the conversion method schematically. The sampling rate is determined by the bandwidth Δf of the analog signal. According to the *sampling theorem* [57]–[59], a bandwidth-limited signal can be fully represented by discrete samples, without any loss of information, provided that the sampling frequency f_s satisfies the *Nyquist criterion* [60], $f_s \geq 2\Delta f$. The first step consists of sampling the analog signal at the right frequency. The sampled values can take any value in the range $0 \leq A \leq A_{max}$, where A_{max} is the maximum amplitude of the given analog signal. Let us assume that A_{max} is divided into M discrete (not necessarily equally spaced) intervals. Each sampled value is quantized to correspond to one of these discrete values. Clearly, this procedure leads to additional noise, known as *quantization noise*, which adds to the noise already present in the analog signal.

The effect of quantization noise can be minimized by choosing the number of discrete levels such that $M > A_{max}/A_N$, where A_N is the root-mean-square noise amplitude of the analog signal. The ratio A_{max}/A_N is called the dynamic range and is related to

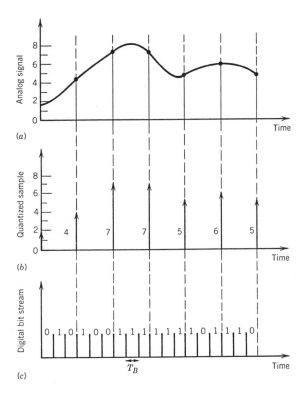

Figure 1.7: Three steps of (a) sampling, (b) quantization, and (c) coding required for converting an analog signal into a binary digital signal.

the *signal-to-noise ratio* (SNR) by the relation

$$\text{SNR} = 20 \log_{10}(A_{\max}/A_N),\qquad\qquad(1.2.1)$$

where SNR is expressed in decibel (dB) units. Any ratio R can be converted into decibels by using the general definition $10 \log_{10} R$ (see Appendix A). Equation (1.2.1) contains a factor of 20 in place of 10 simply because the SNR for electrical signals is defined with respect to the electrical power, whereas A is related to the electric current (or voltage).

The quantized sampled values can be converted into digital format by using a suitable conversion technique. In one scheme, known as *pulse-position modulation*, pulse position within the bit slot is a measure of the sampled value. In another, known as *pulse-duration modulation*, the pulse width is varied from bit to bit in accordance with the sampled value. These techniques are rarely used in practical optical communication systems, since it is difficult to maintain the pulse position or pulse width to high accuracy during propagation inside the fiber. The technique used almost universally, known as *pulse-code modulation* (PCM), is based on a binary scheme in which information is conveyed by the absence or the presence of pulses that are otherwise identical. A binary code is used to convert each sampled value into a string of 1 and 0 bits. The number of bits m needed to code each sample is related to the number of quantized signal levels M by the relation

$$M = 2^m \quad \text{or} \quad m = \log_2 M. \tag{1.2.2}$$

The bit rate associated with the PCM digital signal is thus given by

$$B = mf_s \geq (2\Delta f)\log_2 M, \tag{1.2.3}$$

where the Nyquist criterion, $f_s \geq 2\Delta f$, was used. By noting that $M > A_{\max}/A_N$ and using Eq. (1.2.1) together with $\log_2 10 \approx 3.33$,

$$B > (\Delta f/3)\,\text{SNR}, \tag{1.2.4}$$

where the SNR is expressed in decibel (dB) units.

Equation (1.2.4) provides the minimum bit rate required for digital representation of an analog signal of bandwidth Δf and a specific SNR. When SNR > 30 dB, the required bit rate exceeds $10(\Delta f)$, indicating a considerable increase in the bandwidth requirements of digital signals. Despite this increase, the digital format is almost always used for optical communication systems. This choice is made because of the superior performance of digital transmission systems. Lightwave systems offer such an enormous increase in the system capacity (by a factor $\sim 10^5$) compared with microwave systems that some bandwidth can be traded for improved performance.

As an illustration of Eq. (1.2.4), consider the digital conversion of an audio signal generated in a telephone. The analog audio signal contains frequencies in the range 0.3–3.4 kHz with a bandwidth $\Delta f = 3.1$ kHz and has a SNR of about 30 dB. Equation (1.2.4) indicates that $B > 31$ kb/s. In practice, a digital audio channel operates at 64 kb/s. The analog signal is sampled at intervals of 125 μs (sampling rate $f_s = 8$ kHz), and each sample is represented by 8 bits. The required bit rate for a digital video signal is higher by more than a factor of 1000. The analog television signal has a bandwidth ~ 4 MHz with a SNR of about 50 dB. The minimum bit rate from Eq. (1.2.4) is 66 Mb/s. In practice, a digital video signal requires a bit rate of 100 Mb/s or more unless it is compressed by using a standard format (such as MPEG-2).

1.2.2 Channel Multiplexing

As seen in the preceding discussion, a digital voice channel operates at 64 kb/s. Most fiber-optic communication systems are capable of transmitting at a rate of more than 1 Gb/s. To utilize the system capacity fully, it is necessary to transmit many channels simultaneously through multiplexing. This can be accomplished through *time-division* multiplexing (TDM) or *frequency-division* multiplexing (FDM). In the case of TDM, bits associated with different channels are interleaved in the time domain to form a composite bit stream. For example, the bit slot is about 15 μs for a single voice channel operating at 64 kb/s. Five such channels can be multiplexed through TDM if the bit streams of successive channels are delayed by 3 μs. Figure 1.8(a) shows the resulting bit stream schematically at a composite bit rate of 320 kb/s.

In the case of FDM, the channels are spaced apart in the frequency domain. Each channel is carried by its own carrier wave. The carrier frequencies are spaced more than

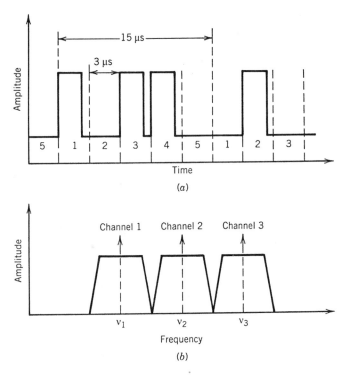

Figure 1.8: (a) Time-division multiplexing of five digital voice channels operating at 64 kb/s; (b) frequency-division multiplexing of three analog signals.

the channel bandwidth so that the channel spectra do not overlap, as seen Fig. 1.8(b). FDM is suitable for both analog and digital signals and is used in broadcasting of radio and television channels. TDM is readily implemented for digital signals and is commonly used for telecommunication networks. It is important to realize that TDM and FDM can be implemented in both the electrical and optical domains; optical FDM is often referred to as WDM. Chapter 8 is devoted to optical-domain multiplexing techniques. This section covers electrical TDM, which is employed universally to multiplex a large number of voice channels into a single electrical bit stream.

The concept of TDM has been used to form *digital hierarchies*. In North America and Japan, the first level corresponds to multiplexing of 24 voice channels with a composite bit rate of 1.544 Mb/s (hierarchy DS-1), whereas in Europe 30 voice channels are multiplexed, resulting in a composite bit rate of 2.048 Mb/s. The bit rate of the multiplexed signal is slightly larger than the simple product of 64 kb/s with the number of channels because of extra control bits that are added for separating (demultiplexing) the channels at the receiver end. The second-level hierarchy is obtained by multiplexing 4 DS-1 TDM channels. This results in a bit rate of 6.312 Mb/s (hierarchy DS-2) for North America or Japan and 8.448 Mb/s for Europe. This procedure is continued to obtain higher-level hierarchies. For example, at the fifth level of hierarchy, the bit rate becomes 565 Mb/s for Europe and 396 Mb/s for Japan.

Table 1.1 SONET/SDH bit rates

SONET	SDH	B (Mb/s)	Channels
OC-1		51.84	672
OC-3	STM-1	155.52	2,016
OC-12	STM-4	622.08	8,064
OC-48	STM-16	2,488.32	32,256
OC-192	STM-64	9,953.28	129,024
OC-768	STM-256	39,813.12	516,096

The lack of an international standard in the telecommunication industry during the 1980s led to the advent of a new standard, first called the *synchronous optical network* (SONET) and later termed the *synchronous digital hierarchy* or SDH [61]–[63]. It defines a synchronous frame structure for transmitting TDM digital signals. The basic building block of the SONET has a bit rate of 51.84 Mb/s. The corresponding optical signal is referred to as OC-1, where OC stands for optical carrier. The basic building block of the SDH has a bit rate of 155.52 Mb/s and is referred to as STM-1, where STM stands for a *synchronous transport module*. A useful feature of the SONET and SDH is that higher levels have a bit rate that is an exact multiple of the basic bit rate. Table 1.1 lists the correspondence between SONET and SDH bit rates for several levels. The SDH provides an international standard that appears to be well adopted. Indeed, lightwave systems operating at the STM-64 level ($B \approx 10$ Gb/s) are available since 1996 [18]. Commercial STM-256 (OC-768) systems operating near 40 Gb/s became available by 2002.

1.2.3 Modulation Formats

The first step in the design of an optical communication system is to decide how the electrical signal would be converted into an optical bit stream. Normally, the output of an optical source such as a semiconductor laser is modulated by applying the electrical signal either directly to the optical source or to an external modulator. There are two choices for the modulation format of the resulting optical bit stream. These are shown in Fig. 1.9 and are known as the *return-to-zero* (RZ) and *nonreturn-to-zero* (NRZ) formats. In the RZ format, each optical pulse representing bit 1 is shorter than the bit slot, and its amplitude returns to zero before the bit duration is over. In the NRZ format, the optical pulse remains on throughout the bit slot and its amplitude does not drop to zero between two or more successive 1 bits. As a result, pulse width varies depending on the bit pattern, whereas it remains the same in the case of RZ format. An advantage of the NRZ format is that the bandwidth associated with the bit stream is smaller than that of the RZ format by about a factor of 2 simply because on–off transitions occur fewer times. However, its use requires tighter control of the pulse width and may lead to bit-pattern-dependent effects if the optical pulse spreads during transmission. The NRZ format is often used in practice because of a smaller signal bandwidth associated with it.

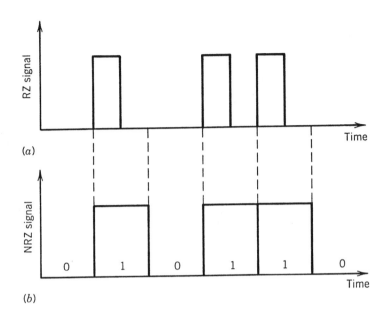

Figure 1.9: Digital bit stream 010110... coded by using (a) return-to-zero (RZ) and (b) nonreturn-to-zero (NRZ) formats.

The use of the RZ format in the optical domain began to attract attention around 1999 after it was found that its use may help the design of high-capacity lightwave systems [64]–[66]. An example of the RZ format is provided by the dispersion-managed soliton systems where a chirped pulse propagates inside the fiber link in a periodic fashion, and the average dispersion is used to counteract the buildup of the nonlinear effects [67]. In an interesting variant of the RZ format, known as the chirped RZ (or CRZ) format, optical pulses in each bit slot are chirped before they are launched into the fiber link but the system is operated in a quasi-linear regime [68]. In other schemes, modulation formats well known in the field of microwave communications are applied to the optical domain. Such formats are known as carrier-suppressed RZ (CSRZ), single-sideband, or vestigial-sideband formats [59]. Such RZ formats benefit from a reduced bandwidth compared to the standard RZ format.

An important issue is related to the choice of the physical variable that is modulated to encode the data on the optical carrier. The optical carrier wave before modulation is of the form

$$\mathbf{E}(t) = \hat{\mathbf{e}}A\cos(\omega_0 t + \phi), \qquad (1.2.5)$$

where $\mathbf{E}$ is the electric field vector, $\hat{\mathbf{e}}$ is the polarization unit vector, A is the amplitude, ω_0 is the carrier frequency, and ϕ is the phase. The spatial dependence of $\mathbf{E}$ is suppressed for simplicity of notation. One may choose to modulate the amplitude A, the frequency ω_0, or the phase ϕ. In the case of analog modulation, the three modulation choices are known as amplitude modulation (AM), frequency modulation (FM), and phase modulation (PM). The same modulation techniques can be applied in the digital case and are called amplitude-shift keying (ASK), frequency-shift keying (FSK), and

Figure 1.10: Generic optical communication system.

phase-shift keying (PSK), depending on whether the amplitude, frequency, or phase of the carrier wave is shifted between the two levels of a binary digital signal. The simplest technique consists of simply changing the signal power between two levels, one of which is set to zero, and is often called *on–off keying* (OOK) to reflect the on–off nature of the resulting optical signal. Most digital lightwave systems employ OOK in combination with PCM.

1.3 Optical Communication Systems

As mentioned earlier, optical communication systems differ in principle from microwave systems only in the frequency range of the carrier wave used to carry the information. The optical carrier frequencies are typically ~ 200 THz, in contrast with the microwave carrier frequencies (~ 1 GHz). An increase in the information capacity of optical communication systems by a factor of up to 10,000 is expected simply because of such high carrier frequencies used for lightwave systems. This increase can be understood by noting that the bandwidth of the modulated carrier can be up to a few percent of the carrier frequency. Taking, for illustration, 1% as the limiting value, optical communication systems have the potential of carrying information at bit rates ~ 1 Tb/s. It is this enormous potential bandwidth of optical communication systems that is the driving force behind the worldwide development and deployment of lightwave systems. Current state-of-the-art systems operate at bit rates ~ 10 Gb/s, indicating that there is considerable room for improvement.

Figure 1.10 shows a generic block diagram of an optical communication system. It consists of a transmitter, a communication channel, and a receiver, the three elements common to all communication systems. Optical communication systems can be classified into two broad categories: *guided* and *unguided*. As the name implies, in the case of guided lightwave systems, the optical beam emitted by the transmitter remains spatially confined. This is realized in practice by using optical fibers, as discussed in Chapter 2. Since all guided optical communication systems currently use optical fibers, the commonly used term for them is fiber-optic communication systems. The term *lightwave system* is also sometimes used for fiber-optic communication systems, although it should generally include both guided and unguided systems.

In the case of unguided optical communication systems, the optical beam emitted by the transmitter spreads in space, similar to the spreading of microwaves. However, unguided optical systems are less suitable for broadcasting applications than microwave systems because optical beams spread mainly in the forward direction (as a result of their short wavelength). Their use generally requires accurate pointing between the transmitter and the receiver. In the case of terrestrial propagation, the signal in unguided systems can deteriorate considerably by scattering within the atmosphere. This problem, of course, disappears in *free-space communications* above the earth atmosphere (e.g., intersatellite communications). Although free-space optical communication systems are needed for certain applications and have been studied extensively [69], most terrestrial applications make use of *fiber-optic communication systems*. This book does not consider unguided optical communication systems.

The application of optical fiber communications is in general possible in any area that requires transfer of information from one place to another. However, fiber-optic communication systems have been developed mostly for telecommunications applications. This is understandable in view of the existing worldwide telephone networks which are used to transmit not only voice signals but also computer data and fax messages. The telecommunication applications can be broadly classified into two categories, *long-haul* and *short-haul*, depending on whether the optical signal is transmitted over relatively long or short distances compared with typical intercity distances (~ 100 km). Long-haul telecommunication systems require high-capacity trunk lines and benefit most by the use of fiber-optic lightwave systems. Indeed, the technology behind optical fiber communication is often driven by long-haul applications. Each successive generation of lightwave systems is capable of operating at higher bit rates and over longer distances. Periodic regeneration of the optical signal by using repeaters is still required for most long-haul systems. However, more than an order-of-magnitude increase in both the repeater spacing and the bit rate compared with those of coaxial systems has made the use of lightwave systems very attractive for long-haul applications. Furthermore, transmission distances of thousands of kilometers can be realized by using optical amplifiers. As shown in Fig. 1.5, a large number of transoceanic lightwave systems have already been installed to create an international fiber-optic network.

Short-haul telecommunication applications cover intracity and local-loop traffic. Such systems typically operate at low bit rates over distances of less than 10 km. The use of single-channel lightwave systems for such applications is not very cost-effective, and multichannel networks with multiple services should be considered. The concept of a broadband integrated-services digital network requires a high-capacity communication system capable of carrying multiple services. The asynchronous transfer mode (ATM) technology also demands high bandwidths. Only fiber-optic communication systems are likely to meet such wideband distribution requirements. Multichannel lightwave systems and their applications in local-area networks are discussed in Chapter 8.

1.4   Lightwave System Components

The generic block diagram of Fig. 1.10 applies to a fiber-optic communication system, the only difference being that the communication channel is an optical fiber cable. The other two components, the optical transmitter and the optical receiver, are designed to meet the needs of such a specific communication channel. In this section we discuss the general issues related to the role of optical fiber as a communication channel and to the design of transmitters and receivers. The objective is to provide an introductory overview, as the three components are discussed in detail in Chapters 2–4.

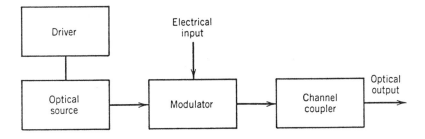

Figure 1.11: Components of an optical transmitter.

1.4.1 Optical Fibers as a Communication Channel

The role of a communication channel is to transport the optical signal from transmitter to receiver without distorting it. Most lightwave systems use optical fibers as the communication channel because silica fibers can transmit light with losses as small as 0.2 dB/km. Even then, optical power reduces to only 1% after 100 km. For this reason, fiber losses remain an important design issue and determines the repeater or amplifier spacing of a long-haul lightwave system. Another important design issue is *fiber dispersion*, which leads to broadening of individual optical pulses with propagation. If optical pulses spread significantly outside their allocated bit slot, the transmitted signal is severely degraded. Eventually, it becomes impossible to recover the original signal with high accuracy. The problem is most severe in the case of multimode fibers, since pulses spread rapidly (typically at a rate of ~ 10 ns/km) because of different speeds associated with different fiber modes. It is for this reason that most optical communication systems use single-mode fibers. Material dispersion (related to the frequency dependence of the refractive index) still leads to pulse broadening (typically < 0.1 ns/km), but it is small enough to be acceptable for most applications and can be reduced further by controlling the spectral width of the optical source. Nevertheless, as discussed in Chapter 2, material dispersion sets the ultimate limit on the bit rate and the transmission distance of fiber-optic communication systems.

1.4.2 Optical Transmitters

The role of an *optical transmitter* is to convert the electrical signal into optical form and to launch the resulting optical signal into the optical fiber. Figure 1.11 shows the block diagram of an optical transmitter. It consists of an optical source, a modulator, and a channel coupler. Semiconductor lasers or light-emitting diodes are used as optical sources because of their compatibility with the optical-fiber communication channel; both are discussed in detail in Chapter 3. The optical signal is generated by modulating the optical carrier wave. Although an external modulator is sometimes used, it can be dispensed with in some cases, since the output of a semiconductor optical source can be modulated directly by varying the injection current. Such a scheme simplifies the transmitter design and is generally cost-effective. The coupler is typically a mi-

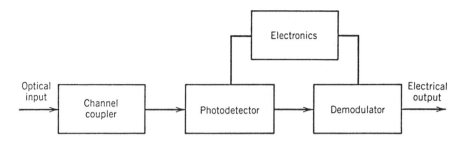

Figure 1.12: Components of an optical receiver.

crolens that focuses the optical signal onto the entrance plane of an optical fiber with
the maximum possible efficiency.

The *launched power* is an important design parameter. One can increase the amplifier (or repeater) spacing by increasing it, but the onset of various *nonlinear effects*
limits how much the input power can be increased. The launched power is often expressed in "dBm" units with 1 mW as the reference level. The general definition is (see
Appendix A)

$$\text{power (dBm)} = 10\log_{10}\left(\frac{\text{power}}{1\ \text{mW}}\right). \tag{1.4.1}$$

Thus, 1 mW is 0 dBm, but 1 μW corresponds to -30 dBm. The launched power is
rather low (< -10 dBm) for light-emitting diodes but semiconductor lasers can launch
powers ~ 10 dBm. As light-emitting diodes are also limited in their modulation capa-
bilities, most lightwave systems use semiconductor lasers as optical sources. The bit
rate of optical transmitters is often limited by electronics rather than by the semiconductor laser itself. With proper design, optical transmitters can be made to operate at
a bit rate of up to 40 Gb/s. Chapter 3 is devoted to a complete description of optical
transmitters.

1.4.3 Optical Receivers

An *optical receiver* converts the optical signal received at the output end of the optical fiber back into the original electrical signal. Figure 1.12 shows the block diagram
of an optical receiver. It consists of a coupler, a photodetector, and a demodulator.
The coupler focuses the received optical signal onto the photodetector. Semiconductor
photodiodes are used as photodetectors because of their compatibility with the whole
system; they are discussed in Chapter 4. The design of the demodulator depends on
the modulation format used by the lightwave system. The use of FSK and PSK formats generally requires heterodyne or homodyne demodulation techniques discussed
in Chapter 10. Most lightwave systems employ a scheme referred to as "intensity
modulation with direct detection" (IM/DD). Demodulation in this case is done by a
decision circuit that identifies bits as 1 or 0, depending on the amplitude of the electric
signal. The accuracy of the decision circuit depends on the SNR of the electrical signal
generated at the photodetector.

The performance of a digital lightwave system is characterized through the *bit-error rate* (BER). Although the BER can be defined as the number of errors made per second, such a definition makes the BER bit-rate dependent. It is customary to define the BER as the average probability of incorrect bit identification. Therefore, a BER of 10^{-6} corresponds to on average one error per million bits. Most lightwave systems specify a BER of 10^{-9} as the operating requirement; some even require a BER as small as 10^{-14}. The error-correction codes are sometimes used to improve the raw BER of a lightwave systems.

An important parameter for any receiver is the *receiver sensitivity*. It is usually defined as the minimum average optical power required to realize a BER of 10^{-9}. Receiver sensitivity depends on the SNR, which in turn depends on various noise sources that corrupt the signal received. Even for a perfect receiver, some noise is introduced by the process of photodetection itself. This is referred to as the *quantum noise* or the *shot noise*, as it has its origin in the particle nature of electrons. Optical receivers operating at the shot-noise limit are called quantum-noise-limited receivers. No practical receiver operates at the quantum-noise limit because of the presence of several other noise sources. Some of the noise sources such as *thermal noise* are internal to the receiver. Others originate at the transmitter or during propagation along the fiber link. For instance, any amplification of the optical signal along the transmission line with the help of optical amplifiers introduces the so-called *amplifier noise* that has its origin in the fundamental process of spontaneous emission. Chromatic dispersion in optical fibers can add additional noise through phenomena such as intersymbol interference and mode-partition noise. The receiver sensitivity is determined by a cumulative effect of all possible noise mechanisms that degrade the SNR at the decision circuit. In general, it also depends on the bit rate as the contribution of some noise sources (e.g., shot noise) increases in proportion to the signal bandwidth. Chapter 4 is devoted to noise and sensitivity issues of optical receivers by considering the SNR and the BER in digital lightwave systems.

Problems

1.1 Calculate the carrier frequency for optical communication systems operating at 0.88, 1.3, and 1.55 μm. What is the photon energy (in eV) in each case?

1.2 Calculate the transmission distance over which the optical power will attenuate by a factor of 10 for three fibers with losses of 0.2, 20, and 2000 dB/km. Assuming that the optical power decreases as $\exp(-\alpha L)$, calculate α (in cm^{-1}) for the three fibers.

1.3 Assume that a digital communication system can be operated at a bit rate of up to 1% of the carrier frequency. How many audio channels at 64 kb/s can be transmitted over a microwave carrier at 5 GHz and an optical carrier at 1.55 μm?

1.4 A 1-hour lecture script is stored on the computer hard disk in the ASCII format. Estimate the total number of bits assuming a delivery rate of 200 words per minute and on average 5 letters per word. How long will it take to transmit the script at a bit rate of 1 Gb/s?

1.5 A 1.55-μm digital communication system operating at 1 Gb/s receives an average power of -40 dBm at the detector. Assuming that 1 and 0 bits are equally likely to occur, calculate the number of photons received within each 1 bit.

1.6 An analog voice signal that can vary over the range 0–50 mA is digitized by sampling it at 8 kHz. The first four sample values are 10, 21, 36, and 16 mA. Write the corresponding digital signal (a string of 1 and 0 bits) by using a 4-bit representation for each sample.

1.7 Sketch the variation of optical power with time for a digital NRZ bit stream 010111101110 by assuming a bit rate of 2.5 Gb/s. What is the duration of the shortest and widest optical pulse?

1.8 A 1.55-μm fiber-optic communication system is transmitting digital signals over 100 km at 2 Gb/s. The transmitter launches 2 mW of average power into the fiber cable, having a net loss of 0.3 dB/km. How many photons are incident on the receiver during a single 1 bit? Assume that 0 bits carry no power, while 1 bits are in the form of a rectangular pulse occupying the entire bit slot (NRZ format).

1.9 A 0.8-μm optical receiver needs at least 1000 photons to detect the 1 bits accurately. What is the maximum possible length of the fiber link for a 100-Mb/s optical communication system designed to transmit -10 dBm of average power? The fiber loss is 2 dB/km at 0.8 μm. Assume the NRZ format and a rectangular pulse shape.

1.10 A 1.3-μm optical transmitter is used to obtain a digital bit stream at a bit rate of 2 Gb/s. Calculate the number of photons contained in a single 1 bit when the average power emitted by the transmitter is 4 mW. Assume that the 0 bits carry no energy.

References

[1] G. J. Holzmann and B. Pehrson, *The Early History of Data Networks*, IEEE Computer Society Press, Los Alamitos, CA, 1995; also available on the Internet at the address http://www.it.kth.se/docs/early_net.

[2] D. Koenig, "Telegraphs and Telegrams in Revolutionary France," *Scientific Monthly*, 431 (1944). See also Chap. 2 of Ref. [1].

[3] A. Jones, *Historical Sketch of the Electrical Telegraph*, Putnam, New York, 1852.

[4] A. G. Bell, U.S. Patent No. 174,465 (1876).

[5] T. H. Maiman, *Nature* **187**, 493 (1960).

[6] W. K. Pratt, *Laser Communication Systems*, Wiley, New York, 1969.

[7] S. E. Miller, *Sci. Am.* **214** (1), 19 (1966).

[8] K. C. Kao and G. A. Hockham, *Proc. IEE* **113**, 1151 (1966); A. Werts, *Onde Electr.* **45**, 967 (1966).

[9] F. P. Kapron, D. B. Keck, and R. D. Maurer, *Appl. Phys. Lett.* **17**, 423 (1970).

[10] I. Hayashi, M. B. Panish, P. W. Foy, and S. Sumski, *Appl. Phys. Lett.* **17**, 109 (1970).

[11] A. E. Willner, Ed., *IEEE J. Sel. Topics Quantum Electron.* **6**, 827 (2000). Several historical articles in this millennium issue cover the development of lasers and optical fibers. See, for example, the articles by Z. Alferov, W. A. Gambling, T. Izawa, D. Keck, H. Kogelnik, and R. H. Rediker.

[12] H. Kogelnik, *IEEE J. Sel. Topics Quantum Electron.* **6**, 1279 (2000).

[13] R. J. Sanferrare, *AT&T Tech. J.* **66**, 95 (1987).

[14] D. Gloge, A. Albanese, C. A. Burrus, E. L. Chinnock, J. A. Copeland, A. G. Dentai, T. P. Lee, T. Li, and K. Ogawa, *Bell Syst. Tech. J.* **59**, 1365 (1980).

[15] J. I. Yamada, S. Machida, and T. Kimura, *Electron. Lett.* **17**, 479 (1981).

[16] T. Miya, Y. Terunuma, T. Hosaka, and T. Miyoshita, *Electron. Lett.* **15**, 106 (1979).

[17] A. H. Gnauck, B. L. Kasper, R. A. Linke, R. W. Dawson, T. L. Koch, T. J. Bridges, E. G. Burkhardt, R. T. Yen, D. P. Wilt, J. C. Campbell, K. C. Nelson, and L. G. Cohen, *J. Lightwave Technol.* **3**, 1032 (1985).

[18] K. Nakagawa, *Trans. IECE Jpn. Pt. J* **78B**, 713 (1995).

[19] R. A. Linke and A. H. Gnauck, *J. Lightwave Technol.* **6**, 1750 (1988); P. S. Henry, *Coherent Lightwave Communications*, IEEE Press, New York, 1990.

[20] N. S. Bergano, J. Aspell, C. R. Davidson, P. R. Trischitta, B. M. Nyman, and F. W. Kerfoot, *Electron. Lett.* **27**, 1889 (1991).

[21] T. Otani, K. Goto, H. Abe, M. Tanaka, H. Yamamoto, and H. Wakabayashi, *Electron. Lett.* **31**, 380 (1995).

[22] J. M. Beaufils, *Opt. Fiber Technol.* **6**, 15 (2000).

[23] T. Welsh, R. Smith, H. Azami, and R. Chrisner, *IEEE Commun. Mag.* **34** (2), 30 (1996).

[24] W. C. Marra and J. Schesser, *IEEE Commun. Mag.* **34** (2), 50 (1996).

[25] N. S. Bergano and H. Kidorf, *Opt. Photon. News* **12** (3), 32 (2001).

[26] K. Fukuchi, T. Kasamatsu, M. Morie, R. Ohhira, T. Ito, K. Sekiya, D. Ogasahara, and T. Ono, Paper PD24, *Proc. Optical Fiber Commun. Conf.*, Optical Society of America, Washington, DC, 2001.

[27] G. Vareille, F. Pitel, and J. F. Marcerou, Paper PD22, *Proc. Optical Fiber Commun. Conf.*, Optical Society of America, Washington, DC, 2001.

[28] G. A. Thomas, B. L. Shraiman, P. F. Glodis, and M. J. Stephan, *Nature* **404**, 262 (2000).

[29] A. Hasegawa and F. Tappert, *Appl. Phys. Lett.* **23**, 142 (1973).

[30] L. F. Mollenauer and K. Smith, *Opt. Lett.* **13**, 675 (1988).

[31] L. F. Mollenauer, *Opt. Photon. News* **11** (4), 15 (1994).

[32] L. F. Mollenauer, P. V. Mamyshev, J. Gripp, M. J. Neubelt, N. Mamysheva, L. Grüner-Nielsen, and T. Veng, *Opt. Lett.* **25**, 704 (2000).

[33] G. Einarrson, *Principles of Lightwave Communication Systems*, Wiley, New York, 1995.

[34] N. Kashima, *Passive Optical Components for Optical Fiber Transmission*, Artec House, Norwood, MA, 1995.

[35] M. M. K. Liu, *Principles and Applications of Optical Communications*, Irwin, Chicago, 1996.

[36] M. Cvijetic, *Coherent and Nonlinear Lightwave Communications*, Artec House, Norwood, MA, 1996.

[37] L. Kazovsky, S. Bendetto, and A. E. Willner, *Optical Fiber Communication Systems*, Artec House, Norwood, MA, 1996.

[38] R. Papannareddy, *Introduction to Lightwave Communication Systems*, Artec House, Norwood, MA, 1997.

[39] K. Nosu, *Optical FDM Network Technologies*, Artec House, Norwood, MA, 1997.

[40] I. P. Kaminow and T. L. Koch, Eds., *Optical Fiber Telecommunications III*, Academic Press, San Diego, CA, 1997.

[41] E. Iannone, F. Matera, A. Mecozzi, and M. Settembre, *Nonlinear Optical Communication Networks*, Wiley, New York, 1998.

[42] G. Lachs, *Fiber Optic Communications: Systems Analysis and Enhancements*, McGraw-Hill, New York, 1998.

[43] A. Borella, G. Cancellieri, and F. Chiaraluce, *WDMA Optical Networks*, Artec House, Norwood, MA, 1998.

[44] R. A. Barry, *Optical Networks and Their Applications*, Optical Society of America, Washington, DC, 1998.

[45] T. E. Stern and K. Bala, *Multiwavelength Optical Networks*, Addison Wesley, Reading, MA, 1999.

[46] R. Sabella and P. Lugli, *High Speed Optical Communications*, Kluwer Academic, Norwell, MA, 1999.

[47] P. J. Wan, *Multichannel Optical Networks*, Kluwer Academic, Norwell, MA, 1999.

[48] A. Bononi, *Optical Networking*, Springer, London, 1999.

[49] G. E. Keiser, *Optical Fiber Communications*, 3rd ed., McGraw-Hill, New York, 2000.

[50] K. M. Sivalingam, *Optical WDM Networks: Principles and Practice*, Kluwer Academic, Norwell, MA, 2000.

[51] J. Franz and V. K. Jain, *Optical Communications: Components and Systems*, CRC Press, Boca Raton, FL, 2000.

[52] B. Ramamurthy, *Design of Optical WDM Networks*, Kluwer Academic, Norwell, MA, 2000.

[53] M. T. Fatehi and M. Wilson, *Optical Networking with WDM*, McGraw-Hill, New York, 2002.

[54] R. Ramaswami and K. Sivarajan, *Optical Networks: A Practical Perspective*, 2nd ed., Morgan Kaufmann Publishers, San Francisco, 2002.

[55] I. P. Kaminow and T. Li, Eds., *Optical Fiber Telecommunications IV*, Academic Press, San Diego, CA, 2002.

[56] M. Schwartz, *Information Transmission, Modulation, and Noise*, 4th ed., McGraw-Hill, New York, 1990.

[57] C. E. Shannon, *Proc. IRE* **37**, 10 (1949).

[58] A. J. Jerri, *Proc. IEEE* **65**, 1565 (1977).

[59] L. W. Couch II, *Modern Communication Systems: Principles and Applications*, 4th ed., Prentice Hall, Upper Saddle River, NJ, 1995.

[60] H. Nyquist, *Trans. AIEE* **47**, 617 (1928).

[61] R. Ballart and Y.-C. Ching, *IEEE Commun. Mag.* **27** (3), 8 (1989).

[62] T. Miki, Jr. and C. A. Siller, Eds., *IEEE Commun. Mag.* **28** (8), 1 (1990).

[63] S. V. Kartalopoulos, *Understanding SONET/SDH and ATM*, IEEE Press, Piscataway, NJ, 1999.

[64] M. I. Hayee and A. E. Willner, *IEEE Photon. Technol. Lett.* **11**, 991 (1999).

[65] R. Ludwig, U. Feiste, E. Dietrich, H. G. Weber, D. Breuer, M. Martin, and F. Kuppers, *Electron. Lett.* **35**, 2216 (1999).

[66] M. Nakazawa, H. Kubota, K. Suzuki, E. Yamada, and A. Sahara, *IEEE J. Sel. Topics Quantum Electron.* **6**, 363 (2000).

[67] G. P. Agrawal, *Applications of Nonlinear Fiber Optics*, Academic Press, San Diego, CA, 2001.

[68] E. A. Golovchenko, A. N. Pilipetskii, and N. A. Bergano, Paper FC3, *Proc. Optical Fiber Commun. Conf.*, Optical Society of America, Washington, DC, 2000.

[69] S. G. Lambert and W. L. Casey, *Laser Communications in Space*, Artec House, Norwood, MA, 1995.

Chapter 2

Optical Fibers

The phenomenon of *total internal reflection*, responsible for guiding of light in optical fibers, has been known since 1854 [1]. Although glass fibers were made in the 1920s [2]–[4], their use became practical only in the 1950s, when the use of a cladding layer led to considerable improvement in their guiding characteristics [5]–[7]. Before 1970, optical fibers were used mainly for medical imaging over short distances [8]. Their use for communication purposes was considered impractical because of high losses (~ 1000 dB/km). However, the situation changed drastically in 1970 when, following an earlier suggestion [9], the loss of optical fibers was reduced to below 20 dB/km [10]. Further progress resulted by 1979 in a loss of only 0.2 dB/km near the 1.55-μm spectral region [11]. The availability of low-loss fibers led to a revolution in the field of lightwave technology and started the era of fiber-optic communications. Several books devoted entirely to optical fibers cover numerous advances made in their design and understanding [12]–[21]. This chapter focuses on the role of optical fibers as a communication channel in lightwave systems. In Section 2.1 we use geometrical-optics description to explain the guiding mechanism and introduce the related basic concepts. Maxwell's equations are used in Section 2.2 to describe wave propagation in optical fibers. The origin of fiber dispersion is discussed in Section 2.3, and Section 2.4 considers limitations on the bit rate and the transmission distance imposed by fiber dispersion. The loss mechanisms in optical fibers are discussed in Section 2.5, and Section 2.6 is devoted to a discussion of the nonlinear effects. The last section covers manufacturing details and includes a discussion of the design of fiber cables.

2.1 Geometrical-Optics Description

In its simplest form an optical fiber consists of a cylindrical core of silica glass surrounded by a cladding whose refractive index is lower than that of the core. Because of an abrupt index change at the core–cladding interface, such fibers are called *step-index fibers*. In a different type of fiber, known as *graded-index fiber*, the refractive index decreases gradually inside the core. Figure 2.1 shows schematically the index profile and the cross section for the two kinds of fibers. Considerable insight in the guiding

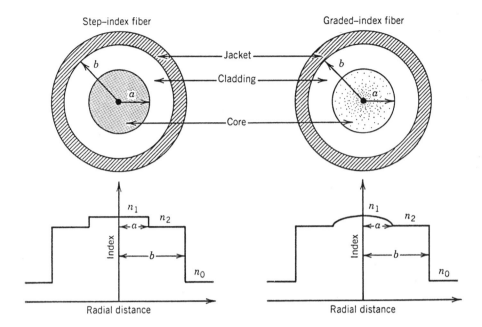

Figure 2.1: Cross section and refractive-index profile for step-index and graded-index fibers.

properties of optical fibers can be gained by using a ray picture based on geometrical optics [22]. The geometrical-optics description, although approximate, is valid when the core radius a is much larger than the light wavelength λ. When the two become comparable, it is necessary to use the wave-propagation theory of Section 2.2.

2.1.1 Step-Index Fibers

Consider the geometry of Fig. 2.2, where a ray making an angle θ_i with the fiber axis is incident at the core center. Because of refraction at the fiber–air interface, the ray bends toward the normal. The angle θ_r of the refracted ray is given by [22]

$$n_0 \sin \theta_i = n_1 \sin \theta_r, \tag{2.1.1}$$

where n_1 and n_0 are the refractive indices of the fiber core and air, respectively. The refracted ray hits the core–cladding interface and is refracted again. However, refraction is possible only for an angle of incidence ϕ such that $\sin \phi < n_2/n_1$. For angles larger than a *critical angle* ϕ_c, defined by [22]

$$\sin \phi_c = n_2/n_1, \tag{2.1.2}$$

where n_2 is the cladding index, the ray experiences total internal reflection at the core–cladding interface. Since such reflections occur throughout the fiber length, all rays with $\phi > \phi_c$ remain confined to the fiber core. This is the basic mechanism behind light confinement in optical fibers.

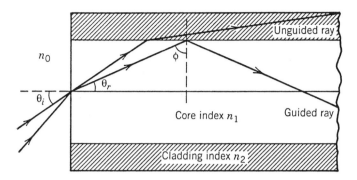

Figure 2.2: Light confinement through total internal reflection in step-index fibers. Rays for which $\phi < \phi_c$ are refracted out of the core.

One can use Eqs. (2.1.1) and (2.1.2) to find the maximum angle that the incident ray should make with the fiber axis to remain confined inside the core. Noting that $\theta_r = \pi/2 - \phi_c$ for such a ray and substituting it in Eq. (2.1.1), we obtain

$$n_0 \sin \theta_i = n_1 \cos \phi_c = (n_1^2 - n_2^2)^{1/2}. \tag{2.1.3}$$

In analogy with lenses, $n_0 \sin \theta_i$ is known as the *numerical aperture* (NA) of the fiber. It represents the light-gathering capacity of an optical fiber. For $n_1 \simeq n_2$ the NA can be approximated by

$$\text{NA} = n_1 (2\Delta)^{1/2}, \qquad \Delta = (n_1 - n_2)/n_1, \tag{2.1.4}$$

where Δ is the fractional index change at the core–cladding interface. Clearly, Δ should be made as large as possible in order to couple maximum light into the fiber. However, such fibers are not useful for the purpose of optical communications because of a phenomenon known as multipath dispersion or *modal dispersion* (the concept of fiber modes is introduced in Section 2.2).

Multipath dispersion can be understood by referring to Fig. 2.2, where different rays travel along paths of different lengths. As a result, these rays disperse in time at the output end of the fiber even if they were coincident at the input end and traveled at the same speed inside the fiber. A short pulse (called an *impulse*) would broaden considerably as a result of different path lengths. One can estimate the extent of pulse broadening simply by considering the shortest and longest ray paths. The shortest path occurs for $\theta_i = 0$ and is just equal to the fiber length L. The longest path occurs for θ_i given by Eq. (2.1.3) and has a length $L/\sin\phi_c$. By taking the velocity of propagation $v = c/n_1$, the time delay is given by

$$\Delta T = \frac{n_1}{c} \left(\frac{L}{\sin \phi_c} - L \right) = \frac{L\, n_1^2}{c\, n_2} \Delta. \tag{2.1.5}$$

The time delay between the two rays taking the shortest and longest paths is a measure of broadening experienced by an impulse launched at the fiber input.

We can relate ΔT to the information-carrying capacity of the fiber measured through the bit rate B. Although a precise relation between B and ΔT depends on many details, such as the pulse shape, it is clear intuitively that ΔT should be less than the allocated bit slot $(T_B = 1/B)$. Thus, an order-of-magnitude estimate of the bit rate is obtained from the condition $B\Delta T < 1$. By using Eq. (2.1.5) we obtain

$$BL < \frac{n_2}{n_1^2} \frac{c}{\Delta}. \tag{2.1.6}$$

This condition provides a rough estimate of a fundamental limitation of step-index fibers. As an illustration, consider an unclad glass fiber with $n_1 = 1.5$ and $n_2 = 1$. The bit rate–distance product of such a fiber is limited to quite small values since $BL < 0.4$ (Mb/s)-km. Considerable improvement occurs for cladded fibers with a small index step. Most fibers for communication applications are designed with $\Delta < 0.01$. As an example, $BL < 100$ (Mb/s)-km for $\Delta = 2 \times 10^{-3}$. Such fibers can communicate data at a bit rate of 10 Mb/s over distances up to 10 km and may be suitable for some local-area networks.

Two remarks are in order concerning the validity of Eq. (2.1.6). First, it is obtained by considering only rays that pass through the fiber axis after each total internal reflection. Such rays are called *meridional rays*. In general, the fiber also supports *skew rays*, which travel at angles oblique to the fiber axis. Skew rays scatter out of the core at bends and irregularities and are not expected to contribute significantly to Eq. (2.1.6). Second, even the oblique meridional rays suffer higher losses than paraxial meridional rays because of scattering. Equation (2.1.6) provides a conservative estimate since all rays are treated equally. The effect of intermodal dispersion can be considerably reduced by using graded-index fibers, which are discussed in the next subsection. It can be eliminated entirely by using the single-mode fibers discussed in Section 2.2.

2.1.2   Graded-Index Fibers

The refractive index of the core in graded-index fibers is not constant but decreases gradually from its maximum value n_1 at the core center to its minimum value n_2 at the core–cladding interface. Most graded-index fibers are designed to have a nearly quadratic decrease and are analyzed by using α-profile, given by

$$n(\rho) = \begin{cases} n_1[1 - \Delta(\rho/a)^\alpha]; & \rho < a, \\ n_1(1 - \Delta) = n_2 \; ; & \rho \geq a, \end{cases} \tag{2.1.7}$$

where a is the core radius. The parameter α determines the index profile. A step-index profile is approached in the limit of large α. A *parabolic-index fiber* corresponds to $\alpha = 2$.

It is easy to understand qualitatively why intermodal or multipath dispersion is reduced for graded-index fibers. Figure 2.3 shows schematically paths for three different rays. Similar to the case of step-index fibers, the path is longer for more oblique rays. However, the ray velocity changes along the path because of variations in the refractive index. More specifically, the ray propagating along the fiber axis takes the shortest path but travels most slowly as the index is largest along this path. Oblique rays have a large part of their path in a medium of lower refractive index, where they travel faster. It is therefore possible for all rays to arrive together at the fiber output by a suitable choice of the refractive-index profile.

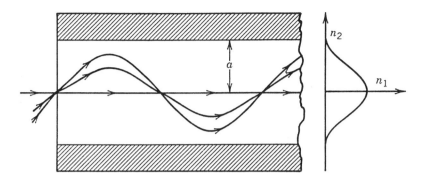

Figure 2.3: Ray trajectories in a graded-index fiber.

Geometrical optics can be used to show that a parabolic-index profile leads to nondispersive pulse propagation within the *paraxial approximation*. The trajectory of a paraxial ray is obtained by solving [22]

$$\frac{d^2\rho}{dz^2} = \frac{1}{n}\frac{dn}{d\rho},$$ (2.1.8)

where ρ is the radial distance of the ray from the axis. By using Eq. (2.1.7) for $\rho < a$ with $\alpha = 2$, Eq. (2.1.8) reduces to an equation of harmonic oscillator and has the general solution

$$\rho = \rho_0 \cos(pz) + (\rho_0'/p)\sin(pz),$$ (2.1.9)

where $p = (2\Delta/a^2)^{1/2}$ and ρ_0 and ρ_0' are the position and the direction of the input ray, respectively. Equation (2.1.9) shows that all rays recover their initial positions and directions at distances $z = 2m\pi/p$, where m is an integer (see Fig. 2.3). Such a complete restoration of the input implies that a parabolic-index fiber does not exhibit intermodal dispersion.

The conclusion above holds only within the paraxial and the geometrical-optics approximations, both of which must be relaxed for practical fibers. Intermodal dispersion in graded-index fibers has been studied extensively by using wave-propagation techniques [13]–[15]. The quantity $\Delta T/L$, where ΔT is the maximum multipath delay in a fiber of length L, is found to vary considerably with α. Figure 2.4 shows this variation for $n_1 = 1.5$ and $\Delta = 0.01$. The minimum dispersion occurs for $\alpha = 2(1-\Delta)$ and depends on Δ as [23]

$$\Delta T/L = n_1\Delta^2/8c.$$ (2.1.10)

The limiting bit rate–distance product is obtained by using the criterion $\Delta T < 1/B$ and is given by

$$BL < 8c/n_1\Delta^2.$$ (2.1.11)

The right scale in Fig. 2.4 shows the BL product as a function of α. Graded-index fibers with a suitably optimized index profile can communicate data at a bit rate of 100 Mb/s over distances up to 100 km. The BL product of such fibers is improved by nearly three orders of magnitude over that of step-index fibers. Indeed, the first generation

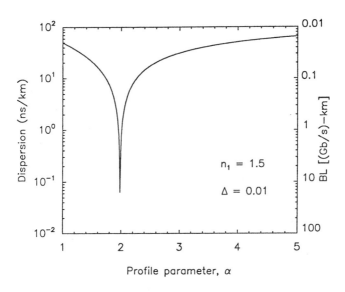

Figure 2.4: Variation of intermodal dispersion $\Delta T/L$ with the profile parameter α for a graded-index fiber. The scale on the right shows the corresponding bit rate–distance product.

of lightwave systems used graded-index fibers. Further improvement is possible only by using single-mode fibers whose core radius is comparable to the light wavelength. Geometrical optics cannot be used for such fibers.

Although graded-index fibers are rarely used for long-haul links, the use of graded-index *plastic* optical fibers for data-link applications has attracted considerable attention during the 1990s [24]–[29]. Such fibers have a relatively large core, resulting in a high numerical aperture and high coupling efficiency but they exhibit high losses (typically exceeding 50 dB/km). The *BL* product of plastic fibers, however, exceeds 2 (Gb/s)-km because of a graded-index profile [24]. As a result, they can be used to transmit data at bit rates > 1 Gb/s over short distances of 1 km or less. In a 1996 demonstration, a 10-Gb/s signal was transmitted over 0.5 km with a bit-error rate of less than 10^{-11} [26]. Graded-index plastic optical fibers provide an ideal solution for transferring data among computers and are becoming increasingly important for Ethernet applications requiring bit rates in excess of 1 Gb/s.

2.2   Wave Propagation (concepts only)

In this section we consider propagation of light in step-index fibers by using Maxwell's equations for electromagnetic waves. These equations are introduced in Section 2.2.1. The concept of fiber modes is discussed in Section 2.2.2, where the fiber is shown to support a finite number of guided modes. Section 2.2.3 focuses on how a step-index fiber can be designed to support only a single mode and discusses the properties of single-mode fibers.

2.2.1 Maxwell's Equations

Like all electromagnetic phenomena, propagation of optical fields in fibers is governed by *Maxwell's equations*. For a nonconducting medium without free charges, these equations take the form [30] (in SI units; see Appendix A)

$$\nabla \times \mathbf{E} = -\partial \mathbf{B}/\partial t, \tag{2.2.1}$$

$$\nabla \times \mathbf{H} = \partial \mathbf{D}/\partial t, \tag{2.2.2}$$

$$\nabla \cdot \mathbf{D} = 0, \tag{2.2.3}$$

$$\nabla \cdot \mathbf{B} = 0, \tag{2.2.4}$$

where $\mathbf{E}$ and $\mathbf{H}$ are the electric and magnetic field vectors, respectively, and $\mathbf{D}$ and $\mathbf{B}$ are the corresponding flux densities. The flux densities are related to the field vectors by the constitutive relations [30]

$$\mathbf{D} = \varepsilon_0 \mathbf{E} + \mathbf{P}, \tag{2.2.5}$$

$$\mathbf{B} = \mu_0 \mathbf{H} + \mathbf{M}, \tag{2.2.6}$$

where ε_0 is the vacuum permittivity, μ_0 is the vacuum permeability, and $\mathbf{P}$ and $\mathbf{M}$ are the induced electric and magnetic polarizations, respectively. For optical fibers $\mathbf{M} = 0$ because of the nonmagnetic nature of silica glass.

Evaluation of the electric polarization $\mathbf{P}$ requires a microscopic quantum-mechanical approach. Although such an approach is essential when the optical frequency is near a medium resonance, a phenomenological relation between $\mathbf{P}$ and $\mathbf{E}$ can be used far from medium resonances. This is the case for optical fibers in the wavelength region 0.5–2 μm, a range that covers the low-loss region of optical fibers that is of interest for fiber-optic communication systems. In general, the relation between $\mathbf{P}$ and $\mathbf{E}$ can be nonlinear. Although the nonlinear effects in optical fibers are of considerable interest [31] and are covered in Section 2.6, they can be ignored in a discussion of fiber modes. $\mathbf{P}$ is then related to $\mathbf{E}$ by the relation

$$\mathbf{P}(\mathbf{r},t) = \varepsilon_0 \int_{-\infty}^{\infty} \chi(\mathbf{r}, t - t') \mathbf{E}(\mathbf{r}, t') \, dt'. \tag{2.2.7}$$

Linear susceptibility χ is, in general, a second-rank tensor but reduces to a scalar for an isotropic medium such as silica glass. Optical fibers become slightly birefringent because of unintentional variations in the core shape or in local strain; such birefringent effects are considered in Section 2.2.3. Equation (2.2.7) assumes a spatially local response. However, it includes the delayed nature of the temporal response, a feature that has important implications for optical fiber communications through chromatic dispersion.

Equations (2.2.1)–(2.2.7) provide a general formalism for studying wave propagation in optical fibers. In practice, it is convenient to use a single field variable $\mathbf{E}$. By taking the curl of Eq. (2.2.1) and using Eqs. (2.2.2), (2.2.5), and (2.2.6), we obtain the wave equation

$$\nabla \times \nabla \times \mathbf{E} = -\frac{1}{c^2} \frac{\partial^2 \mathbf{E}}{\partial t^2} - \mu_0 \frac{\partial^2 \mathbf{P}}{\partial t^2}, \tag{2.2.8}$$

where the speed of light in vacuum is defined as usual by $c = (\mu_0 \varepsilon_0)^{-1/2}$. By introducing the Fourier transform of $\mathbf{E}(\mathbf{r}, t)$ through the relation

$$\tilde{\mathbf{E}}(\mathbf{r}, \omega) = \int_{-\infty}^{\infty} \mathbf{E}(\mathbf{r}, t) \exp(i\omega t)\, dt, \tag{2.2.9}$$

as well as a similar relation for $\mathbf{P}(\mathbf{r}, t)$, and by using Eq. (2.2.7), Eq. (2.2.8) can be written in the frequency domain as

$$\nabla \times \nabla \times \tilde{\mathbf{E}} = -\varepsilon(\mathbf{r}, \omega)(\omega^2/c^2)\tilde{\mathbf{E}}, \tag{2.2.10}$$

where the frequency-dependent *dielectric constant* is defined as

$$\varepsilon(\mathbf{r}, \omega) = 1 + \tilde{\chi}(\mathbf{r}, \omega), \tag{2.2.11}$$

and $\tilde{\chi}(\mathbf{r}, \omega)$ is the Fourier transform of $\chi(\mathbf{r}, t)$. In general, $\varepsilon(\mathbf{r}, \omega)$ is complex. Its real and imaginary parts are related to the *refractive index n* and the *absorption coefficient α* by the definition

$$\varepsilon = (n + i\alpha c/2\omega)^2. \tag{2.2.12}$$

By using Eqs. (2.2.11) and (2.2.12), n and α are related to $\tilde{\chi}$ as

$$n = (1 + \mathrm{Re}\,\tilde{\chi})^{1/2}, \tag{2.2.13}$$
$$\alpha = (\omega/nc)\,\mathrm{Im}\,\tilde{\chi}, \tag{2.2.14}$$

where Re and Im stand for the real and imaginary parts, respectively. Both n and α are frequency dependent. The frequency dependence of n is referred to as *chromatic dispersion* or simply as material dispersion. In Section 2.3, fiber dispersion is shown to limit the performance of fiber-optic communication systems in a fundamental way.

Two further simplifications can be made before solving Eq. (2.2.10). First, ε can be taken to be real and replaced by n^2 because of low optical losses in silica fibers. Second, since $n(\mathbf{r}, \omega)$ is independent of the spatial coordinate $\mathbf{r}$ in both the core and the cladding of a step-index fiber, one can use the identity

$$\nabla \times \nabla \times \tilde{\mathbf{E}} \equiv \nabla(\nabla \cdot \tilde{\mathbf{E}}) - \nabla^2 \tilde{\mathbf{E}} = -\nabla^2 \tilde{\mathbf{E}}, \tag{2.2.15}$$

where we used Eq. (2.2.3) and the relation $\tilde{\mathbf{D}} = \varepsilon \tilde{\mathbf{E}}$ to set $\nabla \cdot \tilde{\mathbf{E}} = 0$. This simplification is made even for graded-index fibers. Equation (2.2.15) then holds approximately as long as the index changes occur over a length scale much longer than the wavelength. By using Eq. (2.2.15) in Eq. (2.2.10), we obtain

$$\nabla^2 \tilde{\mathbf{E}} + n^2(\omega)k_0^2 \tilde{\mathbf{E}} = 0, \tag{2.2.16}$$

where the free-space wave number k_0 is defined as

$$k_0 = \omega/c = 2\pi/\lambda, \tag{2.2.17}$$

and λ is the vacuum wavelength of the optical field oscillating at the frequency ω. Equation (2.2.16) is solved next to obtain the optical modes of step-index fibers.

2.2.2 Fiber Modes

The concept of the mode is a general concept in optics occurring also, for example, in the theory of lasers. An *optical mode* refers to a specific solution of the wave equation (2.2.16) that satisfies the appropriate boundary conditions and has the property that its spatial distribution does not change with propagation. The fiber modes can be classified as guided modes, leaky modes, and radiation modes [14]. As one might expect, signal transmission in fiber-optic communication systems takes place through the guided modes only. The following discussion focuses exclusively on the guided modes of a step-index fiber.

To take advantage of the cylindrical symmetry, Eq. (2.2.16) is written in the cylindrical coordinates ρ, ϕ, and z as

$$\frac{\partial^2 E_z}{\partial \rho^2} + \frac{1}{\rho}\frac{\partial E_z}{\partial \rho} + \frac{1}{\rho^2}\frac{\partial^2 E_z}{\partial \phi^2} + \frac{\partial^2 E_z}{\partial z^2} + n^2 k_0^2 E_z = 0, \tag{2.2.18}$$

where for a step-index fiber of core radius a, the refractive index n is of the form

$$n = \begin{cases} n_1; & \rho \le a, \\ n_2; & \rho > a. \end{cases} \tag{2.2.19}$$

For simplicity of notation, the tilde over $\tilde{E}$ has been dropped and the frequency dependence of all variables is implicitly understood. Equation (2.2.18) is written for the axial component E_z of the electric field vector. Similar equations can be written for the other five components of $\mathbf{E}$ and $\mathbf{H}$. However, it is not necessary to solve all six equations since only two components out of six are independent. It is customary to choose E_z and H_z as the independent components and obtain E_ρ, E_ϕ, H_ρ, and H_ϕ in terms of them. Equation (2.2.18) is easily solved by using the method of separation of variables and writing E_z as

$$E_z(\rho,\phi,z) = F(\rho)\Phi(\phi)Z(z). \tag{2.2.20}$$

By using Eq. (2.2.20) in Eq. (2.2.18), we obtain the three ordinary differential equations:

$$d^2 Z/dz^2 + \beta^2 Z = 0, \tag{2.2.21}$$

$$d^2 \Phi/d\phi^2 + m^2 \Phi = 0, \tag{2.2.22}$$

$$\frac{d^2 F}{d\rho^2} + \frac{1}{\rho}\frac{dF}{d\rho} + \left(n^2 k_0^2 - \beta^2 - \frac{m^2}{\rho^2}\right) F = 0. \tag{2.2.23}$$

Equation (2.2.21) has a solution of the form $Z = \exp(i\beta z)$, where β has the physical significance of the propagation constant. Similarly, Eq. (2.2.22) has a solution $\Phi = \exp(im\phi)$, but the constant m is restricted to take only integer values since the field must be periodic in ϕ with a period of 2π.

Equation (2.2.23) is the well-known differential equation satisfied by the Bessel functions [32]. Its general solution in the core and cladding regions can be written as

$$F(\rho) = \begin{cases} AJ_m(p\rho) + A'Y_m(p\rho); & \rho \le a, \\ CK_m(q\rho) + C'I_m(q\rho); & \rho > a, \end{cases} \tag{2.2.24}$$

where A, A', C, and C' are constants and J_m, Y_m, K_m, and I_m are different kinds of Bessel functions [32]. The parameters p and q are defined by

$$p^2 = n_1^2 k_0^2 - \beta^2, \tag{2.2.25}$$
$$q^2 = \beta^2 - n_2^2 k_0^2. \tag{2.2.26}$$

Considerable simplification occurs when we use the boundary condition that the optical field for a guided mode should be finite at $\rho = 0$ and decay to zero at $\rho = \infty$. Since $Y_m(p\rho)$ has a singularity at $\rho = 0$, $F(0)$ can remain finite only if $A' = 0$. Similarly $F(\rho)$ vanishes at infinity only if $C' = 0$. The general solution of Eq. (2.2.18) is thus of the form

$$E_z = \begin{cases} A J_m(p\rho) \exp(im\phi) \exp(i\beta z); & \rho \le a, \\ C K_m(q\rho) \exp(im\phi) \exp(i\beta z); & \rho > a. \end{cases} \tag{2.2.27}$$

The same method can be used to obtain H_z which also satisfies Eq. (2.2.18). Indeed, the solution is the same but with different constants B and D, that is,

$$H_z = \begin{cases} B J_m(p\rho) \exp(im\phi) \exp(i\beta z); & \rho \le a, \\ D K_m(q\rho) \exp(im\phi) \exp(i\beta z); & \rho > a. \end{cases} \tag{2.2.28}$$

The other four components E_ρ, E_ϕ, H_ρ, and H_ϕ can be expressed in terms of E_z and H_z by using Maxwell's equations. In the core region, we obtain

$$E_\rho = \frac{i}{p^2} \left(\beta \frac{\partial E_z}{\partial \rho} + \mu_0 \frac{\omega}{\rho} \frac{\partial H_z}{\partial \phi} \right), \tag{2.2.29}$$

$$E_\phi = \frac{i}{p^2} \left(\frac{\beta}{\rho} \frac{\partial E_z}{\partial \phi} - \mu_0 \omega \frac{\partial H_z}{\partial \rho} \right), \tag{2.2.30}$$

$$H_\rho = \frac{i}{p^2} \left(\beta \frac{\partial H_z}{\partial \rho} - \varepsilon_0 n^2 \frac{\omega}{\rho} \frac{\partial E_z}{\partial \phi} \right), \tag{2.2.31}$$

$$H_\phi = \frac{i}{p^2} \left(\frac{\beta}{\rho} \frac{\partial H_z}{\partial \phi} + \varepsilon_0 n^2 \omega \frac{\partial E_z}{\partial \rho} \right). \tag{2.2.32}$$

These equations can be used in the cladding region after replacing p^2 by $-q^2$.

Equations (2.2.27)–(2.2.32) express the electromagnetic field in the core and cladding regions of an optical fiber in terms of four constants A, B, C, and D. These constants are determined by applying the boundary condition that the tangential components of $\mathbf{E}$ and $\mathbf{H}$ be continuous across the core–cladding interface. By requiring the continuity of E_z, H_z, E_ϕ, and H_ϕ at $\rho = a$, we obtain a set of four homogeneous equations satisfied by A, B, C, and D [19]. These equations have a nontrivial solution only if the determinant of the coefficient matrix vanishes. After considerable algebraic details, this condition leads us to the following eigenvalue equation [19]–[21]:

$$\left[\frac{J_m'(pa)}{pJ_m(pa)} + \frac{K_m'(qa)}{qK_m(qa)} \right] \left[\frac{J_m'(pa)}{pJ_m(pa)} + \frac{n_2^2}{n_1^2} \frac{K_m'(qa)}{qK_m(qa)} \right]$$
$$= \frac{m^2}{a^2} \left(\frac{1}{p^2} + \frac{1}{q^2} \right) \left(\frac{1}{p^2} + \frac{n_2^2}{n_1^2} \frac{1}{q^2} \right), \tag{2.2.33}$$

where a prime indicates differentiation with respect to the argument.

For a given set of the parameters k_0, a, n_1, and n_2, the eigenvalue equation (2.2.33) can be solved numerically to determine the propagation constant β. In general, it may have multiple solutions for each integer value of m. It is customary to enumerate these solutions in descending numerical order and denote them by β_{mn} for a given m ($n = 1, 2, \ldots.$). Each value β_{mn} corresponds to one possible mode of propagation of the optical field whose spatial distribution is obtained from Eqs. (2.2.27)–(2.2.32). Since the field distribution does not change with propagation except for a phase factor and satisfies all boundary conditions, it is an optical mode of the fiber. In general, both E_z and H_z are nonzero (except for $m = 0$), in contrast with the planar waveguides, for which one of them can be taken to be zero. Fiber modes are therefore referred to as *hybrid modes* and are denoted by HE_{mn} or EH_{mn}, depending on whether H_z or E_z dominates. In the special case $m = 0$, HE_{0n} and EH_{0n} are also denoted by TE_{0n} and TM_{0n}, respectively, since they correspond to transverse-electric ($E_z = 0$) and transverse-magnetic ($H_z = 0$) modes of propagation. A different notation LP_{mn} is sometimes used for weakly guiding fibers [33] for which both E_z and H_z are nearly zero (LP stands for linearly polarized modes).

A mode is uniquely determined by its propagation constant β. It is useful to introduce a quantity $\bar{n} = \beta/k_0$, called the *mode index* or *effective index* and having the physical significance that each fiber mode propagates with an effective refractive index $\bar{n}$ whose value lies in the range $n_1 > \bar{n} > n_2$. A mode ceases to be guided when $\bar{n} \leq n_2$. This can be understood by noting that the optical field of guided modes decays exponentially inside the cladding layer since [32]

$$K_m(q\rho) = (\pi/2q\rho)^{1/2} \exp(-q\rho) \quad \text{for} \quad q\rho \gg 1. \tag{2.2.34}$$

When $\bar{n} \leq n_2$, $q^2 \leq 0$ from Eq. (2.2.26) and the exponential decay does not occur. The mode is said to reach cutoff when q becomes zero or when $\bar{n} = n_2$. From Eq. (2.2.25), $p = k_0(n_1^2 - n_2^2)^{1/2}$ when $q = 0$. A parameter that plays an important role in determining the *cutoff condition* is defined as

$$V = k_0 a (n_1^2 - n_2^2)^{1/2} \approx (2\pi/\lambda) a n_1 \sqrt{2\Delta}. \; = \; \frac{2\pi}{\lambda} a (NA) \tag{2.2.35}$$

It is called the *normalized frequency* ($V \propto \omega$) or simply the V parameter. It is also useful to introduce a normalized propagation constant b as

$$b = \frac{\beta/k_0 - n_2}{n_1 - n_2} = \frac{\bar{n} - n_2}{n_1 - n_2}. \tag{2.2.36}$$

Figure 2.5 shows a plot of b as a function of V for a few low-order fiber modes obtained by solving the eigenvalue equation (2.2.33). A fiber with a large value of V supports many modes. A rough estimate of the number of modes for such a multimode fiber is given by $V^2/2$ [23]. For example, a typical multimode fiber with $a = 25$ μm and $\Delta = 5 \times 10^{-3}$ has $V \simeq 18$ at $\lambda = 1.3$ μm and would support about 162 modes. However, the number of modes decreases rapidly as V is reduced. As seen in Fig. 2.5, a fiber with $V = 5$ supports seven modes. Below a certain value of V all modes except the HE_{11} mode reach cutoff. Such fibers support a single mode and are called single-mode fibers. The properties of single-mode fibers are described next.

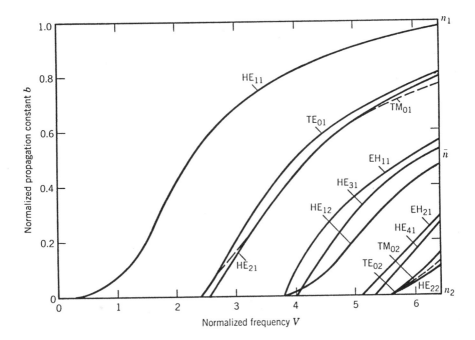

Figure 2.5: Normalized propagation constant b as a function of normalized frequency V for a few low-order fiber modes. The right scale shows the mode index $\bar{n}$. (After Ref. [34]; ©1981 Academic Press; reprinted with permission.)

2.2.3 Single-Mode Fibers

Single-mode fibers support only the HE_{11} mode, also known as the fundamental mode of the fiber. The fiber is designed such that all higher-order modes are cut off at the operating wavelength. As seen in Fig. 2.5, the V parameter determines the number of modes supported by a fiber. The cutoff condition of various modes is also determined by V. The fundamental mode has no cutoff and is always supported by a fiber.

Single-Mode Condition

The *single-mode condition* is determined by the value of V at which the TE_{01} and TM_{01} modes reach cutoff (see Fig. 2.5). The eigenvalue equations for these two modes can be obtained by setting $m = 0$ in Eq. (2.2.33) and are given by

$$pJ_0(pa)K_0'(qa) + qJ_0'(pa)K_0(qa) = 0, \qquad (2.2.37)$$
$$pn_2^2 J_0(pa)K_0'(qa) + qn_1^2 J_0'(pa)K_0(qa) = 0. \qquad (2.2.38)$$

A mode reaches cutoff when $q = 0$. Since $pa = V$ when $q = 0$, the cutoff condition for both modes is simply given by $J_0(V) = 0$. The smallest value of V for which $J_0(V) = 0$ is 2.405. A fiber designed such that $V < 2.405$ supports only the fundamental HE_{11} mode. This is the single-mode condition.

We can use Eq. (2.2.35) to estimate the core radius of single-mode fibers used in lightwave systems. For the operating wavelength range 1.3–1.6 μm, the fiber is generally designed to become single mode for $\lambda > 1.2$ μm. By taking $\lambda = 1.2$ μm, $n_1 = 1.45$, and $\Delta = 5 \times 10^{-3}$, Eq. (2.2.35) shows that $V < 2.405$ for a core radius $a < 3.2$ μm. The required core radius can be increased to about 4 μm by decreasing Δ to 3×10^{-3}. Indeed, most telecommunication fibers are designed with $a \approx 4$ μm.

The mode index $\bar{n}$ at the operating wavelength can be obtained by using Eq. (2.2.36), according to which

$$\bar{n} = n_2 + b(n_1 - n_2) \approx n_2(1 + b\Delta) \tag{2.2.39}$$

and by using Fig. 2.5, which provides b as a function of V for the HE$_{11}$ mode. An analytic approximation for b is [15]

$$b(V) \approx (1.1428 - 0.9960/V)^2 \tag{2.2.40}$$

and is accurate to within 0.2% for V in the range 1.5–2.5.

The field distribution of the fundamental mode is obtained by using Eqs. (2.2.27)–(2.2.32). The axial components E_z and H_z are quite small for $\Delta \ll 1$. Hence, the HE$_{11}$ mode is approximately linearly polarized for weakly guiding fibers. It is also denoted as LP$_{01}$, following an alternative terminology in which all fiber modes are assumed to be linearly polarized [33]. One of the transverse components can be taken as zero for a linearly polarized mode. If we set $E_y = 0$, the E_x component of the electric field for the HE$_{11}$ mode is given by [15]

$$E_x = E_0 \begin{cases} [J_0(p\rho)/J_0(pa)]\exp(i\beta z) ; & \rho \le a, \\ [K_0(q\rho)/K_0(qa)]\exp(i\beta z); & \rho > a, \end{cases} \tag{2.2.41}$$

where E_0 is a constant related to the power carried by the mode. The dominant component of the corresponding magnetic field is given by $H_y = n_2(\varepsilon_0/\mu_0)^{1/2}E_x$. This mode is linearly polarized along the x axis. The same fiber supports another mode linearly polarized along the y axis. In this sense a single-mode fiber actually supports two orthogonally polarized modes that are degenerate and have the same mode index.

Fiber Birefringence

The degenerate nature of the orthogonally polarized modes holds only for an ideal single-mode fiber with a perfectly cylindrical core of uniform diameter. Real fibers exhibit considerable variation in the shape of their core along the fiber length. They may also experience nonuniform stress such that the cylindrical symmetry of the fiber is broken. Degeneracy between the orthogonally polarized fiber modes is removed because of these factors, and the fiber acquires birefringence. The degree of modal birefringence is defined by

$$B_m = |\bar{n}_x - \bar{n}_y|, \tag{2.2.42}$$

where $\bar{n}_x$ and $\bar{n}_y$ are the mode indices for the orthogonally polarized fiber modes. Birefringence leads to a periodic power exchange between the two polarization components. The period, referred to as the *beat length*, is given by

$$L_B = \lambda/B_m. \tag{2.2.43}$$

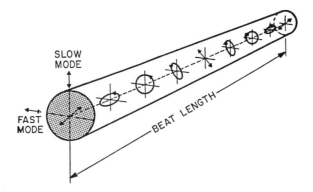

Figure 2.6: State of polarization in a birefringent fiber over one beat length. Input beam is linearly polarized at 45° with respect to the slow and fast axes.

Typically, $B_m \sim 10^{-7}$, and $L_B \sim 10$ m for $\lambda \sim 1\ \mu$m. From a physical viewpoint, linearly polarized light remains linearly polarized only when it is polarized along one of the principal axes. Otherwise, its state of polarization changes along the fiber length from linear to elliptical, and then back to linear, in a periodic manner over the length L_B. Figure 2.6 shows schematically such a periodic change in the state of polarization for a fiber of constant birefringence B. The *fast axis* in this figure corresponds to the axis along which the mode index is smaller. The other axis is called the *slow axis*.

In conventional single-mode fibers, birefringence is not constant along the fiber but changes randomly, both in magnitude and direction, because of variations in the core shape (elliptical rather than circular) and the anisotropic stress acting on the core. As a result, light launched into the fiber with linear polarization quickly reaches a state of arbitrary polarization. Moreover, different frequency components of a pulse acquire different polarization states, resulting in pulse broadening. This phenomenon is called *polarization-mode dispersion* (PMD) and becomes a limiting factor for optical communication systems operating at high bit rates. It is possible to make fibers for which random fluctuations in the core shape and size are not the governing factor in determining the state of polarization. Such fibers are called *polarization-maintaining* fibers. A large amount of birefringence is introduced intentionally in these fibers through design modifications so that small random birefringence fluctuations do not affect the light polarization significantly. Typically, $B_m \sim 10^{-4}$ for such fibers.

Spot Size

Since the field distribution given by Eq. (2.2.41) is cumbersome to use in practice, it is often approximated by a *Gaussian distribution* of the form

$$E_x = A \exp(-\rho^2/w^2) \exp(i\beta z), \qquad (2.2.44)$$

where w is the *field radius* and is referred to as the *spot size*. It is determined by fitting the exact distribution to the Gaussian function or by following a variational procedure [35]. Figure 2.7 shows the dependence of w/a on the V parameter. A comparison

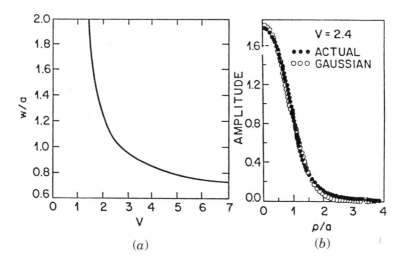

Figure 2.7: (a) Normalized spot size w/a as a function of the V parameter obtained by fitting the fundamental fiber mode to a Gaussian distribution; (b) quality of fit for $V = 2.4$. (After Ref. [35]; ©1978 OSA; reprinted with permission.)

of the actual field distribution with the fitted Gaussian is also shown for $V = 2.4$. The quality of fit is generally quite good for values of V in the neighborhood of 2. The spot size w can be determined from Fig. 2.7. It can also be determined from an analytic approximation accurate to within 1% for $1.2 < V < 2.4$ and given by [35]

$$w/a \approx 0.65 + 1.619V^{-3/2} + 2.879V^{-6}. \tag{2.2.45}$$

The effective core area, defined as $A_{\text{eff}} = \pi w^2$, is an important parameter for optical fibers as it determines how tightly light is confined to the core. It will be seen later that the nonlinear effects are stronger in fibers with smaller values of A_{eff}.

The fraction of the power contained in the core can be obtained by using Eq. (2.2.44) and is given by the *confinement factor*

$$\Gamma = \frac{P_{\text{core}}}{P_{\text{total}}} = \frac{\int_0^a |E_x|^2 \rho \, d\rho}{\int_0^\infty |E_x|^2 \rho \, d\rho} = 1 - \exp\left(-\frac{2a^2}{w^2}\right). \tag{2.2.46}$$

Equations (2.2.45) and (2.2.46) determine the fraction of the mode power contained inside the core for a given value of V. Although nearly 75% of the mode power resides in the core for $V = 2$, this percentage drops down to 20% for $V = 1$. For this reason most telecommunication single-mode fibers are designed to operate in the range $2 < V < 2.4$.

2.3 Dispersion in Single-Mode Fibers

It was seen in Section 2.1 that intermodal dispersion in multimode fibers leads to considerable broadening of short optical pulses (~ 10 ns/km). In the geometrical-optics description, such broadening was attributed to different paths followed by different rays. In the modal description it is related to the different mode indices (or group velocities) associated with different modes. The main advantage of single-mode fibers is that intermodal dispersion is absent simply because the energy of the injected pulse is transported by a single mode. However, pulse broadening does not disappear altogether. The group velocity associated with the fundamental mode is frequency dependent because of chromatic dispersion. As a result, different spectral components of the pulse travel at slightly different group velocities, a phenomenon referred to as *group-velocity dispersion* (GVD), *intramodal dispersion*, or simply *fiber dispersion*. Intramodal dispersion has two contributions, material dispersion and waveguide dispersion. We consider both of them and discuss how GVD limits the performance of lightwave systems employing single-mode fibers.

2.3.1 Group-Velocity Dispersion

Consider a single-mode fiber of length L. A specific spectral component at the frequency ω would arrive at the output end of the fiber after a time delay $T = L/v_g$, where v_g is the *group velocity*, defined as [22]

$$v_g = (d\beta/d\omega)^{-1}. \qquad (2.3.1)$$

By using $\beta = \bar{n}k_0 = \bar{n}\omega/c$ in Eq. (2.3.1), one can show that $v_g = c/\bar{n}_g$, where $\bar{n}_g$ is the *group index* given by

$$\bar{n}_g = \bar{n} + \omega(d\bar{n}/d\omega). \qquad (2.3.2)$$

The frequency dependence of the group velocity leads to pulse broadening simply because different spectral components of the pulse disperse during propagation and do not arrive simultaneously at the fiber output. If $\Delta\omega$ is the spectral width of the pulse, the extent of pulse broadening for a fiber of length L is governed by

$$\Delta T = \frac{dT}{d\omega}\Delta\omega = \frac{d}{d\omega}\left(\frac{L}{v_g}\right)\Delta\omega = L\frac{d^2\beta}{d\omega^2}\Delta\omega = L\beta_2\Delta\omega, \qquad (2.3.3)$$

where Eq. (2.3.1) was used. The parameter $\beta_2 = d^2\beta/d\omega^2$ is known as the GVD parameter. It determines how much an optical pulse would broaden on propagation inside the fiber.

In some optical communication systems, the frequency spread $\Delta\omega$ is determined by the range of wavelengths $\Delta\lambda$ emitted by the optical source. It is customary to use $\Delta\lambda$ in place of $\Delta\omega$. By using $\omega = 2\pi c/\lambda$ and $\Delta\omega = (-2\pi c/\lambda^2)\Delta\lambda$, Eq. (2.3.3) can be written as

$$\Delta T = \frac{d}{d\lambda}\left(\frac{L}{v_g}\right)\Delta\lambda = DL\Delta\lambda, \qquad (2.3.4)$$

↳ spectral width

where

$$D = \frac{d}{d\lambda}\left(\frac{1}{v_g}\right) = -\frac{2\pi c}{\lambda^2}\beta_2. \qquad (2.3.5)$$

D is called the *dispersion parameter* and is expressed in units of ps/(km-nm).

The effect of dispersion on the bit rate B can be estimated by using the criterion $B\Delta T < 1$ in a manner similar to that used in Section 2.1. By using ΔT from Eq. (2.3.4) this condition becomes

$$BL|D|\Delta\lambda < 1. \tag{2.3.6}$$

Equation (2.3.6) provides an order-of-magnitude estimate of the BL product offered by single-mode fibers. The wavelength dependence of D is studied in the next two subsections. For standard silica fibers, D is relatively small in the wavelength region near 1.3 μm [$D \sim 1$ ps/(km-nm)]. For a semiconductor laser, the spectral width $\Delta\lambda$ is 2–4 nm even when the laser operates in several longitudinal modes. The BL product of such lightwave systems can exceed 100 (Gb/s)-km. Indeed, 1.3-μm telecommunication systems typically operate at a bit rate of 2 Gb/s with a repeater spacing of 40–50 km. The BL product of single-mode fibers can exceed 1 (Tb/s)-km when single-mode semiconductor lasers (see Section 3.3) are used to reduce $\Delta\lambda$ below 1 nm.

The dispersion parameter D can vary considerably when the operating wavelength is shifted from 1.3 μm. The wavelength dependence of D is governed by the frequency dependence of the mode index $\bar{n}$. From Eq. (2.3.5), D can be written as

$$D = -\frac{2\pi c}{\lambda^2}\frac{d}{d\omega}\left(\frac{1}{v_g}\right) = -\frac{2\pi}{\lambda^2}\left(2\frac{d\bar{n}}{d\omega} + \omega\frac{d^2\bar{n}}{d\omega^2}\right), \tag{2.3.7}$$

where Eq. (2.3.2) was used. If we substitute $\bar{n}$ from Eq. (2.2.39) and use Eq. (2.2.35), D can be written as the sum of two terms,

$$D = D_M + D_W, \tag{2.3.8}$$

where the *material dispersion* D_M and the *waveguide dispersion* D_W are given by

$$D_M = -\frac{2\pi}{\lambda^2}\frac{dn_{2g}}{d\omega} = \frac{1}{c}\frac{dn_{2g}}{d\lambda}, \tag{2.3.9}$$

$$D_W = -\frac{2\pi\Delta}{\lambda^2}\left[\frac{n_{2g}^2}{n_2\omega}\frac{Vd^2(Vb)}{dV^2} + \frac{dn_{2g}}{d\omega}\frac{d(Vb)}{dV}\right]. \tag{2.3.10}$$

Here n_{2g} is the group index of the cladding material and the parameters V and b are given by Eqs. (2.2.35) and (2.2.36), respectively. In obtaining Eqs. (2.3.8)–(2.3.10) the parameter Δ was assumed to be frequency independent. A third term known as differential material dispersion should be added to Eq. (2.3.8) when $d\Delta/d\omega \neq 0$. Its contribution is, however, negligible in practice.

$V = \frac{2\pi}{\lambda} a n_1 \sqrt{2\Delta}$ (2.2.35)

2.3.2 Material Dispersion

$b = \frac{\bar{n} - n_2}{n_1 - n_2}$ (2.2.36)

Material dispersion occurs because the refractive index of silica, the material used for fiber fabrication, changes with the optical frequency ω. On a fundamental level, the origin of material dispersion is related to the characteristic resonance frequencies at which the material absorbs the electromagnetic radiation. Far from the medium resonances, the refractive index $n(\omega)$ is well approximated by the *Sellmeier equation* [36]

$$n^2(\omega) = 1 + \sum_{j=1}^{M}\frac{B_j\omega_j^2}{\omega_j^2 - \omega^2}, \tag{2.3.11}$$

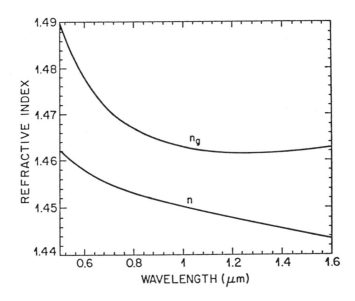

Figure 2.8: Variation of refractive index n and group index n_g with wavelength for fused silica.

where ω_j is the resonance frequency and B_j is the oscillator strength. Here n stands for n_1 or n_2, depending on whether the dispersive properties of the core or the cladding are considered. The sum in Eq. (2.3.11) extends over all material resonances that contribute in the frequency range of interest. In the case of optical fibers, the parameters B_j and ω_j are obtained empirically by fitting the measured dispersion curves to Eq. (2.3.11) with $M = 3$. They depend on the amount of dopants and have been tabulated for several kinds of fibers [12]. For pure silica these parameters are found to be $B_1 = 0.6961663$, $B_2 = 0.4079426$, $B_3 = 0.8974794$, $\lambda_1 = 0.0684043\,\mu m$, $\lambda_2 = 0.1162414\,\mu m$, and $\lambda_3 = 9.896161\,\mu m$, where $\lambda_j = 2\pi c/\omega_j$ with $j = 1$–3 [36]. The group index $n_g = n + \omega(dn/d\omega)$ can be obtained by using these parameter values.

Figure 2.8 shows the wavelength dependence of n and n_g in the range 0.5–1.6 μm for fused silica. Material dispersion D_M is related to the slope of n_g by the relation $D_M = c^{-1}(dn_g/d\lambda)$ [Eq. (2.3.9)]. It turns out that $dn_g/d\lambda = 0$ at $\lambda = 1.276\,\mu m$. This wavelength is referred to as the *zero-dispersion wavelength* λ_{ZD}, since $D_M = 0$ at $\lambda = \lambda_{ZD}$. The dispersion parameter D_M is negative below λ_{ZD} and becomes positive above that. In the wavelength range 1.25–1.66 μm it can be approximated by an empirical relation

$$D_M \approx 122(1 - \lambda_{ZD}/\lambda). \tag{2.3.12}$$

It should be stressed that $\lambda_{ZD} = 1.276\,\mu m$ only for pure silica. It can vary in the range 1.27–1.29 μm for optical fibers whose core and cladding are doped to vary the refractive index. The zero-dispersion wavelength of optical fibers also depends on the core radius a and the index step Δ through the waveguide contribution to the total dispersion.

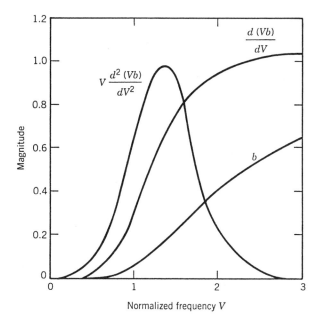

Figure 2.9: Variation of b and its derivatives $d(Vb)/dV$ and $V[d^2(Vb)/dV^2]$ with the V parameter. (After Ref. [33]; ©1971 OSA; reprinted with permission.)

2.3.3 Waveguide Dispersion

The contribution of waveguide dispersion D_W to the dispersion parameter D is given by Eq. (2.3.10) and depends on the V parameter of the fiber. Figure 2.9 shows how $d(Vb)/dV$ and $Vd^2(Vb)/dV^2$ change with V. Since both derivatives are positive, D_W is negative in the entire wavelength range 0–1.6 μm. On the other hand, D_M is negative for wavelengths below λ_{ZD} and becomes positive above that. Figure 2.10 shows D_M, D_W, and their sum $D = D_M + D_W$, for a typical single-mode fiber. The main effect of waveguide dispersion is to shift λ_{ZD} by an amount 30–40 nm so that the total dispersion is zero near 1.31 μm. It also reduces D from its material value D_M in the wavelength range 1.3–1.6 μm that is of interest for optical communication systems. Typical values of D are in the range 15–18 ps/(km-nm) near 1.55 μm. This wavelength region is of considerable interest for lightwave systems, since, as discussed in Section 2.5, the fiber loss is minimum near 1.55 μm. High values of D limit the performance of 1.55-μm lightwave systems.

Since the waveguide contribution D_W depends on fiber parameters such as the core radius a and the index difference Δ, it is possible to design the fiber such that λ_{ZD} is shifted into the vicinity of 1.55 μm [37], [38]. Such fibers are called *dispersion-shifted* fibers. It is also possible to tailor the waveguide contribution such that the total dispersion D is relatively small over a wide wavelength range extending from 1.3 to 1.6 μm [39]–[41]. Such fibers are called *dispersion-flattened* fibers. Figure 2.11 shows typical examples of the wavelength dependence of D for standard (conventional), dispersion-shifted, and dispersion-flattened fibers. The design of dispersion-

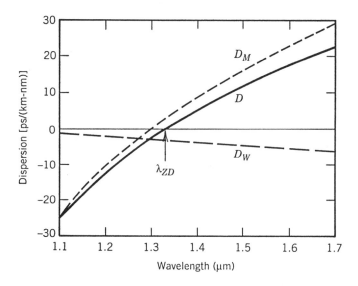

Figure 2.10: Total dispersion D and relative contributions of material dispersion D_M and waveguide dispersion D_W for a conventional single-mode fiber. The zero-dispersion wavelength shifts to a higher value because of the waveguide contribution.

modified fibers involves the use of multiple cladding layers and a tailoring of the refractive-index profile [37]–[43]. Waveguide dispersion can be used to produce *dispersion-decreasing* fibers in which GVD decreases along the fiber length because of axial variations in the core radius. In another kind of fibers, known as the *dispersion-compensating* fibers, GVD is made normal and has a relatively large magnitude. Table 2.1 lists the dispersion characteristics of several commercially available fibers.

2.3.4 Higher-Order Dispersion

It appears from Eq. (2.3.6) that the BL product of a single-mode fiber can be increased indefinitely by operating at the zero-dispersion wavelength λ_{ZD} where $D = 0$. The dispersive effects, however, do not disappear completely at $\lambda = \lambda_{ZD}$. Optical pulses still experience broadening because of higher-order dispersive effects. This feature can be understood by noting that D cannot be made zero at all wavelengths contained within the pulse spectrum centered at λ_{ZD}. Clearly, the wavelength dependence of D will play a role in pulse broadening. Higher-order dispersive effects are governed by the *dispersion slope* $S = dD/d\lambda$. The parameter S is also called a *differential-dispersion* parameter. By using Eq. (2.3.5) it can be written as

$$S = (2\pi c/\lambda^2)^2 \beta_3 + (4\pi c/\lambda^3)\beta_2, \qquad (2.3.13)$$

where $\beta_3 = d\beta_2/d\omega \equiv d^3\beta/d\omega^3$ is the third-order dispersion parameter. At $\lambda = \lambda_{ZD}$, $\beta_2 = 0$, and S is proportional to β_3.

The numerical value of the dispersion slope S plays an important role in the design of modern WDM systems. Since $S > 0$ for most fibers, different channels have slightly

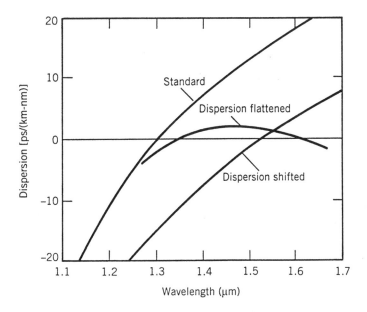

Figure 2.11: Typical wavelength dependence of the dispersion parameter D for standard, dispersion-shifted, and dispersion-flattened fibers.

different GVD values. This feature makes it difficult to compensate dispersion for all channels simultaneously. To solve this problem, new kind of fibers have been developed for which S is either small (reduced-slope fibers) or negative (reverse-dispersion fibers). Table 2.1 lists the values of dispersion slopes for several commercially available fibers.

It may appear from Eq. (2.3.6) that the limiting bit rate of a channel operating at $\lambda = \lambda_{ZD}$ will be infinitely large. However, this is not the case since S or β_3 becomes the limiting factor in that case. We can estimate the limiting bit rate by noting that for a source of spectral width $\Delta\lambda$, the effective value of dispersion parameter becomes $D = S\Delta\lambda$. The limiting bit rate–distance product can now be obtained by using Eq. (2.3.6) with this value of D. The resulting condition becomes

$$BL|S|(\Delta\lambda)^2 < 1. \tag{2.3.14}$$

For a multimode semiconductor laser with $\Delta\lambda = 2$ nm and a dispersion-shifted fiber with $S = 0.05$ ps/(km-nm^2) at $\lambda = 1.55$ μm, the BL product approaches 5 (Tb/s)-km. Further improvement is possible by using single-mode semiconductor lasers.

2.3.5 Polarization-Mode Dispersion

A potential source of pulse broadening is related to fiber birefringence. As discussed in Section 2.2.3, small departures from perfect cylindrical symmetry lead to birefringence because of different mode indices associated with the orthogonally polarized components of the fundamental fiber mode. If the input pulse excites both polarization components, it becomes broader as the two components disperse along the fiber

Table 2.1 Characteristics of several commercial fibers

Fiber Type and Trade Name	A_{eff} (μm^2)	λ_{ZD} (nm)	D (C band) [ps/(km-nm)]	Slope S [ps/(km-nm^2)]
Corning SMF-28	80	1302–1322	16 to 19	0.090
Lucent AllWave	80	1300–1322	17 to 20	0.088
Alcatel ColorLock	80	1300–1320	16 to 19	0.090
Corning Vascade	101	1300–1310	18 to 20	0.060
Lucent TrueWave-RS	50	1470–1490	2.6 to 6	0.050
Corning LEAF	72	1490–1500	2 to 6	0.060
Lucent TrueWave-XL	72	1570–1580	−1.4 to −4.6	0.112
Alcatel TeraLight	65	1440–1450	5.5 to 10	0.058

because of their different group velocities. This phenomenon is called the PMD and has been studied extensively because it limits the performance of modern lightwave systems [44]–[55].

In fibers with constant birefringence (e.g., polarization-maintaining fibers), pulse broadening can be estimated from the time delay ΔT between the two polarization components during propagation of the pulse. For a fiber of length L, ΔT is given by

$$\Delta T = \left| \frac{L}{v_{gx}} - \frac{L}{v_{gy}} \right| = L|\beta_{1x} - \beta_{1y}| = L(\Delta\beta_1), \qquad (2.3.15)$$

where the subscripts x and y identify the two orthogonally polarized modes and $\Delta\beta_1$ is related to the difference in group velocities along the two principal states of polarization [44]. Equation (2.3.1) was used to relate the group velocity v_g to the propagation constant β. Similar to the case of intermodal dispersion discussed in Section 2.1.1, the quantity $\Delta T/L$ is a measure of PMD. For polarization-maintaining fibers, $\Delta T/L$ is quite large (~ 1 ns/km) when the two components are equally excited at the fiber input but can be reduced to zero by launching light along one of the principal axes.

The situation is somewhat different for conventional fibers in which birefringence varies along the fiber in a random fashion. It is intuitively clear that the polarization state of light propagating in fibers with randomly varying birefringence will generally be elliptical and would change randomly along the fiber during propagation. In the case of optical pulses, the polarization state will also be different for different spectral components of the pulse. The final polarization state is not of concern for most lightwave systems as photodetectors used inside optical receivers are insensitive to the state of polarization unless a coherent detection scheme is employed. What affects such systems is not the random polarization state but pulse broadening induced by random changes in the birefringence. This is referred to as *PMD-induced* pulse broadening.

The analytical treatment of PMD is quite complex in general because of its statistical nature. A simple model divides the fiber into a large number of segments. Both the degree of birefringence and the orientation of the principal axes remain constant in each section but change randomly from section to section. In effect, each fiber section can be treated as a phase plate using a Jones matrix [44]. Propagation of each frequency component associated with an optical pulse through the entire fiber length is then governed by a composite Jones matrix obtained by multiplying individual Jones matrices for each fiber section. The composite Jones matrix shows that two principal states of polarization exist for any fiber such that, when a pulse is polarized along them, the polarization state at fiber output is frequency independent to first order, in spite of random changes in fiber birefringence. These states are analogous to the slow and fast axes associated with polarization-maintaining fibers. An optical pulse not polarized along these two principal states splits into two parts which travel at different speeds. The differential group delay ΔT is largest for the two principal states of polarization.

The principal states of polarization provide a convenient basis for calculating the moments of ΔT. The PMD-induced pulse broadening is characterized by the root-mean-square (RMS) value of ΔT, obtained after averaging over random birefringence changes. Several approaches have been used to calculate this average. The variance $\sigma_T^2 \equiv \langle (\Delta T)^2 \rangle$ turns out to be the same in all cases and is given by [46]

$$\sigma_T^2(z) = 2(\Delta\beta_1)^2 l_c^2 [\exp(-z/l_c) + z/l_c - 1], \qquad (2.3.16)$$

where l_c is the correlation length defined as the length over which two polarization components remain correlated; its value can vary over a wide range from 1 m to 1 km for different fibers, typical values being $\sim$ 10 m.

For short distances such that $z \ll l_c$, $\sigma_T = (\Delta\beta_1)z$ from Eq. (2.3.16), as expected for a polarization-maintaining fiber. For distances $z > 1$ km, a good estimate of pulse broadening is obtained using $z \gg l_c$. For a fiber of length L, σ_T in this approximation becomes

$$\sigma_T \approx (\Delta\beta_1)\sqrt{2l_c L} \equiv D_p\sqrt{L}, \qquad (2.3.17)$$

where D_p is the PMD parameter. Measured values of D_p vary from fiber to fiber in the range $D_p = 0.01$–10 ps/$\sqrt{\text{km}}$. Fibers installed during the 1980s have relatively large PMD such that $D_p > 0.1$ ps/$\sqrt{\text{km}}$. In contrast, modern fibers are designed to have low PMD, and typically $D_p < 0.1$ ps/$\sqrt{\text{km}}$ for them. Because of the $\sqrt{L}$ dependence, PMD-induced pulse broadening is relatively small compared with the GVD effects. Indeed, $\sigma_T \sim 1$ ps for fiber lengths $\sim$100 km and can be ignored for pulse widths >10 ps. However, PMD becomes a limiting factor for lightwave systems designed to operate over long distances at high bit rates [48]–[55]. Several schemes have been developed for compensating the PMD effects (see Section 7.9).

Several other factors need to be considered in practice. The derivation of Eq. (2.3.16) assumes that the fiber link has no elements exhibiting polarization-dependent loss or gain. The presence of polarization-dependent losses can induce additional broadening [50]. Also, the effects of second and higher-order PMD become important at high bit rates (40 Gb/s or more) or for systems in which the first-order effects are eliminated using a PMD compensator [54].

2.4 Dispersion-Induced Limitations *(concepts only)*

The discussion of pulse broadening in Section 2.3.1 is based on an intuitive phenomenological approach. It provides a first-order estimate for pulses whose spectral width is dominated by the spectrum of the optical source. In general, the extent of pulse broadening depends on the width and the shape of input pulses [56]. In this section we discuss pulse broadening by using the wave equation (2.2.16).

2.4.1 Basic Propagation Equation

The analysis of fiber modes in Section 2.2.2 showed that each frequency component of the optical field propagates in a single-mode fiber as

$$\tilde{\mathbf{E}}(\mathbf{r}, \omega) = \hat{\mathbf{x}} F(x, y) \tilde{B}(0, \omega) \exp(i\beta z), \tag{2.4.1}$$

where $\hat{\mathbf{x}}$ is the polarization unit vector, $\tilde{B}(0, \omega)$ is the initial amplitude, and β is the propagation constant. The field distribution $F(x,y)$ of the fundamental fiber mode can be approximated by the Gaussian distribution given in Eq. (2.2.44). In general, $F(x,y)$ also depends on ω, but this dependence can be ignored for pulses whose spectral width $\Delta\omega$ is much smaller than ω_0—a condition satisfied by pulses used in lightwave systems. Here ω_0 is the frequency at which the pulse spectrum is centered; it is referred to as the carrier frequency.

Different spectral components of an optical pulse propagate inside the fiber according to the simple relation

$$\tilde{B}(z, \omega) = \tilde{B}(0, \omega) \exp(i\beta z). \tag{2.4.2}$$

The amplitude in the time domain is obtained by taking the inverse Fourier transform and is given by

$$B(z,t) = \frac{1}{2\pi} \int_{-\infty}^{\infty} \tilde{B}(z, \omega) \exp(-i\omega t) \, d\omega. \tag{2.4.3}$$

The initial spectral amplitude $\tilde{B}(0, \omega)$ is just the Fourier transform of the input amplitude $B(0,t)$.

Pulse broadening results from the frequency dependence of β. For pulses for which $\Delta\omega \ll \omega_0$, it is useful to expand $\beta(\omega)$ in a Taylor series around the carrier frequency ω_0 and retain terms up to third order. In this quasi-monochromatic approximation,

$$\beta(\omega) = \bar{n}(\omega) \frac{\omega}{c} \approx \beta_0 + \beta_1(\Delta\omega) + \frac{\beta_2}{2}(\Delta\omega)^2 + \frac{\beta_3}{6}(\Delta\omega)^3, \tag{2.4.4}$$

where $\Delta\omega = \omega - \omega_0$ and $\beta_m = (d^m\beta/d\omega^m)_{\omega=\omega_0}$. From Eq. (2.3.1) $\beta_1 = 1/v_g$, where v_g is the group velocity. The GVD coefficient β_2 is related to the dispersion parameter D by Eq. (2.3.5), whereas β_3 is related to the dispersion slope S through Eq. (2.3.13). We substitute Eqs. (2.4.2) and (2.4.4) in Eq. (2.4.3) and introduce a *slowly varying amplitude* $A(z,t)$ of the pulse envelope as

$$B(z,t) = A(z,t) \exp[i(\beta_0 z - \omega_0 t)]. \tag{2.4.5}$$

The amplitude $A(z,t)$ is found to be given by

$$A(z,t) = \frac{1}{2\pi} \int_{-\infty}^{\infty} d(\Delta\omega) \tilde{A}(0, \Delta\omega)$$
$$\times \exp\left[i\beta_1 z\Delta\omega + \frac{i}{2}\beta_2 z(\Delta\omega)^2 + \frac{i}{6}\beta_3 z(\Delta\omega)^3 - i(\Delta\omega)t\right], \tag{2.4.6}$$

where $\tilde{A}(0,\Delta\omega) \equiv \tilde{B}(0,\omega)$ is the Fourier transform of $A(0,t)$.

By calculating $\partial A/\partial z$ and noting that $\Delta\omega$ is replaced by $i(\partial A/\partial t)$ in the time domain, Eq. (2.4.6) can be written as [31]

$$\frac{\partial A}{\partial z} + \beta_1 \frac{\partial A}{\partial t} + \frac{i\beta_2}{2} \frac{\partial^2 A}{\partial t^2} - \frac{\beta_3}{6} \frac{\partial^3 A}{\partial t^3} = 0. \tag{2.4.7}$$

This is the basic propagation equation that governs pulse evolution inside a single-mode fiber. In the absence of dispersion ($\beta_2 = \beta_3 = 0$), the optical pulse propagates without change in its shape such that $A(z,t) = A(0, t - \beta_1 z)$. Transforming to a reference frame moving with the pulse and introducing the new coordinates

$$t' = t - \beta_1 z \qquad \text{and} \qquad z' = z, \tag{2.4.8}$$

the β_1 term can be eliminated in Eq. (2.4.7) to yield

$$\frac{\partial A}{\partial z'} + \frac{i\beta_2}{2} \frac{\partial^2 A}{\partial t'^2} - \frac{\beta_3}{6} \frac{\partial^3 A}{\partial t'^3} = 0. \tag{2.4.9}$$

For simplicity of notation, we drop the primes over z' and t' in this and the following chapters whenever no confusion is likely to arise.

2.4.2 Chirped Gaussian Pulses

As a simple application of Eq. (2.4.9), let us consider the propagation of chirped Gaussian pulses inside optical fibers by choosing the initial field as

$$A(0,t) = A_0 \exp\left[-\frac{1+iC}{2}\left(\frac{t}{T_0}\right)^2\right], \tag{2.4.10}$$

where A_0 is the peak amplitude. The parameter T_0 represents the half-width at $1/e$ intensity point. It is related to the full-width at half-maximum (FWHM) of the pulse by the relation

$$T_{\text{FWHM}} = 2(\ln 2)^{1/2} T_0 \approx 1.665 T_0. \tag{2.4.11}$$

The parameter C governs the frequency chirp imposed on the pulse. A pulse is said to be chirped if its carrier frequency changes with time. The frequency change is related to the phase derivative and is given by

$$\delta\omega(t) = -\frac{\partial\phi}{\partial t} = \frac{C}{T_0^2} t, \tag{2.4.12}$$

where ϕ is the phase of $A(0,t)$. The time-dependent frequency shift $\delta\omega$ is called the *chirp*. The spectrum of a chirped pulse is broader than that of the unchirped pulse. This can be seen by taking the Fourier transform of Eq. (2.4.10) so that

$$\tilde{A}(0,\omega) = A_0 \left(\frac{2\pi T_0^2}{1+iC}\right)^{1/2} \exp\left[-\frac{\omega^2 T_0^2}{2(1+iC)}\right]. \tag{2.4.13}$$

The spectral half-width (at $1/e$ intensity point) is given by

$$\Delta\omega_0 = (1+C^2)^{1/2}T_0^{-1}. \tag{2.4.14}$$

In the absence of frequency chirp $(C = 0)$, the spectral width satisfies the relation $\Delta\omega_0 T_0 = 1$. Such a pulse has the narrowest spectrum and is called *transform-limited*. The spectral width is enhanced by a factor of $(1+C^2)^{1/2}$ in the presence of linear chirp, as seen in Eq. (2.4.14).

The pulse-propagation equation (2.4.9) can be easily solved in the Fourier domain. Its solution is [see Eq. (2.4.6)]

$$A(z,t) = \frac{1}{2\pi}\int_{-\infty}^{\infty}\tilde{A}(0,\omega)\exp\left(\frac{i}{2}\beta_2 z\omega^2 + \frac{i}{6}\beta_3 z\omega^3 - i\omega t\right)d\omega, \tag{2.4.15}$$

where $\tilde{A}(0,\omega)$ is given by Eq. (2.4.13) for the Gaussian input pulse. Let us first consider the case in which the carrier wavelength is far away from the zero-dispersion wavelength so that the contribution of the β_3 term is negligible. The integration in Eq. (2.4.15) can be performed analytically with the result

$$A(z,t) = \frac{A_0}{\sqrt{Q(z)}}\exp\left[-\frac{(1+iC)t^2}{2T_0^2 Q(z)}\right], \tag{2.4.16}$$

where $Q(z) = 1 + (C-i)\beta_2 z/T_0^2$. This equation shows that a Gaussian pulse remains Gaussian on propagation but its width, chirp, and amplitude change as dictated by the factor $Q(z)$. For example, the chirp at a distance z changes from its initial value C to become $C_1(z) = C + (1+C^2)\beta_2 z/T_0^2$.

Changes in the pulse width with z are quantified through the broadening factor given by

$$\frac{T_1}{T_0} = \left[\left(1+\frac{C\beta_2 z}{T_0^2}\right)^2 + \left(\frac{\beta_2 z}{T_0^2}\right)^2\right]^{1/2}, \tag{2.4.17}$$

where T_1 is the half-width defined similar to T_0. Figure 2.12 shows the broadening factor T_1/T_0 as a function of the propagation distance z/L_D, where $L_D = T_0^2/|\beta_2|$ is the *dispersion length*. An unchirped pulse $(C = 0)$ broadens as $[1 + (z/L_D)^2]^{1/2}$ and its width increases by a factor of $\sqrt{2}$ at $z = L_D$. The chirped pulse, on the other hand, may broaden or compress depending on whether β_2 and C have the same or opposite signs. For $\beta_2 C > 0$ the chirped Gaussian pulse broadens monotonically at a rate faster than the unchirped pulse. For $\beta_2 C < 0$, the pulse width initially decreases and becomes minimum at a distance

$$z_{min} = \left[|C|/(1+C^2)\right]L_D. \tag{2.4.18}$$

The minimum value depends on the chirp parameter as

$$T_1^{min} = T_0/(1+C^2)^{1/2}. \tag{2.4.19}$$

Physically, when $\beta_2 C < 0$, the GVD-induced chirp counteracts the initial chirp, and the effective chirp decreases until it vanishes at $z = z_{min}$.

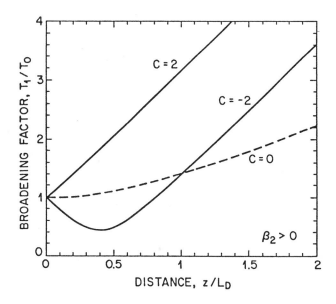

Figure 2.12: Variation of broadening factor with propagated distance for a chirped Gaussian input pulse. Dashed curve corresponds to the case of an unchirped Gaussian pulse. For $\beta_2 < 0$ the same curves are obtained if the sign of the chirp parameter C is reversed.

Equation (2.4.17) can be generalized to include higher-order dispersion governed by β_3 in Eq. (2.4.15). The integral can still be performed in closed form in terms of an Airy function [57]. However, the pulse no longer remains Gaussian on propagation and develops a tail with an oscillatory structure. Such pulses cannot be properly characterized by their FWHM. A proper measure of the pulse width is the RMS width of the pulse defined as

$$\sigma = \left[\langle t^2 \rangle - \langle t \rangle^2 \right]^{1/2}, \tag{2.4.20}$$

where the angle brackets denote averaging with respect to the intensity profile, i.e.,

$$\langle t^m \rangle = \frac{\int_{-\infty}^{\infty} t^m |A(z,t)|^2 \, dt}{\int_{-\infty}^{\infty} |A(z,t)|^2 \, dt}. \tag{2.4.21}$$

The broadening factor defined as σ/σ_0, where σ_0 is the RMS width of the input Gaussian pulse ($\sigma_0 = T_0/\sqrt{2}$) can be calculated following the analysis of Appendix C and is given by [56]

$$\frac{\sigma^2}{\sigma_0^2} = \left(1 + \frac{C\beta_2 L}{2\sigma_0^2} \right)^2 + \left(\frac{\beta_2 L}{2\sigma_0^2} \right)^2 + (1 + C^2)^2 \left(\frac{\beta_3 L}{4\sqrt{2}\sigma_0^3} \right)^2, \tag{2.4.22}$$

where L is the fiber length.

The foregoing discussion assumes that the optical source used to produce the input pulses is nearly monochromatic such that its spectral width satisfies $\Delta\omega_L \ll \Delta\omega_0$ (under continuous-wave, or CW, operation), where $\Delta\omega_0$ is given by Eq. (2.4.14). This condition is not always satisfied in practice. To account for the source spectral width, we must treat the optical field as a stochastic process and consider the coherence properties of the source through the mutual coherence function [22]. Appendix C shows how the broadening factor can be calculated in this case. When the source spectrum is Gaussian with the RMS spectral width σ_ω, the broadening factor is obtained from [56]

$$\frac{\sigma^2}{\sigma_0^2} = \left(1 + \frac{C\beta_2 L}{2\sigma_0^2}\right)^2 + (1 + V_\omega^2)\left(\frac{\beta_2 L}{2\sigma_0^2}\right)^2 + (1 + C^2 + V_\omega^2)^2 \left(\frac{\beta_3 L}{4\sqrt{2}\sigma_0^3}\right)^2, \quad (2.4.23)$$

where V_ω is defined as $V_\omega = 2\sigma_\omega\sigma_0$. Equation (2.4.23) provides an expression for dispersion-induced broadening of Gaussian input pulses under quite general conditions. We use it in the next section to find the limiting bit rate of optical communication systems.

2.4.3 Limitations on the Bit Rate

The limitation imposed on the bit rate by fiber dispersion can be quite different depending on the source spectral width. It is instructive to consider the following two cases separately.

Optical Sources with a Large Spectral Width

This case corresponds to $V_\omega \gg 1$ in Eq. (2.4.23). Consider first the case of a lightwave system operating away from the zero-dispersion wavelength so that the β_3 term can be neglected. The effects of frequency chirp are negligible for sources with a large spectral width. By setting $C = 0$ in Eq. (2.4.23), we obtain

$$\sigma^2 = \sigma_0^2 + (\beta_2 L\sigma_\omega)^2 \equiv \sigma_0^2 + (DL\sigma_\lambda)^2, \quad (2.4.24)$$

where σ_λ is the RMS source spectral width in wavelength units. The output pulse width is thus given by

$$\sigma = (\sigma_0^2 + \sigma_D^2)^{1/2}, \quad (2.4.25)$$

where $\sigma_D \equiv |D|L\sigma_\lambda$ provides a measure of dispersion-induced broadening.

We can relate σ to the bit rate by using the criterion that the broadened pulse should remain inside the allocated bit slot, $T_B = 1/B$, where B is the bit rate. A commonly used criterion is $\sigma \leq T_B/4$; for Gaussian pulses at least 95% of the pulse energy then remains within the bit slot. The limiting bit rate is given by $4B\sigma \leq 1$. In the limit $\sigma_D \gg \sigma_0$, $\sigma \approx \sigma_D = |D|L\sigma_\lambda$, and the condition becomes

$$BL|D|\sigma_\lambda \leq \tfrac{1}{4}. \quad (2.4.26)$$

This condition should be compared with Eq. (2.3.6) obtained heuristically; the two become identical if we interpret $\Delta\lambda$ as $4\sigma_\lambda$ in Eq. (2.3.6).

For a lightwave system operating exactly at the zero-dispersion wavelength, $\beta_2 = 0$ in Eq. (2.4.23). By setting $C = 0$ as before and assuming $V_\omega \gg 1$, Eq. (2.4.23) can be approximated by

$$\sigma^2 = \sigma_0^2 + \tfrac{1}{2}(\beta_3 L\sigma_\omega^2)^2 \equiv \sigma_0^2 + \tfrac{1}{2}(SL\sigma_\lambda^2), \quad (2.4.27)$$

where Eq. (2.3.13) was used to relate β_3 to the dispersion slope S. The output pulse width is thus given by Eq. (2.4.25) but now $\sigma_D \equiv |S|L\sigma_\lambda^2/\sqrt{2}$. As before, we can relate σ to the limiting bit rate by the condition $4B\sigma \leq 1$. When $\sigma_D \gg \sigma_0$, the limitation on the bit rate is governed by

$$BL|S|\sigma_\lambda^2 \leq 1/\sqrt{8}. \tag{2.4.28}$$

This condition should be compared with Eq. (2.3.14) obtained heuristically by using simple physical arguments.

As an example, consider the case of a light-emitting diode (see Section 3.2) for which $\sigma_\lambda \approx 15$ nm. By using $D = 17$ ps/(km-nm) at 1.55 μm, Eq. (2.4.26) yields $BL < 1$ (Gb/s)-km. However, if the system is designed to operate at the zero-dispersion wavelength, BL can be increased to 20 (Gb/s)-km for a typical value $S = 0.08$ ps/(km-nm^2).

Optical Sources with a Small Spectral Width

This case corresponds to $V_\omega \ll 1$ in Eq. (2.4.23). As before, if we neglect the β_3 term and set $C = 0$, Eq. (2.4.23) can be approximated by

$$\sigma^2 = \sigma_0^2 + (\beta_2 L/2\sigma_0)^2 \equiv \sigma_0^2 + \sigma_D^2. \tag{2.4.29}$$

A comparison with Eq. (2.4.25) reveals a major difference between the two cases. In the case of a narrow source spectrum, dispersion-induced broadening depends on the initial width σ_0, whereas it is independent of σ_0 when the spectral width of the optical source dominates. In fact, σ can be minimized by choosing an optimum value of σ_0. The minimum value of σ is found to occur for $\sigma_0 = \sigma_D = (|\beta_2|L/2)^{1/2}$ and is given by $\sigma = (|\beta_2|L)^{1/2}$. The limiting bit rate is obtained by using $4B\sigma \leq 1$ and leads to the condition

$$B\sqrt{|\beta_2|L} \leq \tfrac{1}{4}. \tag{2.4.30}$$

The main difference from Eq. (2.4.26) is that B scales as $L^{-1/2}$ rather than L^{-1}. Figure 2.13 compares the decrease in the bit rate with increasing for $\sigma_\lambda = 0$, 1, and 5 nm L using $D = 16$ ps/(km-nm). Equation (2.4.30) was used in the case $\sigma_\lambda = 0$.

For a lightwave system operating close to the zero-dispersion wavelength, $\beta_2 \approx 0$ in Eq. (2.4.23). Using $V_\omega \ll 1$ and $C = 0$, the pulse width is then given by

$$\sigma^2 = \sigma_0^2 + (\beta_3 L/4\sigma_0^2)^2/2 \equiv \sigma_0^2 + \sigma_D^2. \tag{2.4.31}$$

Similar to the case of Eq. (2.4.29), σ can be minimized by optimizing the input pulse width σ_0. The minimum value of σ occurs for $\sigma_0 = (|\beta_3|L/4)^{1/3}$ and is given by

$$\sigma = (\tfrac{3}{2})^{1/2}(|\beta_3|L/4)^{1/3}. \tag{2.4.32}$$

The limiting bit rate is obtained by using the condition $4B\sigma \leq 1$, or

$$B(|\beta_3|L)^{1/3} \leq 0.324. \tag{2.4.33}$$

The dispersive effects are most forgiving in this case. When $\beta_3 = 0.1$ ps^3/km, the bit rate can be as large as 150 Gb/s for $L = 100$ km. It decreases to only about 70 Gb/s

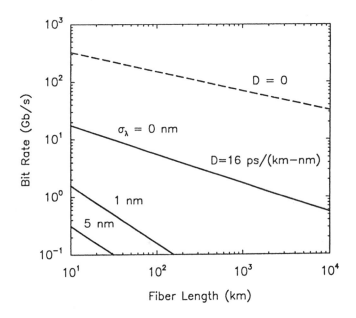

Figure 2.13: Limiting bit rate of single-mode fibers as a function of the fiber length for $\sigma_\lambda = 0$, 1, and 5 nm. The case $\sigma_\lambda = 0$ corresponds to the case of an optical source whose spectral width is much smaller than the bit rate.

even when L increases by a factor of 10 because of the $L^{-1/3}$ dependence of the bit rate on the fiber length. The dashed line in Fig. 2.13 shows this dependence by using Eq. (2.4.33) with $\beta_3 = 0.1$ ps^3/km. Clearly, the performance of a lightwave system can be improved considerably by operating it near the zero-dispersion wavelength of the fiber and using optical sources with a relatively narrow spectral width.

Effect of Frequency Chirp

The input pulse in all preceding cases has been assumed to be an unchirped Gaussian pulse. In practice, optical pulses are often non-Gaussian and may exhibit considerable chirp. A super-Gaussian model has been used to study the bit-rate limitation imposed by fiber dispersion for a NRZ-format bit stream [58]. In this model, Eq. (2.4.10) is replaced by

$$A(0,T) = A_0 \exp\left[-\frac{1+iC}{2}\left(\frac{t}{T_0}\right)^{2m}\right], \tag{2.4.34}$$

where the parameter m controls the pulse shape. Chirped Gaussian pulses correspond to $m = 1$. For large value of m the pulse becomes nearly rectangular with sharp leading and trailing edges. The output pulse shape can be obtained by solving Eq. (2.4.9) numerically. The limiting bit rate–distance product BL is found by requiring that the RMS pulse width does not increase above a tolerable value. Figure 2.14 shows the BL product as a function of the chirp parameter C for Gaussian ($m = 1$) and super-Gaussian ($m = 3$) input pulses. In both cases the fiber length L at which the pulse broadens

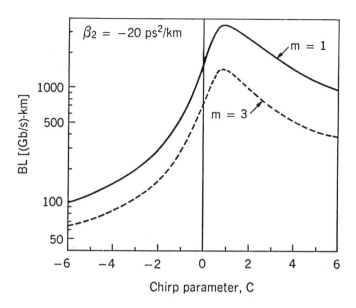

Figure 2.14: Dispersion-limited *BL* product as a function of the chirp parameter for Gaussian (solid curve) and super-Gaussian (dashed curve) input pulses. (After Ref. [58]; ©1986 OSA; reprinted with permission.)

by 20% was obtained for $T_0 = 125$ ps and $\beta_2 = -20$ ps^2/km. As expected, the *BL* product is smaller for super-Gaussian pulses because such pulses broaden more rapidly than Gaussian pulses. The *BL* product is reduced dramatically for negative values of the chirp parameter *C*. This is due to enhanced broadening occurring when $\beta_2 C$ is positive (see Fig. 2.12). Unfortunately, *C* is generally negative for directly modulated semiconductor lasers with a typical value of -6 at 1.55 μm. Since $BL < 100$ (Gb/s)-km under such conditions, fiber dispersion limits the bit rate to about 2 Gb/s for $L = 50$ km. This problem can be overcome by using dispersion-shifted fibers or by using dispersion management (see Chapter 7).

2.4.4 Fiber Bandwidth

The concept of fiber bandwidth originates from the general theory of time-invariant linear systems [59]. If the optical fiber can be treated as a *linear system*, its input and output powers should be related by a general relation

$$P_{\text{out}}(t) = \int_{-\infty}^{\infty} h(t - t') P_{\text{in}}(t') \, dt'. \qquad (2.4.35)$$

For an impulse $P_{\text{in}}(t) = \delta(t)$, where $\delta(t)$ is the delta function, and $P_{\text{out}}(t) = h(t)$. For this reason, $h(t)$ is called the *impulse response* of the linear system. Its Fourier transform,

$$H(f) = \int_{-\infty}^{\infty} h(t) \exp(2\pi i f t) \, dt, \qquad (2.4.36)$$

provides the frequency response and is called the *transfer function*. In general, $|H(f)|$ falls off with increasing f, indicating that the high-frequency components of the input signal are attenuated by the fiber. In effect, the optical fiber acts as a *bandpass filter*. The *fiber bandwidth* $f_{3\,\text{dB}}$ corresponds to the frequency $f = f_{3\,\text{dB}}$ at which $|H(f)|$ is reduced by a factor of 2 or by 3 dB:

$$|H(f_{3\,\text{dB}})/H(0)| = \tfrac{1}{2}. \tag{2.4.37}$$

Note that $f_{3\,\text{dB}}$ is the optical bandwidth of the fiber as the optical power drops by 3 dB at this frequency compared with the zero-frequency response. In the field of electrical communications, the bandwidth of a linear system is defined as the frequency at which electrical power drops by 3 dB.

Optical fibers cannot generally be treated as linear with respect to power, and Eq. (2.4.35) does not hold for them [60]. However, this equation is approximately valid when the source spectral width is much larger than the signal spectral width ($V_\omega \gg 1$). In that case, we can consider propagation of different spectral components independently and add the power carried by them linearly to obtain the output power. For a Gaussian spectrum, the transfer function $H(f)$ is found to be given by [61]

$$H(f) = \left(1 + \frac{if}{f_2}\right)^{-1/2} \exp\left[-\frac{(f/f_1)^2}{2(1 + if/f_2)}\right], \tag{2.4.38}$$

where the parameters f_1 and f_2 are given by

$$f_1 = (2\pi\beta_2 L\sigma_\omega)^{-1} = (2\pi|D|L\sigma_\lambda)^{-1}, \tag{2.4.39}$$

$$f_2 = (2\pi\beta_3 L\sigma_\omega^2)^{-1} = [2\pi(S + 2|D|/\lambda)L\sigma_\lambda^2]^{-1}, \tag{2.4.40}$$

and we used Eqs. (2.3.5) and (2.3.13) to introduce the dispersion parameters D and S.

For lightwave systems operating far away from the zero-dispersion wavelength ($f_1 \ll f_2$), the transfer function is approximately Gaussian. By using Eqs. (2.4.37) and (2.4.38) with $f \ll f_2$, the fiber bandwidth is given by

$$f_{3\,\text{dB}} = (2\ln 2)^{1/2} f_1 \approx 0.188(|D|L\sigma_\lambda)^{-1}. \tag{2.4.41}$$

If we use $\sigma_D = |D|L\sigma_\lambda$ from Eq. (2.4.25), we obtain the relation $f_{3\,\text{dB}}\sigma_D \approx 0.188$ between the fiber bandwidth and dispersion-induced pulse broadening. We can also get a relation between the bandwidth and the bit rate B by using Eqs. (2.4.26) and (2.4.41). The relation is $B \leq 1.33 f_{3\,\text{dB}}$ and shows that the fiber bandwidth is an approximate measure of the maximum possible bit rate of dispersion-limited lightwave systems. In fact, Fig. 2.13 can be used to estimate $f_{3\,\text{dB}}$ and its variation with the fiber length under different operating conditions.

For lightwave systems operating at the zero-dispersion wavelength, the transfer function is obtained from Eq. (2.4.38) by setting $D = 0$. The use of Eq. (2.4.37) then provides the following expression for the fiber bandwidth

$$f_{3\,\text{dB}} = \sqrt{15} f_2 \approx 0.616(SL\sigma_\lambda^2)^{-1}. \tag{2.4.42}$$

The limiting bit rate can be related to $f_{3\,dB}$ by using Eq. (2.4.28) and is given by $B \leq 0.574 f_{3\,dB}$. Again, the fiber bandwidth provides a measure of the dispersion-limited bit rate. As a numerical estimate, consider a 1.55-μm lightwave system employing dispersion-shifted fibers and multimode semiconductor lasers. By using $S = 0.05$ ps/(km-nm^2) and $\sigma_\lambda = 1$ nm as typical values, $f_{3\,dB}L \approx 32$ THz-km. By contrast, the bandwidth–distance product is reduced to 0.1 THz-km for standard fibers with $D = 18$ ps/(km-nm).

2.5 Fiber Losses

Section 2.4 shows that fiber dispersion limits the performance of optical communication systems by broadening optical pulses as they propagate inside the fiber. Fiber losses represent another limiting factor because they reduce the signal power reaching the receiver. As optical receivers need a certain minimum amount of power for recovering the signal accurately, the transmission distance is inherently limited by fiber losses. In fact, the use of silica fibers for optical communications became practical only when losses were reduced to an acceptable level during the 1970s. With the advent of optical amplifiers in the 1990s, transmission distances can exceed several thousands kilometers by compensating accumulated losses periodically. However, low-loss fibers are still required since spacing among amplifiers is set by fiber losses. This section is devoted to a discussion of various loss mechanisms in optical fibers.

2.5.1 Attenuation Coefficient

Under quite general conditions, changes in the average optical power P of a bit stream propagating inside an optical fiber are governed by Beer's law:

$$dP/dz = -\alpha P, \qquad (2.5.1)$$

where α is the attenuation coefficient. Although denoted by the same symbol as the absorption coefficient in Eq. (2.2.12), α in Eq. (2.5.1) includes not only material absorption but also other sources of power attenuation. If P_{in} is the power launched at the input end of a fiber of length L, the output power P_{out} from Eq. (2.5.1) is given by

$$P_{out} = P_{in} \exp(-\alpha L). \qquad (2.5.2)$$

It is customary to express α in units of dB/km by using the relation

$$\alpha\,(dB/km) = -\frac{10}{L} \log_{10}\left(\frac{P_{out}}{P_{in}}\right) \approx 4.343\alpha, \qquad (2.5.3)$$

and refer to it as the fiber-loss parameter.

Fiber losses depend on the wavelength of transmitted light. Figure 2.15 shows the loss spectrum $\alpha(\lambda)$ of a single-mode fiber made in 1979 with 9.4-μm core diameter, $\Delta = 1.9 \times 10^{-3}$, and 1.1-$\mu$m cutoff wavelength [11]. The fiber exhibited a loss of only about 0.2 dB/km in the wavelength region near 1.55 μm, the lowest value first realized in 1979. This value is close to the fundamental limit of about 0.16 dB/km for

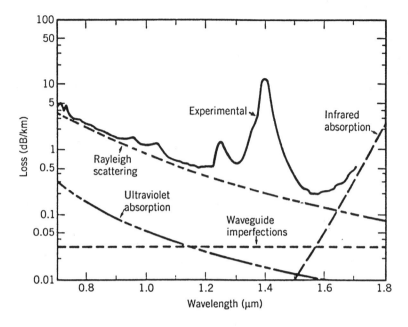

Figure 2.15: Loss spectrum of a single-mode fiber produced in 1979. Wavelength dependence of several fundamental loss mechanisms is also shown. (After Ref. [11]; ©1979 IEE; reprinted with permission.)

silica fibers. The loss spectrum exhibits a strong peak near 1.39 μm and several other smaller peaks. A secondary minimum is found to occur near 1.3 μm, where the fiber loss is below 0.5 dB/km. Since fiber dispersion is also minimum near 1.3 μm, this low-loss window was used for second-generation lightwave systems. Fiber losses are considerably higher for shorter wavelengths and exceed 5 dB/km in the visible region, making it unsuitable for long-haul transmission. Several factors contribute to overall losses; their relative contributions are also shown in Fig. 2.15. The two most important among them are material absorption and Rayleigh scattering.

2.5.2   Material Absorption

Material absorption can be divided into two categories. Intrinsic absorption losses correspond to absorption by fused silica (material used to make fibers) whereas extrinsic absorption is related to losses caused by impurities within silica. Any material absorbs at certain wavelengths corresponding to the electronic and vibrational resonances associated with specific molecules. For silica (SiO_2) molecules, electronic resonances occur in the ultraviolet region ($\lambda < 0.4$ μm), whereas vibrational resonances occur in the infrared region ($\lambda > 7$ μm). Because of the amorphous nature of fused silica, these resonances are in the form of absorption bands whose tails extend into the visible region. Figure 2.15 shows that intrinsic material absorption for silica in the wavelength range 0.8–1.6 μm is below 0.1 dB/km. In fact, it is less than 0.03 dB/km in the 1.3- to

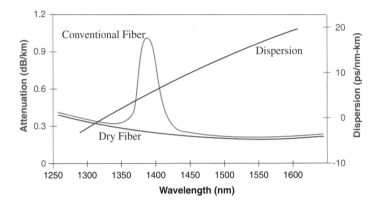

Figure 2.16: Loss and dispersion of the AllWave fiber. Loss of a conventional fiber is shown by the gray line for comparison. (Courtesy Lucent Technologies.)

1.6-μm wavelength window commonly used for lightwave systems.

Extrinsic absorption results from the presence of impurities. Transition-metal impurities such as Fe, Cu, Co, Ni, Mn, and Cr absorb strongly in the wavelength range 0.6–1.6 μm. Their amount should be reduced to below 1 part per billion to obtain a loss level below 1 dB/km. Such high-purity silica can be obtained by using modern techniques. The main source of extrinsic absorption in state-of-the-art silica fibers is the presence of water vapors. A vibrational resonance of the OH ion occurs near 2.73 μm. Its harmonic and combination tones with silica produce absorption at the 1.39-, 1.24-, and 0.95-μm wavelengths. The three spectral peaks seen in Fig. 2.15 occur near these wavelengths and are due to the presence of residual water vapor in silica. Even a concentration of 1 part per million can cause a loss of about 50 dB/km at 1.39 μm. The OH ion concentration is reduced to below 10^{-8} in modern fibers to lower the 1.39-μm peak below 1 dB. In a new kind of fiber, known as the *dry fiber*, the OH ion concentration is reduced to such low levels that the 1.39-μm peak almost disappears [62]. Figure 2.16 shows the loss and dispersion profiles of such a fiber (marketed under the trade name AllWave). Such fibers can be used to transmit WDM signals over the entire 1.30- to1.65-μm wavelength range.

2.5.3 Rayleigh Scattering

Rayleigh scattering is a fundamental loss mechanism arising from local microscopic fluctuations in density. Silica molecules move randomly in the molten state and freeze in place during fiber fabrication. Density fluctuations lead to random fluctuations of the refractive index on a scale smaller than the optical wavelength λ. Light scattering in such a medium is known as *Rayleigh scattering* [22]. The scattering cross section varies as λ^{-4}. As a result, the intrinsic loss of silica fibers from Rayleigh scattering can be written as

$$\alpha_R = C/\lambda^4, \tag{2.5.4}$$

where the constant C is in the range 0.7–0.9 (dB/km)-μm^4, depending on the constituents of the fiber core. These values of C correspond to $\alpha_R = 0.12$–0.16 dB/km at $\lambda = 1.55\ \mu m$, indicating that fiber loss in Fig. 2.15 is dominated by Rayleigh scattering near this wavelength.

The contribution of Rayleigh scattering can be reduced to below 0.01 dB/km for wavelengths longer than 3 μm. Silica fibers cannot be used in this wavelength region, since infrared absorption begins to dominate the fiber loss beyond 1.6 μm. Considerable effort has been directed toward finding other suitable materials with low absorption beyond 2 μm [63]–[66]. Fluorozirconate (ZrF_4) fibers have an intrinsic material absorption of about 0.01 dB/km near 2.55 μm and have the potential for exhibiting loss much smaller than that of silica fibers. State-of-the-art fluoride fibers, however, exhibit a loss of about 1 dB/km because of extrinsic losses. Chalcogenide and polycrystalline fibers exhibit minimum loss in the far-infrared region near 10 μm. The theoretically predicted minimum value of fiber loss for such fibers is below 10^{-3} dB/km because of reduced Rayleigh scattering. However, practical loss levels remain higher than those of silica fibers [66].

2.5.4 Waveguide Imperfections

An ideal single-mode fiber with a perfect cylindrical geometry guides the optical mode without energy leakage into the cladding layer. In practice, imperfections at the core–cladding interface (e.g., random core-radius variations) can lead to additional losses which contribute to the net fiber loss. The physical process behind such losses is *Mie scattering* [22], occurring because of index inhomogeneities on a scale longer than the optical wavelength. Care is generally taken to ensure that the core radius does not vary significantly along the fiber length during manufacture. Such variations can be kept below 1%, and the resulting scattering loss is typically below 0.03 dB/km.

Bends in the fiber constitute another source of scattering loss [67]. The reason can be understood by using the ray picture. Normally, a guided ray hits the core–cladding interface at an angle greater than the critical angle to experience total internal reflection. However, the angle decreases near a bend and may become smaller than the critical angle for tight bends. The ray would then escape out of the fiber. In the mode description, a part of the mode energy is scattered into the cladding layer. The bending loss is proportional to $\exp(-R/R_c)$, where R is the radius of curvature of the fiber bend and $R_c = a/(n_1^2 - n_2^2)$. For single-mode fibers, $R_c = 0.2$–0.4 μm typically, and the bending loss is negligible (< 0.01 dB/km) for bend radius $R > 5$ mm. Since most macroscopic bends exceed $R = 5$ mm, *macrobending losses* are negligible in practice.

A major source of fiber loss, particularly in cable form, is related to the random axial distortions that invariably occur during cabling when the fiber is pressed against a surface that is not perfectly smooth. Such losses are referred to as *microbending losses* and have been studied extensively [68]–[72]. Microbends cause an increase in the fiber loss for both multimode and single-mode fibers and can result in an excessively large loss (~ 100 dB/km) if precautions are not taken to minimize them. For single-mode fibers, microbending losses can be minimized by choosing the V parameter as close to the cutoff value of 2.405 as possible so that mode energy is confined primarily to the core. In practice, the fiber is designed to have V in the range 2.0–2.4 at the operating wavelength. Many other sources of optical loss exist in a fiber cable. These are related to splices and connectors used in forming the fiber link and are often treated as a part of the cable loss; microbending losses can also be included in the total cable loss.

2.6 Nonlinear Optical Effects

The response of any dielectric to light becomes nonlinear for intense electromagnetic fields, and optical fibers are no exception. Even though silica is intrinsically not a highly nonlinear material, the waveguide geometry that confines light to a small cross section over long fiber lengths makes nonlinear effects quite important in the design of modern lightwave systems [31]. We discuss in this section the nonlinear phenomena that are most relevant for fiber-optic communications.

2.6.1 Stimulated Light Scattering

Rayleigh scattering, discussed in Section 2.5.3, is an example of elastic scattering for which the frequency (or the photon energy) of scattered light remains unchanged. By contrast, the frequency of scattered light is shifted downward during inelastic scattering. Two examples of inelastic scattering are *Raman scattering* and *Brillouin scattering* [73]. Both of them can be understood as scattering of a photon to a lower energy photon such that the energy difference appears in the form of a phonon. The main difference between the two is that optical phonons participate in Raman scattering, whereas acoustic phonons participate in Brillouin scattering. Both scattering processes result in a loss of power at the incident frequency. However, their scattering cross sections are sufficiently small that loss is negligible at low power levels.

At high power levels, the nonlinear phenomena of stimulated Raman scattering (SRS) and stimulated Brillouin scattering (SBS) become important. The intensity of the scattered light in both cases grows exponentially once the incident power exceeds a threshold value [74]. SRS and SBS were first observed in optical fibers during the 1970s [75]–[78]. Even though SRS and SBS are quite similar in their origin, different dispersion relations for acoustic and optical phonons lead to the following differences between the two in single-mode fibers [31]: (i) SBS occurs only in the backward direction whereas SRS can occur in both directions; (ii) The scattered light is shifted in frequency by about 10 GHz for SBS but by 13 THz for SRS (this shift is called the Stokes shift); and (iii) the Brillouin gain spectrum is extremely narrow (bandwidth < 100 MHz) compared with the Raman-gain spectrum that extends over 20–30 THz. The origin of these differences lies in a relatively small value of the ratio v_A/c ($\sim 10^{-5}$), where v_A is the acoustic velocity in silica and c is the velocity of light.

Stimulated Brillouin Scattering

The physical process behind Brillouin scattering is the tendency of materials to become compressed in the presence of an electric field—a phenomenon termed electrostriction [73]. For an oscillating electric field at the pump frequency Ω_p, this process generates an acoustic wave at some frequency Ω. Spontaneous Brillouin scattering can be viewed as scattering of the pump wave from this acoustic wave, resulting in creation of a new wave at the pump frequency Ω_s. The scattering process must conserve both the energy and the momentum. The energy conservation requires that the Stokes shift Ω equals $\omega_p - \omega_s$. The momentum conservation requires that the wave vectors satisfy $\mathbf{k}_A = \mathbf{k}_p - \mathbf{k}_s$. Using the dispersion relation $|k_A| = \Omega/v_A$, where v_A is the acoustic velocity, this condition determines the acoustic frequency as [31]

$$\Omega = |k_A|v_A = 2v_A|k_p| \sin(\theta/2), \tag{2.6.1}$$

where $|k_p| \approx |k_s|$ was used and θ represents the angle between the pump and scattered waves. Note that Ω vanishes in the forward direction ($\theta = 0$) and is maximum in the backward direction ($\theta = \pi$). In single-mode fibers, light can travel only in the forward and backward directions. As a result, SBS occurs in the backward direction with a frequency shift $\Omega_B = 2v_A|k_p|$. Using $k_p = 2\pi\bar{n}/\lambda_p$, where λ_p is the pump wavelength, the *Brillouin shift* is given by

$$v_B = \Omega_B/2\pi = 2\bar{n}v_A/\lambda_p, \tag{2.6.2}$$

where $\bar{n}$ is the mode index. Using $v_A = 5.96$ km/s and $\bar{n} = 1.45$ as typical values for silica fibers, $v_B = 11.1$ GHz at $\lambda_p = 1.55$ μm. Equation (2.6.2) shows that v_B scales inversely with the pump wavelength.

Once the scattered wave is generated spontaneously, it beats with the pump and creates a frequency component at the beat frequency $\omega_p - \omega_s$, which is automatically equal to the acoustic frequency Ω. As a result, the beating term acts as source that increases the amplitude of the sound wave, which in turn increases the amplitude of the scattered wave, resulting in a positive feedback loop. SBS has its origin in this positive feedback, which ultimately can transfer all power from the pump to the scattered wave. The feedback process is governed by the following set of two coupled equations [73]:

$$\frac{dI_p}{dz} = -g_B I_p I_s - \alpha_p I_p. \tag{2.6.3}$$

$$-\frac{dI_s}{dz} = +g_B I_p I_s - \alpha_s I_s, \tag{2.6.4}$$

where I_p and I_s are the intensities of the pump and Stokes fields, g_B is the SBS gain, and α_p and α_p account for fiber losses.

The SBS gain g_B is frequency dependent because of a finite damping time T_B of acoustic waves (the lifetime of acoustic phonons). If the acoustic waves decay as $\exp(-t/T_B)$, the Brillouin gain has a Lorentzian spectral profile given by [77]

$$g_B(\Omega) = \frac{g_B(\Omega_B)}{1 + (\Omega - \Omega_B)^2 T_B^2}. \tag{2.6.5}$$

Figure 2.17 shows the Brillouin gain spectra at $\lambda_p = 1.525$ μm for three different kinds of single-mode silica fibers. Both the Brillouin shift v_B and the gain bandwidth Δv_B can vary from fiber to fiber because of the guided nature of light and the presence of dopants in the fiber core. The fiber labeled (a) in Fig. 2.17 has a core of nearly pure silica (germania concentration of about 0.3% per mole). The measured Brillouin

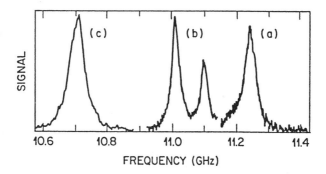

Figure 2.17: Brillouin-gain spectra measured using a 1.525-μm pump for three fibers with different germania doping: (a) silica-core fiber; (b) depressed-cladding fiber; (c) dispersion-shifted fiber. Vertical scale is arbitrary. (After Ref. [78]; ©1986 IEE; reprinted with permission.)

shift $v_B = 11.25$ GHz is in agreement with Eq. (2.6.2). The Brillouin shift is reduced for fibers (b) and (c) of a higher germania concentration in the fiber core. The double-peak structure for fiber (b) results from inhomogeneous distribution of germania within the core. The gain bandwidth in Fig. 2.17 is larger than that expected for bulk silica ($\Delta v_B \approx 17$ MHz at $\lambda_p = 1.525$ μm). A part of the increase is due to the guided nature of acoustic modes in optical fibers. However, most of the increase in bandwidth can be attributed to variations in the core diameter along the fiber length. Because such variations are specific to each fiber, the SBS gain bandwidth is generally different for different fibers and can exceed 100 MHz; typical values are ~50 MHz for λ_p near 1.55 μm.

The peak value of the Brillouin gain in Eq. (2.6.5) occurs for $\Omega = \Omega_B$ and depends on various material parameters such as the density and the elasto-optic coefficient [73]. For silica fibers $g_B \approx 5 \times 10^{-11}$ m/W. The threshold power level for SBS can be estimated by solving Eqs. (2.6.3) and (2.6.4) and finding at what value of I_p, I_s grows from noise to a significant level. The threshold power $P_{th} = I_p A_{eff}$, where A_{eff} is the effective core area, satisfies the condition [74]

$$g_B P_{th} L_{eff} / A_{eff} \approx 21, \qquad (2.6.6)$$

where L_{eff} is the effective interaction length defined as

$$L_{eff} = [1 - \exp(-\alpha L)]/\alpha, \qquad (2.6.7)$$

and α represents fiber losses. For optical communication systems L_{eff} can be approximated by $1/\alpha$ as $\alpha L \gg 1$ in practice. Using $A_{eff} = \pi w^2$, where w is the spot size, P_{th} can be as low as 1 mW depending on the values of w and α [77]. Once the power launched into an optical fiber exceeds the threshold level, most of the light is reflected backward through SBS. Clearly, SBS limits the launched power to a few milliwatts because of its low threshold.

The preceding estimate of P_{th} applies to a narrowband CW beam as it neglects the temporal and spectral characteristics of the incident light. In a lightwave system, the

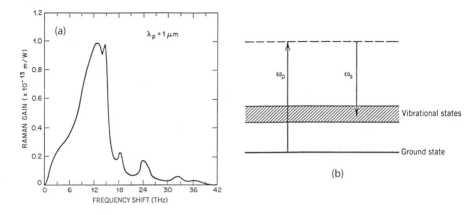

Figure 2.18: (a) Raman gain spectrum of fused silica at $\lambda_p = 1$ μm and (b) energy levels participating in the SRS process. (After Ref. [75]; ©1972 AIP; reprinted with permission.)

signal is in the form of a bit stream. For a single short pulse whose width is much smaller than the phonon lifetime, no SBS is expected to occur. However, for a high-speed bit stream, pulses come at such a fast rate that successive pulses build up the acoustic wave, similar to the case of a CW beam, although the SBS threshold increases. The exact value of the average threshold power depends on the modulation format (RZ versus NRZ) and is typically $\sim$5 mW. It can be increased to 10 mW or more by increasing the bandwidth of the optical carrier to >200 MHz through phase modulation. SBS does not produce interchannel crosstalk in WDM systems because the 10-GHz frequency shift is much smaller than typical channel spacing.

Stimulated Raman Scattering

Spontaneous Raman scattering occurs in optical fibers when a pump wave is scattered by the silica molecules. It can be understood using the energy-level diagram shown in Fig. 2.18(b). Some pump photons give up their energy to create other photons of reduced energy at a lower frequency; the remaining energy is absorbed by silica molecules, which end up in an excited vibrational state. An important difference from Brillouin scattering is that the vibrational energy levels of silica dictate the value of the Raman shift $\Omega_R = \omega_p - \omega_s$. As an acoustic wave is not involved, spontaneous Raman scattering is an isotropic process and occurs in all directions.

Similar to the SBS case, the Raman scattering process becomes stimulated if the pump power exceeds a threshold value. SRS can occur in both the forward and backward directions in optical fibers. Physically speaking, the beating of the pump and with the scattered light in these two directions creates a frequency component at the beat frequency $\omega_p - \omega_s$, which acts as a source that derives molecular oscillations. Since the amplitude of the scattered wave increases in response to these oscillations, a positive feedback loop sets in. In the case of forward SRS, the feedback process is governed by the following set of two coupled equations [31]:

$$\frac{dI_p}{dz} = -g_R I_p I_s - \alpha_p I_p, \qquad (2.6.8)$$

$$\frac{dI_s}{dz} = g_R I_p I_s - \alpha_s I_s, \qquad (2.6.9)$$

where g_R is the SRS gain. In the case of backward SRS, a minus sign is added in front of the derivative in Eq. (2.6.9), and this set of equations becomes identical to the SBS case.

The spectrum of the Raman gain depends on the decay time associated with the excited vibrational state. In the case of a molecular gas or liquid, the decay time is relatively long (~ 1 ns), resulting in a Raman-gain bandwidth of ~ 1 GHz. In the case for optical fibers, the bandwidth exceeds 10 THz. Figure 2.18 shows the Raman-gain spectrum of silica fibers. The broadband and multipeak nature of the spectrum is due to the amorphous nature of glass. More specifically, vibrational energy levels of silica molecules merge together to form a band. As a result, the Stokes frequency ω_s can differ from the pump frequency ω_p over a wide range. The maximum gain occurs when the Raman shift $\Omega_R \equiv \omega_p - \omega_s$ is about 13 THz. Another major peak occurs near 15 THz while minor peaks persist for values of Ω_R as large as 35 THz. The peak value of the Raman gain g_R is about 1×10^{-13} m/W at a wavelength of 1 μm. This value scales linearly with ω_p (or inversely with the pump wavelength λ_p), resulting in $g_R \approx 6 \times 10^{-13}$ m/W at 1.55 μm.

Similar to the case of SBS, the threshold power P_{th} is defined as the incident power at which half of the pump power is transferred to the Stokes field at the output end of a fiber of length L. It is estimated from [74]

$$g_R P_{th} L_{eff}/A_{eff} \approx 16, \qquad (2.6.10)$$

where g_R is the peak value of the Raman gain. As before, L_{eff} can be approximated by $1/\alpha$. If we replace A_{eff} by πw^2, where w is the spot size, P_{th} for SRS is given by

$$P_{th} \approx 16\alpha(\pi w^2)/g_R. \qquad (2.6.11)$$

If we use $\pi w^2 = 50$ μm^2 and $\alpha = 0.2$ dB/km as the representative values, P_{th} is about 570 mW near 1.55 μm. It is important to emphasize that Eq. (2.6.11) provides an order-of-magnitude estimate only as many approximations are made in its derivation. As channel powers in optical communication systems are typically below 10 mW, SRS is not a limiting factor for single-channel lightwave systems. However, it affects the performance of WDM systems considerably; this aspect is covered in Chapter 8.

Both SRS and SBS can be used to advantage while designing optical communication systems because they can amplify an optical signal by transferring energy to it from a pump beam whose wavelength is suitably chosen. SRS is especially useful because of its extremely large bandwidth. Indeed, the Raman gain is used routinely for compensating fiber losses in modern lightwave systems (see Chapter 6).

2.6.2   Nonlinear Phase Modulation

The refractive index of silica was assumed to be power independent in the discussion of fiber modes in Section 2.2. In reality, all materials behave nonlinearly at high intensities and their refractive index increases with intensity. The physical origin of this effect lies in the anharmonic response of electrons to optical fields, resulting in a nonlinear susceptibility [73]. To include nonlinear refraction, we modify the core and cladding indices of a silica fiber as [31]

$$n'_j = n_j + \bar{n}_2(P/A_{\text{eff}}), \qquad j = 1, 2, \tag{2.6.12}$$

where $\bar{n}_2$ is the *nonlinear-index coefficient*, P is the optical power, and A_{eff} is the effective mode area introduced earlier. The numerical value of $\bar{n}_2$ is about 2.6×10^{-20} m^2/W for silica fibers and varies somewhat with dopants used inside the core. Because of this relatively small value, the nonlinear part of the refractive index is quite small ($< 10^{-12}$ at a power level of 1 mW). Nevertheless, it affects modern lightwave systems considerably because of long fiber lengths. In particular, it leads to the phenomena of self- and cross-phase modulations.

Self-Phase Modulation

If we use first-order perturbation theory to see how fiber modes are affected by the nonlinear term in Eq. (2.6.12), we find that the mode shape does not change but the propagation constant becomes power dependent. It can be written as [31]

$$\beta' = \beta + k_0\bar{n}_2 P/A_{\text{eff}} \equiv \beta + \gamma P, \tag{2.6.13}$$

where $\gamma = 2\pi\bar{n}_2/(A_{\text{eff}}\lambda)$ is an important nonlinear parameter with values ranging from 1 to 5 W^{-1}/km depending on the values of A_{eff} and the wavelength. Noting that the optical phase increases linearly with z as seen in Eq. (2.4.1), the γ term produces a nonlinear phase shift given by

$$= \frac{2\pi}{\lambda}\bar{n}_2\frac{P_{in}}{A_{eff}}L_{eff} \quad , \quad L_{eff} = \frac{[1 - \exp(-\alpha L)]}{\alpha}$$

$$\phi_{\text{NL}} = \int_0^L (\beta' - \beta)\,dz = \int_0^L \gamma P(z)\,dz = \gamma P_{\text{in}}L_{\text{eff}}, \tag{2.6.14}$$

where $P(z) = P_{\text{in}}\exp(-\alpha z)$ accounts for fiber losses and L_{eff} is defined in Eq. (2.6.7).

In deriving Eq. (2.6.14) P_{in} was assumed to be constant. In practice, time dependence of P_{in} makes ϕ_{NL} to vary with time. In fact, the optical phase changes with time in exactly the same fashion as the optical signal. Since this nonlinear phase modulation is self-induced, the nonlinear phenomenon responsible for it is called *self-phase modulation* (SPM). It should be clear from Eq. (2.4.12) that SPM leads to frequency chirping of optical pulses. In contrast with the linear chirp considered in Section 2.4, the frequency chirp is proportional to the derivative dP_{in}/dt and depends on the pulse shape. Figure 2.19 shows how chirp varies with time for Gaussian ($m = 1$) and super-Gaussian pulses ($m = 3$). The SPM-induced chirp affects the pulse shape through GVD and often leads to additional pulse broadening [31]. In general, spectral broadening of the pulse induced by SPM [79] increases the signal bandwidth considerably and limits the performance of lightwave systems.

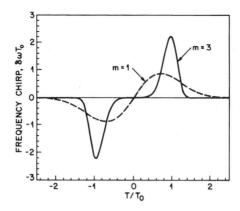

Figure 2.19: SPM-induced frequency chirp for Gaussian (dashed curve) and super-Gaussian (solid curve) pulses.

If fiber losses are compensated periodically using optical amplifiers, ϕ_{NL} in Eq. (2.6.14) should be multiplied by the number of amplifiers N_A because the SPM-induced phase accumulates over multiple amplifiers. To reduce the impact of SPM in lightwave systems, it is necessary that $\phi_{NL} \ll 1$. If we use $\phi_{NL} = 0.1$ as the maximum tolerable value and replace L_{eff} by $1/\alpha$ for long fibers, this condition can be written as a limit on the input peak power as

$$P_{in} < 0.1\alpha/(\gamma N_A). \qquad (2.6.15)$$

For example, if $\gamma = 2$ W^{-1}/km, $N_A = 10$, and $\alpha = 0.2$ dB/km, the input peak power is limited to below 2.2 mW. Clearly, SPM can be a major limiting factor for long-haul lightwave systems.

Cross-Phase Modulation

The intensity dependence of the refractive index in Eq. (2.6.12) can also lead to another nonlinear phenomenon known as *cross-phase modulation* (XPM). It occurs when two or more optical channels are transmitted simultaneously inside an optical fiber using the WDM technique. In such systems, the nonlinear phase shift for a specific channel depends not only on the power of that channel but also on the power of other channels [80]. The phase shift for the jth channel becomes

$$\phi_j^{NL} = \gamma L_{eff} \left(P_j + 2 \sum_{m \neq j} P_m \right), \qquad (2.6.16)$$

where the sum extends over the number of channels. The factor of 2 in Eq. (2.6.16) has its origin in the form of the nonlinear susceptibility [31] and indicates that XPM is twice as effective as SPM for the same amount of power. The total phase shift depends on the powers in all channels and would vary from bit to bit depending on the bit pattern of the neighboring channels. If we assume equal channel powers, the phase shift in the worst case in which all channels simultaneously carry 1 bits and all pulses overlap is given by

$$\phi_j^{\mathrm{NL}} = (\gamma/\alpha)(2M - 1)P_j. \qquad (2.6.17)$$

It is difficult to estimate the impact of XPM on the performance of multichannel lightwave systems. The reason is that the preceding discussion has implicitly assumed that XPM acts in isolation without dispersive effects and is valid only for CW optical beams. In practice, pulses in different channels travel at different speeds. The XPM-induced phase shift can occur only when two pulses overlap in time. For widely separated channels they overlap for such a short time that XPM effects are virtually negligible. On the other hand, pulses in neighboring channels will overlap long enough for XPM effects to accumulate. These arguments show that Eq. (2.6.17) cannot be used to estimate the limiting input power.

A common method for studying the impact of SPM and XPM uses a numerical approach. Equation (2.4.9) can be generalized to include the SPM and XPM effects by adding a nonlinear term. The resulting equation is known as the nonlinear Schrödinger equation and has the form [31]

$$\frac{\partial A}{\partial z} + \frac{i\beta_2}{2}\frac{\partial^2 A}{\partial t^2} = -\frac{\alpha}{2}A + i\gamma|A|^2 A, \qquad (2.6.18)$$

where we neglected the third-order dispersion and added the term containing α to account for fiber losses. This equation is quite useful for designing lightwave systems and will be used in later chapters.

Since the nonlinear parameter γ depends inversely on the effective core area, the impact of fiber nonlinearities can be reduced considerably by enlarging A_{eff}. As seen in Table 2.1, A_{eff} is about 80 $\mu\mathrm{m}^2$ for standard fibers but reduces to 50 $\mu\mathrm{m}^2$ for dispersion-shifted fibers. A new kind of fiber known as large effective-area fiber (LEAF) has been developed for reducing the impact of fiber nonlinearities. The nonlinear effects are not always detrimental for lightwave systems. Numerical solutions of Eq. (2.6.18) show that dispersion-induced broadening of optical pulses is considerably reduced in the case of anomalous dispersion [81]. In fact, an optical pulse can propagate without distortion if the peak power of the pulse is chosen to correspond to a fundamental soliton. Solitons and their use for communication systems are discussed in Chapter 9.

2.6.3   Four-Wave Mixing

The power dependence of the refractive index seen in Eq. (2.6.12) has its origin in the third-order nonlinear susceptibility denoted by $\chi^{(3)}$ [73]. The nonlinear phenomenon, known as *four-wave mixing* (FWM), also originates from $\chi^{(3)}$. If three optical fields with carrier frequencies ω_1, ω_2, and ω_3 copropagate inside the fiber simultaneously, $\chi^{(3)}$ generates a fourth field whose frequency ω_4 is related to other frequencies by a relation $\omega_4 = \omega_1 \pm \omega_2 \pm \omega_3$. Several frequencies corresponding to different plus and minus sign combinations are possible in principle. In practice, most of these combinations do not build up because of a phase-matching requirement [31]. Frequency combinations of the form $\omega_4 = \omega_1 + \omega_2 - \omega_3$ are often troublesome for multichannel communication systems since they can become nearly phase-matched when channel wavelengths lie close to the zero-dispersion wavelength. In fact, the degenerate FWM process for which $\omega_1 = \omega_2$ is often the dominant process and impacts the system performance most.

On a fundamental level, a FWM process can be viewed as a scattering process in which two photons of energies $\hbar\omega_1$ and $\hbar\omega_2$ are destroyed, and their energy appears in the form of two new photons of energies $\hbar\omega_3$ and $\hbar\omega_4$. The *phase-matching condition* then stems from the requirement of momentum conservation. Since all four waves propagate in the same direction, the phase mismatch can be written as

$$\Delta = \beta(\omega_3) + \beta(\omega_4) - \beta(\omega_1) - \beta(\omega_2), \tag{2.6.19}$$

where $\beta(\omega)$ is the propagation constant for an optical field with frequency ω. In the degenerate case, $\omega_2 = \omega_1$, $\omega_3 = \omega_1 + \Omega$, and $\omega_3 = \omega_1 - \Omega$, where Ω represents the channel spacing. Using the Taylor expansion in Eq. (2.4.4), we find that the β_0 and β_1 terms cancel, and the phase mismatch is simply $\Delta = \beta_2\Omega^2$. The FWM process is completely phase matched when $\beta_2 = 0$. When β_2 is small (<1 ps^2/km) and channel spacing is also small ($\Omega < 100$ GHz), this process can still occur and transfer power from each channel to its nearest neighbors. Such a power transfer not only results in the power loss for the channel but also induces interchannel crosstalk that degrades the system performance severely. Modern WDM systems avoid FWM by using the technique of dispersion management in which GVD is kept locally high in each fiber section even though it is low on average (see Chapter 7). Commercial dispersion-shifted fibers are designed with a dispersion of about 4 ps/(km-nm), a value found large enough to suppress FWM.

FWM can also be useful in designing lightwave systems. It is often used for de-multiplexing individual channels when time-division multiplexing is used in the optical domain. It can also be used for wavelength conversion. FWM in optical fibers is sometimes used for generating a spectrally inverted signal through the process of *optical phase conjugation*. As discussed in Chapter 7, this technique is useful for dispersion compensation.

2.7 Fiber Manufacturing

The final section is devoted to the engineering aspects of optical fibers. Manufacturing of fiber cables, suitable for installation in an actual lightwave system, involves sophisticated technology with attention to many practical details. Since such details are available in several texts [12]–[17], the discussion here is intentionally brief.

2.7.1 Design Issues

In its simplest form, a step-index fiber consists of a cylindrical core surrounded by a cladding layer whose index is slightly lower than the core. Both core and cladding use silica as the base material; the difference in the refractive indices is realized by doping the core, or the cladding, or both. Dopants such as GeO_2 and P_2O_5 increase the refractive index of silica and are suitable for the core. On the other hand, dopants such as B_2O_3 and fluorine decrease the refractive index of silica and are suitable for

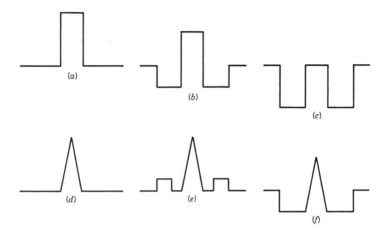

Figure 2.20: Several index profiles used in the design of single-mode fibers. Upper and lower rows correspond to standard and dispersion-shifted fibers, respectively.

the cladding. The major design issues are related to the refractive-index profile, the amount of dopants, and the core and cladding dimensions [82]–[86]. The diameter of the outermost cladding layer has the standard value of 125 μm for all communication-grade fibers.

Figure 2.20 shows typical index profiles that have been used for different kinds of fibers. The top row corresponds to standard fibers which are designed to have minimum dispersion near 1.3 μm with a cutoff wavelength in the range 1.1–1.2 μm. The simplest design [Fig. 2.20(a)] consists of a pure-silica cladding and a core doped with GeO_2 to obtain $\Delta \approx 3 \times 10^{-3}$. A commonly used variation [Fig. 2.20(b)] lowers the cladding index over a region adjacent to the core by doping it with fluorine. It is also possible to have an undoped core by using a design shown in Fig 2.20(c). The fibers of this kind are referred to as doubly clad or *depressed-cladding fibers* [82]. They are also called W fibers, reflecting the shape of the index profile. The bottom row in Fig. 2.20 shows three index profiles used for dispersion-shifted fibers for which the zero-dispersion wavelength is chosen in the range 1.45–1.60 μm (see Table 2.1). A triangular index profile with a depressed or raised cladding is often used for this purpose [83]–[85]. The refractive indices and the thickness of different layers are optimized to design a fiber with desirable dispersion characteristics [86]. Sometimes as many as four cladding layers are used for dispersion-flattened fibers (see Fig. 2.11).

2.7.2   Fabrication Methods

Fabrication of telecommunication-grade silica fibers involves two stages. In the first stage a vapor-deposition method is used to make a *cylindrical preform* with the desired refractive-index profile. The preform is typically 1 m long and 2 cm in diameter and contains core and cladding layers with correct relative dimensions. In the second stage, the preform is drawn into a fiber by using a precision-feed mechanism that feeds the preform into a furnace at the proper speed.

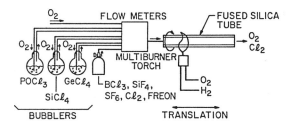

Figure 2.21: MCVD process commonly used for fiber fabrication. (After Ref. [87]; ©1985 Academic Press; reprinted with permission.)

Several methods can be used to make the preform. The three commonly used methods [87]–[89] are modified chemical-vapor deposition (MCVD), outside-vapor deposition (OVD), and vapor-axial deposition (VAD). Figure 2.21 shows a schematic diagram of the MCVD process. In this process, successive layers of SiO_2 are deposited on the inside of a fused silica tube by mixing the vapors of $SiCl_4$ and O_2 at a temperature of about 1800°C. To ensure uniformity, a multiburner torch is moved back and forth across the tube length using an automatic translation stage. The refractive index of the cladding layers is controlled by adding fluorine to the tube. When a sufficient cladding thickness has been deposited, the core is formed by adding the vapors of $GeCl_4$ or $POCl_3$. These vapors react with oxygen to form the dopants GeO_2 and P_2O_5:

$$GeCl_4 + O_2 \rightarrow GeO_2 + 2Cl_2,$$
$$4POCl_3 + 3O_2 \rightarrow 2P_2O_5 + 6Cl_2.$$

The flow rate of $GeCl_4$ or $POCl_3$ determines the amount of dopant and the corresponding increase in the refractive index of the core. A triangular-index core can be fabricated simply by varying the flow rate from layer to layer. When all layers forming the core have been deposited, the torch temperature is raised to collapse the tube into a solid rod of preform.

The MCVD process is also known as the *inner-vapor-deposition method*, as the core and cladding layers are deposited inside a silica tube. In a related process, known as the *plasma-activated chemical vapor deposition* process [90], the chemical reaction is initiated by a microwave plasma. By contrast, in the OVD and VAD processes the core and cladding layers are deposited on the outside of a rotating mandrel by using the technique of *flame hydrolysis*. The mandrel is removed prior to sintering. The porous soot boule is then placed in a sintering furnace to form a glass boule. The central hole allows an efficient way of reducing water vapors through dehydration in a controlled atmosphere of Cl_2–He mixture, although it results in a central dip in the index profile. The dip can be minimized by closing the hole during sintering.

The fiber drawing step is essentially the same irrespective of the process used to make the preform [91]. Figure 2.22 shows the drawing apparatus schematically. The preform is fed into a furnace in a controlled manner where it is heated to a temperature of about 2000°C. The melted preform is drawn into a fiber by using a precision-feed mechanism. The fiber diameter is monitored optically by diffracting light emitted by a laser from the fiber. A change in the diameter changes the diffraction pattern, which

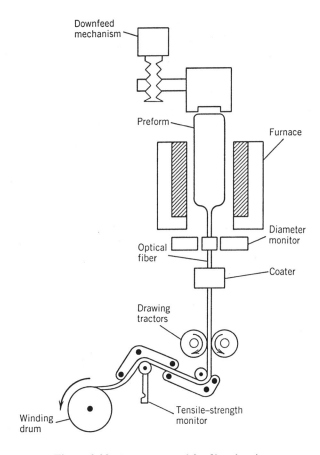

Figure 2.22: Apparatus used for fiber drawing.

in turn changes the photodiode current. This current change acts as a signal for a servocontrol mechanism that adjusts the winding rate of the fiber. The fiber diameter can be kept constant to within 0.1% by this technique. A polymer coating is applied to the fiber during the drawing step. It serves a dual purpose, as it provides mechanical protection and preserves the transmission properties of the fiber. The diameter of the coated fiber is typically 250 μm, although it can be as large as 900 μm when multiple coatings are used. The tensile strength of the fiber is monitored during its winding on the drum. The winding rate is typically 0.2–0.5 m/s. Several hours are required to convert a single preform into a fiber of about 5 km length. This brief discussion is intended to give a general idea. The fabrication of optical fiber generally requires attention to a large number of engineering details discussed in several texts [17].

2.7.3 Cables and Connectors

Cabling of fibers is necessary to protect them from deterioration during transportation and installation [92]. Cable design depends on the type of application. For some

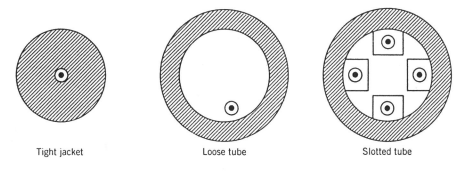

Tight jacket   Loose tube   Slotted tube

Figure 2.23: Typical designs for light-duty fiber cables.

applications it may be enough to buffer the fiber by placing it inside a plastic jacket. For others the cable must be made mechanically strong by using strengthening elements such as steel rods.

A light-duty cable is made by surrounding the fiber by a buffer jacket of hard plastic. Figure 2.23 shows three simple cable designs. A tight jacket can be provided by applying a buffer plastic coating of 0.5–1 mm thickness on top of the primary coating applied during the drawing process. In an alternative approach the fiber lies loosely inside a plastic tube. Microbending losses are nearly eliminated in this loose-tube construction, since the fiber can adjust itself within the tube. This construction can also be used to make multifiber cables by using a slotted tube with a different slot for each fiber.

Heavy-duty cables use steel or a strong polymer such as Kevlar to provide the mechanical strength. Figure 2.24 shows schematically three kinds of cables. In the loose-tube construction, fiberglass rods embedded in polyurethane and a Kevlar jacket provide the necessary mechanical strength (left drawing). The same design can be extended to multifiber cables by placing several loose-tube fibers around a central steel core (middle drawing). When a large number of fibers need to be placed inside a single cable, a ribbon cable is used (right drawing). The ribbon is manufactured by packaging typically 12 fibers between two polyester tapes. Several ribbons are then stacked into a

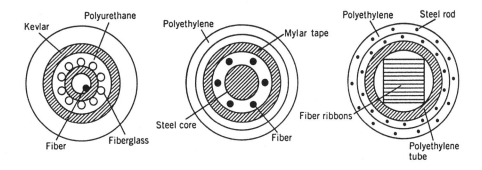

Figure 2.24: Typical designs for heavy-duty fiber cables.

rectangular array which is placed inside a polyethylene tube. The mechanical strength is provided by using steel rods in the two outermost polyethylene jackets. The outer diameter of such fiber cables is about 1–1.5 cm.

Connectors are needed to use optical fibers in an actual communication system. They can be divided into two categories. A permanent joint between two fibers is known as a fiber splice, and a detachable connection between them is realized by using a fiber connector. Connectors are used to link fiber cable with the transmitter (or the receiver), while splices are used to join fiber segments (usually 5–10 km long). The main issue in the use of splices and connectors is related to the loss. Some power is always lost, as the two fiber ends are never perfectly aligned in practice. Splice losses below 0.1 dB are routinely realized by using the technique of fusion splicing [93]. Connector losses are generally larger. State-of-the-art connectors provide an average loss of about 0.3 dB [94]. The technology behind the design of splices and connectors is quite sophisticated. For details, the reader is referred to Ref. [95], a book devoted entirely to this issue.

Problems

2.1 A multimode fiber with a 50-μm core diameter is designed to limit the inter-modal dispersion to 10 ns/km. What is the numerical aperture of this fiber? What is the limiting bit rate for transmission over 10 km at 0.88 μm? Use 1.45 for the refractive index of the cladding.

2.2 Use the ray equation in the paraxial approximation [Eq. (2.1.8)] to prove that intermodal dispersion is zero for a graded-index fiber with a quadratic index profile.

2.3 Use Maxwell's equations to express the field components E_ρ, E_ϕ, H_ρ, and H_ϕ in terms of E_z and H_z and obtain Eqs. (2.2.29)–(2.2.32).

2.4 Derive the eigenvalue equation (2.2.33) by matching the boundary conditions at the core–cladding interface of a step-index fiber.

2.5 A single-mode fiber has an index step $n_1 - n_2 = 0.005$. Calculate the core radius if the fiber has a cutoff wavelength of 1 μm. Estimate the spot size (FWHM) of the fiber mode and the fraction of the mode power inside the core when this fiber is used at 1.3 μm. Use $n_1 = 1.45$ μm.

2.6 A 1.55-μm unchirped Gaussian pulse of 100-ps width (FWHM) is launched into a single-mode fiber. Calculate its FWHM after 50 km if the fiber has a dispersion of 16 ps/(km-nm). Neglect the source spectral width.

2.7 Derive an expression for the confinement factor Γ of single-mode fibers defined as the fraction of the total mode power contained inside the core. Use the Gaussian approximation for the fundamental fiber mode. Estimate Γ for $V = 2$.

2.8 A single-mode fiber is measured to have $\lambda^2(d^2n/d\lambda^2) = 0.02$ at 0.8 μm. Calculate the dispersion parameters β_2 and D.

2.9 Show that a chirped Gaussian pulse is compressed initially inside a single-mode fiber when $\beta_2 C < 0$. Derive expressions for the minimum width and the fiber length at which the minimum occurs.

2.10 Estimate the limiting bit rate for a 60-km single-mode fiber link at 1.3- and 1.55-μm wavelengths assuming transform-limited, 50-ps (FWHM) input pulses. Assume that $\beta_2 = 0$ and -20 ps^2/km and $\beta_3 = 0.1$ ps^3/km and 0 at 1.3- and 1.55-μm wavelengths, respectively. Also assume that $V_\omega \ll 1$.

2.11 A 0.88-μm communication system transmits data over a 10-km single-mode fiber by using 10-ns (FWHM) pulses. Determine the maximum bit rate if the LED has a spectral FWHM of 30 nm. Use $D = -80$ ps/(km-nm).

2.12 Use Eq. (2.4.23) to prove that the bit rate of an optical communication system operating at the zero-dispersion wavelength is limited by $BL|S|\sigma_\lambda^2 < 1/\sqrt{8}$, where $S = dD/d\lambda$ and σ_λ is the RMS spectral width of the Gaussian source spectrum. Assume that $C = 0$ and $V_\omega \gg 1$ in the general expression of the output pulse width.

2.13 Repeat Problem 2.12 for the case of a single-mode semiconductor laser for which $V_\omega \ll 1$ and show that the bit rate is limited by $B(|\beta_3|L)^{1/3} < 0.324$. What is the limiting bit rate for $L = 100$ km if $\beta_3 = 0.1$ ps^3/km?

2.14 An optical communication system is operating with chirped Gaussian input pulses. Assume that $\beta_3 = 0$ and $V_\omega \ll 1$ in Eq. (2.4.23) and obtain a condition on the bit rate in terms of the parameters C, β_2, and L.

2.15 A 1.55-μm optical communication system operating at 5 Gb/s is using Gaussian pulses of width 100 ps (FWHM) chirped such that $C = -6$. What is the dispersion-limited maximum fiber length? How much will it change if the pulses were unchirped? Neglect laser linewidth and assume that $\beta_2 = -20$ ps^2/km.

2.16 A 1.3-μm lightwave system uses a 50-km fiber link and requires at least 0.3 μW at the receiver. The fiber loss is 0.5 dB/km. Fiber is spliced every 5 km and has two connectors of 1-dB loss at both ends. Splice loss is only 0.2 dB. Determine the minimum power that must be launched into the fiber.

2.17 A 1.55-μm continuous-wave signal with 6-dBm power is launched into a fiber with 50-μm^2 effective mode area. After what fiber length would the nonlinear phase shift induced by SPM become 2π? Assume $\bar{n}_2 = 2.6 \times 10^{-20}$ m^2/W and neglect fiber losses.

2.18 Calculate the threshold power for stimulated Brillouin scattering for a 50-km fiber link operating at 1.3 μm and having a loss of 0.5 dB/km. How much does the threshold power change if the operating wavelength is changed to 1.55 μm, where the fiber loss is only 0.2 dB/km? Assume that $A_{\text{eff}} = 50$ μm^2 and $g_B = 5 \times 10^{-11}$ m/W at both wavelengths.

2.19 Calculate the power launched into a 40-km-long single-mode fiber for which the SPM-induced nonlinear phase shift becomes $180°$. Assume $\lambda = 1.55$ μm, $A_{\text{eff}} = 40$ μm^2, $\alpha = 0.2$ dB/km, and $\bar{n}_2 = 2.6 \times 10^{-20}$ m^2/W.

2.20 Find the maximum frequency shift occurring because of the SPM-induced chirp imposed on a Gaussian pulse of 20-ps width (FWHM) and 5-mW peak power after it has propagated 100 km. Use the fiber parameters of the preceding problem but assume $\alpha = 0$.

References

[1] J. Tyndall, *Proc. Roy. Inst.* **1**, 446 (1854).

[2] J. L. Baird, British Patent 285,738 (1927).

[3] C. W. Hansell, U.S. Patent 1,751,584 (1930).

[4] H. Lamm, *Z. Instrumentenk.* **50**, 579 (1930).

[5] A. C. S. van Heel, *Nature* **173**, 39 (1954).

[6] B. I. Hirschowitz, L. E. Curtiss, C. W. Peters, and H. M. Pollard, *Gastro-enterology* **35**, 50 (1958).

[7] N. S. Kapany, *J. Opt. Soc. Am.* **49**, 779 (1959).

[8] N. S. Kapany, *Fiber Optics: Principles and Applications*, Academic Press, San Diego, CA, 1967.

[9] K. C. Kao and G. A. Hockham, *Proc. IEE* **113**, 1151 (1966); A. Werts, *Onde Electr.* **45**, 967 (1966).

[10] F. P. Kapron, D. B. Keck, and R. D. Maurer, *Appl. Phys. Lett.* **17**, 423 (1970).

[11] T. Miya, Y. Terunuma, T. Hosaka, and T. Miyoshita, *Electron. Lett.* **15**, 106 (1979).

[12] M. J. Adams, *An Introduction to Optical Waveguides*, Wiley, New York, 1981.

[13] T. Okoshi, *Optical Fibers*, Academic Press, San Diego, CA, 1982.

[14] A. W. Snyder and J. D. Love, *Optical Waveguide Theory*, Chapman & Hall, London, 1983.

[15] L. B. Jeunhomme, *Single-Mode Fiber Optics*, Marcel Dekker, New York, 1990.

[16] T. Li, Ed., *Optical Fiber Communications*, Vol. 1, Academic Press, San Diego, CA, 1985.

[17] T. Izawa and S. Sudo, *Optical Fibers: Materials and Fabrication*, Kluwer Academic, Boston, 1987.

[18] E. G. Neumann, *Single-Mode Fibers*, Springer, New York, 1988.

[19] D. Marcuse, *Theory of Dielectric Optical Waveguides*, 2nd ed., Academic Press, San Diego, CA, 1991.

[20] G. Cancellieri, *Single-Mode Optical Fibers*, Pergamon Press, Elmsford, NY, 1991.

[21] J. A. Buck, *Fundamentals of Optical Fibers*, Wiley, New York, 1995.

[22] M. Born and E. Wolf, *Principles of Optics*, 7th ed., Cambridge University Press, New York, 1999.

[23] J. Gower, *Optical Communication Systems*, 2nd ed., Prentice Hall, London, 1993.

[24] Y. Koike, T. Ishigure, and E. Nihei, *J. Lightwave Technol.* **13**, 1475 (1995).

[25] T. Ishigure, A. Horibe, E. Nihei, and Y. Koike, *J. Lightwave Technol.* **13**, 1686 (1995).

[26] U. Fiedler, G. Reiner, P. Schnitzer, and K. J. Ebeling, *IEEE Photon. Technol. Lett.* **8**, 746 (1996).

[27] O. Nolen, *Plastic Optical Fibers for Data Communications*, Information Gatekeepers, Boston, 1996.

[28] C. DeCusatis, *Opt. Eng.* **37**, 3082 (1998).

[29] F. Mederer, R. Jager, P. Schnitzer, H. Unold, M. Kicherer, K. J. Ebeling, M. Naritomi, and R. Yoshida, *IEEE Photon. Technol. Lett.* **12**, 201 (2000).

[30] P. Diament, *Wave Transmission and Fiber Optics*, Macmillan, New York, 1990, Chap. 3.

[31] G. P. Agrawal, *Nonlinear Fiber Optics*, 3rd ed., Academic Press, San Diego, CA, 2001.

[32] M. Abramowitz and I. A. Stegun, Eds., *Handbook of Mathematical Functions*, Dover, New York, 1970, Chap. 9.

[33] D. Gloge, *Appl. Opt.* **10**, 2252 (1971); **10**, 2442 (1971).

[34] D. B. Keck, in *Fundamentals of Optical Fiber Communications*, M. K. Barnoski, Ed., Academic Press, San Diego, CA, 1981.

[35] D. Marcuse, *J. Opt. Soc. Am.* **68**, 103 (1978).

[36] I. H. Malitson, *J. Opt. Soc. Am.* **55**, 1205 (1965).

[37] L. G. Cohen, C. Lin, and W. G. French, *Electron. Lett.* **15**, 334 (1979).

[38] C. T. Chang, *Electron. Lett.* **15**, 765 (1979); *Appl. Opt.* **18**, 2516 (1979).

[39] L. G. Cohen, W. L. Mammel, and S. Lumish, *Opt. Lett.* **7**, 183 (1982).

[40] S. J. Jang, L. G. Cohen, W. L. Mammel, and M. A. Shaifi, *Bell Syst. Tech. J.* **61**, 385 (1982).

[41] V. A. Bhagavatula, M. S. Spotz, W. F. Love, and D. B. Keck, *Electron. Lett.* **19**, 317 (1983).

[42] P. Bachamann, D. Leers, H. Wehr, D. V. Wiechert, J. A. van Steenwijk, D. L. A. Tjaden, and E. R. Wehrhahn, *J. Lightwave Technol.* **4**, 858 (1986).

[43] B. J. Ainslie and C. R. Day, *J. Lightwave Technol.* **4**, 967 (1986).

[44] C. D. Poole and J. Nagel, in *Optical Fiber Telecommunications III*, Vol. A, I. P. Kaminow and T. L. Koch, Eds., Academic Press, San Diego, CA, 1997, Chap. 6.

[45] F. Bruyère, *Opt. Fiber Technol.* **2**, 269 (1996).

[46] P. K. A. Wai and C. R. Menyuk, *J. Lightwave Technol.* **14**, 148 (1996).

[47] M. Karlsson, *Opt. Lett.* **23**, 688 (1998).

[48] G. J. Foschini, R. M. Jopson, L. E. Nelson, and H. Kogelnik, *J. Lightwave Technol.* **17**, 1560 (1999).

[49] M. Midrio, *J. Opt. Soc. Am. B* **17**, 169 (2000).

[50] B. Huttner, C. Geiser, and N. Gisin, *IEEE J. Sel. Topics Quantum Electron.* **6**, 317 (2000).

[51] M. Shtaif and A. Mecozzi, *Opt. Lett.* **25**, 707 (2000).

[52] M. Karlsson, J. Brentel, and P. A. Andrekson, *J. Lightwave Technol.* **18**, 941 (2000).

[53] Y. Li and A. Yariv, *J. Opt. Soc. Am. B* **17**, 1821 (2000).

[54] J. M. Fini and H. A. Haus, *IEEE Photon. Technol. Lett.* **13**, 124 (2001).

[55] R. Khosravani and A. E. Willner, *IEEE Photon. Technol. Lett.* **13**, 296 (2001).

[56] D. Marcuse, *Appl. Opt.* **19**, 1653 (1980); **20**, 3573 (1981).

[57] M. Miyagi and S. Nishida, *Appl. Opt.* **18**, 678 (1979); **18**, 2237 (1979).

[58] G. P. Agrawal and M. J. Potasek, *Opt. Lett.* **11**, 318 (1986).

[59] M. Schwartz, *Information, Transmission, Modulation, and Noise*, 4th ed., McGraw-Hill, New York, 1990, Chap. 2.

[60] M. J. Bennett, *IEE Proc.* **130**, Pt. H, 309 (1983).

[61] D. Gloge, K. Ogawa, and L. G. Cohen, *Electron. Lett.* **16**, 366 (1980).

[62] G. A. Thomas, B. L. Shraiman, P. F. Glodis, and M. J. Stephan, *Nature* **404**, 262 (2000).

[63] P. Klocek and G. H. Sigel, Jr., *Infrared Fiber Optics*, Vol. TT2, SPIE, Bellingham, WA, 1989.

[64] T. Katsuyama and H. Matsumura, *Infrared Optical Fibers*, Bristol, Philadelphia, 1989.

[65] J. A. Harrington, Ed., *Infrared Fiber Optics*, SPIE, Bellingham, WA, 1990.

[66] M. F. Churbanov, *J. Non-Cryst. Solids* **184**, 25 (1995).

[67] E. A. J. Marcatili, *Bell Syst. Tech. J.* **48**, 2103 (1969).

[68] W. B. Gardner, *Bell Syst. Tech. J.* **54**, 457 (1975).

[69] D. Marcuse, *Bell Syst. Tech. J.* **55**, 937 (1976).

[70] K. Petermann, *Electron. Lett.* **12**, 107 (1976); *Opt. Quantum Electron.* **9**, 167 (1977).

[71] K. Tanaka, S. Yamada, M. Sumi, and K. Mikoshiba, *Appl. Opt.* **16**, 2391 (1977).

[72] W. A. Gambling, H. Matsumura, and C. M. Rodgal, *Opt. Quantum Electron.* **11**, 43 (1979).

[73] R. W. Boyd, *Nonlinear Optics*, Academic Press, San Diego, CA, 1992.

[74] R. G. Smith, *Appl. Opt.* **11**, 2489 (1972).

[75] R. H. Stolen, E. P. Ippen, and A. R. Tynes, *Appl. Phys. Lett.* **20**, 62 (1972).

[76] E. P. Ippen and R. H. Stolen, *Appl. Phys. Lett.* **21**, 539 (1972).

[77] D. Cotter, *Electron. Lett.* **18**, 495 (1982); *J. Opt. Commun.* **4**, 10 (1983).

[78] R. W. Tkach, A. R. Chraplyvy, and R. M. Derosier, *Electron. Lett.* **22**, 1011 (1986).

[79] R. H. Stolen and C. Lin, *Phys. Rev. A* **17**, 1448 (1978).

[80] A. R. Chraplyvy, D. Marcuse, and P. S. Henry, *J. Lightwave Technol.* **2**, 6 (1984).

[81] M. J. Potasek and G. P. Agrawal, *Electron. Lett.* **22**, 759 (1986).

[82] M. Monerie, *IEEE J. Quantum Electron.* **18**, 535 (1982); *Electron. Lett.* **18**, 642 (1982).

[83] M. A. Saifi, S. J. Jang, L. G. Cohen, and J. Stone, *Opt. Lett.* **7**, 43 (1982).

[84] Y. W. Li, C. D. Hussey, and T. A. Birks, *J. Lightwave Technol.* **11**, 1812 (1993).

[85] R. Lundin, *Appl. Opt.* **32**, 3241 (1993); *Appl. Opt.* **33**, 1011 (1994).

[86] S. P. Survaiya and R. K. Shevgaonkar, *IEEE Photon. Technol. Lett.* **8**, 803 (1996).

[87] S. R. Nagel, J. B. MacChesney, and K. L. Walker, in *Optical Fiber Communications*, Vol. 1, T. Li, Ed., Academic Press, San Diego, CA, 1985, Chap. 1.

[88] A. J. Morrow, A. Sarkar, and P. C. Schultz, in *Optical Fiber Communications*, Vol. 1, T. Li, Ed., Academic Press, San Diego, CA, 1985, Chap. 2.

[89] N. Niizeki, N. Ingaki, and T. Edahiro, in *Optical Fiber Communications*, Vol. 1, T. Li, Ed., Academic Press, San Diego, CA, 1985, Chap. 3.

[90] P. Geittner, H. J. Hagemann, J. Warnier, and H. Wilson, *J. Lightwave Technol.* **4**, 818 (1986).

[91] F. V. DiMarcello, C. R. Kurkjian, and J. C. Williams, in *Optical Fiber Communications*, Vol. 1, T. Li, Ed., Academic Press, San Diego, CA, 1985, Chap. 4.

[92] H. Murata, *Handbook of Optical Fibers and Cables*, Marcel Dekker, New York, 1996.

[93] S. C. Mettler and C. M. Miller, in *Optical Fiber Telecommunications II*, S. E. Miller and I. P. Kaminow, Eds., Academic Press, San Diego, CA, 1988, Chap. 6.

[94] W. C. Young and D. R. Frey, in *Optical Fiber Telecommunications II*, S. E. Miller and I. P. Kaminow, Eds., Academic Press, San Diego, CA, 1988, Chap. 7.

[95] C. M. Miller, S. C. Mettler, and I. A. White, *Optical Fiber Splices and Connectors*, Marcel Dekker, New York, 1986.

Chapter 3

Optical Transmitters

The role of the optical transmitter is to convert an electrical input signal into the corresponding optical signal and then launch it into the optical fiber serving as a communication channel. The major component of optical transmitters is an optical source. Fiber-optic communication systems often use semiconductor optical sources such as light-emitting diodes (LEDs) and semiconductor lasers because of several inherent advantages offered by them. Some of these advantages are compact size, high efficiency, good reliability, right wavelength range, small emissive area compatible with fiber-core dimensions, and possibility of direct modulation at relatively high frequencies. Although the operation of semiconductor lasers was demonstrated as early as 1962, their use became practical only after 1970, when semiconductor lasers operating continuously at room temperature became available [1]. Since then, semiconductor lasers have been developed extensively because of their importance for optical communications. They are also known as laser diodes or injection lasers, and their properties have been discussed in several recent books [2]–[16]. This chapter is devoted to LEDs and semiconductor lasers and their applications in lightwave systems. After introducing the basic concepts in Section 3.1, LEDs are covered in Section 3.2, while Section 3.3 focuses on semiconductor lasers. We describe single-mode semiconductor lasers in Section 3.4 and discuss their operating characteristics in Section 3.5. The design issues related to optical transmitters are covered in Section 3.6.

3.1 Basic Concepts

Under normal conditions, all materials absorb light rather than emit it. The absorption process can be understood by referring to Fig. 3.1, where the energy levels E_1 and E_2 correspond to the ground state and the excited state of atoms of the absorbing medium. If the photon energy $h\nu$ of the incident light of frequency ν is about the same as the energy difference $E_g = E_2 - E_1$, the photon is absorbed by the atom, which ends up in the excited state. Incident light is attenuated as a result of many such absorption events occurring inside the medium.

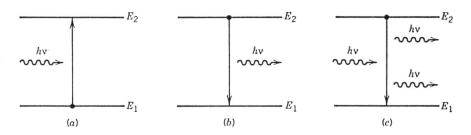

Figure 3.1: Three fundamental processes occurring between the two energy states of an atom: (a) absorption; (b) spontaneous emission; and (c) stimulated emission.

The excited atoms eventually return to their normal "ground" state and emit light in the process. Light emission can occur through two fundamental processes known as *spontaneous emission* and *stimulated emission*. Both are shown schematically in Fig. 3.1. In the case of spontaneous emission, photons are emitted in random directions with no phase relationship among them. Stimulated emission, by contrast, is initiated by an existing photon. The remarkable feature of stimulated emission is that the emitted photon matches the original photon not only in energy (or in frequency), but also in its other characteristics, such as the direction of propagation. All lasers, including semiconductor lasers, emit light through the process of stimulated emission and are said to emit coherent light. In contrast, LEDs emit light through the incoherent process of spontaneous emission.

3.1.1 Emission and Absorption Rates

Before discussing the emission and absorption rates in semiconductors, it is instructive to consider a two-level atomic system interacting with an electromagnetic field through transitions shown in Fig. 3.1. If N_1 and N_2 are the atomic densities in the ground and the excited states, respectively, and $\rho_{ph}(\nu)$ is the spectral density of the electromagnetic energy, the rates of spontaneous emission, stimulated emission, and absorption can be written as [17]

$$R_{\text{spon}} = AN_2, \qquad R_{\text{stim}} = BN_2\rho_{\text{em}}, \qquad R_{\text{abs}} = B'N_1\rho_{\text{em}}, \qquad (3.1.1)$$

where A, B, and B' are constants. In thermal equilibrium, the atomic densities are distributed according to the Boltzmann statistics [18], i.e.,

$$N_2/N_1 = \exp(-E_g/k_BT) \equiv \exp(-h\nu/k_BT), \qquad (3.1.2)$$

where k_B is the Boltzmann constant and T is the absolute temperature. Since N_1 and N_2 do not change with time in thermal equilibrium, the upward and downward transition rates should be equal, or

$$AN_2 + BN_2\rho_{\text{em}} = B'N_1\rho_{\text{em}}. \qquad (3.1.3)$$

By using Eq. (3.1.2) in Eq. (3.1.3), the spectral density ρ_{em} becomes

$$\rho_{\text{em}} = \frac{A/B}{(B'/B)\exp(h\nu/k_BT) - 1}. \qquad (3.1.4)$$

In thermal equilibrium, ρ_{em} should be identical with the spectral density of blackbody radiation given by *Planck's formula* [18]

$$\rho_{em} = \frac{8\pi h \nu^3/c^3}{\exp(h\nu/k_BT) - 1}. \qquad (3.1.5)$$

A comparison of Eqs. (3.1.4) and (3.1.5) provides the relations

$$A = (8\pi h \nu^3/c^3)B; \qquad B' = B. \qquad (3.1.6)$$

These relations were first obtained by Einstein [17]. For this reason, A and B are called *Einstein's coefficients*.

Two important conclusions can be drawn from Eqs. (3.1.1)–(3.1.6). First, R_{spon} can exceed both R_{stim} and R_{abs} considerably if $k_BT > h\nu$. Thermal sources operate in this regime. Second, for radiation in the visible or near-infrared region ($h\nu \sim 1$ eV), spontaneous emission always dominates over stimulated emission in thermal equilibrium at room temperature ($k_BT \approx 25$ meV) because

$$R_{stim}/R_{spon} = [\exp(h\nu/k_BT) - 1]^{-1} \ll 1. \qquad (3.1.7)$$

Thus, all lasers must operate away from thermal equilibrium. This is achieved by pumping lasers with an external energy source.

Even for an atomic system pumped externally, stimulated emission may not be the dominant process since it has to compete with the absorption process. R_{stim} can exceed R_{abs} only when $N_2 > N_1$. This condition is referred to as *population inversion* and is never realized for systems in thermal equilibrium [see Eq. (3.1.2)]. Population inversion is a prerequisite for laser operation. In atomic systems, it is achieved by using three- and four-level pumping schemes [18] such that an external energy source raises the atomic population from the ground state to an excited state lying above the energy state E_2 in Fig. 3.1.

The emission and absorption rates in semiconductors should take into account the energy bands associated with a semiconductor [5]. Figure 3.2 shows the emission process schematically using the simplest band structure, consisting of parabolic conduction and valence bands in the energy–wave-vector space (E–**k** diagram). Spontaneous emission can occur only if the energy state E_2 is occupied by an electron and the energy state E_1 is empty (i.e., occupied by a hole). The occupation probability for electrons in the conduction and valence bands is given by the *Fermi–Dirac distributions* [5]

$$f_c(E_2) = \{1 + \exp[(E_2 - E_{fc})/k_BT]\}^{-1}, \qquad (3.1.8)$$
$$f_v(E_1) = \{1 + \exp[(E_1 - E_{fv})/k_BT]\}^{-1}, \qquad (3.1.9)$$

where E_{fc} and E_{fv} are the Fermi levels. The total spontaneous emission rate at a frequency ω is obtained by summing over all possible transitions between the two bands such that $E_2 - E_1 = E_{em} = \hbar\omega$, where $\omega = 2\pi\nu$, $\hbar = h/2\pi$, and E_{em} is the energy of the emitted photon. The result is

$$R_{spon}(\omega) = \int_{E_c}^{\infty} A(E_1, E_2) f_c(E_2)[1 - f_v(E_1)]\rho_{cv}\, dE_2, \qquad (3.1.10)$$

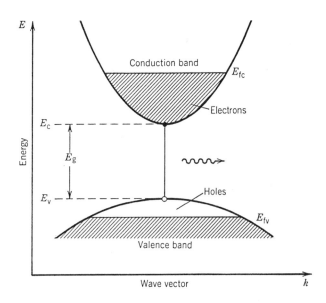

Figure 3.2: Conduction and valence bands of a semiconductor. Electrons in the conduction band and holes in the valence band can recombine and emit a photon through spontaneous emission as well as through stimulated emission.

where ρ_{cv} is the *joint density of states*, defined as the number of states per unit volume per unit energy range, and is given by [18]

$$\rho_{cv} = \frac{(2m_r)^{3/2}}{2\pi^2\hbar^3}(\hbar\omega - E_g)^{1/2}. \tag{3.1.11}$$

In this equation, E_g is the bandgap and m_r is the reduced mass, defined as $m_r = m_c m_v/(m_c + m_v)$, where m_c and m_v are the effective masses of electrons and holes in the conduction and valence bands, respectively. Since ρ_{cv} is independent of E_2 in Eq. (3.1.10), it can be taken outside the integral. By contrast, $A(E_1, E_2)$ generally depends on E_2 and is related to the momentum matrix element in a semiclassical perturbation approach commonly used to calculate it [2].

The stimulated emission and absorption rates can be obtained in a similar manner and are given by

$$R_{\text{stim}}(\omega) = \int_{E_c}^{\infty} B(E_1, E_2) f_c(E_2)[1 - f_v(E_1)]\rho_{cv}\rho_{\text{em}}\,dE_2, \tag{3.1.12}$$

$$R_{\text{abs}}(\omega) = \int_{E_c}^{\infty} B(E_1, E_2) f_v(E_1)[1 - f_c(E_2)]\rho_{cv}\rho_{\text{em}}\,dE_2, \tag{3.1.13}$$

where $\rho_{\text{em}}(\omega)$ is the spectral density of photons introduced in a manner similar to Eq. (3.1.1). The *population-inversion condition* $R_{\text{stim}} > R_{\text{abs}}$ is obtained by comparing Eqs. (3.1.12) and (3.1.13), resulting in $f_c(E_2) > f_v(E_1)$. If we use Eqs. (3.1.8) and (3.1.9), this condition is satisfied when

$$E_{fc} - E_{fv} > E_2 - E_1 > E_g. \tag{3.1.14}$$

Since the minimum value of $E_2 - E_1$ equals E_g, the separation between the Fermi levels must exceed the bandgap for population inversion to occur [19]. In thermal equilibrium, the two Fermi levels coincide ($E_{fc} = E_{fv}$). They can be separated by pumping energy into the semiconductor from an external energy source. The most convenient way for pumping a semiconductor is to use a forward-biased *p–n* junction.

3.1.2 *p–n* Junctions

At the heart of a semiconductor optical source is the *p–n* junction, formed by bringing a *p*-type and an *n*-type semiconductor into contact. Recall that a semiconductor is made *n*-type or *p*-type by doping it with impurities whose atoms have an excess valence electron or one less electron compared with the semiconductor atoms. In the case of *n*-type semiconductor, the excess electrons occupy the conduction-band states, normally empty in undoped (intrinsic) semiconductors. The Fermi level, lying in the middle of the bandgap for intrinsic semiconductors, moves toward the conduction band as the dopant concentration increases. In a heavily doped *n*-type semiconductor, the Fermi level E_{fc} lies inside the conduction band; such semiconductors are said to be degenerate. Similarly, the Fermi level E_{fv} moves toward the valence band for *p*-type semiconductors and lies inside it under heavy doping. In thermal equilibrium, the Fermi level must be continuous across the *p–n* junction. This is achieved through diffusion of electrons and holes across the junction. The charged impurities left behind set up an electric field strong enough to prevent further diffusion of electrons and holds under equilibrium conditions. This field is referred to as the built-in electric field. Figure 3.3(a) shows the energy-band diagram of a *p–n* junction in thermal equilibrium and under forward bias.

When a *p–n* junction is forward biased by applying an external voltage, the built-in electric field is reduced. This reduction results in diffusion of electrons and holes across the junction. An electric current begins to flow as a result of carrier diffusion. The current I increases exponentially with the applied voltage V according to the well-known relation [5]

$$I = I_s[\exp(qV/k_BT) - 1], \qquad (3.1.15)$$

where I_s is the saturation current and depends on the diffusion coefficients associated with electrons and holes. As seen in Fig. 3.3(a), in a region surrounding the junction (known as the depletion width), electrons and holes are present simultaneously when the *p–n* junction is forward biased. These electrons and holes can recombine through spontaneous or stimulated emission and generate light in a semiconductor optical source.

The *p–n* junction shown in Fig. 3.3(a) is called the *homojunction*, since the same semiconductor material is used on both sides of the junction. A problem with the homojunction is that electron–hole recombination occurs over a relatively wide region (~ 1–$10\ \mu m$) determined by the diffusion length of electrons and holes. Since the carriers are not confined to the immediate vicinity of the junction, it is difficult to realize high carrier densities. This carrier-confinement problem can be solved by sandwiching a thin layer between the *p*-type and *n*-type layers such that the bandgap of the sandwiched layer is smaller than the layers surrounding it. The middle layer may or may

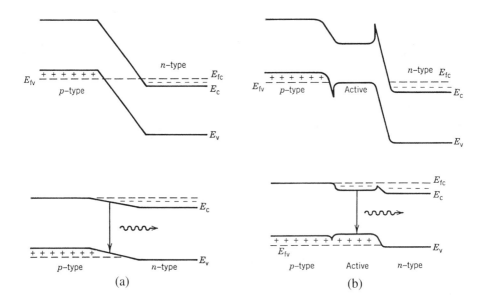

Figure 3.3: Energy-band diagram of (a) homostructure and (b) double-heterostructure p–n junctions in thermal equilibrium (top) and under forward bias (bottom).

not be doped, depending on the device design; its role is to confine the carriers injected inside it under forward bias. The carrier confinement occurs as a result of bandgap discontinuity at the junction between two semiconductors which have the same crystalline structure (the same lattice constant) but different bandgaps. Such junctions are called *heterojunctions*, and such devices are called *double heterostructures*. Since the thickness of the sandwiched layer can be controlled externally (typically, $\sim 0.1 \ \mu$m), high carrier densities can be realized at a given injection current. Figure 3.3(b) shows the energy-band diagram of a double heterostructure with and without forward bias.

The use of a heterostructure geometry for semiconductor optical sources is doubly beneficial. As already mentioned, the bandgap difference between the two semiconductors helps to confine electrons and holes to the middle layer, also called the active layer since light is generated inside it as a result of electron–hole recombination. However, the active layer also has a slightly larger refractive index than the surrounding p-type and n-type cladding layers simply because its bandgap is smaller. As a result of the refractive-index difference, the active layer acts as a dielectric waveguide and supports optical modes whose number can be controlled by changing the active-layer thickness (similar to the modes supported by a fiber core). The main point is that a heterostructure confines the generated light to the active layer because of its higher refractive index. Figure 3.4 illustrates schematically the simultaneous confinement of charge carriers and the optical field to the active region through a heterostructure design. It is this feature that has made semiconductor lasers practical for a wide variety of applications.

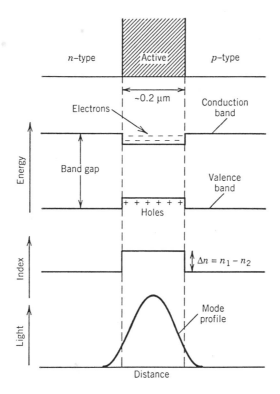

Figure 3.4: Simultaneous confinement of charge carriers and optical field in a double-heterostructure design. The active layer has a lower bandgap and a higher refractive index than those of *p*-type and *n*-type cladding layers.

3.1.3 Nonradiative Recombination

When a *p–n* junction is forward-biased, electrons and holes are injected into the active region, where they recombine to produce light. In any semiconductor, electrons and holes can also recombine nonradiatively. Nonradiative recombination mechanisms include recombination at traps or defects, surface recombination, and the Auger recombination [5]. The last mechanism is especially important for semiconductor lasers emitting light in the wavelength range 1.3–1.6 μm because of a relatively small bandgap of the active layer [2]. In the Auger recombination process, the energy released during electron–hole recombination is given to another electron or hole as kinetic energy rather than producing light.

From the standpoint of device operation, all nonradiative processes are harmful, as they reduce the number of electron–hole pairs that emit light. Their effect is quantified through the *internal quantum efficiency*, defined as

$$\eta_{\text{int}} = \frac{R_{\text{rr}}}{R_{\text{tot}}} = \frac{R_{\text{rr}}}{R_{\text{rr}} + R_{\text{nr}}}, \qquad (3.1.16)$$

where R_{rr} is the radiative recombination rate, R_{nr} is the nonradiative recombination rate, and $R_{tot} \equiv R_{rr} + R_{nr}$ is the total recombination rate. It is customary to introduce the recombination times τ_{rr} and τ_{nr} using $R_{rr} = N/\tau_{rr}$ and $R_{nr} = N/\tau_{nr}$, where N is the carrier density. The internal quantum efficiency is then given by

$$\eta_{int} = \frac{\tau_{nr}}{\tau_{rr} + \tau_{nr}}. \qquad (3.1.17)$$

The radiative and nonradiative recombination times vary from semiconductor to semiconductor. In general, τ_{rr} and τ_{nr} are comparable for direct-bandgap semiconductors, whereas τ_{nr} is a small fraction ($\sim 10^{-5}$) of τ_{rr} for semiconductors with an indirect bandgap. A semiconductor is said to have a direct bandgap if the conduction-band minimum and the valence-band maximum occur for the same value of the electron wave vector (see Fig. 3.2). The probability of radiative recombination is large in such semiconductors, since it is easy to conserve both energy and momentum during electron–hole recombination. By contrast, indirect-bandgap semiconductors require the assistance of a phonon for conserving momentum during electron–hole recombination. This feature reduces the probability of radiative recombination and increases τ_{rr} considerably compared with τ_{nr} in such semiconductors. As evident from Eq. (3.1.17), $\eta_{int} \ll 1$ under such conditions. Typically, $\eta_{int} \sim 10^{-5}$ for Si and Ge, the two semiconductors commonly used for electronic devices. Both are not suitable for optical sources because of their indirect bandgap. For direct-bandgap semiconductors such as GaAs and InP, $\eta_{int} \approx 0.5$ and approaches 1 when stimulated emission dominates.

The radiative recombination rate can be written as $R_{rr} = R_{spon} + R_{stim}$ when radiative recombination occurs through spontaneous as well as stimulated emission. For LEDs, R_{stim} is negligible compared with R_{spon}, and R_{rr} in Eq. (3.1.16) is replaced with R_{spon}. Typically, R_{spon} and R_{nr} are comparable in magnitude, resulting in an internal quantum efficiency of about 50%. However, η_{int} approaches 100% for semiconductor lasers as stimulated emission begins to dominate with an increase in the output power.

It is useful to define a quantity known as the *carrier lifetime* τ_c such that it represents the total recombination time of charged carriers in the absence of stimulated recombination. It is defined by the relation

$$R_{spon} + R_{nr} = N/\tau_c, \qquad (3.1.18)$$

where N is the carrier density. If R_{spon} and R_{nr} vary linearly with N, τ_c becomes a constant. In practice, both of them increase nonlinearly with N such that $R_{spon} + R_{nr} = A_{nr}N + BN^2 + CN^3$, where A_{nr} is the nonradiative coefficient due to recombination at defects or traps, B is the spontaneous radiative recombination coefficient, and C is the Auger coefficient. The carrier lifetime then becomes N dependent and is obtained by using $\tau_c^{-1} = A_{nr} + BN + CN^2$. In spite of its N dependence, the concept of carrier lifetime τ_c is quite useful in practice.

3.1.4 Semiconductor Materials

Almost any semiconductor with a direct bandgap can be used to make a p–n homojunction capable of emitting light through spontaneous emission. The choice is, however, considerably limited in the case of heterostructure devices because their performance

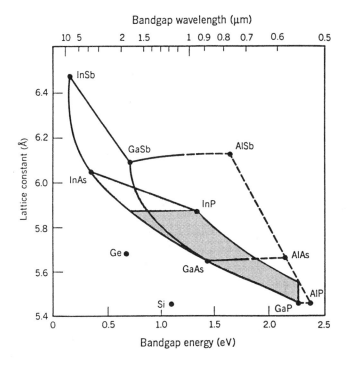

Figure 3.5: Lattice constants and bandgap energies of ternary and quaternary compounds formed by using nine group III–V semiconductors. Shaded area corresponds to possible InGaAsP and AlGaAs structures. Horizontal lines passing through InP and GaAs show the lattice-matched designs. (After Ref. [18]; ©1991 Wiley; reprinted with permission.)

depends on the quality of the heterojunction interface between two semiconductors of different bandgaps. To reduce the formation of lattice defects, the lattice constant of the two materials should match to better than 0.1%. Nature does not provide semiconductors whose lattice constants match to such precision. However, they can be fabricated artificially by forming ternary and quaternary compounds in which a fraction of the lattice sites in a naturally occurring binary semiconductor (e.g., GaAs) is replaced by other elements. In the case of GaAs, a ternary compound $Al_xGa_{1-x}As$ can be made by replacing a fraction x of Ga atoms by Al atoms. The resulting semiconductor has nearly the same lattice constant, but its bandgap increases. The bandgap depends on the fraction x and can be approximated by a simple linear relation [2]

$$E_g(x) = 1.424 + 1.247x \qquad (0 < x < 0.45), \qquad (3.1.19)$$

where E_g is expressed in electron-volt (eV) units.

Figure 3.5 shows the interrelationship between the bandgap E_g and the lattice constant a for several ternary and quaternary compounds. Solid dots represent the binary semiconductors, and lines connecting them corresponds to ternary compounds. The dashed portion of the line indicates that the resulting ternary compound has an indirect bandgap. The area of a closed polygon corresponds to quaternary compounds. The bandgap is not necessarily direct for such semiconductors. The shaded area in Fig. 3.5 represents the ternary and quaternary compounds with a direct bandgap formed by using the elements indium (In), gallium (Ga), arsenic (As), and phosphorus (P).

The horizontal line connecting GaAs and AlAs corresponds to the ternary compound $Al_xGa_{1-x}As$, whose bandgap is direct for values of x up to about 0.45 and is given by Eq. (3.1.19). The active and cladding layers are formed such that x is larger for the cladding layers compared with the value of x for the active layer. The wavelength of the emitted light is determined by the bandgap since the photon energy is approximately equal to the bandgap. By using $E_g \approx h\nu = hc/\lambda$, one finds that $\lambda \approx 0.87 \ \mu m$ for an active layer made of GaAs ($E_g = 1.424$ eV). The wavelength can be reduced to about 0.81 μm by using an active layer with $x = 0.1$. Optical sources based on GaAs typically operate in the range 0.81–0.87 μm and were used in the first generation of fiber-optic communication systems.

As discussed in Chapter 2, it is beneficial to operate lightwave systems in the wavelength range 1.3–1.6 μm, where both dispersion and loss of optical fibers are considerably reduced compared with the 0.85-μm region. InP is the base material for semiconductor optical sources emitting light in this wavelength region. As seen in Fig. 3.5 by the horizontal line passing through InP, the bandgap of InP can be reduced considerably by making the quaternary compound $In_{1-x}Ga_xAs_yP_{1-y}$ while the lattice constant remains matched to InP. The fractions x and y cannot be chosen arbitrarily but are related by $x/y = 0.45$ to ensure matching of the lattice constant. The bandgap of the quaternary compound can be expressed in terms of y only and is well approximated by [2]

$$E_g(y) = 1.35 - 0.72y + 0.12y^2, \tag{3.1.20}$$

where $0 \leq y \leq 1$. The smallest bandgap occurs for $y = 1$. The corresponding ternary compound $In_{0.55}Ga_{0.45}As$ emits light near 1.65 μm ($E_g = 0.75$ eV). By a suitable choice of the mixing fractions x and y, $In_{1-x}Ga_xAs_yP_{1-y}$ sources can be designed to operate in the wide wavelength range 1.0–1.65 μm that includes the region 1.3–1.6 μm important for optical communication systems.

The fabrication of semiconductor optical sources requires epitaxial growth of multiple layers on a base substrate (GaAs or InP). The thickness and composition of each layer need to be controlled precisely. Several epitaxial growth techniques can be used for this purpose. The three primary techniques are known as liquid-phase epitaxy (LPE), vapor-phase epitaxy (VPE), and molecular-beam epitaxy (MBE) depending on whether the constituents of various layers are in the liquid form, vapor form, or in the form of a molecular beam. The VPE technique is also called chemical-vapor deposition. A variant of this technique is metal-organic chemical-vapor deposition (MOCVD), in which metal alkalis are used as the mixing compounds. Details of these techniques are available in the literature [2].

Both the MOCVD and MBE techniques provide an ability to control layer thickness to within 1 nm. In some lasers, the thickness of the active layer is small enough that electrons and holes act as if they are confined to a quantum well. Such confinement leads to quantization of the energy bands into subbands. The main consequence is that the joint density of states ρ_{cv} acquires a staircase-like structure [5]. Such a modification of the density of states affects the gain characteristics considerably and improves the laser performance. Such *quantum-well lasers* have been studied extensively [14]. Often, multiple active layers of thickness 5–10 nm, separated by transparent barrier layers of about 10 nm thickness, are used to improve the device performance. Such lasers are called *multiquantum-well* (MQW) lasers. Another feature that has improved the performance of MQW lasers is the introduction of intentional, but controlled strain within active layers. The use of thin active layers permits a slight mismatch between lattice constants without introducing defects. The resulting strain changes the band structure and improves the laser performance [5]. Such semiconductor lasers are called *strained* MQW lasers. The concept of quantum-well lasers has also been extended to make quantum-wire and quantum-dot lasers in which electrons are confined in more than one dimension [14]. However, such devices were at the research stage in 2001. Most semiconductor lasers deployed in lightwave systems use the MQW design.

3.2   Light-Emitting Diodes

A forward-biased *p–n* junction emits light through spontaneous emission, a pheno-menon referred to as electroluminescence. In its simplest form, an LED is a forward-biased *p–n* homojunction. Radiative recombination of electron–hole pairs in the deple-tion region generates light; some of it escapes from the device and can be coupled into an optical fiber. The emitted light is incoherent with a relatively wide spectral width (30–60 nm) and a relatively large angular spread. In this section we discuss the char-acteristics and the design of LEDs from the standpoint of their application in optical communication systems [20].

3.2.1   Power–Current Characteristics

It is easy to estimate the internal power generated by spontaneous emission. At a given current I the carrier-injection rate is I/q. In the steady state, the rate of electron–hole pairs recombining through radiative and nonradiative processes is equal to the carrier-injection rate I/q. Since the internal quantum efficiency η_{int} determines the fraction of electron–hole pairs that recombine through spontaneous emission, the rate of photon generation is simply $\eta_{int}I/q$. The internal optical power is thus given by

$$P_{int} = \eta_{int}(\hbar\omega/q)I, \qquad (3.2.1)$$

where $\hbar\omega$ is the photon energy, assumed to be nearly the same for all photons. If η_{ext} is the fraction of photons escaping from the device, the emitted power is given by

$$P_e = \eta_{ext}P_{int} = \eta_{ext}\eta_{int}(\hbar\omega/q)I. \qquad (3.2.2)$$

The quantity η_{ext} is called the *external quantum efficiency*. It can be calculated by taking into account internal absorption and the total internal reflection at the semicon-ductor–air interface. As seen in Fig. 3.6, only light emitted within a cone of angle θ_c, where $\theta_c = \sin^{-1}(1/n)$ is the critical angle and n is the refractive index of the semiconductor material, escapes from the LED surface. Internal absorption can be avoided by using heterostructure LEDs in which the cladding layers surrounding the

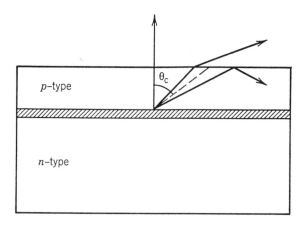

Figure 3.6: Total internal reflection at the output facet of an LED. Only light emitted within a cone of angle θ_c is transmitted, where θ_c is the critical angle for the semiconductor–air interface.

active layer are transparent to the radiation generated. The external quantum efficiency can then be written as

$$\eta_{\text{ext}} = \frac{1}{4\pi} \int_0^{\theta_c} T_f(\theta)(2\pi \sin\theta)\, d\theta, \tag{3.2.3}$$

where we have assumed that the radiation is emitted uniformly in all directions over a solid angle of 4π. The Fresnel transmissivity T_f depends on the incidence angle θ. In the case of normal incidence ($\theta = 0$), $T_f(0) = 4n/(n+1)^2$. If we replace for simplicity $T_f(\theta)$ by $T_f(0)$ in Eq. (3.2.3), η_{ext} is given approximately by

$$\eta_{\text{ext}} = n^{-1}(n+1)^{-2}. \tag{3.2.4}$$

By using Eq. (3.2.4) in Eq. (3.2.2) we obtain the power emitted from one facet (see Fig. 3.6). If we use $n = 3.5$ as a typical value, $\eta_{\text{ext}} = 1.4\%$, indicating that only a small fraction of the internal power becomes the useful output power. A further loss in useful power occurs when the emitted light is coupled into an optical fiber. Because of the incoherent nature of the emitted light, an LED acts as a *Lambertian source* with an angular distribution $S(\theta) = S_0 \cos\theta$, where S_0 is the intensity in the direction $\theta = 0$. The coupling efficiency for such a source [20] is $\eta_c = (\text{NA})^2$. Since the numerical aperture (NA) for optical fibers is typically in the range 0.1–0.3, only a few percent of the emitted power is coupled into the fiber. Normally, the launched power for LEDs is $100\ \mu\text{W}$ or less, even though the internal power can easily exceed 10 mW.

A measure of the LED performance is the total quantum efficiency η_{tot}, defined as the ratio of the emitted optical power P_e to the applied electrical power, $P_{\text{elec}} = V_0 I$, where V_0 is the voltage drop across the device. By using Eq. (3.2.2), η_{tot} is given by

$$\eta_{\text{tot}} = \eta_{\text{ext}}\eta_{\text{int}}(\hbar\omega/qV_0). \tag{3.2.5}$$

Typically, $\hbar\omega \approx qV_0$, and $\eta_{\text{tot}} \approx \eta_{\text{ext}}\eta_{\text{int}}$. The total quantum efficiency η_{tot}, also called the *power-conversion efficiency* or the *wall-plug efficiency*, is a measure of the overall performance of the device.

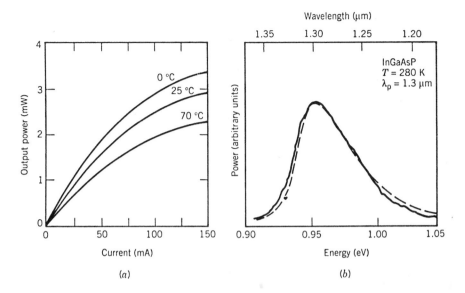

Figure 3.7: (a) Power–current curves at several temperatures; (b) spectrum of the emitted light for a typical 1.3-μm LED. The dashed curve shows the theoretically calculated spectrum. (After Ref. [21]; ©1983 AT&T; reprinted with permission.)

Another quantity sometimes used to characterize the LED performance is the *responsivity* defined as the ratio $R_{LED} = P_e/I$. From Eq. (3.2.2),

$$R_{LED} = \eta_{ext}\eta_{int}(\hbar\omega/q). \qquad (3.2.6)$$

A comparison of Eqs. (3.2.5) and (3.2.6) shows that $R_{LED} = \eta_{tot}V_0$. Typical values of R_{LED} are ~ 0.01 W/A. The responsivity remains constant as long as the linear relation between P_e and I holds. In practice, this linear relationship holds only over a limited current range [21]. Figure 3.7(a) shows the power–current (P–I) curves at several temperatures for a typical 1.3-μm LED. The responsivity of the device decreases at high currents above 80 mA because of bending of the P–I curve. One reason for this decrease is related to the increase in the active-region temperature. The internal quantum efficiency η_{int} is generally temperature dependent because of an increase in the nonradiative recombination rates at high temperatures.

3.2.2 LED Spectrum

As seen in Section 2.3, the spectrum of a light source affects the performance of optical communication systems through fiber dispersion. The LED spectrum is related to the spectrum of spontaneous emission, $R_{spon}(\omega)$, given in Eq. (3.1.10). In general, $R_{spon}(\omega)$ is calculated numerically and depends on many material parameters. However, an approximate expression can be obtained if $A(E_1, E_2)$ is assumed to be nonzero only over a narrow energy range in the vicinity of the photon energy, and the Fermi functions are approximated by their exponential tails under the assumption of weak injection [5]. The result is

$$R_{spon}(\omega) = A_0(\hbar\omega - E_g)^{1/2} \exp[-(\hbar\omega - E_g)/k_BT], \qquad (3.2.7)$$

where A_0 is a constant and E_g is the bandgap. It is easy to deduce that $R_{spon}(\omega)$ peaks when $\hbar\omega = E_g + k_BT/2$ and has a full-width at half-maximum (FWHM) $\Delta\nu \approx 1.8k_BT/h$. At room temperature ($T = 300$ K) the FWHM is about 11 THz. In practice, the spectral width is expressed in nanometers by using $\Delta\nu = (c/\lambda^2)\Delta\lambda$ and increases as λ^2 with an increase in the emission wavelength λ. As a result, $\Delta\lambda$ is larger for In-GaAsP LEDs emitting at 1.3 μm by about a factor of 1.7 compared with GaAs LEDs. Figure 3.7(b) shows the output spectrum of a typical 1.3-μm LED and compares it with the theoretical curve obtained by using Eq. (3.2.7). Because of a large spectral width ($\Delta\lambda = 50$–60 nm), the bit rate–distance product is limited considerably by fiber dispersion when LEDs are used in optical communication systems. LEDs are suitable primarily for local-area-network applications with bit rates of 10–100 Mb/s and transmission distances of a few kilometers.

3.2.3 Modulation Response

The modulation response of LEDs depends on carrier dynamics and is limited by the carrier lifetime τ_c defined by Eq. (3.1.18). It can be determined by using a *rate equation* for the carrier density N. Since electrons and holes are injected in pairs and recombine in pairs, it is enough to consider the rate equation for only one type of charge carrier. The rate equation should include all mechanisms through which electrons appear and disappear inside the active region. For LEDs it takes the simple form (since stimulated emission is negligible)

$$\frac{dN}{dt} = \frac{I}{qV} - \frac{N}{\tau_c}, \qquad (3.2.8)$$

where the last term includes both radiative and nonradiative recombination processes through the carrier lifetime τ_c. Consider sinusoidal modulation of the injected current in the form (the use of complex notation simplifies the math)

$$I(t) = I_b + I_m \exp(i\omega_m t), \qquad (3.2.9)$$

where I_b is the bias current, I_m is the modulation current, and ω_m is the modulation frequency. Since Eq. (3.2.8) is linear, its general solution can be written as

$$N(t) = N_b + N_m \exp(i\omega_m t), \qquad (3.2.10)$$

where $N_b = \tau_c I_b/qV$, V is the volume of active region and N_m is given by

$$N_m(\omega_m) = \frac{\tau_c I_m/qV}{1 + i\omega_m \tau_c}. \qquad (3.2.11)$$

The modulated power P_m is related to $|N_m|$ linearly. One can define the LED transfer function $H(\omega_m)$ as

$$H(\omega_m) = \frac{N_m(\omega_m)}{N_m(0)} = \frac{1}{1 + i\omega_m \tau_c}. \qquad (3.2.12)$$

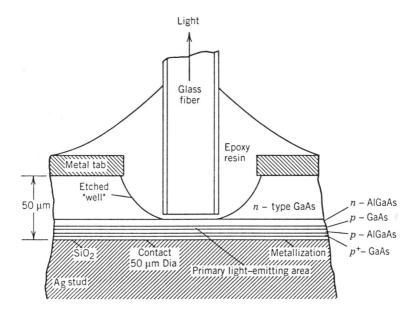

Figure 3.8: Schematic of a surface-emitting LED with a double-heterostructure geometry.

In analogy with the case of optical fibers (see Section 2.4.4), the *3-dB modulation bandwidth* $f_{3\,dB}$ is defined as the modulation frequency at which $|H(\omega_m)|$ is reduced by 3 dB or by a factor of 2. The result is

$$f_{3\,dB} = \sqrt{3}\,(2\pi\tau_c)^{-1}. \qquad (3.2.13)$$

Typically, τ_c is in the range 2–5 ns for InGaAsP LEDs. The corresponding LED modulation bandwidth is in the range 50–140 MHz. Note that Eq. (3.2.13) provides the optical bandwidth because $f_{3\,dB}$ is defined as the frequency at which optical power is reduced by 3 dB. The corresponding electrical bandwidth is the frequency at which $|H(\omega_m)|^2$ is reduced by 3 dB and is given by $(2\pi\tau_c)^{-1}$.

3.2.4 LED Structures

The LED structures can be classified as surface-emitting or edge-emitting, depending on whether the LED emits light from a surface that is parallel to the junction plane or from the edge of the junction region. Both types can be made using either a p–n homo-junction or a heterostructure design in which the active region is surrounded by p- and n-type cladding layers. The heterostructure design leads to superior performance, as it provides a control over the emissive area and eliminates internal absorption because of the transparent cladding layers.

Figure 3.8 shows schematically a surface-emitting LED design referred to as the *Burrus-type* LED [22]. The emissive area of the device is limited to a small region whose lateral dimension is comparable to the fiber-core diameter. The use of a gold stud avoids power loss from the back surface. The coupling efficiency is improved by

etching a well and bringing the fiber close to the emissive area. The power coupled into
the fiber depends on many parameters, such as the numerical aperture of the fiber and
the distance between fiber and LED. The addition of epoxy in the etched well tends
to increase the external quantum efficiency as it reduces the refractive-index mismatch.
Several variations of the basic design exist in the literature. In one variation, a truncated
spherical microlens fabricated inside the etched well is used to couple light into the
fiber [23]. In another variation, the fiber end is itself formed in the form of a spherical
lens [24]. With a proper design, surface-emitting LEDs can couple up to 1% of the
internally generated power into an optical fiber.

The edge-emitting LEDs employ a design commonly used for stripe-geometry
semiconductor lasers (see Section 3.3.3). In fact, a semiconductor laser is converted
into an LED by depositing an antireflection coating on its output facet to suppress lasing
action. Beam divergence of edge-emitting LEDs differs from surface-emitting LEDs
because of waveguiding in the plane perpendicular to the junction. Surface-emitting
LEDs operate as a Lambertian source with angular distribution $S_e(\theta) = S_0 \cos \theta$ in
both directions. The resulting beam divergence has a FWHM of 120° in each direction.
In contrast, edge-emitting LEDs have a divergence of only about 30° in the direction
perpendicular to the junction plane. Considerable light can be coupled into a fiber of
even low numerical aperture (< 0.3) because of reduced divergence and high radiance
at the emitting facet [25]. The modulation bandwidth of edge-emitting LEDs is generally larger (~ 200 MHz) than that of surface-emitting LEDs because of a reduced
carrier lifetime at the same applied current [26]. The choice between the two designs
is dictated, in practice, by a compromise between cost and performance.

In spite of a relatively low output power and a low bandwidth of LEDs compared
with those of lasers, LEDs are useful for low-cost applications requiring data transmission at a bit rate of 100 Mb/s or less over a few kilometers. For this reason, several
new LED structures were developed during the 1990s [27]–[32]. In one design, known
as *resonant-cavity* LED [27], two metal mirrors are fabricated around the epitaxially
grown layers, and the device is bonded to a silicon substrate. In a variant of this idea,
the bottom mirror is fabricated epitaxially by using a stack of alternating layers of two
different semiconductors, while the top mirror consists of a deformable membrane sus-
pended by an air gap [28]. The operating wavelength of such an LED can be tuned over
40 nm by changing the air-gap thickness. In another scheme, several quantum wells
with different compositions and bandgaps are grown to form a MQW structure [29].
Since each quantum well emits light at a different wavelength, such LEDs can have an
extremely broad spectrum (extending over a 500-nm wavelength range) and are useful
for local-area WDM networks.

3.3   Semiconductor Lasers

Semiconductor lasers emit light through stimulated emission. As a result of the fundamental differences between spontaneous and stimulated emission, they are not only
capable of emitting high powers (~ 100 mW), but also have other advantages related
to the coherent nature of emitted light. A relatively narrow angular spread of the output
beam compared with LEDs permits high coupling efficiency ($\sim 50\%$) into single-mode

fibers. A relatively narrow spectral width of emitted light allows operation at high bit rates ($\sim$ 10 Gb/s), since fiber dispersion becomes less critical for such an optical source. Furthermore, semiconductor lasers can be modulated directly at high frequencies (up to 25 GHz) because of a short recombination time associated with stimulated emission. Most fiber-optic communication systems use semiconductor lasers as an optical source because of their superior performance compared with LEDs. In this section the output characteristics of semiconductor lasers are described from the standpoint of their applications in lightwave systems. More details can be found in Refs. [2]–[14], books devoted entirely to semiconductor lasers.

3.3.1 Optical Gain

As discussed in Section 3.1.1, stimulated emission can dominate only if the condition of population inversion is satisfied. For semiconductor lasers this condition is realized by doping the p-type and n-type cladding layers so heavily that the Fermi-level separation exceeds the bandgap [see Eq. (3.1.14)] under forward biasing of the p–n junction. When the injected carrier density in the active layer exceeds a certain value, known as the transparency value, population inversion is realized and the active region exhibits optical gain. An input signal propagating inside the active layer would then amplify as $\exp(gz)$, where g is the *gain coefficient*. One can calculate g by noting that it is proportional to $R_{\text{stim}} - R_{\text{abs}}$, where R_{stim} and R_{abs} are given by Eqs. (3.1.12) and (3.1.13), respectively. In general, g is calculated numerically. Figure 3.9(a) shows the gain calculated for a 1.3-μm InGaAsP active layer at different values of the injected carrier density N. For $N = 1 \times 10^{18}$ cm^{-3}, $g < 0$, as population inversion has not yet occurred. As N increases, g becomes positive over a spectral range that increases with N. The peak value of the gain, g_p, also increases with N, together with a shift of the peak toward higher photon energies. The variation of g_p with N is shown in Fig. 3.9(b). For $N > 1.5 \times 10^{18}$ cm^{-3}, g_p varies almost linearly with N. Figure 3.9 shows that the optical gain in semiconductors increases rapidly once population inversion is realized. It is because of such a high gain that semiconductor lasers can be made with physical dimensions of less than 1 mm.

The nearly linear dependence of g_p on N suggests an empirical approach in which the peak gain is approximated by

$$g_p(N) = \sigma_g(N - N_T), \tag{3.3.1}$$

where N_T is the transparency value of the carrier density and σ_g is the gain cross section; σ_g is also called the *differential gain*. Typical values of N_T and σ_g for InGaAsP lasers are in the range 1.0–1.5$\times 10^{18}$ cm^{-3} and 2–3$\times 10^{-16}$ cm^2, respectively [2]. As seen in Fig. 3.9(b), the approximation (3.3.1) is reasonable in the high-gain region where g_p exceeds 100 cm^{-1}; most semiconductor lasers operate in this region. The use of Eq. (3.3.1) simplifies the analysis considerably, as band-structure details do not appear directly. The parameters σ_g and N_T can be estimated from numerical calculations such as those shown in Fig. 3.9(b) or can be measured experimentally.

Semiconductor lasers with a larger value of σ_g generally perform better, since the same amount of gain can be realized at a lower carrier density or, equivalently, at a

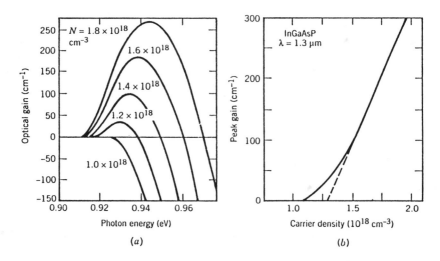

Figure 3.9: (a) Gain spectrum of a 1.3-μm InGaAsP laser at several carrier densities N. (b) Variation of peak gain g_p with N. The dashed line shows the quality of a linear fit in the high-gain region. (After Ref. [2]; ©1993 Van Nostrand Reinhold; reprinted with permission.)

lower injected current. In quantum-well semiconductor lasers, σ_g is typically larger by about a factor of two. The linear approximation in Eq. (3.3.1) for the peak gain can still be used in a limited range. A better approximation replaces Eq. (3.3.1) with $g_p(N) = g_0[1 + \ln(N/N_0)]$, where $g_p = g_0$ at $N = N_0$ and $N_0 = eN_T \approx 2.718N_T$ by using the definition $g_p = 0$ at $N = N_T$ [5].

3.3.2 Feedback and Laser Threshold

The optical gain alone is not enough for laser operation. The other necessary ingredient is *optical feedback*—it converts an amplifier into an oscillator. In most lasers the feedback is provided by placing the gain medium inside a *Fabry–Perot* (FP) cavity formed by using two mirrors. In the case of semiconductor lasers, external mirrors are not required as the two cleaved laser facets act as mirrors whose reflectivity is given by

$$R_m = \left(\frac{n-1}{n+1}\right)^2, \tag{3.3.2}$$

where n is the refractive index of the gain medium. Typically, $n = 3.5$, resulting in 30% facet reflectivity. Even though the FP cavity formed by two cleaved facets is relatively lossy, the gain is large enough that high losses can be tolerated. Figure 3.10 shows the basic structure of a semiconductor laser and the FP cavity associated with it.

The concept of *laser threshold* can be understood by noting that a certain fraction of photons generated by stimulated emission is lost because of cavity losses and needs to be replenished on a continuous basis. If the optical gain is not large enough to compensate for the cavity losses, the photon population cannot build up. Thus, a minimum amount of gain is necessary for the operation of a laser. This amount can be realized

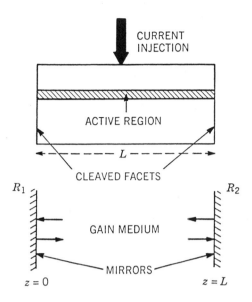

Figure 3.10: Structure of a semiconductor laser and the Fabry–Perot cavity associated with it. The cleaved facets act as partially reflecting mirrors.

only when the laser is pumped above a threshold level. The current needed to reach the threshold is called the *threshold current*.

A simple way to obtain the threshold condition is to study how the amplitude of a plane wave changes during one round trip. Consider a plane wave of amplitude E_0, frequency ω, and wave number $k = n\omega/c$. During one round trip, its amplitude increases by $\exp[(g/2)(2L)]$ because of gain (g is the power gain) and its phase changes by $2kL$, where L is the length of the laser cavity. At the same time, its amplitude changes by $\sqrt{R_1 R_2}\exp(-\alpha_{\text{int}}L)$ because of reflection at the laser facets and because of an internal loss α_{int} that includes free-carrier absorption, scattering, and other possible mechanisms. Here R_1 and R_2 are the reflectivities of the laser facets. Even though $R_1 = R_2$ in most cases, the two reflectivities can be different if laser facets are coated to change their natural reflectivity. In the steady state, the plane wave should remain unchanged after one round trip, i.e.,

$$E_0 \exp(gL)\sqrt{R_1 R_2}\exp(-\alpha_{\text{int}}L)\exp(2ikL) = E_0. \tag{3.3.3}$$

By equating the amplitude and the phase on two sides, we obtain

$$g = \alpha_{\text{int}} + \frac{1}{2L}\ln\left(\frac{1}{R_1 R_2}\right) = \alpha_{\text{int}} + \alpha_{\text{mir}} = \alpha_{\text{cav}}, \tag{3.3.4}$$

$$2kL = 2m\pi \quad \text{or} \quad \nu = \nu_m = mc/2nL, \tag{3.3.5}$$

where $k = 2\pi n\nu/c$ and m is an integer. Equation (3.3.4) shows that the gain g equals total cavity loss α_{cav} at threshold and beyond. It is important to note that g is not the same as the material gain g_m shown in Fig. 3.9. As discussed in Section 3.3.3, the

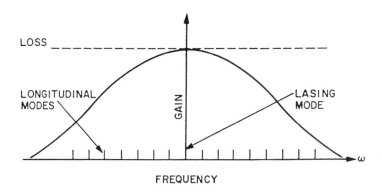

Figure 3.11: Gain and loss profiles in semiconductor lasers. Vertical bars show the location of longitudinal modes. The laser threshold is reached when the gain of the longitudinal mode closest to the gain peak equals loss.

optical mode extends beyond the active layer while the gain exists only inside it. As a result, $g = \Gamma g_m$, where Γ is the confinement factor of the active region with typical values <0.4.

The phase condition in Eq. (3.3.5) shows that the laser frequency ν must match one of the frequencies in the set ν_m, where m is an integer. These frequencies correspond to the *longitudinal modes* and are determined by the optical length nL. The spacing $\Delta\nu_L$ between the longitudinal modes is constant ($\Delta\nu_L = c/2nL$) if the frequency dependence of n is ignored. It is given by $\Delta\nu_L = c/2n_gL$ when material dispersion is included [2]. Here the *group index* n_g is defined as $n_g = n + \omega(dn/d\omega)$. Typically, $\Delta\nu_L = 100$–200 GHz for $L = 200$–400 μm.

A FP semiconductor laser generally emits light in several longitudinal modes of the cavity. As seen in Fig. 3.11, the gain spectrum $g(\omega)$ of semiconductor lasers is wide enough (bandwidth ~ 10 THz) that many longitudinal modes of the FP cavity experience gain simultaneously. The mode closest to the gain peak becomes the dominant mode. Under ideal conditions, the other modes should not reach threshold since their gain always remains less than that of the main mode. In practice, the difference is extremely small (~ 0.1 cm^{-1}) and one or two neighboring modes on each side of the main mode carry a significant portion of the laser power together with the main mode. Such lasers are called multimode semiconductor lasers. Since each mode propagates inside the fiber at a slightly different speed because of group-velocity dispersion, the multimode nature of semiconductor lasers limits the bit-rate–distance product BL to values below 10 (Gb/s)-km for systems operating near 1.55 μm (see Fig. 2.13). The BL product can be increased by designing lasers oscillating in a single longitudinal mode. Such lasers are discussed in Section 3.4.

3.3.3 Laser Structures

The simplest structure of a semiconductor laser consists of a thin active layer (thickness ~ 0.1 μm) sandwiched between p-type and n-type cladding layers of another semi-

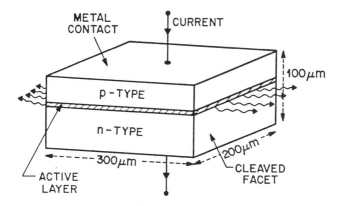

Figure 3.12: A broad-area semiconductor laser. The active layer (hatched region) is sandwiched between *p*-type and *n*-type cladding layers of a higher-bandgap material.

conductor with a higher bandgap. The resulting *p–n* heterojunction is forward-biased through metallic contacts. Such lasers are called *broad-area* semiconductor lasers since the current is injected over a relatively broad area covering the entire width of the laser chip ($\sim 100~\mu$m). Figure 3.12 shows such a structure. The laser light is emitted from the two cleaved facets in the form of an elliptic spot of dimensions $\sim 1 \times 100~\mu$m^2. In the direction perpendicular to the junction plane, the spot size is $\sim 1~\mu$m because of the heterostructure design of the laser. As discussed in Section 3.1.2, the active layer acts as a planar waveguide because its refractive index is larger than that of the surrounding cladding layers ($\Delta n \approx 0.3$). Similar to the case of optical fibers, it supports a certain number of modes, known as the transverse modes. In practice, the active layer is thin enough ($\sim 0.1~\mu$m) that the planar waveguide supports a single transverse mode. However, there is no such light-confinement mechanism in the lateral direction parallel to the junction plane. Consequently, the light generated spreads over the entire width of the laser. Broad-area semiconductor lasers suffer from a number of deficiencies and are rarely used in optical communication systems. The major drawbacks are a relatively high threshold current and a spatial pattern that is highly elliptical and that changes in an uncontrollable manner with the current. These problems can be solved by introducing a mechanism for light confinement in the lateral direction. The resulting semiconductor lasers are classified into two broad categories

Gain-guided semiconductor lasers solve the light-confinement problem by limiting current injection over a narrow stripe. Such lasers are also called *stripe-geometry* semiconductor lasers. Figure 3.13 shows two laser structures schematically. In one approach, a dielectric (SiO$_2$) layer is deposited on top of the *p*-layer with a central opening through which the current is injected [33]. In another, an *n*-type layer is deposited on top of the *p*-layer [34]. Diffusion of Zn over the central region converts the *n*-region into *p*-type. Current flows only through the central region and is blocked elsewhere because of the reverse-biased nature of the *p–n* junction. Many other variations exist [2]. In all designs, current injection over a narrow central stripe ($\sim 5~\mu$m width) leads to a spatially varying distribution of the carrier density (governed by car-

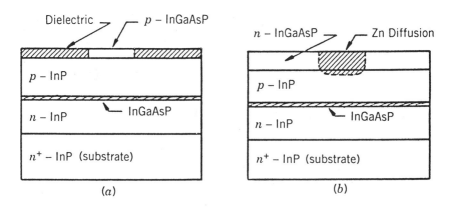

Figure 3.13: Cross section of two stripe-geometry laser structures used to design gain-guided semiconductor lasers and referred to as (a) oxide stripe and (b) junction stripe.

rier diffusion) in the lateral direction. The optical gain also peaks at the center of the stripe. Since the active layer exhibits large absorption losses in the region beyond the central stripe, light is confined to the stripe region. As the confinement of light is aided by gain, such lasers are called gain-guided. Their threshold current is typically in the range 50–100 mA, and light is emitted in the form of an elliptic spot of dimensions $\sim 1 \times 5~\mu m^2$. The major drawback is that the spot size is not stable as the laser power is increased [2]. Such lasers are rarely used in optical communication systems because of mode-stability problems.

The light-confinement problem is solved in the *index-guided* semiconductor lasers by introducing an index step Δn_L in the lateral direction so that a waveguide is formed in a way similar to the waveguide formed in the transverse direction by the heterostructure design. Such lasers can be subclassified as weakly and strongly index-guided semiconductor lasers, depending on the magnitude of Δn_L. Figure 3.14 shows examples of the two kinds of lasers. In a specific design known as the *ridge-waveguide laser*, a ridge is formed by etching parts of the p-layer [2]. A SiO$_2$ layer is then deposited to block the current flow and to induce weak index guiding. Since the refractive index of SiO$_2$ is considerably lower than the central p-region, the effective index of the transverse mode is different in the two regions [35], resulting in an index step $\Delta n_L \sim 0.01$. This index step confines the generated light to the ridge region. The magnitude of the index step is sensitive to many fabrication details, such as the ridge width and the proximity of the SiO$_2$ layer to the active layer. However, the relative simplicity of the ridge-waveguide design and the resulting low cost make such lasers attractive for some applications.

In strongly index-guided semiconductor lasers, the active region of dimensions $\sim 0.1 \times 1~\mu m^2$ is buried on all sides by several layers of lower refractive index. For this reason, such lasers are called *buried heterostructure* (BH) lasers. Several different kinds of BH lasers have been developed. They are known under names such as etched-mesa BH, planar BH, double-channel planar BH, and V-grooved or channeled substrate BH lasers, depending on the fabrication method used to realize the laser structure [2]. They all allow a relatively large index step ($\Delta n_L \sim 0.1$) in the lateral direction and, as

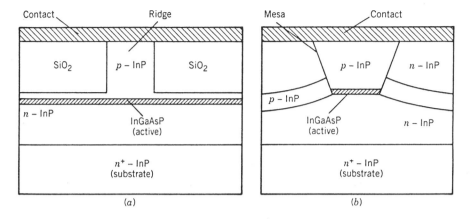

Figure 3.14: Cross section of two index-guided semiconductor lasers: (a) ridge-waveguide structure for weak index guiding; (b) etched-mesa buried heterostructure for strong index guiding.

a result, permit strong mode confinement. Because of a large built-in index step, the spatial distribution of the emitted light is inherently stable, provided that the laser is designed to support a single spatial mode.

As the active region of a BH laser is in the form of a rectangular waveguide, spatial modes can be obtained by following a method similar to that used in Section 2.2 for optical fibers [2]. In practice, a BH laser operates in a single mode if the active-region width is reduced to below 2 μm. The spot size is elliptical with typical dimensions $2 \times 1 \mu$m^2. Because of small spot-size dimensions, the beam diffracts widely in both the lateral and transverse directions. The elliptic spot size and a large divergence angle make it difficult to couple light into the fiber efficiently. Typical coupling efficiencies are in the range 30–50% for most optical transmitters. A spot-size converter is sometimes used to improve the coupling efficiency (see Section 3.6).

3.4 Control of Longitudinal Modes

We have seen that BH semiconductor lasers can be designed to emit light into a single spatial mode by controlling the width and the thickness of the active layer. However, as discussed in Section 3.3.2, such lasers oscillate in several longitudinal modes simultaneously because of a relatively small gain difference (~ 0.1 cm^{-1}) between neighboring modes of the FP cavity. The resulting spectral width (2–4 nm) is acceptable for lightwave systems operating near 1.3 μm at bit rates of up to 1 Gb/s. However, such multimode lasers cannot be used for systems designed to operate near 1.55 μm at high bit rates. The only solution is to design semiconductor lasers [36]–[41] such that they emit light predominantly in a single longitudinal mode (SLM).

The SLM semiconductor lasers are designed such that cavity losses are different for different longitudinal modes of the cavity, in contrast with FP lasers whose losses are mode independent. Figure 3.15 shows the gain and loss profiles schematically for such a laser. The longitudinal mode with the smallest cavity loss reaches threshold first

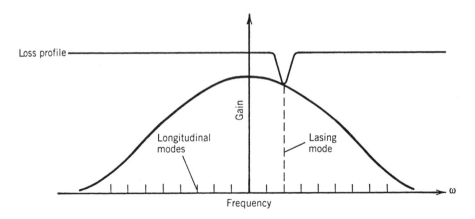

Figure 3.15: Gain and loss profiles for semiconductor lasers oscillating predominantly in a single longitudinal mode.

and becomes the dominant mode. Other neighboring modes are discriminated by their higher losses, which prevent their buildup from spontaneous emission. The power carried by these side modes is usually a small fraction ($< 1\%$) of the total emitted power. The performance of a SLM laser is often characterized by the *mode-suppression ratio* (MSR), defined as [39]

$$\text{MSR} = P_{\text{mm}}/P_{\text{sm}}, \tag{3.4.1}$$

where P_{mm} is the main-mode power and P_{sm} is the power of the most dominant side mode. The MSR should exceed 1000 (or 30 dB) for a good SLM laser.

3.4.1 Distributed Feedback Lasers

Distributed feedback (DFB) semiconductor lasers were developed during the 1980s and are used routinely for WDM lightwave systems [10]–[12]. The feedback in DFB lasers, as the name implies, is not localized at the facets but is distributed throughout the cavity length [41]. This is achieved through an internal built-in grating that leads to a periodic variation of the mode index. Feedback occurs by means of *Bragg diffraction*, a phenomenon that couples the waves propagating in the forward and backward directions. Mode selectivity of the DFB mechanism results from the *Bragg condition*: the coupling occurs only for wavelengths λ_B satisfying

$$\Lambda = m(\lambda_B/2\bar{n}), \tag{3.4.2}$$

where Λ is the grating period, $\bar{n}$ is the average mode index, and the integer m represents the order of Bragg diffraction. The coupling between the forward and backward waves is strongest for the first-order Bragg diffraction ($m = 1$). For a DFB laser operating at $\lambda_B = 1.55$ μm, Λ is about 235 nm if we use $m = 1$ and $\bar{n} = 3.3$ in Eq. (3.4.2). Such gratings can be made by using a holographic technique [2].

From the standpoint of device operation, semiconductor lasers employing the DFB mechanism can be classified into two broad categories: DFB lasers and *distributed*

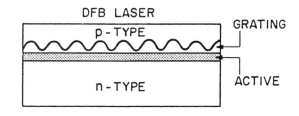

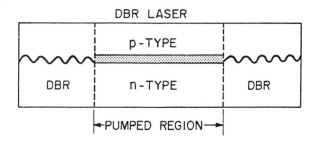

Figure 3.16: DFB and DBR laser structures. The shaded area shows the active region and the wavy line indicates the presence of a Bragg gratin.

Bragg reflector (DBR) lasers. Figure 3.16 shows two kinds of laser structures. Though the feedback occurs throughout the cavity length in DFB lasers, it does not take place inside the active region of a DBR laser. In effect, the end regions of a DBR laser act as mirrors whose reflectivity is maximum for a wavelength λ_B satisfying Eq. (3.4.2). The cavity losses are therefore minimum for the longitudinal mode closest to λ_B and increase substantially for other longitudinal modes (see Fig. 3.15). The MSR is determined by the gain margin defined as the excess gain required by the most dominant side mode to reach threshold. A gain margin of 3–5 cm^{-1} is generally enough to realize an MSR > 30 dB for DFB lasers operating continuously [39]. However, a larger gain margin is needed (> 10 cm^{-1}) when DFB lasers are modulated directly. *Phase-shifted DFB lasers* [38], in which the grating is shifted by $\lambda_B/4$ in the middle of the laser to produce a $\pi/2$ phase shift, are often used, since they are capable of providing much larger gain margin than that of conventional DFB lasers. Another design that has led to improvements in the device performance is known as the *gain-coupled DFB laser* [42]–[44]. In these lasers, both the optical gain and the mode index vary periodically along the cavity length.

Fabrication of DFB semiconductor lasers requires advanced technology with multiple epitaxial growths [41]. The principal difference from FP lasers is that a grating is etched onto one of the cladding layers surrounding the active layer. A thin n-type waveguide layer with a refractive index intermediate to that of active layer and the substrate acts as a grating. The periodic variation of the thickness of the waveguide layer translates into a periodic variation of the mode index $\bar{n}$ along the cavity length and leads to a coupling between the forward and backward propagating waves through Bragg diffraction.

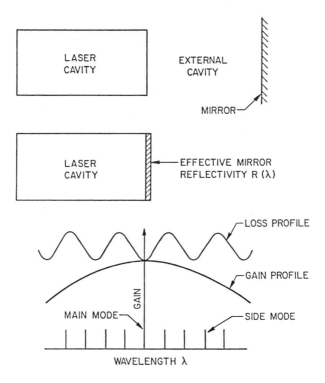

Figure 3.17: Longitudinal-mode selectivity in a coupled-cavity laser. Phase shift in the external cavity makes the effective mirror reflectivity wavelength dependent and results in a periodic loss profile for the laser cavity.

A holographic technique is often used to form a grating with a ~ 0.2-μm periodicity. It works by forming a fringe pattern on a photoresist (deposited on the wafer surface) through interference between two optical beams. In the alternative electron-beam lithographic technique, an electron beam writes the desired pattern on the electron-beam resist. Both methods use chemical etching to form grating corrugations, with the patterned resist acting as a mask. Once the grating has been etched onto the substrate, multiple layers are grown by using an epitaxial growth technique. A second epitaxial regrowth is needed to make a BH device such as that shown in Fig. 3.14(b). Despite the technological complexities, DFB lasers are routinely produced commercially. They are used in nearly all 1.55-μm optical communication systems operating at bit rates of 2.5 Gb/s or more. DFB lasers are reliable enough that they have been used since 1992 in all transoceanic lightwave systems.

3.4.2 Coupled-Cavity Semiconductor Lasers

In a *coupled-cavity* semiconductor laser [2], the SLM operation is realized by coupling the light to an external cavity (see Fig. 3.17). A portion of the reflected light is fed back into the laser cavity. The feedback from the external cavity is not necessarily in phase with the optical field inside the laser cavity because of the phase shift occurring in the external cavity. The in-phase feedback occurs only for those laser modes whose wavelength nearly coincides with one of the longitudinal modes of the external cavity. In effect, the effective reflectivity of the laser facet facing the external cavity becomes wavelength dependent and leads to the loss profile shown in Fig. 3.17. The longitudinal mode that is closest to the gain peak and has the lowest cavity loss becomes the dominant mode.

Several kinds of coupled-cavity schemes have been developed for making SLM laser; Fig. 3.18 shows three among them. A simple scheme couples the light from a semiconductor laser to an external grating [Fig. 3.18(a)]. It is necessary to reduce the natural reflectivity of the cleaved facet facing the grating through an antireflection coating to provide a strong coupling. Such lasers are called *external-cavity* semiconductor lasers and have attracted considerable attention because of their tunability [36]. The wavelength of the SLM selected by the coupled-cavity mechanism can be tuned over a wide range (typically 50 nm) simply by rotating the grating. Wavelength tunability is a desirable feature for lasers used in WDM lightwave systems. A drawback of the laser shown in Fig. 3.18(a) from the system standpoint is its nonmonolithic nature, which makes it difficult to realize the mechanical stability required of optical transmitters.

A monolithic design for coupled-cavity lasers is offered by the cleaved-coupled-cavity laser [37] shown in Fig. 3.18(b). Such lasers are made by cleaving a conventional multimode semiconductor laser in the middle so that the laser is divided into two sections of about the same length but separated by a narrow air gap (width $\sim 1~\mu$m). The reflectivity of cleaved facets ($\sim 30\%$) allows enough coupling between the two sections as long as the gap is not too wide. It is even possible to tune the wavelength of such a laser over a tuning range ~ 20 nm by varying the current injected into one of the cavity sections acting as a mode controller. However, tuning is not continuous, since it corresponds to successive mode hops of about 2 nm.

3.4.3 Tunable Semiconductor Lasers

Modern WDM lightwave systems require single-mode, narrow-linewidth lasers whose wavelength remains fixed over time. DFB lasers satisfy this requirement but their wavelength stability comes at the expense of tunability [9]. The large number of DFB lasers used inside a WDM transmitter make the design and maintenance of such a lightwave system expensive and impractical. The availability of semiconductor lasers whose wavelength can be tuned over a wide range would solve this problem [13].

Multisection DFB and DBR lasers were developed during the 1990s to meet the somewhat conflicting requirements of stability and tunability [45]–[52] and were reaching the commercial stage in 2001. Figure 3.18(c) shows a typical laser structure. It consists of three sections, referred to as the active section, the phase-control section, and the Bragg section. Each section can be biased independently by injecting different amounts of currents. The current injected into the Bragg section is used to change the Bragg wavelength ($\lambda_B = 2n\Lambda$) through carrier-induced changes in the refractive index n. The current injected into the phase-control section is used to change the phase of the feedback from the DBR through carrier-induced index changes in that section. The laser wavelength can be tuned almost continuously over the range 10–15 nm by con-

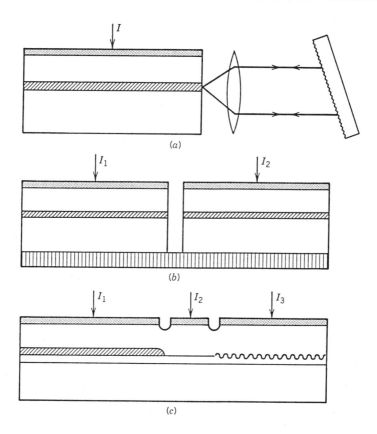

Figure 3.18: Coupled-cavity laser structures: (a) external-cavity laser; (b) cleaved-coupled-cavity laser; (c) multisection DBR laser.

trolling the currents in the phase and Bragg sections. By 1997, such lasers exhibited a tuning range of 17 nm and output powers of up to 100 mW with high reliability [51].

Several other designs of tunable DFB lasers have been developed in recent years. In one scheme, the built-in grating inside a DBR laser is chirped by varying the grating period Λ or the mode index $\bar{n}$ along the cavity length. As seen from Eq. (3.4.2), the Bragg wavelength itself then changes along the cavity length. Since the laser wavelength is determined by the Bragg condition, such a laser can be tuned over a wavelength range determined by the grating chirp. In a simple implementation of the basic idea, the grating period remains uniform, but the waveguide is bent to change the effective mode index $\bar{n}$. Such multisection DFB lasers can be tuned over 5–6 nm while maintaining a single longitudinal mode with high side-mode suppression [47].

In another scheme, a *superstructure grating* is used for the DBR section of a multisection laser [48]–[50]. A superstructure grating consists of an array of gratings (uniform or chirped) separated by a constant distance. As a result, its reflectivity peaks at several wavelengths whose interval is determined by the spacing between the individual gratings forming the array. Such multisection DBR lasers can be tuned discretely over a wavelength range exceeding 100 nm. By controlling the current in the phase-control section, a quasicontinuous tuning range of 40 nm was realized in 1995 with a superstructure grating [48]. The tuning range can be extended considerably by using a four-section device in which another DBR section is added to the left side of the device shown in Fig. 3.18(c). Each DBR section supports its own comb of wavelengths but the spacing in each comb is not the same. The coinciding wavelength in the two combs becomes the output wavelength that can be tuned over a wide range (analogous to the Vernier effect).

In a related approach, the fourth section in Fig. 3.18(c) is added between the gain and phase sections: It consist of a grating-assisted codirectional coupler with a super-structure grating. The coupler has two vertically separated waveguides and selects a single wavelength from the wavelength comb supported by the DBR section with a su-perstructure grating. The largest tuning range of 114 nm was produced in 1995 by this kind of device [49]. Such widely tunable DBR lasers are likely to find applications in many WDM lightwave systems.

3.4.4 Vertical-Cavity Surface-Emitting Lasers

A new class of semiconductor lasers, known as *vertical-cavity surface-emitting lasers* (VCSELs), has emerged during the 1990s with many potential applications [53]–[60]. VCSELs operate in a single longitudinal mode by virtue of an extremely small cav-ity length (~ 1 μm), for which the mode spacing exceeds the gain bandwidth (see Fig. 3.11). They emit light in a direction normal to the active-layer plane in a manner analogous to that of a surface-emitting LED (see Fig. 3.8). Moreover, the emitted light is in the form of a circular beam that can be coupled into a single-node fiber with high efficiency. These properties result in a number of advantages that are leading to rapid adoption of VCSELs for lightwave communications.

As seen in Fig. 3.19, fabrication of VCSELs requires growth of multiple thin lay-ers on a substrate. The active region, in the form of one or several quantum wells, is surrounded by two high-reflectivity ($> 99.5\%$) DBR mirrors that are grown epitaxi-ally on both sides of the active region to form a high-Q microcavity [55]. Each DBR mirror is made by growing many pairs of alternating GaAs and AlAs layers, each $\lambda/4$ thick, where λ is the wavelength emitted by the VCSEL. A wafer-bonding technique is sometimes used for VCSELs operating in the 1.55-μm wavelength region to accommo-date the InGaAsP active region [58]. Chemical etching or a related technique is used to form individual circular disks (each corresponding to one VCSEL) whose diameter can be varied over a wide range (typically 5–20 μm). The entire two-dimensional array of VCSELs can be tested without requiring separation of lasers because of the vertical nature of light emission. As a result, the cost of a VCSEL can be much lower than that of an edge-emitting laser. VCSELs also exhibit a relatively low threshold (~ 1 mA or less). Their only disadvantage is that they cannot emit more than a few milliwatts of power because of a small active volume. For this reason, they are mostly used in local-area and metropolitan-area networks and have virtually replaced LEDs. Early VCSELs were designed to emit near 0.8 μm and operated in multiple transverse modes because of their relatively large diameters (~ 10 μm).

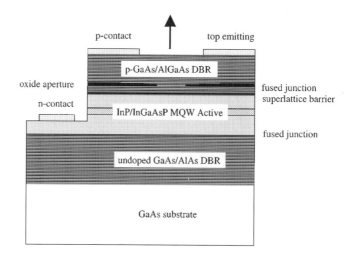

Figure 3.19: Schematic of a 1.55-μm VCSEL made by using the wafer-bonding technique. (After Ref. [58]; ©2000 IEEE; reprinted with permission.)

In recent years, the VCSEL technology have advanced enough that VCSELs can be designed to operate in a wide wavelength range extending from 650 to 1600 nm [55]. Their applications in the 1.3- and 1.55-μm wavelength windows require that VCSELs operate in a single transverse mode. By 2001, several techniques had emerged for controlling the transverse modes of a VCSEL, the most common being the oxide-confinement technique in which an insulating aluminum-oxide layer, acting as a dielectric aperture, confines both the current and the optical mode to a < 3-μm-diameter region. Such VCSELs operate in a single mode with narrow linewidth and can replace a DFB laser in many lightwave applications as long as their low output power is acceptable. They are especially useful for data transfer and local-loop applications because of their low-cost packaging. VCSELs are also well suited for WDM applications for two reasons. First, their wavelengths can be tuned over a wide range ($>$50 nm) using the micro-electro-mechanical system (MEMS) technology [56]. Second, one can make two-dimensional VCSELS arrays such that each laser operates at a different wavelength [60]. WDM sources, containing multiple monolithically integrated lasers, are required for modern lightwave systems.

3.5 Laser Characteristics

The operating characteristics of semiconductor lasers are well described by a set of rate equations that govern the interaction of photons and electrons inside the active region. In this section we use the rate equations to discuss first both the continuous-wave (CW) properties. We then consider small- and large-signal modulation characteristics of single-mode semiconductor lasers. The last two subsections focus on the intensity noise and spectral bandwidth of semiconductor lasers.

3.5.1   CW Characteristics

A rigorous derivation of the rate equations generally starts from Maxwell's equations together with a quantum-mechanical approach for the induced polarization (see Section 2.2). The rate equations can also be written heuristically by considering various physical phenomena through which the number of photons, P, and the number of electrons, N, change with time inside the active region. For a single-mode laser, these equations take the form [2]

$$\frac{dP}{dt} = GP + R_{sp} - \frac{P}{\tau_p}, \tag{3.5.1}$$

$$\frac{dN}{dt} = \frac{I}{q} - \frac{N}{\tau_c} - GP, \tag{3.5.2}$$

where

$$G = \Gamma v_g g_m = G_N(N - N_0). \tag{3.5.3}$$

G is the net rate of stimulated emission and R_{sp} is the rate of spontaneous emission into the lasing mode. Note that R_{sp} is much smaller than the total spontaneous-emission rate in Eq. (3.1.10). The reason is that spontaneous emission occurs in all directions over a wide spectral range (~ 30–40 nm) but only a small fraction of it, propagating along the cavity axis and emitted at the laser frequency, actually contributes to Eq. (3.5.1). In fact, R_{sp} and G are related by $R_{sp} = n_{sp}G$, where n_{sp} is known as the *spontaneous-emission factor* and is about 2 for semiconductor lasers [2]. Although the same notation is used for convenience, the variable N in the rate equations represents the number of electrons rather than the carrier density; the two are related by the active volume V. In Eq. (3.5.3), v_g is the group velocity, Γ is the confinement factor, and g_m is the material gain at the mode frequency. By using Eq. (3.3.1), G varies linearly with N with $G_N = \Gamma v_g \sigma_g / V$ and $N_0 = N_T V$.

The last term in Eq. (3.5.1) takes into account the loss of photons inside the cavity. The parameter τ_p is referred to as the *photon lifetime*. It is related to the *cavity loss* α_{cav} introduced in Eq. (3.3.4) as

$$\tau_p^{-1} = v_g \alpha_{cav} = v_g(\alpha_{mir} + \alpha_{int}). \tag{3.5.4}$$

The three terms in Eq. (3.5.2) indicate the rates at which electrons are created or destroyed inside the active region. This equation is similar to Eq. (3.2.8) except for the addition of the last term, which governs the rate of electron–hole recombination through stimulated emission. The carrier lifetime τ_c includes the loss of electrons due to both spontaneous emission and nonradiative recombination, as indicated in Eq. (3.1.18).

The P–I curve characterizes the emission properties of a semiconductor laser, as it indicates not only the threshold level but also the current that needs to be applied to obtain a certain amount of power. Figure 3.20 shows the P–I curves of a 1.3-μm InGaAsP laser at temperatures in the range 10–130°C. At room temperature, the threshold is reached near 20 mA, and the laser can emit 10 mW of output power from each facet at 100 mA of applied current. The laser performance degrades at high temperatures. The threshold current is found to increase exponentially with temperature, i.e.,

$$I_{th}(T) = I_0 \exp(T/T_0), \tag{3.5.5}$$

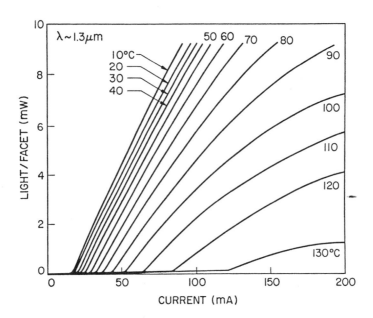

Figure 3.20: *P–I* curves at several temperatures for a 1.3-μm buried heterostructure laser. (After Ref. [2]; ©1993 Van Nostrand Reinhold; reprinted with permission.)

where I_0 is a constant and T_0 is a *characteristic temperature* often used to express the temperature sensitivity of threshold current. For InGaAsP lasers T_0 is typically in the range 50–70 K. By contrast, T_0 exceeds 120 K for GaAs lasers. Because of the temperature sensitivity of InGaAsP lasers, it is often necessary to control their temperature through a built-in thermoelectric cooler.

The rate equations can be used to understand most of the features seen in Fig. 3.20. In the case of CW operation at a constant current I, the time derivatives in Eqs. (3.5.1) and (3.5.2) can be set to zero. The solution takes a particularly simple form if spontaneous emission is neglected by setting $R_{\rm sp} = 0$. For currents such that $G\tau_p < 1$, $P = 0$ and $N = \tau_c I/q$. The threshold is reached at a current for which $G\tau_p = 1$. The carrier population is then clamped to the threshold value $N_{\rm th} = N_0 + (G_N\tau_p)^{-1}$. The threshold current is given by

$$I_{\rm th} = \frac{qN_{\rm th}}{\tau_c} = \frac{q}{\tau_c}\left(N_0 + \frac{1}{G_N\tau_p}\right). \tag{3.5.6}$$

For $I > I_{\rm th}$, the photon number P increases linearly with I as

$$P = (\tau_p/q)(I - I_{\rm th}). \tag{3.5.7}$$

The emitted power P_e is related to P by the relation

$$P_e = \tfrac{1}{2}(v_g\alpha_{\rm mir})\hbar\omega P. \tag{3.5.8}$$

The derivation of Eq. (3.5.8) is intuitively obvious if we note that $v_g\alpha_{\rm mir}$ is the rate at which photons of energy $\hbar\omega$ escape from the two facets. The factor of $\tfrac{1}{2}$ makes P_e

the power emitted from each facet for a FP laser with equal facet reflectivities. For FP lasers with coated facets or for DFB lasers, Eq. (3.5.8) needs to be suitably modified [2]. By using Eqs. (3.5.4) and (3.5.7) in Eq. (3.5.8), the emitted power is given by

$$P_e = \frac{\hbar\omega}{2q} \frac{\eta_{int}\alpha_{mir}}{\alpha_{mir} + \alpha_{int}}(I - I_{th}), \qquad (3.5.9)$$

where the internal quantum efficiency η_{int} is introduced phenomenologically to indicate the fraction of injected electrons that is converted into photons through stimulated emission. In the above-threshold regime, η_{int} is almost 100% for most semiconductor lasers. Equation (3.5.9) should be compared with Eq. (3.2.2) obtained for an LED.

A quantity of practical interest is the slope of the P–I curve for $I > I_{th}$; it is called the *slope efficiency* and is defined as

$$\frac{dP_e}{dI} = \frac{\hbar\omega}{2q}\eta_d \quad \text{with} \quad \eta_d = \frac{\eta_{int}\alpha_{mir}}{\alpha_{mir} + \alpha_{int}}. \qquad (3.5.10)$$

The quantity η_d is called the *differential quantum efficiency*, as it is a measure of the efficiency with which light output increases with an increase in the injected current. One can define the external quantum efficiency η_{ext} as

$$\eta_{ext} = \frac{\text{photon-emission rate}}{\text{electron-injection rate}} = \frac{2P_e/\hbar\omega}{I/q} = \frac{2q}{\hbar\omega}\frac{P_e}{I}. \qquad (3.5.11)$$

By using Eqs. (3.5.9)–(3.5.11), η_{ext} and η_d are found to be related by

$$\eta_{ext} = \eta_d(1 - I_{th}/I). \qquad (3.5.12)$$

Generally, $\eta_{ext} < \eta_d$ but becomes nearly the same for $I \gg I_{th}$. Similar to the case of LEDs, one can define the total quantum efficiency (or wall-plug efficiency) as $\eta_{tot} = 2P_e/(V_0 I)$, where V_0 is the applied voltage. It is related to η_{ext} as

$$\eta_{tot} = \frac{\hbar\omega}{qV_0}\eta_{ext} \approx \frac{E_g}{qV_0}\eta_{ext}, \qquad (3.5.13)$$

where E_g is the bandgap energy. Generally, $\eta_{tot} < \eta_{ext}$ as the applied voltage exceeds E_g/q. For GaAs lasers, η_d can exceed 80% and η_{tot} can approach 50%. The InGaAsP lasers are less efficient with $\eta_d \sim 50\%$ and $\eta_{tot} \sim 20\%$.

The exponential increase in the threshold current with temperature can be understood from Eq. (3.5.6). The carrier lifetime τ_c is generally N dependent because of Auger recombination and decreases with N as N^2. The rate of Auger recombination increases exponentially with temperature and is responsible for the temperature sensitivity of InGaAsP lasers. Figure 3.20 also shows that the slope efficiency decreases with an increase in the output power (bending of the P–I curves). This decrease can be attributed to junction heating occurring under CW operation. It can also result from an increase in internal losses or current leakage at high operating powers. Despite these problems, the performance of DFB lasers improved substantially during the 1990s [10]–[12]. DFB lasers emitting >100 mW of power at room temperature in the 1.55 μm spectral region were fabricated by 1996 using a strained MQW design [61]. Such lasers exhibited < 10 mA threshold current at 20°C and emitted ~ 20 mW of power at 100°C while maintaining a MSR of >40 dB. By 2001, DFB lasers capable of delivering more than 200 mW of power were available commercially.

3.5.2   Small-Signal Modulation

The modulation response of semiconductor lasers is studied by solving the rate equations (3.5.1) and (3.5.2) with a time-dependent current of the form

$$I(t) = I_b + I_m f_p(t), \tag{3.5.14}$$

where I_b is the bias current, I_m is the current, and $f_p(t)$ represents the shape of the current pulse. Two changes are necessary for a realistic description. First, Eq. (3.5.3) for the gain G must be modified to become [2]

$$G = G_N(N - N_0)(1 - \varepsilon_{NL} P), \tag{3.5.15}$$

where ε_{NL} is a nonlinear-gain parameter that leads to a slight reduction in G as P increases. The physical mechanism behind this reduction can be attributed to several phenomena, such as spatial hole burning, spectral hole burning, carrier heating, and two-photon absorption [62]–[65]. Typical values of ε_{NL} are $\sim 10^{-7}$. Equation (3.5.15) is valid for $\varepsilon_{NL} P \ll 1$. The factor $1 - \varepsilon_{NL} P$ should be replaced by $(1 + P/P_s)^{-b}$, where P_s is a material parameter, when the laser power exceeds far above 10 mW. The exponent b equals $\frac{1}{2}$ for spectral hole burning [63] but can vary over the range 0.2–1 because of the contribution of carrier heating [65].

The second change is related to an important property of semiconductor lasers. It turns out that whenever the optical gain changes as a result of changes in the carrier population N, the refractive index also changes. From a physical standpoint, amplitude modulation in semiconductor lasers is always accompanied by phase modulation because of carrier-induced changes in the mode index $\bar{n}$. Phase modulation can be included through the equation [2]

$$\frac{d\phi}{dt} = \frac{1}{2} \beta_c \left[G_N(N - N_0) - \frac{1}{\tau_p} \right], \tag{3.5.16}$$

where β_c is the amplitude-phase coupling parameter, commonly called the *linewidth enhancement factor*, as it leads to an enhancement of the spectral width associated with a single longitudinal mode (see Section 3.5.5). Typical values of β_c for InGaAsP lasers are in the range 4–8, depending on the operating wavelength [66]. Lower values of β_c occur in MQW lasers, especially for strained quantum wells [5].

In general, the nonlinear nature of the rate equations makes it necessary to solve them numerically. A useful analytic solution can be obtained for the case of small-signal modulation in which the laser is biased above threshold ($I_b > I_{th}$) and modulated such that $I_m \ll I_b - I_{th}$. The rate equations can be linearized in that case and solved analytically, using the Fourier-transform technique, for an arbitrary form of $f_p(t)$. The small-signal modulation bandwidth can be obtained by considering the response of semiconductor lasers to sinusoidal modulation at the frequency ω_m so that $f_p(t) = \sin(\omega_m t)$. The laser output is also modulated sinusoidally. The general solution of Eqs. (3.5.1) and (3.5.2) is given by

$$P(t) = P_b + |p_m| \sin(\omega_m t + \theta_m), \tag{3.5.17}$$

$$N(t) = N_b + |n_m| \sin(\omega_m t + \psi_m), \tag{3.5.18}$$

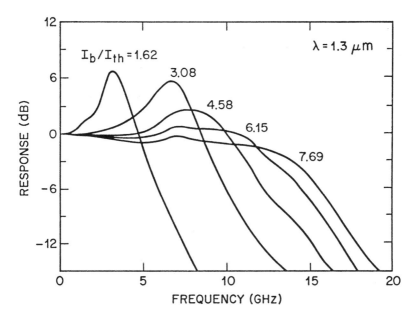

Figure 3.21: Measured (solid curves) and fitted (dashed curves) modulation response of a 1.55-μm DFB laser as a function of modulation frequency at several bias levels. (After Ref. [70]; ©1997 IEEE; reprinted with permission.)

where P_b and N_b are the steady-state values at the bias current I_b, $|p_m|$ and $|n_m|$ are small changes occurring because of current modulation, and θ_m and ψ_m govern the phase lag associated with the small-signal modulation. In particular, $p_m \equiv |p_m| \exp(i\theta_m)$ is given by [2]

$$p_m(\omega_m) = \frac{P_b G_N I_m / q}{(\Omega_R + \omega_m - i\Gamma_R)(\Omega_R - \omega_m + i\Gamma_R)}, \tag{3.5.19}$$

where

$$\Omega_R = [G G_N P_b - (\Gamma_P - \Gamma_N)^2 / 4]^{1/2}, \qquad \Gamma_R = (\Gamma_P + \Gamma_N)/2, \tag{3.5.20}$$

$$\Gamma_P = R_{\rm sp}/P_b + \varepsilon_{\rm NL} G P_b, \qquad \Gamma_N = \tau_c^{-1} + G_N P_b. \tag{3.5.21}$$

Ω_R and Γ_R are the frequency and the damping rate of *relaxation oscillations*. These two parameters play an important role in governing the dynamic response of semiconductor lasers. In particular, the efficiency is reduced when the modulation frequency exceeds Ω_R by a large amount.

Similar to the case of LEDs, one can introduce the *transfer function* as

$$H(\omega_m) = \frac{p_m(\omega_m)}{p_m(0)} = \frac{\Omega_R^2 + \Gamma_R^2}{(\Omega_R + \omega_m - i\Gamma_R)(\Omega_R - \omega_m + i\Gamma_R)}. \tag{3.5.22}$$

The modulation response is flat [$H(\omega_m) \approx 1$] for frequencies such that $\omega_m \ll \Omega_R$, peaks at $\omega_m = \Omega_R$, and then drops sharply for $\omega_m \gg \Omega_R$. These features are observed experimentally for all semiconductor lasers [67]–[70]. Figure 3.21 shows the modulation response of a 1.55-μm DFB laser at several bias levels [70]. The 3-dB modulation bandwidth, $f_{3\,\text{dB}}$, is defined as the frequency at which $|H(\omega_m)|$ is reduced by 3 dB (by a factor of 2) compared with its direct-current (dc) value. Equation (3.5.22) provides the following analytic expression for $f_{3\,\text{dB}}$:

$$f_{3\,\text{dB}} = \frac{1}{2\pi} \left[\Omega_R^2 + \Gamma_R^2 + 2(\Omega_R^4 + \Omega_R^2 \Gamma_R^2 + \Gamma_R^4)^{1/2} \right]^{1/2}. \tag{3.5.23}$$

For most lasers, $\Gamma_R \ll \Omega_R$, and $f_{3\,\text{dB}}$ can be approximated by

$$f_{3\,\text{dB}} \approx \frac{\sqrt{3}\,\Omega_R}{2\pi} \approx \left(\frac{3G_N P_b}{4\pi^2 \tau_p} \right)^{1/2} = \left[\frac{3G_N}{4\pi^2 q}(I_b - I_{\text{th}}) \right]^{1/2}, \tag{3.5.24}$$

where Ω_R was approximated by $(GG_N P_b)^{1/2}$ in Eq. (3.5.21) and G was replaced by $1/\tau_p$ since gain equals loss in the above-threshold regime. The last expression was obtained by using Eq. (3.5.7) at the bias level.

Equation (3.5.24) provides a remarkably simple expression for the modulation bandwidth. It shows that $f_{3\,\text{dB}}$ increases with an increase in the bias level as $\sqrt{P_b}$ or as $(I_b - I_{\text{th}})^{1/2}$. This square-root dependence has been verified for many DFB lasers exhibiting a modulation bandwidth of up to 30 GHz [67]–[70]. Figure 3.21 shows how $f_{3\,\text{dB}}$ can be increased to 24 GHz for a DFB laser by biasing it at 80 mA [70]. A modulation bandwidth of 25 GHz was realized in 1994 for a packaged 1.55-μm InGaAsP laser specifically designed for high-speed response [68].

3.5.3 Large-Signal Modulation

The small-signal analysis, although useful for a qualitative understanding of the modulation response, is not generally applicable to optical communication systems where the laser is typically biased close to threshold and modulated considerably above threshold to obtain optical pulses representing digital bits. In this case of large-signal modulation, the rate equations should be solved numerically. Figure 3.22 shows, as an example, the shape of the emitted optical pulse for a laser biased at $I_b = 1.1 I_{\text{th}}$ and modulated at 2 Gb/s using rectangular current pulses of duration 500 ps and amplitude $I_m = I_{\text{th}}$. The optical pulse does not have sharp leading and trailing edges because of a limited modulation bandwidth and exhibits a rise time ~ 100 ps and a fall time ~ 300 ps. The initial overshoot near the leading edge is a manifestation of relaxation oscillations. Even though the optical pulse is not an exact replica of the applied electrical pulse, deviations are small enough that semiconductor lasers can be used in practice.

As mentioned before, amplitude modulation in semiconductor lasers is accompanied by phase modulation governed by Eq. (3.5.16). A time-varying phase is equivalent to transient changes in the mode frequency from its steady-state value ν_0. Such a pulse is called chirped. The *frequency chirp* $\delta\nu(t)$ is obtained by using Eq. (3.5.16) and is given by

$$\delta\nu(t) = \frac{1}{2\pi} \frac{d\phi}{dt} = \frac{\beta_c}{4\pi} \left[G_N(N - N_0) - \frac{1}{\tau_p} \right]. \tag{3.5.25}$$

The dashed curve in Fig. 3.21 shows the frequency chirp across the optical pulse. The mode frequency shifts toward the blue side near the leading edge and toward the red

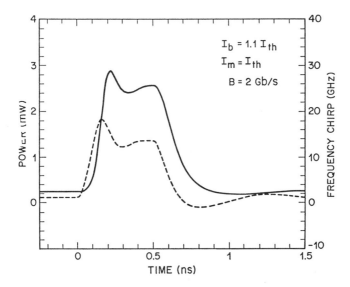

Figure 3.22: Simulated modulation response of a semiconductor laser to 500-ps rectangular current pulses. Solid curve shows the pulse shape and the dashed curve shows the frequency chirp imposed on the pulse ($\beta_c = 5$).

side near the trailing edge of the optical pulse [71]. Such a frequency shift implies that the pulse spectrum is considerably broader than that expected in the absence of frequency chirp.

It was seen in Section 2.4 that the frequency chirp can limit the performance of optical communication systems, especially when $\beta_2 C > 0$, where β_2 is the dispersion parameter and C is the chirp parameter. Even though optical pulses emitted from semiconductors are generally not Gaussian, the analysis of Section 2.4 can be used to study chirp-induced pulse broadening [72] if we identify C with $-\beta_c$ in Eq. (2.4.23). It turns out that 1.55-μm lightwave systems are limited to distances below 100 km even at a bit rate of 2.5 Gb/s because of the frequency chirp [71] when conventional fibers are used ($\beta_2 \approx -20$ ps^2/km). Higher bit rates and longer distances can only be realized by using a dispersion management scheme so that the average dispersion is close to zero (see Chapter 7).

Since frequency chirp is often the limiting factor for lightwave systems operating near 1.55 μm, several methods have been used to reduce its magnitude [73]–[77]. These include pulse-shape tailoring, injection locking, and coupled-cavity schemes. A direct way to reduce the frequency chirp is to design semiconductor lasers with small values of the linewidth enhancement factor β_c. The use of quantum-well design reduces β_c by about a factor of about 2. A further reduction occurs for strained quantum wells [76]. Indeed, $\beta_c \approx 1$ has been measured in *modulation-doped* strained MQW lasers [77]. Such lasers exhibit low chirp under direct modulation. The frequency chirp resulting from current modulation can be avoided altogether if the laser is continuously operated, and an external modulator is used to modulate the laser output [78]. In practice, lightwave systems operating at 10 Gb/s or more use either a monolithically integrated electroabsorption modulator or an external LiNbO$_3$ modulator (see Section 3.6). One can even design a modulator to reverse the sign of chirp such that $\beta_2 C < 0$, resulting in improved system performance.

Lightwave system designed using the RZ format, optical time-division multiplexing, or solitons often require mode-locked lasers that generate short optical pulses (width $\sim$ 10 ps) at a high repetition rate equal to the bit rate. External-cavity semiconductor lasers can be used for this purpose, and are especially practical if a fiber grating is used for an external mirror. An external modulator is still needed to impose the data on the mode-locked pulse train; it blocks pulses in each bit slot corresponding to 0 bits. The gain switching has also been used to generate short pulses from a semiconductor laser. A mode-locked fiber laser can also be used for the same purpose [79].

3.5.4 Relative Intensity Noise

The output of a semiconductor laser exhibits fluctuations in its intensity, phase, and frequency even when the laser is biased at a constant current with negligible current fluctuations. The two fundamental noise mechanisms are *spontaneous emission* and *electron–hole recombination* (shot noise). Noise in semiconductor lasers is dominated by spontaneous emission. Each spontaneously emitted photon adds to the coherent field (established by stimulated emission) a small field component whose phase is random, and thus perturbs both amplitude and phase in a random manner. Moreover, such spontaneous-emission events occur randomly at a high rate ($\sim 10^{12}$ s^{-1}) because of a relatively large value of R_{sp} in semiconductor lasers. The net result is that the intensity and the phase of the emitted light exhibit fluctuations over a time scale as short as 100 ps. Intensity fluctuations lead to a limited *signal-to-noise ratio* (SNR), whereas phase fluctuations lead to a finite spectral linewidth when semiconductor lasers are operated at a constant current. Since such fluctuations can affect the performance of lightwave systems, it is important to estimate their magnitude [80].

The rate equations can be used to study laser noise by adding a noise term, known as the *Langevin force*, to each of them [81]. Equations (3.5.1), (3.5.2), and (3.5.16) then become

$$\frac{dP}{dt} = \left(G - \frac{1}{\tau_p}\right)P + R_{\text{sp}} + F_P(t), \tag{3.5.26}$$

$$\frac{dN}{dt} = \frac{I}{q} - \frac{N}{\tau_c} - GP + F_N(t), \tag{3.5.27}$$

$$\frac{d\phi}{dt} = \frac{1}{2}\beta_c \left[G_N(N - N_0) - \frac{1}{\tau_p}\right] + F_\phi(t), \tag{3.5.28}$$

where $F_P(t)$, $F_N(t)$, and $F_\phi(t)$ are the Langevin forces. They are assumed to be Gaussian random processes with zero mean and to have a correlation function of the form (the *Markoffian approximation*)

$$\langle F_i(t)F_j(t')\rangle = 2D_{ij}\delta(t - t'), \tag{3.5.29}$$

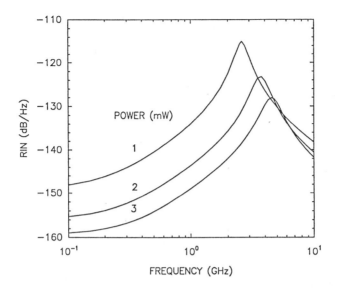

Figure 3.23: RIN spectra at several power levels for a typical 1.55-μm semiconductor laser.

where $i, j = P$, N, or ϕ, angle brackets denote the ensemble average, and D_{ij} is called the *diffusion coefficient*. The dominant contribution to laser noise comes from only two diffusion coefficients $D_{PP} = R_{\rm sp}P$ and $D_{\phi\phi} = R_{\rm sp}/4P$; others can be assumed to be nearly zero [82].

The intensity-autocorrelation function is defined as

$$C_{pp}(\tau) = \langle \delta P(t)\delta P(t+\tau)\rangle / \bar{P}^2, \tag{3.5.30}$$

where $\bar{P} \equiv \langle P\rangle$ is the average value and $\delta P = P - \bar{P}$ represents a small fluctuation. The Fourier transform of $C_{pp}(\tau)$ is known as the *relative-intensity-noise* (RIN) spectrum and is given by

$$\text{RIN}(\omega) = \int_{-\infty}^{\infty} C_{pp}(\tau)\exp(-i\omega t)\,dt. \tag{3.5.31}$$

The RIN can be calculated by linearizing Eqs. (3.5.26) and (3.5.27) in δP and δN, solving the linearized equations in the frequency domain, and performing the average with the help of Eq. (3.5.29). It is given approximately by [2]

$$\text{RIN}(\omega) = \frac{2R_{\rm sp}\{(\Gamma_N^2 + \omega^2) + G_N\bar{P}[G_N\bar{P}(1 + N/\tau_c R_{\rm sp}\bar{P}) - 2\Gamma_N]\}}{\bar{P}[(\Omega_R - \omega)^2 + \Gamma_R^2][(\Omega_R + \omega)^2 + \Gamma_R^2]}, \tag{3.5.32}$$

where Ω_R and Γ_R are the frequency and the damping rate of relaxation oscillations. They are given by Eq. (3.5.21), with P_b replaced by $\bar{P}$.

Figure 3.23 shows the calculated RIN spectra at several power levels for a typical 1.55-μm InGaAsP laser. The RIN is considerably enhanced near the relaxation-oscillation frequency Ω_R but decreases rapidly for $\omega \gg \Omega_R$, since the laser is not able to respond to fluctuations at such high frequencies. In essence, the semiconductor laser acts as a bandpass filter of bandwidth Ω_R to spontaneous-emission fluctuations. At a given frequency, RIN decreases with an increase in the laser power as P^{-3} at low powers, but this behavior changes to P^{-1} dependence at high powers.

The autocorrelation function $C_{pp}(\tau)$ is calculated using Eqs. (3.5.31) and (3.5.32). The calculation shows that $C_{pp}(\tau)$ follows relaxation oscillations and approaches zero for $\tau > \Gamma_R^{-1}$ [83]. This behavior indicates that intensity fluctuations do not remain correlated for times longer than the damping time of relaxation oscillations. The quantity of practical interest is the SNR defined as $\bar{P}/\sigma_p$, where σ_p is the root-mean-square (RMS) noise. From Eq. (3.5.30), SNR $= [C_{pp}(0)]^{-1/2}$. At power levels above a few milliwatts, the SNR exceeds 20 dB and improves linearly with the power as

$$\text{SNR} = \left(\frac{\varepsilon_{\text{NL}}}{R_{\text{sp}}\tau_p}\right)^{1/2}\bar{P}. \tag{3.5.33}$$

The presence of ε_{NL} indicates that the nonlinear form of the gain in Eq. (3.5.15) plays a crucial role. This form needs to be modified at high powers. Indeed, a more accurate treatment shows that the SNR eventually saturates at a value of about 30 dB and becomes power independent [83].

So far, the laser has been assumed to oscillate in a single longitudinal mode. In practice, even DFB lasers are accompanied by one or more side modes. Even though side modes remain suppressed by more than 20 dB on the basis of the average power, their presence can affect the RIN significantly. In particular, the main and side modes can fluctuate in such a way that individual modes exhibit large intensity fluctuations, but the total intensity remains relatively constant. This phenomenon is called *mode-partition noise* (MPN) and occurs due to an anticorrelation between the main and side modes [2]. It manifests through the enhancement of RIN for the main mode by 20 dB or more in the low-frequency range 0–1 GHz; the exact value of the enhancement factor depends on the MSR [84]. In the case of a VCSEL, the MPN involves two transverse modes. [85]. In the absence of fiber dispersion, MPN would be harmless for optical communication systems, as all modes would remain synchronized during transmission and detection. However, in practice all modes do not arrive simultaneously at the receiver because they travel at slightly different speeds. Such a desynchronization not only degrades the SNR of the received signal but also leads to intersymbol interference. The effect of MPN on the system performance is discussed in Section 7.4.3.

3.5.5 Spectral Linewidth

The spectrum of emitted light is related to the field-autocorrelation function $\Gamma_{EE}(\tau)$ through a Fourier-transform relation similar to Eq. (3.5.31), i.e.,

$$S(\omega) = \int_{-\infty}^{\infty} \Gamma_{EE}(t)\exp[-i(\omega - \omega_0)\tau]\,d\tau, \tag{3.5.34}$$

where $\Gamma_{EE}(t) = \langle E^*(t)E(t+\tau)\rangle$ and $E(t) = \sqrt{P}\exp(i\phi)$ is the optical field. If intensity fluctuations are neglected, $\Gamma_{EE}(t)$ is given by

$$\Gamma_{EE}(t) = \langle\exp[i\Delta\phi(t)]\rangle = \exp[-\langle\Delta\phi^2(\tau)\rangle/2], \tag{3.5.35}$$

where the phase fluctuation $\Delta\phi(\tau) = \phi(t+\tau) - \phi(t)$ is taken to be a Gaussian random process. The phase variance $\langle\Delta\phi^2(\tau)\rangle$ can be calculated by linearizing Eqs. (3.5.26)–(3.5.28) and solving the resulting set of linear equations. The result is [82]

$$\langle\Delta\phi^2(\tau)\rangle = \frac{R_{sp}}{2\bar{P}}\left[(1+\beta_c^2 b)\tau + \frac{\beta_c^2 b}{2\Gamma_R\cos\delta}[\cos(3\delta) - e^{-\Gamma_R\tau}\cos(\Omega_R\tau - 3\delta)]\right],$$
(3.5.36)

where

$$b = \Omega_R/(\Omega_R^2 + \Gamma_R^2)^{1/2} \qquad \text{and} \qquad \delta = \tan^{-1}(\Gamma_R/\Omega_R).$$
(3.5.37)

The spectrum is obtained by using Eqs. (3.5.34)–(3.5.36). It is found to consist of a dominant central peak located at ω_0 and multiple satellite peaks located at $\omega = \omega_0 \pm m\Omega_R$, where m is an integer. The amplitude of satellite peaks is typically less than 1% of that of the central peak. The physical origin of the satellite peaks is related to relaxation oscillations, which are responsible for the term proportional to b in Eq. (3.5.36). If this term is neglected, the autocorrelation function $\Gamma_{EE}(\tau)$ decays exponentially with τ. The integral in Eq. (3.5.34) can then be performed analytically, and the spectrum is found to be Lorentzian. The spectral linewidth Δv is defined as the full-width at half-maximum (FWHM) of this Lorentzian line and is given by [82]

$$\Delta v = R_{sp}(1+\beta_c^2)/(4\pi\bar{P}),$$
(3.5.38)

where $b = 1$ was assumed as $\Gamma_R \ll \Omega_R$ under typical operating conditions. The linewidth is enhanced by a factor of $1 + \beta_c^2$ as a result of the amplitude-phase coupling governed by β_c in Eq. (3.5.28); β_c is called the linewidth enhancement factor for this reason.

Equation (3.5.38) shows that Δv should decrease as $\bar{P}^{-1}$ with an increase in the laser power. Such an inverse dependence is observed experimentally at low power levels (< 10 mW) for most semiconductor lasers. However, often the linewidth is found to saturate to a value in the range 1–10 MHz at a power level above 10 mW. Figure 3.24 shows such linewidth-saturation behavior for several 1.55-μm DFB lasers [86]. It also shows that the linewidth can be reduced considerably by using a MQW design for the DFB laser. The reduction is due to a smaller value of the parameter β_c realized by such a design. The linewidth can also be reduced by increasing the cavity length L, since R_{sp} decreases and P increases at a given output power as L is increased. Although not obvious from Eq. (3.5.38), Δv can be shown to vary as L^{-2} when the length dependence of R_{sp} and P is incorporated. As seen in Fig. 3.24, Δv is reduced by about a factor of 4 when the cavity length is doubled. The 800-μm-long MQW-DFB laser is found to exhibit a linewidth as small as 270 kHz at a power output of 13.5 mW [86]. It is further reduced in strained MQW lasers because of relatively low values of β_c, and a value of about 100 kHz has been measured in lasers with $\beta_c \approx 1$ [77]. It should be stressed, however, that the linewidth of most DFB lasers is typically 5–10 MHz when operating at a power level of 10 mW.

Figure 3.24 shows that as the laser power increases, the linewidth not only saturates but begins to rebroaden. Several mechanisms have been invoked to explain such behavior; a few of them are current noise, $1/f$ noise, nonlinear gain, sidemode interaction, and index nonlinearity [87]–[94]. The linewidth of most DFB lasers is small enough that it is not a limiting factor for lightwave systems.

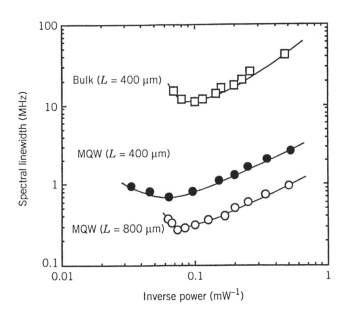

Figure 3.24: Measured linewidth as a function of emitted power for several 1.55-μm DFB lasers. Active layer is 100 nm thick for the bulk laser and 10 nm thick for MQW lasers. (After Ref. [86]; ©1991 IEEE; reprinted with permission.)

3.6 Transmitter Design

So far this chapter has focused on the properties of optical sources. Although an optical source is a major component of optical transmitters, it is not the only component. Other components include a modulator for converting electrical data into optical form (if direct modulation is not used) and an electrical driving circuit for supplying current to the optical source. An external modulator is often used in practice at bit rates of 10 Gb/s or more for avoiding the chirp that is invariably imposed on the directly modulated signal. This section covers the design of optical transmitters with emphasis on the packaging issues [95]–[105].

3.6.1 Source–Fiber Coupling

The design objective for any transmitter is to couple as much light as possible into the optical fiber. In practice, the coupling efficiency depends on the type of optical source (LED versus laser) as well as on the type of fiber (multimode versus single mode). The coupling can be very inefficient when light from an LED is coupled into a single-mode fiber. As discussed briefly in Section 3.2.1, the coupling efficiency for an LED changes with the numerical aperture, and can become < 1% in the case of single-mode fibers. In contrast, the coupling efficiency for edge-emitting lasers is typically 40–50% and can exceed 80% for VCSELs because of their circular spot size. A small piece of fiber (known as a pigtail) is included with the transmitter so that the coupling efficiency can

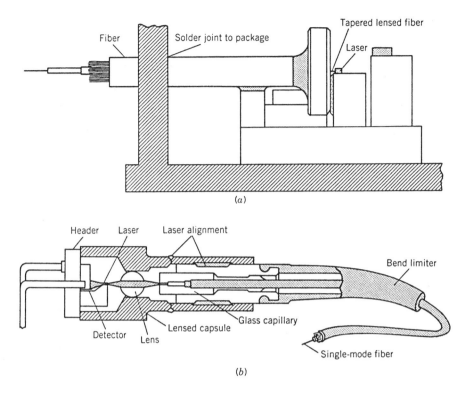

Figure 3.25: Transmitters employing (a) butt-coupling and (b) lens-coupling designs. (After Ref. [97]; ©1989 AT&T; reprinted with permission.)

be maximized during packaging; a splice or connector is used to join the pigtail with the fiber cable.

Two approaches have been used for source–fiber coupling. In one approach, known as direct or *butt coupling*, the fiber is brought close to the source and held in place by epoxy. In the other, known as *lens coupling*, a lens is used to maximize the coupling efficiency. Each approach has its own merits, and the choice generally depends on the design objectives. An important criterion is that the coupling efficiency should not change with time; mechanical stability of the coupling scheme is therefore a necessary requirement.

An example of butt coupling is shown in Fig. 3.25(a), where the fiber is brought in contact with a surface-emitting LED. The coupling efficiency for a fiber of numerical aperture NA is given by [96]

$$n_c = (1 - R_f)(\text{NA})^2, \tag{3.6.1}$$

where R_f is the reflectivity at the fiber front end. R_f is about 4% if an air gap exists between the source and the fiber but can be reduced to nearly zero by placing an index-matching liquid. The coupling efficiency is about 1% for a surface-emitting LED and roughly 10% for an edge-emitting LED. Some improvement is possible in both cases by using fibers that are tapered or have a lensed tip. An external lens also improves the coupling efficiency but only at the expense of reduced mechanical tolerance.

The coupling of a semiconductor laser to a single-mode optical fiber is more efficient than that of an LED. The butt coupling provides only about 10% efficiency, as it makes no attempt to match the mode sizes of the laser and the fiber. Typically, index-guided InGaAsP lasers have a mode size of about 1 μm, whereas the mode size of a single-mode fiber is in the range 6–9 μm. The coupling efficiency can be improved by tapering the fiber end and forming a lens at the fiber tip. Figure 3.25(a) shows such a butt-coupling scheme for a commercial transmitter. The fiber is attached to a jewel, and the jewel is attached to the laser submount by using an epoxy [97]. The fiber tip is aligned with the emitting region of the laser to maximize the coupling efficiency (typically 40%). The use of a lensed fiber can improve the coupling efficiency, and values close to 100% have been realized with an optimum design [98]–[100].

Figure 3.25(b) shows a lens-coupling approach for transmitter design. The coupling efficiency can exceed 70% for such a confocal design in which a sphere is used to collimate the laser light and focus it onto the fiber core. The alignment of the fiber core is less critical for the confocal design because the spot size is magnified to match the fiber's mode size. The mechanical stability of the package is ensured by soldering the fiber into a ferrule which is secured to the body by two sets of laser alignment welds. One set of welds establishes proper axial alignment, while the other set provides transverse alignment.

The laser–fiber coupling issue remains important, and several new schemes have been developed during the 1990s [101]–[105]. In one approach, a *silicon optical bench* is used to align the laser and the fiber [101]. In another, a *silicon micromirror*, fabricated by using the micro-machining technology, is used for optical alignment [102]. In a different approach, a directional coupler is used as the *spot-size converter* for maximizing the coupling efficiency [103]. Coupling efficiencies >80% have been realized by integrating a spot-size converter with semiconductor lasers [105].

An important problem that needs to be addressed in designing an optical transmitter is related to the extreme sensitivity of semiconductor lasers to optical feedback [2]. Even a relatively small amount of feedback ($< 0.1\%$) can destabilize the laser and affect the system performance through phenomena such as linewidth broadening, mode hopping, and RIN enhancement [106]–[110]. Attempts are made to reduce the feedback into the laser cavity by using antireflection coatings. Feedback can also be reduced by cutting the fiber tip at a slight angle so that the reflected light does not hit the active region of the laser. Such precautions are generally enough to reduce the feedback to a tolerable level. However, it becomes necessary to use an *optical isolator* between the laser and the fiber in transmitters designed for more demanding applications. One such application corresponds to lightwave systems operating at high bit rates and requiring a narrow-linewidth DFB laser.

Most optical isolators make use of the *Faraday effect*, which governs the rotation of the plane of polarization of an optical beam in the presence of a magnetic field: The rotation is in the same direction for light propagating parallel or antiparallel to the magnetic field direction. Optical isolators consist of a rod of Faraday material such as yttrium iron garnet (YIG), whose length is chosen to provide 45° rotation. The YIG rod is sandwiched between two polarizers whose axes are tilted by 45° with

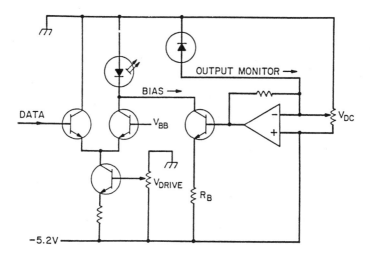

Figure 3.26: Driving circuit for a laser transmitter with feedback control to keep the average optical power constant. A photodiode monitors the output power and provides the control signal. (After Ref. [95]; ©1988 Academic Press; reprinted with permission.)

respect to each other. Light propagating in one direction passes through the second polarizer because of the Faraday rotation. By contrast, light propagating in the opposite direction is blocked by the first polarizer. Desirable characteristics of optical isolators are low insertion loss, high isolation (> 30 dB), compact size, and a wide spectral bandwidth of operation. A very compact isolator can be designed if the lens in Fig. 3.25(b) is replaced by a YIG sphere so that it serves a dual purpose [111]. As light from a semiconductor laser is already polarized, a signal polarizer placed between the YIG sphere and the fiber can reduce the feedback by more than 30 dB.

3.6.2 Driving Circuitry

The purpose of driving circuitry is to provide electrical power to the optical source and to modulate the light output in accordance with the signal that is to be transmitted. Driving circuits are relatively simple for LED transmitters but become increasingly complicated for high-bit-rate optical transmitters employing semiconductor lasers as an optical source [95]. As discussed in Section 3.5.2, semiconductor lasers are biased near threshold and then modulated through an electrical time-dependent signal. Thus the driving circuit is designed to supply a constant bias current as well as modulated electrical signal. Furthermore, a servo loop is often used to keep the average optical power constant.

Figure 3.26 shows a simple driving circuit that controls the average optical power through a feedback mechanism. A photodiode monitors the laser output and generates the control signal that is used to adjust the laser bias level. The rear facet of the laser is generally used for the monitoring purpose (see Fig. 3.25). In some transmitters a front-end tap is used to divert a small fraction of the output power to the detector. The bias-level control is essential, since the laser threshold is sensitive to the operating

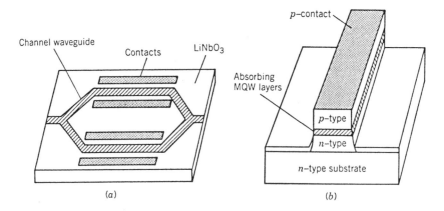

Figure 3.27: Two kinds of external modulators: (a) LiNbO$_3$ modulator in the Mach–Zehnder configuration; (b) semiconductor modulator based on electroabsorption.

temperature. The threshold current also increases with aging of the transmitter because of gradual degradation of the semiconductor laser.

The driving circuit shown in Fig. 3.26 adjusts the bias level dynamically but leaves the modulation current unchanged. Such an approach is acceptable if the slope efficiency of the laser does not change with aging. As discussed in Section 3.5.1 and seen in Fig. 3.20, the slope efficiency of the laser generally decreases with an increase in temperature. A thermoelectric cooler is often used to stabilize the laser temperature. An alternative approach consists of designing driving circuits that use *dual-loop feedback* circuits and adjust both the bias current and the modulation current automatically [112].

3.6.3 Optical Modulators

At bit rates of 10 Gb/s or higher, the frequency chirp imposed by direct modulation becomes large enough that direct modulation of semiconductor lasers is rarely used. For such high-speed transmitters, the laser is biased at a constant current to provide the CW output, and an optical modulator placed next to the laser converts the CW light into a data-coded pulse train with the right modulation format.

Two types of optical modulators developed for lightwave system applications are shown in Fig. 3.27. The electroabsorption modulator makes use of the Franz–Keldysh effect, according to which the bandgap of a semiconductor decreases when an electric field is applied across it. Thus, a transparent semiconductor layer begins to absorb light when its bandgap is reduced electronically by applying an external voltage. An extinction ratio of 15 dB or more can be realized for an applied reverse bias of a few volts at bit rates of up to 40 Gb/s [113]–[120]. Although some chirp is still imposed on coded pulses, it can be made small enough not to be detrimental for the system performance.

An advantage of electroabsorption modulators is that they are made using the same semiconductor material that is used for the laser, and thus the two can be easily integrated on the same chip. Low-chirp transmission at a bit rate of 5 Gb/s was demonstrated as early as 1994 by integrating an electroabsorption modulator with a DBR laser [114]. By 1999, 10-Gb/s optical transmitters with an integrated electroabsorption modulator were available commercially and were used routinely for WDM lightwave systems [119]. By 2001, such integrated modulators exhibited a bandwidth of more than 50 GHz and had the potential of operating at bit rates of up to 100 Gb/s [120]. An electroabsorption modulator can also be used to generate ultrashort pulses suitable for optical time-division multiplexing (OTDM). A DFB laser, integrated monolithically with a MQW modulator, was used as early as 1993 to generate a 20-GHz pulse train [113]. The 7-ps output pulses were nearly transform-limited because of an extremely low chirp associated with the modulator. A 40-GHz train of 1.6 ps pulses was produced in 1999 using an electroabsorption modulator; such pulses can be used for OTDM systems operating at a bit rate of 160 Gb/s [116].

The second category of optical modulators makes use of the $LiNbO_3$ material and a Mach–Zehnder (MZ) interferometer for intensity modulation [121]–[126]. Two titanium-diffused $LiNbO_3$ waveguides form the two arms of a MZ interferometer (see Fig. 3.27). The refractive index of electro-optic materials such as $LiNbO_3$ can be changed by applying an external voltage. In the absence of external voltage, the optical fields in the two arms of the MZ interferometer experience identical phase shifts and interfere constructively. The additional phase shift introduced in one of the arms through voltage-induced index changes destroys the constructive nature of the interference and reduces the transmitted intensity. In particular, no light is transmitted when the phase difference between the two arms equals π, because of destructive interference occurring in that case. As a result, the electrical bit stream applied to the modulator produces an optical replica of the bit stream.

The performance of an external modulator is quantified through the on–off ratio (also called extinction ratio) and the modulation bandwidth. Modern $LiNbO_3$ modulators provide an on–off ratio in excess of 20 and can be modulated at speeds up to 75 GHz [122]. The driving voltage is typically 5 V but can be reduced to below 3 V with a suitable design [125]. $LiNbO_3$ modulators with a bandwidth of 10 GHz were available commercially by 1998, and the bandwidth increased to 40 GHz by 2000 [126].

Other materials can also be used to make external modulators. For example, modulators have been fabricated using electro-optic polymers. Already in 1995 such a modulator exhibited a modulation bandwidth of up to 60 GHz [127]. In a 2001 experiment, a polymeric electro-optic MZ modulator required only 1.8 V for shifting the phase of a 1.55-μm signal by π in one of the arms of the MZ interferometer [128]. The device was only 3 cm long and exhibited about 5-dB chip losses. With further development, such modulators may find applications in lightwave systems.

3.6.4 Optoelectronic Integration

The electrical components used in the driving circuit determine the rate at which the transmitter output can be modulated. For lightwave transmitters operating at bit rates above 1 Gb/s, electrical parasitics associated with various transistors and other components often limit the transmitter performance. The performance of high-speed transmitters can be improved considerably by using monolithic integration of the laser with the driver. Since optical and electrical devices are fabricated on the same chip, such monolithic transmitters are referred to as *optoelectronic integrated-circuit* (OEIC) transmitters. The OEIC approach was first applied to integration of GaAs lasers, since the technology for fabrication of GaAs electrical devices is relatively well established [129]–[131]. The technology for fabrication of InP OEICs evolved rapidly during the 1990s [132]–[136]. A 1.5-μm OEIC transmitter capable of operating at 5 Gb/s was demonstrated in 1988 [132]. By 1995, 10-Gb/s laser transmitters were fabricated by integrating 1.55-μm DFB lasers with field-effect transistors made with the InGaAs/InAlAs material system. Since then, OEIC transmitters with multiple lasers on the same chip have been developed for WDM applications (see Chapter 8).

A related approach to OEIC integrates the semiconductor laser with a photodetector [137]–[139] and/or with a modulator [117]–[120]. The photodetector is generally used for monitoring and stabilizing the output power of the laser. The role of the modulator is to reduce the dynamic chirp occurring when a semiconductor laser is modulated directly (see Section 3.5.2). Photodetectors can be fabricated by using the same material as that used for the laser (see Chapter 4).

The concept of monolithic integration can be extended to build single-chip transmitters by adding all functionality on the same chip. Considerable effort has been directed toward developing such OEICs, often called *photonic integrated circuits* [6], which integrate on the same chip multiple optical components, such as lasers, detectors, modulators, amplifiers, filters, and waveguides [140]–[145]. Such integrated circuits should prove quite beneficial to lightwave technology.

3.6.5 Reliability and Packaging

An optical transmitter should operate reliably over a relatively long period of time (10 years or more) in order to be useful as a major component of lightwave systems. The reliability requirements are quite stringent for undersea lightwave systems, for which repairs and replacement are prohibitively expensive. By far the major reason for failure of optical transmitters is the optical source itself. Considerable testing is performed during assembly and manufacture of transmitters to ensure a reasonable lifetime for the optical source. It is common [95] to quantify the lifetime by a parameter t_F known as *mean time to failure* (MTTF). Its use is based on the assumption of an exponential failure probability $[P_F = \exp(-t/t_F)]$. Typically, t_F should exceed 10^5 hours (about 11 years) for the optical source. Reliability of semiconductor lasers has been studied extensively to ensure their operation under realistic operating conditions [146]–[151].

Both LEDs and semiconductor lasers can stop operating suddenly (catastrophic degradation) or may exhibit a gradual mode of degradation in which the device efficiency degrades with aging [147]. Attempts are made to identify devices that are likely to degrade catastrophically. A common method is to operate the device at high temperatures and high current levels. This technique is referred to as burn-in or *accelerated aging* [146] and is based on the assumption that under high-stress conditions weak devices will fail, while others will stabilize after an initial period of rapid degradation. The change in the operating current at a constant power is used as a measure of device degradation. Figure 3.28 shows the change in the operating current of a 1.3-μm

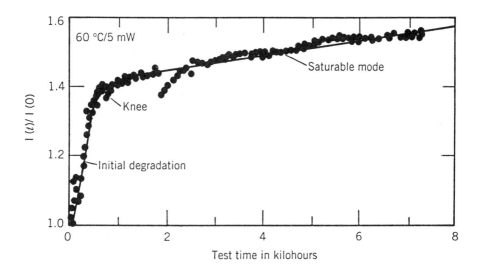

Figure 3.28: Change in current as a function of time for a 1.3-μm InGaAsP laser aged at 60°C with 5 mW of output power. (After Ref. [148]; ©1985 AT&T; reprinted with permission.)

InGaAsP laser aged at 60°C under a constant output power of 5 mW from each facet. The operating current for this laser increases by 40% in the first 400 hours but then stabilizes and increases at a much reduced rate indicative of gradual degradation. The degradation rate can be used to estimate the laser lifetime and the MTTF at the elevated temperature. The MTTF at the normal operating temperature is then extrapolated by using an Arrhenius-type relation $t_F = t_0 \exp(-E_a/k_B T)$, where t_0 is a constant and E_a is the activation energy with a typical value of about 1 eV [147]. Physically, gradual degradation is due to the generation of various kinds of defects (dark-line defects, dark-spot defects) within the active region of the laser or LED [2].

Extensive tests have shown that LEDs are normally more reliable than semiconductor lasers under the same operating conditions. The MTTF for GaAs LEDs easily exceeds 10^6 hours and can be $> 10^7$ hours at 25°C [147]. The MTTF for InGaAsP LEDs is even larger, approaching a value $\sim 10^9$ hours. By contrast, the MTTF for InGaAsP lasers is generally limited to 10^6 hours at 25°C [148]–[150]. Nonetheless, this value is large enough that semiconductor lasers can be used in undersea optical transmitters designed to operate reliably for a period of 25 years. Because of the adverse effect of high temperatures on device reliability, most transmitters use a thermoelectric cooler to maintain the source temperature near 20°C even when the outside temperature may be as high as 80°C.

Even with a reliable optical source, a transmitter may fail in an actual system if the coupling between the source and the fiber degrades with aging. Coupling stability is an important issue in the design of reliable optical transmitters. It depends ultimately on the packaging of transmitters. Although LEDs are often packaged nonhermetically, an hermetic environment is essential for semiconductor lasers. It is common to package the laser separately so that it is isolated from other transmitter components. Figure 3.25 showed two examples of laser packages. In the butt-coupling scheme, an epoxy is used to hold the laser and fiber in place. Coupling stability in this case depends on how epoxy changes with aging of the transmitter. In the lens-coupling scheme, laser welding is used to hold various parts of the assembly together. The laser package becomes a part of the transmitter package, which includes other electrical components associated with the driving circuit. The choice of transmitter package depends on the type of application; a dual-in-line package or a butterfly housing with multiple pins is typically used.

Testing and packaging of optical transmitters are two important parts of the manufacturing process [149], and both of them add considerably to the cost of a transmitter. The development of low-cost packaged transmitters is necessary, especially for local-area and local-loop applications.

Problems

3.1 Show that the external quantum efficiency of a planar LED is given approximately by $\eta_{\text{ext}} = n^{-1}(n+1)^{-2}$, where n is the refractive index of the semiconductor–air interface. Consider Fresnel reflection and total internal reflection at the output facet. Assume that the internal radiation is uniform in all directions.

3.2 Prove that the 3-dB optical bandwidth of a LED is related to the 3-dB electrical bandwidth by the relation $f_{3\,\text{dB}}(\text{optical}) = \sqrt{3} f_{3\,\text{dB}}(\text{electrical})$.

3.3 Find the composition of the quaternary alloy InGaAsP for making semiconductor lasers operating at 1.3- and 1.55-μm wavelengths.

3.4 The active region of a 1.3-μm InGaAsP laser is 250 μm long. Find the active-region gain required for the laser to reach threshold. Assume that the internal loss is 30 cm^{-1}, the mode index is 3.3, and the confinement factor is 0.4.

3.5 Derive the eigenvalue equation for the transverse-electric (TE) modes of a planar waveguide of thickness d and refractive index n_1 sandwiched between two cladding layers of refractive index n_2. (*Hint*: Follow the method of Section 2.2.2 using Cartesian coordinates.)

3.6 Use the result of Problem 3.5 to find the single-mode condition. Use this condition to find the maximum allowed thickness of the active layer for a 1.3-μm semiconductor laser. How does this value change if the laser operates at 1.55 μm? Assume $n_1 = 3.5$ and $n_2 = 3.2$.

3.7 Solve the rate equations in the steady state and obtain the analytic expressions for P and N as a function of the injection current I. Neglect spontaneous emission for simplicity.

3.8 A semiconductor laser is operating continuously at a certain current. Its output power changes slightly because of a transient current fluctuation. Show that the laser power will attain its original value through an oscillatory approach. Obtain the frequency and the damping time of such relaxation oscillations.

3.9 A 250-μm-long InGaAsP laser has an internal loss of 40 cm^{-1}. It operates in a single mode with the modal index 3.3 and the group index 3.4. Calculate the photon lifetime. What is the threshold value of the electron population? Assume that the gain varies as $G = G_N(N - N_0)$ with $G_N = 6 \times 10^3$ s^{-1} and $N_0 = 1 \times 10^8$.

3.10 Determine the threshold current for the semiconductor laser of Problem 3.9 by taking 2 ns as the carrier lifetime. How much power is emitted from one facet when the laser is operated twice above threshold?

3.11 Consider the laser of Problem 3.9 operating twice above threshold. Calculate the differential quantum efficiency and the external quantum efficiency for the laser. What is the device (wall-plug) efficiency if the external voltage is 1.5 V? Assume that the internal quantum efficiency is 90%.

3.12 Calculate the frequency (in GHz units) and the damping time of the relaxation oscillations for the laser of Problem 3.9 operating twice above threshold. Assume that $G_P = -4 \times 10^4$ s^{-1}, where G_P is the derivative of G with respect to P. Also assume that $R_{sp} = 2/\tau_p$.

3.13 Determine the 3-dB modulation bandwidth for the laser of Problem 3.11 biased to operate twice above threshold. What is the corresponding 3-dB electrical bandwidth?

3.14 The threshold current of a semiconductor laser doubles when the operating temperature is increased by 50°C. What is the characteristic temperature of the laser?

3.15 Derive an expression for the 3-dB modulation bandwidth by assuming that the gain G in the rate equations varies with N and P as

$$G(N, P) = G_N(N - N_0)(1 + P/P_s)^{-1/2}.$$

Show that the bandwidth saturates at high operating powers.

3.16 Solve the rate equations (3.5.1) and (3.5.2) numerically by using $I(t) = I_b + I_m f_p(t)$, where $f_p(t)$ represents a rectangular pulse of 200-ps duration. Assume that $I_b/I_{th} = 0.8$, $I_m/I_{th} = 3$, $\tau_p = 3$ ps, $\tau_c = 2$ ns, and $R_{sp} = 2/\tau_p$. Use Eq. (3.5.15) for the gain G with $G_N = 10^4$ s^{-1}, $N_0 = 10^8$, and $\varepsilon_{NL} = 10^{-7}$. Plot the optical pulse shape and the frequency chirp. Why is the optical pulse much shorter than the applied current pulse?

3.17 Complete the derivation of Eq. (3.5.32) for the RIN. How does this expression change if the gain G is assumed of the form of Problem 3.15?

3.18 Calculate the autocorrelation $C_{pp}(\tau)$ by using Eqs. (3.5.31) and (3.5.32). Use it to derive an expression for the SNR of the laser output.

References

[1] Z. Alferov, *IEEE J. Sel. Topics Quantum Electron.* **6**, 832 (2000).

[2] G. P. Agrawal and N. K. Dutta, *Semiconductor Lasers*, 2nd ed., Van Nostrand Reinhold, New York, 1993.

[3] A. R. Adams and Y. Suematsu, *Handbook of Semiconductor Lasers and Photonic Integrated Circuits*, Chapman and Hall, New York, 1994.

[4] N. W. Carlson, *Monolithic Diode Laser Arrays*, Springer, New York, 1994.

[5] S. L. Chuang, *Physics of Optoelectronic Devices*, Wiley, New York, 1995.

[6] L. A. Coldren and S. W. Corzine, *Diode Lasers and Photonic Integrated Circuits*, Wiley, New York, 1995.

[7] G. P. Agrawal, Ed., *Semiconductor Lasers: Past, Present, and Future*, AIP Press, Woodbury, NY, 1995.

[8] T. Igekami, S. Sudo, and S. Sakai, *Frequency Stabilization of Semiconductor Laser Diodes*, Artech House, Norwood, MA, 1995.

[9] M. Ohstu, *Frequency Control of Semiconductor Lasers*, Wiley, New York, 1996.

[10] H. Ghafouri-Shiraz, *Distributed Feedback Laser Diodes*, Wiley, New York, 1995.

[11] G. Morthier and P. Vankwikelberge, *Handbook of Distributed Feedback Laser Diodes*, Artech House, Norwood, MA, 1995.

[12] J. E. Carroll, J. E. Whiteaway, and R. G. Plumb, *Distributed Feedback Semiconductor Lasers*, INSPEC, London, 1998.

[13] M. C. Aman and J. Buus, *Tunable Semiconductor Lasers*, Artech House, Norwood, MA, 1998.

[14] E. Kapon, Ed., *Semiconductor Lasers*, Part I and II, Academic Press, San Diego, CA, 1999.

[15] W. W. Chow and S. W. Koch, *Semiconductor-Laser Fundamentals*, Springer, New York, 1999.

[16] S. Nakamura, S. Pearton, and G. Fasol, *The Blue Laser Diode*, Springer, New York, 2000.

[17] A. Einstein, *Phys. Z.* **18**, 121 (1917).

[18] B. Saleh and M. Teich, *Fundamental of Photonics*, Wiley, New York, 1991, Chaps. 15 and 16.

[19] M. G. A. Bernard and G. Duraffourg, *Phys. Status Solidi* **1**, 699 (1961).

[20] J. Gower, *Optical Communication Systems*, 2nd ed., Prentice-Hall, Upper Saddle River, NJ, 1993.

[21] H. Temkin, G. V. Keramidas, M. A. Pollack, and W. R. Wagner, *J. Appl. Phys.* **52**, 1574 (1981).

[22] C. A. Burrus and R. W. Dawson, *Appl. Phys. Lett.* **17**, 97 (1970).

[23] R. C. Goodfellow, A. C. Carter, I. Griffith, and R. R. Bradley, *IEEE Trans. Electron. Dev.* **26**, 1215 (1979).

[24] O. Wada, S. Yamakoshi, A. Masayuki, Y. Nishitani, and T. Sakurai, *IEEE J. Quantum Electron.* **17**, 174 (1981).

[25] D. Marcuse, *IEEE J. Quantum Electron.* **13**, 819 (1977).

[26] D. Botez and M. Ettenburg, *IEEE Trans. Electron. Dev.* **26**, 1230 (1979).

[27] S. T. Wilkinson, N. M. Jokerst, and R. P. Leavitt, *Appl. Opt.* **34**, 8298 (1995).

[28] M. C. Larson and J. S. Harris, Jr., *IEEE Photon. Technol. Lett.* **7**, 1267 (1995).

[29] I. J. Fritz, J. F. Klem, M. J. Hafich, A. J. Howard, and H. P. Hjalmarson, *IEEE Photon. Technol. Lett.* **7**, 1270 (1995).

[30] T. Whitaker, *Compound Semicond.* **5**, 32 (1999).

[31] P. Bienstman and R Baets, *IEEE J. Quantum Electron.* **36**, 669 (2000).

[32] P. Sipila, M. Saarinen, M. Guina, V. Vilokkinen, M. Toivonen, and M. Pessa, *Semicond. Sci. Technol.* **15**, 418 (2000).

[33] J. C. Dyment, *Appl. Phys. Lett.* **10**, 84 (1967).

[34] K. Oe, S. Ando, and K. Sugiyama, *J. Appl. Phys.* **51**, 43 (1980).

[35] G. P. Agrawal, *J. Lightwave Technol.* **2**, 537 (1984).

[36] R. Wyatt and W. J. Devlin, *Electron. Lett.* **19**, 110 (1983).

[37] W. T. Tsang, in *Semiconductors and Semimetals*, Vol. 22B, W. T. Tsang, Ed., Academic Press, San Diego, CA, 1985, Chap. 4.

[38] S. Akiba, M. Usami, and K. Utaka, *J. Lightwave Technol.* **5**, 1564 (1987).

[39] G. P. Agrawal, in *Progress in Optics*, Vol. 26, E. Wolf, Ed., North-Holland, Amsterdam, 1988, Chap. 3.

[40] J. Buus, *Single Frequency Semiconductor Lasers*, SPIE Press, Bellingham, WA, 1991.

[41] N. Chinone and M. Okai, in *Semiconductor Lasers: Past, Present, and Future*, G. P. Agrawal, Ed., AIP Press, Woodbury, NY, 1995, Chap. 2.

[42] G. P. Li, T. Makino, R. Moor, N. Puetz, K. W. Leong, and H. Lu, *IEEE J. Quantum Electron.* **29**, 1736 (1993).

[43] H. Lu, C. Blaauw, B. Benyon, G. P. Li, and T. Makino, *IEEE J. Sel. Topics Quantum Electron.* **1**, 375 (1995).

[44] J. Hong, C. Blaauw, R. Moore, S. Jatar, and S. Doziba, *IEEE J. Sel. Topics Quantum Electron.* **5**, 441 (1999).

[45] K. Kobayashi and I. Mito, *J. Lightwave Technol.* **6**, 1623 (1988).

[46] T. L. Koch and U. Koren, *J. Lightwave Technol.* **8**, 274 (1990).

[47] H. Hillmer, A. Grabmaier, S. Hansmann, H.-L. Zhu, H. Burkhard, and K. Magari, *IEEE J. Sel. Topics Quantum Electron.* **1**, 356 (1995).

[48] H. Ishii, F. Kano, Y. Tohmori, Y. Kondo, T. Tamamura, and Y. Yoshikuni, *IEEE J. Sel. Topics Quantum Electron.* **1**, 401 (1995).

[49] P.-J. Rigole, S. Nilsson, I. Bäckbom, T. Klinga, J. Wallin, B. Stålnacke, E. Berglind, and B. Stoltz, *IEEE Photon. Technol. Lett.* **7**, 697 (1995); **7**, 1249 (1995).

[50] G. Albert, F. Delorme, S. Grossmaire, S. Slempkes, A. Ougazzaden, and H. Nakajima, *IEEE J. Sel. Topics Quantum Electron.* **3**, 598 (1997).

[51] F. Delorme, G. Albert, P. Boulet, S. Grossmaire, S. Slempkes, and A. Ougazzaden, *IEEE J. Sel. Topics Quantum Electron.* **3**, 607 (1997).

[52] L. Coldren, *IEEE J. Sel. Topics Quantum Electron.* **6**, 988 (2000).

[53] C. J. Chang-Hasnain, in *Semiconductor Lasers: Past, Present, and Future*, G. P. Agrawal, Ed., AIP Press, Woodbury, NY, 1995, Chap. 5.

[54] A. E. Bond, P. D. Dapkus, and J. D. O'Brien, *IEEE J. Sel. Topics Quantum Electron.* **5**, 574 (1999).

[55] C. Wilmsen, H. Temkin, and L.A. Coldren, Eds., *Vertical-Cavity Surface-Emitting Lasers*, Cambridge University Press, New York, 1999.

[56] C. J. Chang-Hasnain, *IEEE J. Sel. Topics Quantum Electron.* **6**, 978 (2000).

[57] K. Iga, *IEEE J. Sel. Topics Quantum Electron.* **6**, 1201 (2000).

[58] A. Karim, S. Björlin, J. Piprek, and J. E. Bowers, *IEEE J. Sel. Topics Quantum Electron.* **6**, 1244 (2000).

[59] H. Li and K. Iga, *Vertical-Cavity Surface-Emitting Laser Devices*, Springer, New York, 2001.

[60] A. Karim, P. Abraham, D. Lofgreen, Y. J. Chiu, J. Piprek, and J. E. Bowers, *Electron. Lett.* **37**, 431 (2001).

[61] T. R. Chen, J. Ungar, J. Iannelli, S. Oh, H. Luong, and N. Bar-Chaim, *Electron. Lett.* **32**, 898 (1996).

[62] G. P. Agrawal, *IEEE J. Quantum Electron.* **23**, 860 (1987).

[63] G. P. Agrawal, *IEEE J. Quantum Electron.* **26**, 1901 (1990).

[64] G. P. Agrawal and G. R. Gray, *Proc. SPIE* **1497**, 444 (1991).

[65] C. Z. Ning and J. V. Moloney, *Appl. Phys. Lett.* **66**, 559 (1995).

[66] M. Osinski and J. Buus, *IEEE J. Quantum Electron.* **23**, 9 (1987).

[67] H. Ishikawa, H. Soda, K. Wakao, K. Kihara, K. Kamite, Y. Kotaki, M. Matsuda, H. Sudo, S. Yamakoshi, S. Isozumi, and H. Imai, *J. Lightwave Technol.* **5**, 848 (1987).

[68] P. A. Morton, T. Tanbun-Ek, R. A. Logan, N. Chand, K. W. Wecht, A. M. Sergent, and P. F. Sciortino, *Electron. Lett.* **30**, 2044 (1994).

[69] E. Goutain, J. C. Renaud, M. Krakowski, D. Rondi, R. Blondeau, D. Decoster, *Electron. Lett.* **32**, 896 (1996).

[70] S. Lindgren, H. Ahlfeldt, L Backlin, L. Forssen, C. Vieider, H. Elderstig, M. Svensson, L. Granlund, et al., *IEEE Photon. Technol. Lett.* **9**, 306 (1997).

[71] R. A. Linke, *Electron. Lett.* **20**, 472 (1984); *IEEE J. Quantum Electron.* **21**, 593 (1985).

[72] G. P. Agrawal and M. J. Potasek, *Opt. Lett.* **11**, 318 (1986).

[73] R. Olshansky and D. Fye, *Electron. Lett.* **20**, 928 (1984).

[74] G. P. Agrawal, *Opt. Lett.* **10**, 10 (1985).

[75] N. A. Olsson, C. H. Henry, R. F. Kazarinov, H. J. Lee, and K. J. Orlowsky, *IEEE J. Quantum Electron.* **24**, 143 (1988).

[76] H. D. Summers and I. H. White, *Electron. Lett.* **30**, 1140 (1994).

[77] F. Kano, T. Yamanaka, N. Yamamoto, H. Mawatan, Y. Tohmori, and Y. Yoshikuni, *IEEE J. Quantum Electron.* **30**, 533 (1994).

[78] D. M. Adams, C. Rolland, N. Puetz, R. S. Moore, F. R. Shepard, H. B. Kim, and S. Bradshaw, *Electron. Lett.* **32**, 485 (1996).

[79] G. P. Agrawal, *Applications of Nonlinear Fiber Optics*, Academic Press, San Diego, CA, 2001.

[80] G. P. Agrawal, *Proc. SPIE* **1376**, 224 (1991).

[81] M. Lax, *Rev. Mod. Phys.* **38**, 541 (1966); *IEEE J. Quantum Electron.* **3**, 37 (1967).

[82] C. H. Henry, *IEEE J. Quantum Electron.* **18**, 259 (1982); **19**, 1391 (1983); *J. Lightwave Technol.* **4**, 298 (1986).

[83] G. P. Agrawal, *Electron. Lett.* **27**, 232 (1991).

[84] G. P. Agrawal, *Phys. Rev. A* **37**, 2488 (1988).

[85] J. Y. Law and G. P. Agrawal, *IEEE Photon. Technol. Lett.* **9**, 437 (1997).

[86] M. Aoki, K. Uomi, T. Tsuchiya, S. Sasaki, M. Okai, and N. Chinone, *IEEE J. Quantum Electron.* **27**, 1782 (1991).

[87] G. P. Agrawal and R. Roy, *Phys. Rev. A* **37**, 2495 (1988).

[88] K. Kikuchi, *Electron. Lett.* **24**, 1001 (1988); *IEEE J. Quantum Electron.* **25**, 684 (1989).

[89] G. P. Agrawal, *IEEE Photon. Technol. Lett.* **1**, 212 (1989).

[90] S. E. Miller, *IEEE J. Quantum Electron.* **24**, 750 (1988); **24**, 1873 (1988).

[91] U. Kruger and K. Petermann, *IEEE J. Quantum Electron.* **24**, 2355 (1988); **26**, 2058 (1990).

[92] G. R. Gray and G. P. Agrawal, *IEEE Photon. Technol. Lett.* **3**, 204 (1991).

[93] G. P. Agrawal, G.-H. Duan, and P. Gallion, *Electron. Lett.* **28**, 1773 (1992).

[94] F. Girardin, G.-H. Duan, and P. Gallion, *IEEE Photon. Technol. Lett.* **8**, 334 (1996).

[95] P. W. Shumate, in *Optical Fiber Telecommunications II*, S. E. Miller and I. P. Kaminow, Eds., Academic Press, San Diego, CA, 1988, Chap. 19.

[96] T. P. Lee, C. A. Burrus, and R. H. Saul, in *Optical Fiber Telecommunications II*, S. E. Miller and I. P. Kaminow, Eds., Academic Press, San Diego, CA, 1988, Chap. 12.

[97] D. S. Alles and K. J. Brady, *AT&T Tech. J.* **68**, 183 (1989).

[98] H. M. Presby and C. A. Edwards, *Electron. Lett.* **28**, 582 (1992).

[99] R. A. Modavis and T. W. Webb, *IEEE Photon. Technol. Lett.* **7**, 798 (1995).

[100] K. Shiraishi, N. Oyama, K. Matsumura, I. Ohisi, and S. Suga, *J. Lightwave Technol.* **13**, 1736 (1995).

[101] P. C. Chen and T. D. Milster, *Laser Diode Chip and Packaging Technology*, Vol. 2610, SPIE Press, Bellingham, WA, 1996.

[102] M. J. Daneman, O. Sologaard, N. C. Tien, K. Y. Lau, and R. S. Muller, *IEEE Photon. Technol. Lett.* **8**, 396 (1996).

[103] B. M. A. Rahman, M. Rajarajan, T. Wongcharoen, and K. T. V. Grattan, *IEEE Photon. Technol. Lett.* **8**, 557 (1996).

[104] I. Moerman, P. P. van Daele, and P. M. Demeester, *IEEE J. Sel. Topics Quantum Electron.* **3**, 1306 (1997).

[105] B. Hubner, G. Vollrath, R. Ries, C. Gréus, H. Janning, E. Ronneberg, E. Kuphal, B. Kempf, R. Gobel, F. Fiedler, R. Zengerle, and H. Burkhard, *IEEE J. Sel. Topics Quantum Electron.* **3**, 1372 (1997).

[106] G. P. Agrawal, *IEEE J. Quantum Electron.* **20**, 468 (1984).

[107] G. P. Agrawal and T. M. Shen, *J. Lightwave Technol.* **4**, 58 (1986).

[108] A. T. Ryan, G. P. Agrawal, G. R. Gray, and E. C. Gage, *IEEE J. Quantum Electron.* **30**, 668 (1994).

[109] G. H. M. van Tartwijk and D. Lenstra, *Quantum Semiclass. Opt.* **7**, 87 (1995).

[110] K. Petermann, *IEEE J. Sel. Topics Quantum Electron.* **1**, 480 (1995).

[111] T. Sugie and M. Saruwatari, *Electron. Lett.* **18**, 1026 (1982).

[112] F. S. Chen, *Electron. Lett.* **16**, 7 (1980).

[113] M. Aoki, M. Suzuki, H. Sano, T. Kawano, T. Ido, T. Taniwatari, K. Uomi, and A. Takai, *IEEE J. Quantum Electron.* **29**, 2088 (1993).

[114] G. Raybon, U. Koren, M. G. Young, B. I. Miller, M. Chien, T. H. Wood, and H. M. Presby, *Electron. Lett.* **30**, 1330 (1994).

[115] T. Tanbun-Ek, P. F. Sciortino, A. M. Sergent, K. W. Wecht, P. Wisk, Y. K. Chen, C. G. Bethea, and S. K. Sputz, *IEEE Photon. Technol. Lett.* **7**, 1019 (1995).

[116] A. D. Ellis, R. J. Manning, I. D. Phillips, and D. Nesset, *Electron. Lett.* **35**, 645 (1999).

[117] H. Takeuchi and H. Yasaka, *NTT Rev.* **12**, 54 (2000).

[118] A. Sano, Y. Miyamoto, K. Yonenaga, and H. Toba, *Electron. Lett.* **36**, 1858 (2000).

[119] Y. Kim, S. K. Kim, J. Lee, Y. Kim, J. Kang, W. Choi, and J. Jeong, *Opt. Fiber Technol.* **7**, 84 (2001).

[120] Y. Akage, K. Kawano, S. Oku, R. Iga, H. Okamoto, Y. Miyamoto, and H. Takeuchi, *Electron. Lett.* **37**, 299 (2001).

[121] L. Thylen, *J. Lightwave Technol.* **6**, 847 (1988).

[122] K. Noguchi, H. Miyazawa, and O. Mitomi, *Electron. Lett.* **30**, 949 (1994).

[123] F. Heismann, S. K. Korotky, and J. J. Veselka, in *Optical Fiber Telecommunications*, Vol. IIIB, I. P. Kaminow and T. L. Loch, Eds., Academic Press, San Diego, CA, 1997, Chap. 8.

[124] K. Noguchi, O. Mitomi, and H. Miyazawa, *J. Lightwave Technol.* **16**, 615 (1998).

[125] E. L. Wooten, K. M. Kissa, A. Yi-Yan, E. J. Murphy, D. A. Lafaw, P. F. Hallemeier, D. Maack, D. V. Attanasio, D. J. Fritz, G. J. McBrien, and D. E. Bossi, *IEEE J. Sel. Topics Quantum Electron.* **6**, 69 (2000).

[126] M. M. Howerton, R. P. Moeller, A. S. Greenblatt, and R. Krahenbuhl, *IEEE Photon. Technol. Lett.* **12**, 792 (2000).

[127] W. Wang, D. Chen, H. R. Fetterman, Y. Shi, W. H. Steier, L. R. Dalton, and P.-M. Chow, *Appl. Phys. Lett.* **67**, 1806 (1995).

[128] H. Zhang, M. C. Oh, A. Szep, W. H. Steier, C. Zhang, L. R. Dalton, H. Erlig, Y. Chang, D. H. Chang, and H. R. Fetterman, *Appl. Phys. Lett.* **78**, 3136 (2001).

[129] O. Wada, T. Sakurai, and T. Nakagami, *IEEE J. Quantum Electron.* **22**, 805 (1986).

[130] T. Horimatsu and M. Sasaki, *J. Lightwave Technol.* **7**, 1613 (1989).

[131] M. Dagenais, R. F. Leheny, H. Temkin, and P. Battacharya, *J. Lightwave Technol.* **8**, 846 (1990).

[132] N. Suzuki, H. Furuyama, Y. Hirayama, M. Morinaga, K. Eguchi, M. Kushibe, M. Funamizu, and M. Nakamura, *Electron. Lett.* **24**, 467 (1988).

[133] K. Pedrotti, *Proc. SPIE* **2149**, 178 (1994).

[134] O. Calliger, A. Clei, D. Robein, R. Azoulay, B. Pierre, S. Biblemont, and C. Kazmierski, *IEE Proc.* **142**, Pt. J, 13 (1995).

[135] R. Pu, C. Duan, and C. W. Wilmsen, *IEEE J. Sel. Topics Quantum Electron.* **5**, 201 (1999).

[136] K. Shuke, T. Yoshida, M. Nakano, A. Kawatani, and Y. Uda, *IEEE J. Sel. Topics Quantum Electron.* **5**, 146 (1999).

[137] K. Sato, I. Katoka, K. Wakita, Y. Kondo, and M. Yamamoto, *Electron. Lett.* **29**, 1087 (1993).

[138] D. Hofstetter, H. P. Zappe, J. E. Epler, and P. Riel, *IEEE Photon. Technol. Lett.* **7**, 1022 (1995).

[139] U. Koren, B. I. Miller, M. G. Young, M. Chien, K. Dreyer, R. Ben-Michael, and R. J. Capik, *IEEE Photon. Technol. Lett.* **8**, 364 (1996).

[140] T. L. Koch and U. Koren, *IEE Proc.* **138**, Pt. J, 139 (1991); *IEEE J. Quantum Electron.* **27**, 641 (1991).

[141] P. J. Williams and A. C. Carter, *GEC J. Res.* **10**, 91 (1993).

[142] R. Matz, J. G. Bauer, P. C. Clemens, G. Heise, H. F. Mahlein, W. Metzger, H. Michel, and G. Schulte-Roth, *IEEE Photon. Technol. Lett.* **6**, 1327 (1994).

[143] Y. Sasaki, Y. Sakata, T. Morimoto, Y. Inomoto, and T. Murakami, *NEC Tech. J.* **48**, 219 (1995).

[144] M. N. Armenise and K.-K. Wong, Eds., *Functional Photonic Integrated Circuits*, SPIE Proc. Series, Vol. 2401, SPIE Press, Bellingham, WA, 1995.

[145] W. Metzger, J. G. Bauer, P. C. Clemens, G. Heise, M. Klein, H. F. Mahlein, R. Matz, H. Michel, and J. Rieger, *Opt. Quantum Electron.* **28**, 51 (1996).

[146] F. R. Nash, W. B. Joyce, R. L. Hartman, E. I. Gordon, and R. W. Dixon, *AT&T Tech. J.* **64**, 671 (1985).

[147] N. K. Dutta and C. L. Zipfel, in *Optical Fiber Telecommunications II*, S. E. Miller and I. P. Kaminow, Eds., Academic Press, San Diego, CA, 1988, Chap. 17.

[148] B. W. Hakki, P. E. Fraley, and T. F. Eltringham, *AT&T Tech. J.* **64**, 771 (1985).

[149] M. Fallahi and S. C. Wang, Eds., *Fabrication, Testing, and Reliability of Semiconductor Lasers*, Vol. 2863, SPIE Press, Bellingham, WA, 1995.

[150] O. Ueda, *Reliability and Degradation of III–V Optical Devices*, Artec House, Boston, 1996.

[151] N. W. Carlson, *IEEE J. Sel. Topics Quantum Electron.* **6**, 615 (2000).

Chapter 4

Optical Receivers

The role of an optical receiver is to convert the optical signal back into electrical form and recover the data transmitted through the lightwave system. Its main component is a photodetector that converts light into electricity through the photoelectric effect. The requirements for a photodetector are similar to those of an optical source. It should have high sensitivity, fast response, low noise, low cost, and high reliability. Its size should be compatible with the fiber-core size. These requirements are best met by photodetectors made of semiconductor materials. This chapter focuses on photodetectors and optical receivers [1]–[9]. We introduce in Section 4.1 the basic concepts behind the photodetection process and discuss in Section 4.2 several kinds of photodetectors commonly used for optical receivers. The components of an optical receiver are described in Section 4.3 with emphasis on the role played by each component. Section 4.4 deals with various noise sources that limit the signal-to-noise ratio in optical receivers. Sections 4.5 and 4.6 are devoted to receiver sensitivity and its degradation under nonideal conditions. The performance of optical receivers in actual transmission experiments is discussed in Section 4.7.

4.1 Basic Concepts

The fundamental mechanism behind the photodetection process is optical absorption. This section introduces basic concepts such as responsivity, quantum efficiency, and bandwidth that are common to all photodetectors and are needed later in this chapter.

4.1.1 Detector Responsivity

Consider the semiconductor slab shown schematically in Fig. 4.1. If the energy $h\nu$ of incident photons exceeds the bandgap energy, an electron–hole pair is generated each time a photon is absorbed by the semiconductor. Under the influence of an electric field set up by an applied voltage, electrons and holes are swept across the semiconductor, resulting in a flow of electric current. The photocurrent I_p is directly proportional to

Incident light

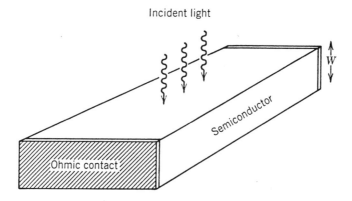

Figure 4.1: A semiconductor slab used as a photodetector.

the incident optical power P_{in}, i.e.,

$$I_p = RP_{in}, \tag{4.1.1}$$

where R is the *responsivity* of the photodetector (in units of A/W).

The responsivity R can be expressed in terms of a fundamental quantity η, called the *quantum efficiency* and defined as

$$\eta = \frac{\text{electron generation rate}}{\text{photon incidence rate}} = \frac{I_p/q}{P_{in}/h\nu} = \frac{h\nu}{q}R, \tag{4.1.2}$$

where Eq. (4.1.1) was used. The responsivity R is thus given by

$$R = \frac{\eta q}{h\nu} \approx \frac{\eta \lambda}{1.24}, \tag{4.1.3}$$

where $\lambda \equiv c/\nu$ is expressed in micrometers. The responsivity of a photodetector increases with the wavelength λ simply because more photons are present for the same optical power. Such a linear dependence on λ is not expected to continue forever because eventually the photon energy becomes too small to generate electrons. In semiconductors, this happens for $h\nu < E_g$, where E_g is the bandgap. The quantum efficiency η then drops to zero.

The dependence of η on λ enters through the absorption coefficient α. If the facets of the semiconductor slab in Fig. 4.1 are assumed to have an antireflection coating, the power transmitted through the slab of width W is $P_{tr} = \exp(-\alpha W)P_{in}$. The absorbed power can be written as

$$P_{abs} = P_{in} - P_{tr} = [1 - \exp(-\alpha W)]P_{in}. \tag{4.1.4}$$

Since each absorbed photon creates an electron–hole pair, the quantum efficiency η is given by

$$\eta = P_{abs}/P_{in} = 1 - \exp(-\alpha W). \tag{4.1.5}$$

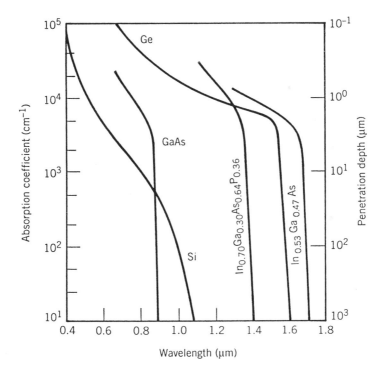

Figure 4.2: Wavelength dependence of the absorption coefficient for several semiconductor materials. (After Ref. [2]; ©1979 Academic Press; reprinted with permission.)

As expected, η becomes zero when $\alpha = 0$. On the other hand, η approaches 1 if $\alpha W \gg 1$.

Figure 4.2 shows the wavelength dependence of α for several semiconductor materials commonly used to make photodetectors for lightwave systems. The wavelength λ_c at which α becomes zero is called the cutoff wavelength, as that material can be used for a photodetector only for $\lambda < \lambda_c$. As seen in Fig. 4.2, indirect-bandgap semiconductors such as Si and Ge can be used to make photodetectors even though the absorption edge is not as sharp as for direct-bandgap materials. Large values of α ($\sim 10^4$ cm^{-1}) can be realized for most semiconductors, and η can approach 100% for $W \sim 10$ μm. This feature illustrates the efficiency of semiconductors for the purpose of photodetection.

4.1.2 Rise Time and Bandwidth

The *bandwidth* of a photodetector is determined by the speed with which it responds to variations in the incident optical power. It is useful to introduce the concept of *rise time* T_r, defined as the time over which the current builds up from 10 to 90% of its final value when the incident optical power is changed abruptly. Clearly, T_r will depend on the time taken by electrons and holes to travel to the electrical contacts. It also depends on the response time of the electrical circuit used to process the photocurrent.

The rise time T_r of a linear electrical circuit is defined as the time during which the response increases from 10 to 90% of its final output value when the input is changed abruptly (a step function). When the input voltage across an RC circuit changes instantaneously from 0 to V_0, the output voltage changes as

$$V_{\text{out}}(t) = V_0[1 - \exp(-t/RC)], \tag{4.1.6}$$

where R is the resistance and C is the capacitance of the RC circuit. The rise time is found to be given by

$$T_r = (\ln 9)RC \approx 2.2\tau_{RC}, \tag{4.1.7}$$

where $\tau_{RC} = RC$ is the time constant of the RC circuit.

The rise time of a photodetector can be written by extending Eq.(4.1.7) as

$$T_r = (\ln 9)(\tau_{\text{tr}} + \tau_{RC}), \tag{4.1.8}$$

where τ_{tr} is the transit time and τ_{RC} is the time constant of the equivalent RC circuit. The transit time is added to τ_{RC} because it takes some time before the carriers are collected after their generation through absorption of photons. The maximum collection time is just equal to the time an electron takes to traverse the absorption region. Clearly, τ_{tr} can be reduced by decreasing W. However, as seen from Eq. (4.1.5), the quantum efficiency η begins to decrease significantly for $\alpha W < 3$. Thus, there is a trade-off between the bandwidth and the responsivity (speed versus sensitivity) of a photodetector. Often, the RC time constant τ_{RC} limits the bandwidth because of electrical parasitics. The numerical values of τ_{tr} and τ_{RC} depend on the detector design and can vary over a wide range.

The bandwidth of a photodetector is defined in a manner analogous to that of a RC circuit and is given by

$$\Delta f = [2\pi(\tau_{\text{tr}} + \tau_{RC})]^{-1}. \tag{4.1.9}$$

As an example, when $\tau_{\text{tr}} = \tau_{RC} = 100$ ps, the bandwidth of the photodetector is below 1 GHz. Clearly, both τ_{tr} and τ_{RC} should be reduced below 10 ps for photodetectors needed for lightwave systems operating at bit rates of 10 Gb/s or more.

Together with the bandwidth and the responsivity, the dark current I_d of a photodetector is the third important parameter. Here, I_d is the current generated in a photodetector in the absence of any optical signal and originates from stray light or from thermally generated electron–hole pairs. For a good photodetector, the dark current should be negligible ($I_d < 10$ nA).

4.2   Common Photodetectors

The semiconductor slab of Fig. 4.1 is useful for illustrating the basic concepts but such a simple device is rarely used in practice. This section focuses on reverse-biased p–n junctions that are commonly used for making optical receivers. Metal–semiconductor–metal (MSM) photodetectors are also discussed briefly.

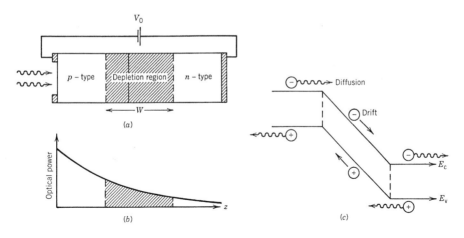

Figure 4.3: (a) A *p–n* photodiode under reverse bias; (b) variation of optical power inside the photodiode; (c) energy-band diagram showing carrier movement through drift and diffusion.

4.2.1 *p–n* Photodiodes

A reverse-biased *p–n* junction consists of a region, known as the *depletion region*, that is essentially devoid of free charge carriers and where a large built-in electric field opposes flow of electrons from the *n*-side to the *p*-side (and of holes from *p* to *n*). When such a *p–n* junction is illuminated with light on one side, say the *p*-side (see Fig. 4.3), electron–hole pairs are created through absorption. Because of the large built-in electric field, electrons and holes generated inside the depletion region accelerate in opposite directions and drift to the *n*- and *p*-sides, respectively. The resulting flow of current is proportional to the incident optical power. Thus a reverse-biased *p–n* junction acts as a photodetector and is referred to as the *p–n* photodiode.

Figure 4.3(a) shows the structure of a *p–n* photodiode. As shown in Fig. 4.3(b), optical power decreases exponentially as the incident light is absorbed inside the depletion region. The electron–hole pairs generated inside the depletion region experience a large electric field and drift rapidly toward the *p*- or *n*-side, depending on the electric charge [Fig. 4.3(c)]. The resulting current flow constitutes the photodiode response to the incident optical power in accordance with Eq. (4.1.1). The responsivity of a photodiode is quite high ($R \sim 1$ A/W) because of a high quantum efficiency.

The bandwidth of a *p–n* photodiode is often limited by the transit time τ_{tr} in Eq. (4.1.9). If W is the width of the depletion region and v_d is the drift velocity, the transit time is given by

$$\tau_{\mathrm{tr}} = W/v_d. \qquad (4.2.1)$$

Typically, $W \sim 10$ μm, $v_d \sim 10^5$ m/s, and $\tau_{\mathrm{tr}} \sim 100$ ps. Both W and v_d can be optimized to minimize τ_{tr}. The depletion-layer width depends on the acceptor and donor concentrations and can be controlled through them. The velocity v_d depends on the applied voltage but attains a maximum value (called the *saturation velocity*) $\sim 10^5$ m/s that depends on the material used for the photodiode. The *RC* time constant τ_{RC} can be

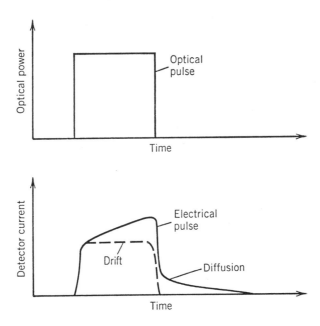

Figure 4.4: Response of a p–n photodiode to a rectangular optical pulse when both drift and diffusion contribute to the detector current.

written as

$$\tau_{RC} = (R_L + R_s)C_p, \tag{4.2.2}$$

where R_L is the external load resistance, R_s is the internal series resistance, and C_p is the parasitic capacitance. Typically, $\tau_{RC} \sim 100$ ps, although lower values are possible with a proper design. Indeed, modern p–n photodiodes are capable of operating at bit rates of up to 40 Gb/s.

The limiting factor for the bandwidth of p–n photodiodes is the presence of a diffusive component in the photocurrent. The physical origin of the diffusive component is related to the absorption of incident light outside the depletion region. Electrons generated in the p-region have to diffuse to the depletion-region boundary before they can drift to the n-side; similarly, holes generated in the n-region must diffuse to the depletion-region boundary. Diffusion is an inherently slow process; carriers take a nanosecond or longer to diffuse over a distance of about 1 μm. Figure 4.4 shows how the presence of a diffusive component can distort the temporal response of a photodiode. The diffusion contribution can be reduced by decreasing the widths of the p- and n-regions and increasing the depletion-region width so that most of the incident optical power is absorbed inside it. This is the approach adopted for p–i–n photodiodes, discussed next.

4.2.2 p–i–n **Photodiodes**

A simple way to increase the depletion-region width is to insert a layer of undoped (or lightly doped) semiconductor material between the p–n junction. Since the middle

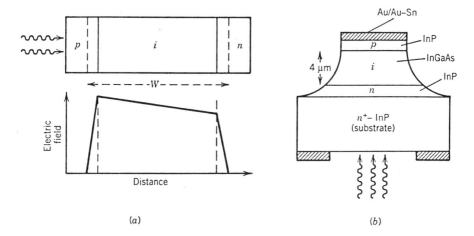

Figure 4.5: (a) A p–i–n photodiode together with the electric-field distribution under reverse bias; (b) design of an InGaAs p–i–n photodiode.

layer consists of nearly intrinsic material, such a structure is referred to as the p–i–n photodiode. Figure 4.5(a) shows the device structure together with the electric-field distribution inside it under reverse-bias operation. Because of its intrinsic nature, the middle i-layer offers a high resistance, and most of the voltage drop occurs across it. As a result, a large electric field exists in the i-layer. In essence, the depletion region extends throughout the i-region, and its width W can be controlled by changing the middle-layer thickness. The main difference from the p–n photodiode is that the drift component of the photocurrent dominates over the diffusion component simply because most of the incident power is absorbed inside the i-region of a p–i–n photodiode.

Since the depletion width W can be tailored in p–i–n photodiodes, a natural question is how large W should be. As discussed in Section 4.1, the optimum value of W depends on a compromise between speed and sensitivity. The responsivity can be increased by increasing W so that the quantum efficiency η approaches 100% [see Eq. (4.1.5)]. However, the response time also increases, as it takes longer for carriers to drift across the depletion region. For indirect-bandgap semiconductors such as Si and Ge, typically W must be in the range 20–50 μm to ensure a reasonable quantum efficiency. The bandwidth of such photodiodes is then limited by a relatively long transit time ($\tau_{tr} > 200$ ps). By contrast, W can be as small as 3–5 μm for photodiodes that use direct-bandgap semiconductors, such as InGaAs. The transit time for such photodiodes is $\tau_{tr} \sim 10$ ps. Such values of τ_{tr} correspond to a detector bandwidth $\Delta f \sim 10$ GHz if we use Eq. (4.1.9) with $\tau_{tr} \gg \tau_{RC}$.

The performance of p–i–n photodiodes can be improved considerably by using a double-heterostructure design. Similar to the case of semiconductor lasers, the middle i-type layer is sandwiched between the p-type and n-type layers of a different semiconductor whose bandgap is chosen such that light is absorbed only in the middle i-layer. A p–i–n photodiode commonly used for lightwave applications uses InGaAs for the middle layer and InP for the surrounding p-type and n-type layers [10]. Figure 4.5(b)

Table 4.1 Characteristics of common p–i–n photodiodes

Parameter	Symbol	Unit	Si	Ge	InGaAs
Wavelength	λ	μm	0.4–1.1	0.8–1.8	1.0–1.7
Responsivity	R	A/W	0.4–0.6	0.5–0.7	0.6–0.9
Quantum efficiency	η	%	75–90	50–55	60–70
Dark current	I_d	nA	1–10	50–500	1–20
Rise time	T_r	ns	0.5–1	0.1–0.5	0.02–0.5
Bandwidth	Δf	GHz	0.3–0.6	0.5–3	1–10
Bias voltage	V_b	V	50–100	6–10	5–6

shows such an InGaAs p–i–n photodiode. Since the bandgap of InP is 1.35 eV, InP is transparent for light whose wavelength exceeds 0.92 μm. By contrast, the bandgap of lattice-matched In$_{1-x}$Ga$_x$As material with $x = 0.47$ is about 0.75 eV (see Section 3.1.4), a value that corresponds to a cutoff wavelength of 1.65 μm. The middle In-GaAs layer thus absorbs strongly in the wavelength region 1.3–1.6 μm. The diffusive component of the detector current is eliminated completely in such a heterostructure photodiode simply because photons are absorbed only inside the depletion region. The front facet is often coated using suitable dielectric layers to minimize reflections. The quantum efficiency η can be made almost 100% by using an InGaAs layer 4–5 μm thick. InGaAs photodiodes are quite useful for lightwave systems and are often used in practice. Table 4.1 lists the operating characteristics of three common p–i–n photo-diodes.

Considerable effort was directed during the 1990s toward developing high-speed p–i–n photodiodes capable of operating at bit rates exceeding 10 Gb/s [10]–[20]. Band-widths of up to 70 GHz were realized as early as 1986 by using a thin absorption layer (< 1 μm) and by reducing the parasitic capacitance C_p with a small size, but only at the expense of a lower quantum efficiency and responsivity [10]. By 1995, p–i–n photodiodes exhibited a bandwidth of 110 GHz for devices designed to reduce τ_{RC} to near 1 ps [15].

Several techniques have been developed to improve the efficiency of high-speed photodiodes. In one approach, a Fabry–Perot (FP) cavity is formed around the p–i–n structure to enhance the quantum efficiency [11]–[14], resulting in a laserlike structure. As discussed in Section 3.3.2, a FP cavity has a set of longitudinal modes at which the internal optical field is resonantly enhanced through constructive interference. As a re-sult, when the incident wavelength is close to a longitudinal mode, such a photodiode exhibits high sensitivity. The wavelength selectivity can even be used to advantage in wavelength-division multiplexing (WDM) applications. A nearly 100% quantum effi-ciency was realized in a photodiode in which one mirror of the FP cavity was formed by using the Bragg reflectivity of a stack of AlGaAs/AlAs layers [12]. This approach was extended to InGaAs photodiodes by inserting a 90-nm-thick InGaAs absorbing layer into a microcavity composed of a GaAs/AlAs Bragg mirror and a dielectric mirror. The device exhibited 94% quantum efficiency at the cavity resonance with a bandwidth of 14 nm [13]. By using an air-bridged metal waveguide together with an undercut mesa

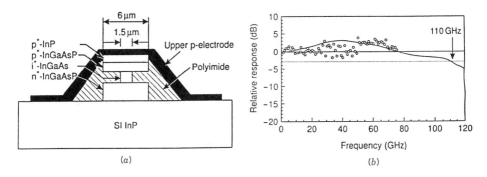

Figure 4.6: (a) Schematic cross section of a mushroom-mesa waveguide photodiode and (b) its measured frequency response. (After Ref. [17]; ©1994 IEEE; reprinted with permission.)

structure, a bandwidth of 120 GHz has been realized [14]. The use of such a structure within a FP cavity should provide a p–i–n photodiode with a high bandwidth and high efficiency.

Another approach to realize efficient high-speed photodiodes makes use of an optical waveguide into which the optical signal is edge coupled [16]–[20]. Such a structure resembles an unpumped semiconductor laser except that various epitaxial layers are optimized differently. In contrast with a semiconductor laser, the waveguide can be made wide to support multiple transverse modes in order to improve the coupling efficiency [16]. Since absorption takes place along the length of the optical waveguide ($\sim 10\ \mu$m), the quantum efficiency can be nearly 100% even for an ultrathin absorption layer. The bandwidth of such *waveguide photodiodes* is limited by τ_{RC} in Eq. (4.1.9), which can be decreased by controlling the waveguide cross-section-area. Indeed, a 50-GHz bandwidth was realized in 1992 for a waveguide photodiode [16].

The bandwidth of waveguide photodiodes can be increased to 110 GHz by adopting a mushroom-mesa waveguide structure [17]. Such a device is shown schematically in Fig. 4.6. In this structure, the width of the i-type absorbing layer was reduced to 1.5 μm while the p- and n-type cladding layers were made 6 μm wide. In this way, both the parasitic capacitance and the internal series resistance were minimized, reducing τ_{RC} to about 1 ps. The frequency response of such a device at the 1.55-μm wavelength is also shown in Fig. 4.6. It was measured by using a spectrum analyzer (circles) as well as taking the Fourier transform of the short-pulse response (solid curve). Clearly, waveguide p–i–n photodiodes can provide both a high responsivity and a large bandwidth. Waveguide photodiodes have been used for 40-Gb/s optical receivers [19] and have the potential for operating at bit rates as high as 100 Gb/s [20].

The performance of waveguide photodiodes can be improved further by adopting an electrode structure designed to support traveling electrical waves with matching impedance to avoid reflections. Such photodiodes are called *traveling-wave photodetectors*. In a GaAs-based implementation of this idea, a bandwidth of 172 GHz with 45% quantum efficiency was realized in a traveling-wave photodetector designed with a 1-μm-wide waveguide [21]. By 2000, such an InP/InGaAs photodetector exhibited a bandwidth of 310 GHz in the 1.55-μm spectral region [22].

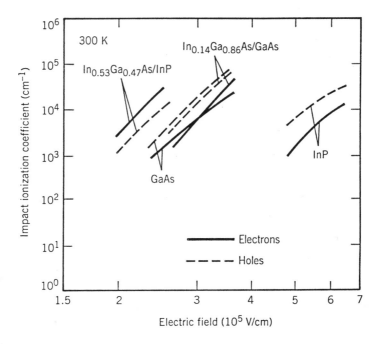

Figure 4.7: Impact-ionization coefficients of several semiconductors as a function of the electric field for electrons (solid line) and holes (dashed line). (After Ref. [24]; ©1977 Elsevier; reprinted with permission.)

4.2.3   Avalanche Photodiodes

All detectors require a certain minimum current to operate reliably. The current requirement translates into a minimum power requirement through $P_{in} = I_p/R$. Detectors with a large responsivity R are preferred since they require less optical power. The responsivity of p–i–n photodiodes is limited by Eq. (4.1.3) and takes its maximum value $R = q/h\nu$ for $\eta = 1$. Avalanche photodiode (APDs) can have much larger values of R, as they are designed to provide an internal current gain in a way similar to photomultiplier tubes. They are used when the amount of optical power that can be spared for the receiver is limited.

The physical phenomenon behind the internal current gain is known as the *impact ionization* [23]. Under certain conditions, an accelerating electron can acquire sufficient energy to generate a new electron–hole pair. In the band picture (see Fig. 3.2) the energetic electron gives a part of its kinetic energy to another electron in the valence band that ends up in the conduction band, leaving behind a hole. The net result of impact ionization is that a single primary electron, generated through absorption of a photon, creates many secondary electrons and holes, all of which contribute to the photodiode current. Of course, the primary hole can also generate secondary electron–hole pairs that contribute to the current. The generation rate is governed by two parameters, α_e and α_h, the *impact-ionization coefficients* of electrons and holes, respectively. Their numerical values depend on the semiconductor material and on the electric field

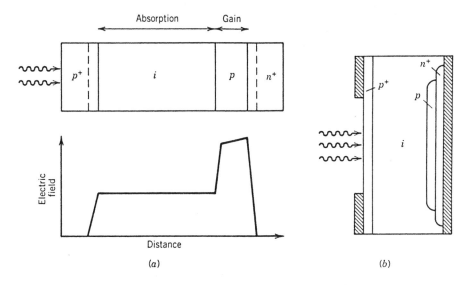

Figure 4.8: (a) An APD together with the electric-field distribution inside various layers under reverse bias; (b) design of a silicon reach-through APD.

that accelerates electrons and holes. Figure 4.7 shows α_e and α_h for several semi-conductors [24]. Values $\sim 1 \times 10^4$ cm^{-1} are obtained for electric fields in the range 2–4×10^5 V/cm. Such large fields can be realized by applying a high voltage (~ 100 V) to the APD.

APDs differ in their design from that of p–i–n photodiodes mainly in one respect: an additional layer is added in which secondary electron–hole pairs are generated through impact ionization. Figure 4.8(a) shows the APD structure together with the variation of electric field in various layers. Under reverse bias, a high electric field exists in the p-type layer sandwiched between i-type and n^+-type layers. This layer is referred to as the *multiplication layer*, since secondary electron–hole pairs are generated here through impact ionization. The i-layer still acts as the depletion region in which most of the incident photons are absorbed and primary electron–hole pairs are generated. Electrons generated in the i-region cross the gain region and generate secondary electron–hole pairs responsible for the current gain.

The current gain for APDs can be calculated by using the two rate equations governing current flow within the multiplication layer [23]:

$$\frac{di_e}{dx} = \alpha_e i_e + \alpha_h i_h, \tag{4.2.3}$$

$$-\frac{di_h}{dx} = \alpha_e i_e + \alpha_h i_h, \tag{4.2.4}$$

where i_e is the electron current and i_h is the hole current. The minus sign in Eq. (4.2.4) is due to the opposite direction of the hole current. The total current,

$$I = i_e(x) + i_h(x), \tag{4.2.5}$$

remains constant at every point inside the multiplication region. If we replace i_h in Eq. (4.2.3) by $I - i_e$, we obtain

$$di_e/dx = (\alpha_e - \alpha_h)i_e + \alpha_h I. \tag{4.2.6}$$

In general, α_e and α_h are x dependent if the electric field across the gain region is nonuniform. The analysis is considerably simplified if we assume a uniform electric field and treat α_e and α_h as constants. We also assume that $\alpha_e > \alpha_h$. The avalanche process is initiated by electrons that enter the gain region of thickness d at $x = 0$. By using the condition $i_h(d) = 0$ (only electrons cross the boundary to enter the n-region), the boundary condition for Eq. (4.2.6) is $i_e(d) = I$. By integrating this equation, the *multiplication factor* defined as $M = i_e(d)/i_e(0)$ is given by

$$M = \frac{1 - k_A}{\exp[-(1 - k_A)\alpha_e d] - k_A}, \tag{4.2.7}$$

where $k_A = \alpha_h/\alpha_e$. The APD gain is quite sensitive to the ratio of the impact-ionization coefficients. When $\alpha_h = 0$ so that only electrons participate in the avalanche process, $M = \exp(\alpha_e d)$, and the APD gain increases exponentially with d. On the other hand, when $\alpha_h = \alpha_e$, so that $k_A = 1$ in Eq. (4.2.7), $M = (1 - \alpha_e d)^{-1}$. The APD gain then becomes infinite for $\alpha_e d = 1$, a condition known as the *avalanche breakdown*. Although higher APD gain can be realized with a smaller gain region when α_e and α_h are comparable, the performance is better in practice for APDs in which either $\alpha_e \gg \alpha_h$ or $\alpha_h \gg \alpha_e$ so that the avalanche process is dominated by only one type of charge carrier. The reason behind this requirement is discussed in Section 4.4, where issues related to the receiver noise are considered.

Because of the current gain, the responsivity of an APD is enhanced by the multiplication factor M and is given by

$$R_{APD} = MR = M(\eta q/h\nu), \tag{4.2.8}$$

where Eq. (4.1.3) was used. It should be mentioned that the avalanche process in APDs is intrinsically noisy and results in a gain factor that fluctuates around an average value. The quantity M in Eq. (4.2.8) refers to the average APD gain. The noise characteristics of APDs are considered in Section 4.4.

The intrinsic bandwidth of an APD depends on the multiplication factor M. This is easily understood by noting that the transit time τ_{tr} for an APD is no longer given by Eq. (4.2.1) but increases considerably simply because generation and collection of secondary electron–hole pairs take additional time. The APD gain decreases at high frequencies because of such an increase in the transit time and limits the bandwidth. The decrease in $M(\omega)$ can be written as [24]

$$M(\omega) = M_0[1 + (\omega \tau_e M_0)^2]^{-1/2}, \tag{4.2.9}$$

where $M_0 = M(0)$ is the low-frequency gain and τ_e is the effective transit time that depends on the ionization coefficient ratio $k_A = \alpha_h/\alpha_e$. For the case $\alpha_h < \alpha_e$, $\tau_e = c_A k_A \tau_{tr}$, where c_A is a constant ($c_A \sim 1$). Assuming that $\tau_{RC} \ll \tau_e$, the APD bandwidth is given approximately by $\Delta f = (2\pi \tau_e M_0)^{-1}$. This relation shows the *trade-off* between

Table 4.2 Characteristics of common APDs

Parameter	Symbol	Unit	Si	Ge	InGaAs
Wavelength	λ	μm	0.4–1.1	0.8–1.8	1.0–1.7
Responsivity	R_{APD}	A/W	80–130	3–30	5–20
APD gain	M	—	100–500	50–200	10–40
k-factor	k_A	—	0.02–0.05	0.7–1.0	0.5–0.7
Dark current	I_d	nA	0.1–1	50–500	1–5
Rise time	T_r	ns	0.1–2	0.5–0.8	0.1–0.5
Bandwidth	Δf	GHz	0.2–1	0.4–0.7	1–10
Bias voltage	V_b	V	200–250	20–40	20–30

the APD gain M_0 and the bandwidth Δf (speed versus sensitivity). It also shows the advantage of using a semiconductor material for which $k_A \ll 1$.

Table 4.2 compares the operating characteristics of Si, Ge, and InGaAs APDs. As $k_A \ll 1$ for Si, silicon APDs can be designed to provide high performance and are useful for lightwave systems operating near 0.8 μm at bit rates $\sim$100 Mb/s. A particularly useful design, shown in Fig. 4.8(b), is known as reach-through APD because the depletion layer reaches to the contact layer through the absorption and multiplication regions. It can provide high gain ($M \approx 100$) with low noise and a relatively large bandwidth. For lightwave systems operating in the wavelength range 1.3–1.6 μm, Ge or InGaAs APDs must be used. The improvement in sensitivity for such APDs is limited to a factor below 10 because of a relatively low APD gain ($M \sim 10$) that must be used to reduce the noise (see Section 4.4.3).

The performance of InGaAs APDs can be improved through suitable design modifications to the basic APD structure shown in Fig. 4.8. The main reason for a relatively poor performance of InGaAs APDs is related to the comparable numerical values of the impact-ionization coefficients α_e and α_h (see Fig. 4.7). As a result, the bandwidth is considerably reduced, and the noise is also relatively high (see Section 4.4). Furthermore, because of a relatively narrow bandgap, InGaAs undergoes tunneling breakdown at electric fields of about 1×10^5 V/cm, a value that is below the threshold for avalanche multiplication. This problem can be solved in heterostructure APDs by using an InP layer for the gain region because quite high electric fields ($> 5 \times 10^5$ V/cm) can exist in InP without tunneling breakdown. Since the absorption region (i-type InGaAs layer) and the multiplication region (n-type InP layer) are separate in such a device, this structure is known as SAM, where SAM stands for *separate absorption and multiplication* regions. As $\alpha_h > \alpha_e$ for InP (see Fig. 4.7), the APD is designed such that holes initiate the avalanche process in an n-type InP layer, and k_A is defined as $k_A = \alpha_e/\alpha_h$. Figure 4.9(a) shows a mesa-type SAM APD structure.

One problem with the SAM APD is related to the large bandgap difference between InP ($E_g = 1.35$ eV) and InGaAs ($E_g = 0.75$ eV). Because of a valence-band step of about 0.4 eV, holes generated in the InGaAs layer are trapped at the heterojunction interface and are considerably slowed before they reach the multiplication region (InP layer). Such an APD has an extremely slow response and a relatively small bandwidth.

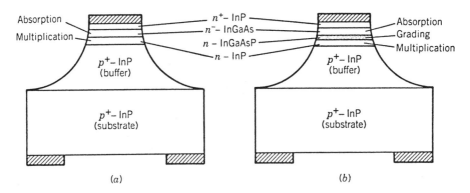

Figure 4.9: Design of (a) SAM and (b) SAGM APDs containing separate absorption, multiplication, and grading regions.

The problem can be solved by using another layer between the absorption and multiplication regions whose bandgap is intermediate to those of InP and InGaAs layers. The quaternary material InGaAsP, the same material used for semiconductor lasers, can be tailored to have a bandgap anywhere in the range 0.75–1.35 eV and is ideal for this purpose. It is even possible to grade the composition of InGaAsP over a region of 10–100 nm thickness. Such APDs are called SAGM APDs, where SAGM indicates *separate absorption, grading, and multiplication* regions [25]. Figure 4.9(b) shows the design of an InGaAs APD with the SAGM structure. The use of an InGaAsP grading layer improves the bandwidth considerably. As early as 1987, a SAGM APD exhibited a gain–bandwidth product $M\Delta f = 70$ GHz for $M > 12$ [26]. This value was increased to 100 GHz in 1991 by using a charge region between the grading and multiplication regions [27]. In such SAGCM APDs, the InP multiplication layer is undoped, while the InP charge layer is heavily n-doped. Holes accelerate in the charge layer because of a strong electric field, but the generation of secondary electron–hole pairs takes place in the undoped InP layer. SAGCM APDs improved considerably during the 1990s [28]–[32]. A gain–bandwidth product of 140 GHz was realized in 2000 using a 0.1-μm-thick multiplication layer that required <20 V across it [32]. Such APDs are quite suitable for making a compact 10-Gb/s APD receiver.

A different approach to the design of high-performance APDs makes use of a superlattice structure [33]–[38]. The major limitation of InGaAs APDs results from comparable values of α_e and α_h. A superlattice design offers the possibility of reducing the ratio $k_A = \alpha_h/\alpha_e$ from its standard value of nearly unity. In one scheme, the absorption and multiplication regions alternate and consist of thin layers ($\sim$10 nm) of semiconductor materials with different bandgaps. This approach was first demonstrated for GaAs/AlGaAs multiquantum-well (MQW) APDs and resulted in a considerable enhancement of the impact-ionization coefficient for electrons [33]. Its use is less successful for the InGaAs/InP material system. Nonetheless, considerable progress has been made through the so-called *staircase* APDs, in which the InGaAsP layer is compositionally graded to form a sawtooth kind of structure in the energy-band diagram that looks like a staircase under reverse bias. Another scheme for making high-speed

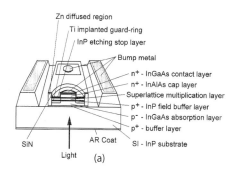
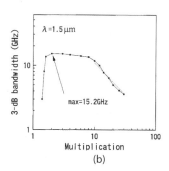

Figure 4.10: (a) Device structure and (b) measured 3-dB bandwidth as a function of M for a superlattice APD. (After Ref. [38]; ©2000 IEEE; reprinted with permission.)

APDs uses alternate layers of InP and InGaAs for the grading region [33]. However, the ratio of the widths of the InP to InGaAs layers varies from zero near the absorbing region to almost infinity near the multiplication region. Since the effective bandgap of a quantum well depends on the quantum-well width (InGaAs layer thickness), a graded "pseudo-quaternary" compound is formed as a result of variation in the layer thickness.

The most successful design for InGaAs APDs uses a superlattice structure for the multiplication region of a SAM APD. A superlattice consists of a periodic structure such that each period is made using two ultrathin ($\sim$10-nm) layers with different bandgaps. In the case of 1.55-μm APDs, alternate layers of InAlGaAs and InAlAs are used, the latter acting as a barrier layer. An InP field-buffer layer often separates the InGaAs absorption region from the superlattice multiplication region. The thickness of this buffer layer is quite critical for the APD performance. For a 52-nm-thick field-buffer layer, the gain–bandwidth product was limited to $M\Delta f = 120$ GHz [34] but increased to 150 GHz when the thickness was reduced to 33.4 nm [37]. These early devices used a mesa structure. During the late 1990s, a planar structure was developed for improving the device reliability [38]. Figure 4.10 shows such a device schematically together with its 3-dB bandwidth measured as a function of the APD gain. The gain–bandwidth product of 110 GHz is large enough for making APDs operating at 10 Gb/s. Indeed, such an APD receiver was used for a 10-Gb/s lightwave system with excellent performance.

The gain–bandwidth limitation of InGaAs APDs results primarily from using the InP material system for the generation of secondary electron–hole pairs. A hybrid approach in which a Si multiplication layer is incorporated next to an InGaAs absorption layer may be useful provided the heterointerface problems can be overcome. In a 1997 experiment, a gain-bandwidth product of more than 300 GHz was realized by using such a hybrid approach [39]. The APD exhibited a 3-dB bandwidth of over 9 GHz for values of M as high as 35 while maintaining a 60% quantum efficiency.

Most APDs use an absorbing layer thick enough (about 1 μm) that the quantum efficiency exceeds 50%. The thickness of the absorbing layer affects the transit time τ_{tr} and the bias voltage V_b. In fact, both of them can be reduced significantly by using a thin absorbing layer ($\sim$0.1 μm), resulting in improved APDs provided that a high quantum efficiency can be maintained. Two approaches have been used to meet these somewhat conflicting design requirements. In one design, a FP cavity is formed to enhance the absorption within a thin layer through multiple round trips. An external quantum efficiency of ~70% and a gain–bandwidth product of 270 GHz were realized in such a 1.55-μm APD using a 60-nm-thick absorbing layer with a 200-nm-thick multiplication layer [40]. In another approach, an optical waveguide is used into which the incident light is edge coupled [41]. Both of these approaches reduce the bias voltage to near 10 V, maintain high efficiency, and reduce the transit time to ~1 ps. Such APDs are suitable for making 10-Gb/s optical receivers.

4.2.4 MSM Photodetectors

In metal–semiconductor–metal (MSM) photodetectors, a semiconductor absorbing layer is sandwiched between two metals, forming a Schottky barrier at each metal–semiconductor interface that prevents flow of electrons from the metal to the semiconductor. Similar to a *p–i–n* photodiode, electron–hole pairs generated through photoabsorption flow toward the metal contacts, resulting in a photocurrent that is a measure of the incident optical power, as indicated in Eq. (4.1.1). For practical reasons, the two metal contacts are made on the same (top) side of the epitaxially grown absorbing layer by using an *interdigited* electrode structure with a finger spacing of about 1 μm [42]. This scheme results in a planar structure with an inherently low parasitic capacitance that allows high-speed operation (up to 300 GHz) of MSM photodetectors. If the light is incident from the electrode side, the responsivity of a MSM photodetector is reduced because of its blockage by the opaque electrodes. This problem can be solved by back illumination if the substrate is transparent to the incident light.

GaAs-based MSM photodetectors were developed throughout the 1980s and exhibit excellent operating characteristics [42]. The development of InGaAs-based MSM photodetectors, suitable for lightwave systems operating in the range 1.3–1.6 μm, started in the late 1980s, with most progress made during the 1990s [43]–[52]. The major problem with InGaAs is its relatively low *Schottky-barrier height* (about 0.2 eV). This problem was solved by introducing a thin layer of InP or InAlAs between the InGaAs layer and the metal contact. Such a layer, called the *barrier-enhancement layer*, improves the performance of InGaAs MSM photodetectors drastically. The use of a 20-nm-thick InAlAs barrier-enhancement layer resulted in 1992 in 1.3-μm MSM photodetectors exhibiting 92% quantum efficiency (through back illumination) with a low dark current [44]. A packaged device had a bandwidth of 4 GHz despite a large 150 μm diameter. If top illumination is desirable for processing or packaging reasons, the responsivity can be enhanced by using semitransparent metal contacts. In one experiment, the responsivity at 1.55 μm increased from 0.4 to 0.7 A/W when the thickness of gold contact was reduced from 100 to 10 nm [45]. In another approach, the structure is separated from the host substrate and bonded to a silicon substrate with the interdigited contact on bottom. Such an "inverted" MSM photodetector then exhibits high responsivity when illuminated from the top [46].

The temporal response of MSM photodetectors is generally different under back and top illuminations [47]. In particular, the bandwidth Δf is larger by about a factor of 2 for top illumination, although the responsivity is reduced because of metal shad-

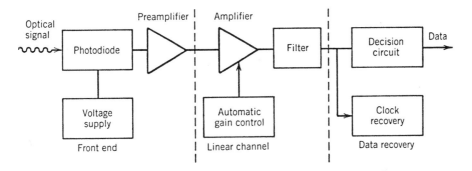

Figure 4.11: Diagram of a digital optical receiver showing various components. Vertical dashed lines group receiver components into three sections.

owing. The performance of a MSM photodetector can be further improved by using a graded superlattice structure. Such devices exhibit a low dark-current density, a responsivity of about 0.6 A/W at 1.3 μm, and a rise time of about 16 ps [50]. In 1998, a 1.55-μm MSM photodetector exhibited a bandwidth of 78 GHz [51]. By 2001, the use of a traveling-wave configuration increased the bandwidth beyond 500 GHz for a GaAs-based device [52]. The planar structure of MSM photodetectors is also suitable for monolithic integration, an issue covered in the next section.

4.3 Receiver Design

The design of an optical receiver depends on the modulation format used by the transmitter. Since most lightwave systems employ the binary intensity modulation, we focus in this chapter on digital optical receivers. Figure 4.11 shows a block diagram of such a receiver. Its components can be arranged into three groups—the front end, the linear channel, and the decision circuit.

4.3.1 Front End

The front end of a receiver consists of a photodiode followed by a preamplifier. The optical signal is coupled onto the photodiode by using a coupling scheme similar to that used for optical transmitters (see Section 3.4.1); butt coupling is often used in practice. The photodiode converts the optical bit stream into an electrical time-varying signal. The role of the preamplifier is to amplify the electrical signal for further processing.

The design of the front end requires a trade-off between speed and sensitivity. Since the input voltage to the preamplifier can be increased by using a large load resistor R_L, a high-impedance front end is often used [see Fig. 4.12(a)]. Furthermore, as discussed in Section 4.4, a large R_L reduces the thermal noise and improves the receiver sensitivity. The main drawback of high-impedance front end is its low bandwidth given by $\Delta f = (2\pi R_L C_T)^{-1}$, where $R_s \ll R_L$ is assumed in Eq. (4.2.2) and $C_T = C_p + C_A$ is the total capacitance, which includes the contributions from the photodiode (C_p) and the transistor used for amplification (C_A). The receiver bandwidth is limited by its slowest

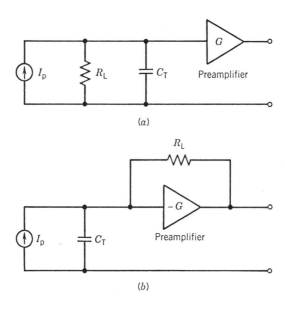

Figure 4.12: Equivalent circuit for (a) high-impedance and (b) transimpedance front ends in optical receivers. The photodiode is modeled as a current source in both cases.

component. A high-impedance front end cannot be used if Δf is considerably less than the bit rate. An equalizer is sometimes used to increase the bandwidth. The equalizer acts as a filter that attenuates low-frequency components of the signal more than the high-frequency components, thereby effectively increasing the front-end bandwidth. If the receiver sensitivity is not of concern, one can simply decrease R_L to increase the bandwidth, resulting in a low-impedance front end.

Transimpedance front ends provide a configuration that has high sensitivity together with a large bandwidth. Its dynamic range is also improved compared with high-impedance front ends. As seen in Fig. 4.12(b), the load resistor is connected as a feedback resistor around an inverting amplifier. Even though R_L is large, the *negative feedback* reduces the effective input impedance by a factor of G, where G is the amplifier gain. The bandwidth is thus enhanced by a factor of G compared with high-impedance front ends. Transimpedance front ends are often used in optical receivers because of their improved characteristics. A major design issue is related to the stability of the feedback loop. More details can be found in Refs. [5]–[9].

4.3.2 Linear Channel

The linear channel in optical receivers consists of a high-gain amplifier (the main amplifier) and a low-pass filter. An equalizer is sometimes included just before the amplifier to correct for the limited bandwidth of the front end. The amplifier gain is controlled automatically to limit the average output voltage to a fixed level irrespective of the incident average optical power at the receiver. The low-pass filter shapes the voltage pulse. Its purpose is to reduce the noise without introducing much *intersymbol interference* (ISI). As discussed in Section 4.4, the receiver noise is proportional to the receiver bandwidth and can be reduced by using a low-pass filter whose bandwidth Δf is smaller than the bit rate. Since other components of the receiver are designed to have a bandwidth larger than the filter bandwidth, the receiver bandwidth is determined by the low-pass filter used in the linear channel. For $\Delta f < B$, the electrical pulse spreads beyond the allocated bit slot. Such a spreading can interfere with the detection of neighboring bits, a phenomenon referred to as ISI.

It is possible to design a low-pass filter in such a way that ISI is minimized [1]. Since the combination of preamplifier, main amplifier, and the filter acts as a linear system (hence the name *linear channel*), the output voltage can be written as

$$V_{\text{out}}(t) = \int_{-\infty}^{\infty} z_T(t - t')I_p(t')\,dt', \tag{4.3.1}$$

where $I_p(t)$ is the photocurrent generated in response to the incident optical power $(I_p = RP_{\text{in}})$. In the frequency domain,

$$\tilde{V}_{\text{out}}(\omega) = Z_T(\omega)\tilde{I}_p(\omega), \tag{4.3.2}$$

where Z_T is the total impedance at the frequency ω and a tilde represents the Fourier transform. Here, $Z_T(\omega)$ is determined by the transfer functions associated with various receiver components and can be written as [3]

$$Z_T(\omega) = G_p(\omega)G_A(\omega)H_F(\omega)/Y_{\text{in}}(\omega), \tag{4.3.3}$$

where $Y_{\text{in}}(\omega)$ is the input admittance and $G_p(\omega)$, $G_A(\omega)$, and $H_F(\omega)$ are transfer functions of the preamplifier, the main amplifier, and the filter. It is useful to isolate the frequency dependence of $\tilde{V}_{\text{out}}(\omega)$ and $\tilde{I}_p(\omega)$ through normalized spectral functions $H_{\text{out}}(\omega)$ and $H_p(\omega)$, which are related to the Fourier transform of the output and input pulse shapes, respectively, and write Eq. (4.3.2) as

$$H_{\text{out}}(\omega) = H_T(\omega)H_p(\omega), \tag{4.3.4}$$

where $H_T(\omega)$ is the total transfer function of the linear channel and is related to the total impedance as $H_T(\omega) = Z_T(\omega)/Z_T(0)$. If the amplifiers have a much larger bandwidth than the low-pass filter, $H_T(\omega)$ can be approximated by $H_F(\omega)$.

The ISI is minimized when $H_{\text{out}}(\omega)$ corresponds to the transfer function of a *raised-cosine filter* and is given by [3]

$$H_{\text{out}}(f) = \begin{cases} \frac{1}{2}[1 + \cos(\pi f/B)], & f < B, \\ 0, & f \geq B, \end{cases} \tag{4.3.5}$$

where $f = \omega/2\pi$ and B is the bit rate. The impulse response, obtained by taking the Fourier transform of $H_{\text{out}}(f)$, is given by

$$h_{\text{out}}(t) = \frac{\sin(2\pi Bt)}{2\pi Bt} \frac{1}{1 - (2Bt)^2}. \tag{4.3.6}$$

The functional form of $h_{\text{out}}(t)$ corresponds to the shape of the voltage pulse $V_{\text{out}}(t)$ received by the decision circuit. At the decision instant $t = 0$, $h_{\text{out}}(t) = 1$, and the

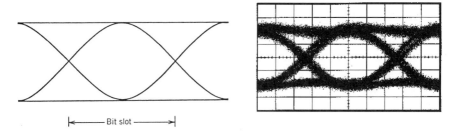

Figure 4.13: Ideal and degraded eye patterns for the NRZ format.

signal is maximum. At the same time, $h_{out}(t) = 0$ for $t = m/B$, where m is an integer. Since $t = m/B$ corresponds to the decision instant of the neighboring bits, the voltage pulse of Eq. (4.3.6) does not interfere with the neighboring bits.

The linear-channel transfer function $H_T(\omega)$ that will result in output pulse shapes of the form (4.3.6) is obtained from Eq. (4.3.4) and is given by

$$H_T(f) = H_{out}(f)/H_p(f). \tag{4.3.7}$$

For an ideal bit stream in the nonreturn-to-zero (NRZ) format (rectangular input pulses of duration $T_B = 1/B$), $H_p(f) = B\sin(\pi f/B)/\pi f$, and $H_T(f)$ becomes

$$H_T(f) = (\pi f/2B)\cot(\pi f/2B). \tag{4.3.8}$$

Equation (4.3.8) determines the frequency response of the linear channel that would produce output pulse shape given by Eq. (4.3.6) under ideal conditions. In practice, the input pulse shape is far from being rectangular. The output pulse shape also deviates from Eq. (4.3.6), and some ISI invariably occurs.

4.3.3 Decision Circuit

The data-recovery section of optical receivers consists of a decision circuit and a clock-recovery circuit. The purpose of the latter is to isolate a spectral component at $f = B$ from the received signal. This component provides information about the bit slot ($T_B = 1/B$) to the decision circuit and helps to synchronize the decision process. In the case of RZ (return-to-zero) format, a spectral component at $f = B$ is present in the received signal; a narrow-bandpass filter such as a surface-acoustic-wave filter can isolate this component easily. Clock recovery is more difficult in the case of NRZ format because the signal received lacks a spectral component at $f = B$. A commonly used technique generates such a component by squaring and rectifying the spectral component at $f = B/2$ that can be obtained by passing the received signal through a high-pass filter.

The decision circuit compares the output from the linear channel to a threshold level, at sampling times determined by the clock-recovery circuit, and decides whether the signal corresponds to bit 1 or bit 0. The best sampling time corresponds to the situation in which the signal level difference between 1 and 0 bits is maximum. It can be determined from the *eye diagram* formed by superposing 2–3-bit-long electrical sequences in the bit stream on top of each other. The resulting pattern is called an eye diagram because of its appearance. Figure 4.13 shows an ideal eye diagram together with a degraded one in which the noise and the timing jitter lead to a partial closing of the eye. The best sampling time corresponds to maximum opening of the eye.

Because of noise inherent in any receiver, there is always a finite probability that a bit would be identified incorrectly by the decision circuit. Digital receivers are designed to operate in such a way that the error probability is quite small (typically $< 10^{-9}$). Issues related to receiver noise and decision errors are discussed in Sections 4.4 and 4.5. The eye diagram provides a visual way of monitoring the receiver performance: Closing of the eye is an indication that the receiver is not performing properly.

4.3.4 Integrated Receivers

All receiver components shown in Fig. 4.11, with the exception of the photodiode, are standard electrical components and can be easily integrated on the same chip by using the integrated-circuit (IC) technology developed for microelectronic devices. Integration is particularly necessary for receivers operating at high bit rates. By 1988, both Si and GaAs IC technologies have been used to make integrated receivers up to a bandwidth of more than 2 GHz [53]. Since then, the bandwidth has been extended to 10 GHz.

Considerable effort has been directed at developing monolithic optical receivers that integrate all components, including the photodetector, on the same chip by using the *optoelectronic integrated-circuit* (OEIC) technology [54]–[74]. Such a complete integration is relatively easy for GaAs receivers, and the technology behind GaAs-based OEICs is quite advanced. The use of MSM photodiodes has proved especially useful as they are structurally compatible with the well-developed *field-effect-transistor* (FET) technology. This technique was used as early as 1986 to demonstrate a four-channel OEIC receiver chip [56].

For lightwave systems operating in the wavelength range 1.3–1.6 μm, InP-based OEIC receivers are needed. Since the IC technology for GaAs is much more mature than for InP, a hybrid approach is sometimes used for InGaAs receivers. In this approach, called *flip-chip OEIC technology* [57], the electronic components are integrated on a GaAs chip, whereas the photodiode is made on top of an InP chip. The two chips are then connected by flipping the InP chip on the GaAs chip, as shown in Fig. 4.14. The advantage of the flip-chip technique is that the photodiode and the electrical components of the receiver can be independently optimized while keeping the parasitics (e.g., effective input capacitance) to a bare minimum.

The InP-based IC technology has advanced considerably during the 1990s, making it possible to develop InGaAs OEIC receivers [58]–[74]. Several kinds of transistors have been used for this purpose. In one approach, a p–i–n photodiode is integrated with the FETs or high-electron-mobility transistors (HEMTs) side by side on an InP substrate [59]–[63]. By 1993, HEMT-based receivers were capable of operating at 10 Gb/s with high sensitivity [62]. The bandwidth of such receivers has been increased to >40 GHz, making it possible to use them at bit rates above 40 Gb/s [63] A waveguide

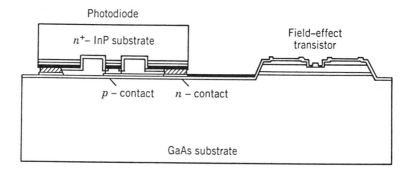

Figure 4.14: Flip-chip OEIC technology for integrated receivers. The InGaAs photodiode is fabricated on an InP substrate and then bonded to the GaAs chip through common electrical contacts. (After Ref. [57]; ©1988 IEE; reprinted with permission.)

p–i–n photodiode has also been integrated with HEMTs to develop a two-channel OEIC receiver.

In another approach [64]–[69], the heterojunction-bipolar transistor (HBT) technology is used to fabricate the p–i–n photodiode within the HBT structure itself through a common-collector configuration. Such transistors are often called *heterojunction phototransistors*. OEIC receivers operating at 5 Gb/s (bandwidth $\Delta f = 3$ GHz) were made by 1993 [64]. By 1995, OEIC receivers making use of the HBT technology exhibited a bandwidth of up to 16 GHz, together with a high gain [66]. Such receivers can be used at bit rates of more than 20 Gb/s. Indeed, a high-sensitivity OEIC receiver module was used in 1995 at a bit rate of 20 Gb/s in a 1.55-μm lightwave system [67]. Even a decision circuit can be integrated within the OEIC receiver by using the HBT technology [68].

A third approach to InP-based OEIC receivers integrates a MSM or a waveguide photodetector with an HEMT amplifier [70]–[73]. By 1995, a bandwidth of 15 GHz was realized for such an OEIC by using modulation-doped FETs [71]. By 2000, such receivers exhibited bandwidths of more than 45 GHz with the use of waveguide photodiodes [73]. Figure 4.15 shows the frequency response together with the epitaxial-layer structure of such an OEIC receiver. This receiver had a bandwidth of 46.5 GHz and exhibited a responsivity of 0.62 A/W in the 1.55-μm wavelength region. It had a clear eye opening at bit rates of up to 50 Gb/s.

Similar to the case of optical transmitters (Section 3.4), packaging of optical receivers is also an important issue [75]–[79]. The fiber–detector coupling issue is quite critical since only a small amount of optical power is typically available at the photodetector. The optical-feedback issue is also important since unintentional reflections fed back into the transmission fiber can affect system performance and should be minimized. In practice, the fiber tip is cut at an angle to reduce the optical feedback. Several different techniques have been used to produce packaged optical receivers capable of operating at bit rates as high as 10 Gb/s. In one approach, an InGaAs APD was bonded to the Si-based IC by using the flip-chip technique [75]. Efficient fiber–APD coupling was realized by using a *slant-ended fiber* and a microlens monolithically fabricated on

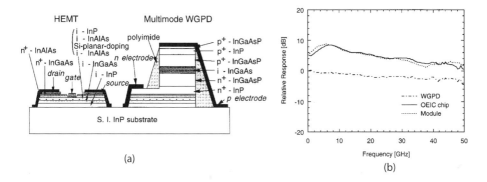

Figure 4.15: (a) Epitaxial-layer structure and (b) frequency response of an OEIC receiver module made using a waveguide photodetector (WGPD). (After Ref. [73]; ©2000 IEEE; reprinted with permission.)

the photodiode. The fiber ferrule was directly laser welded to the package wall with a double-ring structure for mechanical stability. The resulting receiver module withstood shock and vibration tests and had a bandwidth of 10 GHz.

Another hybrid approach makes use of a *planar-lightwave-circuit* platform containing silica waveguides on a silicon substrate. In one experiment, an InP-based OEIC receiver with two channels was flip-chip bonded to the platform [76]. The resulting module could detect two 10-Gb/s channels with negligible crosstalk. GaAs ICs have also been used to fabricate a compact receiver module capable of operating at a bit rate of 10 Gb/s [77]. By 2000, fully packaged 40-Gb/s receivers were available commercially [79]. For local-loop applications, a low-cost package is needed. Such receivers operate at lower bit rates but they should be able to perform well over a wide temperature range extending from -40 to $85\,°C$.

4.4 Receiver Noise

Optical receivers convert incident optical power P_{in} into electric current through a photodiode. The relation $I_p = RP_{in}$ in Eq. (4.1.1) assumes that such a conversion is noise free. However, this is not the case even for a perfect receiver. Two fundamental noise mechanisms, shot noise and thermal noise [80]–[82], lead to fluctuations in the current even when the incident optical signal has a constant power. The relation $I_p = RP_{in}$ still holds if we interpret I_p as the average current. However, electrical noise induced by current fluctuations affects the receiver performance. The objective of this section is to review the noise mechanisms and then discuss the signal-to-nose ratio (SNR) in optical receivers. The *p–i–n* and APD receivers are considered in separate subsections, as the SNR is also affected by the avalanche gain mechanism in APDs.

4.4.1   Noise Mechanisms

The shot noise and thermal noise are the two fundamental noise mechanisms responsible for current fluctuations in all optical receivers even when the incident optical power P_{in} is constant. Of course, additional noise is generated if P_{in} is itself fluctuating because of noise produced by optical amplifiers. This section considers only the noise generated at the receiver; optical noise is discussed in Section 4.6.2.

Shot Noise *(dependent on I_p)*

Shot noise is a manifestation of the fact that an electric current consists of a stream of electrons that are generated at random times. It was first studied by Schottky [83] in 1918 and has been thoroughly investigated since then [80]–[82]. The photodiode current generated in response to a constant optical signal can be written as

$$I(t) = I_p + i_s(t), \tag{4.4.1}$$

where $I_p = RP_{in}$ is the average current and $i_s(t)$ is a current fluctuation related to shot noise. Mathematically, $i_s(t)$ is a stationary random process with *Poisson statistics* (approximated often by Gaussian statistics). The autocorrelation function of $i_s(t)$ is related to the spectral density $S_s(f)$ by the *Wiener–Khinchin theorem* [82]

$$\langle i_s(t)i_s(t+\tau)\rangle = \int_{-\infty}^{\infty} S_s(f)\exp(2\pi i f\tau)\,df, \tag{4.4.2}$$

where angle brackets denote an ensemble average over fluctuations. The spectral density of shot noise is constant and is given by $S_s(f) = qI_p$ (an example of *white noise*). Note that $S_s(f)$ is the *two-sided* spectral density, as negative frequencies are included in Eq. (4.4.2). If only positive frequencies are considered by changing the lower limit of integration to zero, the *one-sided* spectral density becomes $2qI_p$.

The noise variance is obtained by setting $\tau = 0$ in Eq. (4.4.2), i.e.,

$$\sigma_s^2 = \langle i_s^2(t)\rangle = \int_{-\infty}^{\infty} S_s(f)\,df = 2qI_p\Delta f, \tag{4.4.3}$$

where Δf is the *effective noise bandwidth* of the receiver. The actual value of Δf depends on receiver design. It corresponds to the intrinsic photodetector bandwidth if fluctuations in the photocurrent are measured. In practice, a decision circuit may use voltage or some other quantity (e.g., signal integrated over the bit slot). One then has to consider the transfer functions of other receiver components such as the preamplifier and the low-pass filter. It is common to consider current fluctuations and include the total transfer function $H_T(f)$ by modifying Eq. (4.4.3) as

$$\sigma_s^2 = 2qI_p\int_0^{\infty} |H_T(f)|^2\,df = 2qI_p\Delta f, \tag{4.4.4}$$

where $\Delta f = \int_0^{\infty}|H_T(f)|^2 df$ and $H_T(f)$ is given by Eq. (4.3.7). Since the dark current I_d also generates shot noise, its contribution is included in Eq. (4.4.4) by replacing I_p by $I_p + I_d$. The total shot noise is then given by

$$\sigma_s^2 = 2q(I_p+I_d)\Delta f. \tag{4.4.5}$$

The quantity σ_s is the root-mean-square (RMS) value of the noise current induced by shot noise.

Thermal Noise   *(not dependent on I_p)*

At a finite temperature, electrons move randomly in any conductor. Random thermal motion of electrons in a resistor manifests as a fluctuating current even in the absence of an applied voltage. The load resistor in the front end of an optical receiver (see Fig. 4.12) adds such fluctuations to the current generated by the photodiode. This additional noise component is referred to as thermal noise. It is also called *Johnson noise* [84] or *Nyquist noise* [85] after the two scientists who first studied it experimentally and theoretically. Thermal noise can be included by modifying Eq. (4.4.1) as

$$I(t) = I_p + i_s(t) + i_T(t), \tag{4.4.6}$$

where $i_T(t)$ is a current fluctuation induced by thermal noise. Mathematically, $i_T(t)$ is modeled as a stationary Gaussian random process with a spectral density that is frequency independent up to $f \sim 1$ THz (nearly white noise) and is given by

$$S_T(f) = 2k_B T / R_L, \tag{4.4.7}$$

where k_B is the *Boltzmann constant*, T is the absolute temperature, and R_L is the load resistor. As mentioned before, $S_T(f)$ is the two-sided spectral density.

The autocorrelation function of $i_T(t)$ is given by Eq. (4.4.2) if we replace the subscript s by T. The noise variance is obtained by setting $\tau = 0$ and becomes

$$\sigma_T^2 = \langle i_T^2(t) \rangle = \int_{-\infty}^{\infty} S_T(f)\,df = (4k_B T / R_L)\Delta f, \tag{4.4.8}$$

where Δf is the effective noise bandwidth. The same bandwidth appears in the case of both shot and thermal noises. Note that σ_T^2 does not depend on the average current I_p, whereas σ_s^2 does.

Equation (4.4.8) includes thermal noise generated in the load resistor. An actual receiver contains many other electrical components, some of which add additional noise. For example, noise is invariably added by electrical amplifiers. The amount of noise added depends on the front-end design (see Fig. 4.12) and the type of amplifiers used. In particular, the thermal noise is different for field-effect and bipolar transistors. Considerable work has been done to estimate the amplifier noise for different front-end designs [5]. A simple approach accounts for the amplifier noise by introducing a quantity F_n, referred to as the *amplifier noise figure*, and modifying Eq. (4.4.8) as

$$\sigma_T^2 = (4k_B T / R_L)F_n \Delta f. \tag{4.4.9}$$

Physically, F_n represents the factor by which thermal noise is enhanced by various resistors used in pre- and main amplifiers.

The total current noise can be obtained by adding the contributions of shot noise and thermal noise. Since $i_s(t)$ and $i_T(t)$ in Eq. (4.4.6) are independent random processes with approximately Gaussian statistics, the total variance of current fluctuations, $\Delta I = I - I_p = i_s + i_T$, can be obtained simply by adding individual variances. The result is

$$\sigma^2 = \langle (\Delta I)^2 \rangle = \sigma_s^2 + \sigma_T^2 = 2q(I_p + I_d)\Delta f + (4k_B T / R_L) F_n \Delta f. \qquad (4.4.10)$$

Equation (4.4.10) can be used to calculate the SNR of the photocurrent.

4.4.2 *p–i–n* Receivers

The performance of an optical receiver depends on the SNR. The SNR of a receiver with a *p–i–n* photodiode is considered here; APD receivers are discussed in the following subsection. The SNR of any electrical signal is defined as

$$\text{SNR} = \frac{\text{average signal power}}{\text{noise power}} = \frac{I_p^2}{\sigma^2}, \qquad (4.4.11)$$

where we used the fact that electrical power varies as the square of the current. By using Eq. (4.4.10) in Eq. (4.4.11) together with $I_p = RP_{\text{in}}$, the SNR is related to the incident optical power as

$$\text{SNR} = \frac{R^2 P_{\text{in}}^2}{2q(RP_{\text{in}} + I_d)\Delta f + 4(k_B T / R_L) F_n \Delta f}, \qquad (4.4.12)$$

where $R = \eta q / h\nu$ is the responsivity of the *p–i–n* photodiode.

Thermal-Noise Limit

In most cases of practical interest, thermal noise dominates receiver performance ($\sigma_T^2 \gg \sigma_s^2$). Neglecting the shot-noise term in Eq. (4.4.12), the SNR becomes

$$\text{SNR} = \frac{R_L R^2 P_{\text{in}}^2}{4k_B T F_n \Delta f}. \qquad (4.4.13)$$

Thus, the SNR varies as P_{in}^2 in the thermal-noise limit. It can also be improved by increasing the load resistance. As discussed in Section 4.3.1, this is the reason why most receivers use a high-impedance or transimpedance front end. The effect of thermal noise is often quantified through a quantity called the *noise-equivalent power* (NEP). The NEP is defined as the minimum optical power per unit bandwidth required to produce SNR = 1 and is given by

$$\text{NEP} = \frac{P_{\text{in}}}{\sqrt{\Delta f}} = \left(\frac{4k_B T F_n}{R_L R^2} \right)^{1/2} = \frac{h\nu}{\eta q} \left(\frac{4k_B T F_n}{R_L} \right)^{1/2}. \qquad (4.4.14)$$

Another quantity, called *detectivity* and defined as $(\text{NEP})^{-1}$, is also used for this purpose. The advantage of specifying NEP or the detectivity for a *p–i–n* receiver is that it can be used to estimate the optical power needed to obtain a specific value of SNR if the bandwidth Δf is known. Typical values of NEP are in the range 1–10 pW/Hz$^{1/2}$.

Shot-Noise Limit

Consider the opposite limit in which the receiver performance is dominated by shot noise ($\sigma_s^2 \gg \sigma_T^2$). Since σ_s^2 increases linearly with P_{in}, the shot-noise limit can be achieved by making the incident power large. The dark current I_d can be neglected in that situation. Equation (4.4.12) then provides the following expression for SNR:

$$\text{SNR} = \frac{RP_{in}}{2q\Delta f} = \frac{\eta P_{in}}{2h\nu\Delta f}. \tag{4.4.15}$$

The SNR increases linearly with P_{in} in the shot-noise limit and depends only on the quantum efficiency η, the bandwidth Δf, and the photon energy $h\nu$. It can be written in terms of the number of photons N_p contained in the "1" bit. If we use $E_p = P_{in}\int_{-\infty}^{\infty}h_p(t)\,dt = P_{in}/B$ for the pulse energy of a bit of duration $1/B$, where B is the bit rate, and note that $E_p = N_p h\nu$, we can write P_{in} as $P_{in} = N_p h\nu B$. By choosing $\Delta f = B/2$ (a typical value for the bandwidth), the SNR is simply given by ηN_p. In the shot-noise limit, a SNR of 20 dB can be realized if $N_p \approx 100$. By contrast, several thousand photons are required to obtain SNR = 20 dB when thermal noise dominates the receiver. As a reference, for a 1.55-μm receiver operating at 10 Gb/s, $N_p = 100$ when $P_{in} \approx 130$ nW.

4.4.3 APD Receivers

Optical receivers that employ an APD generally provide a higher SNR for the same incident optical power. The improvement is due to the internal gain that increases the photocurrent by a multiplication factor M so that

$$I_p = MRP_{in} = R_{APD}P_{in}, \tag{4.4.16}$$

where R_{APD} is the APD responsivity, enhanced by a factor of M compared with that of p–i–n photodiodes ($R_{APD} = MR$). The SNR should improve by a factor of M^2 if the receiver noise were unaffected by the internal gain mechanism of APDs. Unfortunately, this is not the case, and the SNR improvement is considerably reduced.

Shot-Noise Enhancement

Thermal noise remains the same for APD receivers, as it originates in the electrical components that are not part of the APD. This is not the case for shot noise. The APD gain results from generation of secondary electron–hole pairs through the process of impact ionization. Since such pairs are generated at random times, an additional contribution is added to the shot noise associated with the generation of primary electron–hole pairs. In effect, the multiplication factor itself is a random variable, and M appearing in Eq. (4.4.16) represents the average APD gain. Total shot noise can be calculated by using Eqs. (4.2.3) and (4.2.4) and treating i_e and i_h as random variables [86]. The result is

$$\sigma_s^2 = 2qM^2 F_A(RP_{in} + I_d)\Delta f. \tag{4.4.17}$$

where F_A is the *excess noise factor* of the APD and is given by [86]

$$F_A(M) = k_A M + (1 - k_A)(2 - 1/M). \tag{4.4.18}$$

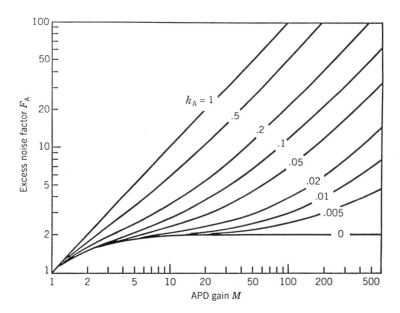

Figure 4.16: Excess noise factor F_A as a function of the average APD gain M for several values of the ionization-coefficient ratio k_A.

The dimensionless parameter $k_A = \alpha_h/\alpha_e$ if $\alpha_h < \alpha_e$ but is defined as $k_A = \alpha_e/\alpha_h$ when $\alpha_h > \alpha_e$. In other words, k_A is in the range $0 < k_A < 1$. Figure 4.16 shows the gain dependence of F_A for several values of k_A. In general, F_A increases with M. However, although F_A is at most 2 for $k_A = 0$, it keeps on increasing linearly ($F_A = M$) when $k_A = 1$. The ratio k_A should be as small as possible for achieving the best performance from an APD [87].

If the avalanche–gain process were noise free ($F_A = 1$), both I_p and σ_s would increase by the same factor M, and the SNR would be unaffected as far as the shot-noise contribution is concerned. In practice, the SNR of APD receivers is worse than that of p–i–n receivers when shot noise dominates because of the excess noise generated inside the APD. It is the dominance of thermal noise in practical receivers that makes APDs attractive. In fact, the SNR of APD receivers can be written as

$$\text{SNR} = \frac{I_p^2}{\sigma_s^2 + \sigma_T^2} = \frac{(MRP_{\text{in}})^2}{2qM^2F_A(RP_{\text{in}} + I_d)\Delta f + 4(k_BT/R_L)F_n\Delta f}, \qquad (4.4.19)$$

where Eqs. (4.4.9), (4.4.16), and (4.4.17) were used. In the thermal-noise limit ($\sigma_s \ll \sigma_T$), the SNR becomes

$$\text{SNR} = (R_LR^2/4k_BTF_n\Delta f)M^2P_{\text{in}}^2 \qquad (4.4.20)$$

and is improved, as expected, by a factor of M^2 compared with that of p–i–n receivers [see Eq. (4.4.13)]. By contrast, in the shot-noise limit ($\sigma_s \gg \sigma_T$), the SNR is given by

$$\text{SNR} = \frac{RP_{\text{in}}}{2qF_A\Delta f} = \frac{\eta P_{\text{in}}}{2h\nu F_A\Delta f} \qquad (4.4.21)$$

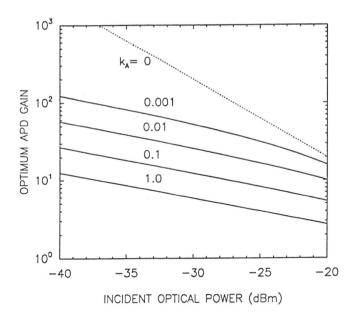

Figure 4.17: Optimum APD gain M_{opt} as a function of the incident optical power P_{in} for several values of k_A. Parameter values corresponding to a typical 1.55-μm InGaAs APD receiver were used.

and is reduced by the excess noise factor F_A compared with that of p–i–n receivers [see Eq. (4.4.15)].

Optimum APD Gain

Equation (4.4.19) shows that for a given P_{in}, the SNR of APD receivers is maximum for an optimum value M_{opt} of the APD gain M. It is easy to show that the SNR is maximum when M_{opt} satisfies the following cubic polynomial:

$$k_A M_{opt}^3 + (1 - k_A)M_{opt} = \frac{4k_B T F_n}{qR_L(RP_{in} + I_d)}. \qquad (4.4.22)$$

The optimum value M_{opt} depends on a large number of the receiver parameters, such as the dark current, the responsivity R, and the ionization-coefficient ratio k_A. However, it is independent of receiver bandwidth. The most notable feature of Eq. (4.4.22) is that M_{opt} decreases with an increase in P_{in}. Figure 4.17 shows the variation of M_{opt} with P_{in} for several values of k_A by using typical parameter values $R_L = 1$ kΩ, $F_n = 2$, $R = 1$ A/W, and $I_d = 2$ nA corresponding to a 1.55-μm InGaAs receiver. The optimum APD gain is quite sensitive to the ionization-coefficient ratio k_A. For $k_A = 0$, M_{opt} decreases inversely with P_{in}, as can readily be inferred from Eq. (4.4.22) by noting that the contribution of I_d is negligible in practice. By contrast, M_{opt} varies as $P_{in}^{-1/3}$ for $k_A = 1$, and this form of dependence appears to hold even for k_A as small as 0.01 as long as $M_{opt} > 10$. In fact, by neglecting the second term in Eq. (4.4.22), M_{opt} is well approximated by

$$M_{\text{opt}} \approx \left[\frac{4k_B T F_n}{k_A q R_L (R P_{\text{in}} + I_d)} \right]^{1/3} \qquad (4.4.23)$$

for k_A in the range 0.01–1. This expression shows the critical role played by the ionization-coefficient ratio k_A. For Si APDs, for which $k_A \ll 1$, M_{opt} can be as large as 100. By contrast, M_{opt} is in the neighborhood of 10 for InGaAs receivers, since $k_A \approx 0.7$. InGaAs APD receivers are nonetheless useful for optical communication systems simply because of their higher sensitivity. Receiver sensitivity is an important issue in the design of lightwave systems and is discussed next.

4.5 Receiver Sensitivity

Among a group of optical receivers, a receiver is said to be more sensitive if it achieves the same performance with less optical power incident on it. The performance criterion for digital receivers is governed by the *bit-error rate* (BER), defined as the probability of incorrect identification of a bit by the decision circuit of the receiver. Hence, a BER of 2×10^{-6} corresponds to on average 2 errors per million bits. A commonly used criterion for digital optical receivers requires the BER to be below 1×10^{-9}. The receiver sensitivity is then defined as the minimum average received power $\bar{P}_{\text{rec}}$ required by the receiver to operate at a BER of 10^{-9}. Since $\bar{P}_{\text{rec}}$ depends on the BER, let us begin by calculating the BER.

4.5.1 Bit-Error Rate

Figure 4.18(a) shows schematically the fluctuating signal received by the decision circuit, which samples it at the decision instant t_D determined through clock recovery. The sampled value I fluctuates from bit to bit around an average value I_1 or I_0, depending on whether the bit corresponds to 1 or 0 in the bit stream. The decision circuit compares the sampled value with a threshold value I_D and calls it bit 1 if $I > I_D$ or bit 0 if $I < I_D$. An error occurs if $I < I_D$ for bit 1 because of receiver noise. An error also occurs if $I > I_D$ for bit 0. Both sources of errors can be included by defining the *error probability* as

$$\text{BER} = p(1)P(0/1) + p(0)P(1/0), \qquad (4.5.1)$$

where $p(1)$ and $p(0)$ are the probabilities of receiving bits 1 and 0, respectively, $P(0/1)$ is the probability of deciding 0 when 1 is received, and $P(1/0)$ is the probability of deciding 1 when 0 is received. Since 1 and 0 bits are equally likely to occur, $p(1) = p(0) = 1/2$, and the BER becomes

$$\text{BER} = \tfrac{1}{2}[P(0/1) + P(1/0)]. \qquad (4.5.2)$$

Figure 4.18(b) shows how $P(0/1)$ and $P(1/0)$ depend on the probability density function $p(I)$ of the sampled value I. The functional form of $p(I)$ depends on the statistics of noise sources responsible for current fluctuations. Thermal noise i_T in Eq. (4.4.6) is well described by Gaussian statistics with zero mean and variance σ_T^2. The

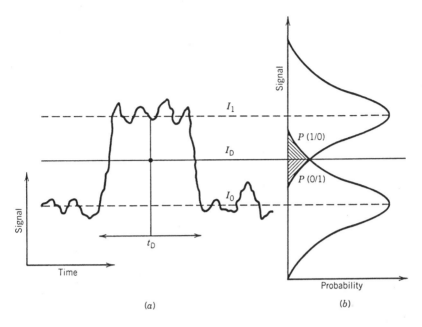

Figure 4.18: (a) Fluctuating signal generated at the receiver. (b) Gaussian probability densities of 1 and 0 bits. The dashed region shows the probability of incorrect identification.

statistics of shot-noise contribution i_s in Eq. (4.4.6) is also approximately Gaussian for p–i–n receivers although that is not the case for APDs [86]–[88]. A common approximation treats i_s as a Gaussian random variable for both p–i–n and APD receivers but with different variance σ_s^2 given by Eqs. (4.4.5) and (4.4.17), respectively. Since the sum of two Gaussian random variables is also a Gaussian random variable, the sampled value I has a Gaussian probability density function with variance $\sigma^2 = \sigma_s^2 + \sigma_T^2$. However, both the average and the variance are different for 1 and 0 bits since I_p in Eq. (4.4.6) equals I_1 or I_0, depending on the bit received. If σ_1^2 and σ_0^2 are the corresponding variances, the conditional probabilities are given by

$$P(0/1) = \frac{1}{\sigma_1\sqrt{2\pi}} \int_{-\infty}^{I_D} \exp\left(-\frac{(I-I_1)^2}{2\sigma_1^2}\right) dI = \frac{1}{2}\,\mathrm{erfc}\left(\frac{I_1-I_D}{\sigma_1\sqrt{2}}\right), \quad (4.5.3)$$

$$P(1/0) = \frac{1}{\sigma_0\sqrt{2\pi}} \int_{I_D}^{\infty} \exp\left(-\frac{(I-I_0)^2}{2\sigma_0^2}\right) dI = \frac{1}{2}\,\mathrm{erfc}\left(\frac{I_D-I_0}{\sigma_0\sqrt{2}}\right), \quad (4.5.4)$$

where erfc stands for the complementary error function, defined as [89]

$$\mathrm{erfc}(x) = \frac{2}{\sqrt{\pi}} \int_{x}^{\infty} \exp(-y^2)\,dy. \quad (4.5.5)$$

By substituting Eqs. (4.5.3) and (4.5.4) in Eq. (4.5.2), the BER is given by

$$\mathrm{BER} = \frac{1}{4}\left[\mathrm{erfc}\left(\frac{I_1-I_D}{\sigma_1\sqrt{2}}\right) + \mathrm{erfc}\left(\frac{I_D-I_0}{\sigma_0\sqrt{2}}\right)\right]. \quad (4.5.6)$$

Equation (4.5.6) shows that the BER depends on the *decision threshold* I_D. In practice, I_D is optimized to minimize the BER. The minimum occurs when I_D is chosen such that

$$\frac{(I_D - I_0)^2}{2\sigma_0^2} = \frac{(I_1 - I_D)^2}{2\sigma_1^2} + \ln\left(\frac{\sigma_1}{\sigma_0}\right). \qquad (4.5.7)$$

The last term in this equation is negligible in most cases of practical interest, and I_D is approximately obtained from

$$(I_D - I_0)/\sigma_0 = (I_1 - I_D)/\sigma_1 \equiv Q. \qquad (4.5.8)$$

An explicit expression for I_D is

$$I_D = \frac{\sigma_0 I_1 + \sigma_1 I_0}{\sigma_0 + \sigma_1}. \qquad (4.5.9)$$

When $\sigma_1 = \sigma_0$, $I_D = (I_1 + I_0)/2$, which corresponds to setting the decision threshold in the middle. This is the situation for most p–i–n receivers whose noise is dominated by thermal noise ($\sigma_T \gg \sigma_s$) and is independent of the average current. By contrast, shot noise is larger for bit 1 than for bit 0, since σ_s^2 varies linearly with the average current. In the case of APD receivers, the BER can be minimized by setting the decision threshold in accordance with Eq. (4.5.9).

The BER with the optimum setting of the decision threshold is obtained by using Eqs. (4.5.6) and (4.5.8) and depends only on the Q parameter as

$$\text{BER} = \frac{1}{2}\operatorname{erfc}\left(\frac{Q}{\sqrt{2}}\right) \approx \frac{\exp(-Q^2/2)}{Q\sqrt{2\pi}}, \qquad (4.5.10)$$

where the parameter Q is obtained from Eqs. (4.5.8) and (4.5.9) and is given by

$$Q = \frac{I_1 - I_0}{\sigma_1 + \sigma_0}. \qquad (4.5.11)$$

The approximate form of BER is obtained by using the asymptotic expansion [89] of $\operatorname{erfc}(Q/\sqrt{2})$ and is reasonably accurate for $Q > 3$. Figure 4.19 shows how the BER varies with the Q parameter. The BER improves as Q increases and becomes lower than 10^{-12} for $Q > 7$. The receiver sensitivity corresponds to the average optical power for which $Q \approx 6$, since BER $\approx 10^{-9}$ when $Q = 6$. The next subsection provides an explicit expression for the receiver sensitivity.

4.5.2   Minimum Received Power

Equation (4.5.10) can be used to calculate the minimum optical power that a receiver needs to operate reliably with a BER below a specified value. For this purpose the Q parameter should be related to the incident optical power. For simplicity, consider the case in which 0 bits carry no optical power so that $P_0 = 0$, and hence $I_0 = 0$. The power P_1 in 1 bits is related to I_1 as

$$I_1 = MRP_1 = 2MR\bar{P}_{\text{rec}}, \qquad (4.5.12)$$

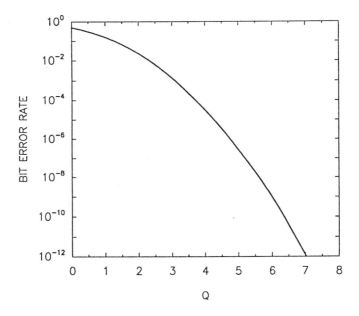

Figure 4.19: Bit-error rate versus the Q parameter.

where $\bar{P}_{\text{rec}}$ is the average received power defined as $\bar{P}_{\text{rec}} = (P_1 + P_0)/2$. The APD gain M is included in Eq. (4.5.12) for generality. The case of p–i–n receivers can be considered by setting $M = 1$.

The RMS noise currents σ_1 and σ_0 include the contributions of both shot noise and thermal noise and can be written as

$$\sigma_1 = (\sigma_s^2 + \sigma_T^2)^{1/2} \quad \text{and} \quad \sigma_0 = \sigma_T, \tag{4.5.13}$$

where σ_s^2 and σ_T^2 are given by Eqs. (4.4.17) and (4.4.9), respectively. Neglecting the contribution of dark current, the noise variances become

$$\sigma_s^2 = 2qM^2 F_A R(2\bar{P}_{\text{rec}})\Delta f, \tag{4.5.14}$$
$$\sigma_T^2 = (4k_B T/R_L)F_n \Delta f. \tag{4.5.15}$$

By using Eqs. (4.5.11)–(4.5.13), the Q parameter is given by

$$Q = \frac{I_1}{\sigma_1 + \sigma_0} = \frac{2MR\bar{P}_{\text{rec}}}{(\sigma_s^2 + \sigma_T^2)^{1/2} + \sigma_T}. \tag{4.5.16}$$

For a specified value of BER, Q is determined from Eq. (4.5.10) and the receiver sensitivity $\bar{P}_{\text{rec}}$ is found from Eq. (4.5.16). A simple analytic expression for $\bar{P}_{\text{rec}}$ is obtained by solving Eq. (4.5.16) for a given value of Q and is given by [3]

$$\bar{P}_{\text{rec}} = \frac{Q}{R}\left(qF_A Q\Delta f + \frac{\sigma_T}{M}\right). \tag{4.5.17}$$

Equation (4.5.17) shows how $\bar{P}_{rec}$ depends on various receiver parameters and how it can be optimized. Consider first the case of a p–i–n receiver by setting $M = 1$. Since thermal noise σ_T generally dominates for such a receiver, $\bar{P}_{rec}$ is given by the simple expression

$$(\bar{P}_{rec})_{pin} \approx Q\sigma_T/R. \tag{4.5.18}$$

From Eq. (4.5.15), σ_T^2 depends not only on receiver parameters such as R_L and F_n but also on the bit rate through the receiver bandwidth Δf (typically, $\Delta f = B/2$). Thus, $\bar{P}_{rec}$ increases as $\sqrt{B}$ in the thermal-noise limit. As an example, consider a 1.55-μm p–i–n receiver with $R = 1$ A/W. If we use $\sigma_T = 100$ nA as a typical value and $Q = 6$ corresponding to a BER of 10^{-9}, the receiver sensitivity is given by $\bar{P}_{rec} = 0.6\ \mu$W or -32.2 dBm.

Equation (4.5.17) shows how receiver sensitivity improves with the use of APD receivers. If thermal noise remains dominant, $\bar{P}_{rec}$ is reduced by a factor of M, and the received sensitivity is improved by the same factor. However, shot noise increases considerably for APD, and Eq. (4.5.17) should be used in the general case in which shot-noise and thermal-noise contributions are comparable. Similar to the case of SNR discussed in Section 4.4.3, the receiver sensitivity can be optimized by adjusting the APD gain M. By using F_A from Eq. (4.4.18) in Eq. (4.5.17), it is easy to verify that $\bar{P}_{rec}$ is minimum for an optimum value of M given by [3]

$$M_{opt} = k_A^{-1/2} \left(\frac{\sigma_T}{Qq\Delta f} + k_A - 1 \right)^{1/2} \approx \left(\frac{\sigma_T}{k_A Qq\Delta f} \right)^{1/2}, \tag{4.5.19}$$

and the minimum value is given by

$$(\bar{P}_{rec})_{APD} = (2q\Delta f/R)Q^2(k_A M_{opt} + 1 - k_A). \tag{4.5.20}$$

The improvement in receiver sensitivity obtained by the use of an APD can be estimated by comparing Eqs. (4.5.18) and (4.5.20). It depends on the ionization-coefficient ratio k_A and is larger for APDs with a smaller value of k_A. For InGaAs APD receivers, the sensitivity is typically improved by 6–8 dB; such an improvement is sometimes called the APD advantage. Note that $\bar{P}_{rec}$ for APD receivers increases linearly with the bit rate B ($\Delta f \approx B/2$), in contrast with its $\sqrt{B}$ dependence for p–i–n receivers. The linear dependence of $\bar{P}_{rec}$ on B is a general feature of shot-noise-limited receivers. For an ideal receiver for which $\sigma_T = 0$, the receiver sensitivity is obtained by setting $M = 1$ in Eq. (4.5.17) and is given by

$$(\bar{P}_{rec})_{ideal} = (q\Delta f/R)Q^2. \tag{4.5.21}$$

A comparison of Eqs. (4.5.20) and (4.5.21) shows sensitivity degradation caused by the excess-noise factor in APD receivers.

Alternative measures of receiver sensitivity are sometimes used. For example, the BER can be related to the SNR and to the average number of photons N_p contained within the "1" bit. In the thermal-noise limit $\sigma_0 \approx \sigma_1$. By using $I_0 = 0$, Eq. (4.5.11) provides $Q = I_1/2\sigma_1$. As SNR $= I_1^2/\sigma_1^2$, it is related to Q by the simple relation SNR $= 4Q^2$. Since $Q = 6$ for a BER of 10^{-9}, the SNR must be at least 144 or 21.6 dB for achieving BER $\leq 10^{-9}$. The required value of SNR changes in the shot-noise limit. In the absence of thermal noise, $\sigma_0 \approx 0$, since shot noise is negligible for the "0" bit if the dark-current contribution is neglected. Since $Q = I_1/\sigma_1 = (\text{SNR})^{1/2}$ in the shot-noise limit, an SNR of 36 or 15.6 dB is enough to obtain BER $= 1 \times 10^{-9}$. It was shown in Section 4.4.2 that SNR $\approx \eta N_p$ [see Eq. (4.4.15) and the following discussion] in the shot-noise limit. By using $Q = (\eta N_p)^{1/2}$ in Eq. (4.5.10), the BER is given by

$$\text{BER} = \tfrac{1}{2} \text{erfc}\left(\sqrt{\eta N_p/2}\right). \qquad (4.5.22)$$

For a receiver with 100% quantum efficiency ($\eta = 1$), BER $= 1 \times 10^{-9}$ when $N_p = 36$. In practice, most optical receivers require $N_p \sim 1000$ to achieve a BER of 10^{-9}, as their performance is severely limited by thermal noise.

4.5.3 Quantum Limit of Photodetection

The BER expression (4.5.22) obtained in the shot-noise limit is not totally accurate, since its derivation is based on the Gaussian approximation for the receiver noise statistics. For an ideal detector (no thermal noise, no dark current, and 100% quantum efficiency), $\sigma_0 = 0$, as shot noise vanishes in the absence of incident power, and thus the decision threshold can be set quite close to the 0-level signal. Indeed, for such an ideal receiver, 1 bits can be identified without error as long as even one photon is detected. An error is made only if a 1 bit fails to produce even a single electron–hole pair. For such a small number of photons and electrons, shot-noise statistics cannot be approximated by a Gaussian distribution, and the exact Poisson statistics should be used. If N_p is the average number of photons in each 1 bit, the probability of generating m electron–hole pairs is given by the Poisson distribution [90]

$$P_m = \exp(-N_p)N_p^m/m!. \qquad (4.5.23)$$

The BER can be calculated by using Eqs. (4.5.2) and (4.5.23). The probability $P(1/0)$ that a 1 is identified when 0 is received is zero since no electron–hole pair is generated when $N_p = 0$. The probability $P(0/1)$ is obtained by setting $m = 0$ in Eq. (4.5.23), since a 0 is decided in that case even though 1 is received. Since $P(0/1) = \exp(-N_p)$, the BER is given by the simple expression

$$\text{BER} = \exp(-N_p)/2. \qquad (4.5.24)$$

For BER $< 10^{-9}$, $\bar{N}_p$ must exceed 20. Since this requirement is a direct result of quantum fluctuations associated with the incoming light, it is referred to as the quantum limit. Each 1 bit must contain at least 20 photons to be detected with a BER $< 10^{-9}$. This requirement can be converted into power by using $P_1 = N_p h\nu B$, where B is the bit rate and $h\nu$ the photon energy. The receiver sensitivity, defined as $\bar{P}_{\text{rec}} = (P_1 + P_0)/2 = P_1/2$, is given by

$$\bar{P}_{\text{rec}} = N_p h\nu B/2 = \bar{N}_p h\nu B. \qquad (4.5.25)$$

The quantity $\bar{N}_p$ expresses the receiver sensitivity in terms of the average number of photons/bit and is related to N_p as $\bar{N}_p = N_p/2$ when 0 bits carry no energy. Its use as a measure of receiver sensitivity is quite common. In the quantum limit $\bar{N}_p = 10$. The power can be calculated from Eq. (4.5.25). For example, for a 1.55-μm receiver ($h\nu = 0.8$ eV), $\bar{P}_{rec} = 13$ nW or -48.9 dBm at $B = 10$ Gb/s. Most receivers operate away from the quantum limit by 20 dB or more. This is equivalent to saying that $\bar{N}_p$ typically exceeds 1000 photons in practical receivers.

4.6   Sensitivity Degradation

The sensitivity analysis in Section 4.5 is based on the consideration of receiver noise only. In particular, the analysis assumes that the optical signal incident on the receiver consists of an ideal bit stream such that 1 bits consist of an optical pulse of constant energy while no energy is contained in 0 bits. In practice, the optical signal emitted by a transmitter deviates from this ideal situation. Moreover, it can be degraded during its transmission through the fiber link. An example of such degradation is provided by the noise added at optical amplifiers. The minimum average optical power required by the receiver increases because of such nonideal conditions. This increase in the average received power is referred to as the *power penalty*. In this section we focus on the sources of power penalties that can lead to sensitivity degradation even without signal transmission through the fiber. The transmission-related power-penalty mechanisms are discussed in Chapter 7.

4.6.1   Extinction Ratio

A simple source of a power penalty is related to the energy carried by 0 bits. Some power is emitted by most transmitters even in the off state. In the case of semiconductor lasers, the off-state power P_0 depends on the bias current I_b and the threshold current I_{th}. If $I_b < I_{th}$, the power emitted during 0 bits is due to spontaneous emission, and generally $P_0 \ll P_1$, where P_1 is the on-state power. By contrast, P_0 can be a significant fraction of P_1 if the laser is biased close to but above threshold. The *extinction ratio* is defined as

$$r_{ex} = P_0/P_1. \tag{4.6.1}$$

The power penalty can be obtained by using Eq. (4.5.11). For a *p–i–n* receiver $I_1 = RP_1$ and $I_0 = RP_0$, where R is the responsivity (the APD gain can be included by replacing R with MR). By using the definition $\bar{P}_{rec} = (P_1 + P_0)/2$ for the receiver sensitivity, the parameter Q is given by

$$Q = \left(\frac{1 - r_{ex}}{1 + r_{ex}}\right) \frac{2R\bar{P}_{rec}}{\sigma_1 + \sigma_0}. \tag{4.6.2}$$

In general, σ_1 and σ_0 depend on $\bar{P}_{rec}$ because of the dependence of the shot-noise contribution on the received optical signal. However, both of them can be approximated by the thermal noise σ_T when receiver performance is dominated by thermal noise. By using $\sigma_1 \approx \sigma_0 \approx \sigma_T$ in Eq. (4.6.2), $\bar{P}_{rec}$ is given by

$$\bar{P}_{rec}(r_{ex}) = \left(\frac{1 + r_{ex}}{1 - r_{ex}}\right) \frac{\sigma_T Q}{R}. \tag{4.6.3}$$

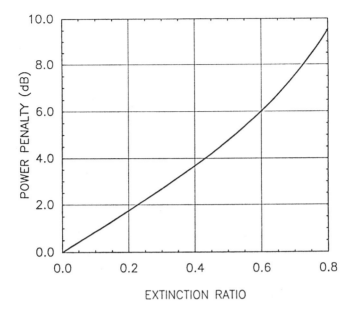

Figure 4.20: Power penalty versus the extinction ratio r_{ex}.

This equation shows that $\bar{P}_{rec}$ increases when $r_{ex} \neq 0$. The power penalty is defined as the ratio $\delta_{ex} = \bar{P}_{rec}(r_{ex})/\bar{P}_{rec}(0)$. It is commonly expressed in decibel (dB) units by using

$$\delta_{ex} = 10 \log_{10}\left(\frac{\bar{P}_{rec}(r_{ex})}{\bar{P}_{rec}(0)}\right) = 10 \log_{10}\left(\frac{1 + r_{ex}}{1 - r_{ex}}\right). \tag{4.6.4}$$

Figure 4.20 shows how the power penalty increases with r_{ex}. A 1-dB penalty occurs for $r_{ex} = 0.12$ and increases to 4.8 dB for $r_{ex} = 0.5$. In practice, for lasers biased below threshold, r_{ex} is typically below 0.05, and the corresponding power penalty (<0.4 dB) is negligible. Nonetheless, it can become significant if the semiconductor laser is biased above threshold. An expression for $\bar{P}_{rec}(r_{ex})$ can be obtained [3] for APD receivers by including the APD gain and the shot-noise contribution to σ_0 and σ_1 in Eq. (4.6.2). The optimum APD gain is lower than that in Eq. (4.5.19) when $r_{ex} \neq 0$. The sensitivity is also reduced because of the lower optimum gain. Normally, the power penalty for an APD receiver is larger by about a factor of 2 for the same value of r_{ex}.

4.6.2 Intensity Noise

The noise analysis of Section 4.4 is based on the assumption that the optical power incident on the receiver does not fluctuate. In practice, light emitted by any transmitter exhibits power fluctuations. Such fluctuations, called intensity noise, were discussed in Section 3.3.8 in the context of semiconductor lasers. The optical receiver converts power fluctuations into current fluctuations which add to those resulting from shot noise and thermal noise. As a result, the receiver SNR is degraded and is lower than that given by Eq. (4.4.19). An exact analysis is complicated, as it involves the calculation of photocurrent statistics [91]. A simple approach consists of adding a third term to the current variance given by Eq. (4.4.10), so that

$$\sigma^2 = \sigma_s^2 + \sigma_T^2 + \sigma_I^2, \tag{4.6.5}$$

where

$$\sigma_I = R\langle(\Delta P_{\text{in}}^2)\rangle^{1/2} = RP_{\text{in}}r_I. \tag{4.6.6}$$

The parameter r_I, defined as $r_I = \langle(\Delta P_{\text{in}}^2)\rangle^{1/2}/P_{\text{in}}$, is a measure of the noise level of the incident optical signal. It is related to the *relative intensity noise* (RIN) of the transmitter as

$$r_I^2 = \frac{1}{2\pi} \int_{-\infty}^{\infty} \text{RIN}(\omega)\, d\omega, \tag{4.6.7}$$

where $\text{RIN}(\omega)$ is given by Eq. (3.5.32). As discussed in Section 3.5.4, r_I is simply the inverse of the SNR of light emitted by the transmitter. Typically, the transmitter SNR is better than 20 dB, and $r_I < 0.01$.

As a result of the dependence of σ_0 and σ_1 on the parameter r_I, the parameter Q in Eq. (4.5.11) is reduced in the presence of intensity noise, Since Q should be maintained to the same value to maintain the BER, it is necessary to increase the received power. This is the origin of the power penalty induced by intensity noise. To simplify the following analysis, the extinction ratio is assumed to be zero, so that $I_0 = 0$ and $\sigma_0 = \sigma_T$. By using $I_1 = RP_1 = 2R\bar{P}_{\text{rec}}$ and Eq. (4.6.5) for σ_1, Q is given by

$$Q = \frac{2R\bar{P}_{\text{rec}}}{(\sigma_T^2 + \sigma_s^2 + \sigma_I^2)^{1/2} + \sigma_T}, \tag{4.6.8}$$

where

$$\sigma_s = (4qR\bar{P}_{\text{rec}}\Delta f)^{1/2}, \qquad \sigma_I = 2r_I R\bar{P}_{\text{rec}}, \tag{4.6.9}$$

and σ_T is given by Eq. (4.4.9). Equation (4.6.8) is easily solved to obtain the following expression for the receiver sensitivity:

$$\bar{P}_{\text{rec}}(r_I) = \frac{Q\sigma_T + Q^2 q\Delta f}{R(1 - r_I^2 Q^2)}. \tag{4.6.10}$$

The power penalty, defined as the increase in $\bar{P}_{\text{rec}}$ when $r_I \neq 0$, is given by

$$\delta_I = 10\log_{10}[\bar{P}_{\text{rec}}(r_I)/\bar{P}_{\text{rec}}(0)] = -10\log_{10}(1 - r_I^2 Q^2). \tag{4.6.11}$$

Figure 4.21 shows the power penalty as a function of r_I for maintaining $Q = 6$ corresponding to a BER of 10^{-9}. The penalty is negligible for $r_I < 0.01$ as δ_I is below 0.02 dB. Since this is the case for most optical transmitters, the effect of transmitter noise is negligible in practice. The power penalty is almost 2 dB for $r_I = 0.1$ and becomes infinite when $r_I = Q^{-1} = 0.167$. An infinite power penalty implies that the receiver cannot operate at the specific BER even if the received optical power is increased indefinitely. In the BER diagram shown in Fig. 4.19, an infinite power penalty corresponds to a saturation of the BER curve above the 10^{-9} level, a feature referred to as the BER floor. In this respect, the effect of intensity noise is qualitatively different

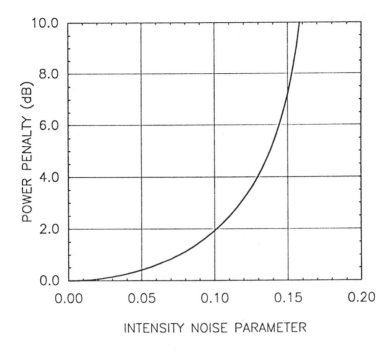

Figure 4.21: Power penalty versus the intensity noise parameter r_I.

than the extinction ratio, for which the power penalty remains finite for all values of r_{ex} such that $r_{ex} < 1$.

The preceding analysis assumes that the intensity noise at the receiver is the same as at the transmitter. This is not typically the case when the optical signal propagates through a fiber link. The intensity noise added by in-line optical amplifiers often becomes a limiting factor for most long-haul lightwave systems (see Chapter 5). When a multimode semiconductor laser is used, fiber dispersion can lead to degradation of the receiver sensitivity through the mode-partition noise. Another phenomenon that can enhance intensity noise is optical feedback from parasitic reflections occurring all along the fiber link. Such transmission-induced power-penalty mechanisms are considered in Chapter 7.

4.6.3 Timing Jitter

The calculation of receiver sensitivity in Section 4.5 is based on the assumption that the signal is sampled at the peak of the voltage pulse. In practice, the decision instant is determined by the clock-recovery circuit (see Fig. 4.11). Because of the noisy nature of the input to the clock-recovery circuit, the sampling time fluctuates from bit to bit. Such fluctuations are called *timing jitter* [92]–[95]. The SNR is degraded because fluctuations in the sampling time lead to additional fluctuations in the signal. This can be understood by noting that if the bit is not sampled at the bit center, the sampled value is reduced by an amount that depends on the timing jitter Δt. Since Δt is a random variable, the reduction in the sampled value is also random. The SNR is reduced as a result of such additional fluctuations, and the receiver performance is degraded. The SNR can be maintained by increasing the received optical power. This increase is the power penalty induced by timing jitter.

To simplify the following analysis, let us consider a p–i–n receiver dominated by thermal noise σ_T and assume a zero extinction ratio. By using $I_0 = 0$ in Eq. (4.5.11), the parameter Q is given by

$$Q = \frac{I_1 - \langle \Delta i_j \rangle}{(\sigma_T^2 + \sigma_j^2)^{1/2} + \sigma_T}, \tag{4.6.12}$$

where $\langle \Delta i_j \rangle$ is the average value and σ_j is the RMS value of the current fluctuation Δi_j induced by timing jitter Δt. If $h_{\text{out}}(t)$ governs the shape of the current pulse,

$$\Delta i_j = I_1 [h_{\text{out}}(0) - h_{\text{out}}(\Delta t)], \tag{4.6.13}$$

where the ideal sampling instant is taken to be $t = 0$.

Clearly, σ_j depends on the shape of the signal pulse at the decision current. A simple choice [92] corresponds to $h_{\text{out}}(t) = \cos^2(\pi B t/2)$, where B is the bit rate. Here Eq. (4.3.6) is used as many optical receivers are designed to provide that pulse shape. Since Δt is likely to be much smaller than the bit period $T_B = 1/B$, it can be approximated as

$$\Delta i_j = (2\pi^2/3 - 4)(B\Delta t)^2 I_1 \tag{4.6.14}$$

by assuming that $B\Delta t \ll 1$. This approximation provides a reasonable estimate of the power penalty as long as the penalty is not too large [92]. This is expected to be the case in practice. To calculate σ_j, the probability density function of the timing jitter Δt is assumed to be Gaussian, so that

$$p(\Delta t) = \frac{1}{\tau_j \sqrt{2\pi}} \exp\left(-\frac{\Delta t^2}{2\tau_j^2}\right), \tag{4.6.15}$$

where τ_j is the RMS value (standard deviation) of Δt. The probability density of Δi_j can be obtained by using Eqs. (4.6.14) and (4.6.15) and noting that Δi_j is proportional to $(\Delta t)^2$. The result is

$$p(\Delta i_j) = \frac{1}{\sqrt{\pi b \Delta i_j I_1}} \exp\left(-\frac{\Delta i_j}{b I_1}\right), \tag{4.6.16}$$

where

$$b = (4\pi^2/3 - 8)(B\tau_j)^2. \tag{4.6.17}$$

Equation (4.6.16) is used to calculate $\langle \Delta i_j \rangle$ and $\sigma_j = \langle (\Delta i_j)^2 \rangle^{1/2}$. The integration over Δi_j is easily done to obtain

$$\langle \Delta i_j \rangle = b I_1/2, \qquad \sigma_j = b I_1/\sqrt{2}. \tag{4.6.18}$$

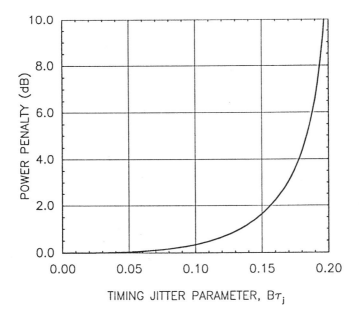

Figure 4.22: Power penalty versus the timing jitter parameter $B\tau_j$.

By using Eqs. (4.6.12) and (4.6.18) and noting that $I_1 = 2R\bar{P}_{rec}$, where R is the responsivity, the receiver sensitivity is given by

$$\bar{P}_{rec}(b) = \left(\frac{\sigma_T Q}{R}\right) \frac{1 - b/2}{(1 - b/2)^2 - b^2 Q^2/2}. \tag{4.6.19}$$

The power penalty, defined as the increase in $\bar{P}_{rec}$, is given by

$$\delta_j = 10 \log_{10}\left(\frac{\bar{P}_{rec}(b)}{\bar{P}_{rec}(0)}\right) = 10 \log_{10}\left(\frac{1 - b/2}{(1 - b/2)^2 - b^2 Q^2/2}\right). \tag{4.6.20}$$

Figure 4.22 shows how the power penalty varies with the parameter $B\tau_j$, which has the physical significance of the fraction of the bit period over which the decision time fluctuates (one standard deviation). The power penalty is negligible for $B\tau_j < 0.1$ but increases rapidly beyond $B\tau_j = 0.1$. A 2-dB penalty occurs for $B\tau_j = 0.16$. Similar to the case of intensity noise, the jitter-induced penalty becomes infinite beyond $B\tau_j = 0.2$. The exact value of $B\tau_j$ at which the penalty becomes infinite depends on the model used to calculate the jitter-induced power penalty. Equation (4.6.20) is obtained by using a specific pulse shape and a specific jitter distribution. It is also based on the use of Eqs. (4.5.10) and (4.6.12), which assumes Gaussian statistics for the receiver current. As evident from Eq. (4.6.16), jitter-induced current fluctuations are not Gaussian in nature. A more accurate calculation shows that Eq. (4.6.20) underestimates the power penalty [94]. The qualitative behavior, however, remains the same. In general, the RMS value of the timing jitter should be below 10% of the bit period for a negligible power penalty. A similar conclusion holds for APD receivers, for which the penalty is generally larger [95].

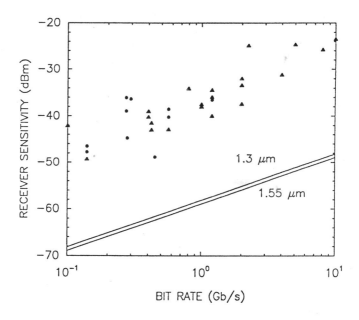

Figure 4.23: Measured receiver sensitivities versus the bit rate for p–i–n (circles) and APD (triangles) receivers in transmission experiments near 1.3- and 1.55-μm wavelengths. The quantum limit of receiver sensitivity is also shown for comparison (solid lines).

4.7 Receiver Performance

The receiver performance is characterized by measuring the BER as a function of the average optical power received. The average optical power corresponding to a BER of 10^{-9} is a measure of receiver sensitivity. Figure 4.23 shows the receiver sensitivity measured in various transmission experiments [96]–[107] by sending a long sequence of pseudorandom bits (typical sequence length $2^{15} - 1$) over a single-mode fiber and then detecting it by using either a p–i–n or an APD receiver. The experiments were performed at the 1.3- or 1.55-μm wavelength, and the bit rate varied from 100 MHz to 10 GHz. The theoretical quantum limit at these two wavelengths is also shown in Fig. 4.23 by using Eq. (4.5.25). A direct comparison shows that the measured receiver sensitivities are worse by 20 dB or more compared with the quantum limit. Most of the degradation is due to the thermal noise that is unavoidable at room temperature and generally dominates the shot noise. Some degradation is due to fiber dispersion, which leads to power penalties; sources of such penalties are discussed in the following chapter.

The dispersion-induced sensitivity degradation depends on both the bit rate B and the fiber length L and increases with BL. This is the reason why the sensitivity degradation from the quantum limit is larger (25–30 dB) for systems operating at high bit rates. The receiver sensitivity at 10 Gb/s is typically worse than -25 dBm [107]. It can be improved by 5–6 dB by using APD receivers. In terms of the number of photons/bit, APD receivers require nearly 1000 photons/bit compared with the quantum

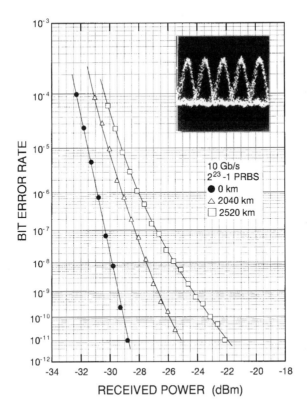

Figure 4.24: BER curves measured for three fiber-link lengths in a 1.55-μm transmission experiment at 10 Gb/s. Inset shows an example of the eye diagram at the receiver. (After Ref. [110]; ©2000 IEEE; reprinted with permission.)

limit of 10 photons/bit. The receiver performance is generally better for shorter wavelengths in the region near 0.85 μm, where silicon APDs can be used; they perform satisfactorily with about 400 photons/bit; an experiment in 1976 achieved a sensitivity of only 187 photons/bit [108]. It is possible to improve the receiver sensitivity by using coding schemes. A sensitivity of 180 photons/bit was realized in a 1.55-μm system experiment [109] after 305 km of transmission at 140 Mb/s.

It is possible to isolate the extent of sensitivity degradation occurring as a result of signal propagation inside the optical fiber. The common procedure is to perform a separate measurement of the receiver sensitivity by connecting the transmitter and receiver directly, without the intermediate fiber. Figure 4.24 shows the results of such a measurement for a 1.55-μm field experiment in which the RZ-format signal consisting of a pseudorandom bit stream in the form of solitons (sequence length $2^{23} - 1$) was propagated over more than 2000 km of fiber [110]. In the absence of fiber (0-km curve), a BER of 10^{-9} is realized for -29.5 dBm of received power. However, the launched signal is degraded considerably during transmission, resulting in about a 3-dB penalty for a 2040-km fiber link. The power penalty increases rapidly with further propagation. In fact, the increasing curvature of BER curves indicates that the BER of 10^{-9} would be unreachable after a distance of 2600 km. This behavior is typical of most lightwave systems. The eye diagram seen in Fig. 4.24 is qualitatively different than that appearing in Fig. 4.13. This difference is related to the use of the RZ format.

The performance of an optical receiver in actual lightwave systems may change with time. Since it is not possible to measure the BER directly for a system in operation, an alternative is needed to monitor system performance. As discussed in Section 4.3.3, the eye diagram is best suited for this purpose; closing of the eye is a measure of degradation in receiver performance and is associated with a corresponding increase in the BER. Figures 4.13 and 4.24 show examples of the eye diagrams for lightwave systems making use of the NRZ and RZ formats, respectively. The eye is wide open in the absence of optical fiber but becomes partially closed when the signal is transmitted through a long fiber link. Closing of the eye is due to amplifier noise, fiber dispersion, and various nonlinear effects, all of which lead to considerable distortion of optical pulses as they propagate through the fiber. The continuous monitoring of the eye pattern is common in actual systems as a measure of receiver performance.

The performance of optical receivers operating in the wavelength range 1.3–1.6 μm is severely limited by thermal noise, as seen clearly from the data in Fig. 4.23. The use of APD receivers improves the situation, but to a limited extent only, because of the excess noise factor associated with InGaAs APDs. Most receivers operate away from the quantum limit by 20 dB or more. The effect of thermal noise can be considerably reduced by using coherent-detection techniques in which the received signal is mixed coherently with the output of a narrow-linewidth laser. The receiver performance can also be improved by amplifying the optical signal before it is incident on the photodetector. We turn to optical amplifiers in the next chapter.

Problems

4.1 Calculate the responsivity of a p–i–n photodiode at 1.3 and 1.55 μm if the quantum efficiency is 80%. Why is the photodiode more responsive at 1.55 μm?

4.2 Photons at a rate of 10^{10}/s are incident on an APD with responsivity of 6 A/W. Calculate the quantum efficiency and the photocurrent at the operating wavelength of 1.5 μm for an APD gain of 10.

4.3 Show by solving Eqs. (4.2.3) and (4.2.4) that the multiplication factor M is given by Eq. (4.2.7) for an APD in which electrons initiate the avalanche process. Treat α_e and α_h as constants.

4.4 Draw a block diagram of a digital optical receiver showing its various components. Explain the function of each component. How is the signal used by the decision circuit related to the incident optical power?

4.5 The raised-cosine pulse shape of Eq. (4.3.6) can be generalized to generate a family of such pulses by defining

$$h_{\text{out}}(t) = \frac{\sin(\pi Bt)}{\pi Bt} \frac{\cos(\pi \beta Bt)}{1 - (2\beta Bt)^2},$$

where the parameter β varies between 0 and 1. Derive an expression for the transfer function $H_{out}(f)$ given by the Fourier transform of $h_{out}(t)$. Plot $h_{out}(t)$ and $H_{out}(f)$ for $\beta = 0, 0.5$, and 1.

4.6 Consider a 0.8-μm receiver with a silicon p–i–n photodiode. Assume 20 MHz bandwidth, 65% quantum efficiency, 1 nA dark current, 8 pF junction capacitance, and 3 dB amplifier noise figure. The receiver is illuminated with 5 μW of optical power. Determine the RMS noise currents due to shot noise, thermal noise, and amplifier noise. Also calculate the SNR.

4.7 The receiver of Problem 4.6 is used in a digital communication system that requires a SNR of at least 20 dB for satisfactory performance. What is the minimum received power when the detection is limited by (a) shot noise and (b) thermal noise? Also calculate the noise-equivalent power in the two cases.

4.8 The excess noise factor of avalanche photodiodes is often approximated by M^x instead of Eq. (4.4.18). Find the range of M for which Eq. (4.4.18) can be approximated within 10% by $F_A(M) = M^x$ by choosing $x = 0.3$ for Si, 0.7 for InGaAs, and 1.0 for Ge. Use $k_A = 0.02$ for Si, 0.35 for InGaAs, and 1.0 for Ge.

4.9 Derive Eq. (4.4.22). Plot M_{opt} versus k_A by solving the cubic polynomial on a computer by using $R_L = 1$ kΩ, $F_n = 2$, $R = 1$ A/W, $P_{in} = 1$ μW, and $I_d = 2$ nA. Compare the results with the approximate analytic solution given by Eq. (4.4.23) and comment on its validity.

4.10 Derive an expression for the optimum value of M for which the SNR becomes maximum by using $F_A(M) = M^x$ in Eq. (4.4.19).

4.11 Prove that the bit-error rate (BER) given by Eq. (4.5.6) is minimum when the decision threshold is set close to a value given by Eq. (4.5.9).

4.12 A 1.3-μm digital receiver is operating at 100 Mb/s and has an effective noise bandwidth of 60 MHz. The p–i–n photodiode has negligible dark current and 90% quantum efficiency. The load resistance is 100 Ω and the amplifier noise figure is 3 dB. Calculate the receiver sensitivity corresponding to a BER of 10^{-9}. How much does it change if the receiver is designed to operate reliably up to a BER of 10^{-12}?

4.13 Calculate the receiver sensitivity (at a BER of 10^{-9}) for the receiver in Problem 4.12 in the shot-noise and thermal-noise limits. How many photons are incident during bit 1 in the two limits if the optical pulse can be approximated by a square pulse?

4.14 Derive an expression for the optimum gain M_{opt} of an APD receiver that would maximize the receiver sensitivity by taking the excess-noise factor as M^x. Plot M_{opt} as a function of x for $\sigma_T = 0.2$ mA and $\Delta f = 1$ GHz and estimate its value for InGaAs APDs (see Problem 4.8).

4.15 Derive an expression for the sensitivity of an APD receiver by taking into account a finite extinction ratio for the general case in which both shot noise and thermal noise contribute to the receiver sensitivity. You can neglect the dark current.

4.16 Derive an expression for the intensity-noise-induced power penalty of a *p–i–n* receiver by taking into account a finite extinction ratio. Shot-noise and intensity-noise contributions can both be neglected compared with the thermal noise in the off state but not in the on state.

4.17 Use the result of Problem 4.16 to plot the power penalty as a function of the intensity-noise parameter r_I [see Eq. (4.6.6) for its definition] for several values of the extinction ratio. When does the power penalty become infinite? Explain the meaning of an infinite power penalty.

4.18 Derive an expression for the timing-jitter-induced power penalty by assuming a parabolic pulse shape $I(t) = I_p(1 - B^2 t^2)$ and a Gaussian jitter distribution with a standard deviation τ (RMS value). You can assume that the receiver performance is dominated by thermal noise. Calculate the tolerable value of $B\tau$ that would keep the power penalty below 1 dB.

References

[1] S. D. Personick, *Bell Syst. Tech. J.* **52**, 843 (1973); **52**, 875 (1973).

[2] T. P. Lee and T. Li, in *Optical Fiber Telecommunications I*, S. E. Miller and A. G. Chynoweth, Eds., Academic Press, San Diego, CA, 1979, Chap. 18.

[3] R. G. Smith and S. D. Personick, in *Semiconductor Devices for Optical Communications*, H. Kressel, Ed., Springer, New York, 1980.

[4] S. R. Forrest, in *Optical Fiber Telecommunications II*, S. E. Miller and I. P. Kaminow, Eds., Academic Press, San Diego, CA, 1988, Chap. 14.

[5] B. L. Kasper, in *Optical Fiber Telecommunications II*, S. E. Miller and I. P. Kaminow, Eds., Academic Press, San Diego, CA, 1988, Chap. 18.

[6] S. B. Alexander, *Optical Communication Receiver Design*, Vol. TT22, SPIE Press, Bellingham, WA, 1995.

[7] R. J. Keyes, *Optical and Infrared Detectors*, Springer, New York, 1997.

[8] G. J. Brown, Ed., *Photodetectors Materials & Devices III*, SPIE Press, Bellingham, WA, 1998.

[9] M. J. Digonnet, Ed., *Optical Devices for Fiber Communication*, SPIE Press, Bellingham, WA, 1999.

[10] R. S. Tucker, A. J. Taylor, C. A. Burrus, G. Eisenstein, and J. M. Westfield, *Electron. Lett.* **22**, 917 (1986).

[11] K. Kishino, S. Ünlü, J. I. Chyi, J. Reed, L. Arsenault, and H. Morkoç, *IEEE J. Quantum Electron.* **27**, 2025 (1991).

[12] C. C. Barron, C. J. Mahon, B. J. Thibeault, G. Wang, W. Jiang, L. A. Coldren, and J. E. Bowers, *Electron. Lett.* **30**, 1796 (1994).

[13] I.-H. Tan, J. Dudley, D. I. Babić, D. A. Cohen, B. D. Young, E. L. Hu, J. E. Bowers, B. I. Miller, U. Koren, and M. G. Young, *IEEE Photon. Technol. Lett.* **6**, 811 (1994).

[14] I.-H. Tan, C.-K. Sun, K. S. Giboney, J. E. Bowers E. L. Hu, B. I. Miller, and R. J. Kapik, *IEEE Photon. Technol. Lett.* **7**, 1477 (1995).

[15] Y.-G. Wey, K. S. Giboney, J. E. Bowers, M. J. Rodwell, P. Silvestre, P. Thiagarajan, and G. Robinson, *J. Lightwave Technol.* **13**, 1490 (1995).

[16] K. Kato, S. Hata, K. Kwano, J. Yoshida, and A. Kozen, *IEEE J. Quantum Electron.* **28**, 2728 (1992).

[17] K. Kato, A. Kozen, Y. Muramoto, Y. Itaya, N. Nagatsuma, and M. Yaita, *IEEE Photon. Technol. Lett.* **6**, 719 (1994).

[18] K. Kato and Y. Akatsu, *Opt. Quantum Electron.* **28**, 557 (1996).

[19] T. Takeuchi, T. Nakata, K. Fukuchi, K. Makita, and K. Taguchi, *IEICE Trans. Electron.* **E82C**, 1502 (1999).

[20] K. Kato, *IEEE Trans. Microwave Theory Tech.* **47**, 1265 (1999).

[21] K. S. Giboney, R. L. Nagarajan, T. E. Reynolds, S. T. Allen, R. P. Mirin, M. J. W. Rodwell, and J. E. Bowers, *IEEE Photon. Technol. Lett.* **7**, 412 (1995).

[22] H. Ito, T. Furuta, S. Kodama, and T. Ishibashi, *Electron. Lett.* **36**, 1809 (2000).

[23] G. E. Stillman and C. M. Wolfe, in *Semiconductors and Semimetals*, Vol. 12, R. K. Willardson and A. C. Beer, Eds., Academic Press, San Diego, CA, 1977, pp. 291–393.

[24] H. Melchior, in *Laser Handbook*, Vol. 1, F. T. Arecchi and E. O. Schulz-Dubois, Eds., North-Holland, Amsterdam, 1972, pp. 725–835.

[25] J. C. Campbell, A. G. Dentai, W. S. Holden, and B. L. Kasper, *Electron. Lett.* **19**, 818 (1983).

[26] B. L. Kasper and J. C. Campbell, *J. Lightwave Technol.* **5**, 1351 (1987).

[27] L. E. Tarof, *Electron. Lett.* **27**, 34 (1991).

[28] L. E. Tarof, J. Yu, R. Bruce, D. G. Knight, T. Baird, and B. Oosterbrink, *IEEE Photon. Technol. Lett.* **5**, 672 (1993).

[29] J. Yu, L. E. Tarof, R. Bruce, D. G. Knight, K. Visvanatha, and T. Baird, *IEEE Photon. Technol. Lett.* **6**, 632 (1994).

[30] C. L. F. Ma, M. J. Deen, and L. E. Tarof, *IEEE J. Quantum Electron.* **31**, 2078 (1995).

[31] K. A. Anselm, H. Nie, C. Lenox, P. Yuan, G. Kinsey, J. C. Campbell, B. G. Streetman, *IEEE J. Quantum Electron.* **34**, 482 (1998).

[32] T. Nakata, I. Watanabe, K. Makita, and T. Torikai, *Electron. Lett.* **36**, 1807 (2000).

[33] F. Capasso, in *Semiconductor and Semimetals*, Vol. 22D, W. T. Tsang, Ed., Academic Press, San Diego, CA, 1985, pp. 1–172.

[34] I. Watanabe, S. Sugou, H. Ishikawa, T. Anan, K. Makita, M. Tsuji, and K. Taguchi, *IEEE Photon. Technol. Lett.* **5**, 675 (1993).

[35] T. Kagawa, Y. Kawamura, and H. Iwamura, *IEEE J. Quantum Electron.* **28**, 1419 (1992); *IEEE J. Quantum Electron.* **29**, 1387 (1993).

[36] S. Hanatani, H. Nakamura, S. Tanaka, T. Ido, and C. Notsu, *Microwave Opt. Tech. Lett.* **7**, 103 (1994).

[37] I. Watanabe, M. Tsuji, K. Makita, and K. Taguchi, *IEEE Photon. Technol. Lett.* **8**, 269 (1996).

[38] I. Watanabe, T. Nakata, M. Tsuji, K. Makita, T. Torikai, and K. Taguchi, *J. Lightwave Technol.* **18**, 2200 (2000).

[39] A. R. Hawkins, W. Wu, P. Abraham, K. Streubel, and J. E. Bowers, *Appl. Phys. Lett.* **70**, 303 (1997).

[40] C. Lenox, H. Nie, P. Yuan, G. Kinsey, A. L. Homles, B. G. Streetman, and J. C. Campbell, *IEEE Photon. Technol. Lett.* **11**, 1162 (1999).

[41] T. Nakata, T. Takeuchi, I. Watanabe, K. Makita, and T. Torikai, *Electron. Lett.* **36**, 2033 (2000).

[42] J. Burm, K. I. Litvin, D. W. Woodard, W. J. Schaff, P. Mandeville, M. A. Jaspan, M. M. Gitin, and L. F. Eastman, *IEEE J. Quantum Electron.* **31**, 1504 (1995).

[43] J. B. D. Soole and H. Schumacher, *IEEE J. Quantum Electron.* **27**, 737 (1991).

[44] J. H. Kim, H. T. Griem, R. A. Friedman, E. Y. Chan, and S. Roy, *IEEE Photon. Technol. Lett.* **4**, 1241 (1992).

[45] R.-H. Yuang, J.-I. Chyi, Y.-J. Chan, W. Lin, and Y.-K. Tu, *IEEE Photon. Technol. Lett.* **7**, 1333 (1995).

[46] O. Vendier, N. M. Jokerst, and R. P. Leavitt, *IEEE Photon. Technol. Lett.* **8**, 266 (1996).

[47] M. C. Hargis, S. E. Ralph, J. Woodall, D. McInturff, A. J. Negri, and P. O. Haugsjaa, *IEEE Photon. Technol. Lett.* **8**, 110 (1996).

[48] W. A. Wohlmuth, P. Fay, C. Caneau, and I. Adesida, *Electron. Lett.* **32**, 249 (1996).

[49] A. Bartels, E. Peiner, G.-P. Tang, R. Klockenbrink, H.-H. Wehmann, and A. Schlachetzki, *IEEE Photon. Technol. Lett.* **8**, 670 (1996).

[50] Y. G. Zhang, A. Z. Li, and J. X. Chen, *IEEE Photon. Technol. Lett.* **8**, 830 (1996).

[51] E. Droge, E. H. Bottcher, S. Kollakowski, A. Strittmatter, D. Bimberg, O. Reimann, and R. Steingruber, *Electron. Lett.* **34**, 2241 (1998).

[52] J. W. Shi, K. G. Gan, Y. J. Chiu, Y. H. Chen, C. K. Sun, Y. J. Yang, and J. E. Bowers, *IEEE Photon. Technol. Lett.* **16**, 623 (2001).

[53] R. G. Swartz, in *Optical Fiber Telecommunications II*, S. E. Miller and I. P. Kaminow, Eds., Academic Press, San Diego, CA, 1988, Chap. 20.

[54] K. Kobayashi, in *Optical Fiber Telecommunications II*, S. E. Miller and I. P. Kaminow, Eds., Academic Press, San Diego, CA, 1988, Chap. 11.

[55] T. Horimatsu and M. Sasaki, *J. Lightwave Technol.* **7**, 1612 (1989).

[56] O. Wada, H. Hamaguchi, M. Makiuchi, T. Kumai, M. Ito, K. Nakai, T. Horimatsu, and T. Sakurai, *J. Lightwave Technol.* **4**, 1694 (1986).

[57] M. Makiuchi, H. Hamaguchi, T. Kumai, O. Aoki, Y. Oikawa, and O. Wada, *Electron. Lett.* **24**, 995 (1988).

[58] K. Matsuda, M. Kubo, K. Ohnaka, and J. Shibata, *IEEE Trans. Electron. Dev.* **35**, 1284 (1988).

[59] H. Yano, K. Aga, H. Kamei, G. Sasaki, and H. Hayashi, *J. Lightwave Technol.* **8**, 1328 (1990).

[60] H. Hayashi, H. Yano, K. Aga, M. Murata, H. Kamei, and G. Sasaki, *IEE Proc.* **138**, Pt. J, 164 (1991).

[61] H. Yano, G. Sasaki, N. Nishiyama, M. Murata, and H. Hayashi, *IEEE Trans. Electron. Dev.* **39**, 2254 (1992).

[62] Y. Akatsu, M. Miyugawa, Y. Miyamoto, Y. Kobayashi, and Y. Akahori, *IEEE Photon. Technol. Lett.* **5**, 163 (1993).

[63] K. Takahata, Y. Muramoto, H. Fukano, K. Kato, A. Kozen, O. Nakajima, and Y. Matsuoka, *IEEE Photon. Technol. Lett.* **10**, 1150 (1998).

[64] S. Chandrasekhar, L. M. Lunardi, A. H. Gnauck, R. A. Hamm, and G. J. Qua, *IEEE Photon. Technol. Lett.* **5**, 1316 (1993).

[65] E. Sano, M. Yoneyama, H. Nakajima, and Y. Matsuoka, *J. Lightwave Technol.* **12**, 638 (1994).

[66] H. Kamitsuna, *J. Lightwave Technol.* **13**, 2301 (1995).

[67] L. M. Lunardi, S. Chandrasekhar, C. A. Burrus, and R. A. Hamm, *IEEE Photon. Technol. Lett.* **7**, 1201 (1995).

[68] M. Yoneyama, E. Sano, S. Yamahata, and Y. Matsuoka, *IEEE Photon. Technol. Lett.* **8**, 272 (1996).

[69] E. Sano, K. Kurishima, and S. Yamahata, *Electron. Lett.* **33**, 159 (1997).

[70] W. P. Hong, G. K. Chang, R. Bhat, C. K. Nguyen, and M. Koza, *IEEE Photon. Technol. Lett.* **3**, 156 (1991).

[71] P. Fay, W. Wohlmuth, C. Caneau, and I. Adesida, *Electron. Lett.* **31**, 755 (1995).

[72] G. G. Mekonnen, W. Schlaak, H. G. Bach, R. Steingruber, A. Seeger, T. Enger, W. Passenberg, A. Umbach, C. Schramm, G. Unterborsch, and S. van Waasen, *IEEE Photon. Technol. Lett.* **11**, 257 (1999).

[73] K. Takahata, Y. Muramoto, H. Fukano, K. Kato, A. Kozen, S. Kimura, Y. Imai, Y. Miyamoto, O. Nakajima, and Y. Matsuoka, *IEEE J. Sel. Topics Quantum Electron.* **6**, 31 (2000).

[74] N. Shimizu, K. Murata, A. Hirano, Y. Miyamoto, H. Kitabayashi, Y. Umeda, T. Akeyoshi, T. Furuta, and N. Watanabe, *Electron. Lett.* **36**, 1220 (2000).

[75] Y. Oikawa, H. Kuwatsuka, T. Yamamoto, T. Ihara, H. Hamano, and T. Minami, *J. Lightwave Technol.* **12**, 343 (1994).

[76] T. Ohyama, S. Mino, Y. Akahori, M. Yanagisawa, T. Hashimoto, Y. Yamada, Y. Muramoto, and T. Tsunetsugu, *Electron. Lett.* **32**, 845 (1996).

[77] Y. Kobayashi, Y. Akatsu, K. Nakagawa, H. Kikuchi, and Y. Imai, *IEEE Trans. Microwave Theory Tech.* **43**, 1916 (1995).

[78] K. Emura, *Solid-State Electron.* **43**, 1613 (1999).

[79] M. Bitter, R. Bauknecht, W. Hunziker, and H. Melchior, *IEEE Photon. Technol. Lett.* **12**, 74 (2000).

[80] W. R. Bennett, *Electrical Noise*, McGraw-Hill, New York, 1960.

[81] D. K. C. MacDonald, *Noise and Fluctuations: An Introduction*, Wiley, New York, 1962.

[82] F. N. H. Robinson, *Noise and Fluctuations in Electronic Devices and Circuits*, Oxford University Press, Oxford, 1974.

[83] W. Schottky, *Ann. Phys.* **57**, 541 (1918).

[84] J. B. Johnson, *Phys. Rev.* **32**, 97 (1928).

[85] H. Nyquist, *Phys. Rev.* **32**, 110 (1928).

[86] R. J. McIntyre, *IEEE Trans. Electron. Dev.* **13**, 164 (1966); **19**, 703 (1972).

[87] P. P. Webb, R. J. McIntyre, and J. Conradi, *RCA Rev.* **35**, 235 (1974).

[88] P. Balaban, *Bell Syst. Tech. J.* **55**, 745 (1976).

[89] M. Abramowitz and I. A. Stegun, Eds., *Handbook of Mathematical Functions*, Dover, New York, 1970.

[90] B. E. A. Saleh and M. Teich, *Fundamentals of Photonics*, Wiley, New York, 1991.

[91] L. Mandel and E. Wolf, *Optical Coherence and Quantum Optics*, Cambride University Press, New York, 1995.

[92] G. P. Agrawal and T. M. Shen, *Electron. Lett.* **22**, 450 (1986).

[93] J. J. O'Reilly, J. R. F. DaRocha, and K. Schumacher, *IEE Proc.* **132**, Pt. J, 309 (1985).

[94] K. Schumacher and J. J. O'Reilly, *Electron. Lett.* **23**, 718 (1987).

[95] T. M. Shen, *Electron. Lett.* **22**, 1043 (1986).

[96] T. P. Lee, C. A. Burrus, A. G. Dentai, and K. Ogawa, *Electron. Lett.* **16**, 155 (1980).

[97] D. R. Smith, R. C. Hooper, P. P. Smyth, and D. Wake, *Electron. Lett.* **18**, 453 (1982).

[98] J. Yamada, A. Kawana, T. Miya, H. Nagai, and T. Kimura, *IEEE J. Quantum Electron.* **18**, 1537 (1982).

[99] M. C. Brain, P. P. Smyth, D. R. Smith, B. R. White, and P. J. Chidgey, *Electron. Lett.* **20**, 894 (1984).

[100] M. L. Snodgrass and R. Klinman, *J. Lightwave Technol.* **2**, 968 (1984).

[101] S. D. Walker and L. C. Blank, *Electron. Lett.* **20**, 808 (1984).

[102] C. Y. Chen, B. L. Kasper, H. M. Cox, and J. K. Plourde, *Appl. Phys. Lett.* **46**, 379 (1985).

[103] B. L. Kasper, J. C. Campbell, A. H. Gnauck, A. G. Dentai, and J. R. Talman, *Electron. Lett.* **21**, 982 (1985).

[104] B. L. Kasper, J. C. Campbell, J. R. Talman, A. H. Gnauck, J. E. Bowers, and W. S. Holden, *J. Lightwave Technol.* **5**, 344 (1987).

[105] R. Heidemann, U. Scholz, and B. Wedding, *Electron. Lett.* **23**, 1030 (1987).

[106] M. Shikada, S. Fujita, N. Henmi, I. Takano, I. Mito, K. Taguchi, and K. Minemura, *J. Lightwave Technol.* **5**, 1488 (1987).

[107] S. Fujita, M. Kitamura, T. Torikai, N. Henmi, H. Yamada, T. Suzaki, I. Takano, and M. Shikada, *Electron. Lett.* **25**, 702 (1989).

[108] P. K. Runge, *IEEE Trans. Commun.* **24**, 413 (1976).

[109] L. Pophillat and A. Levasseur, *Electron. Lett.* **27**, 535 (1991).

[110] M. Nakazawa, H. Kubota, K. Suzuki, E. Yamada, and A. Sahara, *IEEE J. Sel. Topics Quantum Electron.* **6**, 363 (2000).

Chapter 5

Lightwave Systems

The preceding three chapters focused on the three main components of a fiber-optic communication system—optical fibers, optical transmitters, and optical receivers. In this chapter we consider the issues related to system design and performance when the three components are put together to form a practical lightwave system. Section 5.1 provides an overview of various system architectures. The design guidelines for fiber-optic communication systems are discussed in Section 5.2 by considering the effects of fiber losses and group-velocity dispersion. The power and the rise-time budgets are also described in this section. Section 5.3 focuses on long-haul systems for which the nonlinear effects become quite important. This section also covers various terrestrial and undersea lightwave systems that have been developed since 1977 when the first field trial was completed in Chicago. Issues related to system performance are treated in Section 5.4 with emphasis on performance degradation occurring as a result of signal transmission through the optical fiber. The physical mechanisms that can lead to power penalty in actual lightwave systems include modal noise, mode-partition noise, source spectral width, frequency chirp, and reflection feedback; each of them is discussed in separate subsections. In Section 5.5 we emphasize the importance of computer-aided design for lightwave systems.

5.1 System Architectures

From an architectural standpoint, fiber-optic communication systems can be classified into three broad categories—point-to-point links, distribution networks, and local-area networks [1]–[7]. This section focuses on the main characteristics of these three system architectures.

5.1.1 Point-to-Point Links

Point-to-point links constitute the simplest kind of lightwave systems. Their role is to transport information, available in the form of a digital bit stream, from one place to another as accurately as possible. The link length can vary from less than a kilometer

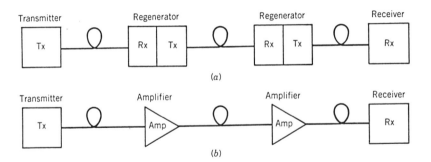

Figure 5.1: Point-to-point fiber links with periodic loss compensation through (a) regenerators and (b) optical amplifiers. A regenerator consists of a receiver followed by a transmitter.

(short haul) to thousands of kilometers (long haul), depending on the specific application. For example, optical data links are used to connect computers and terminals within the same building or between two buildings with a relatively short transmission distance (<10 km). The low loss and the wide bandwidth of optical fibers are not of primary importance for such data links; fibers are used mainly because of their other advantages, such as immunity to electromagnetic interference. In contrast, undersea lightwave systems are used for high-speed transmission across continents with a link length of several thousands of kilometers. Low losses and a large bandwidth of optical fibers are important factors in the design of transoceanic systems from the standpoint of reducing the overall operating cost.

When the link length exceeds a certain value, in the range 20–100 km depending on the operating wavelength, it becomes necessary to compensate for fiber losses, as the signal would otherwise become too weak to be detected reliably. Figure 5.1 shows two schemes used commonly for loss compensation. Until 1990, optoelectronic repeaters, called *regenerators* because they regenerate the optical signal, were used exclusively. As seen in Fig. 5.1(a), a regenerator is nothing but a receiver–transmitter pair that detects the incoming optical signal, recovers the electrical bit stream, and then converts it back into optical form by modulating an optical source. Fiber losses can also be compensated by using optical amplifiers, which amplify the optical bit stream directly without requiring conversion of the signal to the electric domain. The advent of optical amplifiers around 1990 revolutionized the development of fiber-optic communication systems [8]–[10]. Amplifiers are especially valuable for wavelength-division multiplexed (WDM) lightwave systems as they can amplify many channels simultaneously; Chapter 6 is devoted to them.

Optical amplifiers solve the loss problem but they add noise (see Chapter 6) and worsen the impact of fiber dispersion and nonlinearity since signal degradation keeps on accumulating over multiple amplification stages. Indeed, periodically amplified lightwave systems are often limited by fiber dispersion unless dispersion-compensation techniques (discussed in Chapter 7) are used. Optoelectronic repeaters do not suffer from this problem as they regenerate the original bit stream and thus effectively compensate for all sources of signal degradation automatically. An optical regenerator should perform the same three functions—reamplification, reshaping, and retiming (the 3Rs)—to replace an optoelectronic repeater. Although considerable research effort is being directed toward developing such all-optical regenerators [11], most terrestrial systems use a combination of the two techniques shown in Fig. 5.1 and place an optoelectronic regenerator after a certain number of optical amplifiers. Until 2000, the regenerator spacing was in the range of 600–800 km. Since then, ultralong-haul systems have been developed that are capable of transmitting optical signals over 3000 km or more without using a regenerator [12].

The spacing L between regenerators or optical amplifiers (see Fig. 5.1), often called the *repeater spacing*, is a major design parameter simply because the system cost reduces as L increases. However, as discussed in Section 2.4, the distance L depends on the bit rate B because of fiber dispersion. The bit rate–distance product, BL, is generally used as a measure of the system performance for point-to-point links. The BL product depends on the operating wavelength, since both fiber losses and fiber dispersion are wavelength dependent. The first three generations of lightwave systems correspond to three different operating wavelengths near 0.85, 1.3, and 1.55 μm. Whereas the BL product was $\sim$1 (Gb/s)-km for the first-generation systems operating near 0.85 μm, it becomes $\sim$1 (Tb/s)-km for the third-generation systems operating near 1.55 μm and can exceed 100 (Tb/s)-km for the fourth-generation systems.

5.1.2 Distribution Networks

Many applications of optical communication systems require that information is not only transmitted but is also distributed to a group of subscribers. Examples include local-loop distribution of telephone services and broadcast of multiple video channels over cable television (CATV, short for common-antenna television). Considerable effort is directed toward the integration of audio and video services through a broadband *integrated-services digital network* (ISDN). Such a network has the ability to distribute a wide range of services, including telephone, facsimile, computer data, and video broadcasts. Transmission distances are relatively short ($L < 50$ km), but the bit rate can be as high as 10 Gb/s for a broadband ISDN.

Figure 5.2 shows two topologies for distribution networks. In the case of *hub topology*, channel distribution takes place at central locations (or hubs), where an automated cross-connect facility switches channels in the electrical domain. Such networks are called *metropolitan-area networks* (MANs) as hubs are typically located in major cities [13]. The role of fiber is similar to the case of point-to-point links. Since the fiber bandwidth is generally much larger than that required by a single hub office, several offices can share a single fiber headed for the main hub. Telephone networks employ hub topology for distribution of audio channels within a city. A concern for the hub topology is related to its reliability—outage of a single fiber cable can affect the service to a large portion of the network. Additional point-to-point links can be used to guard against such a possibility by connecting important hub locations directly.

In the case of *bus topology*, a single fiber cable carries the multichannel optical signal throughout the area of service. Distribution is done by using optical taps, which divert a small fraction of the optical power to each subscriber. A simple CATV application of bus topology consists of distributing multiple video channels within a city. The use of optical fiber permits distribution of a large number of channels (100 or more)

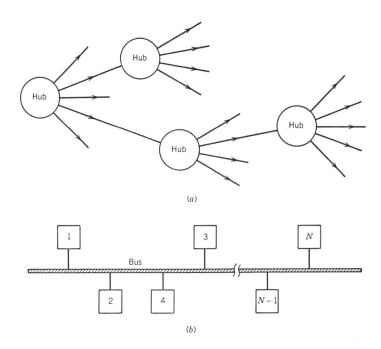

Figure 5.2: (a) Hub topology and (b) bus topology for distribution networks.

because of its large bandwidth compared with coaxial cables. The advent of *high-definition television* (HDTV) also requires lightwave transmission because of a large bandwidth (about 100 Mb/s) of each video channel unless a compression technique (such as MPEG-2, or 2nd recommendation of the motion-picture entertainment group) is used.

A problem with the bus topology is that the signal loss increases exponentially with the number of taps and limits the number of subscribers served by a single optical bus. Even when fiber losses are neglected, the power available at the Nth tap is given by [1]

$$P_N = P_T C[(1 - \delta)(1 - C)]^{N-1}, \tag{5.1.1}$$

where P_T is the transmitted power, C is the fraction of power coupled out at each tap, and δ accounts for insertion losses, assumed to be the same at each tap. The derivation of Eq. (5.1.1) is left as an exercise for the reader. If we use $\delta = 0.05$, $C = 0.05$, $P_T = 1$ mW, and $P_N = 0.1$ μW as illustrative values, N should not exceed 60. A solution to this problem is offered by optical amplifiers which can boost the optical power of the bus periodically and thus permit distribution to a large number of subscribers as long as the effects of fiber dispersion remain negligible.

5.1.3 Local-Area Networks

Many applications of fiber-optic communication technology require networks in which a large number of users within a local area (e.g., a university campus) are intercon-

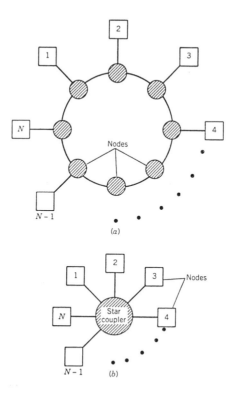

Figure 5.3: (a) Ring topology and (b) star topology for local-area networks.

nected in such a way that any user can access the network randomly to transmit data to any other user [14]–[16]. Such networks are called *local-area networks* (LANs). Optical-access networks used in a local subscriber loop also fall in this category [17]. Since the transmission distances are relatively short (<10 km), fiber losses are not of much concern for LAN applications. The major motivation behind the use of optical fibers is the large bandwidth offered by fiber-optic communication systems.

The main difference between MANs and LANs is related to the random access offered to multiple users of a LAN. The system architecture plays an important role for LANs, since the establishment of predefined protocol rules is a necessity in such an environment. Three commonly used topologies are known as bus, ring, and star configurations. The bus topology is similar to that shown in Fig. 5.2(b). A well-known example of bus topology is provided by the *Ethernet*, a network protocol used to connect multiple computers and used by the *Internet*. The Ethernet operates at speeds up to 1 Gb/s by using a protocol based on *carrier-sense multiple access* (CSMA) with collision detection. Although the Ethernet LAN architecture has proven to be quite successful when coaxial cables are used for the bus, a number of difficulties arise when optical fibers are used. A major limitation is related to the losses occurring at each tap, which limits the number of users [see Eq. (5.1.1)].

Figure 5.3 shows the ring and star topologies for LAN applications. In the ring topology [18], consecutive nodes are connected by point-to-point links to form a closed ring. Each node can transmit and receive the data by using a transmitter–receiver pair, which also acts as a repeater. A token (a predefined bit sequence) is passed around the ring. Each node monitors the bit stream to listen for its own address and to receive the data. It can also transmit by appending the data to an empty token. The use of ring topology for fiber-optic LANs has been commercialized with the standardized interface known as the fiber distributed data interface, FDDI for short [18]. The FDDI operates at 100 Mb/s by using multimode fibers and 1.3-μm transmitters based on light-emitting diodes (LEDs). It is designed to provide backbone services such as the interconnection of lower-speed LANs or mainframe computers.

In the *star topology*, all nodes are connected through point-to-point links to a central node called a hub, or simply a star. Such LANs are further subclassified as *active-star* or *passive-star* networks, depending on whether the central node is an active or passive device. In the active-star configuration, all incoming optical signals are converted to the electrical domain through optical receivers. The electrical signal is then distributed to drive individual node transmitters. Switching operations can also be performed at the central node since distribution takes place in the electrical domain. In the passive-star configuration, distribution takes place in the optical domain through devices such as directional couplers. Since the input from one node is distributed to many output nodes, the power transmitted to each node depends on the number of users. Similar to the case of bus topology, the number of users supported by passive-star LANs is limited by the distribution losses. For an ideal $N \times N$ star coupler, the power reaching each node is simply P_T/N (if we neglect transmission losses) since the transmitted power P_T is divided equally among N users. For a passive star composed of directional couplers (see Section 8.2.4), the power is further reduced because of insertion losses and can be written as [1]

$$P_N = (P_T/N)(1 - \delta)^{\log_2 N}, \tag{5.1.2}$$

where δ is the insertion loss of each directional coupler. If we use $\delta = 0.05$, $P_T = 1$ mW, and $P_N = 0.1$ μW as illustrative values, N can be as large as 500. This value of N should be compared with $N = 60$ obtained for the case of bus topology by using Eq. (5.1.1). A relatively large value of N makes star topology attractive for LAN applications. The remainder of this chapter focuses on the design and performance of point-to-point links, which constitute a basic element of all communication systems, including LANs, MANS, and other distribution networks.

5.2   Design Guidelines

The design of fiber-optic communication systems requires a clear understanding of the limitations imposed by the loss, dispersion, and nonlinearity of the fiber. Since fiber properties are wavelength dependent, the choice of operating wavelength is a major design issue. In this section we discuss how the bit rate and the transmission distance of a single-channel system are limited by fiber loss and dispersion; Chapter 8 is devoted to multichannel systems. We also consider the power and rise-time budgets and illustrate them through specific examples [5]. The power budget is also called the link budget, and the rise-time budget is sometimes referred to as the bandwidth budget.

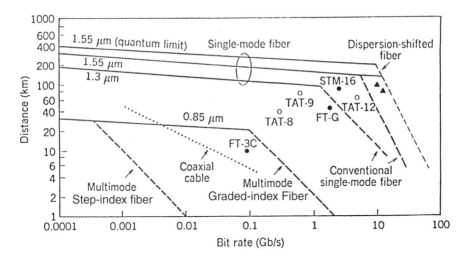

Figure 5.4: Loss (solid lines) and dispersion (dashed lines) limits on transmission distance L as a function of bit rate B for the three wavelength windows. The dotted line corresponds to coaxial cables. Circles denote commercial lightwave systems; triangles show laboratory experiments. (After Ref. [1]; ©1988 Academic Press; reprinted with permission.)

5.2.1 Loss-Limited Lightwave Systems

Except for some short-haul fiber links, fiber losses play an important role in the system design. Consider an optical transmitter that is capable of launching an average power $\bar{P}_{tr}$. If the signal is detected by a receiver that requires a minimum average power $\bar{P}_{rec}$ at the bit rate B, the maximum transmission distance is limited by

$$L = \frac{10}{\alpha_f} \log_{10}\left(\frac{\bar{P}_{tr}}{\bar{P}_{rec}}\right), \tag{5.2.1}$$

where α_f is the net loss (in dB/km) of the fiber cable, including splice and connector losses. The bit-rate dependence of L arises from the linear dependence of $\bar{P}_{rec}$ on the bit rate B. Noting that $\bar{P}_{rec} = \bar{N}_p h\nu B$, where $h\nu$ is the photon energy and $\bar{N}_p$ is the average number of photons/bit required by the receiver [see Eq. (4.5.24)], the distance L decreases logarithmically as B increases at a given operating wavelength.

The solid lines in Fig. 5.4 show the dependence of L on B for three common operating wavelengths of 0.85, 1.3, and 1.55 μm by using $\alpha_f = 2.5, 0.4$, and 0.25 dB/km, respectively. The transmitted power is taken to be $\bar{P}_{tr} = 1$ mW at the three wavelengths, whereas $\bar{N}_p = 300$ at $\lambda = 0.85$ μm and $\bar{N}_p = 500$ at 1.3 and 1.55 μm. The smallest value of L occurs for first-generation systems operating at 0.85 μm because of relatively large fiber losses near that wavelength. The repeater spacing of such systems is limited to 10–25 km, depending on the bit rate and the exact value of the loss parameter. In contrast, a repeater spacing of more than 100 km is possible for lightwave systems operating near 1.55 μm.

It is interesting to compare the loss limit of 0.85-μm lightwave systems with that of electrical communication systems based on coaxial cables. The dotted line in Fig.

5.4 shows the bit-rate dependence of L for coaxial cables by assuming that the loss increases as $\sqrt{B}$. The transmission distance is larger for coaxial cables at small bit rates ($B < 5$ Mb/s), but fiber-optic systems take over at bit rates in excess of 5 Mb/s. Since a longer transmission distance translates into a smaller number of repeaters in a long-haul point-to-point link, fiber-optic communication systems offer an economic advantage when the operating bit rate exceeds 10 Mb/s.

The system requirements typically specified in advance are the bit rate B and the transmission distance L. The performance criterion is specified through the bit-error rate (BER), a typical requirement being BER $< 10^{-9}$. The first decision of the system designer concerns the choice of the operating wavelength. As a practical matter, the cost of components is lowest near 0.85 μm and increases as wavelength shifts toward 1.3 and 1.55 μm. Figure 5.4 can be quite helpful in determining the appropriate operating wavelength. Generally speaking, a fiber-optic link can operate near 0.85 μm if $B < 200$ Mb/s and $L < 20$ km. This is the case for many LAN applications. On the other hand, the operating wavelength is by necessity in the 1.55-μm region for long-haul lightwave systems operating at bit rates in excess of 2 Gb/s. The curves shown in Fig. 5.4 provide only a guide to the system design. Many other issues need to be addressed while designing a realistic fiber-optic communication system. Among them are the choice of the operating wavelength, selection of appropriate transmitters, receivers, and fibers, compatibility of various components, issue of cost versus performance, and system reliability and upgradability concerns.

5.2.2   Dispersion-Limited Lightwave Systems

In Section 2.4 we discussed how fiber dispersion limits the bit rate–distance product BL because of pulse broadening. When the dispersion-limited transmission distance is shorter than the loss-limited distance of Eq. (5.2.1), the system is said to be dispersion-limited. The dashed lines in Fig. 5.4 show the dispersion-limited transmission distance as a function of the bit rate. Since the physical mechanisms leading to dispersion limitation can be different for different operating wavelengths, let us examine each case separately.

Consider first the case of 0.85-μm lightwave systems, which often use multimode fibers to minimize the system cost. As discussed in Section 2.1, the most limiting factor for multimode fibers is intermodal dispersion. In the case of step-index multimode fibers, Eq. (2.1.6) provides an approximate upper bound on the BL product. A slightly more restrictive condition $BL = c/(2n_1\Delta)$ is plotted in Fig. 5.4 by using typical values $n_1 = 1.46$ and $\Delta = 0.01$. Even at a low bit rate of 1 Mb/s, such multimode systems are dispersion-limited, and their transmission distance is limited to below 10 km. For this reason, multimode step-index fibers are rarely used in the design of fiber-optic communication systems. Considerable improvement can be realized by using graded-index fibers for which intermodal dispersion limits the BL product to values given by Eq. (2.1.11). The condition $BL = 2c/(n_1\Delta^2)$ is plotted in Fig. 5.4 and shows that 0.85-μm lightwave systems are loss-limited, rather than dispersion-limited, for bit rates up to 100 Mb/s when graded-index fibers are used. The first generation of terrestrial telecommunication systems took advantage of such an improvement and used graded-index fibers. The first commercial system became available in 1980 and operated at a bit rate of 45 Mb/s with a repeater spacing of less than 10 km.

The second generation of lightwave systems used primarily single-mode fibers near the minimum-dispersion wavelength occurring at about 1.31 μm. The most limiting factor for such systems is dispersion-induced pulse broadening dominated by a relatively large source spectral width. As discussed in Section 2.4.3, the *BL* product is then limited by [see Eq. (2.4.26)]

$$BL \leq (4|D|\sigma_\lambda)^{-1}, \tag{5.2.2}$$

where σ_λ is the root-mean-square (RMS) width of the source spectrum. The actual value of $|D|$ depends on how close the operating wavelength is to the zero-dispersion wavelength of the fiber and is typically ~ 1 ps/(km-nm). Figure 5.4 shows the dispersion limit for 1.3-μm lightwave systems by choosing $|D|\sigma_\lambda = 2$ ps/km so that $BL \leq 125$ (Gb/s)-km. As seen there, such systems are generally loss-limited for bit rates up to 1 Gb/s but become dispersion-limited at higher bit rates.

Third- and fourth-generation lightwave systems operate near 1.55 μm to take advantage of the smallest fiber losses occurring in this wavelength region. However, fiber dispersion becomes a major problem for such systems since $D \approx 16$ ps/(km-nm) near 1.55 μm for standard silica fibers. Semiconductor lasers operating in a single longitudinal mode provide a solution to this problem. The ultimate limit is then given by [see Eq. (2.4.30)]

$$B^2 L < (16|\beta_2|)^{-1}, \tag{5.2.3}$$

where β_2 is related to D as in Eq. (2.3.5). Figure 5.4 shows this limit by choosing $B^2 L = 4000$ (Gb/s)^2-km. As seen there, such 1.55-μm systems become dispersion-limited only for $B > 5$ Gb/s. In practice, the frequency chirp imposed on the optical pulse during direct modulation provides a much more severe limitation. The effect of frequency chirp on system performance is discussed in Section 5.4.4. Qualitatively speaking, the frequency chirp manifests through a broadening of the pulse spectrum. If we use Eq. (5.2.2) with $D = 16$ ps/(km-nm) and $\sigma_\lambda = 0.1$ nm, the *BL* product is limited to 150 (Gb/s)-km. As a result, the frequency chirp limits the transmission distance to 75 km at $B = 2$ Gb/s, even though loss-limited distance exceeds 150 km. The frequency-chirp problem is often solved by using an external modulator for systems operating at bit rates >5 Gb/s.

A solution to the dispersion problem is offered by *dispersion-shifted fibers* for which dispersion and loss both are minimum near 1.55 μm. Figure 5.4 shows the improvement by using Eq. (5.2.3) with $|\beta_2| = 2$ ps^2/km. Such systems can be operated at 20 Gb/s with a repeater spacing of about 80 km. Further improvement is possible by operating the lightwave system very close to the zero-dispersion wavelength, a task that requires careful matching of the laser wavelength to the zero-dispersion wavelength and is not always feasible because of variations in the dispersive properties of the fiber along the transmission link. In practice, the frequency chirp makes it difficult to achieve even the limit indicated in Fig. 5.4. By 1989, two laboratory experiments had demonstrated transmission over 81 km at 11 Gb/s [19] and over 100 km at 10 Gb/s [20] by using low-chirp semiconductor lasers together with dispersion-shifted fibers. The triangles in Fig. 5.4 show that such systems operate quite close to the fundamental limits set by fiber dispersion. Transmission over longer distances requires the use of dispersion-management techniques discussed in Chapter 7.

5.2.3 Power Budget

The purpose of the *power budget* is to ensure that enough power will reach the receiver to maintain reliable performance during the entire system lifetime. The minimum average power required by the receiver is the receiver sensitivity $\bar{P}_{rec}$ (see Section 4.4). The average launch power $\bar{P}_{tr}$ is generally known for any transmitter. The power budget takes an especially simple form in decibel units with optical powers expressed in dBm units (see Appendix A). More specifically,

$$\bar{P}_{tr} = \bar{P}_{rec} + C_L + M_s, \tag{5.2.4}$$

where C_L is the total channel loss and M_s is the *system margin*. The purpose of system margin is to allocate a certain amount of power to additional sources of power penalty that may develop during system lifetime because of component degradation or other unforeseen events. A system margin of 4–6 dB is typically allocated during the design process.

The channel loss C_L should take into account all possible sources of power loss, including connector and splice losses. If α_f is the fiber loss in decibels per kilometer, C_L can be written as

$$C_L = \alpha_f L + \alpha_{con} + \alpha_{splice}, \tag{5.2.5}$$

where α_{con} and α_{splice} account for the connector and splice losses throughout the fiber link. Sometimes splice loss is included within the specified loss of the fiber cable. The connector loss α_{con} includes connectors at the transmitter and receiver ends but must include other connectors if used within the fiber link.

Equations (5.2.4) and (5.2.5) can be used to estimate the maximum transmission distance for a given choice of the components. As an illustration, consider the design of a fiber link operating at 100 Mb/s and requiring a maximum transmission distance of 8 km. As seen in Fig. 5.4, such a system can be designed to operate near 0.85 μm provided that a graded-index multimode fiber is used for the optical cable. The operation near 0.85 μm is desirable from the economic standpoint. Once the operating wavelength is selected, a decision must be made about the appropriate transmitters and receivers. The GaAs transmitter can use a semiconductor laser or an LED as an optical source. Similarly, the receiver can be designed to use either a *p–i–n* or an avalanche photodiode. Keeping the low cost in mind, let us choose a *p–i–n* receiver and assume that it requires 2500 photons/bit on average to operate reliably with a BER below 10^{-9}. Using the relation $\bar{P}_{rec} = \bar{N}_p h\nu B$ with $\bar{N}_p = 2500$ and $B = 100$ Mb/s, the receiver sensitivity is given by $\bar{P}_{rec} = -42$ dBm. The average launch power for LED and laser-based transmitters is typically 50 μW and 1 mW, respectively.

Table 5.1 shows the power budget for the two transmitters by assuming that the splice loss is included within the cable loss. The transmission distance L is limited to 6 km in the case of LED-based transmitters. If the system specification is 8 km, a more expensive laser-based transmitter must be used. The alternative is to use an avalanche photodiode (APD) receiver. If the receiver sensitivity improves by more than 7 dB

Table 5.1 Power budget of a 0.85-μm lightwave system

Quantity	Symbol	Laser	LED
Transmitter power	$\bar{P}_{tr}$	0 dBm	-13 dBm
Receiver sensitivity	$\bar{P}_{rec}$	-42 dBm	-42 dBm
System margin	M_s	6 dB	6 dB
Available channel loss	C_L	36 dB	23 dB
Connector loss	α_{con}	2 dB	2 dB
Fiber cable loss	α_f	3.5 dB/km	3.5 dB/km
Maximum fiber length	L	9.7 km	6 km

when an APD is used in place of a *p–i–n* photodiode, the transmission distance can be increased to 8 km even for an LED-based transmitter. Economic considerations would then dictate the choice between the laser-based transmitters and APD receivers.

5.2.4 Rise-Time Budget

The purpose of the *rise-time budget* is to ensure that the system is able to operate properly at the intended bit rate. Even if the bandwidth of the individual system components exceeds the bit rate, it is still possible that the total system may not be able to operate at that bit rate. The concept of rise time is used to allocate the bandwidth among various components. The rise time T_r of a linear system is defined as the time during which the response increases from 10 to 90% of its final output value when the input is changed abruptly. Figure 5.5 illustrates the concept graphically.

An inverse relationship exists between the bandwidth Δf and the rise time T_r associated with a linear system. This relationship can be understood by considering a simple *RC* circuit as an example of the linear system. When the input voltage across an *RC* circuit changes instantaneously from 0 to V_0, the output voltage changes as

$$V_{out}(t) = V_0[1 - \exp(-t/RC)], \tag{5.2.6}$$

where R is the resistance and C is the capacitance of the *RC* circuit. The rise time is found to be given by

$$T_r = (\ln 9)RC \approx 2.2RC. \tag{5.2.7}$$

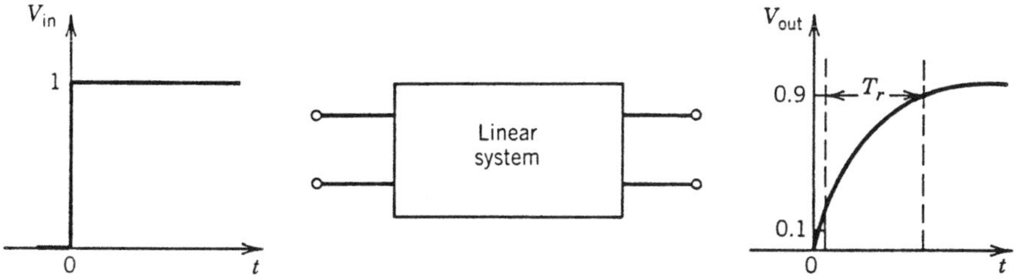

Figure 5.5: Rise time T_r associated with a bandwidth-limited linear system.

The transfer function $H(f)$ of the *RC* circuit is obtained by taking the Fourier transform of Eq. (5.2.6) and is of the form

$$H(f) = (1 + i2\pi f RC)^{-1}. \qquad (5.2.8)$$

The bandwidth Δf of the *RC* circuit corresponds to the frequency at which $|H(f)|^2 = 1/2$ and is given by the well-known expression $\Delta f = (2\pi RC)^{-1}$. By using Eq. (5.2.7), Δf and T_r are related as

$$T_r = \frac{2.2}{2\pi \Delta f} = \frac{0.35}{\Delta f}. \qquad (5.2.9)$$

The inverse relationship between the rise time and the bandwidth is expected to hold for any linear system. However, the product $T_r \Delta f$ would generally be different than 0.35. One can use $T_r \Delta f = 0.35$ in the design of optical communication systems as a conservative guideline. The relationship between the bandwidth Δf and the bit rate B depends on the digital format. In the case of return-to-zero (RZ) format (see Section 1.2), $\Delta f = B$ and $BT_r = 0.35$. By contrast, $\Delta f \approx B/2$ for the nonreturn-to-zero (NRZ) format and $BT_r = 0.7$. In both cases, the specified bit rate imposes an upper limit on the maximum rise time that can be tolerated. The communication system must be designed to ensure that T_r is below this maximum value, i.e.,

$$T_r \leq \begin{cases} 0.35/B & \text{for RZ format,} \\ 0.70/B & \text{for NRZ format.} \end{cases} \qquad (5.2.10)$$

The three components of fiber-optic communication systems have individual rise times. The total rise time of the whole system is related to the individual component rise times approximately as [21]

$$T_r^2 = T_{\text{tr}}^2 + T_{\text{fiber}}^2 + T_{\text{rec}}^2, \qquad (5.2.11)$$

where T_{tr}, T_{fiber}, and T_{rec} are the rise times associated with the transmitter, fiber, and receiver, respectively. The rise times of the transmitter and the receiver are generally known to the system designer. The transmitter rise time T_{tr} is determined primarily by the electronic components of the driving circuit and the electrical parasitics associated with the optical source. Typically, T_{tr} is a few nanoseconds for LED-based transmitters but can be shorter than 0.1 ns for laser-based transmitters. The receiver rise time T_{rec} is determined primarily by the 3-dB electrical bandwidth of the receiver front end. Equation (5.2.9) can be used to estimate T_{rec} if the front-end bandwidth is specified.

The fiber rise time T_{fiber} should in general include the contributions of both the intermodal dispersion and group-velocity dispersion (GVD) through the relation

$$T_{\text{fiber}}^2 = T_{\text{modal}}^2 + T_{\text{GVD}}^2. \qquad (5.2.12)$$

For single-mode fibers, $T_{\text{modal}} = 0$ and $T_{\text{fiber}} = T_{\text{GVD}}$. In principle, one can use the concept of fiber bandwidth discussed in Section 2.4.4 and relate T_{fiber} to the 3-dB fiber bandwidth $f_{3\,\text{dB}}$ through a relation similar to Eq. (5.2.9). In practice it is not easy to calculate $f_{3\,\text{dB}}$, especially in the case of modal dispersion. The reason is that a fiber link consists of many concatenated fiber sections (typical length 5 km), which may have different dispersion characteristics. Furthermore, mode mixing occurring at splices and connectors tends to average out the propagation delay associated with different modes of a multimode fiber. A statistical approach is often necessary to estimate the fiber bandwidth and the corresponding rise time [22]–[25].

In a phenomenological approach, T_{modal} can be approximated by the time delay ΔT given by Eq. (2.1.5) in the absence of mode mixing, i.e.,

$$T_{\text{modal}} \approx (n_1 \Delta / c) L, \qquad (5.2.13)$$

where $n_1 \approx n_2$ was used. For graded-index fibers, Eq. (2.1.10) is used in place of Eq. (2.1.5), resulting in $T_{\text{modal}} \approx (n_1 \Delta^2 / 8c) L$. In both cases, the effect of mode mixing is included by changing the linear dependence on L by a sublinear dependence L^q, where q has a value in the range 0.5–1, depending on the extent of mode mixing. A reasonable estimate based on the experimental data is $q = 0.7$. The contribution T_{GVD} can also be approximated by ΔT given by Eq. (2.3.4), so that

$$T_{\text{GVD}} \approx |D| L \Delta \lambda, \qquad (5.2.14)$$

where $\Delta \lambda$ is the spectral width of the optical source (taken as a full width at half maximum). The dispersion parameter D may change along the fiber link if different sections have different dispersion characteristics; an average value should be used in Eq. (5.2.14) in that case.

As an illustration of the rise-time budget, consider a 1.3-μm lightwave system designed to operate at 1 Gb/s over a single-mode fiber with a repeater spacing of 50 km. The rise times for the transmitter and the receiver have been specified as $T_{\text{tr}} = 0.25$ ns and $T_{\text{rec}} = 0.35$ ns. The source spectral width is specified as $\Delta \lambda = 3$ nm, whereas the average value of D is 2 ps/(km-nm) at the operating wavelength. From Eq. (5.2.14), $T_{\text{GVD}} = 0.3$ ns for a link length $L = 50$ km. Modal dispersion does not occur in single-mode fibers. Hence $T_{\text{modal}} = 0$ and $T_{\text{fiber}} = 0.3$ ns. The system rise time is estimated by using Eq. (5.2.11) and is found to be $T_r = 0.524$ ns. The use of Eq. (5.2.10) indicates that such a system cannot be operated at 1 Gb/s when the RZ format is employed for the optical bit stream. However, it would operate properly if digital format is changed to the NRZ format. If the use of RZ format is a prerequisite, the designer must choose different transmitters and receivers to meet the rise-time budget requirement. The NRZ format is often used as it permits a larger system rise time at the same bit rate.

5.3 Long-Haul Systems

With the advent of optical amplifiers, fiber losses can be compensated by inserting amplifiers periodically along a long-haul fiber link (see Fig. 5.1). At the same time, the effects of fiber dispersion (GVD) can be reduced by using dispersion management (see Chapter 7). Since neither the fiber loss nor the GVD is then a limiting factor, one may ask how many in-line amplifiers can be cascaded in series, and what limits the total link length. This topic is covered in Chapter 6 in the context of erbium-doped fiber amplifiers. Here we focus on the factors that limit the performance of amplified fiber links and provide a few design guidelines. The section also outlines the progress realized in the development of terrestrial and undersea lightwave systems since 1977 when the first field trial was completed.

5.3.1 Performance-Limiting Factors

The most important consideration in designing a periodically amplified fiber link is related to the *nonlinear effects* occurring inside all optical fibers [26] (see Section 2.6). For single-channel lightwave systems, the dominant nonlinear phenomenon that limits the system performance is *self-phase modulation* (SPM). When optoelectronic regenerators are used, the SPM effects accumulate only over one repeater spacing (typically <100 km) and are of little concern if the launch power satisfies Eq. (2.6.15) or the condition $P_{in} \ll 22$ mW when $N_A = 1$. In contrast, the SPM effects accumulate over long lengths ($\sim$1000 km) when in-line amplifiers are used periodically for loss compensation. A rough estimate of the limitation imposed by the SPM is again obtained from Eq. (2.6.15). This equation predicts that the peak power should be below 2.2 mW for 10 cascaded amplifiers when the nonlinear parameter $\gamma = 2$ W^{-1}/km. The condition on the average power depends on the modulation format and the shape of optical pulses. It is nonetheless clear that the average power should be reduced to below 1 mW for SPM effects to remain negligible for a lightwave system designed to operate over a distance of more than 1000 km. The limiting value of the average power also depends on the type of fiber in which light is propagating through the effective core area A_{eff}. The SPM effects are most dominant inside dispersion-compensating fibers for which A_{eff} is typically close to 20 μm^2.

The forgoing discussion of the SPM-induced limitations is too simplistic to be accurate since it completely ignores the role of fiber dispersion. In fact, as the dispersive and nonlinear effects act on the optical signal simultaneously, their mutual interplay becomes quite important [26]. The effect of SPM on pulses propagating inside an optical fiber can be included by using the nonlinear Schrödinger (NLS) equation of Section 2.6. This equation is of the form [see Eq. (2.6.18)]

$$\frac{\partial A}{\partial z} + \frac{i\beta_2}{2}\frac{\partial^2 A}{\partial t^2} = -\frac{\alpha}{2}A + i\gamma|A|^2 A, \tag{5.3.1}$$

where fiber losses are included through the α term. This term can also include periodic amplification of the signal by treating α as a function of z. The NLS equation is used routinely for designing modern lightwave systems.

Because of the nonlinear nature of Eq. (5.3.1), it should be solved numerically in general. A numerical approach has indeed been adopted (see Appendix E) since the early 1990s for quantifying the impact of SPM on the performance of long-haul lightwave systems [27]–[35]. The use of a large-effective-area fiber (LEAF) helps by reducing the nonlinear parameter γ defined as $\gamma = 2\pi n_2/(\lambda A_{eff})$. Appropriate chirping of input pulses can also be beneficial for reducing the SPM effects. This feature has led to the adoption of a new modulation format known as the chirped RZ or CRZ format. Numerical simulations show that, in general, the launch power must be optimized to a value that depends on many design parameters such as the bit rate, total link length, and amplifier spacing. In one study, the optimum launch power was found to be about 1 mW for a 5-Gb/s signal transmitted over 9000 km with 40-km amplifier spacing [31].

The combined effects of GVD and SPM also depend on the sign of the dispersion parameter β_2. In the case of anomalous dispersion ($\beta_2 < 0$), the nonlinear phenomenon of *modulation instability* [26] can affect the system performance drastically [32]. This problem can be overcome by using a combination of fibers with normal and anomalous GVD such that the average dispersion over the entire fiber link is "normal." However, a new kind of modulation instability, referred to as *sideband instability* [36], can occur in both the normal and anomalous GVD regions. It has its origin in the periodic variation of the signal power along the fiber link when equally spaced optical amplifiers are used to compensate for fiber losses. Since the quantity $\gamma|A|^2$ in Eq. (5.3.1) is then a periodic function of z, the resulting nonlinear-index grating can initiate a four-wave-mixing process that generates sidebands in the signal spectrum. It can be avoided by making the amplifier spacing nonuniform.

Another factor that plays a crucial role is the noise added by optical amplifiers. Similar to the case of electronic amplifiers (see Section 4.4), the noise of optical amplifiers is quantified through an amplifier noise figure F_n (see Chapter 6). The nonlinear interaction between the amplified spontaneous emission and the signal can lead to a large spectral broadening through the nonlinear phenomena such as cross-phase modulation and four-wave mixing [37]. Because the noise has a much larger bandwidth than the signal, its impact can be reduced by using optical filters. Numerical simulations indeed show a considerable improvement when optical filters are used after every in-line amplifier [31].

The polarization effects that are totally negligible in the traditional "nonamplified" lightwave systems become of concern for long-haul systems with in-line amplifiers. The polarization-mode dispersion (PMD) issue has been discussed in Section 2.3.5. In addition to PMD, optical amplifiers can also induce polarization-dependent gain and loss [30]. Although the PMD effects must be considered during system design, their impact depends on the design parameters such as the bit rate and the transmission distance. For bit rates as high as 10-Gb/s, the PMD effects can be reduced to an acceptable level with a proper design. However, PMD becomes of major concern for 40-Gb/s systems for which the bit slot is only 25 ps wide. The use of a PMD-compensation technique appears to be necessary at such high bit rates.

The fourth generation of lightwave systems began in 1995 when lightwave systems employing amplifiers first became available commercially. Of course, the laboratory demonstrations began as early as 1989. Many experiments used a recirculating fiber loop to demonstrate system feasibility as it was not practical to use long lengths of fiber in a laboratory setting. Already in 1991, an experiment showed the possibility of data transmission over 21,000 km at 2.5 Gb/s, and over 14,300 km at 5 Gb/s, by using the recirculating-loop configuration [38]. In a system trial carried out in 1995 by using actual submarine cables and repeaters [39], a 5.3-Gb/s signal was transmitted over 11,300 km with 60 km of amplifier spacing. This system trial led to the deployment of a commercial transpacific cable (TPC–5) that began operating in 1996.

The bit rate of fourth-generation systems was extended to 10 Gb/s beginning in 1992. As early as 1995, a 10-Gb/s signal was transmitted over 6480 km with 90-km amplifier spacing [40]. With a further increase in the distance, the SNR decreased below the value needed to maintain the BER below 10^{-9}. One may think that the performance should improve by operating close to the zero-dispersion wavelength of the

Table 5.2 Terrestrial lightwave systems

System	Year	λ (μm)	B (Mb/s)	L (km)	Voice Channels
FT–3	1980	0.85	45	< 10	672
FT–3C	1983	0.85	90	< 15	1,344
FT–3X	1984	1.30	180	< 25	2,688
FT–G	1985	1.30	417	< 40	6,048
FT–G-1.7	1987	1.30	1,668	< 46	24,192
STM–16	1991	1.55	2,488	< 85	32,256
STM–64	1996	1.55	9,953	< 90	129,024
STM–256	2002	1.55	39,813	< 90	516,096

fiber. However, an experiment, performed under such conditions, achieved a distance of only 6000 km at 10 Gb/s even with 40-km amplifier spacing [41], and the situation became worse when the RZ modulation format was used. Starting in 1999, the single-channel bit rate was pushed toward 40 Gb/s in several experiments [42]–[44]. The design of 40-Gb/s lightwave systems requires the use of several new ideas including the CRZ format, dispersion management with GVD-slope compensation, and distributed Raman amplification. Even then, the combined effects of the higher-order dispersion, PMD, and SPM degrade the system performance considerably at a bit rate of 40 Gb/s.

5.3.2 Terrestrial Lightwave Systems

An important application of fiber-optic communication links is for enhancing the capacity of telecommunication networks worldwide. Indeed, it is this application that started the field of optical fiber communications in 1977 and has propelled it since then by demanding systems with higher and higher capacities. Here we focus on the status of commercial systems by considering the terrestrial and undersea systems separately.

After a successful Chicago field trial in 1977, terrestrial lightwave systems became available commercially beginning in 1980 [45]–[47]. Table 5.2 lists the operating characteristics of several terrestrial systems developed since then. The first-generation systems operated near 0.85 μm and used multimode graded-index fibers as the transmission medium. As seen in Fig. 5.4, the BL product of such systems is limited to 2 (Gb/s)-km. A commercial lightwave system (FT–3C) operating at 90 Mb/s with a repeater spacing of about 12 km realized a BL product of nearly 1 (Gb/s)-km; it is shown by a filled circle in Fig. 5.4. The operating wavelength moved to 1.3 μm in second-generation lightwave systems to take advantage of low fiber losses and low dispersion near this wavelength. The BL product of 1.3-μm lightwave systems is limited to about 100 (Gb/s)-km when a multimode semiconductor laser is used inside the transmitter. In 1987, a commercial 1.3-μm lightwave system provided data transmission at 1.7 Gb/s with a repeater spacing of about 45 km. A filled circle in Fig. 5.4 shows that this system operates quite close to the dispersion limit.

The third generation of lightwave systems became available commercially in 1991. They operate near 1.55 μm at bit rates in excess of 2 Gb/s, typically at 2.488 Gb/s, corresponding to the OC-48 level of the synchronized optical network (SONET) [or the STS–16 level of the synchronous digital hierarchy (SDH)] specifications. The switch to the 1.55-μm wavelength helps to increase the loss-limited transmission distance to more than 100 km because of fiber losses of less than 0.25 dB/km in this wavelength region. However, the repeater spacing was limited to below 100 km because of the high GVD of standard telecommunication fibers. In fact, the deployment of third-generation lightwave systems was possible only after the development of distributed feedback (DFB) semiconductor lasers, which reduce the impact of fiber dispersion by reducing the source spectral width to below 100 MHz (see Section 2.4).

The fourth generation of lightwave systems appeared around 1996. Such systems operate in the 1.55-μm region at a bit rate as high as 40 Gb/s by using dispersion-shifted fibers in combination with optical amplifiers. However, more than 50 million kilometers of the standard telecommunication fiber is already installed in the world-wide telephone network. Economic reasons dictate that the fourth generation of lightwave systems make use of this existing base. Two approaches are being used to solve the dispersion problem. First, several dispersion-management schemes (discussed in Chapter 7) make it possible to extend the bit rate to 10 Gb/s while maintaining an amplifier spacing of up to 100 km. Second, several 10-Gb/s signals can be transmitted simultaneously by using the WDM technique discussed in Chapter 8. Moreover, if the WDM technique is combined with dispersion management, the total transmission distance can approach several thousand kilometers provided that fiber losses are compensated periodically by using optical amplifiers. Such WDM lightwave systems were deployed commercially worldwide beginning in 1996 and allowed a system capacity of 1.6 Tb/s by 2000 for the 160-channel commercial WDM systems.

The fifth generation of lightwave systems was just beginning to emerge in 2001. The bit rate of each channel in this generation of WDM systems is 40 Gb/s (corresponding to the STM-256 or OC-768 level). Several new techniques developed in recent years make it possible to transmit a 40-Gb/s optical signal over long distances. New fibers known as reverse-dispersion fibers have been developed with a negative GVD slope. Their use in combination with tunable dispersion-compensating techniques can compensate the GVD for all channels simultaneously. The PMD compensators help to reduce the PMD-induced degradation of the signal. The use of Raman amplification helps to reduce the noise and improves the signal-to-noise ratio (SNR) at the receiver. The use of a forward-error-correction technique helps to increase the transmission distance by reducing the required SNR. The number of WDM channels can be increased by using the L and S bands located on the long- and short-wavelength sides of the conventional C band occupying the 1530–1570-nm spectral region. In one 3-Tb/s experiment, 77 channels, each operating at 42.7-Gb/s, were transmitted over 1200 km by using the C and L bands simultaneously [48]. In another experiment, the system capacity was extended to 10.2 Tb/s by transmitting 256 channels over 100 km at 42.7 Gb/s per channel using only the C and L bands, resulting in a spectral efficiency of 1.28 (b/s)/Hz [49]. The bit rate was 42.7 Gb/s in both of these experiments because of the overhead associated with the forward-error-correction technique. The highest capacity achieved in 2001 was 11 Tb/s and was realized by transmitting 273 channels

Table 5.3 Commercial transatlantic lightwave systems

System	Year	Capacity (Gb/s)	L (km)	Comments
TAT–8	1988	0.28	70	1.3 μm, multimode lasers
TAT–9	1991	0.56	80	1.55 μm, DFB lasers
TAT–10/11	1993	0.56	80	1.55 μm, DFB lasers
TAT–12/13	1996	5.00	50	1.55 μm, optical amplifiers
AC–1	1998	80.0	50	1.55 μm, WDM with amplifiers
TAT–14	2001	1280	50	1.55 μm, dense WDM
AC–2	2001	1280	50	1.55 μm, dense WDM
360Atlantic-1	2001	1920	50	1.55 μm, dense WDM
Tycom	2001	2560	50	1.55 μm, dense WDM
FLAG Atlantic-1	2001	4800	50	1.55 μm, dense WDM

over 117 km at 40 Gb/s per channel while using all three bands simultaneously [50].

5.3.3   Undersea Lightwave Systems

Undersea or submarine transmission systems are used for intercontinental communications and are capable of providing a network spanning the whole earth [51]–[53]. Figure 1.5 shows several undersea systems deployed worldwide. Reliability is of major concern for such systems as repairs are expensive. Generally, undersea systems are designed for a 25-year service life, with at most three failures during operation. Table 5.3 lists the main characteristics of several transatlantic fiber-optic cable systems. The first undersea fiber-optic cable (TAT–8) was a second-generation system. It was installed in 1988 in the Atlantic Ocean for operation at a bit rate of 280 Mb/s with a repeater spacing of up to 70 km. The system design was on the conservative side, mainly to ensure reliability. The same technology was used for the first transpacific lightwave system (TPC–3), which became operational in 1989.

By 1990 the third-generation lightwave systems had been developed. The TAT–9 submarine system used this technology in 1991; it was designed to operate near 1.55 μm at a bit rate of 560 Mb/s with a repeater spacing of about 80 km. The increasing traffic across the Atlantic Ocean led to the deployment of the TAT–10 and TAT–11 lightwave systems by 1993 with the same technology. The advent of optical amplifiers prompted their use in the next generation of undersea systems, and the TAT–12 submarine fiber-optic cable became operational by 1996. This fourth-generation system employed optical amplifiers in place of optoelectronic regenerators and operated at a bit rate of 5.3 Gb/s with an amplifier spacing of about 50 km. The bit rate is slightly larger than the STM-32-level bit rate of 5 Gb/s because of the overhead associated with the forward-error-correction technique. As discussed earlier, the design of such lightwave systems is much more complex than that of previous undersea systems because of the cumulative effects of fiber dispersion and nonlinearity, which must be controlled over long distances. The transmitter power and the dispersion profile along the link must be optimized to combat such effects. Even then, amplifier spacing is typically limited to 50 km, and the use of an error-correction scheme is essential to ensure a bit-error rate of $< 2 \times 10^{-11}$.

A second category of undersea lightwave systems requires repeaterless transmission over several hundred kilometers [52]. Such systems are used for interisland communication or for looping a shoreline such that the signal is regenerated on the shore periodically after a few hundred kilometers of undersea transmission. The dispersive and nonlinear effects are of less concern for such systems than for transoceanic lightwave systems, but fiber losses become a major issue. The reason is easily appreciated by noting that the cable loss exceeds 100 dB over a distance of 500 km even under the best operating conditions. In the 1990s several laboratory experiments demonstrated repeaterless transmission at 2.5 Gb/s over more than 500 km by using two in-line amplifiers that were pumped remotely from the transmitter and receiver ends with high-power pump lasers. Another amplifier at the transmitter boosted the launched power to close to 100 mW.

Such high input powers exceed the threshold level for stimulated Brillouin scattering (SBS), a nonlinear phenomenon discussed in Section 2.6. The suppression of SBS is often realized by modulating the phase of the optical carrier such that the carrier linewidth is broadened to 200 MHz or more from its initial value of <10 MHz [54]. Directly modulated DFB lasers can also be used for this purpose. In a 1996 experiment. a 2.5-Gb/s signal was transmitted over 465 km by direct modulation of a DFB laser [55]. Chirping of the modulated signal broadened the spectrum enough that an external phase modulator was not required provided that the launched power was kept below 100 mW. The bit rate of repeaterless undersea systems can be increased to 10 Gb/s by employing the same techniques used at 2.5 Gb/s. In a 1996 experiment [56], the 10-Gb/s signal was transmitted over 442 km by using two remotely pumped in-line amplifiers. Two external modulators were used, one for SBS suppression and another for signal generation. In a 1998 experiment, a 40-Gb/s signal was transmitted over 240 km using the RZ format and an alternating polarization format [57]. These results indicate that undersea lightwave systems looping a shoreline can operate at 10 Gb/s or more with only shore-based electronics [58].

The use of the WDM technique in combination with optical amplifiers, dispersion management, and error correction has revolutionized the design of submarine fiber-optic systems. In 1998, a submarine cable known as Atlantic-Crossing 1 (AC–1) with a capacity of 80 Gb/s was deployed using the WDM technology. An identically designed system (Pacific-Crossing 1 or PC–1) crossed the Pacific Ocean. The use of dense WDM, in combination with multiple fiber pairs per cable, resulted in systems with much larger capacities. By 2001, several systems with a capacity of >1 Tb/s became operational across the Atlantic Ocean (see Table 5.3). These systems employ a ring configuration and cross the Atlantic Ocean twice to ensure fault tolerance. The "360Atlantic" submarine system can operate at speeds up to 1.92 Tb/s and spans a total distance of 11,700 km. Another system, known as FLAG Atlantic-1, is capable of carrying traffic at speeds up to 4.8 Tb/s as it employs six fiber pairs. A global network, spanning 250,000 km and capable of operating at 3.2 Tb/s using 80 channels (at 10 Gb/s) over 4 fibers, was under development in 2001 [53]. Such a submarine network can transmit nearly 40 million voice channels simultaneously, a capacity that should be

contrasted with the TAT–8 capacity of 8000 channels in 1988, which in turn should be
compared to the 48-channel capacity of TAT–1 in 1959.

5.4   Sources of Power Penalty

The sensitivity of the optical receiver in a realistic lightwave system is affected by
several physical phenomena which, in combination with fiber dispersion, degrade the
SNR at the decision circuit. Among the phenomena that degrade the receiver sensitivity
are modal noise, dispersion broadening and intersymbol interference, mode-partition
noise, frequency chirp, and reflection feedback. In this section we discuss how the
system performance is affected by fiber dispersion by considering the extent of power
penalty resulting from these phenomena.

5.4.1   Modal Noise

Modal noise is associated with multimode fibers and was studied extensively during the
1980s [59]–[72]. Its origin can be understood as follows. Interference among various
propagating modes in a multimode fiber creates a *speckle pattern* at the photodetector.
The nonuniform intensity distribution associated with the speckle pattern is harmless
in itself, as the receiver performance is governed by the total power integrated over
the detector area. However, if the speckle pattern fluctuates with time, it will lead to
fluctuations in the received power that would degrade the SNR. Such fluctuations are
referred to as *modal noise*. They invariably occur in multimode fiber links because
of mechanical disturbances such as vibrations and microbends. In addition, splices
and connectors act as spatial filters. Any temporal changes in spatial filtering translate
into speckle fluctuations and enhancement of the modal noise. Modal noise is strongly
affected by the source spectral bandwidth Δv since mode interference occurs only if
the coherence time ($T_c \approx 1/\Delta v$) is longer than the intermodal delay time ΔT given by
Eq. (2.1.5). For LED-based transmitters Δv is large enough ($\Delta v \sim 5$ THz) that this
condition is not satisfied. Most lightwave systems that use multimode fibers also use
LEDs to avoid the modal-noise problem.

Modal noise becomes a serious problem when semiconductor lasers are used in
combination with multimode fibers. Attempts have been made to estimate the extent
of sensitivity degradation induced by modal noise [61]–[63] by calculating the BER
after adding modal noise to the other sources of receiver noise. Figure 5.6 shows the
power penalty at a BER of 10^{-12} calculated for a 1.3-μm lightwave system operating at
140 Mb/s. The graded-index fiber has a 50-μm core diameter and supports 146 modes.
The power penalty depends on the mode-selective coupling loss occurring at splices
and connectors. It also depends on the longitudinal-mode spectrum of the semiconductor laser. As expected, power penalty decreases as the number of longitudinal modes
increases because of a reduction in the coherence time of the emitted light.

Modal noise can also occur in single-mode systems if short sections of fiber are
installed between two connectors or splices during repair or normal maintenance [63]–
[66]. A higher-order mode can be excited at the fiber discontinuity occurring at the
first splice and then converted back to the fundamental mode at the second connector

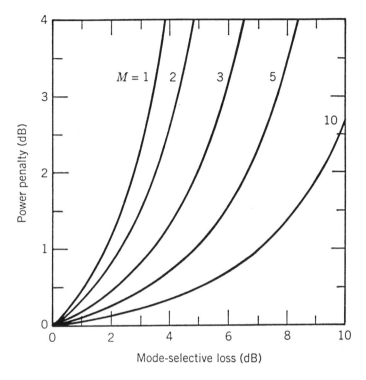

Figure 5.6: Modal-noise power penalty versus mode-selective loss. The parameter M is defined as the total number of longitudinal modes whose power exceeds 10% of the peak power. (After Ref. [61]; ©1986 IEEE; reprinted with permission.)

or splice. Since a higher-order mode cannot propagate far from its excitation point, this problem can be avoided by ensuring that the spacing between two connectors or splices exceeds 2 m. Generally speaking, modal noise is not a problem for properly designed and maintained single-mode fiber-optic communication systems.

With the development of the vertical-cavity surface-emitting laser (VCSEL), the modal-noise issue has resurfaced in recent years [67]–[71]. The use of such lasers in short-haul optical data links, making use of multimode fibers (even those made of plastic), is of considerable interest because of the high bandwidth associated with VCSELs. Indeed, rates of several gigabits per second have been demonstrated in laboratory experiments with plastic-cladded multimode fibers [73]. However, VCSELs have a long coherence length as they oscillate in a single longitudinal mode. In a 1994 experiment the BER measurements showed an error floor at a level of 10^{-7} even when the mode-selective loss was only 1 dB [68]. This problem can be avoided to some extent by using larger-diameter VCSELs which oscillate in several transverse modes and thus have a shorter coherence length. Computer models are generally used to estimate the power penalty for optical data links under realistic operating conditions [70]. Analytic tools such as the saddle-point method can also provide a reasonable estimate of the BER [71].

5.4.2 Dispersive Pulse Broadening

The use of single-mode fibers for lightwave systems nearly avoids the problem of inter-modal dispersion and the associated modal noise. The group-velocity dispersion still limits the bit rate–distance product BL by broadening optical pulses beyond their allocated bit slot; Eq. (5.2.2) provides the limiting BL product and shows how it depends on the source spectral width σ_λ. Dispersion-induced pulse broadening can also decrease the receiver sensitivity. In this subsection we discuss the power penalty associated with such a decrease in receiver sensitivity.

Dispersion-induced pulse broadening affects the receiver performance in two ways. First, a part of the pulse energy spreads beyond the allocated bit slot and leads to intersymbol interference (ISI). In practice, the system is designed to minimize the effect of ISI (see Section 4.3.2). Second, the pulse energy within the bit slot is reduced when the optical pulse broadens. Such a decrease in the pulse energy reduces the SNR at the decision circuit. Since the SNR should remain constant to maintain the system performance, the receiver requires more average power. This is the origin of dispersion-induced power penalty δ_d. An exact calculation of δ_d is difficult, as it depends on many details, such as the extent of pulse shaping at the receiver. A rough estimate is obtained by following the analysis of Section 2.4.2, where broadening of Gaussian pulses is discussed. Equation (2.4.16) shows that the optical pulse remains Gaussian, but its peak power is reduced by a pulse-broadening factor given by Eq. (2.4.17). If we define the power penalty δ_d as the increase (in dB) in the received power that would compensate the peak-power reduction, δ_d is given by

$$\delta_d = 10 \log_{10} f_b, \tag{5.4.1}$$

where f_b is the pulse broadening factor. When pulse broadening is due mainly to a wide source spectrum at the transmitter, the broadening factor f_b is given by Eq. (2.4.24), i.e.,

$$f_b = \sigma/\sigma_0 = [1 + (DL\sigma_\lambda/\sigma_0)^2]^{1/2}, \tag{5.4.2}$$

where σ_0 is the RMS width of the optical pulse at the fiber input and σ_λ is the RMS width of the source spectrum assumed to be Gaussian.

Equations (5.4.1) and (5.4.2) can be used to estimate the dispersion penalty for lightwave systems that use single-mode fiber together with a multimode laser or an LED. The ISI is minimized when the bit rate B is such that $4B\sigma \leq 1$, as little pulse energy spreads beyond the bit slot ($T_B = 1/B$). By using $\sigma = (4B)^{-1}$, Eq. (5.4.2) can be written as

$$f_b^2 = 1 + (4BLD\sigma_\lambda f_b)^2. \tag{5.4.3}$$

By solving this equation for f_b and substituting it in Eq. (5.4.1), the power penalty is given by

$$\delta_d = -5 \log_{10}[1 - (4BLD\sigma_\lambda)^2]. \tag{5.4.4}$$

Figure 5.7 shows the power penalty as a function of the dimensionless parameter combination $BLD\sigma_\lambda$. Although the power penalty is negligible ($\delta_d = 0.38$ dB) for $BLD\sigma_\lambda = 0.1$, it increases to 2.2 dB when $BLD\sigma_\lambda = 0.2$ and becomes infinite when $BLD\sigma_\lambda = 0.25$. The BL product, shown in Fig. 5.4, is truly limiting, since receiver

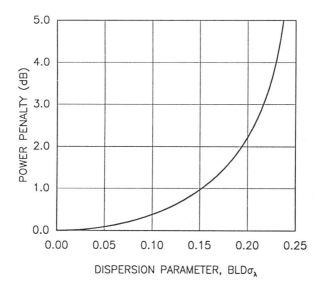

Figure 5.7: Dispersion-induced power penalty for a Gaussian pulse as a function of $BLD\sigma_\lambda$. Source spectrum is also assumed to be Gaussian with an RMS width σ_λ.

sensitivity degrades severely when a system is designed to approach it. Most lightwave systems are designed such that $BLD\sigma_\lambda < 0.2$, so that the dispersion penalty is below 2 dB. It should be stressed that Eq. (5.4.4) provides a rough estimate only as its derivation is based on several simplifying assumptions, such as a Gaussian pulse shape and a Gaussian source spectrum. These assumptions are not always satisfied in practice. Moreover, it is based on the condition $4B\sigma = 1$, so that the pulse remains nearly confined within the bit slot. It is possible to design a system such that the pulse spreads outside the bit slot but ISI is reduced through pulse shaping at the receiver.

5.4.3 Mode-Partition Noise

As discussed in Section 3.5.4, multimode semiconductor lasers exhibit *mode-partition noise* (MPN), a phenomenon occurring because of an anticorrelation among pairs of longitudinal modes. In particular, various longitudinal modes fluctuate in such a way that individual modes exhibit large intensity fluctuations even though the total intensity remains relatively constant. MPN would be harmless in the absence of fiber dispersion, as all modes would remain synchronized during transmission and detection. In practice, different modes become unsynchronized, since they travel at slightly different speeds inside the fiber because of group-velocity dispersion. As a result of such desynchronization, the receiver current exhibits additional fluctuations, and the SNR at the decision circuit becomes worse than that expected in the absence of MPN. A power penalty must be paid to improve the SNR to the same value that is necessary to achieve the required BER (see Section 4.5). The effect of MPN on system performance has been studied extensively for both multimode semiconductor lasers [74]–[83] and nearly single-mode lasers [84]–[98].

In the case of multimode semiconductor lasers, the power penalty can be calculated by following an approach similar to that of Section 4.6.2 and is given by [74]

$$\delta_{mpn} = -5 \log_{10}(1 - Q^2 r_{mpn}^2), \tag{5.4.5}$$

where r_{mpn} is the relative noise level of the received power in the presence of MPN. A simple model for estimating the parameter r_{mpn} assumes that laser modes fluctuate in such a way that the total power remains constant under CW operation [75]. It also assumes that the average mode power is distributed according to a Gaussian distribution of RMS width σ_λ and that the pulse shape at the decision circuit of the receiver is described by a cosine function [74]. Different laser modes are assumed to have the same cross-correlation coefficient γ_{cc}, i.e.,

$$\gamma_{cc} = \frac{\langle P_i P_j \rangle}{\langle P_i \rangle \langle P_j \rangle} \tag{5.4.6}$$

for all i and j such that $i \neq j$. The angular brackets denote an average over power fluctuations associated with mode partitioning. A straightforward calculation shows that r_{mpn} is given by [78]

$$r_{mpn} = (k/\sqrt{2})\{1 - \exp[-(\pi BLD\sigma_\lambda)^2]\}, \tag{5.4.7}$$

where the mode-partition coefficient k is related to γ_{cc} as $k = \sqrt{1 - \gamma_{cc}}$. The model assumes that mode partition can be quantified in terms of a single parameter k with values in the range 0–1. The numerical value of k is difficult to estimate and is likely to vary from laser to laser. Experimental measurements suggest that the values of k are in the range 0.6–0.8 and vary for different mode pairs [75], [80].

Equations (5.4.5) and (5.4.7) can be used to calculate the MPN-induced power penalty. Figure 5.8 shows the power penalty at a BER of 10^{-9} ($Q = 6$) as a function of the normalized dispersion parameter $BLD\sigma_\lambda$ for several values of the mode-partition coefficient k. For a given value of k, the variation of power penalty is similar to that shown in Fig. 5.7; δ_{mpn} increases rapidly with an increase in $BLD\sigma_\lambda$ and becomes infinite when $BLD\sigma_\lambda$ reaches a critical value. For $k > 0.5$, the MPN-induced power penalty is larger than the penalty occurring due to dispersion-induced pulse broadening (see Fig. 5.7). However, it can be reduced to a negligible level ($\delta_{mpn} < 0.5$ dB) by designing the optical communication system such that $BLD\sigma_\lambda < 0.1$. As an example, consider a 1.3-μm lightwave system. If we assume that the operating wavelength is matched to the zero-dispersion wavelength to within 10 nm, $D \approx 1$ ps/(km-nm). A typical value of σ_λ for multimode semiconductor lasers is 2 nm. The MPN-induced power penalty would be negligible if the BL product were below 50 (Gb/s)-km. At $B = 2$ Gb/s the transmission distance is then limited to 25 km. The situation becomes worse for 1.55-μm lightwave systems for which $D \approx 16$ ps/(km-nm) unless dispersion-shifted fibers are used. In general, the MPN-induced power penalty is quite sensitive to the spectral bandwidth of the multimode laser and can be reduced by reducing the bandwidth. In one study [83], a reduction in the carrier lifetime from 340 to 130 ps, realized by p-doping of the active layer, reduced the bandwidth of 1.3-μm semiconductor lasers by only 40% (from 5.6 to 3.4 nm), but the power penalty decreased from an infinite value (BER floor above 10^{-9} level) to a mere 0.5 dB.

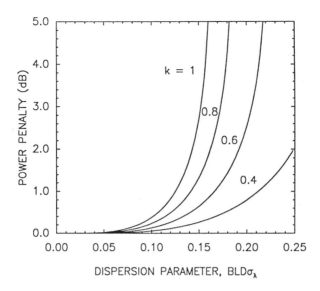

Figure 5.8: MPN-induced Power penalty versus $BLD\sigma_\lambda$ for a multimode semiconductor laser of RMS spectral width σ_λ. Different curves correspond to different values of the mode-partition coefficient k.

One may think that MPN can be avoided completely by using DFB lasers designed to oscillate in a single longitudinal mode. Unfortunately, this is not necessarily the case [88]–[91]. The reason is that the main mode of any DFB laser is accompanied by several side modes of much smaller amplitudes. The single-mode nature of DFB lasers is quantified through the *mode-suppression ratio* (MSR), defined as the ratio of the main-mode power P_m to the power P_s of the most dominant side mode. Clearly, the effect of MPN on system performance would depend on the MSR. Attempts have therefore been made to estimate the dependence of the MPN-induced power penalty on the MSR [84]–[98].

A major difference between the multimode and nearly single-mode semiconductor lasers is related to the statistics associated with mode-partition fluctuations. In a multimode laser, both main and side modes are above threshold and their fluctuations are well described by a Gaussian probability density function. By contrast, side modes in a DFB semiconductor laser are typically below threshold, and the optical power associated with them follows an exponential distribution given by [84]

$$p(P_s) = \bar{P}_s^{-1} \exp[-(P_s/\bar{P}_s)], \qquad (5.4.8)$$

where $\bar{P}_s$ is the average value of the random variable P_s.

The effect of side-mode fluctuations on system performance can be appreciated by considering an ideal receiver. Let us assume that the relative delay $\Delta T = DL\Delta\lambda$ between the main and side modes is large enough that the side mode appears outside the bit slot (i.e., $\Delta T > 1/B$ or $BLD\Delta\lambda_L > 1$, where $\Delta\lambda_L$ is the mode spacing). The decision circuit of the receiver would make an error for 0 bits if the side-mode power P_s were to exceed the decision threshold set at $\bar{P}_m/2$, where $\bar{P}_m$ is the average main-mode power. Furthermore, the two modes are anticorrelated in such a way that the main-mode power drops below $\bar{P}_m/2$ whenever side-mode power exceeds $\bar{P}_m/2$, so that the total power remains nearly constant [85]. Thus, an error would occur even for "1" bits whenever $P_s > \bar{P}_m/2$. Since the two terms in Eq. (4.5.2) make equal contributions, the BER is given by [84]

$$\text{BER} = \int_{\bar{P}_m/2}^{\infty} p(P_s)\,dP_s = \exp\left(-\frac{\bar{P}_m}{2\bar{P}_s}\right) = \exp\left(-\frac{R_{ms}}{2}\right). \qquad (5.4.9)$$

The BER depends on the MSR defined as $R_{ms} = \bar{P}_m/\bar{P}_s$ and exceeds 10^{-9} when MSR $<$ 42.

To calculate the MPN-induced power penalty in the presence of receiver noise, one should follow the analysis in Section 4.5.1 and add an additional noise term that accounts for side-mode fluctuations. For a p–i–n receiver the BER is found to be [85]

$$\text{BER} = \frac{1}{2}\text{erfc}\left(\frac{Q}{\sqrt{2}}\right) + \exp\left(-\frac{R_{ms}}{2} + \frac{R_{ms}^2}{4Q^2}\right)\left[1 - \frac{1}{2}\text{erfc}\left(\frac{Q}{\sqrt{2}} - \frac{R_{ms}}{Q\sqrt{2}}\right)\right], \quad (5.4.10)$$

where the parameter Q is defined by Eq. (4.5.10). In the limit of an infinite MSR, Eq. (5.4.10) reduces to Eq. (4.5.9). For a noise-free receiver ($Q = \infty$), Eq. (5.4.10) reduces to Eq. (5.4.9). Figure 5.9 shows the BER versus the power penalty at a BER of 10^{-9} as a function of MSR. As expected, the power penalty becomes infinite for MSR values below 42, since the 10^{-9} BER cannot be realized irrespective of the power received. The penalty can be reduced to a negligible level (<0.1 dB) for MSR values in excess of 100 (20 dB).

The experimental measurements of the BER in several transmission experiments show that a BER floor above the 10^{-9} level can occur even for DFB lasers which exhibit a MSR in excess of 30 dB under continuous-wave (CW) operation [88]–[91]. The reason behind the failure of apparently good lasers is related to the possibility of side-mode excitation under transient conditions occurring when the laser is repeatedly turned on and off to generate the bit stream. When the laser is biased below threshold and modulated at a high bit rate ($B \geq 1$ Gb/s), the probability of side-mode excitation above $\bar{P}_m/2$ is much higher than that predicted by Eq. (5.4.8). Considerable attention has been paid to calculate, both analytically and numerically, the probability of the transient excitation of side modes and its dependence on various device parameters [87]–[98]. An important device parameter is found to be the *gain margin* between the main and side modes. The gain margin should exceed a critical value which depends on the bit rate. The critical value is about 5–6 cm^{-1} at 500 Mb/s [88] but can exceed 15 cm^{-1} at high bit rates, depending on the bias and modulation currents [93]. The bias current plays a critical role. Numerical simulations show that the best performance is achieved when the DFB laser is biased close to but slightly below threshold to avoid the bit-pattern effects [98]. Moreover, the effects of MPN are independent of the bit rate as long as the gain margin exceeds a certain value. The required value of gain margin exceeds 25 cm^{-1} for the 5-GHz modulation frequency. Phase-shifted DFB lasers have a large built-in gain margin and have been developed for this purpose.

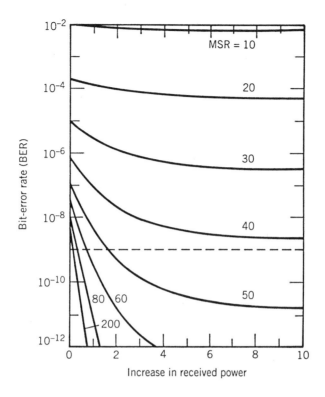

Figure 5.9: Effect of MPN on bit-error rate of DFB lasers for several values of MSR. Intersection of the dashed line with the solid curves provides MPN-induced power penalty. (After Ref. [85]; ©1985 IEEE; reprinted with permission.)

5.4.4 Frequency Chirping

Frequency chirping is an important phenomenon that is known to limit the performance of 1.55-μm lightwave systems even when a DFB laser with a large MSR is used to generate the digital bit stream [99]–[112]. As discussed in Section 3.5.3, intensity modulation in semiconductor lasers is invariably accompanied by phase modulation because of the carrier-induced change in the refractive index governed by the linewidth enhancement factor. Optical pulses with a time-dependent phase shift are called chirped. As a result of the frequency chirp imposed on an optical pulse, its spectrum is considerably broadened. Such spectral broadening affects the pulse shape at the fiber output because of fiber dispersion and degrades system performance.

An exact calculation of the chirp-induced power penalty δ_c is difficult because frequency chirp depends on both the shape and the width of the optical pulse [101]–[104]. For nearly rectangular pulses, experimental measurements of time-resolved pulse spectra show that frequency chirp occurs mainly near the leading and trailing edges such that the leading edge shifts toward the blue while the trailing edge shifts toward the red. Because of the spectral shift, the power contained in the chirped portion of the pulse moves out of the bit slot when the pulse propagates inside the optical fiber. Such

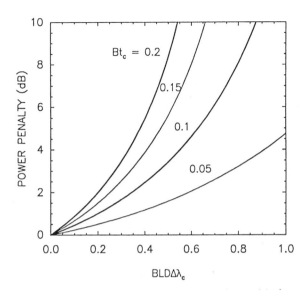

Figure 5.10: Chirp-induced power penalty as a function of $BLD\Delta\lambda_c$ for several values of the parameter Bt_c, where $\Delta\lambda_c$ is the wavelength shift occurring because of frequency chirp and t_c is the duration of such a wavelength shift.

a power loss decreases the SNR at the receiver and results in power penalty. In a simple model the chirp-induced power penalty is given by [100]

$$\delta_c = -10\,\log_{10}(1 - 4BLD\Delta\lambda_c), \qquad (5.4.11)$$

where $\Delta\lambda_c$ is the spectral shift associated with frequency chirping. This equation applies as long as $LD\Delta\lambda_c < t_c$, where t_c is the chirp duration. Typically, t_c is 100–200 ps, depending on the relaxation-oscillation frequency, since chirping lasts for about one-half of the relaxation-oscillation period. By the time $LD\Delta\lambda_c$ equals t_c, the power penalty stops increasing because all the chirped power has left the bit interval. For $LD\Delta\lambda_c > t_c$, the product $LD\Delta\lambda_c$ in Eq. (5.4.11) should be replaced by t_c.

The model above is overly simplistic, as it does not take into account pulse shaping at the receiver. A more accurate calculation based on raised-cosine filtering (see Section 4.3.2) leads to the following expression [107]:

$$\delta_c = -20\,\log_{10}\{1 - (4\pi^2/3 - 8)B^2 L_D\Delta\lambda_c t_c[1 + (2B/3)(LD\Delta\lambda_c - t_c)]\}. \qquad (5.4.12)$$

The receiver is assumed to contain a p–i–n photodiode. The penalty is larger for an APD, depending on the excess-noise factor of the APD. Figure 5.10 shows the power penalty δ_c as a function of the parameter combination $BLD\Delta\lambda_c$ for several values of the parameter Bt_c, which is a measure of the fraction of the bit period over which chirping occurs. As expected, δ_c increases with both the chirp $\Delta\lambda_c$ and the chirp duration t_c. The power penalty can be kept below 1 dB if the system is designed such that $BLD\Delta\lambda_c < 0.1$ and $Bt_c < 0.2$. A shortcoming of this model is that $\Delta\lambda_c$ and t_c appear as free

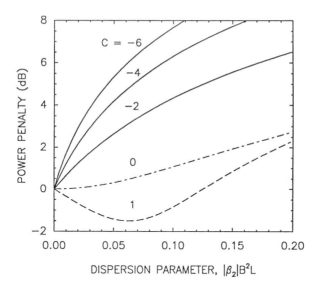

Figure 5.11: Chirp-induced power penalty as a function of $|\beta_2|B^2L$ for several values of the chirp parameter C. The Gaussian optical pulse is assumed to be linearly chirped over its entire width.

parameters and must be determined for each laser through experimental measurements of the frequency chirp. In practice, $\Delta\lambda_c$ itself depends on the bit rate B and increases with it.

For lightwave systems operating at high bit rates ($B > 2$ Gb/s), the bit duration is generally shorter than the total duration $2t_c$ over which chirping is assumed to occur in the foregoing model. The frequency chirp in that case increases almost linearly over the entire pulse width (or bit slot). A similar situation occurs even at low bit rates if the optical pulses do not contain sharp leading and trailing edges but have long rise and fall times (Gaussian-like shape rather than a rectangular shape). If we assume a Gaussian pulse shape and a linear chirp, the analysis of Section 2.4.2 can be used to estimate the chirp-induced power penalty. Equation (2.4.16) shows that the chirped Gaussian pulse remains Gaussian but its peak power decreases because of dispersion-induced pulse broadening. Defining the power penalty as the increase (in dB) in the received power that would compensate the peak-power reduction, δ_c is given by

$$\delta_c = 10\log_{10} f_b, \qquad (5.4.13)$$

where f_b is the broadening factor given by Eq. (2.4.22) with $\beta_3 = 0$. The RMS width σ_0 of the input pulse should be such that $4\sigma_0 \leq 1/B$. Choosing the worst-case condition $\sigma_0 = 1/4B$, the power penalty is given by

$$\delta_c = 5\log_{10}[(1 + 8C\beta_2B^2L)^2 + (8\beta_2B^2L)^2]. \qquad (5.4.14)$$

Figure 5.11 shows the chirp-induced power penalty as a function of $|\beta_2|B^2L$ for several values of the chirp parameter C. The parameter β_2 is taken to be negative,

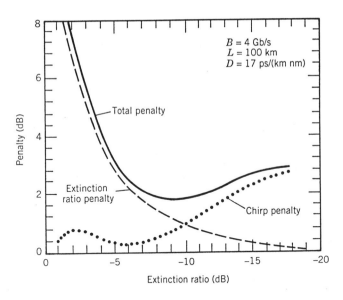

Figure 5.12: Power penalty as a function of the extinction ratio. (After Ref. [105]; ©1987 IEEE; reprinted with permission.)

as is the case for 1.55-μm lightwave systems. The $C = 0$ curve corresponds to the case of a chirp-free pulse. The power penalty is negligible (<0.1 dB) in this ideal case as long as $|\beta_2|B^2L < 0.05$. However, the penalty can exceed 5 dB if the pulses transmitted are chirped such that $C = -6$. To keep the penalty below 0.1 dB, the system should be designed with $|\beta_2|B^2L < 0.002$. For $|\beta_2| = 20$ ps^2/km, B^2L is limited to 100 (Gb/s)^2-km. Interestingly, system performance is improved for positive values of C since the optical pulse then goes through an initial compression phase (see Section 2.4). Unfortunately, C is negative for semiconductor lasers; it can be approximated by $-\beta_c$, where β_c is the linewidth enhancement factor with positive values of 2–6.

It is important to stress that the analytic results shown in Figs. 5.10 and 5.11 provide only a rough estimate of the power penalty. In practice, the chirp-induced power penalty depends on many system parameters. For instance, several system experiments have shown that the effect of chirp can be reduced by biasing the semiconductor laser above threshold [103]. However, above-threshold biasing increases that extinction ratio r_{ex}, defined in Eq. (4.6.1) as $r_{ex} = P_0/P_1$, where P_0 and P_1 are the powers received for bit 0 and bit 1, respectively. As discussed in Section 4.6.1, an increase in r_{ex} decreases the receiver sensitivity and leads to its own power penalty. Clearly, r_{ex} cannot be increased indefinitely in an attempt to reduce the chirp penalty. The total system performance can be optimized by designing the system so that it operates with an optimum value of r_{ex} that takes into account the trade-off between the chirp and the extinction ratio. Numerical simulations are often used to understand such trade-offs in actual lightwave systems [110]–[113]. Figure 5.12 shows the power penalty as a function of the extinction ratio r_{ex} by simulating numerically the performance of a 1.55-μm lightwave system transmitting at 4 Gb/s over a 100-km-long fiber. The total penalty can be reduced below 2 dB by operating the system with an extinction ratio of about 0.1. The optimum values of r_{ex} and the total penalty are sensitive to many other laser parameters such as the active-region width. A semiconductor laser with a wider active region is found to have a larger chirp penalty [105]. The physical phenomenon behind this width dependence appears to be the nonlinear gain [see Eq. (3.3.40)] and the associated damping of relaxation oscillations. In general, rapid damping of relaxation oscillations decreases the effect of frequency chirp and improves system performance [113].

The origin of chirp in semiconductor lasers is related to carrier-induced index changes governed by the linewidth enhancement factor β_c. The frequency chirp would be absent for a laser with $\beta_c = 0$. Unfortunately, β_c cannot be made zero for semiconductor lasers, although it can be reduced by adopting a multiquantum-well (MQW) design [114]–[118]. The use of a MQW active region reduces β_c by about a factor of 2. In one 1.55-μm experiment [120], the 10-Gb/s signal could be transmitted over 60–70 km, despite the high dispersion of standard telecommunication fiber, by biasing the laser above threshold. The MQW DFB laser used in the experiment had $\beta_c \approx 3$. A further reduction in β_c occurs for strained quantum wells [118]. Indeed, $\beta_c \approx 1$ has been measured in modulation-doped strained MQW lasers [119]. Such lasers exhibit low chirp under direct modulation at bit rates as high as 10 Gb/s.

An alternative scheme eliminates the laser-chirp problem completely by operating the laser continuously and using an external modulator to generate the bit stream. This approach has become practical with the development of optical transmitters in which a modulator is integrated monolithically with a DFB laser (see Section 3.6.4). The chirp parameter C is close to zero in such transmitters. As shown by the $C = 0$ curve in Fig. 5.11, the dispersion penalty is below 2 dB in that case even when $|\beta_2|B^2L$ is close to 0.2. Moreover, an external modulator can be used to modulate the phase of the optical carrier in such a way that $\beta_2C < 0$ in Eq. (5.4.14). As seen in Fig. 5.11, the chirp-induced power penalty becomes negative over a certain range of $|\beta_2|B^2L$, implying that such frequency chirping is beneficial to combat the effects of dispersion. In a 1996 experiment [121], the 10-Gb/s signal was transmitted penalty free over 100 km of standard telecommunication fiber by using a modulator-integrated transmitter such that C was effectively positive. By using $\beta_2 \approx -20$ ps^2/km, it is easy to verify that $|\beta_2|B^2L = 0.2$ for this experiment, a value that would have produced a power penalty of more than 8 dB if the DFB laser were modulated directly.

5.4.5 Reflection Feedback and Noise

In most fiber-optic communication systems, some light is invariably reflected back because of refractive-index discontinuities occurring at splices, connectors, and fiber ends. The effects of such unintentional feedback have been studied extensively [122]–[140] because they can degrade the performance of lightwave systems considerably. Even a relatively small amount of optical feedback affects the operation of semiconductor lasers [126] and can lead to excess noise in the transmitter output. Even when an isolator is used between the transmitter and the fiber, multiple reflections between splices and connectors can generate additional intensity noise and degrade receiver performance [128]. This subsection is devoted to the effect of reflection-induced noise on receiver sensitivity.

Most reflections in a fiber link originate at glass–air interfaces whose reflectivity can be estimated by using $R_f = (n_f - 1)^2/(n_f + 1)^2$, where n_f is the refractive index of the fiber material. For silica fibers $R_f = 3.6\%$ (-14.4 dB) if we use $n_f = 1.47$. This value increases to 5.3% for polished fiber ends since polishing can create a thin surface layer with a refractive index of about 1.6. In the case of multiple reflections occurring between two splices or connectors, the reflection feedback can increase considerably because the two reflecting surfaces act as mirrors of a Fabry–Perot interferometer. When the resonance condition is satisfied, the reflectivity increases to 14% for unpolished surfaces and to over 22% for polished surfaces. Clearly, a considerable fraction of the signal transmitted can be reflected back unless precautions are taken to reduce the optical feedback. A common technique for reducing reflection feedback is to use index-matching oil or gel near glass–air interfaces. Sometimes the tip of the fiber is curved or cut at an angle so that the reflected light deviates from the fiber axis. Reflection feedback can be reduced to below 0.1% by such techniques.

Semiconductor lasers are extremely sensitive to optical feedback [133]; their operating characteristics can be affected by feedback as small as -80 dB [126]. The most dramatic effect of feedback is on the laser linewidth, which can narrow or broaden by several order of magnitude, depending on the exact location of the surface where feedback originates [122]. The reason behind such a sensitivity is related to the fact that the phase of the reflected light can perturb the laser phase significantly even for relatively weak feedback levels. Such feedback-induced phase changes are detrimental mainly for coherent communication systems. The performance of direct-detection lightwave systems is affected by intensity noise rather than phase noise.

Optical feedback can increase the intensity noise significantly. Several experiments have shown a feedback-induced enhancement of the intensity noise occurring at frequencies corresponding to multiples of the external-cavity mode spacing [123]–[125]. In fact, there are several mechanisms through which the relative intensity noise (RIN) of a semiconductor laser can be enhanced by the external optical feedback. In a simple model [127], the feedback-induced enhancement of the intensity noise is attributed to the onset of multiple, closely spaced, external-cavity longitudinal modes whose spacing is determined by the distance between the laser output facet and the glass–air interface where feedback originates. The number and the amplitudes of the external-cavity modes depend on the amount of feedback. In this model, the RIN enhancement is due to intensity fluctuations of the feedback-generated side modes. Another source of RIN enhancement has its origin in the feedback-induced chaos in semiconductor lasers. Numerical simulations of the rate equations show that the RIN can be enhanced by 20 dB or more when the feedback level exceeds a certain value [134]. Even though the feedback-induced chaos is deterministic in nature, it manifests as an apparent RIN increase.

Experimental measurements of the RIN and the BER in the presence of optical feedback confirm that the feedback-induced RIN enhancement leads to a power penalty in lightwave systems [137]–[140]. Figure 5.13 shows the results of the BER measurements for a VCSEL operating at 958 nm. Such a laser operates in a single longitudinal mode because of an ultrashort cavity length (~ 1 μm) and exhibits a RIN near -130 dB/Hz in the absence of reflection feedback. However, the RIN increases by as much as 20 dB when the feedback exceeds the -30-dB level. The BER measurements

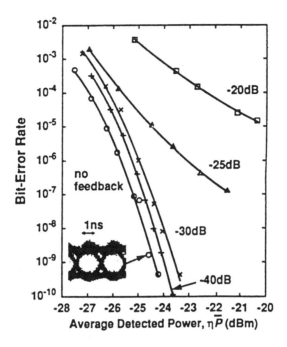

Figure 5.13: Experimentally measured BER at 500 Mb/s for a VCSEL under optical feedback. The BER is measured at several feedback levels. (After Ref. [139]; ©1993 IEEE; reprinted with permission.)

at a bit rate of 500 Mb/s show a power penalty of 0.8 dB at a BER of 10^{-9} for -30-dB feedback, and the penalty increases rapidly at higher feedback levels [139].

The power penalty can be calculated by following the analysis of Section 4.6.2 and is given by

$$\delta_{\text{ref}} = -10 \log_{10}(1 - r_{\text{eff}}^2 Q^2), \tag{5.4.15}$$

where r_{eff} is the effective intensity noise over the receiver bandwidth Δf and is obtained from

$$r_{\text{eff}}^2 = \frac{1}{2\pi} \int_{-\infty}^{\infty} \text{RIN}(\omega)\, d\omega = 2(\text{RIN})\Delta f. \tag{5.4.16}$$

In the case of feedback-induced external-cavity modes, r_{eff} can be calculated by using a simple model and is found to be [127]

$$r_{\text{eff}}^2 \approx r_I^2 + N/(\text{MSR})^2, \tag{5.4.17}$$

where r_I is the relative noise level in the absence of reflection feedback, N is the number of external-cavity modes, and MSR is the factor by which the external-cavity modes remain suppressed. Figure 5.14 shows the reflection-noise power penalty as a function of MSR for several values of N by choosing $r_I = 0.01$. The penalty is negligible in the absence of feedback ($N = 0$). However, it increases with an increase in N and a decrease in MSR. In fact, the penalty becomes infinite when MSR is reduced below a critical

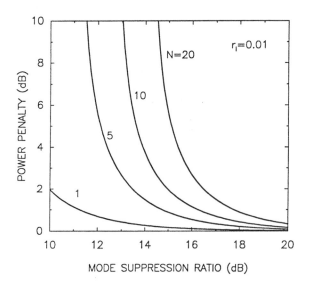

Figure 5.14: Feedback-induced power penalty as a function of MSR for several values of N and $r_f = 0.01$. Reflection feedback into the laser is assumed to generate N side modes of the same amplitude.

value. Thus, reflection feedback can degrade system performance to the extent that the system cannot achieve the desired BER despite an indefinite increase in the power received. Such reflection-induced BER floors have been observed experimentally [125] and indicate the severe impact of reflection noise on the performance of lightwave systems. An example of the reflection-induced BER floor is seen in Fig. 5.13, where the BER remains above 10^{-9} for feedback levels in excess of -25 dB. Generally speaking, most lightwave systems operate satisfactorily when the reflection feedback is below -30 dB. In practice, the problem can be nearly eliminated by using an optical isolator within the transmitter module.

Even when an isolator is used, reflection noise can be a problem for lightwave systems. In long-haul fiber links making use of optical amplifiers, fiber dispersion can convert the phase noise to intensity noise, leading to performance degradation [130]. Similarly, two reflecting surfaces anywhere along the fiber link act as a Fabry–Perot interferometer which can convert phase noise into intensity noise [128]. Such a conversion can be understood by noting that multiple reflections inside a Fabry–Perot interferometer lead to a phase-dependent term in the transmitted intensity which fluctuates in response to phase fluctuations. As a result, the RIN of the signal incident on the receiver is higher than that occurring in the absence of reflection feedback. Most of the RIN enhancement occurs over a narrow frequency band whose spectral width is governed by the laser linewidth (~ 100 MHz). Since the total noise is obtained by integrating over the receiver bandwidth, it can affect system performance considerably at bit rates larger than the laser linewidth. The power penalty can still be calculated by using Eq. (5.4.15). A simple model that includes only two reflections between the reflecting interfaces shows that r_{eff} is proportional to $(R_1 R_2)^{1/2}$, where R_1 and R_2 are the

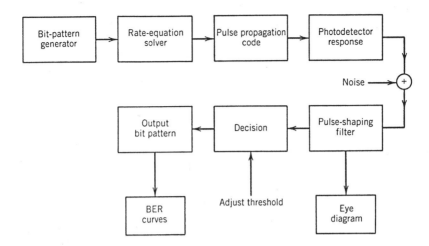

Figure 5.15: Steps involved in computer modeling of fiber-optic communication systems.

reflectivities of the two interfaces [128]. Figure 4.19 can be used to estimate the power penalty. It shows that power penalty can become infinite and lead to BER floors when $r_{\rm eff}$ exceeds 0.2. Such BER floors have been observed experimentally [128]. They can be avoided only by eliminating or reducing parasitic reflections along the entire fiber link. It is therefore necessary to employ connectors and splices that reduce reflections through the use of index matching or other techniques.

5.5 Computer-Aided Design

The design of a fiber-optic communication system involves optimization of a large number of parameters associated with transmitters, optical fibers, in-line amplifiers, and receivers. The design aspects discussed in Section 5.2 are too simple to provide the optimized values for all system parameters. The power and the rise-time budgets are only useful for obtaining a conservative estimate of the transmission distance (or repeater spacing) and the bit rate. The system margin in Eq. (5.2.4) is used as a vehicle to include various sources of power penalties discussed in Section 5.4. Such a simple approach fails for modern high-capacity systems designed to operate over long distances using optical amplifiers.

An alternative approach uses computer simulations and provides a much more realistic modeling of fiber-optic communication systems [141]–[156]. The computer-aided design techniques are capable of optimizing the whole system and can provide the optimum values of various system parameters such that the design objectives are met at a minimum cost. Figure 5.15 illustrates the various steps involved in the simulation process. The approach consists of generating an optical bit pattern at the transmitter, transmitting it through the fiber link, detecting it at the receiver, and then analyzing it through the tools such as the eye diagram and the Q factor.

Each step in the block diagram shown in Fig. 5.15 can be carried out numerically by using the material given in Chapters 2–4. The input to the optical transmitter is a pseudorandom sequence of electrical pulses which represent 1 and 0 bits. The length N of the pseudorandom bit sequence determines the computing time and should be chosen judiciously. Typically, $N = 2^M$, where M is in the range 6–10. The optical bit stream can be obtained by solving the rate equations that govern the modulation response of semiconductor lasers (see Section 3.5). The equations governing the modulation response should be used if an external modulator is used. Chirping is automatically included in both cases. Deformation of the optical bit stream during its transmission through the optical fiber is calculated by solving the NLS equation (5.3.1). The noise added by optical amplifiers should be included at the location of each amplifier.

The optical signal is converted into the electrical domain at the receiver. The shot and thermal noise is adding through a fluctuating term with Gaussian statistics. The electrical bit stream is shaped by passing it through a filter whose bandwidth is also a design parameter. An eye diagram is constructed using the filtered bit stream. The effect of varying system parameters can be studied by monitoring the eye degradation or by calculating the Q parameter given in Eq. (4.5.11). Such an approach can be used to obtain the power penalty associated with various mechanisms discussed in Section 5.4. It can also be used to investigate trade-offs that would optimize the overall system performance. An example is shown in Fig. 5.12, where the dependence of the calculated system penalty on the frequency chirp and extinction ratio is found. Numerical simulations reveal the existence of an optimum extinction ratio for which the system penalty is minimum.

Computer-aided design has another important role to play. A long-haul lightwave system may contain many repeaters, both optical and electrical. Transmitters, receivers, and amplifiers used at repeaters, although chosen to satisfy nominal specifications, are never identical. Similarly, fiber cables are constructed by splicing many different pieces (typical length 4–8 km) which have slightly different loss and dispersion characteristics. The net result is that many system parameters vary around their nominal values. For example, the dispersion parameter D, responsible not only for pulse broadening but also for other sources of power penalty, can vary significantly in different sections of the fiber link because of variations in the zero-dispersion wavelength and the transmitter wavelength. A statistical approach is often used to estimate the effect of such inherent variations on the performance of a realistic lightwave system [146]–[150]. The idea behind such an approach is that it is extremely unlikely that all system parameters would take their worst-case values at the same time. Thus, repeater spacing can be increased well above its worst-case value if the system is designed to operate reliably at the specific bit rate with a high probability (say 99.9%).

The importance of computer-aided design for fiber-optic communication systems became apparent during the 1990s when the dispersive and nonlinear effects in optical fibers became of paramount concern with increasing bit rates and transmission distances. All modern lightwave systems are designed using numerical simulations, and several software packages are available commercially. Appendix E provides details on the simulation package available on the CD-ROM included with this book (Courtesy OptiWave Corporation). The reader is encouraged to use it for a better understanding of the material covered in this book.

Problems

5.1 A distribution network uses an optical bus to distribute the signal to 10 users. Each optical tap couples 10% of the power to the user and has 1-dB insertion loss. Assuming that the station 1 transmits 1 mW of power over the optical bus, calculate the power received by the stations 8, 9, and 10.

5.2 A cable-television operator uses an optical bus to distribute the video signal to its subscribers. Each receiver needs a minimum of 100 nW to operate satisfactorily. Optical taps couple 5% of the power to each subscriber. Assuming 0.5 dB insertion loss for each tap and 1 mW transmitter power, estimate the number of subscribers that can be added to the optical bus?

5.3 A star network uses directional couplers with 0.5-dB insertion loss to distribute data to its subscribers. If each receiver requires a minimum of 100 nW and each transmitter is capable of emitting 0.5 mW, calculate the maximum number of subscribers served by the network.

5.4 Make the power budget and calculate the maximum transmission distance for a 1.3-μm lightwave system operating at 100 Mb/s and using an LED for launching 0.1 mW of average power into the fiber. Assume 1-dB/km fiber loss, 0.2-dB splice loss every 2 km, 1-dB connector loss at each end of fiber link, and 100-nW receiver sensitivity. Allow 6-dB system margin.

5.5 A 1.3-μm long-haul lightwave system is designed to operate at 1.5 Gb/s. It is capable of coupling 1 mW of average power into the fiber. The 0.5-dB/km fiber-cable loss includes splice losses. The connectors at each end have 1-dB losses. The InGaAs p–i–n receiver has a sensitivity of 250 nW. Make the power budget and estimate the repeater spacing.

5.6 Prove that the rise time T_r and the 3-dB bandwidth Δf of a RC circuit are related by $T_r \Delta f = 0.35$.

5.7 Consider a super-Gaussian optical pulse with the power distribution

$$P(t) = P_0 \exp[-(t/T_0)^{2m}],$$

where the parameter m controls the pulse shape. Derive an expression for the rise time T_r of such a pulse. Calculate the ratio T_r/T_{FWHM}, where T_{FWHM} is the full width at half maximum, and show that for a Gaussian pulse ($m = 1$) this ratio equals 0.716.

5.8 Prove that for a Gaussian optical pulse, the rise time T_r and the 3-dB optical bandwidth Δf are related by $T_r \Delta f = 0.316$.

5.9 Make the rise-time budget for a 0.85-μm, 10-km fiber link designed to operate at 50 Mb/s. The LED transmitter and the Si p–i–n receiver have rise times of 10 and 15 ns, respectively. The graded-index fiber has a core index of 1.46, $\Delta = 0.01$, and $D = 80$ ps/(km-nm). The LED spectral width is 50 nm. Can the system be designed to operate with the NRZ format?

5.10 A 1.3-μm lightwave system is designed to operate at 1.7 Gb/s with a repeater spacing of 45 km. The single-mode fiber has a dispersion slope of 0.1 ps/(km-nm^2) in the vicinity of the zero-dispersion wavelength occurring at 1.308 μm. Calculate the wavelength range of multimode semiconductor lasers for which the mode-partition-noise power penalty remains below 1 dB. Assume that the RMS spectral width of the laser is 2 nm and the mode-partition coefficient $k = 0.7$.

5.11 Generalize Eq. (5.4.5) for the case of APD receivers by including the excess-noise factor in the form $F(M) = M^x$.

5.12 Consider a 1.55-μm lightwave system operating at 1 Gb/s by using multimode semiconductor lasers of 2 nm (RMS) spectral width. Calculate the maximum transmission distance that would keep the mode-partition-noise power penalty below 2 dB. Use $k = 0.8$ for the mode-partition coefficient.

5.13 Follow the rate-equation analysis of Section 3.3.8 (see also Ref. [84]) to prove that the side-mode power P_s follows an exponential probability density function given by Eq. (5.4.8).

5.14 Use Eq. (5.4.14) to determine the maximum transmission distance for a 1.55-μm lightwave system operating at 4 Gb/s such that the chirp-induced power penalty is below 1 dB. Assume that $C = -6$ for the single-mode semiconductor laser and $\beta_2 = -20$ ps^2/km for the single-mode fiber.

5.15 Repeat Problem 5.14 for the case of 8-Gb/s bit rate.

5.16 Use the results of Problem 4.16 to obtain an expression of the reflection-induced power penalty in the case of a finite extinction ratio r_{ex}. Reproduce the penalty curves shown in Fig. 5.13 for the case $r_{ex} = 0.1$.

5.17 Consider a Fabry–Perot interferometer with two surfaces of reflectivity R_1 and R_2. Follow the analysis of Ref. [128] to derive an expression of the relative intensity noise RIN(ω) of the transmitted light as a function of the linewidth of the incident light. Assume that R_1 and R_2 are small enough that it is enough to consider only a single reflection at each surface.

5.18 Follow the analysis of Ref. [142] to obtain an expression for the total receiver noise by including thermal noise, shot noise, intensity noise, mode-partition noise, chirp noise, and reflection noise.

References

[1] P. S. Henry, R. A. Linke, and A. H. Gnauck, in *Optical Fiber Telecommunications II*, S. E. Miller and I. P. Kaminow, Eds., Academic Press, San Diego, CA, 1988, Chap. 21.

[2] S. E. Miller and I. P. Kaminow, Eds., *Optical Fiber Telecommunications II*, Academic Press, San Diego, CA, 1988, Chaps. 22–25.

[3] P. E. Green, Jr., *Fiber-Optic Networks*, Prentice Hall, Upper Saddle River, NJ, 1993.

[4] I. P. Kaminow and T. L. Koch, Eds., *Optical Fiber Telecommunications III*, Academic Press, San Diego, CA, 1997.

[5] G. E. Keiser, *Optical Fiber Communications*, 3rd ed., McGraw-Hill, New York, 2000.

[6] R. Ramaswami and K. Sivarajan, *Optical Networks: A Practical Perspective*, 2nd ed., Morgan Kaufmann Publishers, San Francisco, 2002.

[7] I. P. Kaminow and T. Li, Eds., *Optical Fiber Telecommunications IV*, Academic Press, San Diego, CA, 2002.

[8] T. Li, *Proc. IEEE* **81**, 1568 (1993); R. Giles and T. Li, *Proc. IEEE* **84**, 870 (1996).

[9] E. Desurvire, *Erbium-Doped Fiber Amplifiers*, Wiley, New York, 1994.

[10] P. C. Becker, N. A. Olsson, and J. R. Simpson, *Erbium-Doped Fiber Amplifiers: Fundamentals and Technology*, Academic Press, San Diego, CA, 1999.

[11] O. Leclerc, B. Lavigne, D. Chiaroni, and E. Desurvire, in *Optical Fiber Telecommunications IV*, Vol. A, I. P. Kaminow and T. Li, Eds., Academic Press, San Diego, CA, 2002.

[12] J. Zyskind and R. Barry, in *Optical Fiber Telecommunications IV*, Vol. B, I. P. Kaminow and T. Li, Eds., Academic Press, San Diego, CA, 2002.

[13] N. Ghani, J.-Y. Pan, and X. Cheng, in *Optical Fiber Telecommunications IV*, Vol. B, I. P. Kaminow and T. Li, Eds., Academic Press, San Diego, CA, 2002.

[14] E. G. Rawson, Ed., *Selected Papers on Fiber-Optic Local-Area Networks*, SPIE, Bellingham, WA, 1994.

[15] D. W. Smith, *Optical Network Technology*, Chapman & Hall, New York, 1995.

[16] P. E. Green, Jr., *IEEE J. Sel. Areas Commun.* **14**, 764 (1996).

[17] E. Harstead and P. H. Van Heyningen, in *Optical Fiber Telecommunications IV*, Vol. B, I. P. Kaminow and T. Li, Eds., Academic Press, San Diego, CA, 2002.

[18] F. E. Ross, *IEEE J. Sel. Areas Commun.* **7**, 1043 (1989).

[19] J. L. Gimlett, M. Z. Iqbal, J. Young, L. Curtis, R. Spicer, and N. K. Cheung, *Electron. Lett.* **25**, 596 (1989).

[20] S. Fujita, M. Kitamura, T. Torikai, N. Henmi, H. Yamada, T. Suzaki, I. Takano, and M. Shikada, *Electron. Lett.* **25**, 702 (1989).

[21] C. Kleekamp and B. Metcalf, *Designer's Guide to Fiber Optics*, Cahners, Boston, 1978.

[22] M. Eve, *Opt. Quantum Electron.* **10**, 45 (1978).

[23] P. M. Rodhe, *J. Lightwave Technol.* **3**, 145 (1985).

[24] D. A. Nolan, R. M. Hawk, and D. B. Keck, *J. Lightwave Technol.* **5**, 1727 (1987).

[25] R. D. de la Iglesia and E. T. Azpitarte, *J. Lightwave Technol.* **5**, 1768 (1987).

[26] G. P. Agrawal, *Nonlinear Fiber Optics*, 3rd ed., Academic Press, San Diego, CA, 2001.

[27] J. P. Hamaide, P. Emplit, and J. M. Gabriagues, *Electron. Lett.* **26**, 1451 (1990).

[28] A. Mecozzi, *J. Opt. Soc. Am. B* **11**, 462 (1994).

[29] A. Naka and S. Saito, *J. Lightwave Technol.* **12**, 280 (1994).

[30] E. Lichtman, *J. Lightwave Technol.* **13**, 898 (1995).

[31] F. Matera and M. Settembre, *J. Lightwave Technol.* **14**, 1 (1996); *Opt. Fiber Technol.* **4**, 34 (1998).

[32] N. Kikuchi and S. Sasaki, *Electron. Lett.* **32**, 570 (1996).

[33] D. Breuer, K. Obermann, and K. Petermann, *IEEE Photon. Technol. Lett.* **10**, 1793 (1998).

[34] F. M. Madani and K. Kikuchi, *J. Lightwave Technol.* **17**, 1326 (1999).

[35] A. Sano, Y. Miyamoto, S. Kuwahara, and H. Toba, *J. Lightwave Technol.* **18**, 1519 (2000).

[36] F. Matera, A. Mecozzi, M. Romagnoli, and M. Settembre, *Opt. Lett.* **18**, 1499 (1993).

[37] C. Lorattanasane and K. Kikuchi, *IEEE J. Quantum Electron.* **33**, 1084 (1997).

[38] N. S. Bergano, J. Aspell, C. R. Davidson, P. R. Trischitta, B. M. Nyman, and F. W. Kerfoot, *Electron. Lett.* **27**, 1889 (1991).

[39] T. Otani, K. Goto, H. Abe, M. Tanaka, H. Yamamoto, and H. Wakabayashi, *Electron. Lett.* **31**, 380 (1995).

[40] M. Murakami, T. Takahashi, M. Aoyama, M. Amemiya, M. Sumida, N. Ohkawa, Y. Fukuda, T. Imai, and M. Aiki, *Electron. Lett.* **31**, 814 (1995).

[41] T. Matsuda, A. Naka, and S. Saito, *Electron. Lett.* **32**, 229 (1996).

[42] T. N. Nielsen, A. J. Stentz, K. Rottwitt, D. S. Vengsarkar, Z. J. Chen, P. B. Hansen, J. H. Park, K. S. Feder, S. Cabot, et al., *IEEE Photon. Technol. Lett.* **12**, 1079 (2000).

[43] T. Zhu, W. S. Lee, and C. Scahill, *Electron. Lett.* **37**, 15 (2001).

[44] T. Matsuda, M. Murakami, and T. Imai, *Electron. Lett.* **37**, 237 (2001).

[45] R. J. Sanferrare, *AT&T Tech. J.* **66** (1), 95 (1987).

[46] C. Fan and L. Clark, *Opt. Photon. News* **6** (2), 26 (1995).

[47] I. Jacobs, *Opt. Photon. News* **6** (2), 19 (1995).

[48] B. Zhu, L. Leng, L. E. Nelson, Y. Qian, L. Cowsar, S. Stulz, C. Doerr, L. Stulz, S. Chandrasekhar, et al., *Electron. Lett.* **37**, 844 (2001).

[49] S. Bigo, E. Lach, Y. Frignac, D. Hamoir, P. Sillard, W. Idler, S. Gauchard, A. Bertaina, S. Borne, et al., *Electron. Lett.* **37**, 448 (2001).

[50] K. Fukuchi, T. Kasamatsu, M. Morie, R. Ohhira, T. Ito, K. Sekiya, D. Ogasahara, and T. Ono, Proc. Optical Fiber Commun. Conf., Paper PD24, Optical Society of America, Washington, DC, 2001.

[51] J. M. Sipress, Special issue, *AT&T Tech. J.* **73** (1), 4 (1995).

[52] E. K. Stafford, J. Mariano, and M. M. Sanders, *AT&T Tech. J.* **73** (1), 47 (1995).

[53] N. Bergano, in *Optical Fiber Telecommunications IV*, Vol. B, I. P. Kaminow and T. Li, Eds., Academic Press, San Diego, CA, 2002.

[54] P. B. Hansen, L. Eskildsen, S. G. Grubb, A. M. Vengsarkar, S. K. Korotky, T. A. Strasser, J. E. J. Alphonsus, J. J. Veselka, D. J. DiGiovanni, et al., *Electron. Lett.* **31**, 1460 (1995).

[55] L. Eskildsen, P. B. Hansen, S. G. Grubb, A. M. Vengsarkar, T. A. Strasser, J. E. J. Alphonsus, D. J. DiGiovanni, D. W. Peckham, D. Truxal, and W. Y. Cheung, *IEEE Photon. Technol. Lett.* **8**, 724 (1996).

[56] P. B. Hansen, L. Eskildsen, S. G. Grubb, A. M. Vengsarkar, S. K. Korotky, T. A. Strasser, J. E. J. Alphonsus, J. J. Veselka, D. J. DiGiovanni, D. W. Peckham, and D. Truxal,, *Electron. Lett.* **32**, 1018 (1996).

[57] K. I. Suzuki, N. Ohkawa, M. Murakami, and K. Aida, *Electron. Lett.* **34**, 799 (1998).

[58] N. Ohkawa, T. Takahashi, Y. Miyajima, and M. Aiki, *IEICE Trans. Commun.* **E81B**, 586 (1998).

[59] P. E. Couch and R. E. Epworth, *J. Lightwave Technol.* **1**, 591 (1983).

[60] T. Kanada, *J. Lightwave Technol.* **2**, 11 (1984).

[61] A. M. J. Koonen, *IEEE J. Sel. Areas Commun.* **4**, 1515 (1986).

[62] P. Chan and T. T. Tjhung, *J. Lightwave Technol.* **7**, 1285 (1989).

[63] P. M. Shankar, *J. Opt. Commun.* **10**, 19 (1989).

[64] G. A. Olson and R. M. Fortenberry, *Fiber Integ. Opt.* **9**, 237 (1990).

[65] J. C. Goodwin and P. J. Vella, *J. Lightwave Technol.* **9**, 954 (1991).

[66] C. M. Olsen, *J. Lightwave Technol.* **9**, 1742 (1991).

[67] K. Abe, Y. Lacroix, L. Bonnell, and Z. Jakubczyk, *J. Lightwave Technol.* **10**, 401 (1992).

[68] D. M. Kuchta and C. J. Mahon, *IEEE Photon. Technol. Lett.* **6**, 288 (1994).

[69] C. M. Olsen and D. M. Kuchta, *Fiber Integ. Opt.* **14**, 121 (1995).

[70] R. J. S. Bates, D. M. Kuchta, and K. P. Jackson, *Opt. Quantum Electron.* **27**, 203 (1995).

[71] C.-L. Ho, *J. Lightwave Technol.* **17**, 1820 (1999).

[72] G. C. Papen and G. M. Murphy, *J. Lightwave Technol.* **13**, 817 (1995).

[73] H. Kosaka, A. K. Dutta, K. Kurihara, Y. Sugimoto, and K. Kasahara, *IEEE Photon. Technol. Lett.* **7**, 926 (1995).

[74] K. Ogawa, *IEEE J. Quantum Electron.* **18**, 849 (1982).

[75] K. Ogawa, in *Semiconductors and Semimetals*, vol. 22C, W. T. Tsang, Ed., Academic Press, San Diego, CA, 1985, pp. 299–330.

[76] W. R. Throssell, *J. Lightwave Technol.* **4**, 948 (1986).

[77] J. C. Campbell, *J. Lightwave Technol.* **6**, 564 (1988).

[78] G. P. Agrawal, P. J. Anthony, and T. M. Shen, *J. Lightwave Technol.* **6**, 620 (1988).

[79] C. M. Olsen, K. E. Stubkjaer, and H. Olesen, *J. Lightwave Technol.* **7**, 657 (1989).

[80] M. Mori, Y. Ohkuma, and N. Yamaguchi, *J. Lightwave Technol.* **7**, 1125 (1989).

[81] W. Jiang, R. Feng, and P. Ye, *Opt. Quantum Electron.* **22**, 23 (1990).

[82] R. S. Fyath and J. J. O'Reilly, *IEE Proc.* **137**, Pt. J, 230 (1990).

[83] W.-H. Cheng and A.-K. Chu, *IEEE Photon. Technol. Lett.* **8**, 611 (1996).

[84] C. H. Henry, P. S. Henry, and M. Lax, *J. Lightwave Technol.* **2**, 209 (1984).

[85] R. A. Linke, B. L. Kasper, C. A. Burrus, I. P. Kaminow, J. S. Ko, and T. P. Lee, *J. Lightwave Technol.* **3**, 706 (1985).

[86] E. E. Basch, R. F. Kearns, and T. G. Brown, *J. Lightwave Technol.* **4**, 516 (1986).

[87] J. C. Cartledge, *J. Lightwave Technol.* **6**, 626 (1988).

[88] N. Henmi, Y. Koizumi, M. Yamaguchi, M. Shikada, and I. Mito, *J. Lightwave Technol.* **6**, 636 (1988).

[89] M. M. Choy, P. L. Liu, and S. Sasaki, *Appl. Phys. Lett.* **52**, 1762 (1988).

[90] P. L. Liu and M. M. Choy, *IEEE J. Quantum Electron.* **25**, 854 (1989); *IEEE J. Quantum Electron.* **25**, 1767 (1989).

[91] D. A. Fishmen, *J. Lightwave Technol.* **8**, 634 (1990).

[92] S. E. Miller, *IEEE J. Quantum Electron.* **25**, 1771 (1989); **26**, 242 (1990).

[93] A. Mecozzi, A. Sapia, P. Spano, and G. P. Agrawal, *IEEE J. Quantum Electron.* **27**, 332 (1991).

[94] K. Kiasaleh and L. S. Tamil, *Opt. Commun.* **82**, 41 (1991).

[95] T. B. Anderson and B. R. Clarke, *IEEE J. Quantum Electron.* **29**, 3 (1993).

[96] A. Valle, P. Colet, L. Pesquera, and M. San Miguel, *IEE Proc.* **140**, Pt. J, 237 (1993).

[97] H. Wu and H. Chang, *IEEE J. Quantum Electron.* **29**, 2154 (1993).

[98] A. Valle, C. R. Mirasso, and L. Pesquera, *IEEE J. Quantum Electron.* **31**, 876 (1995).

[99] D. A. Frisch and I. D. Henning, *Electron. Lett.* **20**, 631 (1984).

[100] R. A. Linke, *Electron. Lett.* **20**, 472 (1984); *IEEE J. Quantum Electron.* **21**, 593 (1985).

[101] T. L. Koch and J. E. Bowers, *Electron. Lett.* **20**, 1038 (1984).

[102] F. Koyama and Y. Suematsu, *IEEE J. Quantum Electron.* **21**, 292 (1985).

[103] A. H. Gnauck, B. L. Kasper, R. A. Linke, R. W. Dawson, T. L. Koch, T. J. Bridges, E. G. Burkhardt, R. T. Yen, D. P. Wilt, J. C. Campbell, K. C. Nelson, and L. G. Cohen, *J. Lightwave Technol.* **3**, 1032 (1985).

[104] G. P. Agrawal and M. J. Potasek, *Opt. Lett.* **11**, 318 (1986).

[105] P. J. Corvini and T. L. Koch, *J. Lightwave Technol.* **5**, 1591 (1987).

[106] J. J. O'Reilly and H. J. A. da Silva, *Electron. Lett.* **23**, 992 (1987).

[107] S. Yamamoto, M. Kuwazuru, H. Wakabayashi, and Y. Iwamoto, *J. Lightwave Technol.* **5**, 1518 (1987).

[108] D. A. Atlas, A. F. Elrefaie, M. B. Romeiser, and D. G. Daut, *Opt. Lett.* **13**, 1035 (1988).

[109] K. Hagimoto and K. Aida, *J. Lightwave Technol.* **6**, 1678 (1988).

[110] H. J. A. da Silva, R. S. Fyath, and J. J. O'Reilly, *IEE Proc.* **136**, Pt. J, 209 (1989).

[111] J. C. Cartledge and G. S. Burley, *J. Lightwave Technol.* **7**, 568 (1989).

[112] J. C. Cartledge and M. Z. Iqbal, *IEEE Photon. Technol. Lett.* **1**, 346 (1989).

[113] G. P. Agrawal and T. M. Shen, *Electron. Lett.* **22**, 1087 (1986).

[114] K. Uomi, S. Sasaki, T. Tsuchiya, H. Nakano, and N. Chinone, *IEEE Photon. Technol. Lett.* **2**, 229 (1990).

[115] S. Kakimoto, Y. Nakajima, Y. Sakakibara, H. Watanabe, A. Takemoto, and N. Yoshida, *IEEE J. Quantum Electron.* **26**, 1460 (1990).

[116] K. Uomi, T. Tsuchiya, H. Nakano, M. Suzuki, and N. Chinone, *IEEE J. Quantum Electron.* **27**, 1705 (1991).

[117] M. Blez, D. Mathoorasing, C. Kazmierski, M. Quillec, M. Gilleron, J. Landreau, and H. Nakajima, *IEEE J. Quantum Electron.* **29**, 1676 (1993).

[118] H. D. Summers and I. H. White, *Electron. Lett.* **30**, 1140 (1994).

[119] F. Kano, T. Yamanaka, N. Yamamoto, H. Mawatan, Y. Tohmori, and Y. Yoshikuni, *IEEE J. Quantum Electron.* **30**, 533 (1994).

[120] S. Mohredik, H. Burkhard, F. Steinhagen, H. Hilmer, R. Lösch, W. Scalapp, and R. Göbel, *IEEE Photon. Technol. Lett.* **7**, 1357 (1996).

[121] K. Morito, R. Sahara, K. Sato, and Y. Kotaki, *IEEE Photon. Technol. Lett.* **8**, 431 (1996).

[122] G. P. Agrawal, *IEEE J. Quantum Electron.* **20**, 468 (1984).

[123] G. P. Agrawal, N. A. Olsson, and N. K. Dutta, *Appl. Phys. Lett.* **45**, 597 (1984).

[124] T. Fujita, S. Ishizuka, K. Fujito, H. Serizawa, and H. Sato, *IEEE J. Quantum Electron.* **20**, 492 (1984).

[125] N. A. Olsson, W. T. Tsang, H. Temkin, N. K. Dutta, and R. A. Logan, *J. Lightwave Technol.* **3**, 215 (1985).

[126] R. W. Tkach and A. R. Chraplyvy, *J. Lightwave Technol.* **4**, 1655 (1986).

[127] G. P. Agrawal and T. M. Shen, *J. Lightwave Technol.* **4**, 58 (1986).

[128] J. L. Gimlett and N. K. Cheung, *J. Lightwave Technol.* **7**, 888 (1989).

[129] M. Tur and E. L. Goldstein, *J. Lightwave Technol.* **7**, 2055 (1989).

[130] S. Yamamoto, N. Edagawa, H. Taga, Y. Yoshida, and H. Wakabayashi, *J. Lightwave Technol.* **8**, 1716 (1990).

[131] G. P. Agrawal, *Proc. SPIE* **1376**, 224 (1991).

[132] K. Petermann, *Laser Diode Modulation and Noise*, 2nd ed., Kluwer Academic, Dordrecht, The Netherlands, 1991.

[133] G. P. Agrawal and N. K. Dutta, *Semiconductor Lasers*, 2nd ed., Van Nostrand Reinhold, New York, 1993.

[134] A. T. Ryan, G. P. Agrawal, G. R. Gray, and E. C. Gage, *IEEE J. Quantum Electron.* **30**, 668 (1994).

[135] K. Petermann, *IEEE J. Sel. Topics Quantum Electron.* **1**, 480 (1995).

[136] R. S. Fyath and R. S. A. Waily, *Int. J. Optoelectron.* **10**, 195 (1995).

[137] M. Shikada, S. Takano, S. Fujita, I. Mito, and K. Minemura, *J. Lightwave Technol.* **6**, 655 (1988).

[138] R. Heidemann, *J. Lightwave Technol.* **6**, 1693 (1988).

[139] K.-P. Ho, J. D. Walker, and J. M. Kahn, *IEEE Photon. Technol. Lett.* **5**, 892 (1993).

[140] U. Fiedler, E. Zeeb, G. Reiner, G. Jost, and K. J. Ebeling, *Electron. Lett.* **30**, 1898 (1994).

[141] D. G. Duff, *IEEE J. Sel. Areas Commun.* **2**, 171 (1984).

[142] T. M. Shen and G. P. Agrawal, *J. Lightwave Technol.* **5**, 653 (1987).

[143] K. S. Shanmugan, *IEEE J. Sel. Areas Commun.* **6**, 5 (1988).

[144] W. H. Tranter and C. R. Ryan, *IEEE J. Sel. Areas Commun.* **6**, 13 (1988).

[145] A. F. Elrefaie, J. K. Townsend, M. B. Romeiser, and K. S. Shanmugan, *IEEE J. Sel. Areas Commun.* **6**, 94 (1988).

[146] R. D. de la Iglesia, *J. Lightwave Technol.* **4**, 767 (1986).

[147] T. J. Batten, A. J. Gibbs, and G. Nicholson, *J. Lightwave Technol.* **7**, 209 (1989).

[148] K. Kikushima and K. Hogari, *J. Lightwave Technol.* **8**, 11 (1990).

[149] M. K. Moaveni and M. Shafi, *J. Lightwave Technol.* **8**, 1064 (1990).

[150] M. C. Jeruchim, P. Balaban, and K. S. Shamugan, *Simulation of Communication Systems*, Plenum Press, New York, 1992.

[151] K. Hinton and T. Stephens, *IEEE J. Sel. Areas Commun.* **11**, 380 (1993).

[152] K. B. Letaief, *IEEE Trans. Commun.* **43**, Pt. 1, 240 (1995).

[153] A. J. Lowery, *IEEE Spectrum* **34** (4), 26 (1997).

[154] A. J. Lowery and P. C. R. Gurney, *Appl. Opt.* **37**, 6066 (1998).

[155] A. J. Lowery, O. Lenzmann, I. Koltchanov, R. Moosburger, R. Freund, A. Richter, S. Georgi, D. Breuer, and H. Hamster, *IEEE J. Sel. Topics Quantum Electron.* **6**, 282 (2000).

[156] F. Matera and M. Settembre, *IEEE J. Sel. Topics Quantum Electron.* **6**, 308 (2000).

Chapter 6

Optical Amplifiers

As seen in Chapter 5, the transmission distance of any fiber-optic communication system is eventually limited by fiber losses. For long-haul systems, the loss limitation has traditionally been overcome using optoelectronic repeaters in which the optical signal is first converted into an electric current and then regenerated using a transmitter. Such regenerators become quite complex and expensive for wavelength-division multiplexed (WDM) lightwave systems. An alternative approach to loss management makes use of optical amplifiers, which amplify the optical signal directly without requiring its conversion to the electric domain. Several kinds of optical amplifiers were developed during the 1980s, and the use of optical amplifiers for long-haul lightwave systems became widespread during the 1990s. By 1996, optical amplifiers were a part of the fiber-optic cables laid across the Atlantic and Pacific oceans. This chapter is devoted to optical amplifiers. In Section 6.1 we discuss general concepts common to all optical amplifiers. Semiconductor optical amplifiers are considered in Section 6.2, while Section 6.3 focuses on Raman amplifiers. Section 6.4 is devoted to fiber amplifiers made by doping the fiber core with a rare-earth element. The emphasis is on the erbium-doped fiber amplifiers, used almost exclusively for 1.55-μm lightwave systems. System applications of optical amplifiers are discussed in Section 6.5.

6.1 Basic Concepts

Most optical amplifiers amplify incident light through stimulated emission, the same mechanism that is used by lasers (see Section 3.1). Indeed, an optical amplifier is nothing but a laser without feedback. Its main ingredient is the *optical gain* realized when the amplifier is pumped (optically or electrically) to achieve *population inversion*. The optical gain, in general, depends not only on the frequency (or wavelength) of the incident signal, but also on the local beam intensity at any point inside the amplifier. Details of the frequency and intensity dependence of the optical gain depend on the amplifier medium. To illustrate the general concepts, let us consider the case in which the gain medium is modeled as a homogeneously broadened two-level system. The *gain coefficient* of such a medium can be written as [1]

$$g(\omega) = \frac{g_0}{1 + (\omega - \omega_0)^2 T_2^2 + P/P_s}, \quad (6.1.1)$$

where g_0 is the peak value of the gain, ω is the optical frequency of the incident signal, ω_0 is the atomic transition frequency, and P is the optical power of the signal being amplified. The saturation power P_s depends on gain-medium parameters such as the fluorescence time T_1 and the transition cross section; its expression for different kinds of amplifiers is given in the following sections. The parameter T_2 in Eq. (6.1.1), known as the *dipole relaxation time*, is typically quite small (<1 ps). The fluorescence time T_1, also called the *population relaxation time*, varies in the range 100 ps–10 ms, depending on the gain medium. Equation (6.1.1) can be used to discuss important characteristics of optical amplifiers, such as the gain bandwidth, amplification factor, and output saturation power.

6.1.1 Gain Spectrum and Bandwidth

Consider the unsaturated regime in which $P/P_s \ll 1$ throughout the amplifier. By neglecting the term P/P_s in Eq. (6.1.1), the gain coefficient becomes

$$g(\omega) = \frac{g_0}{1 + (\omega - \omega_0)^2 T_2^2}. \quad (6.1.2)$$

This equation shows that the gain is maximum when the incident frequency ω coincides with the atomic transition frequency ω_0. The gain reduction for $\omega \neq \omega_0$ is governed by a Lorentzian profile that is a characteristic of homogeneously broadened two-level systems [1]. As discussed later, the gain spectrum of actual amplifiers can deviate considerably from the Lorentzian profile. The gain bandwidth is defined as the full width at half maximum (FWHM) of the gain spectrum $g(\omega)$. For the Lorentzian spectrum, the gain bandwidth is given by $\Delta \omega_g = 2/T_2$, or by

$$\Delta v_g = \frac{\Delta \omega_g}{2\pi} = \frac{1}{\pi T_2}. \quad (6.1.3)$$

As an example, $\Delta v_g \sim 5$ THz for semiconductor optical amplifiers for which $T_2 \sim 60$ fs. Amplifiers with a relatively large bandwidth are preferred for optical communication systems because the gain is then nearly constant over the entire bandwidth of even a multichannel signal.

The concept of *amplifier bandwidth* is commonly used in place of the gain bandwidth. The difference becomes clear when one considers the amplifier gain G, known as the *amplification factor* and defined as

$$G = P_{\text{out}}/P_{\text{in}}, \quad (6.1.4)$$

where P_{in} and P_{out} are the input and output powers of the continuous-wave (CW) signal being amplified. We can obtain an expression for G by using

$$\frac{dP}{dz} = gP, \quad (6.1.5)$$

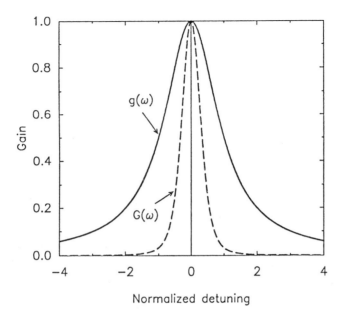

Figure 6.1: Lorentzian gain profile $g(\omega)$ and the corresponding amplifier-gain spectrum $G(\omega)$ for a two-level gain medium.

where $P(z)$ is the optical power at a distance z from the input end. A straightforward integration with the initial condition $P(0) = P_{in}$ shows that the signal power grows exponentially as

$$P(z) = P_{in} \exp(gz). \tag{6.1.6}$$

By noting that $P(L) = P_{out}$ and using Eq. (6.1.4), the amplification factor for an amplifier of length L is given by

$$G(\omega) = \exp[g(\omega)L], \tag{6.1.7}$$

where the frequency dependence of both G and g is shown explicitly. Both the amplifier gain $G(\omega)$ and the gain coefficient $g(\omega)$ are maximum when $\omega = \omega_0$ and decrease with the signal detuning $\omega - \omega_0$. However, $G(\omega)$ decreases much faster than $g(\omega)$. The amplifier bandwidth Δv_A is defined as the FWHM of $G(\omega)$ and is related to the gain bandwidth Δv_g as

$$\Delta v_A = \Delta v_g \left[\frac{\ln 2}{\ln(G_0/2)} \right]^{1/2}, \tag{6.1.8}$$

where $G_0 = \exp(g_0 L)$. Figure 6.1 shows the gain profile $g(\omega)$ and the amplification factor $G(\omega)$ by plotting g/g_0 and G/G_0 as a function of $(\omega - \omega_0)T_2$. The amplifier bandwidth is smaller than the gain bandwidth, and the difference depends on the amplifier gain itself.

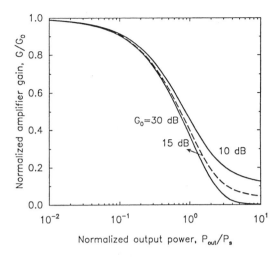

Figure 6.2: Saturated amplifier gain G as a function of the output power (normalized to the saturation power) for several values of the unsaturated amplifier gain G_0.

6.1.2   Gain Saturation

The origin of gain saturation lies in the power dependence of the $g(\omega)$ in Eq. (6.1.1). Since g is reduced when P becomes comparable to P_s, the amplification factor G decreases with an increase in the signal power. This phenomenon is called gain saturation. Consider the case in which incident signal frequency is exactly tuned to the gain peak ($\omega = \omega_0$). The detuning effects can be incorporated in a straightforward manner. By substituting g from Eq. (6.1.1) in Eq. (6.1.5), we obtain

$$\frac{dP}{dz} = \frac{g_0 P}{1 + P/P_s}. \tag{6.1.9}$$

This equation can easily be integrated over the amplifier length. By using the initial condition $P(0) = P_{\text{in}}$ together with $P(L) = P_{\text{out}} = GP_{\text{in}}$, we obtain the following implicit relation for the large-signal amplifier gain:

$$G = G_0 \exp\left(-\frac{G-1}{G}\frac{P_{\text{out}}}{P_s}\right). \tag{6.1.10}$$

Equation (6.1.10) shows that the amplification factor G decreases from its unsaturated value G_0 when P_{out} becomes comparable to P_s. Figure 6.2 shows the saturation characteristics by plotting G as a function of P_{out}/P_s for several values of G_0. A quantity of practical interest is the output saturation power P_{out}^s, defined as the output power for which the amplifier gain G is reduced by a factor of 2 (or by 3 dB) from its unsaturated value G_0. By using $G = G_0/2$ in Eq. (6.1.10),

$$P_{\text{out}}^s = \frac{G_0 \ln 2}{G_0 - 2} P_s. \tag{6.1.11}$$

Here, P_{out}^s is smaller than P_s by about 30%. Indeed, by noting that $G_0 \gg 2$ in practice ($G_0 = 1000$ for 30-dB amplifier gain), $P_{out}^s \approx (\ln 2)P_s \approx 0.69 P_s$. As seen in Fig. 6.2, P_{out}^s becomes nearly independent of G_0 for $G_0 > 20$ dB.

6.1.3 Amplifier Noise

All amplifiers degrade the signal-to-noise ratio (SNR) of the amplified signal because of spontaneous emission that adds noise to the signal during its amplification. The SNR degradation is quantified through a parameter F_n, called the *amplifier noise figure* in analogy with the electronic amplifiers (see Section 4.4.1) and defined as [2]

$$F_n = \frac{(\text{SNR})_{in}}{(\text{SNR})_{out}}, \qquad (6.1.12)$$

where SNR refers to the electric power generated when the optical signal is converted into an electric current. In general, F_n depends on several detector parameters that govern thermal noise associated with the detector (see Section 4.4.1). A simple expression for F_n can be obtained by considering an ideal detector whose performance is limited by shot noise only [2].

Consider an amplifier with the gain G such that the output and input powers are related by $P_{out} = GP_{in}$. The SNR of the input signal is given by

$$(\text{SNR})_{in} = \frac{\langle I \rangle^2}{\sigma_s^2} = \frac{(RP_{in})^2}{2q(RP_{in})\Delta f} = \frac{P_{in}}{2h\nu\Delta f}, \qquad (6.1.13)$$

where $\langle I \rangle = RP_{in}$ is the average photocurrent, $R = q/h\nu$ is the responsivity of an ideal photodetector with unit quantum efficiency (see Section 4.1), and

$$\sigma_s^2 = 2q(RP_{in})\Delta f \qquad (6.1.14)$$

is obtained from Eq. (4.4.5) for the shot noise by setting the dark current $I_d = 0$. Here Δf is the detector bandwidth. To evaluate the SNR of the amplified signal, one should add the contribution of spontaneous emission to the receiver noise.

The spectral density of spontaneous-emission-induced noise is nearly constant (white noise) and can be written as [2]

$$S_{sp}(\nu) = (G-1)n_{sp}h\nu, \qquad (6.1.15)$$

where ν is the optical frequency. The parameter n_{sp} is called the *spontaneous-emission factor* (or the population-inversion factor) and is given by

$$n_{sp} = N_2/(N_2 - N_1), \qquad (6.1.16)$$

where N_1 and N_2 are the atomic populations for the ground and excited states, respectively. The effect of spontaneous emission is to add fluctuations to the amplified signal; these are converted to current fluctuations during the photodetection process.

It turns out that the dominant contribution to the receiver noise comes from the beating of spontaneous emission with the signal [2]. The spontaneously emitted radiation mixes with the amplified signal and produces the current $I = R|\sqrt{G}E_{in} + E_{sp}|^2$ at the photodetector of responsivity R. Noting that E_{in} and E_{sp} oscillate at different frequencies with a random phase difference, it is easy to see that the beating of spontaneous emission with the signal will produce a noise current $\Delta I = 2R(GP_{in})^{1/2}|E_{sp}|\cos\theta$, where θ is a rapidly varying random phase. Averaging over the phase, and neglecting all other noise sources, the variance of the photocurrent can be written as

$$\sigma^2 \approx 4(RGP_{in})(RS_{sp})\Delta f, \qquad (6.1.17)$$

where $\cos^2\theta$ was replaced by its average value $\frac{1}{2}$. The SNR of the amplified signal is thus given by

$$(\text{SNR})_{out} = \frac{\langle I \rangle^2}{\sigma^2} = \frac{(RGP_{in})^2}{\sigma^2} \approx \frac{GP_{in}}{4S_{sp}\Delta f}. \qquad (6.1.18)$$

The amplifier noise figure can now be obtained by substituting Eqs. (6.1.13) and (6.1.18) in Eq. (6.1.12). If we also use Eq. (6.1.15) for S_{sp},

$$F_n = 2n_{sp}(G-1)/G \approx 2n_{sp}. \qquad (6.1.19)$$

This equation shows that the SNR of the amplified signal is degraded by 3 dB even for an ideal amplifier for which $n_{sp} = 1$. For most practical amplifiers, F_n exceeds 3 dB and can be as large as 6–8 dB. For its application in optical communication systems, an optical amplifier should have F_n as low as possible.

6.1.4 Amplifier Applications

Optical amplifiers can serve several purposes in the design of fiber-optic communication systems: three common applications are shown schematically in Fig. 6.3. The most important application for long-haul systems consists of using amplifiers as in-line amplifiers which replace electronic regenerators (see Section 5.1). Many optical amplifiers can be cascaded in the form of a periodic chain as long as the system performance is not limited by the cumulative effects of fiber dispersion, fiber nonlinearity, and amplifier noise. The use of optical amplifiers is particularly attractive for WDM lightwave systems as all channels can be amplified simultaneously.

   Another way to use optical amplifiers is to increase the transmitter power by placing an amplifier just after the transmitter. Such amplifiers are called power amplifiers or *power boosters*, as their main purpose is to boost the power transmitted. A power amplifier can increase the transmission distance by 100 km or more depending on the amplifier gain and fiber losses. Transmission distance can also be increased by putting an amplifier just before the receiver to boost the received power. Such amplifiers are called *optical preamplifiers* and are commonly used to improve the receiver sensitivity. Another application of optical amplifiers is to use them for compensating distribution losses in local-area networks. As discussed in Section 5.1, distribution losses often limit the number of nodes in a network. Many other applications of optical amplifiers are discussed in Chapter 8 devoted to WDM lightwave systems.

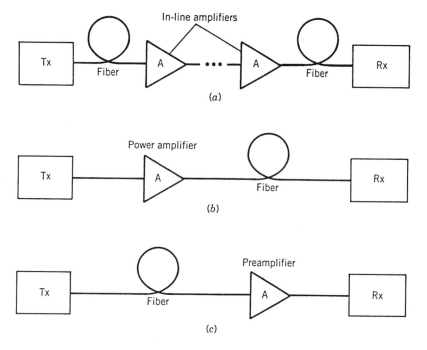

Figure 6.3: Three possible applications of optical amplifiers in lightwave systems: (a) as in-line amplifiers; (b) as a booster of transmitter power; (c) as a preamplifier to the receiver.

6.2   Semiconductor Optical Amplifiers

All lasers act as amplifiers close to but before reaching threshold, and semiconductor lasers are no exception. Indeed, research on semiconductor optical amplifiers (SOAs) started soon after the invention of semiconductor lasers in 1962. However, it was only during the 1980s that SOAs were developed for practical applications, motivated largely by their potential applications in lightwave systems [3]–[8]. In this section we discuss the amplification characteristics of SOAs and their applications.

6.2.1   Amplifier Design

The amplifier characteristics discussed in Section 6.1 were for an optical amplifier without feedback. Such amplifiers are called *traveling-wave* (TW) amplifiers to emphasize that the amplified signal travels in the forward direction only. Semiconductor lasers experience a relatively large feedback because of reflections occurring at the cleaved facets (32% reflectivity). They can be used as amplifiers when biased below threshold, but multiple reflections at the facets must be included by considering a Fabry–Perot (FP) cavity. Such amplifiers are called *FP amplifiers*. The amplification factor is obtained by using the standard theory of FP interferometers and is given by [4]

$$G_{FP}(v) = \frac{(1-R_1)(1-R_2)G(v)}{(1-G\sqrt{R_1R_2})^2 + 4G\sqrt{R_1R_2}\sin^2[\pi(v-v_m)/\Delta v_L]}, \tag{6.2.1}$$

where R_1 and R_2 are the facet reflectivities, ν_m represents the cavity-resonance frequencies [see Eq. (3.3.5)], and $\Delta \nu_L$ is the longitudinal-mode spacing, also known as the free spectral range of the FP cavity. The single-pass amplification factor G corresponds to that of a TW amplifier and is given by Eq. (6.1.7) when gain saturation is negligible. Indeed, G_{FP} reduces to G when $R_1 = R_2 = 0$.

As evident from Eq. (6.2.1), $G_{FP}(\nu)$ peaks whenever ν coincides with one of the cavity-resonance frequencies and drops sharply in between them. The amplifier bandwidth is thus determined by the sharpness of the cavity resonance. One can calculate the amplifier bandwidth from the detuning $\nu - \nu_m$ for which G_{FP} drops by 3 dB from its peak value. The result is given by

$$\Delta \nu_A = \frac{2\Delta \nu_L}{\pi} \sin^{-1} \left(\frac{1 - G\sqrt{R_1 R_2}}{(4G\sqrt{R_1 R_2})^{1/2}} \right). \qquad (6.2.2)$$

To achieve a large amplification factor, $G\sqrt{R_1 R_2}$ should be quite close to 1. As seen from Eq. (6.2.2), the amplifier bandwidth is then a small fraction of the free spectral range of the FP cavity (typically, $\Delta \nu_L \sim 100$ GHz and $\Delta \nu_A < 10$ GHz). Such a small bandwidth makes FP amplifiers unsuitable for most lightwave system applications.

TW-type SOAs can be made if the reflection feedback from the end facets is suppressed. A simple way to reduce the reflectivity is to coat the facets with an *antireflection coating*. However, it turns out that the reflectivity must be extremely small ($<0.1\%$) for the SOA to act as a TW amplifier. Furthermore, the minimum reflectivity depends on the amplifier gain itself. One can estimate the tolerable value of the facet reflectivity by considering the maximum and minimum values of G_{FP} from Eq. (6.2.1) near a cavity resonance. It is easy to verify that their ratio is given by

$$\Delta G = \frac{G_{FP}^{max}}{G_{FP}^{min}} = \left(\frac{1 + G\sqrt{R_1 R_2}}{1 - G\sqrt{R_1 R_2}} \right)^2. \qquad (6.2.3)$$

If ΔG exceeds 3 dB, the amplifier bandwidth is set by the cavity resonances rather than by the gain spectrum. To keep $\Delta G < 2$, the facet reflectivities should satisfy the condition

$$G\sqrt{R_1 R_2} < 0.17. \qquad (6.2.4)$$

It is customary to characterize the SOA as a TW amplifier when Eq. (6.2.4) is satisfied. A SOA designed to provide a 30-dB amplification factor ($G = 1000$) should have facet reflectivities such that $\sqrt{R_1 R_2} < 1.7 \times 10^{-4}$.

Considerable effort is required to produce antireflection coatings with reflectivities less than 0.1%. Even then, it is difficult to obtain low facet reflectivities in a predictable and regular manner. For this reason, alternative techniques have been developed to reduce the reflection feedback in SOAs. In one method, the active-region stripe is tilted from the facet normal, as shown in Fig. 6.4(a). Such a structure is referred to as the *angled-facet* or *tilted-stripe* structure [9]. The reflected beam at the facet is physically separated from the forward beam because of the angled facet. Some feedback can still occur, as the optical mode spreads beyond the active region in all semiconductor laser devices. In practice, the combination of an antireflection coating and the tilted stripe can produce reflectivities below 10^{-3} (as small as 10^{-4} with design optimization). In

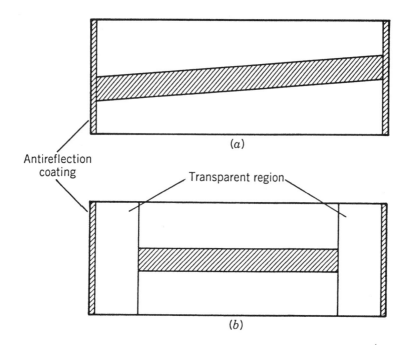

Figure 6.4: (a) Tilted-stripe and (b) buried-facet structures for nearly TW semiconductor optical amplifiers.

an alternative scheme [10] a transparent region is inserted between the active-layer ends and the facets [see Fig. 6.4(b)]. The optical beam spreads in this window region before arriving at the semiconductor–air interface. The reflected beam spreads even further on the return trip and does not couple much light into the thin active layer. Such a structure is called buried-facet or window-facet structure and has provided reflectivities as small as 10^{-4} when used in combination with antireflection coatings.

6.2.2 Amplifier Characteristics

The amplification factor of SOAs is given by Eq. (6.2.1). Its frequency dependence results mainly from the frequency dependence of $G(v)$ when condition (6.2.4) is satisfied. The measured amplifier gain exhibits ripples reflecting the effects of residual facet reflectivities. Figure 6.5 shows the wavelength dependence of the amplifier gain measured for a SOA with the facet reflectivities of about 4×10^{-4}. Condition (6.2.4) is well satisfied as $G\sqrt{R_1 R_2} \approx 0.04$ for this amplifier. Gain ripples were negligibly small as the SOA operated in a nearly TW mode. The 3-dB amplifier bandwidth is about 70 nm because of a relatively broad gain spectrum of SOAs (see Section 3.3.1).

To discuss gain saturation, consider the peak gain and assume that it increases linearly with the carrier population N as (see Section 3.3.1)

$$g(N) = (\Gamma \sigma_g / V)(N - N_0), \tag{6.2.5}$$

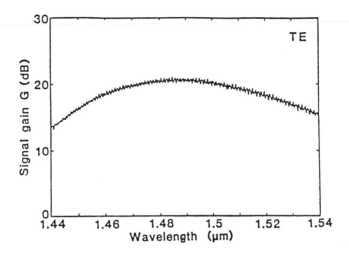

Figure 6.5: Amplifier gain versus signal wavelength for a semiconductor optical amplifier whose facets are coated to reduce reflectivity to about 0.04%. (After Ref. [3]; ©1987 IEEE; reprinted with permission.)

where Γ is the confinement factor, σ_g is the differential gain, V is the active volume, and N_0 is the value of N required at transparency. The gain has been reduced by Γ to account for spreading of the waveguide mode outside the gain region of SOAs. The carrier population N changes with the injection current I and the signal power P as indicated in Eq. (3.5.2). Expressing the photon number in terms of the optical power, this equation can be written as

$$\frac{dN}{dt} = \frac{I}{q} - \frac{N}{\tau_c} - \frac{\sigma_g(N - N_0)}{\sigma_m h\nu} P, \tag{6.2.6}$$

where τ_c is the carrier lifetime and σ_m is the cross-sectional area of the waveguide mode. In the case of a CW beam, or pulses much longer than τ_c, the steady-state value of N can be obtained by setting $dN/dt = 0$ in Eq. (6.2.6). When the solution is substituted in Eq. (6.2.5), the optical gain is found to saturate as

$$g = \frac{g_0}{1 + P/P_s}, \tag{6.2.7}$$

where the small-signal gain g_0 is given by

$$g_0 = (\Gamma\sigma_g/V)(I\tau_c/q - N_0), \tag{6.2.8}$$

and the saturation power P_s is defined as

$$P_s = h\nu\sigma_m/(\sigma_g\tau_c). \tag{6.2.9}$$

A comparison of Eqs. (6.1.1) and (6.2.7) shows that the SOA gain saturates in the same way as that for a two-level system. Thus, the output saturation power P_{out}^s is obtained from Eq. (6.1.11) with P_s given by Eq. (6.2.9). Typical values of P_{out}^s are in the range 5–10 mW.

The noise figure F_n of SOAs is larger than the minimum value of 3 dB for several reasons. The dominant contribution comes from the spontaneous-emission factor n_{sp}. For SOAs, n_{sp} is obtained from Eq. (6.1.16) by replacing N_2 and N_1 by N and N_0, respectively. An additional contribution results from internal losses (such as free-carrier absorption or scattering loss) which reduce the available gain from g to $g - \alpha_{int}$. By using Eq. (6.1.19) and including this additional contribution, the noise figure can be written as [6]

$$F_n = 2\left(\frac{N}{N - N_0}\right)\left(\frac{g}{g - \alpha_{int}}\right). \qquad (6.2.10)$$

Residual facet reflectivities increase F_n by an additional factor that can be approximated by $1 + R_1 G$, where R_1 is the reflectivity of the input facet [6]. In most TW amplifiers, $R_1 G \ll 1$, and this contribution can be neglected. Typical values of F_n for SOAs are in the range 5–7 dB.

An undesirable characteristic of SOAs is their *polarization sensitivity*. The amplifier gain G differs for the transverse electric and magnetic (TE, TM) modes by as much as 5–8 dB simply because both G and σ_g are different for the two orthogonally polarized modes. This feature makes the amplifier gain sensitive to the polarization state of the input beam, a property undesirable for lightwave systems in which the state of polarization changes with propagation along the fiber (unless polarization-maintaining fibers are used). Several schemes have been devised to reduce the polarization sensitivity [10]–[15]. In one scheme, the amplifier is designed such that the width and the thickness of the active region are comparable. A gain difference of less than 1.3 dB between TE and TM polarizations has been realized by making the active layer 0.26 μm thick and 0.4 μm wide [10]. Another scheme makes use of a large-optical-cavity structure; a gain difference of less than 1 dB has been obtained with such a structure [11].

Several other schemes reduce the polarization sensitivity by using two amplifiers or two passes through the same amplifier. Figure 6.6 shows three such configurations. In Fig. 6.6(a), the TE-polarized signal in one amplifier becomes TM polarized in the second amplifier, and vice versa. If both amplifiers have identical gain characteristics, the twin-amplifier configuration provides signal gain that is independent of the signal polarization. A drawback of the *series configuration* is that residual facet reflectivities lead to mutual coupling between the two amplifiers. In the *parallel configuration* shown in Fig. 6.6(b) the incident signal is split into a TE- and a TM-polarized signal, each of which is amplified by separate amplifiers. The amplified TE and TM signals are then combined to produce the amplified signal with the same polarization as that of the input beam [12]. The *double-pass configuration* of Fig. 6.6(c) passes the signal through the same amplifier twice, but the polarization is rotated by 90° between the two passes [13]. Since the amplified signal propagates in the backward direction, a 3-dB fiber coupler is needed to separate it from the incident signal. Despite a 6-dB loss occurring at the fiber coupler (3 dB for the input signal and 3 dB for the amplified signal) this configuration provides high gain from a single amplifier, as the same amplifier supplies gain on the two passes.

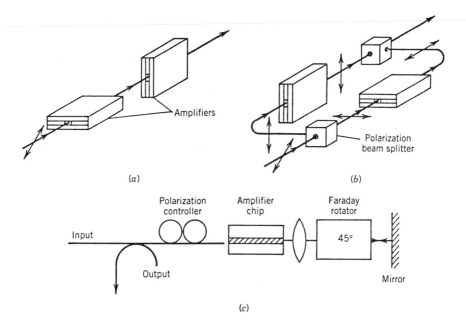

Figure 6.6: Three configurations used to reduce the polarization sensitivity of semiconductor optical amplifiers: (a) twin amplifiers in series; (b) twin amplifiers in parallel; and (c) double pass through a single amplifier.

6.2.3 Pulse Amplification

One can adapt the formulation developed in Section 2.4 for pulse propagation in optical fibers to the case of SOAs by making a few changes. The dispersive effects are not important for SOAs because of negligible material dispersion and a short amplifier length (<1 mm in most cases). The amplifier gain can be included by adding the term $gA/2$ on the right side of Eq. (2.4.7). By setting $\beta_2 = \beta_3 = 0$, the amplitude $A(z,t)$ of the pulse envelope then evolves as [18]

$$\frac{\partial A}{\partial z} + \frac{1}{v_g}\frac{\partial A}{\partial t} = \frac{1}{2}(1 - i\beta_c)gA, \tag{6.2.11}$$

where carrier-induced index changes are included through the linewidth enhancement factor β_c (see Section 3.5.2). The time dependence of g is governed by Eqs. (6.2.5) and (6.2.6). The two equations can be combined to yield

$$\frac{\partial g}{\partial t} = \frac{g_0 - g}{\tau_c} - \frac{g|A|^2}{E_{\text{sat}}}, \tag{6.2.12}$$

where the saturation energy E_{sat} is defined as

$$E_{\text{sat}} = h\nu(\sigma_m/\sigma_g), \tag{6.2.13}$$

and g_0 is given by Eq. (6.2.8). Typically $E_{\text{sat}} \sim 1$ pJ.

Equations (6.2.11) and (6.2.12) govern amplification of optical pulses in SOAs. They can be solved analytically for pulses whose duration is short compared with the carrier lifetime ($\tau_p \ll \tau_c$). The first term on the right side of Eq. (6.2.12) can then be neglected during pulse amplification. By introducing the reduced time $\tau = t - z/v_g$ together with $A = \sqrt{P}\exp(i\phi)$, Eqs. (6.2.11) and (6.2.12) can be written as [18]

$$\frac{\partial P}{\partial z} = g(z, \tau)P(z, \tau),\tag{6.2.14}$$

$$\frac{\partial \phi}{\partial z} = -\tfrac{1}{2}\beta_c g(z, \tau),\tag{6.2.15}$$

$$\frac{\partial g}{\partial \tau} = -g(z, \tau)P(z, \tau)/E_{\text{sat}}.\tag{6.2.16}$$

Equation (6.2.14) can easily be integrated over the amplifier length L to yield

$$P_{\text{out}}(\tau) = P_{\text{in}}(\tau)\exp[h(\tau)],\tag{6.2.17}$$

where $P_{\text{in}}(\tau)$ is the input power and $h(\tau)$ is the total integrated gain defined as

$$h(\tau) = \int_0^L g(z, \tau)\,dz.\tag{6.2.18}$$

If Eq. (6.2.16) is integrated over the amplifier length after replacing gP by $\partial P/\partial z$, $h(\tau)$ satisfies [18]

$$\frac{dh}{d\tau} = -\frac{1}{E_{\text{sat}}}[P_{\text{out}}(\tau) - P_{\text{in}}(\tau)] = -\frac{P_{\text{in}}(\tau)}{E_{\text{sat}}}(e^h - 1).\tag{6.2.19}$$

Equation (6.2.19) can easily be solved to obtain $h(\tau)$. The amplification factor $G(\tau)$ is related to $h(\tau)$ as $G = \exp(h)$ and is given by [1]

$$G(\tau) = \frac{G_0}{G_0 - (G_0 - 1)\exp[-E_0(\tau)/E_{\text{sat}}]},\tag{6.2.20}$$

where G_0 is the unsaturated amplifier gain and $E_0(\tau) = \int_{-\infty}^{\tau} P_{\text{in}}(\tau)\,d\tau$ is the partial energy of the input pulse defined such that $E_0(\infty)$ equals the input pulse energy E_{in}.

The solution (6.2.20) shows that the amplifier gain is different for different parts of the pulse. The leading edge experiences the full gain G_0 as the amplifier is not yet saturated. The trailing edge experiences the least gain since the whole pulse has saturated the amplifier gain. The final value of $G(\tau)$ after passage of the pulse is obtained from Eq. (6.2.20) by replacing $E_0(\tau)$ by E_{in}. The intermediate values of the gain depend on the pulse shape. Figure 6.7 shows the shape dependence of $G(\tau)$ for super-Gaussian input pulses by using

$$P_{\text{in}}(t) = P_0\exp[-(\tau/\tau_p)^{2m}],\tag{6.2.21}$$

where m is the shape parameter. The input pulse is Gaussian for $m = 1$ but becomes nearly rectangular as m increases. For comparison purposes, the input energy is held constant for different pulse shapes by choosing $E_{\text{in}}/E_{\text{sat}} = 0.1$. The shape dependence of the amplification factor $G(\tau)$ implies that the output pulse is distorted, and distortion is itself shape dependent.

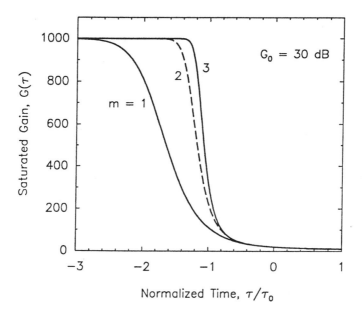

Figure 6.7: Time-dependent amplification factor for super-Gaussian input pulses of input energy such that $E_{in}/E_{sat} = 0.1$. The unsaturated value G_0 is 30 dB in all cases. The input pulse is Gaussian for $m = 1$ but becomes nearly rectangular as m increases.

As seen from Eq. (6.2.15), gain saturation leads to a time-dependent phase shift across the pulse. This phase shift is found by integrating Eq. (6.2.15) over the amplifier length and is given by

$$\phi(\tau) = -\tfrac{1}{2}\beta_c \int_0^L g(z,\tau)\,dz = -\tfrac{1}{2}\beta_c h(\tau) = -\tfrac{1}{2}\beta_c \ln[G(\tau)]. \qquad (6.2.22)$$

Since the pulse modulates its own phase through gain saturation, this phenomenon is referred to as *saturation-induced* self-phase modulation [18]. The frequency chirp is related to the phase derivative as

$$\Delta\nu_c = -\frac{1}{2\pi}\frac{d\phi}{d\tau} = \frac{\beta_c}{4\pi}\frac{dh}{d\tau} = -\frac{\beta_c P_{in}(\tau)}{4\pi E_{sat}}[G(\tau) - 1], \qquad (6.2.23)$$

where Eq. (6.2.19) was used. Figure 6.8 shows the chirp profiles for several input pulse energies when a Gaussian pulse is amplified in a SOA with 30-dB unsaturated gain. The frequency chirp is larger for more energetic pulses simply because gain saturation sets in earlier for such pulses.

Self-phase modulation and the associated frequency chirp can affect lightwave systems considerably. The spectrum of the amplified pulse becomes considerably broad and contains several peaks of different amplitudes [18]. The dominant peak is shifted toward the red side and is broader than the input spectrum. It is also accompanied by one or more satellite peaks. Figure 6.9 shows the expected shape and spectrum of amplified pulses when a Gaussian pulse of energy such that $E_{in}/E_{sat} = 0.1$ is amplified

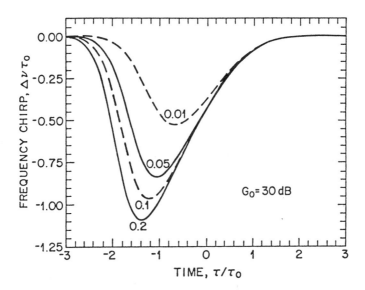

Figure 6.8: Frequency chirp imposed across the amplified pulse for several values of E_{in}/E_{sat}. A Gaussian input pulse is assumed together with $G_0 = 30$ dB and $\beta_c = 5$. (After Ref. [19]; ©1989 IEEE; reprinted with permission.)

by a SOA. The temporal and spectral changes depend on amplifier gain and are quite significant for $G_0 = 30$ dB. The experiments performed by using picosecond pulses from mode-locked semiconductor lasers confirm this behavior [18]. In particular, the spectrum of amplified pulses is found to be shifted toward the red side by 50–100 GHz, depending on the amplifier gain. Spectral distortion in combination with the frequency chirp would affect the transmission characteristics when amplified pulses are propagated through optical amplifiers.

It turns out that the frequency chirp imposed by the SOA is opposite in nature compared with that imposed by directly modulated semiconductor lasers. If we also note that the chirp is nearly linear over a considerable portion of the amplified pulse (see Fig. 6.8), it is easy to understand that the amplified pulse would pass through an initial compression stage when it propagates in the anomalous-dispersion region of optical fibers (see Section 2.4.2). Such a compression was observed in an experiment [19] in which 40-ps optical pulses were first amplified in a 1.52-μm SOA and then propagated through 18 km of single-mode fiber with $\beta_2 = -18$ ps^2/km. This compression mechanism can be used to design fiber-optic communication systems in which in-line SOAs are used to compensate simultaneously for both fiber loss and dispersion by operating SOAs in the saturation region so that they impose frequency chirp on the amplified pulse. The basic concept was demonstrated in 1989 in an experiment [20] in which a 16-Gb/s signal was transmitted over 70 km by using an SOA. In the absence of the SOA or when the SOA was operated in the unsaturated regime, the system was dispersion limited to the extent that the signal could not be transmitted over more than 20 km.

The preceding analysis considers a single pulse. In a lightwave system, the signal

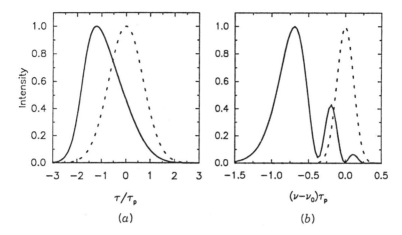

Figure 6.9: (a) Shape and (b) spectrum at the output of a semiconductor optical amplifier with $G_0 = 30$ dB and $\beta_c = 5$ for a Gaussian input pulse of energy $E_{in}/E_{sat} = 0.1$. The dashed curves show for comparison the shape and spectrum of the input pulse.

consists of a random sequence of 1 and 0 bits. If the energy of each 1 bit is large enough to saturate the gain partially, the following bit will experience less gain. The gain will recover partially if the bit 1 is preceded by one or more 0 bits. In effect, the gain of each bit in an SOA depends on the bit pattern. This phenomenon becomes quite problematic for WDM systems in which several pulse trains pass through the amplifier simultaneously. It is possible to implement a gain-control mechanism that keeps the amplifier gain pinned at a constant value. The basic idea is to make the SOA oscillate at a controlled wavelength outside the range of interest (typically below 1.52 μm). Since the gain remains clamped at the threshold value for a laser, the signal is amplified by the same factor for all pulses.

6.2.4 System Applications

The use of SOAs as a preamplifier to the receiver is attractive since it permits mono-lithic integration of the SOA with the receiver. As seen in Fig. 6.3(c), in this application the signal is optically amplified before it falls on the receiver. The preamplifier boosts the signal to such a high level that the receiver performance is limited by shot noise rather than by thermal noise. The basic idea is similar to the case of avalanche pho-todiodes (APDs), which amplify the signal in the electrical domain. However, just as APDs add additional noise (see Section 4.4.3), preamplifiers also degrade the SNR through spontaneous-emission noise. A relatively large noise figure of SOAs ($F_n = 5$–7 dB) makes them less than ideal as a preamplifier. Nonetheless, they can improve the receiver sensitivity considerably. SOAs can also be used as power amplifiers to boost the transmitter power. It is, however, difficult to achieve powers in excess of 10 mW because of a relatively small value of the output saturation power (~ 5 mW).

SOAs were used as in-line amplifiers in several system experiments before 1990. In a 1988 experiment, a signal at 1 Gb/s was transmitted over 313 km by using four cascaded SOAs [21]. SOAs have also been employed to overcome distribution losses in the local-area network (LAN) applications. In one experiment, an SOA was used as a dual-function device [22]. It amplified five channels, but at the same time the SOA was used to monitor the network performance through a baseband control channel. The 100-Mb/s baseband control signal modulated the carrier density of the amplifier, which in turn produced a corresponding electric signal that was used for monitoring.

Although SOAs can be used to amplify several channels simultaneously, they suffer from a fundamental problem related to their relatively fast response. Ideally, the signal in each channel should be amplified by the same amount. In practice, several nonlinear phenomena in SOAs induce *interchannel crosstalk*, an undesirable feature that should be minimized for practical lightwave systems. Two such nonlinear phenomena are *cross-gain saturation* and *four-wave mixing* (FWM). Both of them originate from the stimulated recombination term in Eq. (6.2.6). In the case of multichannel amplification, the power P in this equation is replaced with

$$P = \frac{1}{2} \left| \sum_{j=1}^{M} A_j \exp(-i\omega_j t) + \text{c.c.} \right|^2, \qquad (6.2.24)$$

where c.c. stands for the complex conjugate, M is the number of channels, A_j is the amplitude, and ω_j is the carrier frequency of the jth channel. Because of the coherent addition of individual channel fields, Eq. (6.2.24) contains time-dependent terms resulting from beating of the signal in different channels, i.e.,

$$P = \sum_{j=1}^{M} P_j + \sum_{j=1}^{M} \sum_{k \neq j}^{M} 2\sqrt{P_j P_k} \cos(\Omega_{jk} t + \phi_j - \phi_k), \qquad (6.2.25)$$

where $A_j = \sqrt{P_j} \exp(i\phi_j)$ was assumed together with $\Omega_{jk} = \omega_j - \omega_k$. When Eq. (6.2.25) is substituted in Eq. (6.2.6), the carrier population is also found to oscillate at the beat frequency Ω_{jk}. Since the gain and the refractive index both depend on N, they are also modulated at the frequency Ω_{jk}; such a modulation creates gain and index gratings, which induce interchannel crosstalk by scattering a part of the signal from one channel to another. This phenomenon can also be viewed as FWM [16].

The origin of cross-gain saturation is also evident from Eq. (6.2.25). The first term on the right side shows that the power P in Eq. (6.2.7) should be replaced by the total power in all channels. Thus, the gain of a specific channel is saturated not only by its own power but also by the power of neighboring channels, a phenomenon known as cross-gain saturation. It is undesirable in WDM systems since the amplifier gain changes with time depending on the bit pattern of neighboring channels. As a result, the amplified signal appears to fluctuate more or less randomly. Such fluctuations degrade the effective SNR at the receiver. The interchannel crosstalk occurs regardless of the channel spacing. It can be avoided only by reducing the channel powers to low enough values that the SOA operates in the unsaturated regime. Interchannel crosstalk induced by FWM occurs for all WDM lightwave systems irrespective of the modulation format used [23]–[26]. Its impact is most severe for coherent systems because of a relatively small channel spacing [25]. FWM can occur even for widely spaced channels through intraband nonlinearities [17] occurring at fast time scales (<1 ps).

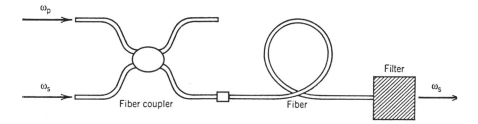

Figure 6.10: Schematic of a fiber-based Raman amplifier in the forward-pumping configuration.

It is clear that SOAs suffer from several drawbacks which make their use as in-line amplifiers impractical. A few among them are polarization sensitivity, interchannel crosstalk, and large coupling losses. The unsuitability of SOAs led to a search for alternative amplifiers during the 1980s, and two types of fiber-based amplifiers using the Raman effect and rare-earth dopants were developed. The following two sections are devoted to these two types of amplifiers. It should be stressed that SOAs have found many other applications. They can be used for wavelength conversion and can act as a fast switch for wavelength routing in WDM networks. They are also being pursued for metropolitan-area networks as a low-cost alternative to fiber amplifiers.

6.3 Raman Amplifiers

A fiber-based Raman amplifier uses *stimulated Raman scattering* (SRS) occurring in silica fibers when an intense pump beam propagates through it [27]–[29]. The main features of SRS have been discussed in Sections 2.6. SRS differs from stimulated emission in one fundamental aspect. Whereas in the case of stimulated emission an incident photon stimulates emission of another identical photon without losing its energy, in the case of SRS the incident pump photon gives up its energy to create another photon of reduced energy at a lower frequency (inelastic scattering); the remaining energy is absorbed by the medium in the form of molecular vibrations (optical phonons). Thus, Raman amplifiers must be pumped optically to provide gain. Figure 6.10 shows how a fiber can be used as a Raman amplifier. The pump and signal beams at frequencies ω_p and ω_s are injected into the fiber through a fiber coupler. The energy is transferred from the pump beam to the signal beam through SRS as the two beams copropagate inside the fiber. The pump and signal beams counterpropagate in the backward-pumping configuration commonly used in practice.

6.3.1 Raman Gain and Bandwidth

The Raman-gain spectrum of silica fibers is shown in Figure 2.18; its broadband nature is a consequence of the amorphous nature of glass. The Raman-gain coefficient g_R is related to the optical gain $g(z)$ as $g = g_R I_p(z)$, where I_p is the pump intensity. In terms of the pump power P_p, the gain can be written as

$$g(\omega) = g_R(\omega)(P_p/a_p), \qquad (6.3.1)$$

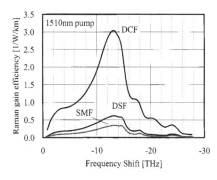 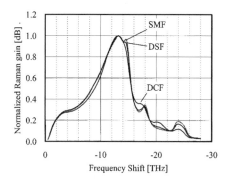

Figure 6.11: Raman-gain spectra (ratio g_R/a_p) for standard (SMF), dispersion-shifted (DSF) and dispersion-compensating (DCF) fibers. Normalized gain profiles are also shown. (After Ref. [30]; ©2001 IEEE; reprinted with permission.)

where a_p is the cross-sectional area of the pump beam inside the fiber. Since a_p can vary considerably for different types of fibers, the ratio g_R/a_p is a measure of the Raman-gain efficiency [30]. This ratio is plotted in Fig. 6.11 for three different fibers. A dispersion-compensating fiber (DCF) can be 8 times more efficient than a standard silica fiber (SMF) because of its smaller core diameter. The frequency dependence of the Raman gain is almost the same for the three kinds of fibers as evident from the normalized gain spectra shown in Fig. 6.11. The gain peaks at a Stokes shift of about 13.2 THz. The gain bandwidth Δv_g is about 6 THz if we define it as the FWHM of the dominant peak in Fig. 6.11.

The large bandwidth of fiber Raman amplifiers makes them attractive for fiber-optic communication applications. However, a relatively large pump power is required to realize a large amplification factor. For example, if we use Eq. (6.1.7) by assuming operation in the unsaturated region, $gL \approx 6.7$ is required for $G = 30$ dB. By using $g_R = 6 \times 10^{-14}$ m/W at the gain peak at 1.55 μm and $a_p = 50$ μm^2, the required pump power is more than 5 W for 1-km-long fiber. The required power can be reduced for longer fibers, but then fiber losses must be included. In the following section we discuss the theory of Raman amplifiers including both fiber losses and pump depletion.

6.3.2 Amplifier Characteristics

It is necessary to include the effects of fiber losses because of a long fiber length required for Raman amplifiers. Variations in the pump and signal powers along the amplifier length can be studied by solving the two coupled equations given in Section 2.6.1. In the case of forward pumping, these equations take the form

$$dP_s/dz = -\alpha_s P_s + (g_R/a_p)P_p P_s, \qquad (6.3.2)$$

$$dP_p/dz = -\alpha_p P_p - (\omega_p/\omega_s)(g_R/a_p)P_s P_p, \qquad (6.3.3)$$

where α_s and α_p represent fiber losses at the signal and pump frequencies ω_s and ω_p, respectively. The factor ω_p/ω_s results from different energies of pump and signal photons and disappears if these equations are written in terms of photon numbers.

Consider first the case of small-signal amplification for which pump depletion can be neglected [the last term in Eq. (6.3.3)]. Substituting $P_p(z) = P_p(0) \exp(-\alpha_p z)$ in Eq. (6.3.2), the signal power at the output of an amplifier of length L is given by

$$P_s(L) = P_s(0) \exp(g_R P_0 L_{\text{eff}}/a_p - \alpha_s L), \qquad (6.3.4)$$

where $P_0 = P_p(0)$ is the input pump power and L_{eff} is defined as

$$L_{\text{eff}} = [1 - \exp(-\alpha_p L)]/\alpha_p. \qquad (6.3.5)$$

Because of fiber losses at the pump wavelength, the effective length of the amplifier is less than the actual length L; $L_{\text{eff}} \approx 1/\alpha_p$ for $\alpha_p L \gg 1$. Since $P_s(L) = P_s(0) \exp(-\alpha_s L)$ in the absence of Raman amplification, the amplifier gain is given by

$$G_A = \frac{P_s(L)}{P_s(0) \exp(-\alpha_s L)} = \exp(g_0 L), \qquad (6.3.6)$$

where the small-signal gain g_0 is defined as

$$g_0 = g_R \left(\frac{P_0}{a_p} \right) \left(\frac{L_{\text{eff}}}{L} \right) \approx \frac{g_R P_0}{a_p \alpha_p L}. \qquad (6.3.7)$$

The last relation holds for $\alpha_p L \gg 1$. The amplification factor G_A becomes length independent for large values of $\alpha_p L$. Figure 6.12 shows variations of G_A with P_0 for several values of input signal powers for a 1.3-km-long Raman amplifier operating at 1.064 μm and pumped at 1.017 μm. The amplification factor increases exponentially with P_0 initially but then starts to deviate for $P_0 > 1$ W because of gain saturation. Deviations become larger with an increase in $P_s(0)$ as gain saturation sets in earlier along the amplifier length. The solid lines in Fig. 6.12 are obtained by solving Eqs. (6.3.2) and (6.3.3) numerically to include pump depletion.

The origin of gain saturation in Raman amplifiers is quite different from SOAs. Since the pump supplies energy for signal amplification, it begins to deplete as the signal power P_s increases. A decrease in the pump power P_p reduces the optical gain as seen from Eq. (6.3.1). This reduction in gain is referred to as gain saturation. An approximate expression for the saturated amplifier gain G_s can be obtained assuming $\alpha_s = \alpha_p$ in Eqs. (6.3.2) and (6.3.3). The result is given by [29]

$$G_s = \frac{1 + r_0}{r_0 + G_A^{-(1+r_0)}}, \qquad r_0 = \frac{\omega_p}{\omega_s} \frac{P_s(0)}{P_p(0)}. \qquad (6.3.8)$$

Figure 6.13 shows the saturation characteristics by plotting G_s/G_A as a function of $G_A r_0$ for several values of G_A. The amplifier gain is reduced by 3 dB when $G_A r_0 \approx 1$. This condition is satisfied when the power of the amplified signal becomes comparable to the input pump power P_0. In fact, P_0 is a good measure of the saturation power. Since typically $P_0 \sim 1$ W, the saturation power of fiber Raman amplifiers is much larger than that of SOAs. As typical channel powers in a WDM system are ~ 1 mW, Raman amplifiers operate in the unsaturated or linear regime, and Eq. (6.3.7) can be used in place of Eq. (6.3.8)

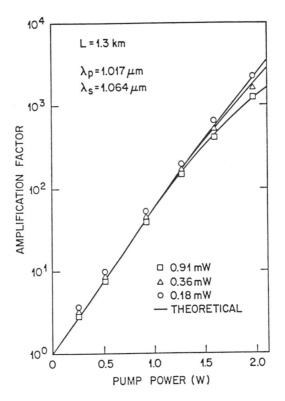

Figure 6.12: Variation of amplifier gain G_0 with pump power P_0 in a 1.3-km-long Raman amplifier for three values of the input power. Solid lines show the theoretical prediction. (After Ref. [31]; ©1981 Elsevier; reprinted with permission.)

Noise in Raman amplifiers stems from spontaneous Raman scattering. It can be included in Eq. (6.3.2) by replacing P_s in the last term with $P_s + P_{sp}$, where $P_{sp} = 2n_{sp}h\nu_s\Delta\nu_R$ is the total spontaneous Raman power over the entire Raman-gain bandwidth $\Delta\nu_R$. The factor of 2 accounts for the two polarization directions. The factor $n_{sp}(\Omega)$ equals $[1 - \exp(-\hbar\Omega_s/k_BT)]^{-1}$, where k_BT is the thermal energy at room temperature (about 25 meV). In general, the added noise is much smaller for Raman amplifiers because of the distributed nature of the amplification.

6.3.3 Amplifier Performance

As seen in Fig. 6.12, Raman amplifiers can provide 20-dB gain at a pump power of about 1 W. For the optimum performance, the frequency difference between the pump and signal beams should correspond to the peak of the Raman gain in Fig. 6.11 (occurring at about 13 THz). In the near-infrared region, the most practical pump source is a diode-pumped Nd:YAG laser operating at 1.06 μm. For such a pump laser, the maximum gain occurs for signal wavelengths near 1.12 μm. However, the wavelengths of most interest for fiber-optic communication systems are near 1.3 and 1.5 μm. A

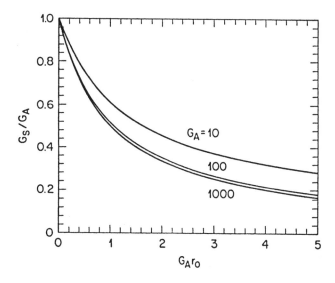

Figure 6.13: Gain–saturation characteristics of Raman amplifiers for several values of the unsaturated amplifier gain G_A.

Nd:YAG laser can still be used if a higher-order Stokes line, generated through cascaded SRS, is used as a pump. For instance, the third-order Stokes line at 1.24 μm can act as a pump for amplifying the 1.3-μm signal. Amplifier gains of up to 20 dB were measured in 1984 with this technique [32]. An early application of Raman amplifiers was as a preamplifier for improving the receiver sensitivity [33].

The broad bandwidth of Raman amplifiers is useful for amplifying several channels simultaneously. As early as 1988 [34], signals from three DFB semiconductor lasers operating in the range 1.57–1.58 μm were amplified simultaneously using a 1.47-μm pump. This experiment used a semiconductor laser as a pump source. An amplifier gain of 5 dB was realized at a pump power of only 60 mW. In another interesting experiment [35], a Raman amplifier was pumped by a 1.55-μm semiconductor laser whose output was amplified using an erbium-doped fiber amplifier. The 140-ns pump pulses had 1.4 W peak power at the 1-kHz repetition rate and were capable of amplifying 1.66-μm signal pulses by more than 23 dB through SRS in a 20-km-long dispersion-shifted fiber. The 200 mW peak power of 1.66-μm pulses was large enough for their use for optical time-domain reflection measurements commonly used for supervising and maintaining fiber-optic networks [36].

The use of Raman amplifiers in the 1.3-μm spectral region has also attracted attention [37]–[40]. However, a 1.24-μm pump laser is not readily available. Cascaded SRS can be used to generate the 1.24-μm pump light. In one approach, three pairs of fiber gratings are inserted within the fiber used for Raman amplification [37]. The Bragg wavelengths of these gratings are chosen such that they form three cavities for three Raman lasers operating at wavelengths 1.117, 1.175, and 1.24 μm that correspond to first-, second-, and third-order Stokes lines of the 1.06-μm pump. All three lasers are pumped by using a diode-pumped Nd-fiber laser through cascaded SRS. The 1.24-μm laser then pumps the Raman amplifier and amplifies a 1.3-μm signal. The same idea of cascaded SRS was used to obtain 39-dB gain at 1.3 μm by using WDM couplers in place of fiber gratings [38]. Such 1.3-μm Raman amplifiers exhibit high gains with a low noise figure (about 4 dB) and are also suitable as an optical preamplifier for high-speed optical receivers. In a 1996 experiment, such a receiver yielded the sensitivity of 151 photons/bit at a bit rate of 10 Gb/s [39]. The 1.3-μm Raman amplifiers can also be used to upgrade the capacity of existing fiber links from 2.5 to 10 Gb/s [40].

Raman amplifiers are called lumped or distributed depending on their design. In the lumped case, a discrete device is made by spooling 1–2 km of a especially prepared fiber that has been doped with Ge or phosphorus for enhancing the Raman gain. The fiber is pumped at a wavelength near 1.45 μm for amplification of 1.55-μm signals. In the case of distributed Raman amplification, the same fiber that is used for signal transmission is also used for signal amplification. The pump light is often injected in the backward direction and provides gain over relatively long lengths (>20 km). The main drawback in both cases from the system standpoint is that high-power lasers are required for pumping. Early experiments often used a tunable color-center laser as a pump; such lasers are too bulky for system applications. For this reason, Raman amplifiers were rarely used during the 1990s after erbium-doped fiber amplifiers became available. The situation changed by 2000 with the availability of compact high-power semiconductor and fiber lasers.

The phenomenon that limits the performance of distributed Raman amplifiers most turns out to be Rayleigh scattering [41]–[45]. As discussed in Section 2.5, Rayleigh scattering occurs in all fibers and is the fundamental loss mechanism for them. A small part of light is always backscattered because of this phenomenon. Normally, this Rayleigh backscattering is negligible. However, it can be amplified over long lengths in fibers with distributed gain and affects the system performance in two ways. First, a part of backward propagating noise appears in the forward direction, enhancing the overall noise. Second, double Rayleigh scattering of the signal creates a crosstalk component in the forward direction. It is this Rayleigh crosstalk, amplified by the distributed Raman gain, that becomes the major source of power penalty. The fraction of signal power propagating in the forward direction after double Rayleigh scattering is the Rayleigh crosstalk. This fraction is given by [43]

$$f_{\text{DRS}} = r_s^2 \int_0^z dz_1\, G^{-2}(z_1) \int_{z_1}^L G^2(z_2)\, dz_2, \tag{6.3.9}$$

where $r_s \sim 10^{-4}$ km^{-1} is the Rayleigh scattering coefficient and $G(z)$ is the Raman gain at a distance z in the backward-pumping configuration for an amplifier of length L. The crosstalk level can exceed 1% (-20-dB crosstalk) for $L > 80$ km and $G(L) > 10$. Since this crosstalk accumulates over multiple amplifiers, it can lead to large power penalties for undersea lightwave systems with long lengths.

Raman amplifiers can work at any wavelength as long as the pump wavelength is suitably chosen. This property, coupled with their wide bandwidth, makes Raman amplifiers quite suitable for WDM systems. An undesirable feature is that the Raman gain is somewhat polarization sensitive. In general, the gain is maximum when the signal and pump are polarized along the same direction but is reduced when they are orthogonally polarized. The polarization problem can be solved by pumping a Raman amplifier with two orthogonally polarized lasers. Another requirement for WDM systems is that the gain spectrum be relatively uniform over the entire signal bandwidth so that all channels experience the same gain. In practice, the gain spectrum is flattened by using several pumps at different wavelengths. Each pump creates the gain that mimics the spectrum shown in Fig. 6.11. The superposition of several such spectra then creates relatively flat gain over a wide spectral region. Bandwidths of more than 100 nm have been realized using multiple pump lasers [46]–[48] .

The design of broadband Raman amplifiers suitable for WDM applications requires consideration of several factors. The most important among them is the inclusion of pump–pump interactions. In general, multiple pump beams are also affected by the Raman gain, and some power from each short-wavelength pump is invariably transferred to long-wavelength pumps. An appropriate model that includes pump interactions, Rayleigh backscattering, and spontaneous Raman scattering considers each frequency component separately and solves the following set of coupled equations [48]:

$$\frac{dP_f(v)}{dz} = \int_{\mu>v} g_R(\mu - v)a_\mu^{-1}[P_f(\mu) + P_b(\mu)][P_f(v) + 2hvn_{sp}(\mu - v)]d\mu$$

$$- \int_{\mu<v} g_R(v - \mu)a_v^{-1}[P_f(\mu) + P_b(\mu)][P_f(v) + 2hvn_{sp}(v - \mu)]d\mu,$$

$$- \alpha(v)P_f(v) + r_sP_b(v) \tag{6.3.10}$$

where μ and v denote optical frequencies, $n_{sp}(\Omega) = [1 - \exp(-\hbar\Omega/k_BT)]^{-1}$, and the subscripts f and b denote forward- and backward-propagating waves, respectively. In this equation, the first and second terms account for the Raman-induced power transfer into and out of each frequency band. Fiber losses and Rayleigh backscattering are included through the third and fourth terms, respectively. The noise induced by spontaneous Raman scattering is included by the temperature-dependent factor in the two integrals. A similar equation can be written for the backward-propagating waves.

To design broadband Raman amplifiers, the entire set of such equations is solved numerically to find the channel gains, and input pump powers are adjusted until the gain is nearly the same for all channels. Figure 6.14 shows an example of the gain spectrum measured for a Raman amplifier made by pumping a 25-km-long dispersion-shifted fiber with 12 diode lasers. The frequencies and power levels of the pump lasers, required to achieve a nearly flat gain profile, are also shown. Notice that all power levels are under 100 mW. The amplifier provides about 10.5 dB gain over an 80-nm bandwidth with a ripple of less than 0.1 dB. Such an amplifier is suitable for dense WDM systems covering both the C and L bands. Several experiments have used broadband Raman amplifiers to demonstrate transmission over long distances at high bit rates. In one 3-Tb/s experiment, 77 channels, each operating at 42.7 Gb/s, were transmitted over 1200 km by using the C and L bands simultaneously [49].

Several other nonlinear processes can provide gain inside silica fibers. An example is provided by the parametric gain resulting from FWM [29]. The resulting fiber amplifier is called a *parametric amplifier* and can have a gain bandwidth larger than 100 nm. Parametric amplifiers require a large pump power (typically >1 W) that may be reduced using fibers with high nonlinearities. They also generate a phase-conjugated

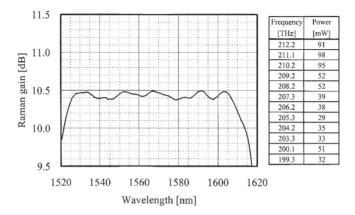

Figure 6.14: Measured gain profile of a Raman amplifier with nearly flat gain over an 80-nm bandwidth. Pump frequencies and powers used are shown on the right. (After Ref. [30]; ©2001 IEEE; reprinted with permission.)

signal that can be useful for dispersion compensation (see Section 7.7). Fiber amplifiers can also be made using stimulated Brillouin scattering (SBS) in place of SRS [29]. The operating mechanism behind Brillouin amplifiers is essentially the same as that for fiber Raman amplifiers in the sense that both amplifiers are pumped backward and provide gain through a scattering process. Despite this formal similarity, Brillouin amplifiers are rarely used in practice because their gain bandwidth is typically below 100 MHz. Moreover, as the Stokes shift for SBS is ~ 10 GHz, pump and signal wavelengths nearly coincide. These features render Brillouin amplifiers unsuitable for WDM lightwave systems although they can be exploited for other applications.

6.4 Erbium-Doped Fiber Amplifiers

An important class of fiber amplifiers makes use of *rare-earth elements* as a gain medium by doping the fiber core during the manufacturing process (see Section 2.7). Although doped-fiber amplifiers were studied as early as 1964 [50], their use became practical only 25 years later, after the fabrication and characterization techniques were perfected [51]. Amplifier properties such as the operating wavelength and the gain bandwidth are determined by the dopants rather than by the silica fiber, which plays the role of a host medium. Many different rare-earth elements, such as erbium, holmium, neodymium, samarium, thulium, and ytterbium, can be used to realize fiber amplifiers operating at different wavelengths in the range 0.5–3.5 μm. Erbium-doped fiber amplifiers (EDFAs) have attracted the most attention because they operate in the wavelength region near 1.55 μm [52]–[56]. Their deployment in WDM systems after 1995 revolutionized the field of fiber-optic communications and led to lightwave systems with capacities exceeding 1 Tb/s. This section focuses on the main characteristics of EDFAs.

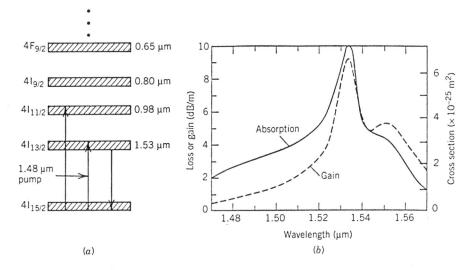

Figure 6.15: (a) Energy-level diagram of erbium ions in silica fibers; (b) absorption and gain spectra of an EDFA whose core was codoped with germania. (After Ref. [64]; ©1991 IEEE; reprinted with permission.)

6.4.1 Pumping Requirements

The design of an EDFA looks similar to that shown in Fig. 6.10 with the main difference that the fiber core contains erbium ions (Er^{3+}). Pumping at a suitable wavelength provides gain through population inversion. The gain spectrum depends on the pumping scheme as well as on the presence of other dopants, such as germania and alumina, within the fiber core. The amorphous nature of silica broadens the energy levels of Er^{3+} into bands. Figure 6.15(a) shows a few energy levels of Er^{3+} in silica glasses. Many transitions can be used to pump an EDFA. Early experiments used the visible radiation emitted from argon-ion, Nd:YAG, or dye lasers even though such pumping schemes are relatively inefficient. From a practical standpoint the use of semiconductor lasers is preferred.

Efficient EDFA pumping is possible using semiconductor lasers operating near 0.98- and 1.48-μm wavelengths. Indeed, the development of such pump lasers was fueled with the advent of EDFAs. It is possible to realize 30-dB gain with only 10–15 mW of absorbed pump power. Efficiencies as high as 11 dB/mW were achieved by 1990 with 0.98-μm pumping [57]. The pumping transition $^4I_{15/2} \rightarrow {}^4I_{9/2}$ can use high-power GaAs lasers, and the *pumping efficiency* of about 1 dB/mW has been obtained at 820 nm [58]. The required pump power can be reduced by using silica fibers doped with aluminum and phosphorus or by using *fluorophosphate fibers* [59]. With the availability of visible semiconductor lasers, EDFAs can also be pumped in the wavelength range 0.6–0.7 μm. In one experiment [60], 33-dB gain was realized at 27 mW of pump power obtained from an AlGaInP laser operating at 670 nm. The pumping efficiency was as high as 3 dB/mW at low pump powers. Most EDFAs use 980-nm pump lasers as such lasers are commercially available and can provide more than 100 mW of pump power. Pumping at 1480 nm requires longer fibers and higher powers because it uses the tail of the absorption band shown in Fig. 6.15(b).

EDFAs can be designed to operate in such a way that the pump and signal beams propagate in opposite directions, a configuration referred to as backward pumping to distinguish it from the forward-pumping configuration shown in Fig. 6.10. The performance is nearly the same in the two pumping configurations when the signal power is small enough for the amplifier to remain unsaturated. In the saturation regime, the power-conversion efficiency is generally better in the backward-pumping configuration [61], mainly because of the important role played by the amplified spontaneous emission (ASE). In the bidirectional pumping configuration, the amplifier is pumped in both directions simultaneously by using two semiconductor lasers located at the two fiber ends. This configuration requires two pump lasers but has the advantage that the population inversion, and hence the small-signal gain, is relatively uniform along the entire amplifier length.

6.4.2   Gain Spectrum

The gain spectrum shown in Fig. 6.15 is the most important feature of an EDFA as it determines the amplification of individual channels when a WDM signal is amplified. The shape of the gain spectrum is affected considerably by the amorphous nature of silica and by the presence of other codopants within the fiber core such as germania and alumina [62]–[64]. The gain spectrum of erbium ions alone is homogeneously broadened; its bandwidth is determined by the dipole relaxation time T_2 in accordance with Eq. (6.1.2). However, the spectrum is considerably broadened in the presence of randomly located silica molecules. Structural disorders lead to inhomogeneous broadening of the gain spectrum, whereas *Stark splitting* of various energy levels is responsible for homogeneous broadening. Mathematically, the gain $g(\omega)$ in Eq. (6.1.2) should be averaged over the distribution of atomic transition frequencies ω_0 such that the effective gain is given by

$$g_{\text{eff}}(\omega) = \int_{-\infty}^{\infty} g(\omega, \omega_0) f(\omega_0) \, d\omega_0, \qquad (6.4.1)$$

where $f(\omega_0)$ is the distribution function whose form also depends on the presence of other dopants within the fiber core.

Figure 6.15(b) shows the gain and absorption spectra of an EDFA whose core was doped with germania [64]. The gain spectrum is quite broad and has a double-peak structure. The addition of alumina to the fiber core broadens the gain spectrum even more. Attempts have been made to isolate the contributions of homogeneous and inhomogeneous broadening through measurements of *spectral hole burning*. For germania-doped EDFAs the contributions of homogeneous and inhomogeneous broadening are relatively small [63]. In contrast, the gain spectrum of aluminosilicate glasses has roughly equal contributions from homogeneous and inhomogeneous broadening mechanisms. The gain bandwidth of such EDFAs typically exceeds 35 nm.

The gain spectrum of EDFAs can vary from amplifier to amplifier even when core composition is the same because it also depends on the amplifier length. The reason is that the gain depends on both the absorption and emission cross sections having different spectral characteristics. The local inversion or local gain varies along the fiber

length because of pump power variations. The total gain is obtained by integrating over
the amplifier length. This feature can be used to realize EDFAs that provide amplification in the L band covering the spectral region 1570–1610 nm. The wavelength range
over which an EDFA can provide nearly constant gain is of primary interest for WDM
systems. This issue is discussed later in this section.

6.4.3 Simple Theory

The gain of an EDFA depends on a large number of device parameters such as erbium-
ion concentration, amplifier length, core radius, and pump power [64]–[68]. A three-
level rate-equation model commonly used for lasers [1] can be adapted for EDFAs. It
is sometimes necessary to add a fourth level to include the *excited-state absorption*. In
general, the resulting equations must be solved numerically.

Considerable insight can be gained by using a simple two-level model that is valid
when ASE and excited-state absorption are negligible. The model assumes that the
top level of the three-level system remains nearly empty because of a rapid transfer
of the pumped population to the excited state. It is, however, important to take into
account the different emission and absorption cross sections for the pump and signal
fields. The population densities of the two states, N_1 and N_2, satisfy the following two
rate equations [55]:

$$\frac{\partial N_2}{\partial t} = (\sigma_p^a N_1 - \sigma_p^e N_2)\phi_p + (\sigma_s^a N_1 - \sigma_s^e N_2)\phi_s - \frac{N_2}{T_1}, \tag{6.4.2}$$

$$\frac{\partial N_1}{\partial t} = (\sigma_p^e N_2 - \sigma_p^a N_1)\phi_p + (\sigma_s^e N_2 - \sigma_s^a N_1)\phi_s + \frac{N_2}{T_1}, \tag{6.4.3}$$

where σ_j^a and σ_j^e are the absorption and emission cross sections at the frequency ω_j
with $j = p, s$. Further, T_1 is the spontaneous lifetime of the excited state (about 10 ms
for EDFAs). The quantities ϕ_p and ϕ_s represent the photon flux for the pump and
signal waves, defined such that $\phi_j = P_j/(a_j h\nu_j)$, where P_j is the optical power, σ_j is
the transition cross section at the frequency ν_j, and a_j is the cross-sectional area of the
fiber mode for $j = p, s$.

The pump and signal powers vary along the amplifier length because of absorption,
stimulated emission, and spontaneous emission. If the contribution of spontaneous
emission is neglected, P_s and P_p satisfy the simple equations

$$\frac{\partial P_s}{\partial z} = \Gamma_s(\sigma_s^e N_2 - \sigma_s^a N_1)P_s - \alpha P_s, \tag{6.4.4}$$

$$s\frac{\partial P_p}{\partial z} = \Gamma_p(\sigma_p^e N_2 - \sigma_p^a N_1)P_p - \alpha' P_p, \tag{6.4.5}$$

where α and α' take into account fiber losses at the signal and pump wavelengths,
respectively. These losses can be neglected for typical amplifier lengths of 10–20 m.
However, they must be included in the case of distributed amplification discussed later.
The confinement factors Γ_s and Γ_p account for the fact that the doped region within the
core provides the gain for the entire fiber mode. The parameter $s = \pm 1$ in Eq. (6.4.5)

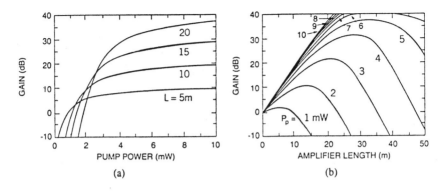

Figure 6.16: Small-signal gain as a function of (a) pump power and (b) amplifier length for an EDFA assumed to be pumped at 1.48 μm. (After Ref. [64]; ©1991 IEEE; reprinted with permission.)

depending on the direction of pump propagation; $s = -1$ in the case of a backward-propagating pump.

Equations (6.4.2)–(6.4.5) can be solved analytically, in spite of their complexity, after some justifiable approximations [65]. For lumped amplifiers, the fiber length is short enough that both α and α' can be set to zero. Noting that $N_1 + N_2 = N_t$ where N_t is the total ion density, only one equation, say Eq. (6.4.2) for N_2, need be solved. Noting again that the absorption and stimulated-emission terms in the field and population equations are related, the steady-state solution of Eq. (6.4.2), obtained by setting the time derivative to zero, can be written as

$$N_2(z) = -\frac{T_1}{a_d h v_s}\frac{\partial P_s}{\partial z} - \frac{sT_1}{a_d h v_p}\frac{\partial P_p}{\partial z}, \qquad (6.4.6)$$

where $a_d = \Gamma_s a_s = \Gamma_p a_p$ is the cross-sectional area of the doped portion of the fiber core. Substituting this solution into Eqs. (6.4.4) and (6.4.5) and integrating them over the fiber length, the powers P_s and P_p at the fiber output can be obtained in an analytical form. This model has been extended to include the ASE propagation in both the forward and backward directions [68].

The total amplifier gain G for an EDFA of length L is obtained using

$$G = \Gamma_s \exp\left[\int_0^L (\sigma_s^e N_2 - \sigma_s^a N_1)\, dz\right], \qquad (6.4.7)$$

where $N_1 = N_t - N_2$ and N_2 is given by Eq. (6.4.6). Figure 6.16 shows the small-signal gain at 1.55 μm as a function of the pump power and the amplifier length by using typical parameter values. For a given amplifier length L, the amplifier gain initially increases exponentially with the pump power, but the increase becomes much smaller when the pump power exceeds a certain value [corresponding to the "knee" in Fig. 6.16(a)]. For a given pump power, the amplifier gain becomes maximum at an optimum value of L and drops sharply when L exceeds this optimum value. The reason is that the latter portion of the amplifier remains unpumped and absorbs the amplified signal.

Since the optimum value of L depends on the pump power P_p, it is necessary to choose both L and P_p appropriately. Figure 6.16(b) shows that a 35-dB gain can be realized at a pump power of 5 mW for $L = 30$ m and 1.48-μm pumping. It is possible to design amplifiers such that high gain is obtained for amplifier lengths as short as a few meters. The qualitative features shown in Fig. 6.16 are observed in all EDFAs; the agreement between theory and experiment is generally quite good [67]. The saturation characteristics of EDFAs are similar to those shown in Figs. 6.13 for Raman amplifiers. In general, the output saturation power is smaller than the output pump power expected in the absence of signal. It can vary over a wide range depending on the EDFA design, with typical values ~10 mW. For this reason the output power levels of EDFAs are generally limited to below 100 mW, although powers as high as 250 mW have been obtained with a proper design [69].

The foregoing analysis assumes that both pump and signal waves are in the form of CW beams. In practice, EDFAs are pumped by using CW semiconductor lasers, but the signal is in the form of a pulse train (containing a random sequence of 1 and 0 bits), and the duration of individual pulses is inversely related to the bit rate. The question is whether all pulses experience the same gain or not. As discussed in Section 6.2, the gain of each pulse depends on the preceding bit pattern for SOAs because an SOA can respond on time scales of 100 ps or so. Fortunately, the gain remains constant with time in an EDFA for even microsecond-long pulses. The reason is related to a relatively large value of the fluorescence time associated with the excited erbium ions ($T_1 \sim 10$ ms). When the time scale of signal-power variations is much shorter than T_1, erbium ions are unable to follow such fast variations. As single-pulse energies are typically much below the saturation energy (~10 μJ), EDFAs respond to the average power. As a result, gain saturation is governed by the average signal power, and amplifier gain does not vary from pulse to pulse even for a WDM signal.

In some applications such as packet-switched networks, signal power may vary on a time scale comparable to T_1. Amplifier gain in that case is likely to become time dependent, an undesirable feature from the standpoint of system performance. A gain-control mechanism that keeps the amplifier gain pinned at a constant value consists of making the EDFA oscillate at a controlled wavelength outside the range of interest (typically below 1.5 μm). Since the gain remains clamped at the threshold value for a laser, the signal is amplified by the same factor despite variations in the signal power. In one implementation of this scheme, an EDFA was forced to oscillate at 1.48 μm by fabricating two fiber Bragg gratings acting as high-reflectivity mirrors at the two ends of the amplifier [70].

6.4.4 Amplifier Noise

Amplifier noise is the ultimate limiting factor for system applications [71]–[74]. For a lumped EDFA, the impact of ASE is quantified through the noise figure F_n given by $F_n = 2n_{sp}$. The spontaneous emission factor n_{sp} depends on the relative populations N_1 and N_2 of the ground and excited states as $n_{sp} = N_2/(N_2 - N_1)$. Since EDFAs operate on the basis of a three-level pumping scheme, $N_1 \neq 0$ and $n_{sp} > 1$. Thus, the noise figure of EDFAs is expected to be larger than the ideal value of 3 dB.

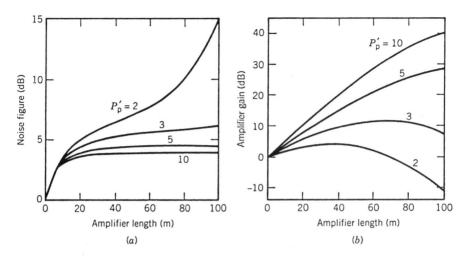

Figure 6.17: (a) noise figure and (b) amplifier gain as a function of the length for several pumping levels. (After Ref. [74]; ©1990 IEE; reprinted with permission.)

The spontaneous-emission factor can be calculated for an EDFA by using the rate-equation model discussed earlier. However, one should take into account the fact that both N_1 and N_2 vary along the fiber length because of their dependence on the pump and signal powers; hence $n_{\rm sp}$ should be averaged along the amplifier length. As a result, the noise figure depends both on the amplifier length L and the pump power P_p, just as the amplifier gain does. Figure 6.17(a) shows the variation of F_n with the amplifier length for several values of $P_p/P_p^{\rm sat}$ when a 1.53-μm signal is amplified with an input power of 1 mW. The amplifier gain under the same conditions is also shown in Fig. 6.17(b). The results show that a noise figure close to 3 dB can be obtained for a high-gain amplifier pumped such that $P_p \gg P_p^{\rm sat}$ [71].

The experimental results confirm that F_n close to 3 dB is possible in EDFAs. A noise figure of 3.2 dB was measured in a 30-m-long EDFA pumped at 0.98 μm with 11 mW of power [72]. A similar value was found for another EDFA pumped with only 5.8 mW of pump power at 0.98 μm [73]. In general, it is difficult to achieve high gain, low noise, and high pumping efficiency simultaneously. The main limitation is imposed by the ASE traveling backward toward the pump and depleting the pump power. Incorporation of an internal isolator alleviates this problem to a large extent. In one implementation, 51-dB gain was realized with a 3.1-dB noise figure at a pump power of only 48 mW [75].

The measured values of F_n are generally larger for EDFAs pumped at 1.48 μm. A noise figure of 4.1 dB was obtained for a 60-m-long EDFA when pumped at 1.48 μm with 24 mW of pump power [72]. The reason for a larger noise figure for 1.48-μm pumped EDFAs can be understood from Fig. 6.17(a), which shows that the pump level and the excited level lie within the same band for 1.48-μm pumping. It is difficult to achieve complete population inversion ($N_1 \approx 0$) under such conditions. It is nonetheless possible to realize $F_n < 3.5$ dB for pumping wavelengths near 1.46 μm.

Relatively low noise levels of EDFAs make them an ideal choice for WDM lightwave systems. In spite of low noise, the performance of long-haul fiber-optic communication systems employing multiple EDFAs is often limited by the amplifier noise. The noise problem is particularly severe when the system operates in the anomalous-dispersion region of the fiber because a nonlinear phenomenon known as the modulation instability [29] enhances the amplifier noise [76] and degrades the signal spectrum [77]. Amplifier noise also introduces timing jitter. These issue are discussed later in this chapter.

6.4.5 Multichannel Amplification

The bandwidth of EDFAs is large enough that they have proven to be the optical amplifier of choice for WDM applications. The gain provided by them is nearly polarization insensitive. Moreover, the interchannel crosstalk that cripples SOAs because of the carrier-density modulation occurring at the channel spacing does not occur in EDFAs. The reason is related to the relatively large value of T_1 (about 10 ms) compared with the carrier lifetime in SOAs (<1 ns). The sluggish response of EDFAs ensures that the gain cannot be modulated at frequencies much larger than 10 kHz.

A second source of interchannel crosstalk is cross-gain saturation occurring because the gain of a specific channel is saturated not only by its own power (self-saturation) but also by the power of neighboring channels. This mechanism of crosstalk is common to all optical amplifiers including EDFAs [78]–[80]. It can be avoided by operating the amplifier in the unsaturated regime. Experimental results support this conclusion. In a 1989 experiment [78], negligible power penalty was observed when an EDFA was used to amplify two channels operating at 2 Gb/s and separated by 2 nm as long as the channel powers were low enough to avoid the gain saturation.

The main practical limitation of an EDFA stems from the spectral nonuniformity of the amplifier gain. Even though the gain spectrum of an EDFA is relatively broad, as seen in Fig. 6.15, the gain is far from uniform (or flat) over a wide wavelength range. As a result, different channels of a WDM signal are amplified by different amounts. This problem becomes quite severe in long-haul systems employing a cascaded chain of EDFAs. The reason is that small variations in the amplifier gain for individual channels grow exponentially over a chain of in-line amplifiers if the gain spectrum is the same for all amplifiers. Even a 0.2-dB gain difference grows to 20 dB over a chain of 100 in-line amplifiers, making channel powers vary by a factor of 100, an unacceptable variation range in practice. To amplify all channels by nearly the same amount, the double-peak nature of the EDFA gain spectrum forces one to pack all channels near one of the gain peaks. In a simple approach, input powers of different channels were adjusted to reduce power variations at the receiver to an acceptable level [81]. This technique may work for a small number of channels but becomes unsuitable for dense WDM systems.

The entire bandwidth of 35–40 nm can be used if the gain spectrum is flattened by introducing wavelength-selective losses through an optical filter. The basic idea behind gain flattening is quite simple. If an optical filter whose transmission losses mimic the gain profile (high in the high-gain region and low in the low-gain region) is inserted after the doped fiber, the output power will become constant for all channels.

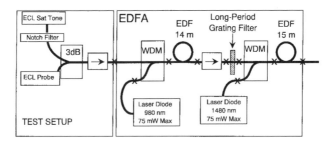

Figure 6.18: Schematic of an EDFA designed to provide uniform gain over the 1530–1570-nm bandwidth using an optical filter containing several long-period fiber gratings. The two-stage design helps to reduce the noise level. (After Ref. [85]; ©1997 IEEE; reprinted with permission.)

Although fabrication of such a filter is not simple, several gain-flattening techniques have been developed [55]. For example, thin-film interference filters, Mach–Zehnder filters, acousto-optic filters, and long-period fiber gratings have been used for flattening the gain profile and equalizing channel gains [82]–[84].

The gain-flattening techniques can be divided into active and passive categories. Most filter-based methods are passive in the sense that channel gains cannot be adjusted in a dynamic fashion. The location of the optical filter itself requires some thought because of high losses associated with it. Placing it before the amplifier increases the noise while placing it after the amplifier reduces the output power. Often a two-stage configuration shown in Fig. 6.18 is used. The second stage acts as a power amplifier while the noise figure is mostly determined by the first stage whose noise is relatively low because of its low gain. A combination of several long-period fiber gratings acting as the optical filter in the middle of two stages resulted by 1977 in an EDFA whose gain was flat to within 1 dB over the 40-nm bandwidth in the wavelength range of 1530–1570 nm [85].

Ideally, an optical amplifier should provide the same gain for all channels under all possible operating conditions. This is not the case in general. For instance, if the number of channels being transmitted changes, the gain of each channel will change since it depends on the total signal power because of gain saturation. The active control of channel gains is thus desirable for WDM applications. Many techniques have been developed for this purpose. The most commonly used technique stabilizes the gain dynamically by incorporating within the amplifier a laser that operates outside the used bandwidth. Such devices are called gain-clamped EDFAs (as their gain is clamped by a built-in laser) and have been studied extensively [86]–[91].

WDM lightwave systems capable of transmitting more than 80 channels appeared by 1998. Such systems use the C and L bands simultaneously and need uniform amplifier gain over a bandwidth exceeding 60 nm. Moreover, the use of the L band requires optical amplifiers capable of providing gain in the wavelength range 1570–1610 nm. It turns out that EDFAs can provide gain over this wavelength range, with a suitable design. An L-band EDFA requires long fiber lengths ($>$100 m) to keep the inversion level relatively low. Figure 6.19 shows an L-band amplifier with a two-stage

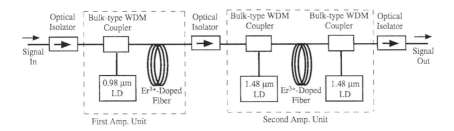

Figure 6.19: Schematic of an L-band EDFA providing uniform gain over the 1570–1610-nm bandwidth with a two-stage design. (After Ref. [92]; ©1999 IEEE; reprinted with permission.)

design [92]. The first stage is pumped at 980 nm and acts as a traditional EDFA (fiber length 20–30 m) capable of providing gain in the range 1530–1570 nm. In contrast, the second stage has 200-m-long doped fiber and is pumped bidirectionally using 1480-nm lasers. An optical isolator between the two stages passes the ASE from the first stage to the second stage (necessary for pumping the second stage) but blocks the backward-propagating ASE from entering the first stage. Such cascaded, two-stage amplifiers can provide flat gain over a wide bandwidth while maintaining a relatively low noise level. As early as 1996, flat gain to within 0.5 dB was realized over the wavelength range of 1544–1561 nm [93]. The second EDFA was codoped with ytterbium and phosphorus and was optimized such that it acted as a power amplifier. Since then, EDFAs providing flat gain over the entire C and L bands have been made [55]. Raman amplification can also be used for the L band. Combining Raman amplification with one or two EDFAs, uniform gain can be realized over a 75-nm bandwidth covering the C and L bands [94].

A parallel configuration has also been developed for EDFAs capable of amplifying over the C and L bands simultaneously [95]. In this approach, the incoming WDM signal is split into two branches, which amplify the C-band and L-band signals separately using an optimized EDFA in each branch. The two-arm design has produced a relatively uniform gain of 24 dB over a bandwidth as large as 80 nm when pumped with 980-nm semiconductor lasers while maintaining a noise figure of about 6 dB [55]. The two-arm or two-stage amplifiers are complex devices and contain multiple components, such as optical filters and isolators, within them for optimizing the amplifier performance. An alternative approach to broadband EDFAs uses a *fluoride* fiber in place of silica fibers as the host medium in which erbium ions are doped. Gain flatness over a 76-nm bandwidth has been realized by doping a *tellurite* fiber with erbium ions [96]. Although such EDFAs are simpler in design compared with multistage amplifiers, they suffer from the splicing difficulties because of the use of nonsilica glasses.

Starting in 2001, high-capacity lightwave systems began to use the short-wavelength region—the so-called S band—extending from 1470 to 1520 nm [97]. Erbium ions cannot provide gain in this spectral band. Thulium-doped fiber amplifiers have been developed for this purpose, and they are capable of providing flat gain in the wavelength range 1480–1510 nm when pumped using 1420-nm and 1560-nm semiconductor lasers [98]. Both lasers are needed to reach the $^3F_4$ state of thulium ions. The gain is realized on the $^3F_4 \rightarrow {}^3H_4$ transition. Raman amplification can also be used for the S band, and such amplifiers were under development in 2001.

6.4.6 Distributed-Gain Amplifiers

Most EDFAs provide 20–25 dB amplification over a length ∼10 m through a relatively high density of dopants (∼500 parts per million). Since such EDFAs compensate for losses accumulated over 80–100 km in a relatively short distance of 10–20 m, they are referred to as the lumped amplifiers. Similar to the case of Raman amplification, fiber losses can also be compensated through *distributed amplification*. In this approach, the transmission fiber itself is lightly doped (dopant density ∼50 parts per billion) to provide the gain distributed over the entire fiber length such that it compensates for fiber losses locally. Such an approach results in a virtually transparent fiber at a specific wavelength when the fiber is pumped using the bidirectional pumping configuration. The scheme is similar to that discussed in Section 6.3 for distributed Raman amplifiers, except that the dopants provide the gain instead of the nonlinear phenomenon of SRS. Although considerable research has been done on distributed EDFAs [99]–[106], this scheme has not yet been used commercially as it requires special fibers.

Ideally, one would like to compensate for fiber losses in such a way that the signal power does not change at all during propagation. Such a performance is, however, never realized in practice as the pump power is not uniform along the fiber length because of fiber losses at the pump wavelength. Pumping at 980 nm is ruled out because fiber losses exceed 1 dB/km at that wavelength. The optimal pumping wavelength for distributed EDFAs is 1.48 μm, where losses are about 0.25 dB/km. If we include pump absorption by dopants, total pump losses typically exceed 0.4 dB/km, resulting in losses of 10 dB for a fiber length of only 25 km. If the fiber is pumped unidirectionally by injecting the pump beam from one end, nonuniform pumping leads to large variations in the signal power. A bidirectional pumping configuration is therefore used in which the fiber is pumped from both ends by using two 1.48-μm lasers. In general, variations in the signal power due to nonuniform pumping can be kept small for a relatively short fiber length of 10–15 km [99]. For practical reasons, it is important to increase the fiber length close to 50 km or more so that the pumping stations could be spaced that far apart. In a 1995 experiment, 11.5-ps pulses were transmitted over 93.4 km of a distributed EDFA by injecting up to 90 mW of pump power from each end [104]. The signal power was estimated to vary by a factor of more than 10 because of the nonuniform pumping.

The performance of distributed EDFAs depends on the signal wavelength since both the noise figure and the pump power required to achieve transparency change with the signal wavelength [103]. In a 1996 experiment, a 40-Gb/s return-to-zero (RZ) signal was transmitted over 68 km by using 7.8-ps optical pulses [105]. The ASE noise added to the signal is expected to be smaller than lumped EDFAs as the gain is relatively small all along the fiber. Computer simulations show that the use of distributed amplification for nonreturn-to-zero (NRZ) systems has the potential of doubling the pump-station spacing in comparison with the spacing for the lumped amplifiers [106]. For long fiber lengths, one should consider the effect of SRS in distributed EDFAs (pumped at 1.48 μm) because the pump-signal wavelength difference lies within the Raman-gain bandwidth, and the signal experiences not only the gain provided by the dopants but also the gain provided by SRS. The SRS increases the net gain and reduces the noise figure for a given amount of pump power.

6.5 System Applications

Fiber amplifiers have become an integral part of almost all fiber-optic communication systems installed after 1995 because of their excellent amplification characteristics such as low insertion loss, high gain, large bandwidth, low noise, and low crosstalk. In this section we first consider the use of EDFAs as preamplifiers at the receiver end and then focus on the design issues for long-haul systems employing a cascaded chain of optical amplifiers.

6.5.1 Optical Preamplification

Optical amplifiers are routinely used for improving the sensitivity of optical receivers by preamplifying the optical signal before it falls on the photodetector. Preamplification of the optical signal makes it strong enough that thermal noise becomes negligible compared with the noise induced by the preamplifier. As a result, the receiver sensitivity can be improved by 10–20 dB using an EDFA as a preamplifier [107]–[112]. In a 1990 experiment [107], only 152 photons/bit were needed for a lightwave system operating at bit rates in the range 0.6–2.5 Gb/s. In another experiment [110], a receiver sensitivity of -37.2 dBm (147 photons/bit) was achieved at the bit rate of 10 Gb/s. It is even possible to use two preamplifiers in series; the receiver sensitivity improved by 18.8 dB with this technique [109]. An experiment in 1992 demonstrated a record sensitivity of -38.8 dBm (102 photons/bit) at 10 Gb/s by using two EDFAs [111]. Sensitivity degradation was limited to below 1.2 dB when the signal was transmitted over 45 km of dispersion-shifted fiber.

To calculate the receiver sensitivity, we need to include all sources of current noise at the receiver. The most important performance issue in designing optical preamplifiers is the contamination of the amplified signal by the ASE. Because of the incoherent nature of spontaneous emission, the amplified signal is noisier than the input signal. Following Sections 4.4.1 and 6.1.3, the photocurrent generated at the detector can be written as

$$I = R|\sqrt{G}E_s + E_{sp}|^2 + i_s + i_T, \qquad (6.5.1)$$

where R is the photodetector responsivity, G is the amplifier gain, E_s is the signal field, E_{sp} is the optical field associated with the ASE, and i_s and i_T are current fluctuations induced by the shot noise and thermal noise, respectively, within the receiver. The average value of the current consists of

$$\bar{I} = R(GP_s + P_{sp}), \qquad (6.5.2)$$

where $P_s = |E_s|^2$ is the optical signal before its preamplification, and P_{sp} is the ASE noise power added to the signal with the magnitude

$$P_{sp} = |E_{sp}|^2 = S_{sp}\Delta v_{sp}. \qquad (6.5.3)$$

The spectral density S_{sp} is given by Eq. (6.1.15) and Δv_{sp} is the effective bandwidth of spontaneous emission set by the amplifier bandwidth or the filter bandwidth if an optical filter is placed after the amplifier. Notice that E_{sp} in Eq. (6.5.1) includes only the component of ASE that is copolarized with the signal as the orthogonally polarized component cannot beat with the signal.

The current noise ΔI consists of fluctuations originating from the shot noise, thermal noise, and ASE noise. The ASE-induced current noise has its origin in the beating of E_s with E_{sp} and the beating of E_{sp} with itself. To understand this beating phenomenon more clearly, notice that the ASE field E_{sp} is broadband and can be written in the form

$$E_{sp} = \int \sqrt{S_{sp}} \exp(\phi_n - i\omega_n t)\, d\omega_n, \tag{6.5.4}$$

where ϕ_n is the phase of the noise-spectral component at the frequency ω_n, and the integral extends over the entire bandwidth of the amplifier (or optical filter). Using $E_s = \sqrt{P_s} \exp(\phi_s - i\omega_s t)$, the interference term in Eq. (6.5.1) consists of two parts and leads to current fluctuations of the form

$$i_{sig-sp} = 2R \int (GP_s S_{sp})^{1/2} \cos \theta_1\, d\omega_n, \quad i_{sp-sp} = 2R \iint S_{sp} \cos \theta_2\, d\omega_n\, d\omega_n', \tag{6.5.5}$$

where $\theta_1 = (\omega_s - \omega_n)t + \phi_n - \phi_s$ and $\theta_2 = (\omega_n - \omega_n')t + \phi_n' - \phi_n$ are two rapidly varying random phases. These two contributions to current noise are due to the beating of E_s with E_{sp} and the beating of E_{sp} with itself, respectively. Averaging over the random phases, the total variance $\sigma^2 = \langle (\Delta I)^2 \rangle$ of current fluctuations can be written as [5]

$$\sigma^2 = \sigma_T^2 + \sigma_s^2 + \sigma_{sig-sp}^2 + \sigma_{sp-sp}^2, \tag{6.5.6}$$

where σ_T^2 is the thermal noise and the remaining three terms are [113]

$$\sigma_s^2 = 2q[R(GP_s + P_{sp})]\Delta f, \tag{6.5.7}$$

$$\sigma_{sig-sp}^2 = 4R^2 GP_s S_{sp}\Delta f, \tag{6.5.8}$$

$$\sigma_{sp-sp}^2 = 4R^2 S_{sp}^2 \Delta\nu_{opt}\Delta f, \tag{6.5.9}$$

where $\Delta\nu_{opt}$ is the bandwidth of the optical filter and Δf is the electrical noise bandwidth of the receiver. The shot-noise term σ_s^2 is the same as in Section 4.4.1 except that P_{sp} has been added to GP_s to account for the shot noise generated by spontaneous emission.

The BER can be obtained by following the analysis of Section 4.5.1. As before, it is given by

$$\text{BER} = \tfrac{1}{2}\text{erfc}(Q/\sqrt{2}), \tag{6.5.10}$$

with the Q parameter

$$Q = \frac{I_1 - I_0}{\sigma_1 + \sigma_0} = \frac{RG(2\bar{P}_{rec})}{\sigma_1 + \sigma_0}. \tag{6.5.11}$$

Equation (6.5.11) is obtained by assuming zero extinction ratio ($I_0 = 0$) so that $I_1 = RGP_1 = RG(2\bar{P}_{rec})$, where $\bar{P}_{rec}$ is the receiver sensitivity for a given value of BER ($Q = 6$ for BER $= 10^{-9}$). The RMS noise currents σ_1 and σ_0 are obtained from Eqs. (6.5.6)–(6.5.9) by setting $P_s = P_1 = 2\bar{P}_{rec}$ and $P_s = 0$, respectively.

The analysis can be simplified considerably by comparing the magnitude of various terms in Eqs. (6.5.6). For this purpose it is useful to substitute S_{sp} from Eq. (6.1.15), use $R = \eta q/h\nu$ and Eq. (6.1.19), and write Eqs. (6.5.7)–(6.5.9) in terms of the amplifier noise figure F_n as

$$\sigma_s^2 = 2q^2\eta GP_s\Delta f/h\nu, \qquad (6.5.12)$$

$$\sigma_{\text{sig-sp}}^2 = 2(q\eta G)^2 F_n P_s\Delta f/h\nu, \qquad (6.5.13)$$

$$\sigma_{\text{sp-sp}}^2 = (q\eta GF_n)^2 \Delta\nu_{\text{opt}}\Delta f, \qquad (6.5.14)$$

where the RP_{sp} term was neglected in Eq. (6.5.7) as it contributes negligibly to the shot noise. A comparison of Eqs. (6.5.12) and (6.5.13) shows that σ_s^2 can be neglected in comparison with $\sigma_{\text{sig-sp}}^2$ as it is smaller by a large factor ηGF_n. The thermal noise σ_T^2 can also be neglected in comparison with the dominant terms. The noise currents σ_1 and σ_0 are thus well approximated by

$$\sigma_1 = (\sigma_{\text{sig-sp}}^2 + \sigma_{\text{sp-sp}}^2)^{1/2}, \qquad \sigma_0 = \sigma_{\text{sp-sp}}. \qquad (6.5.15)$$

The receiver sensitivity is obtained by substituting Eq. (6.5.15) in Eq. (6.5.11), using Eqs. (6.5.13) and (6.5.14) with $P_s = 2\bar{P}_{\text{rec}}$, and solving for $\bar{P}_{\text{rec}}$. The result is

$$\bar{P}_{\text{rec}} = h\nu F_n\Delta f[Q^2 + Q(\Delta\nu_{\text{opt}}/\Delta f)^{1/2}]. \qquad (6.5.16)$$

The receiver sensitivity can also be written in terms of the average number of photons/bit, $\bar{N}_p$, by using $\bar{P}_{\text{rec}} = \bar{N}_p h\nu B$. Taking $\Delta f = B/2$ as a typical value of the receiver bandwidth, $\bar{N}_p$ is given by

$$\bar{N}_p = \tfrac{1}{2}F_n[Q^2 + Q(2\Delta\nu_{\text{opt}}/B)^{1/2}]. \qquad (6.5.17)$$

Equation (6.5.17) is a remarkably simple expression for the receiver sensitivity. It shows clearly why amplifiers with a small noise figure must be used; the receiver sensitivity degrades as F_n increases. It also shows how optical filters can improve the receiver sensitivity by reducing $\Delta\nu_{\text{opt}}$. Figure 6.20 shows $\bar{N}_p$ as a function of $\Delta\nu_{\text{opt}}/B$ for several values of the noise figure F_n by using $Q = 6$, a value required to achieve a BER of 10^{-9}. The minimum optical bandwidth is equal to the bit rate to avoid blocking the signal. The minimum value of F_n is 2 for an ideal amplifier (see Section 8.1.3). Thus, by using $Q = 6$, the best receiver sensitivity from Eq. (6.5.17) is $\bar{N}_p = 44.5$ photons/bit. This value should be compared with $\bar{N}_p = 10$ for an ideal receiver (see Section 4.5.3) operating in the quantum-noise limit. Of course, $\bar{N}_p = 10$ is never realized in practice because of thermal noise; typically, $\bar{N}_p$ exceeds 1000 for p–i–n receivers without optical amplifiers. The analysis of this section shows that $\bar{N}_p < 100$ can be realized when optical amplifiers are used to preamplify the signal received despite the degradation caused by spontaneous-emission noise. The effect of a finite laser linewidth on the receiver sensitivity has also been included with similar conclusions [114].

Improvements in the receiver sensitivity, realized with an EDFA acting as a preamplifier, can be used to increase the transmission distance of point-to-point fiber links used for intercity and interisland communications. Another EDFA acting as a power booster is often used to increase the launched power to levels as high as 100 mW. In a 1992 experiment, a 2.5-Gb/s signal was transmitted over 318 km by such a technique [115]. Bit rate was later increased to 5 Gb/s in an experiment [116] that used two

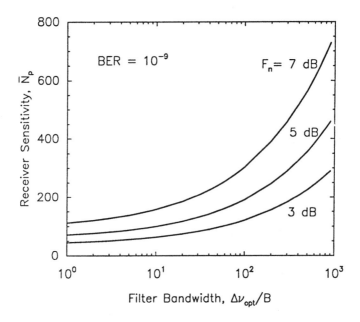

Figure 6.20: Receiver sensitivity versus optical-filter bandwidth for several values of the noise figure F_n when an optical amplifier is used for preamplification of the received signal.

EDFAs to boost the signal power from -8 to 15.5 dBm (about 35 mW). This power level is large enough that SBS becomes a problem. SBS can be suppressed through phase modulation of the optical carrier that broadens the carrier linewidth to 200 MHz or more. Direct modulation of lasers also helps through frequency chirping that broadens the signal spectrum. In a 1996 experiment, a 10-Gb/s signal was transmitted over 442 km using two remotely pumped in-line amplifiers [117].

6.5.2 Noise Accumulation in Long-Haul Systems

Optical amplifiers are often cascaded to overcome fiber losses in a long-haul lightwave system. The buildup of amplifier-induced noise is the most critical factor for such systems. There are two reasons behind it. First, in a cascaded chain of optical amplifiers (see Fig. 5.1), the ASE accumulates over many amplifiers and degrades the optical SNR as the number of amplifiers increases [118]–[121]. Second, as the level of ASE grows, it begins to saturate optical amplifiers and reduce the gain of amplifiers located further down the fiber link. The net result is that the signal level drops further while the ASE level increases. Clearly, if the number of amplifiers is large, the SNR will degrade so much at the receiver that the BER will become unacceptable. Numerical simulations show that the system is self-regulating in the sense that the total power obtained by adding the signal and ASE powers remains relatively constant. Figure 6.21 shows this self-regulating behavior for a cascaded chain of 100 amplifiers with 100-km spacing and 35-dB small-signal gain. The power launched by the transmitter is 1 mW. The other parameters are $P_{out}^s = 8$ mW, $n_{sp} = 1.3$, and $G_0 \exp(-\alpha L_A) = 3$, where L_A is the

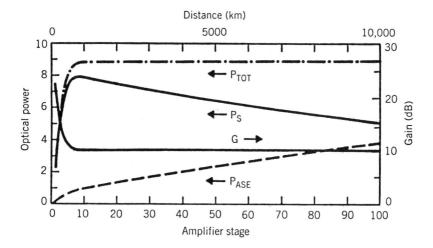

Figure 6.21: Variation of the signal power P_s and the ASE power P_{ASE} along a cascaded chain of optical amplifiers. The total power P_{TOT} becomes nearly constant after a few amplifiers. (After Ref. [119]; ©1991 IEEE; reprinted with permission.)

amplifier spacing. The signal and ASE powers become comparable after 10,000 km, indicating the SNR problem at the receiver.

To estimate the SNR associated with a long-haul lightwave system, we assume that all amplifiers are spaced apart by a constant distance L_A, and the amplifier gain $G \equiv \exp(\alpha L_A)$ is just large enough to compensate for fiber losses in each fiber section. The total ASE power for a chain of N_A amplifiers is then obtained by multiplying Eq. (6.5.3) with N_A and is given by

$$P_{sp} = 2N_A S_{sp}\Delta v_{opt} = 2n_{sp}hv_0 N_A(G-1)\Delta v_{opt}, \tag{6.5.18}$$

where the factor of 2 accounts for the unpolarized nature of ASE. We can use this equation to find the optical SNR using $SNR_{opt} = P_{in}/P_{sp}$. However, optical SNR is not the quantity that determines the receiver performance. As discussed earlier, the electrical SNR is dominated by the signal-spontaneous beat noise generated at the photodetector. If we include only this dominant contribution, the electrical SNR is related to optical SNR as

$$SNR_{el} = \frac{R^2 P_{in}^2}{N_A \sigma_{sig-sp}^2} = \frac{\Delta v_{opt}}{2\Delta f} SNR_{opt} \tag{6.5.19}$$

if we use Eq. (6.5.8) with $G = 1$ assuming no net amplification of the input signal.

We can now evaluate the impact of multiple amplifiers. Clearly, the electrical SNR can become quite small for large values of G and N_A. For a fixed system length L_T, the number of amplifiers depends on the amplifier spacing L_A and can be reduced by increasing it. However, a longer amplifier spacing will force one to increase the gain of each amplifier since $G = \exp(\alpha L_A)$. Noting that $N_A = L_T/L_A = \alpha L_T/\ln G$, we find that SNR_{el} scales with G as $\ln G/(G-1)$ and can be increased by lowering the gain of each amplifier. In practice, the amplifier spacing L_A cannot be made too small because

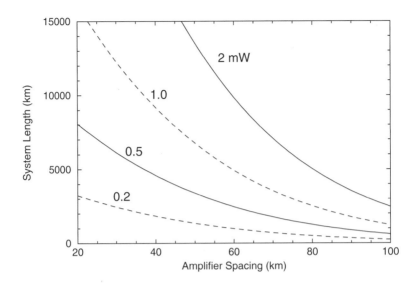

Figure 6.22: Maximum ASE-limited system length for a 10-Gb/s channel as a function of amplifier spacing L_A for several values of input powers.

of cost considerations. To estimate the optimum value of L_A, Fig. 6.22 shows the total system length L_T as a function of L_A for several values of input powers P_{in} assuming that an electrical SNR of 20 dB is required for the system to function properly and using $\alpha = 0.2$ dB/km, $n_{sp} = 1.6$ (noise figure 5 dB), and $\Delta f = 10$ GHz. The main point to note is that amplifier spacing becomes smaller as the system length increases. Typically, L_A is kept near 50 km for undersea systems but can be increased to 80 km or so for terrestrial systems with link lengths under 3000 km. Although amplifier spacing can be improved by increasing the input power P_{in}, in practice, the maximum power that can be launched is limited by the onset of various nonlinear effects. We turn to this issue next.

6.5.3   ASE-Induced Timing Jitter

The amplifier noise can also induce timing jitter in the bit stream by shifting optical pulses from their original time slot in a random fashion. Such jitter was first studied in 1986 in the context of solitons and is called the Gordon–Haus jitter [122]. It was later recognized that timing jitter can occur with any transmission format [NRZ, RZ, or chirped RZ (CRZ)] and imposes a fundamental limitation on all long-haul systems designed with a cascaded chain of optical amplifiers [123]–[126].

The physical origin of ASE-induced jitter can be understood by noting that optical amplifiers affect not only the amplitude but also the phase of the amplified signal as apparent from Eq. (6.5.32) or Eq. (6.5.34). Time-dependent variations in the optical phase lead to a change in the signal frequency by a small amount. Since the group velocity depends on the frequency because of dispersion, the speed at which a soliton propagates through the fiber is affected by each amplifier in a random fashion. Such random speed changes produce random shifts in the pulse position at the receiver and are responsible for the timing jitter.

Timing jitter induced by the ASE noise can be calculated using the *moment method*. According to this method, changes in the pulse position q and the frequency Ω along the link length are calculated using [125]

$$q(z) = \frac{1}{E} \int_{-\infty}^{\infty} t |A(z,t)|^2 \, dt, \tag{6.5.20}$$

$$\Omega(z) = \frac{i}{2E} \int_{-\infty}^{\infty} \left(A^* \frac{\partial A}{\partial t} - A \frac{\partial A^*}{\partial t} \right) dt, \tag{6.5.21}$$

where $E = \int_{-\infty}^{\infty} |A|^2 \, dt$ represents the pulse energy.

The NLS equation can be used to find how T and W evolve along the fiber link. Differentiating Eqs. (6.5.20) and (6.5.21) with respect to z and using Eq. (6.5.31), we obtain [126]

$$\frac{d\Omega}{dz} = \sum_i \delta\Omega_i \, \delta(z - z_i), \tag{6.5.22}$$

$$\frac{dq}{dz} = \beta_2 \Omega + \sum_i \delta q_i \, \delta(z - z_i), \tag{6.5.23}$$

where $\delta\Omega_i$ and δq_i are the random frequency and position changes imparted by noise at the ith amplifier and the sum is over the total number N_A of amplifiers. These equations show that frequency fluctuations induced by an amplifier become temporal fluctuations because of GVD; no jitter occurs when $\beta_2 = 0$.

Equations (6.5.22) and (6.5.23) can be integrated in a straightforward manner. For a cascaded chain of N_A amplifiers with spacing L_A, the pulse position at the last amplifier is given by

$$q_f = \sum_{n=1}^{N_A} \delta q_n + \bar{\beta}_2 L_A \sum_{n=1}^{N_A} \sum_{i=1}^{n-1} \delta\Omega_i, \tag{6.5.24}$$

where $\bar{\beta}_2$ is the average value of the GVD. Timing jitter is calculated from this equation using $\sigma_t^2 = \langle q_f^2 \rangle - \langle q_f \rangle^2$ together with $\langle q_f \rangle = 0$. The average can be performed by noting that fluctuations at two different amplifiers are not correlated. However, the timing jitter depends not only on the variances of position and frequency fluctuations but also on the cross-correlation function $\langle \delta q \, \delta\Omega \rangle$ at the same amplifier. These quantities depend on the pulse amplitude $A(z_i, t)$ at the amplifier location z_i (see Section 9.5).

Consider a low-power lightwave system employing the CRZ format and assume that the input pulse is in the form of a chirped Gaussian pulse. As seen in Section 2.4, the pulse maintains its Gaussian shape on propagation such that

$$A(z,t) = a \exp[i\phi - i\Omega(t - q) - (1 + iC)(t - q)^2 / 2T^2], \tag{6.5.25}$$

where the amplitude a, phase ϕ, frequency Ω, position q, chirp C, and width T all are functions of z. The variances and cross-correlation of δq_i and Ω_i at the location of the ith amplifier are found to be [126]

$$\langle (\delta\Omega)^2 \rangle = (S_{\text{sp}}/E_0)[(1 + C_i^2)/T_i^2], \tag{6.5.26}$$

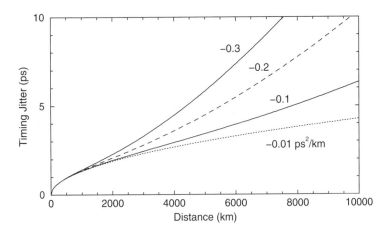

Figure 6.23: ASE-induced timing jitter as a function of system length for several values of the average dispersion $\bar{\beta}_2$.

$$\langle(\delta q)^2\rangle = (S_{sp}/E_0)T_i^2, \qquad \langle\delta\Omega\delta q\rangle = (S_{sp}/E_0)C_i, \qquad (6.5.27)$$

where E_0 is the input pulse energy and C_i and T_i are the chirp and width at $z = z_i$. These quantities can be calculated easily using the theory of Section 2.4. Note that the ratio $(1+C_i^2)/T_i^2$ is related to the spectral width that does not change if the nonlinear effects are negligible. It can be replaced by T_m^{-2}, where T_m is the minimum width occurring when the pulse is unchirped.

Many lightwave systems employ the postcompensation technique in which a fiber is placed at the end of the last amplifier to reduce the net accumulated dispersion (see Section 7.4). Using Eqs. (6.5.24)–(6.5.27), the timing jitter for a CRZ system employing postcompensation is found to be [126]

$$\sigma_t^2 = (S_{sp}/E_0)T_m^2\left[N_A + N_A(N_Ad + C_0 + d_f)^2\right], \qquad (6.5.28)$$

where C_0 is the input chirp, $d = \bar{\beta}_2L_A/T_m^2$, and $d_f = \beta_{2f}L_f/T_m^2$ for a postcompensation fiber of length L_f and dispersion β_{2f}. Several points are noteworthy. First, if postcompensation is not used ($d_f = 0$), the dominant term in Eq. (6.5.28) varies as $N_A^3d^2$. This is the general feature of the ASE jitter resulting from frequency fluctuations [122]. Second, if the average dispersion of the fiber link is zero, the cubic term vanishes, and the jitter increases only linearly with N_A. Third, the smallest value of the jitter occurs when $N_Ad + C_0 + d_f = 0$. This condition corresponds to zero net dispersion over the entire link, including the fiber used to chirp the pulse initially.

The average dispersion of the fiber link can lead to considerable timing jitter in CRZ systems when postcompensation is not used. Figure 6.23 shows the timing jitter as a function of the total system length $L_T = N_AL_A$ for a 10-Gb/s system using four values of $\bar{\beta}_2$ with $T_m = 30$ ps, $L_A = 50$ km, $C_0 = 0.2$, and $S_{sp}/E_0 = 10^{-4}$. The ASE-induced jitter becomes a significant fraction of the pulse width for values of $|\bar{\beta}_2|$ as small as 0.2 ps²/km because of the cubic dependence of σ_t^2 on the system length L_T. Such jitter would lead to large power penalties, as discussed in Section 4.6.3, if left uncontrolled. The tolerable value of the jitter can be estimated assuming the Gaussian statistics for q so that

$$p(q) = (2\pi\sigma_t^2)^{-1/2} \exp(-q^2/2\sigma_t^2). \qquad (6.5.29)$$

The BER can be calculated following the method of Section 4.5. If we assume that an error occurs whenever the pulse has moved out of the bit slot, we need to find the accumulated probability for $|q|$ to exceed $T_B/2$, where $T_B \equiv 1/B$ is the bit slot. This probability is found to be

$$\text{BER} = 2 \int_{T_B/2}^{\infty} p(q)\,dq = \text{erfc}\left(\frac{T_B}{2\sqrt{2}\sigma_t}\right) \approx \frac{4\sigma_t}{\sqrt{2\pi}T_B} \exp\left(-\frac{T_B^2}{8\sigma_t^2}\right), \qquad (6.5.30)$$

where erfc stands for the complimentary error function defined in Eq. (4.5.5). To reduce the BER below 10^{-9} for σ_t/T_B should be less than 8% of the bit slot, resulting in a tolerable value of the jitter of 8 ps for 10-Gb/s systems and only 2 ps for 40-Gb/s systems. Clearly, the average dispersion of a fiber link should nearly vanish if the system is designed not to be limited by the ASE-induced jitter. This can be accomplished through dispersion management discussed in Chapter 7.

6.5.4 Accumulated Dispersive and Nonlinear Effects

Many single-channel experiments performed during the early 1990s demonstrated the benefits of in-line amplifiers for increasing the transmission distance of point-to-point fiber links [127]–[132]. These experiments showed that fiber dispersion becomes the limiting factor in periodically amplified long-haul systems. Indeed, the experiments were possible only because the system was operated close to the zero-dispersion wavelength of the fiber link. Moreover, the zero-dispersion wavelength varied along the link in such a way that the total dispersion over the entire link length was quite small at the operating wavelength of 1.55 μm. By 1992, the total system length could be increased to beyond 10,000 km using such dispersion-management techniques. In a 1992 experiment [130], a 2.5-Gb/s signal was transmitted over 10,073 km using 199 EDFAs. An effective transmission distance of 21,000 km at 2.5 Gb/s and of 14,300 km at 5 Gb/s was demonstrated using a recirculating fiber loop [133].

A crude estimate of dispersion-limited L_T can be obtained if the input power is low enough that one can neglect the nonlinear effects during signal transmission. Since amplifiers compensate only for fiber losses, dispersion limitations discussed in Section 5.2.2 and shown in Fig. 5.4 apply for each channel of a WDM system if L is replaced by L_T. From Eq. (5.2.3), the dispersion limit for systems making use of standard fibers ($\beta_2 \approx -20$ ps^2/km at 1.55 μm) is $B^2 L_T < 3000$ (Gb/s)^2-km: The distance is limited to below 30 km at 10 Gb/s for such fibers. An increase by a factor of 20 can be realized by using dispersion-shifted fibers. To extend the distance to beyond 5000 km at 10 Gb/s, the average GVD along the link should be smaller than $\bar{\beta}_2 = -0.1$ ps^2/km.

The preceding estimate is crude since it does not include the impact of the nonlinear effects. Even though power levels are relatively modest for each channel, the nonlinear effects can become quite important because of their accumulation over long distances [29]. Moreover, amplifier noise often forces one to increase the channel power to more than 1 mW in order to maintain a high SNR (or a high Q factor). The accumulation of the nonlinear effects then limits the system length L_T [134]–[147]. For single-channel systems, the most dominant nonlinear phenomenon that limits the system performance is self-phase modulation (SPM). An estimate of the power limitation imposed by the SPM can be obtained from Eq. (2.6.15). In general, the condition $\phi_{NL} \ll 1$ limits the total link length to $L_T \ll L_{NL}$, where the *nonlinear length* is defined as $L_{NL} = (\gamma \bar{P})^{-1}$. Typically, $\gamma \sim 1$ W^{-1}/km, and the link length is limited to below 1000 km even for $\bar{P} = 1$ mW.

The estimate of the SPM-limited distance is too simplistic to be accurate since it completely ignores the role of fiber dispersion. In fact, since the dispersive and nonlinear effects act on the optical signal simultaneously, their mutual interplay becomes quite important. As discussed in Section 5.3, it is necessary to solve the nonlinear Schrödinger equation

$$\frac{\partial A}{\partial z} + \frac{i\beta_2}{2}\frac{\partial^2 A}{\partial t^2} = i\gamma|A|^2 A - \frac{\alpha}{2}A \qquad (6.5.31)$$

numerically, while including the gain and ASE noise at the location of each amplifier. Such an approach is indeed used to quantify the impact of nonlinear effects on the performance of periodically amplified lightwave systems [134]–[148]. A common technique solves Eq. (6.5.31) in each fiber segment using the split-step Fourier method [56]. At each optical amplifier, the noise is added using

$$A_{\text{out}}(t) = \sqrt{G}A_{\text{in}}(t) + a_n(t), \qquad (6.5.32)$$

where G is the amplification factor. The spontaneous-emission noise field a_n added by the amplifier vanishes on average but its second moment satisfies

$$\langle a_n(t)a_n(t')\rangle = S_{\text{sp}}\delta(t-t'), \qquad (6.5.33)$$

where the noise spectral density S_{sp} is given by Eq. (6.1.15).

In practice, Eq. (6.5.32) is often implemented in the frequency domain as

$$\tilde{A}_{\text{out}}(\nu) = \sqrt{G}\tilde{A}_{\text{in}}(\nu) + \tilde{a}_n(\nu), \qquad (6.5.34)$$

where a tilde represents the Fourier transform. The noise $\tilde{a}_n(\nu)$ is assumed to be frequency independent (white noise) over the whole amplifier bandwidth, or the filter bandwidth if an optical filter is used after each amplifier. Mathematically, $\tilde{a}_n(\nu)$ is a complex Gaussian random variable whose real and imaginary parts have the spectral density $S_{\text{sp}}/2$. The system performance is quantified through the Q factor as defined in Eq. (4.5.10) and related directly to the BER through Eq. (4.5.9).

As an example of the numerical results, the curve (a) in Fig. 6.24 shows variations of the Q factor with the average input power for a NRZ, single-channel lightwave system designed to operate at 5 Gb/s over 9000 km of constant-dispersion fibers [$D = 1$ ps/(km-nm)] with 40-km amplifier spacing [134]. Since $Q < 6$ for all input powers, such a system cannot operate reliably in the absence of in-line filters ($Q > 6$ is required for a BER of $< 10^{-9}$). An optical filter of 150-GHz bandwidth, inserted after every amplifier, reduces the ASE for the curve (b). In the presence of optical filters, $Q > 6$ can be realized only at a specific value of the average input power (about 0.5 mW). This

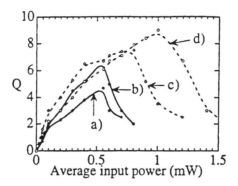

Figure 6.24: Q factor as a function of the average input power for a 9000-km fiber link: (a) 5-Gb/s system; (b) improvement realized with 150-GHz optical filters; (c) 6-Gb/s operation near the zero-dispersion wavelength; (d) 10-Gb/s system with dispersion management. (After Ref. [134]; ©1996 IEEE; reprinted with permission.)

behavior can be understood by noting that as the input power increases, the system performance improves initially because of a better SNR but becomes worse at high input powers as the nonlinear effects (SPM) begins to dominate.

The role of dispersion can be minimized either by operating close to the zero-dispersion wavelength of the fiber or by using a dispersion management technique in which the fiber GVD alternates its sign in such a way that the average dispersion is close to zero (see Chapter 7). In both cases, the GVD parameter β_2 fluctuates because of unintentional variations in the zero-dispersion wavelength of various fiber segments. The curve (c) in Fig. 6.24 is drawn for a 6-Gb/s system for the case of a Gaussian distribution of β_2 with a standard deviation of 0.3 ps^2/km. The filter bandwidth is taken to be 60 GHz [134]. The curve (d) shows the dispersion-managed case for a 10-Gb/s system with a filter bandwidth of 50 GHz. All other parameters remain the same. Clearly, system performance can improve considerably with dispersion management, although the input pump power needs to be optimized in each case.

6.5.5 WDM-Related Impairments

The advantages of EDFAs for WDM systems were demonstrated as early as 1990 [149]–[154]. In a 1993 experiment, four channels were transmitted over 1500 km using 22 cascaded amplifiers [150]. By 1996, 55 channels, spaced apart by 0.8 nm and each operating at 20 Gb/s, were transmitted over 150 km by using two in-line amplifiers, resulting in a total bit rate of 1.1 Tb/s and the BL product of 165 (Tb/s)-km [151]. For submarine applications, one needs to transmit a large number of channels over a distance of more than 5000 km. Such systems employ a large number of cascaded amplifiers and are affected most severely by the amplifier noise. Already in 1996, transmission at 100 Gb/s (20 channels at 5 Gb/s) over a distance of 9100 km was possible using the polarization-scrambling and forward-error correction techniques [152]. By 2001, transmission at 2.4 Tb/s (120 channels at 20 Gb/s) over 6200 km has been realized within the C band using EDFAs every 50 km [154]. The adjacent channels were orthogonally polarized for reducing the nonlinear effects resulting from a relatively small channel spacing of 42 GHz. This technique is referred to as polarization multiplexing and is quite useful for WDM systems.

The two major nonlinear phenomena affecting the performance of WDM systems are the cross-phase modulation (XPM) and four-wave mixing (FWM). FWM can be avoided by using dispersion management such that the GVD is locally high all along the fiber but quite small on average. The SPM and XPM then become the most limiting factors for WDM systems. The XPM effects within an EDFA are normally negligible because of a small length of doped fiber used. The situation changes for the L-band amplifiers, which operate in the 1570- to 1610-nm wavelength region and require fiber lengths in excess of 100 m. The effective core area of doped fibers used in such amplifiers is relatively small, resulting in larger values of the nonlinear parameter γ and enhanced XPM-induced phase shifts. As a result, the XPM can lead to considerable power fluctuations within an L-band amplifier [155]–[160]. A new feature is that such XPM effects are independent of the channel spacing and can occur over the entire bandwidth of the amplifier [156]. The reason for this behavior is that all XPM effects occur before pulses walk off because of group-velocity mismatch. The effects of FWM are also enhanced in L-band amplifiers because of their long lengths [161].

Problems

6.1 The Lorentzian gain profile of an optical amplifier has a FWHM of 1 THz. Calculate the amplifier bandwidths when it is operated to provide 20- and 30-dB gain. Neglect gain saturation.

6.2 An optical amplifier can amplify a 1-μW signal to the 1-mW level. What is the output power when a 1-mW signal is incident on the same amplifier? Assume that the saturation power is 10 mW.

6.3 Explain the concept of noise figure for an optical amplifier. Why does the SNR of the amplified signal degrade by 3 dB even for an ideal amplifier?

6.4 A 250-μm-long semiconductor laser is used as an FP amplifier by biasing it below threshold. Calculate the amplifier bandwidth by assuming 32% reflectivity for both facets and 30-dB peak gain. The group index $n_g = 4$. How much does the bandwidth change when both facets are coated to reduce the facet reflectivities to 1%?

6.5 Complete the derivation of Eq. (6.2.3) starting from Eq. (6.2.1). What should be the facet reflectivities to ensure traveling-wave operation of a semiconductor optical amplifier designed to provide 20-dB gain. Assume that $R_1 = 2R_2$.

6.6 A semiconductor optical amplifier is used to amplify two channels separated by 1 GHz. Each channel can be amplified by 30 dB in isolation. What are the channel gains when both channels are amplified simultaneously? Assume that $P_{in}/P_s = 10^{-3}$, $\tau_c = 0.5$ ns, and $\beta_c = 5$.

6.7 Integrate Eq. (6.2.19) to obtain the time-dependent saturated gain given by Eq. (6.2.20). Plot $G(\tau)$ for a 10-ps square pulse using $G_0 = 30$ dB and $E_s = 10$ pJ.

6.8 Explain why semiconductor optical amplifiers impose a chirp on the pulse during amplification. Derive an expression for the imposed chirp when a Gaussian pulse is incident on the amplifier. Use Eq. (6.2.21) with $m = 1$ for the input pulse.

6.9 Discuss the origin of gain saturation in fiber Raman amplifiers. Solve Eqs. (6.3.2) and (6.3.3) with $\alpha_s = \alpha_p$ and derive Eq. (6.3.8) for the saturated gain.

6.10 A Raman amplifier is pumped in the backward direction using 1 W of power. Find the output power when a 1-μW signal is injected into the 5-km-long amplifier. Assume losses of 0.2 and 0.25 dB/km at the signal and pump wavelengths, respectively, $A_{\text{eff}} = 50~\mu\text{m}^2$, and $g_R = 6 \times 10^{-14}$ m/W. Neglect gain saturation.

6.11 Explain the gain mechanism in EDFAs. Use Eqs. (6.4.2) and (6.4.3) to derive an expression for the small-signal gain in the steady state.

6.12 Discuss how EDFAs can be used to provide gain in the L band. How can you use them to provide amplification over the both C and L bands?

6.13 Starting from Eq. (6.5.11), derive Eq. (6.5.16) for the sensitivity of a direct-detection receiver when an EDFA is used as a preamplifier.

6.14 Calculate the receiver sensitivity at a BER of 10^{-9} and 10^{-12} by using Eq. (6.5.16). Assume that the receiver operates at 1.55 μm with 3-GHz bandwidth. The preamplifier has a noise figure of 4 dB, and a 1-nm optical filter is installed between the preamplifier and the detector.

6.15 Calculate the optical SNR at the output end of a 4000-km lightwave system designed using 50 EDFAs with 4.5-dB noise figure. Assume a fiber-cable loss of 0.25 dB/km at 1.55 μm. A 2-nm-bandwidth optical filter is inserted after every amplifier to reduce the noise.

6.16 Find the electrical SNR for the system of the preceding problem for a receiver of 8-GHz bandwidth.

6.17 Why does ASE induce timing jitter in lightwave systems? How would you design the system to reduce the jitter?

6.18 Use the moment method to find an expression for the timing jitter for a lightwave system employing the CRZ format.

References

[1] A. E. Siegman, *Lasers*, University Science Books, Mill Valley, CA, 1986.

[2] A. Yariv, *Opt. Lett.* **15**, 1064 (1990); H. Kogelnik and A. Yariv, *Proc. IEEE* **52**, 165 (1964).

[3] T. Saitoh and T. Mukai, *IEEE J. Quantum Electron.* **23**, 1010 (1987).

[4] M. J. O'Mahony, *J. Lightwave Technol.* **6**, 531 (1988).

[5] N. A. Olsson, *J. Lightwave Technol.* **7**, 1071 (1989).

[6] T. Saitoh and T. Mukai, in *Coherence, Amplification, and Quantum Effects in Semiconductor Lasers*, Y. Yamamoto, Ed., Wiley, New York, 1991, Chap. 7.

[7] G. P. Agrawal, *Semiconductor Lasers*, Van Nostrand Reinhold, New York, 1993.

[8] G.-H. Duan, in *Semiconductor Lasers: Past, Present, and Future*, G. P. Agrawal, Ed., AIP Press, Woodbury, NY, 1995, Chap. 10.

[9] C. E. Zah, J. S. Osinski, C. Caneau, S. G. Menocal, L. A. Reith, J. Salzman, F. K. Shokoohi, and T. P. Lee, *Electron. Lett.* **23**, 990 (1987).

[10] I. Cha, M. Kitamura, H. Honmou, and I. Mito, *Electron. Lett.* **25**, 1241 (1989).

[11] S. Cole, D. M. Cooper, W. J. Devlin, A. D. Ellis, D. J. Elton, J. J. Isaak, G. Sherlock, P. C. Spurdens, and W. A. Stallard, *Electron. Lett.* **25**, 314 (1989).

[12] G. Großkopf, R. Ludwig, R. G. Waarts, and H. G. Weber, *Electron. Lett.* **23**, 1387 (1987).

[13] N. A. Olsson, *Electron. Lett.* **24**, 1075 (1988).

[14] M. Sumida, *Electron. Lett.* **25**, 1913 (1989).

[15] M. Koga and T. Mutsumoto, *J. Lightwave Technol.* **9**, 284 (1991).

[16] G. P. Agrawal, *Opt. Lett.* **12**, 260 (1987).

[17] G. P. Agrawal, *Appl. Phys. Lett.* **51**, 302 (1987); *J. Opt. Soc. Am. B* **5**, 147 (1988).

[18] G. P. Agrawal and N. A. Olsson, *IEEE J. Quantum Electron.* **25**, 2297 (1989).

[19] G. P. Agrawal and N. A. Olsson, *Opt. Lett.* **14**, 500 (1989).

[20] N. A. Olsson, G. P. Agrawal, and K. W. Wecht, *Electron. Lett.* **25**, 603 (1989).

[21] N. A. Olsson, M. G. Öberg, L. A. Koszi, and G. J. Przybylek, *Electron. Lett.* **24**, 36 (1988).

[22] K. T. Koai, R. Olshansky, and P. M. Hill, *IEEE Photon. Technol. Lett.* **2**, 926 (1990).

[23] G. P. Agrawal, *Electron. Lett.* **23**, 1175 (1987).

[24] I. M. I. Habbab and G. P. Agrawal, *J. Lightwave Technol.* **7**, 1351 (1989).

[25] S. Ryu, K. Mochizuki, and H. Wakabayashi, *J. Lightwave Technol.* **7**, 1525 (1989).

[26] G. P. Agrawal and I. M. I. Habbab, *IEEE J. Quantum Electron.* **26**, 501 (1990).

[27] R. H. Stolen, E. P. Ippen, and A. R. Tynes, *Appl. Phys. Lett.* **20**, 62 (1972).

[28] R. H. Stolen, *Proc. IEEE* **68**, 1232 (1980).

[29] G. P. Agrawal, *Nonlinear Fiber Optics*, 3rd ed., Academic Press, San Diego, CA, 2001.

[30] S. Namiki and Y. Emori, *IEEE J. Sel. Topics Quantum Electron.* **7**, 3 (2001).

[31] M. Ikeda, *Opt. Commun.* **39**, 148 (1981).

[32] M. Nakazawa, M. Tokuda, Y. Negishi, and N. Uchida, *J. Opt. Soc. Am. B* **1** 80 (1984).

[33] J. Hegarty, N. A. Olsson, and L. Goldner, *Electron. Lett.* **21**, 290 (1985).

[34] M. L. Dakss and P. Melman, *IEE Proc.* **135**, Pt. J, 95 (1988).

[35] T. Horiguchi, T. Sato, and Y. Koyamada, *IEEE Photon. Technol. Lett.* **4**, 64 (1992).

[36] T. Sato, T. Horiguchi, Y. Koyamada, and I. Sankawa, *IEEE Photon. Technol. Lett.* **4**, 923 (1992).

[37] S. G. Grubb and A. J. Stentz, *Laser Focus World* **32** (2), 127 (1996).

[38] S. V. Chernikov, Y. Zhu, R. Kashyap, and J. R. Taylor, *Electron. Lett.* **31**, 472 (1995).

[39] P. B. Hansen, A. J. Stentz, L. Eskilden, S. G. Grubb, T. A. Strasser, and J. R. Pedrazzani, *Electron. Lett.* **27**, 2164 (1996).

[40] P. B. Hansen, L. Eskilden, S. G. Grubb, A. J. Stentz, T. A. Strasser, J. Judkins, J. J. De-Marco, R. Pedrazzani, and D. J. DiGiovanni, *IEEE Photon. Technol. Lett.* **9**, 262 (1997).

[41] P. Wan and J. Conradi, *J. Lightwave Technol.* **14**, 288 (1996).

[42] P. B. Hansen, L. Eskilden, A. J. Stentz, T. A. Strasser, J. Judkins, J. J. DeMarco, R. Pedrazzani, and D. J. DiGiovanni, *IEEE Photon. Technol. Lett.* **10**, 159 (1998).

[43] M. Nissov, K. Rottwitt, H. D. Kidorf, and M. X. Ma, *Electron. Lett.* **35**, 997 (1999).

[44] S. Chinn, *IEEE Photon. Technol. Lett.* **11**, 1632 (1999).

[45] S. A. E. Lewis, S. V. Chernikov, and J. R. Taylor, *IEEE Photon. Technol. Lett.* **12**, 528 (2000).

[46] Y. Emori, K. Tanaka, and S. Namiki, *Electron. Lett.* **35**, 1355 (1999).

[47] S. A. E. Lewis, S. V. Chernikov, and J. R. Taylor, *Electron. Lett.* **35**, 1761 (1999).

[48] H. D. Kidorf, K. Rottwitt, M. Nissov, M. X. Ma, and E. Rabarijaona, *IEEE Photon. Technol. Lett.* **12**, 530 (1999).

[49] B. Zhu, L. Leng, L. E. Nelson, Y. Qian, L. Cowsar, S. Stulz, C. Doerr, L. Stulz, S. Chandrasekhar, et al., *Electron. Lett.* **37**, 844 (2001).

[50] C. J. Koester and E. Snitzer, *Appl. Opt.* **3**, 1182 (1964).

[51] S. B. Poole, D. N. Payne, R. J. Mears, M. E. Fermann, and R. E. Laming, *J. Lightwave Technol.* **4**, 870 (1986).

[52] M. J. F. Digonnet, Ed., *Rare-Earth Doped Fiber Lasers and Amplifiers*, Marcel Dekker, New York, 1993.

[53] A. Bjarklev, *Optical Fiber Amplifiers: Design and System Applications*, Artech House, Boston, 1993.

[54] E. Desurvire, *Erbium-Doped Fiber Amplifiers*, Wiley, New York, 1994.

[55] P. C. Becker, N. A. Olsson, and J. R. Simpson, Academic Press, San Diego, CA, 1999.

[56] G. P. Agrawal, *Applications of Nonlinear Fiber Optics*, Academic Press, San Diego, CA, 2001.

[57] M. Shimizu, M. Yamada, H. Horiguchi, T. Takeshita, and M. Okayasu, *Electron. Lett.* **26**, 1641 (1990).

[58] M. Nakazawa, Y. Kimura, E. Yoshida, and K. Suzuki, *Electron. Lett.* **26**, 1936 (1990).

[59] B. Pederson, A. Bjarklev, H. Vendeltorp-Pommer, and J. H. Povlesen, *Opt. Commun.* **81**, 23 (1991).

[60] M. Horiguchi, K. Yoshino, M. Shimizu, and M. Yamada, *Electron. Lett.* **29**, 593 (1993).

[61] R. I. Laming, J. E. Townsend, D. N. Payne, F. Meli, G. Grasso, and E. J. Tarbox, *IEEE Photon. Technol. Lett.* **3**, 253 (1991).

[62] W. J. Miniscalco, *J. Lightwave Technol.* **9**, 234 (1991).

[63] J. L. Zyskind, E. Desurvire, J. W. Sulhoff, and D. J. DiGiovanni, *IEEE Photon. Technol. Lett.* **2**, 869 (1990).

[64] C. R. Giles and E. Desurvire, *J. Lightwave Technol.* **9**, 271 (1991).

[65] A. A. M. Saleh, R. M. Jopson, J. D. Evankow, and J. Aspell, *IEEE Photon. Technol. Lett.* **2**, 714 (1990).

[66] B. Pedersen, A. Bjarklev, O. Lumholt, and J. H. Povlsen, *IEEE Photon. Technol. Lett.* **3**, 548 (1991).

[67] K. Nakagawa, S. Nishi, K. Aida, and E. Yoneda, *J. Lightwave Technol.* **9**, 198 (1991).

[68] R. M. Jopson and A. A. M. Saleh, *Proc. SPIE* **1581**, 114 (1992).

[69] S. G. Grubb, W. H. Humer, R. S. Cannon, S. W. Nendetta, K. L. Sweeney, P. A. Leilabady, M. R. Keur, J. G. Kwasegroch, T. C. Munks, and P. W. Anthony, *Electron. Lett.* **28**, 1275 (1992).

[70] E. Delevaque, T. Georges, J. F. Bayon, M. Monerie, P. Niay, and P. Benarge, *Electron. Lett.* **29**, 1112 (1993).

[71] R. Olshansky, *Electron. Lett.* **24**, 1363 (1988).

[72] M. Yamada, M. Shimizu, M. Okayasu, T. Takeshita, M. Horiguchi, Y. Tachikawa, and E. Sugita, *IEEE Photon. Technol. Lett.* **2**, 205 (1990).

[73] R. I. Laming and D. N. Payne, *IEEE Photon. Technol. Lett.* **2**, 418 (1990).

[74] K. Kikuchi, *Electron. Lett.* **26**, 1851 (1990).

[75] R. I. Laming, M. N. Zervas, and D. N. Payne, *IEEE Photon. Technol. Lett.* **4**, 1345 (1992).

[76] K. Kikuchi, *IEEE Photon. Technol. Lett.* **5**, 221 (1993).

[77] M. Murakami and S. Saito, *IEEE Photon. Technol. Lett.* **4**, 1269 (1992).

[78] E. Desurvire, C. R. Giles, and J. R. Simpson, *J. Lightwave Technol.* **7**, 2095 (1989).

[79] K. Inoue, H. Toba, N. Shibata, K. Iwatsuki, A. Takada, and M. Shimizu, *Electron. Lett.* **25**, 594 (1989).

[80] C. R. Giles, E. Desurvire, and J. R. Simpson, *Opt. Lett.* **14**, 880 (1990).

[81] A. R. Chraplyvy, R. W. Tkach, K. C. Reichmann, P. D. Magill, and J. A. Nagel, *IEEE Photon. Technol. Lett.* **5**, 428 (1993).

[82] K. Inoue, T. Korninaro, and H. Toba, *IEEE Photon. Technol. Lett.* **3**, 718 (1991).

[83] S. H. Yun, B. W. Lee, H. K. Kim, and B. Y. Kim, *IEEE Photon. Technol. Lett.* **11**, 1229 (1999).

[84] R. Kashyap, *Fiber Bragg Gratings*, Academic Press, San Diego, CA, 1999.

[85] P. F. Wysocki, J. B. Judkins, R. P. Espindola, M. Andrejco, A. M. Vengsarkar, and K. Walker, *IEEE Photon. Technol. Lett.* **9**, 1343 (1997).

[86] X. Y. Zhao, J. Bryce, and R. Minasian, *IEEE J. Sel. Topics Quantum Electron.* **3**, 1008 (1997).

[87] R. H. Richards, J. L. Jackel, and M. A. Ali, *IEEE J. Sel. Topics Quantum Electron.* **3**, 1027 (1997).

[88] G. Luo, J. L. Zyskind, J. A Nagel, and M. A. Ali, *J. Lightwave Technol.* **16**, 527 (1998).

[89] M. Karasek and J. A. Valles, *J. Lightwave Technol.* **16**, 1795 (1998).

[90] A. Bononi and L. Barbieri, *J. Lightwave Technol.* **17**, 1229 (1999).

[91] M. Karasek, M. Menif, and R. A. Rusch, *J. Lightwave Technol.* **19**, 933 (2001).

[92] H. Ono, M. Yamada, T. Kanamori, S. Sudo, and Y. Ohishi, *J. Lightwave Technol.* **17**, 490 (1999).

[93] P. F. Wysocki, N. Park, and D. DiGiovanni, *Opt. Lett.* **21**, 1744 (1996).

[94] M. Masuda and S. Kawai, *IEEE Photon. Technol. Lett.* **11**, 647 (1999).

[95] M. Yamada, H. Ono, T. Kanamori, S. Sudo, and Y. Ohishi, *Electron. Lett.* **33**, 710 (1997).

[96] M. Yamada, A. Mori, K. Kobayashi, H. Ono, T. Kanamori, K. Oikawa, Y. Nishida, and Y. Ohishi, *IEEE Photon. Technol. Lett.* **10**, 1244 (1998).

[97] K. Fukuchi, T. Kasamatsu, M. Morie, R. Ohhira, T. Ito, K. Sekiya, D. Ogasahara, and T. Ono, Paper PD24, *Proc. Optical Fiber Commun. Conf.*, Optical Society of America, Washington, DC, 2001.

[98] T. Kasamatsu, Y. Yano, and T. Ono, *IEEE Photon. Technol. Lett.* **13**, 433 (2001).

[99] D. L. Williams, S. T. Davey, D. M. Spirit, and B. L. Ainslie, *Electron. Lett.* **26**, 517 (1990).

[100] G. R. Walker, D. M. Spirit, D. L. Williams, and S. T. Davey, *Electron. Lett.* **27**, 1390 (1991)

[101] D. N. Chen and E. Desurvire, *IEEE Photon. Technol. Lett.* **4**, 52 (1992).

[102] K. Rottwitt, J. H. Povlsen, A. Bjarklev, O. Lumholt, B. Pedersen, and T. Rasmussen, *IEEE Photon. Technol. Lett.* **4**, 714 (1992); **5**, 218 (1993).

[103] K. Rottwitt, J. H. Povlsen, and A. Bjarklev, *J. Lightwave Technol.* **11**, 2105 (1993).

[104] C. Lester, K. Bertilsson, K. Rottwitt, P. A. Andrekson, M. A. Newhouse, and A. J. Antos, *Electron. Lett.* **31**, 219 (1995).

[105] A. Altuncu, L. Noel, W. A. Pender, A. S. Siddiqui, T. Widdowson, A. D. Ellis, M. A. Newhouse, A. J. Antos, G. Kar, and P. W. Chu, *Electron. Lett.* **32**, 233 (1996).

[106] M. Nissov, H. N. Poulsen, R. J. Pedersen, B. F. Jørgensen, M. A. Newhouse, and A. J. Antos, *Electron. Lett.* **32**, 1905 (1996).

[107] P. P. Smyth, R. Wyatt, A. Fidler, P. Eardley, A. Sayles, and S. Graig-Ryan, *Electron. Lett.* **26**, 1604 (1990).

[108] R. C. Steele and G. R. Walker, *IEEE Photon. Technol. Lett.* **2**, 753 (1990).

[109] T. L. Blair and H. Nakano, *Electron. Lett.* **27**, 835 (1991).

[110] T. Saito, Y. Sunohara, K. Fukagai, S. Ishikawa, N. Henmi, S. Fujita, and Y. Aoki, *IEEE Photon. Technol. Lett.* **3**, 551 (1991).

[111] A. H. Gnauck and C. R. Giles, *IEEE Photon. Technol. Lett.* **4**, 80 (1992).

[112] F. F. Röhl and R. W. Ayre, *IEEE Photon. Technol. Lett.* **5**, 358 (1993).

[113] R. C. Steele, G. R. Walker, and N. G. Walker, *IEEE Photon. Technol. Lett.* **3**, 545 (1991).

[114] O. K. Tonguz and L. G. Kazovsky, *J. Lightwave Technol.* **9**, 174 (1991).

[115] Y. K. Park, S. W. Granlund, T. W. Cline, L. D. Tzeng, J. S. French, J.-M. P. Delavaux, R. E. Tench, S. K. Korotky, J. J. Veselka, and D. J. DiGiovanni, *IEEE Photon. Technol. Lett.* **4**, 179 (1992).

[116] Y. K. Park, O. Mizuhara, L. D. Tzeng, J.-M. P. Delavaux, T. V. Nguyen, M. L. Kao, P. D. Yates, and J. Stone, *IEEE Photon. Technol. Lett.* **5**, 79 (1993).

[117] P. B. Hansen, L. Eskildsen, S. G. Grubb, A. M. Vengsarkar, S. K. Korotky, T. A. Strasser, J. E. J. Alphonsus, J. J. Veselka, D. J. DiGiovanni, et al., *Electron. Lett.* **32**, 1018 (1996).

[118] Y. Yamamoto and T. Mukai, *Opt. Quantum Electron.* **21**, S1 (1989).

[119] C. R. Giles and E. Desurvire, *J. Lightwave Technol.* **9**, 147 (1991).

[120] G. R. Walker, N. G. Walker, R. C. Steele, M. J. Creaner, and M. C. Brain, *J. Lightwave Technol.* **9**, 182 (1991).

[121] S. Ryu, S. Yamamoto, H. Taga, N. Edagawa, Y. Yoshida, and H. Wakabayashi, *J. Lightwave Technol.* **9**, 251 (1991).

[122] J. P. Gordon and H. A. Haus, *Opt. Lett.* **11**, 665 (1986).

[123] R. J. Essiambre and G. P. Agrawal, *J. Opt. Soc. Am. B* **14**, 314 (1997).

[124] E. Iannone, F. Matera, A. Mecozzi, and M. Settembre, *Nonlinear Optical Communication Networks*, Wiley, New York, 1998, Chap. 5.

[125] V. S. Grigoryan, C. R. Menyuk, and R. M. Mu, *J. Lightwave Technol.* **17**, 1347 (1999).

[126] J. Santhanam, C. J. McKinstrie, T. I. Lakoba, and G. P. Agrawal, *Opt. Lett.* **26** 1131 (2001).

[127] N. Edagawa, Y. Toshida, H. Taga, S. Yamamoto, K. Mochizuchi, and H. Wakabayashi, *Electron. Lett.* **26**, 66 (1990).

[128] S. Saito, T. Imai, and T. Ito, *J. Lightwave Technol.* **9**, 161 (1991).

[129] S. Saito, *J. Lightwave Technol.* **10**, 1117 (1992).

[130] T. Imai, M. Murakami, T. Fukuda, M. Aiki, and T. Ito, *Electron. Lett.* **28**, 1484 (1992).

[131] M. Murakami, T. Kataoka, T. Imai, K. Hagimoto, and M. Aiki, *Electron. Lett.* **28**, 2254 (1992).

[132] H. Taga, M. Suzuki, N. Edagawa, Y. Yoshida, S. Yamamoto, S. Akiba, and H. Wakabayashi, *Electron. Lett.* **28**, 2247 (1992).

[133] N. S. Bergano, J. Aspell, C. R. Davidson, P. R. Trischitta, B. M. Nyman, and F. W. Kerfoot, *Electron. Lett.* **27**, 1889 (1991).

[134] F. Matera and M. Settembre, *J. Lightwave Technol.* **14**, 1 (1996).

[135] X. Y. Zou, M. I. Hayee, S. M. Hwang, and A. E. Willner, *J. Lightwave Technol.* **14**, 1144 (1996).

[136] K. Inser and K. Petermann, *IEEE Photon. Technol. Lett.* **8**, 443 (1996).

[137] D. Breuer and K. Petermann, *IEEE Photon. Technol. Lett.* **9**, 398 (1997).

[138] F. Forghieri, P. R. Prucnal, R. W. Tkach, and A. R. Chraplyvy, *IEEE Photon. Technol. Lett.* **9**, 1035 (1997).

[139] F. Matera and M. Settembre, *Fiber Integ. Opt.* **15**, 89 (1996); *J. Opt. Commun.* **17**, 1 (1996); *Opt. Fiber Technol.* **4**, 34 (1998).

[140] S. Bigo, A. Bertaina, M. W. Chbat, S. Gurib, J. Da Loura, J. C. Jacquinot, J. Hervo, P. Bousselet, S. Borne, D. Bayart, L. Gasca, and J. L. Beylat, *IEEE Photon. Technol. Lett.* **10**, 1045 (1998).

[141] D. Breuer, K. Obermann, and K. Petermann, *IEEE Photon. Technol. Lett.* **10**, 1793 (1998).

[142] M. I. Hayee and A. E. Willner, *IEEE Photon. Technol. Lett.* **11**, 991 (1999).

[143] A. Sahara, H. Kubota, and M. Nakazawa, *Opt. Commun.* **160**, 139 (1999).

[144] R. Lebref, A. Ciani, F. Matera, and M. Tamburrini, *Fiber Integ. Opt.* **18**, 245 (1999).

[145] F. M. Madani and K. Kikuchi, *J. Lightwave Technol.* **17**, 1326 (1999).

[146] J. Kani, M. Jinno, T. Sakamoto, S. Aisawa, M. Fukui, K. Hattori, and K. Oguchi, *J. Lightwave Technol.* **17**, 2249 (1999).

[147] C. M. Weinert, R. Ludwig, W. Pieper, H. G. Weber, D. Breuer, K. Petermann, and F. Kuppers, *J. Lightwave Technol.* **17**, 2276 (1999).

[148] G. P. Agrawal, *Applications of Nonlinear Fiber Optics*, Academic Press, San Diego, CA, 2001.

[149] H. Taga, Y. Yoshida, S. Yamamoto, and H. Wakabayashi, *Electron. Lett.* **26**, 500 (1990); *J. Lightwave Technol.* **14**, 1287 (1996).

[150] H. Toga, N. Edagawa, Y. Yoshida, S. Yamamoto, and H. Wakabayashi, *Electron. Lett.* **29**, 485 (1993).

[151] H. Onaka, H. Miyata, G. Ishikawa, K. Otsuka, H. Ooi, Y. Kai, S. Kinoshita, M. Seino, H. Nishimoto, and T. Chikama, Paper PD19, *Proc. Optical Fiber Commun. Conf.*, Optical Society of America, Washington, DC, 1996.

[152] N. S. Bergano and C. R. Davidson, *J. Lightwave Technol.* **14**, 1287 (1996).

[153] O. Gautheron, G. Bassier, V. Letellier, G. Grandpierre, and P. Bollaert, *Electron. Lett.* **32**, 1019 (1996).

[154] J. X. Cai, M. Nissov, A. N. Pilipetskii, A.J. Lucero, C. . Davidson, D. Foursa, H. Kidorf, M. A. Mills, R. Menges, P. C. Corbett, D. Sutton, and N. S. Bergano, Paper PD20, *Proc. Optical Fiber Commun. Conf.*, Optical Society of America, Washington, DC, 2001.

[155] M. Shtaif, M. Eiselt, R. W. Tkach, R. H. Stolen, and A. H. Gnauck, *IEEE Photon. Technol. Lett.* **10**, 1796 (1998).

[156] M. Eiselt, M. Shtaif, R. W. Tkach, F. A. Flood, S. Ten, and D. Butler, *IEEE Photon. Technol. Lett.* **11**, 1575 (1999).

[157] G. J. Pendock, S. Y. Park, A. K. Srivastava, S. Radic, J. W. Sulhoff, C. L. Wolf, K. Kantor, and Y. Sun, *IEEE Photon. Technol. Lett.* **11**, 1578 (1999).

[158] R. Q. Hui, K. R. Demarest, and C. T. Allen, *J. Lightwave Technol.* **17**, 1018 (1999).

[159] S. Reichel and R. Zengerle, *J. Lightwave Technol.* **17**, 1152 (1999).

[160] A. Sano, Y. Miyamoto, S. Kuwahara, and H. Toba, *J. Lightwave Technol.* **18**, 1519 (2000).

[161] S. Radic, G. J. Pendock, A. K. Srivastava, P. Wysocki, and A. Chraplyvy, *J. Lightwave Technol.* **19**, 636 (2001).

Chapter 7

Dispersion Management

It should be clear from Chapter 6 that with the advent of optical amplifiers, fiber losses are no longer a major limiting factor for optical communication systems. Indeed, modern lightwave systems are often limited by the dispersive and nonlinear effects rather than fiber losses. In some sense, optical amplifiers solve the loss problem but, at the same time, worsen the dispersion problem since, in contrast with electronic regenerators, an optical amplifier does not restore the amplified signal to its original state. As a result, dispersion-induced degradation of the transmitted signal accumulates over multiple amplifiers. For this reason, several dispersion-management schemes were developed during the 1990s to address the dispersion problem [1]. In this chapter we review these techniques with emphasis on the underlying physics and the improvement realized in practice. In Section 7.1 we explain why dispersion management is needed. Sections 7.2 and 7.3 are devoted to the methods used at the transmitter or receiver for managing the dispersion. In Sections 7.4–7.6 we consider the use of several high-dispersion optical elements along the fiber link. The technique of optical phase conjugation, also known as midspan spectral inversion, is discussed in Section 7.7. Section 7.8 is devoted to dispersion management in long-haul systems. Section 7.9 focuses on high-capacity systems by considering broadband, tunable, and higher-order compensation techniques. Polarization-mode dispersion (PMD) compensation is also discussed in this section.

7.1 Need for Dispersion Management

In Section 2.4 we have discussed the limitations imposed on the system performance by dispersion-induced pulse broadening. As shown by the dashed line in Fig. 2.13, the group-velocity dispersion (GVD) effects can be minimized using a narrow-linewidth laser and operating close to the zero-dispersion wavelength λ_{ZD} of the fiber. However, it is not always practical to match the operating wavelength λ with λ_{ZD}. An example is provided by the third-generation terrestrial systems operating near 1.55 μm and using optical transmitters containing a distributed feedback (DFB) laser. Such systems generally use the existing fiber-cable network installed during the 1980s and consisting of more than 50 million kilometers of the "standard" single-mode fiber with $\lambda_{ZD} \approx 1.31$ μm. Since the dispersion parameter $D \approx 16$ ps/(km-nm) in the 1.55-μm region of such fibers, the GVD severely limits the performance when the bit rate exceeds 2 Gb/s (see Fig. 2.13). For a directly modulated DFB laser, we can use Eq. (2.4.26) for estimating the maximum transmission distance so that

$$L < (4B|D|s_\lambda)^{-1}, \tag{7.1.1}$$

where s_λ is the root-mean-square (RMS) width of the pulse spectrum broadened considerably by frequency chirping (see Section 3.5.3). Using $D = 16$ ps/(km-nm) and $s_\lambda = 0.15$ nm in Eq. (7.1.1), lightwave systems operating at 2.5 Gb/s are limited to $L \approx 42$ km. Indeed, such systems use electronic regenerators, spaced apart by about 40 km, and cannot benefit from the availability of optical amplifiers. Furthermore, their bit rate cannot be increased beyond 2.5 Gb/s because the regenerator spacing becomes too small to be feasible economically.

System performance can be improved considerably by using an external modulator and thus avoiding spectral broadening induced by frequency chirping. This option has become practical with the commercialization of transmitters containing DFB lasers with a monolithically integrated modulator. The $s_\lambda = 0$ line in Fig. 2.13 provides the dispersion limit when such transmitters are used with the standard fibers. The limiting transmission distance is then obtained from Eq. (2.4.31) and is given by

$$L < (16|\beta_2|B^2)^{-1}, \qquad D = -\frac{2\pi c}{\lambda^2}\beta_2 \tag{7.1.2}$$

where β_2 is the GVD coefficient related to D by Eq. (2.3.5). If we use a typical value $\beta_2 = -20$ ps^2/km at 1.55 μm, $L < 500$ km at 2.5 Gb/s. Although improved considerably compared with the case of directly modulated DFB lasers, this dispersion limit becomes of concern when in-line amplifiers are used for loss compensation. Moreover, if the bit rate is increased to 10 Gb/s, the GVD-limited transmission distance drops to 30 km, a value so low that optical amplifiers cannot be used in designing such lightwave systems. It is evident from Eq. (7.1.2) that the relatively large GVD of standard single-mode fibers severely limits the performance of 1.55-μm systems designed to use the existing telecommunication network at a bit rate of 10 Gb/s or more.

A dispersion-management scheme attempts to solve this practical problem. The basic idea behind all such schemes is quite simple and can be understood by using the pulse-propagation equation derived in Section 2.4.1 and written as

$$\frac{\partial A}{\partial z} + \frac{i\beta_2}{2}\frac{\partial^2 A}{\partial t^2} - \frac{\beta_3}{6}\frac{\partial^3 A}{\partial t^3} = 0, \tag{7.1.3}$$

where A is the pulse-envelope amplitude. The effects of third-order dispersion are included by the β_3 term. In practice, this term can be neglected when $|\beta_2|$ exceeds 0.1 ps^2/km. Equation (7.1.3) has been solved in Section 2.4.2, and the solution is given by Eq. (2.4.15). In the specific case of $\beta_3 = 0$ the solution becomes

$$A(z,t) = \frac{1}{2\pi}\int_{-\infty}^{\infty}\tilde{A}(0,\omega)\exp\left(\frac{i}{2}\beta_2 z\omega^2 - i\omega t\right)d\omega, \tag{7.1.4}$$

where $\tilde{A}(0,\omega)$ is the Fourier transform of $A(0,t)$.

Dispersion-induced degradation of the optical signal is caused by the phase factor $\exp(i\beta_2 z\omega^2/2)$, acquired by spectral components of the pulse during its propagation in the fiber. All dispersion-management schemes attempt to cancel this phase factor so that the input signal can be restored. Actual implementation can be carried out at the transmitter, at the receiver, or along the fiber link. In the following sections we consider the three cases separately.

7.2 Precompensation Schemes (transmitter)

This approach to dispersion management modifies the characteristics of input pulses at the transmitter before they are launched into the fiber link. The underlying idea can be understood from Eq. (7.1.4). It consists of changing the spectral amplitude $\tilde{A}(0,\omega)$ of the input pulse in such a way that GVD-induced degradation is eliminated, or at least reduced substantially. Clearly, if the spectral amplitude is changed as

$$\tilde{A}(0,\omega) \rightarrow \tilde{A}(0,\omega)\exp(-i\omega^2\beta_2 L/2), \qquad (7.2.1)$$

where L is the fiber length, GVD will be compensated exactly, and the pulse will retain its shape at the fiber output. Unfortunately, it is not easy to implement Eq. (7.2.1) in practice. In a simple approach, the input pulse is chirped suitably to minimize the GVD-induced pulse broadening. Since the frequency chirp is applied at the transmitter before propagation of the pulse, this scheme is called the prechirp technique.

7.2.1 Prechirp Technique

A simple way to understand the role of prechirping is based on the theory presented in Section 2.4 where propagation of chirped Gaussian pulses in optical fibers is discussed. The input amplitude is taken to be

$$A(0,t) = A_0\exp\left[-\frac{1+iC}{2}\left(\frac{t}{T_0}\right)^2\right], \qquad (7.2.2)$$

where C is the chirp parameter. As seen in Fig. 2.12, for values of C such that $\beta_2 C < 0$, the input pulse initially compresses in a dispersive fiber. Thus, a suitably chirped pulse can propagate over longer distances before it broadens outside its allocated bit slot. As a rough estimate of the improvement, consider the case in which pulse broadening by a factor of up to $\sqrt{2}$ is acceptable. By using Eq. (2.4.17) with $T_1/T_0 = \sqrt{2}$, the transmission distance is given by

$\left. \frac{T_1}{T_0} = \left[\left(1 + \frac{C\beta_2 z}{T_0^2}\right)^2 + \left(\frac{\beta_2 z}{T_0^2}\right)^2\right]^{1/2} \right.$

$$L = \frac{C+\sqrt{1+2C^2}}{1+C^2}L_D, \qquad (7.2.3)$$

where $L_D = T_0^2/|\beta_2|$ is the dispersion length. For unchirped Gaussian pulses, $C = 0$ and $L = L_D$. However, L increases by 36% for $C = 1$. Note also that $L < L_D$ for large values of C. In fact, the maximum improvement by a factor of $\sqrt{2}$ occurs for

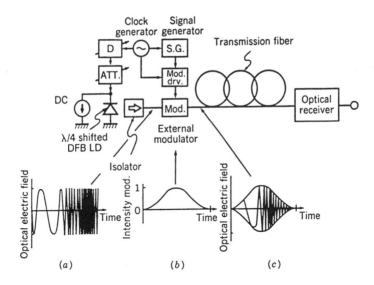

Figure 7.1: Schematic of the prechirp technique used for dispersion compensation: (a) FM output of the DFB laser; (b) pulse shape produced by external modulator; and (c) prechirped pulse used for signal transmission. (After Ref. [9]; ©1994 IEEE; reprinted with permission.)

$C = 1/\sqrt{2}$. These features clearly illustrate that the prechirp technique requires careful optimization. Even though the pulse shape is rarely Gaussian in practice, the prechirp technique can increase the transmission distance by a factor of about 2 when used with care. As early as 1986, a super-Gaussian model [2] suitable for nonreturn-to-zero (NRZ) transmission predicted such an improvement, a feature also evident in Fig. 2.14, which shows the results of numerical simulations for chirped super-Gaussian pulses.

The prechirp technique was considered during the 1980s in the context of directly modulated semiconductor lasers [2]–[5]. Such lasers chirp the pulse automatically through the carrier-induced index changes governed by the linewidth enhancement factor β_c (see Section 3.5.3). Unfortunately, the chirp parameter C is negative ($C = -\beta_c$) for directly modulated semiconductor lasers. Since β_2 in the 1.55-μm wavelength region is also negative for standard fibers, the condition $\beta_2 C < 0$ is not satisfied. In fact, as seen in Fig. 2.12, the chirp induced during direct modulation increases GVD-induced pulse broadening, thereby reducing the transmission distance drastically. Several schemes during the 1980s considered the possibility of shaping the current pulse appropriately in such a way that the transmission distance improved over that realized without current-pulse shaping [3]–[5].

In the case of external modulation, optical pulses are nearly chirp-free. The prechirp technique in this case imposes a frequency chirp with a positive value of the chirp parameter C so that the condition $\beta_2 C < 0$ is satisfied. Several schemes have been proposed for this purpose [6]–[12]. In a simple approach shown schematically in Fig. 7.1, the frequency of the DFB laser is first frequency modulated (FM) before the laser output is passed to an external modulator for amplitude modulation (AM). The resulting optical signal exhibits simultaneous AM and FM [9]. From a practical standpoint, FM of the optical carrier can be realized by modulating the current injected into the DFB laser by a small amount (~ 1 mA). Although such a direct modulation of the DFB laser also modulates the optical power sinusoidally, the magnitude is small enough that it does not interfere with the detection process.

It is clear from Fig. 7.1 that FM of the optical carrier, followed by external AM, generates a signal that consists of chirped pulses. The amount of chirp can be determined as follows. Assuming that the pulse shape is Gaussian, the optical signal can be written as

$$E(0,t) = A_0 \exp(-t^2/T_0^2) \exp[-i\omega_0(1 + \delta \sin \omega_m t)t], \qquad (7.2.4)$$

where the carrier frequency ω_0 of the pulse is modulated sinusoidally at the frequency ω_m with a modulation depth δ. Near the pulse center, $\sin(\omega_m t) \approx \omega_m t$, and Eq. (7.2.4) becomes

$$E(0,t) \approx A_0 \exp\left[-\frac{1+iC}{2}\left(\frac{t}{T_0}\right)^2\right]\exp(-i\omega_0 t), \qquad (7.2.5)$$

where the chirp parameter C is given by

$$C = 2\delta\omega_m\omega_0 T_0^2. \qquad (7.2.6)$$

Both the sign and magnitude of the chirp parameter C can be controlled by changing the FM parameters δ and ω_m.

Phase modulation of the optical carrier also leads to a positive chirp, as can be verified by replacing Eq. (7.2.4) with

$$E(0,t) = A_0 \exp(-t^2/T_0^2) \exp[-i\omega_0 t + i\delta \cos(\omega_m t)] \qquad (7.2.7)$$

and using $\cos x \approx 1 - x^2/2$. An advantage of the phase-modulation technique is that the external modulator itself can modulate the carrier phase. The simplest solution is to employ an external modulator whose refractive index can be changed electronically in such a way that it imposes a frequency chirp with $C > 0$ [6]. As early as 1991, a 5-Gb/s signal was transmitted over 256 km [7] using a LiNbO$_3$ modulator such that values of C were in the range 0.6–0.8. These experimental values are in agreement with the Gaussian-pulse theory on which Eq. (7.2.3) is based. Other types of semiconductor modulators, such as an electroabsorption modulator [8] or a Mach–Zehnder (MZ) modulator [10], can also chirp the optical pulse with $C > 0$, and have indeed been used to demonstrate transmission beyond the dispersion limit [11]. With the development of DFB lasers containing a monolithically integrated electroabsorption modulator, the implementation of the prechirp technique has become quite practical. In a 1996 experiment, a 10-Gb/s NRZ signal was transmitted over 100 km of standard fiber using such a transmitter [12].

7.2.2 Novel Coding Techniques

Simultaneous AM and FM of the optical signal is not essential for dispersion compensation. In a different approach, referred to as *dispersion-supported transmission*, the frequency-shift keying (FSK) format is used for signal transmission [13]–[17]. The FSK signal is generated by switching the laser wavelength by a constant amount $\Delta\lambda$

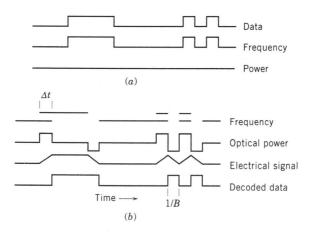

Figure 7.2: Dispersion compensation using FSK coding: (a) Optical frequency and power of the transmitted signal; (b) frequency and power of the received signal and the electrically decoded data. (After Ref. [13]; ©1994 IEEE; reprinted with permission.)

between 1 and 0 bits while leaving the power unchanged (see Chapter 10). During propagation inside the fiber, the two wavelengths travel at slightly different speeds. The time delay between the 1 and 0 bits is determined by the wavelength shift $\Delta\lambda$ and is given by $\Delta T = DL\Delta\lambda$, as shown in Eq. (2.3.4). The wavelength shift $\Delta\lambda$ is chosen such that $\Delta T = 1/B$. Figure 7.2 shows schematically how the one-bit delay produces a three-level optical signal at the receiver. In essence, because of fiber dispersion, the FSK signal is converted into a signal whose amplitude is modulated. The signal can be decoded at the receiver by using an electrical integrator in combination with a decision circuit [13].

Several transmission experiments have shown the usefulness of the dispersion-supported transmission scheme [13]–[15]. All of these experiments were concerned with increasing the transmission distance of a 1.55-μm lightwave system operating at 10 Gb/s or more over the standard fibers. In 1994, transmission of a 10-Gb/s signal over 253 km of standard fiber was realized [13]. By 1998, in a 40-Gb/s field trial, the signal was transmitted over 86 km of standard fiber [15]. These values should be compared with the prediction of Eq. (7.1.2). Clearly, the transmission distance can be improved by a large factor by using the FSK technique when the system is properly designed [17].

Another approach for increasing the transmission distance consists of transmitting an optical signal whose bandwidth at a given bit rate is smaller compared with that of the standard on–off coding technique. One scheme makes use of the *duobinary coding*, which can reduce the signal bandwidth by 50% [18]. In the simplest duobinary scheme, the two successive bits in the digital bit stream are summed, forming a three-level duobinary code at half the bit rate. Since the GVD-induced degradation depends on the signal bandwidth, the transmission distance should improve for a reduced-bandwidth signal. This is indeed found to be the case experimentally [19]–[24].

In a 1994 experiment designed to compare the binary and duobinary schemes, a

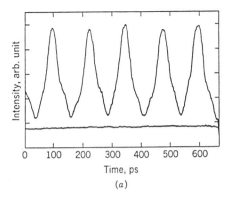 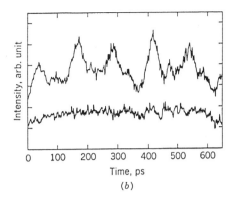

Figure 7.3: Streak-camera traces of the 16-Gb/s signal transmitted over 70 km of standard fiber (a) with and (b) without SOA-induced chirp. Bottom trace shows the background level in each case. (After Ref. [26]; ©1989 IEE; reprinted with permission.)

10-Gb/s signal could be transmitted over distances 30 to 40 km longer by replacing binary coding with duobinary coding [19]. The duobinary scheme can be combined with the prechirping technique. Indeed, transmission of a 10-Gb/s signal over 160 km of a standard fiber has been realized by combining duobinary coding with an external modulator capable of producing a frequency chirp with $C > 0$ [19]. Since chirping increases the signal bandwidth, it is hard to understand why it would help. It appears that phase reversals occurring in practice when a duobinary signal is generated are primarily responsible for improvement realized with duobinary coding [20]. A new dispersion-management scheme, called the *phase-shaped* binary transmission, has been proposed to take advantage of phase reversals [21]. The use of duobinary transmission increases signal-to-noise requirements and requires decoding at the receiver. Despite these shortcomings, it is useful for upgrading the existing terrestrial lightwave systems to bit rates of 10 Gb/s and more [22]–[24].

7.2.3 Nonlinear Prechirp Techniques

A simple nonlinear prechirp technique, demonstrated in 1989, amplifies the transmitter output using a semiconductor optical amplifier (SOA) operating in the gain-saturation regime [25]–[29]. As discussed in Section 6.2.4, gain saturation leads to time-dependent variations in the carrier density, which, in turn, chirp the amplified pulse through carrier-induced variations in the refractive index. The amount of chirp is given by Eq. (6.2.23) and depends on the input pulse shape. As seen in Fig. 6.8, the chirp is nearly linear over most of the pulse. The SOA not only amplifies the pulse but also chirps it such that the chirp parameter $C > 0$. Because of this chirp, the input pulse can be compressed in a fiber with $\beta_2 < 0$. Such a compression was observed in an experiment in which 40-ps input pulses were compressed to 23 ps when they were propagated over 18 km of standard fiber [25].

The potential of this technique for dispersion compensation was demonstrated in a 1989 experiment by transmitting a 16-Gb/s signal, obtained from a mode-locked external-cavity semiconductor laser, over 70 km of fiber [26]. Figure 7.3 compares the streak-camera traces of the signal obtained with and without dispersion compensation. From Eq. (7.1.2), in the absence of amplifier-induced chirp, the transmission distance at 16 Gb/s is limited by GVD to about 14 km for a fiber with $D = 15$ ps/(km-nm). The use of the amplifier in the gain-saturation regime increased the transmission distance fivefold, a feature that makes this approach to dispersion compensation quite attractive. It has an added benefit that it can compensate for the coupling and insertion losses that invariably occur in a transmitter by amplifying the signal before it is launched into the optical fiber. Moreover, this technique can be used for simultaneous compensation of fiber losses and GVD if SOAs are used as in-line amplifiers [29].

A nonlinear medium can also be used to prechirp the pulse. As discussed in Section 2.6, the intensity-dependent refractive index chirps an optical pulse through the phenomenon of self-phase modulation (SPM). Thus, a simple prechirp technique consists of passing the transmitter output through a fiber of suitable length before launching it into the fiber link. Using Eq. (2.6.13), the optical signal at the fiber input is given by

$$A(0,t) = \sqrt{P(t)}\exp[i\gamma L_m P(t)],\qquad(7.2.8)$$

where $P(t)$ is the power of the pulse, L_m is the length of the nonlinear medium, and γ is the nonlinear parameter. In the case of Gaussian pulses for which $P(t) = P_0\exp(-t^2/T_0^2)$, the chirp is nearly linear, and Eq. (7.2.8) can be approximated by

$$A(0,t) \approx \sqrt{P_0}\exp\left[-\frac{1+iC}{2}\left(\frac{t}{T_0}\right)^2\right]\exp(-i\gamma L_m P_0),\qquad(7.2.9)$$

where the chirp parameter is given by $C = 2\gamma L_m P_0$. For $\gamma > 0$, the chirp parameter C is positive, and is thus suitable for dispersion compensation.

Since $\gamma > 0$ for silica fibers, the transmission fiber itself can be used for chirping the pulse. This approach was suggested in a 1986 study [30]. It takes advantage of higher-order solitons which pass through a stage of initial compression (see Chapter 9). Figure 7.4 shows the GVD-limited transmission distance as a function of the average launch power for 4- and 8-Gb/s lightwave systems. It indicates the possibility of doubling the transmission distance by optimizing the average power of the input signal to about 3 mW.

7.3 Postcompensation Techniques (receiver)

Electronic techniques can be used for compensation of GVD within the receiver. The philosophy behind this approach is that even though the optical signal has been degraded by GVD, one may be able to equalize the effects of dispersion electronically if the fiber acts as a *linear system*. It is relatively easy to compensate for dispersion if a heterodyne receiver is used for signal detection (see Section 10.1). A heterodyne receiver first converts the optical signal into a microwave signal at the intermediate frequency ω_{IF} while preserving both the amplitude and phase information. A microwave bandpass filter whose impulse response is governed by the transfer function

$$H(\omega) = \exp[-i(\omega - \omega_{IF})^2\beta_2 L/2],\qquad(7.3.1)$$

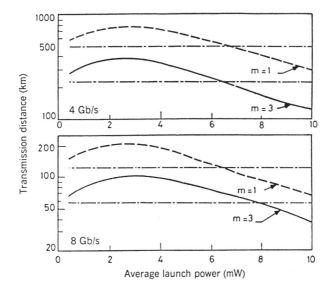

Figure 7.4: Dispersion-limited transmission distance as a function of launch power for Gaussian ($m = 1$) and super-Gaussian ($m = 3$) pulses at bit rates of 4 and 8 Gb/s. Horizontal lines correspond to the linear case. (After Ref. [30]; ©1986 IEE; reprinted with permission.)

where L is the fiber length, should restore to its original form the signal received. This conclusion follows from the standard theory of linear systems (see Section 4.3.2) by using Eq. (7.1.4) with $z = L$. This technique is most practical for dispersion compensation in coherent lightwave systems [31]. In a 1992 transmission experiment, a 31.5-cm-long *microstrip line* was used for dispersion equalization [32]. Its use made it possible to transmit the 8-Gb/s signal over 188 km of standard fiber having a dispersion of 18.5 ps/(km-nm). In a 1993 experiment, the technique was extended to homodyne detection using *single-sideband transmission* [33], and the 6-Gb/s signal could be recovered at the receiver after propagating over 270 km of standard fiber. Microstrip lines can be designed to compensate for GVD acquired over fiber lengths as long as 4900 km for a lightwave system operating at a bit rate of 2.5 Gb/s [34].

As discussed in Chapter 10, use of a coherent receiver is often not practical. An electronic dispersion equalizer is much more practical for a direct-detection receiver. A linear electronic circuit cannot compensate GVD in this case. The problem lies in the fact that all phase information is lost during direct detection as a photodetector responds to optical intensity only (see Chapter 4). As a result, no linear equalization technique can recover a signal that has spread outside its allocated bit slot. Nevertheless, several nonlinear equalization techniques have been developed that permit recovery of the degraded signal [35]–[38]. In one method, the decision threshold, normally kept fixed at the center of the eye diagram (see Section 4.3.3), is varied depending on the preceding bits. In another, the decision about a given bit is made after examining the analog waveform over a multiple-bit interval surrounding the bit in question [35]. The main difficulty with all such techniques is that they require electronic logic circuits, which must operate at the bit rate and whose complexity increases exponentially with the number of bits over which an optical pulse has spread because of GVD-induced pulse broadening. Consequently, electronic equalization is generally limited to low bit rates and to transmission distances of only a few dispersion lengths.

An optoelectronic equalization technique based on a *transversal filter* has also been proposed [39]. In this technique, a power splitter at the receiver splits the received optical signal into several branches. Fiber-optic delay lines introduce variable delays in different branches. The optical signal in each branch is converted into photocurrent by using variable-sensitivity photodetectors, and the summed photocurrent is used by the decision circuit. The technique can extend the transmission distance by about a factor of 3 for a lightwave system operating at 5 Gb/s.

7.4 Dispersion-Compensating Fibers *(fiber)*

The preceding techniques may extend the transmission distance of a dispersion-limited system by a factor of 2 or so but are unsuitable for long-haul systems for which GVD must be compensated along the transmission line in a periodic fashion. What one needs for such systems is an all-optical, fiber-based, dispersion-management technique [40]. A special kind of fiber, known as the *dispersion-compensating fiber* (DCF), has been developed for this purpose [41]–[44]. The use of DCF provides an all-optical technique that is capable of compensating the fiber GVD completely if the average optical power is kept low enough that the nonlinear effects inside optical fibers are negligible. It takes advantage of the linear nature of Eq. (7.1.3).

To understand the physics behind this dispersion-management technique, consider the situation in which each optical pulse propagates through two fiber segments, the second of which is the DCF. Using Eq. (7.1.4) for each fiber section consecutively, we obtain

$$A(L,t) = \frac{1}{2\pi} \int_{-\infty}^{\infty} \tilde{A}(0,\omega) \exp\left[\frac{i}{2}\omega^2(\beta_{21}L_1 + \beta_{22}L_2) - i\omega t\right] d\omega, \qquad (7.4.1)$$

where $L = L_1 + L_2$ and β_{2j} is the GVD parameter for the fiber segment of length L_j ($j = 1, 2$). If the DCF is chosen such that the ω^2 phase term vanishes, the pulse will recover its original shape at the end of DCF. The condition for perfect dispersion compensation is thus $\beta_{21}L_1 + \beta_{22}L_2 = 0$, or

$$D_1 L_1 + D_2 L_2 = 0. \qquad (7.4.2)$$

Equation (7.4.2) shows that the DCF must have normal GVD at 1.55 μm ($D_2 < 0$) because $D_1 > 0$ for standard telecommunication fibers. Moreover, its length should be chosen to satisfy

$$L_2 = -(D_1/D_2)L_1. \qquad (7.4.3)$$

For practical reasons, L_2 should be as small as possible. This is possible only if the DCF has a large negative value of D_2.

Although the idea of using a DCF has been around since 1980 [40], it was only after the advent of optical amplifiers around the 1990 that the development of DCFs accelerated in pace. There are two basic approaches to designing DCFs. In one approach, the DCF supports a single mode, but it is designed with a relatively small value of the fiber parameter V given in Eq. (2.2.35). As discussed in Section 2.2.3 and seen in Fig. 2.7, the fundamental mode is weakly confined for $V \approx 1$. As a large fraction of the mode propagates inside the cladding layer, where the refractive index is smaller, the waveguide contribution to the GVD is quite different and results in values of $D \sim -100$ ps/(km-nm). A depressed-cladding design is often used in practice for making DCFs [41]–[44]. Unfortunately, DCFs also exhibit relatively high losses because of increase in bending losses ($\alpha = 0.4$–0.6 dB/km). The ratio $|D|/\alpha$ is often used as a figure of merit M for characterizing various DCFs [41]. By 1997, DCFs with $M > 250$ ps/(nm-dB) have been fabricated.

A practical solution for upgrading the terrestrial lightwave systems making use of the existing standard fibers consists of adding a DCF module (with 6–8 km of DCF) to optical amplifiers spaced apart by 60–80 km. The DCF compensates GVD while the amplifier takes care of fiber losses. This scheme is quite attractive but suffers from two problems. First, insertion losses of a DCF module typically exceed 5 dB. Insertion losses can be compensated by increasing the amplifier gain but only at the expense of enhanced amplified spontaneous emission (ASE) noise. Second, because of a relatively small mode diameter of DCFs, the effective mode area is only $\sim 20~\mu\mathrm{m}^2$. As the optical intensity is larger inside a DCF at a given input power, the nonlinear effects are considerably enhanced [44].

The problems associated with a DCF can be solved to a large extent by using a *two-mode fiber* designed with values of V such that the higher-order mode is near cutoff ($V \approx 2.5$). Such fibers have almost the same loss as the single-mode fiber but can be designed such that the dispersion parameter D for the higher-order mode has large negative values [45]–[48]. Indeed, values of D as large as -770 ps/(km-nm) have been measured for elliptical-core fibers [45]. A 1-km length of such a DCF can compensate the GVD for a 40-km-long fiber link, adding relatively little to the total link loss.

The use of a two-mode DCF requires a mode-conversion device capable of converting the energy from the fundamental mode to the higher-order mode supported by the DCF. Several such all-fiber devices have been developed [49]–[51]. The all-fiber nature of the mode-conversion device is important from the standpoint of compatibility with the fiber network. Moreover, such an approach reduces the insertion loss. Additional requirements on a mode converter are that it should be polarization insensitive and should operate over a broad bandwidth. Almost all practical mode-conversion devices use a two-mode fiber with a fiber grating that provides coupling between the two modes. The grating period Λ is chosen to match the mode-index difference $\delta\bar{n}$ of the two modes ($\Lambda = \lambda/\delta\bar{n}$) and is typically $\sim 100~\mu\mathrm{m}$. Such gratings are called long-period fiber gratings [51]. Figure 7.5 shows schematically a two-mode DCF with two long-period gratings. The measured dispersion characteristics of this DCF are also shown [47]. The parameter D has a value of -420 ps/(km-nm) at 1550 nm and changes considerably with wavelength. This is an important feature that allows for broadband dispersion compensation [48]. In general, DCFs are designed such that $|D|$ increases with wavelength. The wavelength dependence of D plays an important role for wavelength-division multiplexed (WDM) systems. This issue is discussed later in Section 7.9.

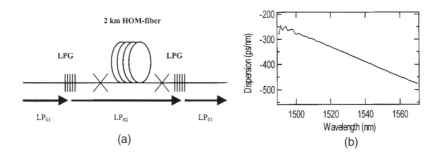

Figure 7.5: (a) Schematic of a DCF made using a higher-order mode (HOM) fiber and two long-period gratings (LPGs). (b) Dispersion spectrum of the DCF. (After Ref. [47]; ©2001 IEEE; reprinted with permission.)

7.5   Optical Filters *(concepts)*

A shortcoming of DCFs is that a relatively long length (> 5 km) is required to compensate the GVD acquired over 50 km of standard fiber. This adds considerably to the link loss, especially in the case of long-haul applications. For this reason, several other all-optical schemes have been developed for dispersion management. Most of them can be classified under the category of *optical equalizing filters*. Interferometric filters are considered in this section while the next section is devoted to fiber gratings.

The function of optical filters is easily understood from Eq. (7.1.4). Since the GVD affects the optical signal through the spectral phase $\exp(i\beta_2 z\omega^2/2)$, it is evident that an optical filter whose transfer function cancels this phase will restore the signal. Unfortunately, no optical filter (except for an optical fiber) has a transfer function suitable for compensating the GVD exactly. Nevertheless, several optical filters have provided partial GVD compensation by mimicking the ideal transfer function. Consider an optical filter with the transfer function $H(\omega)$. If this filter is placed after a fiber of length L, the filtered optical signal can be written using Eq. (7.1.4) as

$$A(L,t) = \frac{1}{2\pi}\int_{-\infty}^{\infty}\tilde{A}(0,\omega)H(\omega)\exp\left(\frac{i}{2}\beta_2 L\omega^2 - i\omega t\right)d\omega, \qquad (7.5.1)$$

By expanding the phase of $H(\omega)$ in a Taylor series and retaining up to the quadratic term,

$$H(\omega) = |H(\omega)|\exp[i\phi(\omega)] \approx |H(\omega)|\exp[i(\phi_0 + \phi_1\omega + \tfrac{1}{2}\phi_2\omega^2)], \qquad (7.5.2)$$

where $\phi_m = d^m\phi/d\omega^m (m = 0,1,\dots)$ is evaluated at the optical carrier frequency ω_0. The constant phase ϕ_0 and the time delay ϕ_1 do not affect the pulse shape and can be ignored. The spectral phase introduced by the fiber can be compensated by choosing an optical filter such that $\phi_2 = -\beta_2 L$. The pulse will recover perfectly only if $|H(\omega)| = 1$ and the cubic and higher-order terms in the Taylor expansion in Eq. (7.5.2) are negligible. Figure 7.6 shows schematically how such an optical filter can be combined with optical amplifiers such that both fiber losses and GVD can be compensated simultaneously. Moreover, the optical filter can also reduce the amplifier noise if its bandwidth is much smaller than the amplifier bandwidth.

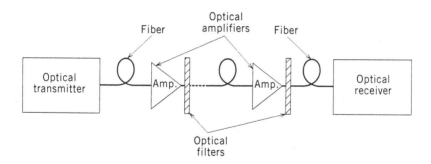

Figure 7.6: Dispersion management in a long-haul fiber link using optical filters after each amplifier. Filters compensate for GVD and also reduce amplifier noise.

Optical filters can be made using an interferometer which, by its very nature, is sensitive to the frequency of the input light and acts as an optical filter because of its frequency-dependent transmission characteristics. A simple example is provided by the Fabry–Perot (FP) interferometer encountered in Sections 3.3.2 and 6.2.1 in the context of a laser cavity. In fact, the transmission spectrum $|H_{FP}|^2$ of a FP interferometer can be obtained from Eq. (6.2.1) by setting $G = 1$ if losses per pass are negligible. For dispersion compensation, we need the frequency dependence of the phase of the transfer function $H(\omega)$, which can be obtained by considering multiple round trips between the two mirrors. A reflective FP interferometer, known as the *Gires–Tournois interferometer*, is designed with a back mirror that is 100% reflective. Its transfer function is given by [52]

$$H_{FP}(\omega) = H_0 \frac{1 + r\exp(-i\omega T)}{1 + r\exp(i\omega T)},\tag{7.5.3}$$

where the constant H_0 takes into account all losses, $|r|^2$ is the front-mirror reflectivity, and T is the round-trip time within the FP cavity. Since $|H_{FP}(\omega)|$ is frequency independent, only the spectral phase is modified by the FP filter. However, the phase $\phi(\omega)$ of $H_{FP}(\omega)$ is far from ideal. It is a periodic function that peaks at the FP resonances (longitudinal-mode frequencies of Section 3.3.2). In the vicinity of each peak, a spectral region exists in which the phase variation is nearly quadratic. By expanding $\phi(\omega)$ in a Taylor series, ϕ_2 is given by

$$\phi_2 = 2T^2 r(1-r)/(1+r)^3.\tag{7.5.4}$$

As an example, for a 2-cm-long FP cavity with $r = 0.8$, $\phi_2 \approx 2200$ ps^2. Such a filter can compensate the GVD acquired over 110 km of standard fiber. In a 1991 experiment [53], such an all-fiber device was used to transmit a 8-Gb/s signal over 130 km of standard fiber. The relatively high insertion loss of 8 dB was compensated by using an optical amplifier. A loss of 6 dB was due to a 3-dB fiber coupler used to separate the reflected signal from the incident signal. This amount can be reduced to about 1 dB using an optical circulator, a three-port device that transfers power one port to another in a circular fashion. Even then, relatively high losses and narrow bandwidths of FP filters limit their use in practical lightwave systems.

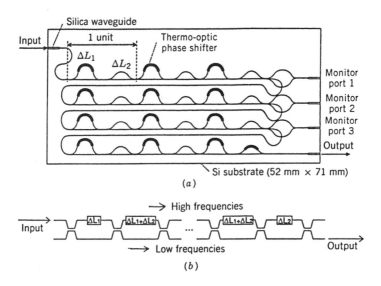

Figure 7.7: (a) A planar lightwave circuit made using of a chain of Mach–Zehnder interferometers; (b) unfolded view of the device. (After Ref. [56]; ©1996 IEEE; reprinted with permission.)

A Mach–Zehnder (MZ) interferometer can also act as an optical filter. An all-fiber MZ interferometer can be constructed by connecting two 3-dB directional couplers in series, as shown schematically in Fig. 7.7(b). The first coupler splits the input signal into two equal parts, which acquire different phase shifts if arm lengths are different, before they interfere at the second coupler. The signal may exit from either of the two output ports depending on its frequency and the arm lengths. It is easy to show that the transfer function for the bar port is given by [54]

$$H_{MZ}(\omega) = \tfrac{1}{2}[1 + \exp(i\omega\tau)], \tag{7.5.5}$$

where τ is the extra delay in the longer arm of the MZ interferometer.

A single MZ interferometer does not act as an optical equalizer but a cascaded chain of several MZ interferometers forms an excellent equalizing filter [55]. Such filters have been fabricated in the form of a *planar lightwave circuit* by using silica waveguides [56]. Figure 7.7(a) shows the device schematically. The device is 52×71 mm^2 in size and exhibits a chip loss of 8 dB. It consists of 12 couplers with asymmetric arm lengths that are cascaded in series. A chromium heater is deposited on one arm of each MZ interferometer to provide thermo-optic control of the optical phase. The main advantage of such a device is that its dispersion-equalization characteristics can be controlled by changing the arm lengths and the number of MZ interferometers.

The operation of the MZ filter can be understood from the unfolded view shown in Fig. 7.7(b). The device is designed such that the higher-frequency components propagate in the longer arm of the MZ interferometers. As a result, they experience more delay than the lower-frequency components taking the shorter route. The relative delay introduced by such a device is just the opposite of that introduced by an optical fiber in the anomalous-dispersion regime. The transfer function $H(\omega)$ can be obtained analytically and is used to optimize the device design and performance [57]. In a 1994 implementation [58], a planar lightwave circuit with only five MZ interferometers provided a relative delay of 836 ps/nm. Such a device is only a few centimeters long, but it is capable of compensating for 50 km of fiber dispersion. Its main limitations are a relatively narrow bandwidth (~ 10 GHz) and sensitivity to input polarization. However, it acts as a programmable optical filter whose GVD as well as the operating wavelength can be adjusted. In one device, the GVD could be varied from -1006 to 834 ps/nm [59].

7.6 Fiber Bragg Gratings

A fiber Bragg grating acts as an optical filter because of the existence of a *stop band*, the frequency region in which most of the incident light is reflected back [51]. The stop band is centered at the *Bragg wavelength* $\lambda_B = 2\bar{n}\Lambda$, where Λ is the grating period and $\bar{n}$ is the average mode index. The periodic nature of index variations couples the forward- and backward-propagating waves at wavelengths close to the Bragg wavelength and, as a result, provides frequency-dependent reflectivity to the incident signal over a bandwidth determined by the grating strength. In essence, a fiber grating acts as a reflection filter. Although the use of such gratings for dispersion compensation was proposed in the 1980s [60], it was only during the 1990s that fabrication technology advanced enough to make their use practical.

7.6.1 Uniform-Period Gratings

We first consider the simplest type of grating in which the refractive index along the length varies periodically as $n(z) = \bar{n} + n_g \cos(2\pi z/\Lambda)$, where n_g is the modulation depth (typically $\sim 10^{-4}$). Bragg gratings are analyzed using the *coupled-mode equations* that describe the coupling between the forward- and backward-propagating waves at a given frequency ω and are written as [51]

$$dA_f/dz = i\delta A_f + i\kappa A_b, \tag{7.6.1}$$

$$dA_b/dz = -i\delta A_b - i\kappa A_f, \tag{7.6.2}$$

where A_f and A_b are the spectral amplitudes of the two waves and

$$\delta = \frac{2\pi}{\lambda_0} - \frac{2\pi}{\lambda_B}, \qquad \kappa = \frac{\pi n_g \Gamma}{\lambda_B}. \tag{7.6.3}$$

Here δ is the detuning from the Bragg wavelength, κ is the *coupling coefficient*, and the confinement factor Γ is defined as in Eq. (2.2.50).

The coupled-mode equations can be solved analytically owing to their linear nature. The transfer function of the grating, acting as a reflective filter, is found to be [54]

$$H(\omega) = r(\omega) = \frac{A_b(0)}{A_f(0)} = \frac{i\kappa \sin(qL_g)}{q\cos(qL_g) - i\delta \sin(qL_g)}, \tag{7.6.4}$$

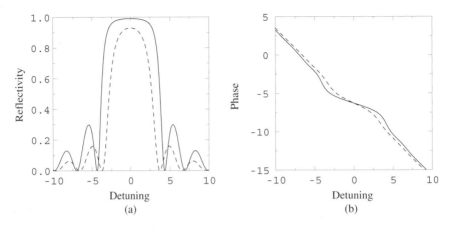

Figure 7.8: (a) Magnitude and (b) phase of the reflectivity plotted as a function of detuning δL_g for a uniform fiber grating with $\kappa L_g = 2$ (dashed curve) or $\kappa L_g = 3$ (solid curve).

where $q^2 = \delta^2 - \kappa^2$ and L_g is the grating length. Figure 7.8 shows the reflectivity $|H(\omega)|^2$ and the phase of $H(\omega)$ for $\kappa L_g = 2$ and 3. The grating reflectivity becomes nearly 100% within the stop band for $\kappa L_g = 3$. However, as the phase is nearly linear in that region, the grating-induced dispersion exists only outside the stop band. Noting that the propagation constant $\beta = \beta_B \pm q$, where the choice of sign depends on the sign of δ, and expanding β in a Taylor series as was done in Eq. (2.4.4) for fibers, the dispersion parameters of a fiber grating are given by [54]

$$\beta_2^g = -\frac{\text{sgn}(\delta)\kappa^2/v_g^2}{(\delta^2 - \kappa^2)^{3/2}}, \qquad \beta_3^g = \frac{3|\delta|\kappa^2/v_g^3}{(\delta^2 - \kappa^2)^{5/2}}, \qquad (7.6.5)$$

where v_g is the group velocity of the pulse with the carrier frequency $\omega_0 = 2\pi c/\lambda_0$.

Figure 7.9 shows how β_2^g varies with the detuning parameter δ for values of κ in the range 1 to 10 cm^{-1}. The grating-induced GVD depends on the sign of detuning δ. The GVD is anomalous on the high-frequency or "blue" side of the stop band where δ is positive and the carrier frequency exceeds the Bragg frequency. In contrast, GVD becomes normal ($\beta_2^g > 0$) on the low-frequency or "red" side of the stop band. The red side can be used for compensating the anomalous GVD of standard fibers. Since β_2^g can exceed 1000 ps^2/cm, a single 2-cm-long grating can be used for compensating the GVD of 100-km fiber. However, the third-order dispersion of the grating, reduced transmission, and rapid variations of $|H(\omega)|$ close to the bandgap make use of uniform fiber gratings for dispersion compensation far from being practical.

The problem can be solved by using the *apodization technique* in which the index change n_g is made nonuniform across the grating, resulting in z-dependent κ. In practice, such an apodization occurs naturally when an ultraviolet Gaussian beam is used to write the grating holographically [51]. For such gratings, κ peaks in the center and tapers down to zero at both ends. A better approach consists of making a grating such that κ varies linearly over the entire length of the fiber grating. In a 1996 experiment [61], such an 11-cm-long grating was used to compensate the GVD acquired by a 10-Gb/s

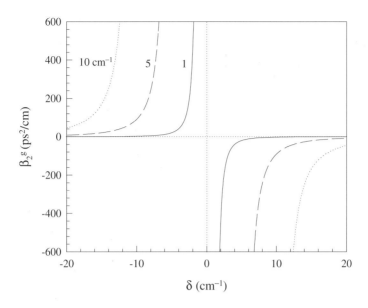

Figure 7.9: Grating-induced GVD plotted as a function of δ for several values of the coupling coefficient κ.

signal transmitted over 100 km of standard fiber. The coupling coefficient $\kappa(z)$ varied smoothly from 0 to 6 cm^{-1} over the grating length. Figure 7.10 shows the transmission characteristics of this grating, calculated by solving the coupled-mode equations numerically. The solid curve shows the group delay related to the phase derivative $d\phi/d\omega$ in Eq. (7.5.2). In a 0.1-nm-wide wavelength region near 1544.2 nm, the group delay varies almost linearly at a rate of about 2000 ps/nm, indicating that the grating can compensate for the GVD acquired over 100 km of standard fiber while providing more than 50% transmission to the incident light. Indeed, such a grating compensated GVD over 106 km for a 10-Gb/s signal with only a 2-dB power penalty at a bit-error rate (BER) of 10^{-9} [61]. In the absence of the grating, the penalty was infinitely large because of the existence of a BER floor.

Tapering of the coupling coefficient along the grating length can also be used for dispersion compensation when the signal wavelength lies within the stop band and the grating acts as a reflection filter. Numerical solutions of the coupled-mode equations for a uniform-period grating for which $\kappa(z)$ varies linearly from 0 to 12 cm^{-1} over the 12-cm length show that the V-shaped group-delay profile, centered at the Bragg wavelength, can be used for dispersion compensation if the wavelength of the incident signal is offset from the center of the stop band such that the signal spectrum sees a linear variation of the group delay. Such a 8.1-cm-long grating was capable of compensating the GVD acquired over 257 km of standard fiber by a 10-Gb/s signal [62]. Although uniform gratings have been used for dispersion compensation [61]–[64], they suffer from a relatively narrow stop band (typically < 0.1 nm) and cannot be used at high bit rates.

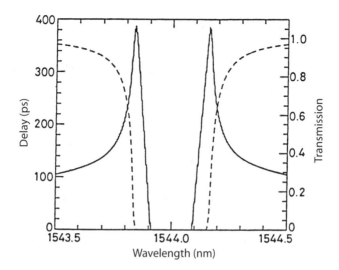

Figure 7.10: Transmittivity (dashed curve) and time delay (solid curve) as a function of wavelength for a uniform-pitch grating for which $\kappa(z)$ varies linearly from 0 to 6 cm^{-1} over the 11-cm length. (After Ref. [61]; ©1996 IEE; reprinted with permission.)

7.6.2 Chirped Fiber Gratings

Chirped fiber gratings have a relatively broad stop band and were proposed for dispersion compensation as early as 1987 [65]. The optical period $\bar{n}\Lambda$ in a chirped grating is not constant but changes over its length [51]. Since the Bragg wavelength ($\lambda_B = 2\bar{n}\Lambda$) also varies along the grating length, different frequency components of an incident optical pulse are reflected at different points, depending on where the Bragg condition is satisfied locally. In essence, the stop band of a chirped fiber grating results from overlapping of many mini stop bands, each shifted as the Bragg wavelength shifts along the grating. The resulting stop band can be as wide as a few nanometers.

It is easy to understand the operation of a chirped fiber grating from Fig. 7.11, where the low-frequency components of a pulse are delayed more because of increasing optical period (and the Bragg wavelength). This situation corresponds to anomalous GVD. The same grating can provide normal GVD if it is flipped (or if the light is incident from the right). Thus, the optical period $\bar{n}\Lambda$ of the grating should decrease for it to provide normal GVD. From this simple picture, the dispersion parameter D_g of a chirped grating of length L_g can be determined by using the relation $T_R = D_g L_g \Delta\lambda$, where T_R is the round-trip time inside the grating and $\Delta\lambda$ is the difference in the Bragg wavelengths at the two ends of the grating. Since $T_R = 2\bar{n}L_g/c$, the grating dispersion is given by a remarkably simple expression,

$$D_g = 2\bar{n}/c(\Delta\lambda). \tag{7.6.6}$$

As an example, $D_g \approx 5 \times 10^7$ ps/(km-nm) for a grating bandwidth $\Delta\lambda = 0.2$ nm. Because of such large values of D_g, a 10-cm-long chirped grating can compensate for the GVD acquired over 300 km of standard fiber.

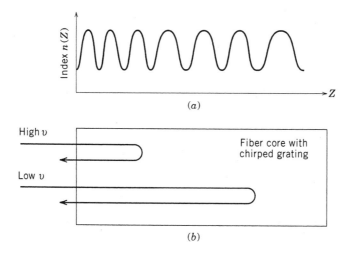

Figure 7.11: Dispersion compensation by a linearly chirped fiber grating: (a) index profile $n(z)$ along the grating length; (b) reflection of low and high frequencies at different locations within the grating because of variations in the Bragg wavelength.

Chirped fiber gratings have been fabricated by using several different methods [51]. It is important to note that it is the optical period $\bar{n}\Lambda$ that needs to be varied along the grating (z axis), and thus chirping can be induced either by varying the physical grating period Λ or by changing the effective mode index $\bar{n}$ along z. In the commonly used *dual-beam holographic technique*, the fringe spacing of the interference pattern is made nonuniform by using dissimilar curvatures for the interfering wavefronts [66], resulting in Λ variations. In practice, cylindrical lenses are used in one or both arms of the interferometer. In a *double-exposure technique* [67], a moving mask is used to vary $\bar{n}$ along z during the first exposure. A uniform-period grating is then written over the same section of the fiber by using the *phase-mask technique*. Many other variations are possible. For example, chirped fiber gratings have been fabricated by tilting or stretching the fiber, by using strain or temperature gradients, and by stitching together multiple uniform sections.

The potential of chirped fiber gratings for dispersion compensation was demonstrated during the 1990s in several transmission experiments [68]–[73]. In 1994, GVD compensation over 160 km of standard fiber at 10 and 20 Gb/s was realized [69]. In 1995, a 12-cm-long chirped grating was used to compensate GVD over 270 km of fiber at 10 Gb/s [70]. Later, the transmission distance was increased to 400 km using a 10-cm-long apodized chirped fiber grating [71]. This is a remarkable performance by an optical filter that is only 10 cm long. Note also from Eq. (7.1.2) that the transmission distance is limited to only 20 km in the absence of dispersion compensation.

Figure 7.12 shows the measured reflectivity and the group delay (related to the phase derivative $d\phi/d\omega$) as a function of the wavelength for the 10-cm-long grating with a bandwidth $\Delta\lambda = 0.12$ nm chosen to ensure that the 10-Gb/s signal fits within the stop band of the grating. For such a grating, the period Λ changes by only 0.008% over its entire length. Perfect dispersion compensation occurs in the spectral range

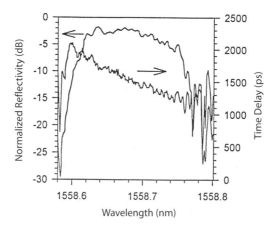

Figure 7.12: Measured reflectivity and time delay for a linearly chirped fiber grating with a bandwidth of 0.12 nm. (After Ref. [73]; ©1996 IEEE; reprinted with permission.)

over which $d\phi/d\omega$ varies linearly. The slope of the group delay (about 5000 ps/nm) is a measure of the dispersion-compensation capability of the grating. Such a grating can recover the 10-Gb/s signal by compensating the GVD acquired over 400 km of the standard fiber. The chirped grating should be apodized in such a way that the coupling coefficient peaks in the middle but vanishes at the grating ends. The apodization is essential to remove the ripples that occur for gratings with a constant κ.

It is clear from Eq. (7.6.6) that D_g of a chirped grating is ultimately limited by the bandwidth $\Delta\lambda$ over which GVD compensation is required, which in turn is determined by the bit rate B. Further increase in the transmission distance at a given bit rate is possible only if the signal bandwidth is reduced or a prechirp technique is used at the transmitter. In a 1996 system trial [72], prechirping of the 10-Gb/s optical signal was combined with the two chirped fiber gratings, cascaded in series, to increase the transmission distance to 537 km. The bandwidth-reduction technique can also be combined with the grating. As discussed in Section 7.3, a duobinary coding scheme can reduce the bandwidth by up to 50%. In a 1996 experiment, the transmission distance of a 10-Gb/s signal was extended to 700 km by using a 10-cm-long chirped grating in combination with a phase-alternating duobinary scheme [73]. The grating bandwidth was reduced to 0.073 nm, too narrow for the 10-Gb/s signal but wide enough for the reduced-bandwidth duobinary signal.

The main limitation of chirped fiber gratings is that they work as a reflection filter. A 3-dB fiber coupler is sometimes used to separate the reflected signal from the incident one. However, its use imposes a 6-dB loss that adds to other insertion losses. An optical circulator can reduce insertion losses to below 2 dB and is often used in practice. Several other techniques can be used. Two or more fiber gratings can be combined to form a transmission filter, which provides dispersion compensation with relatively low insertion losses [74]. A single grating can be converted into a transmission filter by introducing a phase shift in the middle of the grating [75]. A Moiré grating, formed by superimposing two chirped gratings formed on the same piece of fiber, also has a

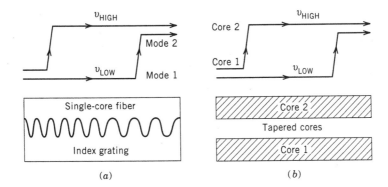

Figure 7.13: Schematic illustration of dispersion compensation by two fiber-based transmission filters: (a) chirped dual-mode coupler; (b) tapered dual-core fiber. (After Ref. [77]; ©1994 IEEE; reprinted with permission.)

transmission peak within its stop band [76]. The bandwidth of such transmission filters is relatively small.

7.6.3 Chirped Mode Couplers

This subsection focuses on two fiber devices that can act as a transmission filter suitable for dispersion compensation. A chirped mode coupler is an all-fiber device designed using the concept of chirped distributed resonant coupling [77]. Figure 7.13 shows the operation of two such devices schematically. The basic idea behind a chirped mode coupler is quite simple [78]. Rather than coupling the forward and backward propagating waves of the same mode (as is done in a fiber grating), the chirped grating couples the two spatial modes of a dual-mode fiber. Such a device is similar to the mode converter discussed in Section 7.4 in the context of a DCF except that the grating period is varied linearly over the fiber length. The signal is transferred from the fundamental mode to a higher-order mode by the grating, but different frequency components travel different lengths before being transferred because of the chirped nature of the grating that couples the two modes. If the grating period increases along the coupler length, the coupler can compensate for the fiber GVD. The signal remains propagating in the forward direction, but ends up in a higher-order mode of the coupler. A uniform-grating mode converter can be used to reconvert the signal back into the fundamental mode.

A variant of the same idea uses the coupling between the fundamental modes of a dual-core fiber with dissimilar cores [79]. If the two cores are close enough, evanescent-wave coupling between the modes leads to a transfer of energy from one core to another, similar to the case of a directional coupler. When the spacing between the cores is linearly tapered, such a transfer takes place at different points along the fiber, depending on the frequency of the propagating signal. Thus, a dual-core fiber with the linearly tapered core spacing can compensate for fiber GVD. Such a device keeps the signal propagating in the forward direction, although it is physically transferred to the neighboring core. This scheme can also be implemented in the form of a compact device by using semiconductor waveguides since the supermodes of two coupled waveguides exhibit a large amount of GVD that is also tunable [80].

7.7 Optical Phase Conjugation

Although the use of optical phase conjugation (OPC) for dispersion compensation was proposed in 1979 [81], it was only in 1993 that the OPC technique was implemented experimentally; it has attracted considerable attention since then [82]–[103]. In contrast with other optical schemes discussed in this chapter, the OPC is a nonlinear optical technique. This section describes the principle behind it and discusses its implementation in practical lightwave systems.

7.7.1 Principle of Operation

The simplest way to understand how OPC can compensate the GVD is to take the complex conjugate of Eq. (7.1.3) and obtain

$$\frac{\partial A^*}{\partial z} - \frac{i\beta_2}{2}\frac{\partial^2 A^*}{\partial t^2} - \frac{\beta_3}{6}\frac{\partial^3 A^*}{\partial t^3} = 0. \tag{7.7.1}$$

A comparison of Eqs. (7.1.3) and (7.7.1) shows that the phase-conjugated field A^* propagates with the sign reversed for the GVD parameter β_2. This observation suggests immediately that, if the optical field is phase-conjugated in the middle of the fiber link, the dispersion acquired over the first half will be exactly compensated in the second-half section of the link. Since the β_3 term does not change sign on phase conjugation, OPC cannot compensate for the third-order dispersion. In fact, it is easy to show, by keeping the higher-order terms in the Taylor expansion in Eq. (2.4.4), that OPC compensates for all even-order dispersion terms while leaving the odd-order terms unaffected.

The effectiveness of midspan OPC for dispersion compensation can also be verified by using Eq. (7.1.4). The optical field just before OPC is obtained by using $z = L/2$ in this equation. The propagation of the phase-conjugated field A^* in the second-half section then yields

$$A^*(L,t) = \frac{1}{2\pi}\int_{-\infty}^{\infty}\tilde{A}^*\left(\frac{L}{2},\omega\right)\exp\left(\frac{i}{4}\beta_2 L\omega^2 - i\omega t\right)d\omega, \tag{7.7.2}$$

where $\tilde{A}^*(L/2,\omega)$ is the Fourier transform of $A^*(L/2,t)$ and is given by

$$\tilde{A}^*(L/2,\omega) = \tilde{A}^*(0,-\omega)\exp(-i\omega^2\beta_2 L/4). \tag{7.7.3}$$

By substituting Eq. (7.7.3) in Eq. (7.7.2), one finds that $A(L,t) = A^*(0,t)$. Thus, except for a phase reversal induced by the OPC, the input field is completely recovered, and the pulse shape is restored to its input form. Since the signal spectrum after OPC becomes the mirror image of the input spectrum, the OPC technique is also referred to as *midspan spectral inversion*.

7.7.2 Compensation of Self-Phase Modulation

As discussed in Section 2.6, the nonlinear phenomenon of SPM leads to fiber-induced chirping of the transmitted signal. Section 7.3 indicated that this chirp can be used to advantage with a proper design. Optical solitons also use the SPM to their advantage (see Chapter 9). However, in most lightwave systems, the SPM-induced nonlinear effects degrade the signal quality, especially when the signal is propagated over long distances using multiple optical amplifiers (see Section 6.5).

The OPC technique differs from all other dispersion-compensation schemes in one important way: Under certain conditions, it can compensate simultaneously for both the GVD and SPM. This feature of OPC was noted in the early 1980s [104] and has been studied extensively after 1993 [97]. It is easy to show that both the GVD and SPM are compensated perfectly in the absence of fiber losses. Pulse propagation in a lossy fiber is governed by Eq. (5.3.1) or by

$$\frac{\partial A}{\partial z} + \frac{i\beta_2}{2}\frac{\partial^2 A}{\partial t^2} = i\gamma|A|^2 A - \frac{\alpha}{2}A, \tag{7.7.4}$$

where the β_3 term is neglected and α accounts for the fiber losses. When $\alpha = 0$, A^* satisfies the same equation when we take the complex conjugate of Eq. (7.7.4) and change z to $-z$. As a result, midspan OPC can compensate for SPM and GVD simultaneously.

Fiber losses destroy this important property of midspan OPC. The reason is intuitively obvious if we note that the SPM-induced phase shift is power dependent. As a result, much larger phase shifts are induced in the first-half of the link than the second half, and OPC cannot compensate for the nonlinear effects. Equation (7.7.4) can be used to study the impact of fiber losses. By making the substitution

$$A(z,t) = B(z,t)\exp(-\alpha z/2), \tag{7.7.5}$$

Eq. (7.7.4) can be written as

$$\frac{\partial B}{\partial z} + \frac{i\beta_2}{2}\frac{\partial^2 B}{\partial t^2} = i\gamma(z)|B|^2 B, \tag{7.7.6}$$

where $\gamma(z) = \gamma\exp(-\alpha z)$. The effect of fiber losses is mathematically equivalent to the loss-free case but with a z-dependent nonlinear parameter. By taking the complex conjugate of Eq. (7.7.6) and changing z to $-z$, it is easy to see that perfect SPM compensation can occur only if $\gamma(z) = \gamma(L - z)$. This condition cannot be satisfied when $\alpha \neq 0$.

One may think that the problem can be solved by amplifying the signal after OPC so that the signal power becomes equal to the input power before it is launched in the second-half section of the fiber link. Although such an approach reduces the impact of SPM, it does not lead to perfect compensation of it. The reason can be understood by noting that propagation of a phase-conjugated signal is equivalent to propagating a *time-reversed* signal [105]. Thus, perfect SPM compensation can occur only if the power variations are symmetric around the midspan point where the OPC is performed so that $\gamma(z) = \gamma(L - z)$ in Eq. (7.7.6). Optical amplification does not satisfy this property. One can come close to SPM compensation if the signal is amplified often enough that the power does not vary by a large amount during each amplification stage. This approach is, however, not practical because it requires closely spaced amplifiers.

Perfect compensation of both GVD and SPM can be realized by using *dispersion-decreasing fibers* for which β_2 decreases along the fiber length. To see how such a scheme can be implemented, assume that β_2 in Eq. (7.7.6) is a function of z. By making the transformation

$$\xi = \int_0^z \gamma(z)\, dz, \qquad (7.7.7)$$

Eq. (7.7.6) can be written as [97]

$$\frac{\partial B}{\partial \xi} + \frac{i}{2} b(\xi) \frac{\partial^2 B}{\partial t^2} = i|B|^2 B, \qquad (7.7.8)$$

where $b(\xi) = \beta_2(\xi)/\gamma(\xi)$. Both GVD and SPM are compensated if $b(\xi) = b(\xi_L - \xi)$, where ξ_L is the value of ξ at $z = L$. This condition is automatically satisfied when the dispersion decreases in exactly the same way as $\gamma(z)$ so that $\beta_2(\xi) = \gamma(\xi)$ and $b(\xi) = 1$. Since fiber losses make $\gamma(z)$ to decrease exponentially as $\exp(-\alpha z)$, both GVD and SPM can be compensated exactly in a dispersion-decreasing fiber whose GVD decreases as $\exp(-\alpha z)$. This approach is quite general and applies even when in-line amplifiers are used.

7.7.3   Phase-Conjugated Signal

The implementation of the midspan OPC technique requires a nonlinear optical element that generates the phase-conjugated signal. The most commonly used method makes use of *four-wave mixing* (FWM) in a nonlinear medium. Since the optical fiber itself is a nonlinear medium, a simple approach is to use a few-kilometer-long fiber especially designed to maximize the FWM efficiency.

The FWM phenomenon in optical fibers has been studied extensively [106]. Its use requires injection of a pump beam at a frequency ω_p that is shifted from the signal frequency ω_s by a small amount (~ 0.5 THz). The fiber nonlinearity generates the phase-conjugated signal at the frequency $\omega_c = 2\omega_p - \omega_s$ provided that the *phase-matching condition* $k_c = 2k_p - k_s$ is approximately satisfied, where $k_j = n(\omega_j)\omega_c/c$ is the wave number for the optical field of frequency ω_j. The phase-matching condition can be approximately satisfied if the zero-dispersion wavelength of the fiber is chosen to coincide with the pump wavelength. This was the approach adopted in the 1993 experiments in which the potential of OPC for dispersion compensation was first demonstrated. In one experiment [82], the 1546-nm signal was phase conjugated by using FWM in a 23-km-long fiber with pumping at 1549 nm. The 6-Gb/s signal was transmitted over 152 km of standard fiber in a coherent transmission experiment employing the FSK format. In another experiment [83], a 10-Gb/s signal was transmitted over 360 km. The midspan OPC was performed in a 21-km-long fiber by using a pump laser whose wavelength was tuned exactly to the zero-dispersion wavelength of the fiber. The pump and signal wavelengths differed by 3.8 nm. Figure 7.14 shows the

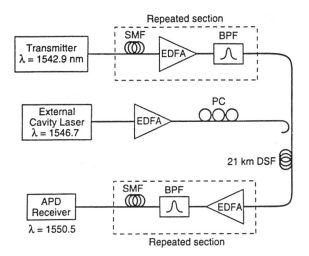

Figure 7.14: Experimental setup for dispersion compensation through midspan spectral inversion in a 21-km-long dispersion-shifted fiber. (After Ref. [83]; ©1993 IEEE; reprinted with permission.)

experimental setup. A bandpass filter (BPF) is used to separate the phase-conjugated signal from the pump.

Several factors need to be considered while implementing the midspan OPC technique in practice. First, since the signal wavelength changes from ω_s to $\omega_c = 2\omega_p - \omega_s$ at the phase conjugator, the GVD parameter β_2 becomes different in the second-half section. As a result, perfect compensation occurs only if the phase conjugator is slightly offset from the midpoint of the fiber link. The exact location L_p can be determined by using the condition $\beta_2(\omega_s)L_p = \beta_2(\omega_c)(L - L_p)$, where L is the total link length. By expanding $\beta_2(\omega_c)$ in a Taylor series around the signal frequency ω_s, L_p is found to be

$$\frac{L_p}{L} = \frac{\beta_2 + \delta_c \beta_3}{2\beta_2 + \delta_c \beta_3}, \tag{7.7.9}$$

where $\delta_c = \omega_c - \omega_s$ is the frequency shift of the signal induced by the OPC technique. For a typical wavelength shift of 6 nm, the phase-conjugator location changes by about 1%. The effect of residual dispersion and SPM in the phase-conjugation fiber itself can also affect the placement of phase conjugator [94].

A second factor that needs to be addressed is that the FWM process in optical fibers is polarization sensitive. As signal polarization is not controlled in optical fibers, it varies at the OPC in a random fashion. Such random variations affect the FWM efficiency and make the standard FWM technique unsuitable for practical purposes. Fortunately, the FWM scheme can be modified to make it polarization insensitive. In one approach, two orthogonally polarized pump beams at different wavelengths, located symmetrically on the opposite sides of the zero-dispersion wavelength λ_{ZD} of the fiber, are used [85]. This scheme has another advantage: the phase-conjugate wave can be generated at the frequency of the signal itself by choosing λ_{ZD} such that it coincides with the signal frequency. This feature follows from the relation $\omega_c = \omega_{p1} + \omega_{p2} - \omega_s$, where $\omega_{p1} \neq \omega_{p2}$. Polarization insensitivity of OPC can also be realized by using a single pump in combination with a fiber grating and an *orthoconjugate mirror* [90], but the device works in the reflective mode and requires the separation of the conjugate wave from the signal by using a 3-dB coupler or an optical circulator.

The relatively low efficiency of the OPC process in optical fibers is also of some concern. Typically, the conversion efficiency η_c is below 1%, making it necessary to amplify the phase-conjugated signal [83]. Effectively, the insertion loss of the phase conjugator exceeds 20 dB. However, the FWM process is not inherently a low-efficiency process and, in principle, it can even provide net gain [106]. Indeed, the analysis of the FWM equations shows that η_c increases considerably by increasing the pump power while decreasing the signal power; it can even exceed 100% by optimizing the power levels and the pump-signal wavelength difference [92]. High pump powers are often avoided because of the onset of stimulated Brillouin scattering (SBS). However, SBS can be suppressed by modulating the pump at a frequency ~ 100 MHz. In a 1994 experiment, 35% conversion efficiency was realized by using this technique [107].

The FWM process in a semiconductor optical amplifier (SOA) has also been used to generate the phase-conjugated signal for dispersion compensation. This approach was first used in a 1993 experiment to demonstrate transmission of a 2.5-Gb/s signal, obtained through direct modulation of a semiconductor laser, over 100 km of standard fiber [84]. Later, in a 1995 experiment the same approach was used for transmitting a 40-Gb/s signal over 200 km of standard fiber [93]. The possibility of highly nondegenerate FWM inside SOAs was suggested in 1987, and this technique is used extensively in the context of wavelength conversion [108]. Its main advantage is that the phase-conjugated signal can be generated in a device of 1-mm length. The conversion efficiency is also typically higher than that of FWM in an optical fiber because of amplification, although this advantage is offset by the relatively large coupling losses resulting from the need to couple the signal back into the fiber. By a proper choice of the pump-signal detuning, conversion efficiencies of more than 100% (net gain for the phase-conjugated signal) have been realized for FWM in SOAs [109].

A periodically poled LiNbO$_3$ waveguide has also been used to make a wideband spectral inverter [102]. The phase-conjugated signal is generated using cascaded second-order nonlinear processes, which are quasi-phase-matched through periodic poling of the crystal. Such an OPC device exhibited only 7-dB insertion losses and was capable of compensating dispersion of four 10-Gb/s channels simultaneously over 150 km of standard fiber. The system potential of the OPC technique was demonstrated in a 1999 field trial in which a FWM-based phase conjugator was used to compensate the GVD of a 40-Gb/s signal over 140 km of standard fiber [100]. In the absence of OPC, the 40-Gb/s signal cannot be transmitted over more than 7 km as deduced from Eq. (7.1.2).

Most of the experimental work on dispersion compensation has considered transmission distances of several hundred kilometers. For long-haul applications, one may ask whether the OPC technique can compensate the GVD acquired over thousands of kilometers in fiber links which use amplifiers periodically for loss compensation. This question has been studied mainly through numerical simulations. In one set of simulations, a 10-Gb/s signal could be transmitted over 6000 km when the average launch power was kept below 3 mW to reduce the effects of fiber nonlinearity [95]. In another study, the amplifier spacing was found to play an important role; transmission over 9000 km was feasible by keeping the amplifiers 40 km apart [98]. The choice of the operating wavelength with respect to the zero-dispersion wavelength was also critical. In the anomalous-dispersion region ($\beta_2 < 0$), the periodic variation of the signal power along the fiber link can lead to the generation of additional sidebands through the phenomenon of *modulation instability* [110]. This instability can be avoided if the dispersion parameter is relatively large [$D > 10$ ps/(km-nm)]. This is the case for standard fibers near 1.55 μm. It should be remarked that the maximum transmission distance depends critically on many factors, such as the FWM efficiency, the input power, and the amplifier spacing, and may decrease to below 3000 km, depending on the operating parameters [96].

The use of OPC for long-haul lightwave systems requires periodic use of optical amplifiers and phase conjugators. These two optical elements can be combined into one by using *parametric amplifiers*, which not only generate the phase-conjugated signal through the FWM process but also amplify it. The analysis of such a long-haul system shows that 20- to 30-ps input pulses can travel over thousands of kilometers despite a high GVD; the total transmission distance can exceed 15,000 km for dispersion-shifted fibers with $\beta_2 = -2$ ps^2/km near 1.55 μm [111]. The phase-conjugation technique is not used in practice as parametric amplifiers are not yet available commercially. The next section focuses on the techniques commonly used for dispersion management in long-haul systems.

7.8 Long-Haul Lightwave Systems

This chapter has so far focused on lightwave systems in which dispersion management helps to extend the transmission distance from a value of $\sim$10 km to a few hundred kilometers. The important question is how dispersion management can be used for long-haul systems for which transmission distance is several thousand kilometers. If the optical signal is regenerated electronically every 100–200 km, all techniques discussed in this chapter should work well since the nonlinear effects do not accumulate over long lengths. In contrast, if the signal is maintained in the optical domain over the entire link by using periodic amplification, the nonlinear effects such as SPM, cross-phase modulation (XPM), and FWM [106] would limit the system ultimately. Indeed, the impact of nonlinear effects on the performance of dispersion-managed systems has been a subject of intense study [112]–[137]. In this section we focus on long-haul lightwave systems in which the loss and dispersion-management schemes are used simultaneously.

7.8.1 Periodic Dispersion Maps

In the absence of the nonlinear effects, total GVD accumulated over thousands of kilometers can be compensated at the receiver end without degrading the system performance. The reason is that each optical pulse recovers its original position within the bit slot for a linear system (except for the amplifier-induced timing jitter) even if it was

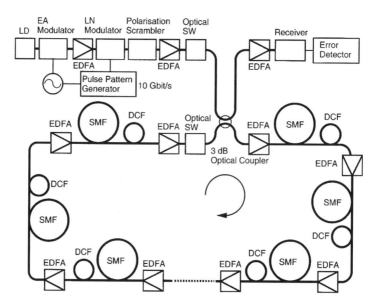

Figure 7.15: Recirculating fiber loop used to demonstrate transmission of a 10-Gb/s signal over 10,000 km of standard fiber using a DCF periodically. Components used include laser diode (LD), electroabsorption (EA) modulator, optical switch (SW), fiber amplifier (EDFA), single-mode fiber (SMF), and DCF. (After Ref. [135]; ©2000 IEEE; reprinted with permission.)

spread over several bit slots before the GVD was compensated. This is not the case when nonlinear effects cannot be neglected. The nonlinear interaction among optical pulses of the same channel (intrachannel effects), and among pulses of neighboring channels in a WDM system (interchannel effects), degrade the signal quality to the extent that the GVD compensation at the receiver alone fails to work for long-haul systems.

A simple solution is provided by the technique of *periodic* dispersion management. The underlying idea is quite simple and consists of mixing fibers with positive and negative GVDs in a periodic fashion such that the total dispersion over each period is close to zero. The simplest scheme uses just two fibers of opposite dispersions and lengths with the average dispersion

$$\bar{D} = (D_1 L_1 + D_2 L_2)/L_m, \qquad (7.8.1)$$

where D_j is the dispersion of the fiber section of length L_j ($j = 1, 2$) and $L_m = L_1 + L_2$ is the period of dispersion map, also referred to as the map period. If $\bar{D}$ is nearly zero, dispersion is compensated over each map period. The length L_m is a free design parameter that can be chosen to meet the system-performance requirements. In practice, it is common to choose L_m to be equal to the amplifier spacing L_A as this choice simplifies the system design. Typically $L_m = L_A \approx 80$ km for terrestrial lightwave systems but is reduced to about 50 km for submarine systems.

Because of cost considerations, most laboratory experiments use a fiber loop in which the optical signal is forced to recirculate many times to simulate a long-haul lightwave system. Figure 7.15 shows such a recirculating fiber loop schematically. It was used to demonstrate transmission of a 10-Gb/s signal over a distance of up to 10,000 km over standard fibers with periodic loss and dispersion management [135]. Two optical switches determine how long a pseudorandom bit stream circulates inside the loop before it reaches the receiver. The loop length and the number of round trips determine the total transmission distance. The loop length is typically 300–500 km. The length of DCF is chosen in accordance with Eq. (7.8.1) and is set to $L_2 = -D_1 L_1 / D_2$ for complete compensation ($\bar{D} = 0$). An optical bandpass filter is also inserted inside the loop to reduce the effects of amplifier noise.

7.8.2 Simple Theory

The major nonlinear phenomenon affecting the performance of a single-channel system is SPM. As before, the propagation of an optical bit stream inside a dispersion-managed system is governed by the nonlinear Schrödinger (NLS) equation [Eq. (7.7.4)]:

$$i\frac{\partial A}{\partial z} - \frac{\beta_2}{2}\frac{\partial^2 A}{\partial t^2} + \gamma|A|^2 A = -\frac{i\alpha}{2}A, \tag{7.8.2}$$

with the main difference that β_2, γ, and α are now periodic functions of z because of their different values in two or more fiber sections used to form the dispersion map. Loss compensation at lumped amplifiers can be included by changing the loss parameter suitably at the amplifier locations.

In general, Eq. (7.8.2) is solved numerically to study the performance of dispersion-managed systems [120]–[129]. It is useful to eliminate the last term in this equation with the transformation [see Eq. (7.7.5)]

$$A(z,t) = B(z,t)\exp\left[-\frac{1}{2}\int_0^z \alpha(z)\,dz\right]. \tag{7.8.3}$$

Equation (7.8.2) then takes the form

$$i\frac{\partial B}{\partial z} - \frac{\beta_2(z)}{2}\frac{\partial^2 B}{\partial t^2} + \bar{\gamma}(z)|B|^2 B = 0, \tag{7.8.4}$$

where power variations along the dispersion-managed fiber link are included through a periodically varying nonlinear parameter $\bar{\gamma}(z) = \gamma \exp[-\int_0^z \alpha(z)\,dz]$.

Considerable insight into the design of a dispersion-managed system can be gained by solving Eq. (7.8.4) with a variational approach [123]. Its use is based on the observation that a chirped Gaussian pulse maintains its functional form in the linear case ($\gamma = 0$) although its amplitude, width, and chirp change with propagation (see Section 2.4). Since the nonlinear effects are relatively weak locally in each fiber section compared with the dispersive effects, the pulse shape is likely to retain its Gaussian shape. One can thus assume that the pulse evolves along the fiber in the form of a chirped Gaussian pulse such that

$$B(z,t) = a\exp[-(1+iC)t^2/2T^2 + i\phi], \tag{7.8.5}$$

where a is the amplitude, T is the width, C is the chirp, and ϕ is the phase. All four parameters vary with z. The variational method is useful to find the z dependence of these parameters. It makes use of the fact that Eq. (7.8.4) can be derived from the Euler–Lagrange equation using the following Lagrangian density:

$$L_{den} = \frac{i}{2}\left(B\frac{\partial B^*}{\partial z} - B^*\frac{\partial B}{\partial z}\right) + \frac{1}{2}\left[\bar{\gamma}(z)|B|^4 - \beta_2(z)\left|\frac{\partial B}{\partial t}\right|^2\right]. \tag{7.8.6}$$

Following the variational method, we can find the evolution equations for the four parameters a, T, C, and ϕ. The phase equation can be ignored as it is not coupled to the other three equations. The amplitude equation can be integrated to find that the combination $a^2 T$ does not vary with z and is related to the input pulse energy E_0 as $a^2 T = \sqrt{\pi}E_0$. Thus, one only needs to solve the following two coupled equations:

$$\frac{dT}{dz} = \frac{\beta_2 C}{T}, \tag{7.8.7}$$

$$\frac{dC}{dz} = \frac{\bar{\gamma}E_0}{\sqrt{2\pi}T} + (1+C^2)\frac{\beta_2}{T^2}. \tag{7.8.8}$$

Consider first the linear case by setting $\gamma = 0$. Noting that the ratio $(1+C^2)/T^2$ is related to the spectral width of the pulse that remains constant in a linear medium, we can replace it by its initial value $(1+C_0^2)/T_0^2$, where T_0 and C_0 are the width and the chirp of input pulses before they are launched into the dispersion-managed fiber link. Equations (7.8.7) and (7.8.8) can now be solved analytically and have the following general solution:

$$T^2(z) = T_0^2 + 2\int_0^z \beta_2(z)C(z)\,dz, \qquad C(z) = C_0 + \frac{1+C_0^2}{T_0^2}\int_0^z \beta_2(z)\,dz. \tag{7.8.9}$$

This solution looks complicated but is is easy to perform the integrations for a two-section dispersion map. In fact, the values of T and C at the end of the first map period $(z = L_m)$ are given by

$$T_1 = T_0[(1+C_0 d)^2 + d^2]^{1/2}, \qquad C_1 = C_0 + (1+C_0^2)d, \tag{7.8.10}$$

where $d = \bar{\beta}_2 L_m/T_0^2$ and $\bar{\beta}_2$ is the average GVD value. This is exactly what one would expect from the theory of Section 2.4. It is easy to see that when $\bar{\beta}_2 = 0$, both T and C return to their input values at the end of each map period, as they should for a linear medium. When the average GVD of the dispersion-managed link is not zero, T and C change after each map period, and pulse evolution is not periodic.

When the nonlinear term is not negligible, the pulse parameters do not return to their input values for perfect GVD compensation $(d = 0)$. It was noted in several experiments that the nonlinear system performs best when GVD compensation is only 90–95% so that some residual dispersion remains after each map period. In fact, if the input pulse is initially chirped such that $\bar{\beta}_2 C < 0$, the pulse at the end of the fiber link may be shorter than the input pulse. This behavior is expected for a linear system (see Section 2.4) and follows from Eq. (7.8.10) for $C_0 d < 0$. It also persists for weakly nonlinear systems. This observation has led to the adoption of the CRZ (chirped RZ) format for dispersion-managed fiber links. If the dispersion map is made such that the pulse broadens in the first section and compresses in the second section, the impact of the nonlinear effects can be reduced significantly. The reason is as follows: The pulse peak power is reduced considerably in the first section because of rapid broadening of chirped pulses, while in the second section it is lower because of the accumulated fiber losses. Such dispersion-managed links are called *quasi-linear* transmission links [127]. The solution given in Eq. (7.8.9) applies reasonably well for such links. As optical pulses spread considerably outside their assigned bit slot over a considerable fraction of each map period, their overlapping can degrade the system performance when the nonlinear effects are not negligible. These effects are considered in the next section.

If the input peak power is so large that a quasi-linear situation cannot be realized, one must solve Eqs. (7.8.7) and (7.8.8) with the nonlinear term included. No analytic solution is possible in this case. However, one can find periodic solutions of these equations numerically by imposing the periodic boundary conditions

$$T(L_m) = T_0, \qquad C(L_m) = C_0, \tag{7.8.11}$$

which ensure that the pulse recovers its initial shape at the end of each map period. Such pulses propagate through the dispersion-managed link in a periodic fashion and are called dispersion-managed solitons because they exhibit soliton-like features. This case is discussed in Chapter 9.

7.8.3 Intrachannel Nonlinear Effects

The nonlinear effects play an important role in dispersion-managed systems, especially because they are enhanced within the DCF because of its reduced effective core area. Placement of the amplifier after the DCF helps since the signal is then weak enough that the nonlinear effects are less important in spite of the small effective area of DCFs. The optimization of system performance using different dispersion maps has been a subject of intense study. In a 1994 experiment, a 1000-km-long fiber loop containing 31 fiber amplifiers was used to study three different dispersion maps [112]. The maximum transmission distance of 12,000 km was realized for the case in which short sections of normal GVD fibers were used to compensate for the anomalous GVD of long sections. In a 1995 experiment, a 80-Gb/s signal, obtained by multiplexing eight 10-Gb/s channels with 0.8-nm channel spacing, was propagated inside a recirculating fiber loop [114]. The total transmission distance was limited to 1171 km because of various nonlinear effects.

Perfect compensation of GVD in each map period is not the best solution in the presence of nonlinear effects. A numerical approach is often used to optimize the design of dispersion-managed systems [115]–[124]. In general, local GVD should be kept relatively large to suppress the nonlinear effects, while minimizing the average dispersion for all channels. In a 1998 experiment, a 40-Gb/s signal was transmitted over 2000 km of standard fiber using a novel dispersion map [125]. The distance could be increased to 16,500 km at a lower bit rate of 10 Gb/s by placing an optical amplifier right after the DCF within the recirculating fiber loop [126]. Since the nonlinear effects played an important role, these experiments are thought to be making use of soliton properties (see Chapter 9). The main limitation stems from a large pulse broadening in the standard-fiber section of the dispersion map, resulting in the nonlinear interaction between the neighboring overlapping pulses. Such nonlinear effects have been studied extensively [127]–[134] and are referred to as the *intrachannel* effects to distinguish them from the *interchannel* nonlinear effects that occur when pulses in two neighboring channels at different wavelengths overlap in the time domain (see Section 8.3).

The origin of intrachannel nonlinear effects can be seen from Eq. (7.8.4) by considering three neighboring pulses and writing the total field as $B = B_1 + B_2 + B_3$. Equation (7.8.4) then reduces to the following set of three coupled NLS equations [132]:

$$i\frac{\partial B_1}{\partial z} - \frac{\beta_2}{2}\frac{\partial^2 B_1}{\partial t^2} + \bar{\gamma}[(|B_1|^2 + 2|B_2|^2 + 2|B_3|^2)B_1 + B_2^2 B_3^*] = 0, \quad (7.8.12)$$

$$i\frac{\partial B_2}{\partial z} - \frac{\beta_2}{2}\frac{\partial^2 B_2}{\partial t^2} + \bar{\gamma}[(|B_2|^2 + 2|B_1|^2 + 2|B_3|^2)B_2 + 2B_1 B_2^* B_3] = 0, \quad (7.8.13)$$

$$i\frac{\partial B_3}{\partial z} - \frac{\beta_2}{2}\frac{\partial^2 B_3}{\partial t^2} + \bar{\gamma}[(|B_3|^2 + 2|B_1|^2 + 2|B_2|^2)B_3 + B_2^2 B_1^*] = 0. \quad (7.8.14)$$

The first nonlinear term corresponds to SPM. The next two terms result from XPM induced by the other two pulses. The last term is FWM-like. Although it is common to refer to its effect as intrachannel FWM, it is somewhat of a misnomer because all three pulses have the same wavelength. Nevertheless, this term can create new pulses in the time domain, in analogy with FWM that creates new waves in the spectral domain. Such pulses are referred to as ghost pulses [128]. The ghost pulses can impact the system performance considerably if they fall within the 0-bit time slots [134].

The intrachannel XPM affects only the phase but the phase shift is time dependent. The resulting frequency chirp leads to timing jitter through fiber dispersion [130]. The impact of intrachannel XPM and FWM on the system performance depends on the choice of the dispersion map among other things [127]. In general, the optimization of a dispersion-managed systems depends on many design parameters such as the launch power, amplifier spacing, and the location of DCFs [129]. In a 2000 experiment, a 40-Gb/s signal was transmitted over transoceanic distances, in spite of its use of standard fibers, using the in-line synchronous modulation method originally proposed for solitons [137]. Pseudolinear transmission of a 320-Gb/s channel has also been demonstrated over 200 km of fiber whose dispersion of 5.7 ps/(km-nm) was compensated using DCFs [138].

7.9 High-Capacity Systems *(concepts)*

Modern WDM lightwave systems use a large number of channels to realize a system capacity of more than 1 Tb/s. For such systems, the dispersion-management technique should be compatible with the broad bandwidth occupied by the multichannel signal. In this section we discuss the dispersion-management issues relevant for high-capacity systems.

7.9.1 Broadband Dispersion Compensation

As discussed in Chapter 8, a WDM signal typically occupies a bandwidth of 30 nm or more, although it is bunched in spectral packets of bandwidth ~ 0.1 nm (depending on the bit rate of individual channels). For 10-Gb/s channels, the third-order dispersion does not play an important role as relatively wide (> 10 ps) optical pulses are used for individual channels. However, because of the wavelength dependence of β_2, or the dispersion parameter D, the accumulated dispersion will be different for each channel. Any dispersion-management scheme should compensate the GVD of all channels simultaneously to be effective in practice. Several different methods have been used for dispersion compensation in WDM systems. One can use either a single broadband fiber grating or multiple fiber gratings with their stop bands tuned to individual channels. Alternatively, one can take advantage of the periodic nature of the WDM spectrum, and use an optical filter with periodic transmission peaks. A common approach consists of extending the DCF approach to WDM systems by designing the DCF appropriately.

Consider first the case of fiber gratings [51]. A chirped fiber grating can have a stop band as wide as10 nm if it is made long enough. Such a grating can be used in a WDM system if the number of channels is small enough (typically <10) that the total signal bandwidth fits inside its stop band. In a 1999 experiment, a 6-nm-bandwidth chirped grating was used for a four-channel WDM system, each channel operating at 40 Gb/s [139]. When the WDM-signal bandwidth is much larger than that, one can use several cascaded chirped gratings in series such that each grating reflects one channel and compensates its dispersion [140]–[144]. The advantage of this technique is that the gratings can be tailored to match the GVD of each channel. Figure 7.16 shows the cascaded-grating scheme schematically for a four-channel WDM system [143]. Every 80 km, a set of four gratings compensates the GVD for all channels while two optical amplifiers take care of all losses. The gratings opened the "closed eye" almost completely in this experiment. By 2000, this approach was applied to a 32-channel WDM system with 18-nm bandwidth [144]. Six chirped gratings, each with 6-nm-wide stop band, were cascaded to compensate GVD for all channels simultaneously.

The multiple-gratings approach becomes cumbersome when the number of channels is so large that the signal bandwidth exceeds 30 nm. A FP filter has multiple transmission peaks, spaced apart periodically by the free spectral range of the filter. Such a filter can compensate the GVD of all channels if (i) all channels are spaced apart equally and (ii) the free spectral range of the filter is matched to the channel spacing. It is difficult to design FP filters with a large amount of dispersion. A new kind of fiber grating, referred to as the *sampled* fiber grating, has been developed to solve this problem [145]–[147]. Such a grating has multiple stop bands and is relatively easy to fabricate. Rather than making a single long grating, multiple short-length gratings are written with uniform spacing among them. (Each short section is a sample, hence the name "sampled" grating.) The wavelength spacing among multiple reflectivity peaks is determined by the sample period and is controllable during the fabrication process. Moreover, if each sample is chirped, the dispersion characteristics of each reflectivity peak are governed by the amount of chirp introduced. Such a grating was first used in 1995 to demonstrate simultaneous compensation of fiber dispersion over 240 km for two 10-Gb/s channels [145]. A 1999 experiment used a sampled grating for a four-

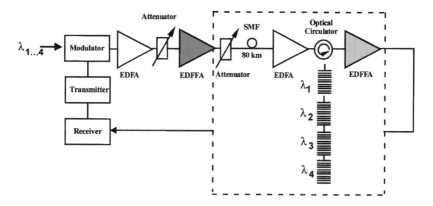

Figure 7.16: Cascaded gratings used for dispersion compensation in a WDM system. (After Ref. [143]; ©1999 IEEE; reprinted with permission.)

channel WDM system [147]. As the number of channels increases, it becomes more and more difficult to compensate the GVD of all channels at the same time.

The use of *negative-slope* DCFs offers the simplest solution to dispersion management in high-capacity WDM systems with a large number of channels. Indeed, such DCFs were developed and commercialized during the 1990s and are employed in virtually all dense WDM systems [148]–[159]. The need of a negative dispersion slope can be understood from the condition (7.4.2) obtained in Section 7.4 for a single channel. This condition should be satisfied for all channels, i.e.,

$$D_1(\lambda_n)L_1 + D_2(\lambda_n)L_2 = 0, \tag{7.9.1}$$

where λ_n is the wavelength of the nth channel. Because of a finite positive value of the dispersion slope S, or the third-order dispersion β_3 (see Section 2.3.4), D_1 increases with wavelength for both the standard and dispersion-shifted fibers (see Fig. 2.11). As a result the accumulated dispersion D_1L_1 is different for each channel. If the same DCF has to work for all channels, its dispersion slope should be negative and has a value such that Eq. (7.9.1) is approximately satisfied for all channels.

Writing $D_j(\lambda_n) = D_j + S_j(\lambda_n - \lambda_c)$ in Eq. (7.9.1), where D_j $(j = 1, 2)$ is the value at the wavelength λ_c of the central channel, the dispersion slope of the DCF should be

$$S_2 = -S_1(L_1/L_2) = S_1(D_2/D_1), \tag{7.9.2}$$

where we used the condition (7.4.2) for the central channel. This equation shows that the ratio S/D, called the *relative dispersion slope*, should be the same for both fibers used to form the dispersion map [44]. For standard fibers with $D \approx 16$ ps/(km-nm) and $S \approx 0.05$ ps/(km-nm^2), this ratio is about 0.003 nm^{-1}. Thus, for a DCF with $D \approx -100$ ps/(km-nm), the dispersion slope should be about -0.3 ps/(km-nm^2). Such DCFs have been made and are available commercially. In the case of dispersion-shifted fibers, the ratio S/D can exceed 0.02 nm^{-1}. It is hard to manufacture DCFs with such large values of the relative dispersion slope, although two-mode DCFs can provide values as large as 0.01 nm^{-1} (see Fig. 7.5). In their place, *reverse-dispersion* fibers have been developed for which the signs of both D and S are reversed compared with the conventional dispersion-shifted fibers. Dispersion map in this case is made using roughly equal lengths of the two types of fibers.

Many experiments during the 1990s demonstrated the usefulness of DCFs for WDM systems. In a 1995 experiment [148], 8 channels with 1.6-nm spacing, each operating at 20 Gb/s, were transmitted over 232 km of standard fiber by using multiple DCFs. The residual dispersion for each channel was relatively small ($\sim$ 100 ps/nm for the entire span) since all channels were compensated simultaneously by the DCFs. In a 2001 experiment, broadband DCFs were used to transmit a 1-Tb/s WDM signal (101 channels, each operating at 10 Gb/s) over 9000 km [158]. The highest capacity of 11 Tb/s was also realized using the reverse-dispersion fibers in an experiment [159] that transmitted 273 channels, each operating at 40 Gb/s, over the C, L, and S bands simultaneously (resulting in the total bandwidth of more than 100 nm).

7.9.2 Tunable Dispersion Compensation

It is difficult to attain full GVD compensation for all channels in a WDM system. A small amount of residual dispersion remains and often becomes of concern for long-haul systems. In many laboratory experiments, a postcompensation technique is adopted in which the residual dispersion for individual channels is compensated by adding adjustable lengths of a DCF (or a fiber grating) at the receiver end (dispersion trimming). This technique is not suitable for commercial WDM systems for several reasons. First, the exact amount of channel-dependent residual dispersion is not always known because of uncontrollable variations in fiber GVD in the fiber segments forming the transmission path. Second, even the path length may change in reconfigurable optical networks. Third, as the single-channel bit rate increases toward 40 Gb/s, the tolerable value of the residual dispersion becomes so small that even temperature-induced changes in GVD become of concern. For these reasons, the best approach may be to adopt a tunable dispersion-compensation scheme that allows the GVD control for each channel in a dynamic fashion.

Several techniques for tunable dispersion compensation have been developed and used for system experiments [160]–[167]. Most of them make use of a fiber Bragg grating whose dispersion is tuned by changing the grating period $\bar{n}\Lambda$. In one scheme, the grating is made with a nonlinear chirp (Bragg wavelength increases nonlinearly along the grating length) that can be changed by stretching the grating with a piezoelectric transducer [160]. In another approach, the grating is made with either no chirp or with a linear chirp and a temperature gradient is used to produce a controllable chirp [164]. In both cases, the stress- or temperature-induced changes in the mode index $\bar{n}$ change the local Bragg wavelength as $\lambda_B(z) = 2\bar{n}(z)\Lambda(z)$. For such a grating, Eq. (7.6.6) is replaced with

$$D_g(\lambda) = \frac{d\tau_g}{d\lambda} = \frac{2}{c} \frac{d}{d\lambda} \left(\int_0^{L_g} \bar{n}(z)\,dz \right), \qquad (7.9.3)$$

where τ_g is the group delay and L_g is the grating length. The value of D_g at any wavelength can be changed by changing the mode index $\bar{n}$ (through heating or stretching), resulting in tunable dispersion characteristics for the Bragg grating.

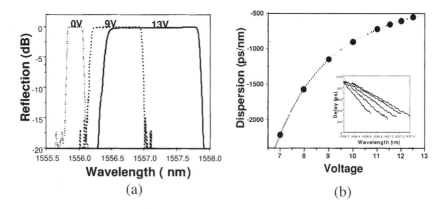

Figure 7.17: (a) Reflection spectrum and (b) total GVD as a function of voltage for a fiber grating with temperature gradient. Inset shows $\tau_g(\lambda)$ at several voltages. (After Ref. [164]; ©2000 IEEE; reprinted with permission.)

Distributed heating of the Bragg grating requires a thin-film heater deposited on the outer surface of the fiber with an intracore grating [164]. The film thickness changes along the grating length and creates a temperature gradient through nonuniform heating when a voltage is applied across the film. A segmented thin-film heater can also be used for this purpose [167]. Figure 7.17 shows the reflection spectra of a 8-cm-long grating at three voltage levels together with the total dispersion $D_g L_g$ as a function of voltage. The inset shows $\tau_g(\lambda)$ for several values of the applied voltage. The grating is initially unchirped and has a narrow stop band that shifts and broadens as the grating is chirped through nonuniform heating. Physically, the Bragg wavelength λ_B changes along the grating because the optical period $\bar{n}(z)\Lambda$ becomes z dependent when a temperature gradient is established along the grating. The total dispersion $D_g L_g$ can be changed in the range -500 to -2200 ps/nm by this approach. Such gratings can be used to provide tunable dispersion for 10-Gb/s systems.

When the bit rate becomes 40 Gb/s or more, it is necessary to chirp the grating so that the stop band is wide enough for passing the signal spectrum. The use of a nonlinear chirp then provides an additional control over the device [160]. Such chirped gratings have been made and used to provide tunable dispersion compensation at bit rates as high as 160 Gb/s. Figure 7.18 shows the measured receiver sensitivities in the 160-Gb/s experiment as a function of the residual (preset) dispersion with and without the chirped grating with tunable dispersion [165]. In the absence of the grating, the minimum sensitivity occurs around 91 ps/nm because the DCF used provided this value of constant dispersion. A power penalty of 4 dB occurred when the residual GVD changed by as little as 8 ps/nm. It was reduced to below 0.5 dB with tunable dispersion compensation. The eye diagrams for a residual dispersion of 110 ps/nm show that the system becomes inoperable without the grating but the eye remains wide open when tunable dispersion compensation is employed. This experiment used 2-ps optical pulses as the bit slot is only 6.25 ps wide at the 160-Gb/s bit rate. The effects of third-order dispersion becomes important for such short pulses. We turn to this issue next.

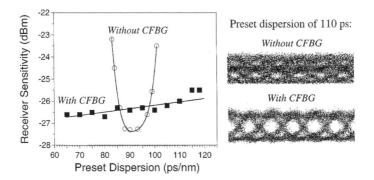

Figure 7.18: Receiver sensitivities measured in a 160-Gb/s experiment as a function of the preset dispersion with (squares) and without (circles) a chirped fiber–Bragg grating (CFBG). The improvement in the eye diagram is shown for 110 ps/nm on the right. (After Ref. [165]; ©2000 IEEE; reprinted with permission.)

7.9.3 Higher-Order Dispersion Management

When the bit rate of a single channel exceeds 40 Gb/s (through the used of time-division multiplexing, for example), the third- and higher-order dispersive effects begin to influence the optical signal. For example, the bit slot at a bit rate of 100 Gb/s is only 10 ps wide, and an RZ optical signal would consist of pulses of width <5 ps. Equation (2.4.34) can be used to estimate the maximum transmission distance L, limited by the third-order dispersion β_3, when only second-order dispersion is compensated. The result is

$$L \leq 0.034(|\beta_3|B^3)^{-1}. \tag{7.9.4}$$

This limitation is shown in Fig. 2.13 by the dashed line. At a bit rate of 200 Gb/s, L is limited to about 50 km and drops to only 3.4 km at 500 Gb/s if we use a typical value $\beta_3 = 0.08$ ps^3/km. Clearly, it is essential to use techniques that compensate for both the second- and third-order dispersion simultaneously when the single-channel bit rate exceeds 100 Gb/s, and several techniques have been developed for this purpose [168]–[180].

The simplest solution to third-order dispersion compensation is provided by DCFs designed to have a negative dispersion slope so that both β_2 and β_3 have opposite signs, in comparison with the standard fibers. The necessary conditions for designing such fibers can be obtained by solving Eq. (7.1.3) using the Fourier-transform method. For a fiber link containing two different fibers of lengths L_1 and L_2, the conditions for dispersion compensation become

$$\beta_{21}L_1 + \beta_{22}L_2 = 0 \quad \text{and} \quad \beta_{31}L_1 + \beta_{32}L_2 = 0, \tag{7.9.5}$$

where β_{2j} and β_{3j} are second- and third-order dispersion parameters for the fiber of length L_j. The first condition is the same as Eq. (7.4.2). By using Eq. (7.4.3), the second condition can be used to find the third-order dispersion parameter for the DCF:

$$\beta_{32} = (\beta_{22}/\beta_{21})\beta_{31} = -(L_1/L_2)\beta_{31}. \tag{7.9.6}$$

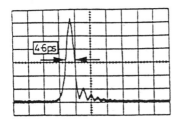 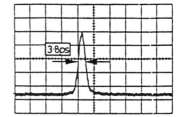

Figure 7.19: Pulse shapes after a 2.6-ps input pulse propagated over 300 km of dispersion-shifted fiber ($\beta_2 = 0$). Left and right traces compare the improvement realized by compensating the third-order dispersion. (After Ref. [169]; ©1996 IEE; reprinted with permission.)

This requirement is nearly the same as that obtained earlier in Eq. (7.9.2) for DCFs used in WDM systems because β_3 is related to the dispersion slope S through Eq. (2.3.13).

For a single-channel system, the signal bandwidth is small enough even at bit rates of 500 Gb/s that it is sufficient to satisfy Eq. (7.9.5) over a 4-nm bandwidth. This requirement is easily met by an optical filter or a chirped fiber grating [51]. Consider the case of optical filters first. Planar lightwave circuits based on multiple MZ interferometric filters (see Section 7.5) have proved quite successful because of the programmable nature of such filters. In one experiment [169], such a filter was designed to have a dispersion slope of -15.8 ps/nm^2 over a 170-GHz bandwidth. It was used to compensate third-order dispersion over 300 km of a dispersion-shifted fiber with $\beta_3 \approx 0.05$ ps/(km-nm^2) at the operating wavelength. Figure 7.19 compares the pulse shapes at the fiber output observed with and without β_3 compensation when a 2.6-ps pulse was transmitted over 300 km of such a fiber. The equalizer eliminates the long oscillatory tail and reduces the width of the main peak from 4.6 to 3.8 ps. The increase in the pulse width from its input value of 2.6 ps is attributed to polarization-mode dispersion (PMD), a topic covered later.

Chirped fiber gratings are often preferred in practice because of their all-fiber nature. Long fiber gratings (~ 1 m) were developed by 1997 for this purpose [170]. In 1998, a nonlinearly chirped fiber grating was capable of compensating the third-order dispersion over 6 nm for distances as long as 60 km [171]. Cascading of several chirped gratings can provide a dispersion compensator that has arbitrary dispersion characteristics and is capable for compensating dispersion to all higher orders [172]. An arrayed-waveguide grating [173] or a sampled fiber grating [174] can also compensate for second- and third-order dispersion simultaneously. Although a sampled fiber grating chirped nonlinearly can provide tunable dispersion for several channels simultaneously [177], its bandwidth is still limited. An arrayed-waveguide grating in combination with a spatial phase filter can provide dispersion-slope compensation over a bandwidth as large as 8 THz and should be suitable for 40-Gb/s multichannel systems [178]. The feasibility of transmitting of a 100-Gb/s signal over 10,000 km has also been investigated using midway optical phase conjugation in combination with third-order dispersion compensation [179].

Several single-channel experiments have explored the possibility of transmitting a single channel at bit rates of more than 200 Gb/s [181]–[183]. Assuming that a 2-ps bit slot is required for a RZ system making use of 1-ps pulses, transmission at bit rates as high as 500 Gb/s appears to be feasible using DCFs or chirped fiber gratings designed to provide compensation of β_3 over a 4-nm bandwidth. In a 1996 experiment [181], a 400-Gb/s signal was transmitted by managing the fiber dispersion and transmitting 0.98-ps pulses inside a 2.5-ps time slot. Without compensation of the third-order dispersion, the pulse broadened to 2.3 ps after 40 km and exhibited a long oscillatory tail extending over 5–6 ps, a characteristic feature of the third-order dispersion [106]. With partial compensation of third-order dispersion, the oscillatory tail disappeared, and the pulse width reduced to 1.6 ps, making it possible to recover the 400-Gb/s data with high accuracy. Optical pulses as short as 0.4 ps were used in 1998 to realize a bit rate of 640 Gb/s [182]. In a 2001 experiment, the bit rate was extended to 1.28 Tb/s by transmitting 380-fs pulses over 70 km of fiber [183]. Propagation of such short pulses requires compensation of second- third- and fourth-order dispersion simultaneously. It turns out that if sinusoidal phase modulation of the right kind is applied to the linearly chirped pulse before it is transmitted through a GVD-compensated fiber, it can compensate for both the third- and fourth-order dispersion.

7.9.4 PMD Compensation

As discussed in Section 2.3.5, PMD leads to broadening of optical pulses because of random variations in the birefringence of an optical fiber along its length. This broadening is in addition to GVD-induced pulse broadening. The use of dispersion management can eliminate GVD-induced broadening but does not affect the PMD-induced broadening. For this reason, PMD has become a major source of concern for modern dispersion-managed systems [184]–[196].

Before considering the techniques used for PMD compensation, we provide an order-of-magnitude estimate of the system length in uncompensated systems. Equation (2.3.17) shows that the RMS pulse broadening for a link of length L is given by $\sigma_T \equiv \langle (\Delta T)^2 \rangle^{1/2} = D_p \sqrt{L}$, where D_p is the PMD parameter and ΔT is the relative delay along the two principal states of polarization (PSPs). It is important to note that σ_T denotes an average value. The instantaneous value of ΔT fluctuates with time over a wide range because of temperature and other environmental factors [192]. If ΔT exceeds the bit slot even for a short time interval, the system will stop functioning properly; this is referred to as fading or *outage* in analogy with a similar effect occurring in radio systems [184].

The performance of a PMD-limited system is quantified using the concept of the outage probability, which should be below a prescribed value (often set near 10^{-5} or 5 minutes/year) for the system [184]. This probability can be calculated noting that ΔT follows a Maxwellian distribution. In general, the RMS value σ_T should only be a small fraction of the bit slot T_B at a certain bit rate $B \equiv 1/T_B$. The exact value of this fraction varies in the range 0.1–0.15 depending on the modulation format (RZ, CRZ, or NRZ) and details of the input pulse. Using 10% as a conservative criterion, the system length and the bit rate should satisfy the condition

$$B^2 L < (10 D_p)^{-2}. \tag{7.9.7}$$

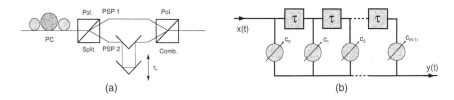

Figure 7.20: Schematic illustration of (a) optical and (b) electrical PMD compensators. (After Ref. [210]; ©2000 Elsevier; reprinted with permission.)

Consider a few relevant examples. In the case of "old" fiber links installed using standard fibers, the condition (7.9.7) becomes $B^2L < 10^4$ (Gb/s)^2-km if we use $D_p = 1$ ps/$\sqrt{\text{km}}$ as a representative value. Such fibers require PMD compensation at $B = 10$ Gb/s if the link length exceeds even 100 km. In contrast, modern fibers have typically $D_p < 0.1$ ps/$\sqrt{\text{km}}$. For systems designed using such fibers, B^2L can exceed 10^6 (Gb/s)^2-km. As a result, PMD compensation is not necessary at 10 Gb/s but may be required at 40 Gb/s if the link length exceeds 600 km. It should be stressed that these numbers represent only an order-of-magnitude estimate. A more accurate estimate can be obtained following the PMD theory developed in recent years [192]–[195].

The preceding discussion shows that PMD limits the system performance when the single-channel bit rate is extended to beyond 10 Gb/s. Several techniques have been developed for PMD compensation in dispersion-managed lightwave systems [197]–[214]; they can be classified as being optical or electrical. Figure 7.20 shows the basic idea behind the electrical and optical PMD compensation schemes. An electrical PMD equalizer corrects for the PMD effects within the receiver using a transversal filter. The filter splits the electrical signal $x(t)$ into a number of branches using multiple tapped delay lines and then combines the output as

$$y(t) = \sum_{m=0}^{N-1} c_m x(t - m\tau), \qquad (7.9.8)$$

where N is the total number of taps, τ is the delay time, and c_m is the tap weight for the mth tap. Tap weights are adjusted in a dynamic fashion using a control algorithm in such a way that the system performance is improved [210]. The error signal for the control electronics is often based on the closing of the "eye" at the receiver. Such an electrical technique cannot eliminate the PMD effects completely as it does not consider the PMD-induced delay between the two PSPs. On the positive side, it corrects for all sources of degradation that lead to eye closing.

An optical PMD compensator also makes use of a delay line. It can be inserted periodically all along the fiber link (at the amplifier locations, for example) or just before the receiver. Typically, the PMD-distorted signal is separated into two components along the PSPs using a polarization controller (PC) followed by a polarization beam splitter; the two components are combined after introducing an adjustable delay in one branch through a variable delay line (see Fig. 7.20). A feedback loop is still needed to obtain an error signal that is used to adjust the polarization controller in response to the environmental changes in the fiber PSPs. The success of this technique depends on

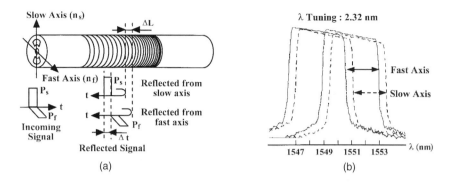

Figure 7.21: Tunable PMD compensation provided by a birefringent chirped fiber grating. (a) The origin of differential group delay; (b) stop-band shift induced by stretching the grating. (After Ref. [160]; ©1999 IEEE; reprinted with permission.)

the ratio L/L_{PMD} for a fiber of length L, where $L_{PMD} = (T_0/D_p)^2$ is the PMD length for pulses of width T_0 [202]. Considerable improvement is expected as long as this ratio does not exceed 4. Because L_{PMD} is close to 10,000 km for $D_p \approx 0.1$ ps/$\sqrt{km}$ and $T_0 = 10$ ps, such a PMD compensator can work over transoceanic distances for 10-Gb/s systems.

Several other all-optical techniques can be used for PMD compensation [205]. For example, a LiNbO$_3$-based Soleil–Babinet compensator can provide endless polarization control. Other devices include ferroelectric liquid crystals, twisted polarization-maintaining fibers, optical all-pass filters [209], and birefringent chirped fiber gratings [204]. Figure 7.21 shows how a grating-based PMD compensator works. Because of a large birefringence, the two field components polarized along the slow and fast axis have different Bragg wavelengths and see slightly shifted stop bands. As a result, they are reflected at different places within the grating and experience a differential group delay that can compensate for the PMD-induced group delay. The delay is wavelength dependent because of the chirped nature of the grating. Moreover, it can be tuned over several nanometers by stretching the grating [160]. Such a device can provide tunable PMD compensation and is suited for WDM systems.

It should be stressed that optical PMD compensators shown in Figs. 7.20 and 7.21 remove only the first-order PMD effects. At high bit rates, optical pulses are short enough and their spectrum becomes wide enough that the PSPs cannot be assumed to remain constant over the whole pulse spectrum. Higher-order PMD effects have become of concern with the advent of 40-Gb/s lightwave systems, and techniques for compensating them have been proposed [207].

The effectiveness of first-order PMD compensation can be judged by considering how much PMD-induced pulse broadening is reduced by such a compensator. An analytical theory of PMD compensation shows that the average or expected value of the broadening factor, defined as $b^2 = \sigma^2/\sigma_0^2$, is given by the following expression for an unchirped Gaussian pulse of width T_0 [206]:

$$b_c^2 = b_u^2 + 2x/3 - 4[(1 + 2x/3)^{1/2} - 1], \tag{7.9.9}$$

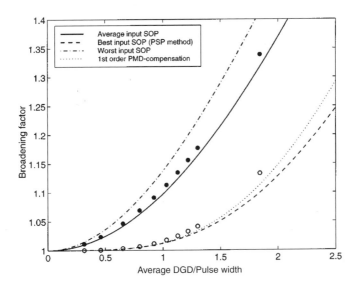

Figure 7.22: Pulse broadening factor as a function of average DGD in four cases. The dotted curve shows the improvement realized using a first-order PMD compensator. Filled and empty circles show the results of numerical simulations (After Ref. [206]; ©2000 IEEE; reprinted with permission.)

where $x = \langle(\Delta T)^2\rangle/4T_0^2$, ΔT is the differential group delay along the PSPs, and b_u^2 is the value before PMD compensation:

$$b_u^2 = 1 + x - \tfrac{1}{2}[(1 + 4x/3)^{1/2} - 1]. \qquad (7.9.10)$$

Figure 7.22 shows the broadening factors b_u (solid line) and b_c (dotted line) as a function of $\langle\Delta T\rangle/T_0$. For comparison, the worst and best cases corresponding to the two specific choices of the input state of polarization (SOP) are also shown.

Figure 7.22 can be used to estimate the improvement realized by a first-order PMD compensator. As discussed earlier, the average DGD should not exceed about 10% of the bit slot in uncompensated systems for keeping the outage probability below 10^{-5}. Thus, the tolerable value of PMD-induced pulse broadening is close to $b = 1.02$. From Eqs. (7.9.9) and (7.9.10) it is easy to show that this value can be maintained in PMD-compensated systems even when σ_T exceeds 30%. Thus, a first-order PMD compensator can increase the tolerable value of DGD by more than a factor of 3. The net result is a huge increase in the transmission distance of PMD-compensated systems. One should note that a single PMD compensator cannot be used for all WDM channels. Rather, a separate PMD compensator is required for each channel. This fact makes PMD compensation along the fiber link a costly proposition for WDM systems. An optical compensator just before the receiver or an electrical PMD equalizer built into the receiver provides the most practical solution; both were being pursued in 2001 for commercial applications.

Problems

7.1 What is the dispersion-limited transmission distance for a 1.55-μm lightwave system making use of direct modulation at 10 Gb/s? Assume that frequency chirping broadens the Gaussian-shape pulse spectrum by a factor of 6 from its transform-limited width. Use $D = 17$ ps/(km-nm) for fiber dispersion.

7.2 How much improvement in the dispersion-limited transmission distance is expected if an external modulator is used in place of direct modulation for the lightwave system of Problem 7.1?

7.3 Solve Eq. (7.1.3) by using the Fourier transform method. Use the solution to find an analytic expression for the pulse shape after a Gaussian input pulse has propagated to $z = L$ in a fiber with $\beta_2 = 0$.

7.4 Use the result obtained in Problem 7.3 to plot the pulse shape after a Gaussian pulse with a full-width at half-maximum (FWHM) of 1 ps is transmitted over 20 km of dispersion-shifted fiber with $\beta_2 = 0$ and $\beta_3 = 0.08$ ps^3/km. How would the pulse shape change if the sign of β_3 is inverted?

7.5 Use Eqs. (7.1.4) and (7.2.2) to plot the pulse shapes for $C = -1, 0$, and 1 when 50-ps (FWHM) chirped Gaussian pulses are transmitted over 100 km of standard fiber with $D = 16$ ps/(km-nm). Compare the three cases and comment on their relative merits.

7.6 The prechirp technique is used for dispersion compensation in a 10-Gb/s lightwave system operating at 1.55 μm and transmitting the 1 bits as chirped Gaussian pulses of 40 ps width (FWHM). Pulse broadening by up to 50% can be tolerated. What is the optimum value of the chirp parameter C, and how far can the signal be transmitted for this optimum value? Use $D = 17$ ps/(km-nm).

7.7 The prechirp technique in Problem 7.6 is implemented through frequency modulation of the optical carrier. Determine the modulation frequency for a maximum change of 10% from the average value.

7.8 Repeat Problem 7.7 for the case in which the prechirp technique is implemented through sinusoidal modulation of the carrier phase.

7.9 The transfer function of an optical filter is given by

$$H(\omega) = \exp[-(1 + ib)\omega^2/\omega_f^2].$$

What is the impulse response of this filter? Use Eq. (7.5.1) to find the pulse shape at the filter output when a Gaussian pulse is launched at the fiber input. How would you optimize the filter to minimize the effect of fiber dispersion?

7.10 Use the result obtained in Problem 7.9 to compare the pulse shapes before and after the filter when 30-ps (FWHM) Gaussian pulses are propagated over 100 km of fiber with $\beta_2 = -20$ ps^2/km. Assume that the filter bandwidth is the same as the pulse spectral width and that the filter parameter b is optimized. What is the optimum value of b?

7.11 Derive Eq. (7.5.3) by considering multiple round trips inside a FP filter whose back mirror is 100% reflecting.

7.12 Solve Eqs. (7.6.1) and (7.6.2) and show that the transfer function of a Bragg grating is indeed given by Eq. (7.6.4).

7.13 Write a computer program to solve Eqs. (7.6.1) and (7.6.2) for chirped fiber gratings such that both δ and κ vary with z. Use it to plot the amplitude and phase of the reflectivity of a grating in which the period varies linearly by 0.01% over the 10-cm length. Assume $\kappa L = 4$ and the Bragg wavelength of 1.55 μm at the input end of the grating.

7.14 Use the dispersion relation $q^2 = \delta^2 - \kappa^2$ of a Bragg grating to show that the second- and third-order dispersion parameters of the grating are given by Eq. (7.6.5).

7.15 Explain how a chirped fiber grating compensates for GVD. Derive an expression for the GVD parameter of such a grating when the grating period varies linearly by $\Delta\Lambda$ over the grating length L.

7.16 Explain how midspan OPC compensates for fiber dispersion. Show that the OPC process inverts the signal spectrum.

7.17 Prove that both SPM and GVD can be compensated through midspan OPC only if the fiber loss $\alpha = 0$. Show also that simultaneous compensation of SPM and GVD can occur when $\alpha \neq 0$ if GVD decreases along the fiber length. What is the optimum GVD profile of such a fiber?

7.18 Prove that the phase conjugator should be located at a distance given in Eq. (7.7.9) when the frequency ω_c of the phase-conjugated field does not coincide with the signal frequency ω_s.

7.19 Derive the variational equations for the pulse width and chirp using the Lagrangian density given in Eq. (7.8.6).

7.20 Solve the variational equations (7.8.7) and (7.8.8) after setting $\gamma = 0$ and find the pulse width and chirp after one map period in terms of their initial values.

References

[1] A. Gnauck and R. Jopson, in *Optical Fiber Telecommunications III*, Vol. A, I. P. Kaminow and T. L. Koch, Eds., Academic Press, San Diego, CA, 1997, Chap. 7.

[2] G. P. Agrawal and M. J. Potasek, *Opt. Lett.* **11**, 318 (1986).

[3] R. Olshansky and D. Fye, *Electron. Lett.* **20**, 80 (1984).

[4] L. Bickers and L. D. Westbrook, *Electron. Lett.* **21**, 103 (1985).

[5] T. L. Koch and R. C. Alferness, *J. Lightwave Technol.* **3**, 800 (1985).

[6] F. Koyoma and K. Iga, *J. Lightwave Technol.* **6**, 87 (1988).

[7] A. H. Gnauck, S. K. Korotky, J. J. Veselka, J. Nagel, C. T. Kemmerer, W. J. Minford, and D. T. Moser, *IEEE Photon. Technol. Lett.* **3**, 916 (1991).

[8] E. Devaux, Y. Sorel, and J. F. Kerdiles, *J. Lightwave Technol.* **11**, 1937 (1993).

[9] N. Henmi, T. Saito, and T. Ishida, *J. Lightwave Technol.* **12**, 1706 (1994).

[10] J. C. Cartledge, H. Debrégeas, and C. Rolland, *IEEE Photon. Technol. Lett.* **7**, 224 (1995).

[11] J. A. J. Fells, M. A. Gibbon, I. H. White, G. H. B. Thompson, R. V. Penty, C. J. Armistead, E. M. Kinber, D. J. Moule, and E. J. Thrush, *Electron. Lett.* **30**, 1168 (1994).

[12] K. Morito, R. Sahara, K. Sato, and Y. Kotaki, *IEEE Photon. Technol. Lett.* **8**, 431 (1996).

[13] B. Wedding, B. Franz, and B. Junginger, *J. Lightwave Technol.* **12**, 1720 (1994).

[14] B. Wedding, K. Koffers, B. Franz, D. Mathoorasing, C. Kazmierski, P. P. Monteiro, and J. N. Matos, *Electron. Lett.* **31**, 566 (1995).

[15] W. Idler, B. Franz, D. Schlump, B. Wedding, and A. J. Ramos, *Electron. Lett.* **35**, 2425 (1998).

[16] K. Perlicki and J. Siuzdak, *Opt. Quantum Electron.* **31**, 243 (1999).

[17] J. A. V. Morgado and A. V. T. Cartaxo, *IEE Proc. Optoelect.* **148**, 107 (2001).

[18] M. Schwartz, *Information, Transmission, Modulation, and Noise*, 4th ed., McGraw-Hill, New York, 1990, Sec. 3.10.

[19] G. May, A. Solheim, and J. Conradi, *IEEE Photon. Technol. Lett.* **6**, 648 (1994).

[20] D. Penninckx, L. Pierre, J.-P. Thiery, B. Clesca, M. Chbat, and J.-L. Beylat, *Electron. Lett.* **32**, 1023 (1996).

[21] D. Penninckx, M. Chbat, L. Pierre, and J.-P. Thiery, *IEEE Photon. Technol. Lett.* **9**, 259 (1997).

[22] K. Yonenaga and S. Kuwano, *J. Lightwave Technol.* **15**, 1530 (1997).

[23] T. Ono, Y. Yano, K. Fukuchi, T. Ito, H. Yamazaki, M. Yamaguchi, and K. Emura, *J. Lightwave Technol.* **16**, 788 (1998).

[24] W. Kaiser, T. Wuth, M. Wichers, and W. Rosenkranz, *IEEE Photon. Technol. Lett.* **13**, 884 (2001).

[25] G. P. Agrawal and N. A. Olsson, *Opt. Lett.* **14**, 500 (1989).

[26] N. A. Olsson, G. P. Agrawal, and K. W. Wecht, *Electron. Lett.* **25**, 603 (1989).

[27] N. A. Olsson and G. P. Agrawal, *Appl. Phys. Lett.* **55**, 13 (1989).

[28] G. P. Agrawal and N. A. Olsson, *IEEE J. Quantum Electron.* **25**, 2297 (1989).

[29] G. P. Agrawal and N. A. Olsson, U.S. Patent 4,979,234 (1990).

[30] M. J. Potasek and G. P. Agrawal, *Electron. Lett.* **22**, 759 (1986).

[31] K. Iwashita and N. Takachio, *J. Lightwave Technol.* **8**, 367 (1990).

[32] N. Takachio, S. Norimatsu, and K. Iwashita, *IEEE Photon. Technol. Lett.* **4**, 278 (1992).

[33] K. Yonenaga and N. Takachio, *IEEE Photon. Technol. Lett.* **5**, 949 (1993).

[34] S. Yamazaki, T. Ono, and T. Ogata, *J. Lightwave Technol.* **11**, 603 (1993).

[35] J. H. Winters and R. D. Gitlin, *IEEE Trans. Commun.* **38**, 1439 (1990).

[36] J. H. Winters, *J. Lightwave Technol.* **8**, 1487 (1990).

[37] J. C. Cartledge, R. G. McKay, and M. C. Nowell, *J. Lightwave Technol.* **10**, 1105 (1992).

[38] J. H. Winters, *Proc. SPIE* **1787**, 346 (1992).

[39] R. I. MacDonald, *IEEE Photon. Technol. Lett.* **6**, 565 (1994).

[40] C. Lin, H. Kogelnik, and L. G. Cohen, *Opt. Lett.* **5**, 476 (1980).

[41] A. J. Antos and D. K. Smith, *J. Lightwave Technol.* **12**, 1739 (1994).

[42] M. Onishi, T. Kashiwada, Y. Ishiguro, Y. Koyano, M. Nishimura, and H. Kanamori, *Fiber Integ. Opt.* **16**, 277 (1997).

[43] J. Liu, Y. L. Lam, Y. C. Chan, Y. Zhou, and J. Yao, *Fiber Integ. Opt.* **18**, 63 (1999).

[44] L. Gruner-Nielsen, S. N. Knudsen, B. Edvold, T. Veng, D. Magnussen, C. C. Larsen, and H. Damsgaard, *Opt. Fiber Technol.* **6**, 164 (2000).

[45] C. D. Poole, J. M. Wiesenfeld, D. J. DiGiovanni, and A. M. Vengsarkar, *J. Lightwave Technol.* **12** 1746 (1994).

[46] M. Eguchi, M Koshiba, and Y. Tsuji, *J. Lightwave Technol.* **14**, 2387 (1996).

[47] S. Ramachandran, B. Mikkelsen, L. C. Cowsar, M. F. Yan, G. Raybon, L. Boivin, M. Fishteyn, W. A. Reed, P. Wisk, et al., *IEEE Photon. Technol. Lett.* **13**, 632 (2001).

[48] M. Eguchi, *J. Opt. Soc. Am. B* **18**, 737 (2001).

[49] R. C. Youngquist, J. L. Brooks, and H. J. Shaw, *Opt. Lett.* **9**, 177 (1984).

[50] C. D. Poole, C. D. Townsend, and K. T. Nelson, *J. Lightwave Technol.* **9**, 598 (1991).

[51] R. Kashyap, *Fiber Bragg Gratings*, Academic Press, San Diego, CA, 1999.

[52] L. J. Cimini, L. J. Greenstein, and A. A. M. Saleh, *J. Lightwave Technol.* **8**, 649 (1990).

[53] A. H. Gnauck, C. R. Giles, L. J. Cimini, J. Stone, L. W. Stulz, S. K. Korotoky, and J. J. Veselka, *IEEE Photon. Technol. Lett.* **3**, 1147 (1991).

[54] G. P. Agrawal, *Applications of Nonlinear Fiber Optics*, Academic Press, San Diego, CA, 2001.

[55] T. Ozeki, *Opt. Lett.* **17**, 375 (1992).

[56] K. Takiguchi, K. Okamoto, S. Suzuki, and Y. Ohmori, *IEEE Photon. Technol. Lett.* **6**, 86 (1994).

[57] M. Sharma, H. Ibe, and T. Ozeki, *J. Lightwave Technol.* **12**, 1759 (1994).

[58] K. Takiguchi, K. Okamoto, and K. Moriwaki, *IEEE Photon. Technol. Lett.* **6**, 561 (1994).

[59] K. Takiguchi, K. Okamoto, and K.. Moriwaki, *J. Lightwave Technol.* **14**, 2003 (1996).

[60] D. K. W. Lam, B. K. Garside, and K. O. Hill, *Opt. Lett.* **7**, 291 (1982).

[61] B. J. Eggleton, T. Stephens, P. A. Krug, G. Dhosi, Z. Brodzeli, and F. Ouellette, *Electron. Lett.* **32**, 1610 (1996).

[62] T. Stephens, P. A. Krug, Z. Brodzeli, G. Dhosi, F. Ouellette, and L. Poladian, *Electron. Lett.* **32**, 1599 (1996).

[63] N. M. Litchinister, B. J. Eggleton, and D. B. Pearson, *J. Lightwave Technol.* **15**, 1303 (1997).

[64] K. Hinton, *J. Lightwave Technol.* **15**, 1411 (1997).

[65] F. Ouellette, *Opt. Lett.* **12**, 622, 1987.

[66] M. C. Farries, K. Sugden, D. C. J. Reid, I. Bennion, A. Molony, and M. J. Goodwin, *Electron. Lett.* **30**, 891 (1994).

[67] K. O. Hill, F. Bilodeau, B. Malo, T. Kitagawa, S. Thériault, D. C. Johnson, and J. Albert, *Opt. Lett.* **19**, 1314 (1994).

[68] K. O. Hill, S. Thériault, B. Malo, F. Bilodeau, T. Kitagawa, D. C. Johnson, J. Albert, K. Takiguchi, T. Kataoka, and K. Hagimoto, *Electron. Lett.* **30**, 1755 (1994).

[69] D. Garthe, R. E. Epworth, W. S. Lee, A. Hadjifotiou, C. P. Chew, T. Bricheno, A. Fielding, H. N. Rourke, S. R. Baker, et al., *Electron. Lett.* **30**, 2159 (1994).

[70] P. A. Krug, T. Stephens, G. Yoffe, F. Ouellette, P. Hill, and G. Dhosi, *Electron. Lett.* **31**, 1091 (1995).

[71] W. H. Loh, R. I. Laming, X. Gu, M. N. Zervas, M. J. Cole, T. Widdowson, and A. D. Ellis, *Electron. Lett.* **31**, 2203 (1995).

[72] W. H. Loh, R. I. Laming, N. Robinson, A. Cavaciuti, F. Vaninetti, C. J. Anderson, M. N. Zervas, and M. J. Cole, *IEEE Photon. Technol. Lett.* **8**, 944 (1996).

[73] W. H. Loh, R. I. Laming, A. D. Ellis, and D. Atkinson, *IEEE Photon. Technol. Lett.* **8**, 1258 (1996).

[74] S. V. Chernikov, J. R. Taylor, and R. Kashyap, *Opt. Lett.* **20**, 1586 (1995).

[75] G. P. Agrawal and S. Radic, *IEEE Photon. Technol. Lett.* **6**, 995 (1994).

[76] L. Zhang, K. Sugden, I. Bennion, and A. Molony, *Electron. Lett.* **31**, 477 (1995).

[77] F. Ouellette, J.-F. Cliche, and S. Gagnon, *J. Lightwave Technol.* **12**, 1278 (1994).

[78] F. Ouellette, *Opt. Lett.* **16**, 303 (1991).

[79] F. Ouellette, *Electron. Lett.* **27**, 1668 (1991).

[80] U. Peschel, T. Peschel, and F. Lederer, *Appl. Phys. Lett.* **67**, 2111 (1995).

[81] A. Yariv, D. Fekete, and D. M. Pepper, *Opt. Lett.* **4**, 52 (1979).

[82] S. Watanabe, N. Saito, and T. Chikama, *IEEE Photon. Technol. Lett.* **5**, 92 (1993).

[83] R. M. Jopson, A. H. Gnauck, and R. M. Derosier, *IEEE Photon. Technol. Lett.* **5**, 663 (1993).

[84] M. C. Tatham, G. Sherlock, and L. D. Westbrook, *Electron. Lett.* **29**, 1851 (1993).

[85] R. M. Jopson and R. E. Tench, *Electron. Lett.* **29**, 2216 (1993).

[86] S. Watanabe, T. Chikama, G. Ishikawa, T. Terahara, and H. Kuwahara, *IEEE Photon. Technol. Lett.* **5**, 1241 (1993).

[87] K. Kikuchi and C. Lorattanasane, *IEEE Photon. Technol. Lett.* **6**, 104 (1994).

[88] S. Watanabe, G. Ishikawa, T. Naito, and T. Chikama, *J. Lightwave Technol.* **12**, 2139 (1994).

[89] W. Wu, P. Yeh, and S. Chi, *IEEE Photon. Technol. Lett.* **6**, 1448 (1994).

[90] C. R. Giles, V. Mizrahi, and T. Erdogan, *IEEE Photon. Technol. Lett.* **7**, 126 (1995).

[91] M. E. Marhic, N. Kagi, T.-K. Chiang, and L. G. Kazovsky, *Opt. Lett.* **20**, 863 (1995).

[92] S. Wabnitz, *IEEE Photon. Technol. Lett.* **7**, 652 (1995).

[93] A. D. Ellis, M. C. Tatham, D. A. O. Davies, D. Nesser, D. G. Moodie, and G. Sherlock, *Electron. Lett.* **31**, 299 (1995).

[94] M. Yu, G. P. Agrawal, and C. J. McKinstrie, *IEEE Photon. Technol. Lett.* **7**, 932 (1995).

[95] X. Zhang, F. Ebskamp, and B. F. Jorgensen, *IEEE Photon. Technol. Lett.* **7**, 819 (1995).

[96] X. Zhang and B. F. Jorgensen, *Electron. Lett.* **32**, 753 (1996).

[97] S. Watanabe and M. Shirasaki, *J. Lightwave Technol.* **14**, 243 (1996).

[98] C. Lorattanasane and K. Kikuchi, *J. Lightwave Technol.* **15**, 948 (1997).

[99] S. F. Wen, *J. Lightwave Technol.* **15**, 1061 (1997).

[100] S. Y. Set, R. Girardi, E. Riccardi, B. E. Olsson, M. Puleo, M. Ibsen, R. I. Laming, P. A. Andrekson, F. Cisternino, and H. Geiger, *Electron. Lett.* **35**, 581 (1999).

[101] H. C. Lim, F. Futami, K. Taira, and K. Kikuchi, *IEEE Photon. Technol. Lett.* **11**, 1405 (1999).

[102] M. H. Chou, I. Brener, G. Lenz, R. Scotti, E. E. Chaban, J. Shmulovich, D. Philen, S. Kosinski, K. R. Parameswaran, and M. M. Fejer, *IEEE Photon. Technol. Lett.* **12**, 82 (2000).

[103] H. C. Lim and K. Kikuchi, *IEEE Photon. Technol. Lett.* **13**, 481 (2001).

[104] R. A. Fisher, B. R. Suydam, and D. Yevick, *Opt. Lett.* **8**, 611 (1983).

[105] R. A. Fisher, Ed., *Optical Phase Conjugation*, Academic Press, San Diego, CA, 1983.

[106] G. P. Agrawal, *Nonlinear Fiber Optics*, 3rd ed., Academic Press, San Diego, CA, 2001.

[107] S. Watanabe and T. Chikama, *Electron. Lett.* **30**, 163 (1994).

[108] G. P. Agrawal, in *Semiconductor Lasers: Past, Present, Future*, G. P. Agrawal, Ed., AIP Press, Woodbury, NY, 1995, Chap. 8.

[109] A. D'Ottavi, F. Martelli, P. Spano, A. Mecozzi, and S. Scotti, *Appl. Phys. Lett.* **68**, 2186 (1996).

[110] F. Matera, A. Mecozzi, M. Romagnoli, and M. Settembre, *Opt. Lett.* **18**, 1499 (1993).

[111] R. D. Li, P. Kumar, W. L. Kath, and J. N. Kutz, *IEEE Photon. Technol. Lett.* **5**, 669 (1993).

[112] H. Taga, S. Yamamoto, N. Edagawa, Y. Yoshida, S. Akiba, and H. Wakabayashi, *J. Lightwave Technol.* **12**, 1616 (1994).

[113] N. Kikuchi, S. Sasaki, and K. Sekine, *Electron. Lett.* **31**, 375 (1995).

[114] S. Sekine, N. Kikuchi, S. Sasaki, and Y. Uchida, *Electron. Lett.* **31**, 1080 (1995).

[115] A. Naka and S. Saito, *J. Lightwave Technol.* **13**, 862 (1995).

[116] N. Kikuchi and S. Sasaki, *J. Lightwave Technol.* **13**, 868 (1995); *Electron. Lett.* **32**, 570 (1996).

[117] J. Nakagawa and K. Hotate, *J. Opt. Commun.* **16**, 202 (1995).

[118] F. Matera and M. Settembre, *J. Lightwave Technol.* **14**, 1 (1996).

[119] M. E. Marhic, N. Kagi, T.-K. Chiang, and L. G. Kazovsky, *IEEE Photon. Technol. Lett.* **8**, 145 (1996).

[120] D. Breuer and K. Petermann, *IEEE Photon. Technol. Lett.* **9**, 398 (1997).

[121] D. Breuer, K. Jurgensen, F. Kuppers, A. Mattheus, I. Gabitov, and S. K. Turitsyn, *Opt. Commun.* **140**, 15 (1997).

[122] F. Forghieri, P. R. Prucnal, R. W. Tkach, and A. R. Chraplyvy, *IEEE Photon. Technol. Lett.* **9**, 1035 (1997).

[123] S. K. Turitsyn and E. G. Shapiro, *Opt. Fiber Technol.* **4**, 151 (1998).

[124] D. Breuer, K. Obermann, and K. Petermann, *IEEE Photon. Technol. Lett.* **10**, 1793 (1998).

[125] D. S. Govan, W. Forysiak, and N. J. Doran, *Opt. Lett.* **23**, 1523 (1998).

[126] I. S. Penketh, P. Harper, S. B. Alleston, A. M. Niculae, I. Bennion I, and N. J. Doran, *Opt. Lett.* **24**, 802 (1999).

[127] R. J. Essiambre, G. Raybon, and B. Mikkelsen, in *Optical Fiber Telecommunications IV*, Vol. B, I. P. Kaminow and T. Li, Eds., Academeic Press, San Diego, CA, 2002.

[128] P. V. Mamyshev and N. A. Mamysheva, *Opt. Lett.* **24**, 1454 (1999).

[129] M. Zitelli, F. Matera, and M. Settembre, *J. Lightwave Technol.* **17**, 2498 (1999).

[130] A. Mecozzi, C. B. Clausen, and M. Shtaif, *IEEE Photon. Technol. Lett.* **12**, 392 (2000); *IEEE Photon. Technol. Lett.* **12**, 1633 (2000).

[131] R. I. Killey, H. J. Thiele, V. Mikhailov, and P. Bayvel, *IEEE Photon. Technol. Lett.* **12**, 1624 (2000).

[132] S. Kumar, *IEEE Photon. Technol. Lett.* **13**, 800 (2001).

[133] M. J. Ablowitz and T. Hirooka, *Opt. Lett.* **25**, 1750 (2000); *IEEE Photon. Technol. Lett.* **13**, 1082 (2001).

[134] P. Johannisson, D. Anderson, A. Berntson, and J. Mårtensson, *Opt. Lett.* **26**, 1227 (2001).

[135] M. Nakazawa, H. Kubota, K. Suzuki, E. Yamada, and A. Sahara, *IEEE J. Sel. Topics Quantum Electron.* **6**, 363 (2000).

[136] M. Murakami, T. Matsuda, H. Maeda, and T. Imai, *J. Lightwave Technol.* **18**, 1197 (2000).

[137] A. Sahara, T. Inui, T. Komukai, H. Kubota, and M. Nakazawa, *J. Lightwave Technol.* **18**, 1364 (2000).

[138] B. Mikkelsen, G. Raybon, R. J. Essiambre, A. J. Stentz, T. N. Nielsen, D. W. Peckham, L. Hsu, L. Gruner-Nielsen, K. Dreyer, and J. E. Johnson, *IEEE Photon. Technol. Lett.* **12**, 1400 (2000).

[139] A. H. Gnauck, J. M. Wiesenfeld, L. D. Garrett, R. M. Derosier, F. Forghieri, V. Gusmeroli, and D. Scarano, *IEEE Photon. Technol. Lett.* **11**, 1503 (1999).

[140] L. D. Garrett, A. H. Gnauck, F. Forghieri, V. Gusmeroli, and D. Scarano, *IEEE Photon. Technol. Lett.* **11**, 484 (1999).

[141] K. Hinton, *Opt. Fiber Technol.* **5**, 145 (1999).

[142] X. F. Chen, Z. B. Ma, C. Guang, Y. Meng, and X. L. Yang, *Microwave Opt. Tech. Lett.* **23**, 352 (1999).

[143] I. Riant, S. Gurib, J. Gourhant, P. Sansonetti, C. Bungarzeanu, and R. Kashyap, *IEEE J. Sel. Topics Quantum Electron.* **5**, 1312 (1999).

[144] L. D. Garrett, A. H. Gnauck, R. W. Tkach, B. Agogliati, L. Arcangeli, D. Scarano, V. Gusmeroli, C. Tosetti, G. Di Maio, and F. Forghieri, *IEEE Photon. Technol. Lett.* **12**, 356 (2000).

[145] F. Ouellette, P. A. Krug, T. Stephens, G. Dhosi, and B. Eggleton, *Electron. Lett.* **31**, 899 (1995).

[146] M. Ibsen, M. K. Durkin, M. J. Cole, and R. I. Laming, *IEEE Photon. Technol. Lett.* **10**, 842 (1998).

[147] M. Ibsen, A. Fu, H. Geiger, and R. I. Laming, *Electron. Lett.* **35**, 982 (1999).

[148] R. W. Tkach, R. M. Derosier, A. H. Gnauck, A. M. Vengsarkar, D. W. Peckham, J. J. Zyskind, J. W. Sulhoff, and A. R. Chraplyvy, *IEEE Photon. Technol. Lett.* **7**, 1369 (1995).

[149] H. Taga, M. Suzuki, N. Edagawa, S. Yamamoto, and S. Akiba, *IEEE J. Quantum Electron.* **34**, 2055 (1998).

[150] M. I. Hayee and A. E. Willner, *IEEE Photon. Technol. Lett.* **11**, 991 (1999).

[151] A. Sahara, H. Kubota, and M. Nakazawa, *Opt. Commun.* **160**, 139 (1999).

[152] F. M. Madani and K. Kikuchi, *J. Lightwave Technol.* **17**, 1326 (1999).

[153] J. Kani, M. Jinno, T. Sakamoto, S. Aisawa, M. Fukui, K. Hattori, and K. Oguchi, *J. Lightwave Technol.* **17**, 2249 (1999).

[154] C. M. Weinert, R. Ludwig, W. Pieper, H. G. Weber, D. Breuer, K. Petermann, and F. Kuppers, *J. Lightwave Technol.* **17**, 2276 (1999).

[155] C. D. Chen, I. Kim, O. Mizuhara, T. V. Nguyen, K. Ogawa, R. E. Tench, L. D. Tzeng, and P. D. Yeates, *Electron. Lett.* **35**, 648 (1999).

[156] M. R. C. Caputo and M. E. Gouvea, *Opt. Commun.* **178**, 323 (2000).

[157] H. S. Chung, H. Kim, S. E. Jin, E. S. Son, D. W. Kim, K. M. Lee, H. Y. Park, and Y. C. Chung, *IEEE Photon. Technol. Lett.* **12**, 1397 (2000).

[158] B. Bakhshi, M. F. Arend, M. Vaa, W. W. Patterson, R. L. Maybach, and N. S. Bergano, Paper PD2, *Proc. Optical Fiber Commun. Conf.*, Optical Society of America, Washington, DC, 2001.

[159] K. Fukuchi, T. Kasamatsu, M. Morie, R. Ohhira, T. Ito, K. Sekiya, D. Ogasahara, and T. Ono, Paper PD24, *Proc. Optical Fiber Commun. Conf.*, Optical Society of America, Washington, DC, 2001.

[160] A. E. Willner, K. M. Feng, J. Cai, S. Lee, J. Peng, and H. Sun, *IEEE J. Sel. Topics Quantum Electron.* **5**, 1298 (1999).

[161] M. J. Erro, M. A. G. Laso, D. Benito, M. J. Garde, and M. A. Muriel, *IEEE J. Sel. Topics Quantum Electron.* **5**, 1332 (1999).

[162] C. K. Madsen, G. Lenz, A. J. Bruce, M. A. Cappuzzo, L. T. Gomez, and R. E. Scotti, *IEEE Photon. Technol. Lett.* **11**, 1623 (1999).

[163] C. K. Madsen, J. A. Walker, J. E. Ford, K. W. Goossen, T. N. Nielsen, and G. Lenz, *IEEE Photon. Technol. Lett.* **12**, 651 (2000).

[164] B. J. Eggleton, A. Ahuja, P. S.Westbrook, J. A. Rogers, P. Kuo, T. N. Nielsen, and B. Mikkelsen, *J. Lightwave Technol.* **18**, 1418 (2000).

[165] B. J. Eggleton, B. Mikkelsen, G. Raybon, A. Ahuja, J. A. Rogers,P. S. Westbrook, T. N. Nielsen, S. Stulz, and K. Dreyer, *IEEE Photon. Technol. Lett.* **12**, 1022 (2000).

[166] T. Inui, T. Komukai, and M. Nakazawa, *Opt. Commun.* **190**, 1 (2001).

[167] S. Matsumoto, T. Ohira, M. Takabayashi, K. Yoshiara, and T. Sugihara, *IEEE Photon. Technol. Lett.* **13**, 827 (2001).

[168] M. Onishi, T. Kashiwada, Y. Koyano, Y. Ishiguro, M. Nishimura, and H. Kanamori, *Electron. Lett.* **32**, 2344 (1996).

[169] K. Takiguchi, S. Kawanishi, H. Takara, K. Okamoto, and Y. Ohmori, *Electron. Lett.* **32**, 755 (1996).

[170] M. Durkin, M. Ibsen, M. J. Cole, and R. I. Laming, *Electron. Lett.* **33**, 1891 (1997).

[171] T. Komukai and M. Nakazawa, *Opt. Commun.* **154**, 5 (1998).

[172] T. Komukai, T. Inui, and M. Nakazawa, *IEEE J. Quantum Electron.* **36**, 409 (2000).

[173] H. Tsuda, K. Okamoto, T. Ishii, K. Naganuma, Y. Inoue, H. Takenouchi, and T. Kurokawa, *IEEE Photon. Technol. Lett.* **11**, 569 (1999).

[174] W. H. Loh, F. Q. Zhou, and J. J. Pan, *IEEE Photon. Technol. Lett.* **11**, 1280 (1999).

[175] M. J. Erro, M. A. G. Laso, D. Benito, M. J. Garde, and M. A. Muriel, *Fiber Integ. Opt.* **19**, 367 (2000).

[176] A. H. Gnauck, L. D. Garrett, Y. Danziger, U. Levy, and M. Tur, *Electron. Lett.* **36**, 1946 (2000).

[177] Y. Xie, S. Lee, Z. Pan, J. X. Cai, A. E. Willner,V. Grubsky, D. S. Starodubov, E. Salik, and J. Feinberg, *IEEE Photon. Technol. Lett.* **12**, 1417 (2000).

[178] H. Takenouchi, T. Ishii, and T. Goh, *Electron. Lett.* **37**, 777 (2001).

[179] P. Kaewplung, R. Angkaew, and K. Kikuchi, *IEEE Photon. Technol. Lett.* **13**, 293 (2001).

[180] M. Kato, N. Yoshizawa, T. Sugie, and K. Okamoto, *IEEE Photon. Technol. Lett.* **13**, 463 (2001).

[181] S. Kawanishi, H. Takara, T. Morioka, O. Kamatani, K. Takiguchi, T. Kitoh, and M. Saruwatari, *Electron. Lett.* **32**, 916 (1996).

[182] M. Nakazawa, E. Yoshida, T. Yamamoto, E. Yamada, and A. Sahara, *Electron. Lett.* **34**, 907 (1998).

[183] T. Yamamoto and M. Nakazawa, *Opt. Lett.* **26**, 647 (2001).

[184] C. D. Poole and J. Nagel, in *Optical Fiber Telecommunications III*, Vol. A, I. P. Kaminow and T. L. Koch, Eds., Academic Press, San Diego, CA, 1997, Chap. 6.

[185] F. Bruyère, *Opt. Fiber Technol.* **2**, 269 (1996).

[186] P. K. A. Wai and C. R. Menyuk, *J. Lightwave Technol.* **14**, 148 (1996).

[187] M. Karlsson, *Opt. Lett.* **23**, 688 (1998).

[188] G. J. Foschini, R. M. Jopson, L. E. Nelson, and H. Kogelnik, *J. Lightwave Technol.* **17**, 1560 (1999).

[189] M. Midrio, *J. Opt. Soc. Am. B* **17**, 169 (2000).

[190] B. Huttner, C. Geiser, and N. Gisin, *IEEE J. Sel. Topics Quantum Electron.* **6**, 317 (2000).

[191] M. Shtaif and A. Mecozzi, *Opt. Lett.* **25**, 707 (2000).

[192] M. Karlsson, J. Brentel, and P. A. Andrekson, *J. Lightwave Technol.* **18**, 941 (2000).

[193] Y. Li and A. Yariv, *J. Opt. Soc. Am. B* **17**, 1821 (2000).

[194] J. M. Fini and H. A. Haus, *IEEE Photon. Technol. Lett.* **13**, 124 (2001).

[195] R. Khosravani and A. E. Willner, *IEEE Photon. Technol. Lett.* **13**, 296 (2001).

[196] H. Sunnerud, M. Karlsson, and P. A. Andrekson, *IEEE Photon. Technol. Lett.* **13**, 448 (2001).

[197] J. H. Winters and M. A. Santoro, *IEEE Photon. Technol. Lett.* **2**, 591 (1990).

[198] T. Ono, S. Yamamzaki, H. Shimizu, and K. Emura, *J. Lightwave Technol.* **12**, 891 (1994).

[199] T. Takahashi, T. Imai, and M. Aiki, *Electron. Lett.* **30**, 348 (1994).

[200] B. W. Hakki, *IEEE Photon. Technol. Lett.* **9**, 121 (1997).

[201] C. Francia, F. Bruyère, J. P. Thiéry, and D. Penninckx, *Electron. Lett.* **35**, 414 (1999).

[202] D. Mahgerefteh and C. R. Menyuk, *IEEE Photon. Technol. Lett.* **11**, 340 (1999).

[203] D. A. Watley, K. S. Farley, B. J. Shaw, W. S. Lee, G. Bordogna, A. P. Hadjifotiou, and R. E. Epworth, *Electron. Lett.* **35**, 1094 (1999).

[204] S. Lee, R. Khosravani, J. Peng, V. Grubsky, D. S. Starodubov, A. E. Willner, and J. Feinberg, *IEEE Photon. Technol. Lett.* **11**, 1277 (1999).

[205] R. Noé, D. Sandel, M. Yoshida-Dierolf, S. Hinz, V. Mirvoda, A. Schopflin, C. Gungener, E. Gottwald, C. Scheerer, G. Fischer, T. Weyrauch, and W. Haase, *J. Lightwave Technol.* **17**, 1602 (1999).

[206] H. Sunnerud, M. Karlsson, and P. A. Andrekson, *IEEE Photon. Technol. Lett.* **12**, 50 (2000).

[207] M. Shtaif, A. Mecozzi, M. Tur, and J. Nagel, *IEEE Photon. Technol. Lett.* **12**, 434 (2000).

[208] H. Y. Pua, K. Peddanarappagari, B. Zhu, C. Allen, K. Demarest, and R. Hui, *J. Lightwave Technol.* **18**, 832 (2000).

[209] C. K. Madsen, *Opt. Lett.* **25**, 878 (2000).

[210] T. Merker, N. Hahnenkamp, and P. Meissner, *Opt. Commun.* **182**, 135 (2000).

[211] L. Möller, *IEEE Photon. Technol. Lett.* **12**, 1258 (2000).

[212] S. Sarkimukka, A. Djupsjobacka, A. Gavler, and G Jacobsen, *J. Lightwave Technol.* **18**, 1374 (2000).

[213] D. Sobiski, D. Pikula, J. Smith, C. Henning, D. Chowdhury, E. Murphy, E. Kolltveit, F. Annunziata, *Electron. Lett.* **37**, 46 (2001).

[214] Q. Yu, L. S. Yan, Y. Xie, M. Hauer, and A. E. Willner, *IEEE Photon. Technol. Lett.* **13**, 863 (2001).

Chapter 8

Multichannel Systems

In principle, the capacity of optical communication systems can exceed 10 Tb/s because of a large frequency associated with the optical carrier. In practice, however, the bit rate was limited to 10 Gb/s or less until 1995 because of the limitations imposed by the dispersive and nonlinear effects and by the speed of electronic components. Since then, transmission of multiple optical channels over the same fiber has provided a simple way for extending the system capacity to beyond 1 Tb/s. Channel multiplexing can be done in the time or the frequency domain through time-division multiplexing (TDM) and frequency-division multiplexing (FDM), respectively. The TDM and FDM techniques can also be used in the electrical domain (see Section 1.2.2). To make the distinction explicit, it is common to refer to the two optical-domain techniques as *optical* TDM (OTDM) and *wavelength-division multiplexing* (WDM), respectively. The development of such multichannel systems attracted considerable attention during the 1990s. In fact, WDM lightwave systems were available commercially by 1996.

This chapter is organized as follows. Sections 8.1–8.3 are devoted to WDM lightwave systems by considering in different sections the architectural aspects of such systems, the optical components needed for their implementation, and the performance issues such as interchannel crosstalk. In Section 8.4 we focus on the basic concepts behind OTDM systems and issues related to their practical implementation. Subcarrier multiplexing, a scheme in which FDM is implemented in the microwave domain, is discussed in Section 8.5. The technique of code-division multiplexing is the focus of Section 8.6.

8.1   WDM Lightwave Systems

WDM corresponds to the scheme in which multiple optical carriers at different wavelengths are modulated by using independent electrical bit streams (which may themselves use TDM and FDM techniques in the electrical domain) and are then transmitted over the same fiber. The optical signal at the receiver is demultiplexed into separate channels by using an optical technique. WDM has the potential for exploiting the large bandwidth offered by optical fibers. For example, hundreds of 10-Gb/s channels can

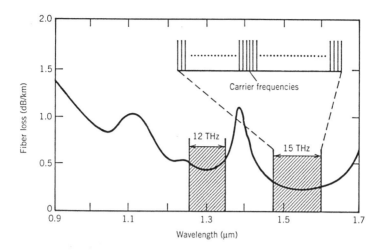

Figure 8.1: Low-loss transmission windows of silica fibers in the wavelength regions near 1.3 and 1.55 μm. The inset shows the WDM technique schematically.

be transmitted over the same fiber when channel spacing is reduced to below 100 GHz. Figure 8.1 shows the low-loss transmission windows of optical fibers centered near 1.3 and 1.55 μm. If the OH peak can be eliminated using "dry" fibers, the total capacity of a WDM system can ultimately exceed 30 Tb/s.

The concept of WDM has been pursued since the first commercial lightwave system became available in 1980. In its simplest form, WDM was used to transmit two channels in different transmission windows of an optical fiber. For example, an existing 1.3-μm lightwave system can be upgraded in capacity by adding another channel near 1.55 μm, resulting in a channel spacing of 250 nm. Considerable attention was directed during the 1980s toward reducing the channel spacing, and multichannel systems with a channel spacing of less than 0.1 nm had been demonstrated by 1990 [1]–[4]. However, it was during the decade of the 1990s that WDM systems were developed most aggressively [5]–[12]. Commercial WDM systems first appeared around 1995, and their total capacity exceeded 1.6 Tb/s by the year 2000. Several laboratory experiments demonstrated in 2001 a system capacity of more than 10 Tb/s although the transmission distance was limited to below 200 km. Clearly, the advent of WDM has led to a virtual revolution in designing lightwave systems. This section focuses on WDM systems by classifying them into three categories introduced in Section 5.1.

8.1.1 High-Capacity Point-to-Point Links

For long-haul fiber links forming the backbone or the core of a telecommunication network, the role of WDM is simply to increase the total bit rate [14]. Figure 8.2 shows schematically such a point-to-point, high-capacity, WDM link. The output of several transmitters, each operating at its own carrier frequency (or wavelength), is multiplexed together. The multiplexed signal is launched into the optical fiber for transmission to the other end, where a demultiplexer sends each channel to its own receiver. When N

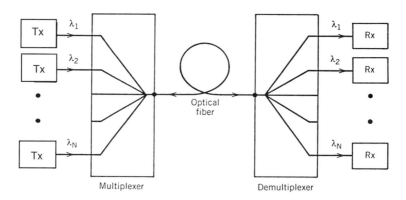

Figure 8.2: Multichannel point-to-point fiber link. Separate transmitter-receiver pairs are used to send and receive the signal at different wavelengths

channels at bit rates B_1, B_2,..., and B_N are transmitted simultaneously over a fiber of length L, the total bit rate–distance product, BL, becomes

$$BL = (B_1 + B_2 + \cdots + B_N)L. \tag{8.1.1}$$

For equal bit rates, the system capacity is enhanced by a factor of N. An early experiment in 1985 demonstrated the BL product of 1.37 (Tb/s)-km by transmitting 10 channels at 2 Gb/s over 68.3 km of standard fiber with a channel spacing of 1.35 nm [3].

The ultimate capacity of WDM fiber links depends on how closely channels can be packed in the wavelength domain. The minimum channel spacing is limited by interchannel crosstalk, an issue covered in Section 8.3. Typically, channel spacing Δv_{ch} should exceed $2B$ at the bit rate B. This requirement wastes considerable bandwidth. It is common to introduce a measure of the *spectral efficiency* of a WDM system as $\eta_s = B/\Delta v_{ch}$. Attempts are made to make η_s as large as possible.

The channel frequencies (or wavelengths) of WDM systems have been standardized by the International Telecommunication Union (ITU) on a 100-GHz grid in the frequency range 186–196 THz (covering the C and L bands in the wavelength range 1530–1612 nm). For this reason, channel spacing for most commercial WDM systems is 100 GHz (0.8 nm at 1552 nm). This value leads to only 10% spectral efficiency at the bit rate of 10 Gb/s. More recently, ITU has specified WDM channels with a frequency spacing of 50 GHz. The use of this channel spacing in combination with the bit rate of 40 Gb/s has the potential of increasing the spectral efficiency to 80%. WDM systems were moving in that direction in 2001.

What is the ultimate capacity of WDM systems? The low-loss region of the state-of-the-art "dry" fibers (e.g, fibers with reduced OH-absorption near 1.4 μm) extends over 300 nm in the wavelength region covering 1.3–1.6 μm (see Fig. 8.1). The minimum channel spacing can be as small as 50 GHz or 0.4 nm for 40-Gb/s channels. Since 750 channels can be accommodated over the 300-nm bandwidth, the resulting effective bit rate can be as large as 30 Tb/s. If we assume that the WDM signal can be transmitted over 1000 km by using optical amplifiers with dispersion management, the effective BL product may exceed 30,000 (Tb/s)-km with the use of WDM technology.

Table 8.1 High-capacity WDM transmission experiments

Channels N	Bit Rate B (Gb/s)	Capacity NB (Tb/s)	Distance L (km)	NBL Product [(Pb/s)-km]
120	20	2.40	6200	14.88
132	20	2.64	120	0.317
160	20	3.20	1500	4.80
82	40	3.28	300	0.984
256	40	10.24	100	1.024
273	40	10.92	117	1.278

This should be contrasted with the third-generation commercial lightwave systems, which transmitted a single channel over 80 km or so at a bit rate of up to 2.5 Gb/s, resulting in BL values of at most 0.2 (Tb/s)-km. Clearly, the use of WDM has the potential of improving the performance of modern lightwave systems by a factor of more than 100,000.

In practice, many factors limit the use of the entire low-loss window. As seen in Chapter 6, most optical amplifiers have a finite bandwidth. The number of channels is often limited by the bandwidth over which amplifiers can provide nearly uniform gain. The bandwidth of erbium-doped fiber amplifiers is limited to 40 nm even with the use of gain-flattening techniques (see Section 6.4). The use of Raman amplification has extended the bandwidth to near 100 nm. Among other factors that limit the number of channels are (i) stability and tunability of distributed feedback (DFB) semiconductor lasers, (ii) signal degradation during transmission because of various nonlinear effects, and (iii) interchannel crosstalk during demultiplexing. High-capacity WDM fiber links require many high-performance components, such as transmitters integrating multiple DFB lasers, channel multiplexers and demultiplexers with add-drop capability, and large-bandwidth constant-gain amplifiers.

Experimental results on WDM systems can be divided into two groups based on whether the transmission distance is ~ 100 km or exceeds 1000 km. Since the 1985 experiment in which ten 2-Gb/s channels were transmitted over 68 km [3], both the number of channels and the bit rate of individual channels have increased considerably. A capacity of 340 Gb/s was demonstrated in 1995 by transmitting 17 channels, each operating at 20 Gb/s, over 150 km [15]. This was followed within a year by several experiments that realized a capacity of 1 Tb/s. By 2001, the capacity of WDM systems exceeded 10 Tb/s in several laboratory experiments. In one experiment, 273 channels, spaced 0.4-nm apart and each operating at 40 Gb/s, were transmitted over 117 km using three in-line amplifiers, resulting in a total bit rate of 11 Tb/s and a BL product of 1300 (Tb/s)-km [16]. Table 8.1 lists several WDM transmission experiments in which the system capacity exceeded 2 Tb/s.

The second group of WDM experiments is concerned with transmission distance of more than 5000 km for submarine applications. In a 1996 experiment, 100-Gb/s transmission (20 channels at 5 Gb/s) over 9100 km was realized using the polarization-scrambling and forward-error-correction techniques [17]. The number of channels was later increased to 32, resulting in a 160-Gb/s transmission over 9300 km [18]. In a 2001 experiment, a 2.4-Tb/s WDM signal (120 channels, each operating at 20 Gb/s) was transmitted over 6200 km, resulting in a *NBL* product of almost 15 (Pb/s)-km (see Table 8.1). This should be compared with the first fiber-optic cable laid across the Atlantic ocean (TAT-8); it operated at 0.27 Gb/s with $NBL \approx 1.5$ (Tb/s)-km. The use of WDM had improved the capacity of undersea systems by a factor of 10,000 by 2001.

On the commercial side, WDM systems with a capacity of 40 Gb/s (16 channels at 2.5 Gb/s or 4 channels at 10 Gb/s) were available in 1996. The 16-channel system covered a wavelength range of about 12 nm in the 1.55-μm region with a channel spacing of 0.8 nm. WDM fiber links operating at 160 Gb/s (16 channels at 10 Gb/s) appeared in 1998. By 2001, WDM systems with a capacity of 1.6 Tb/s (realized by multiplexing 160 channels, each operating at 10 Gb/s) were available. Moreover, systems with a 6.4-Tb/s capacity were in the development stage (160 channels at 40 Gb/s). This should be contrasted with the 10-Gb/s capacity of the third-generation systems available before the advent of the WDM technique. The use of WDM had improved the capacity of commercial terrestrial systems by a factor of more than 6000 by 2001.

8.1.2   Wide-Area and Metro-Area Networks

Optical networks, as discussed in Section 5.1, are used to connect a large group of users spread over a geographical area. They can be classified as a local-area network (LAN), metropolitan-area network (MAN), or a wide-area network (WAN) depending on the area they cover [6]–[11]. All three types of networks can benefit from the WDM technology. They can be designed using the hub, ring, or star topology. A ring topology is most practical for MANs and WANs, while the star topology is commonly used for LANs. At the LAN level, a broadcast star is used to combine multiple channels. At the next level, several LANs are connected to a MAN by using passive wavelength routing. At the highest level, several MANs connect to a WAN whose nodes are interconnected in a mesh topology. At the WAN level, the network makes extensive use of switches and wavelength-shifting devices so that it is dynamically configurable.

Consider first a WAN covering a wide area (e.g, a country). Historically, telecommunication and computer networks (such as the Internet) occupying the entire U.S. geographical region have used a hub topology shown schematically in Fig. 8.3. Such networks are often called mesh networks [19]. Hubs or nodes located in large metropolitan areas house electronic switches, which connect any two nodes either by creating a "virtual circuit" between them or by using *packet switching* through protocols such as TCP/IP (transmission control protocol/Internet protocol) and *asynchronous transfer mode* (ATM). With the advent of WDM during the 1990s, the nodes were connected through point-to-point WDM links, but the switching was being done electronically even in 2001. Such transport networks are termed "opaque" networks because they require optical-to-electronic conversion. As a result, neither the bit rate nor the modulation format can be changed without changing the switching equipment.

An all-optical network in which a WDM signal can pass through multiple nodes (possibly modified by adding or dropping certain channels) is called optically "transparent." Transparent WDM networks are desirable as they do not require demultiplexing and optical-to-electronic conversion of all WDM channels. As a result, they are

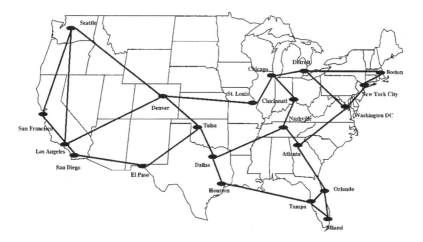

Figure 8.3: An example of a wide-area network in the form of several interconnected SONET rings. (After Ref. [19]; ©2000 IEEE; reproduced with permission.)

not limited by the electronic-speed bottleneck and may help in reducing the cost of installing and maintaining the network. The nodes in a transparent WDM network (see Fig. 8.3) switch channels using optical cross-connects. Such devices were still in their infancy in 2001.

An alternative topology implements a regional WDM network in the form of several interconnected rings. Figure 8.4 shows such a scheme schematically [20]. The feeder ring connects to the backbone of the network through an egress node. This ring employs four fibers to ensure robustness. Two of the fibers are used to route the data in the clockwise and counterclockwise directions. The other two fibers are called protection fibers and are used in case a point-to-point link fails (self-healing). The feeder ring supplies data to several other rings through access nodes. An add–drop multiplexer can be used at all nodes to drop and to add individual WDM channels. Dropped channels can be distributed to users using bus, tree, or ring networks. Notice that nodes are not always directly connected and require data transfer at multiple hubs. Such networks are called multihop networks.

Metro networks or MANs connect several central offices within a metropolitan area. The ring topology is also used for such networks. The main difference from the ring shown in Fig. 8.4 stems from the scaling and cost considerations. The traffic flows in a metro ring at a modest bit rate compared with a WAN ring forming the backbone of a nationwide network. Typically, each channel operates at 2.5 Gb/s. To reduce the cost, a coarse WDM technique is used (in place of dense WDM common in the backbone rings) by using a channel spacing in the 2- to 10-nm range. Moreover, often just two fibers are used inside the ring, one for carrying the data and the other for protecting against a failure. Most metro networks were using electrical switching in 2001 although optical switching is the ultimate goal. In a test-bed implementation of an optically switched metro network, called the *multiwavelength optical network* (MONET), several sites within the Washington, DC, area of the United States were connected us-

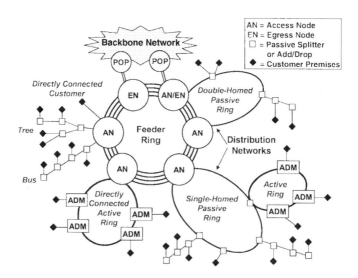

Figure 8.4: A WDM network with a feeder ring connected to several local distribution networks. (After Ref. [20]; ©1999 IEEE; reproduced with permission.)

ing a set of eight standard wavelengths in the 1.55-μm region with a channel spacing of 200 GHz [21]. MONET incorporated diverse switching technologies [synchronous digital hierarchy (SDH), asynchronous transfer mode (ATM), etc.] into an all-optical ring network using cross-connect switches based on the LiNbO$_3$ technology.

8.1.3 Multiple-Access WDM Networks

Multiple-access networks offer a random bidirectional access to each subscriber. Each user can receive and transmit information to any other user of the network at all times. Telephone networks provide one example; they are known as subscriber loop, local-loop, or access networks. Another example is provided by the Internet used for connecting multiple computers. In 2001, both the local-loop and computer networks were using electrical techniques to provide bidirectional access through circuit or packet switching. The main limitation of such techniques is that each node on the network must be capable of processing the entire network traffic. Since it is difficult to achieve electronic processing speeds in excess of 10 Gb/s, such networks are inherently limited by the electronics.

The use of WDM permits a novel approach in which the channel wavelength itself can be used for switching, routing, or distributing each channel to its destination, resulting in an all-optical network. Since wavelength is used for multiple access, such a WDM approach is referred to as *wavelength-division multiple access* (WDMA). A considerable amount of research and development work was done during the 1990s for developing WDMA networks [22]–[26]. Broadly speaking, WDMA networks can be classified into two categories, called single-hop and multihop all-optical networks [6]. Every node is directly connected to all other nodes in a single-hop network, resulting in a fully connected network. In contrast, multihop networks are only partially con-

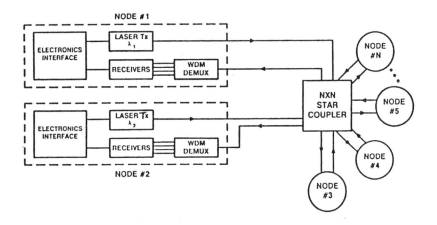

Figure 8.5: Schematic of the Lambdanet with N nodes. Each node consists of one transmitter and N receivers. (After Ref. [28]; ©1990 IEEE; reprinted with permission.)

nected such that an optical signal sent by one node may require several hops through intermediate nodes before reaching its destination. In each category, transmitters and receivers can have their operating frequencies either fixed or tunable.

Several architectures can be used for all-optical multihop networks [6]–[11]. Hypercube architecture provides one example—it has been used for interconnecting multiple processors in a supercomputer [27]. The hypercube configuration can be easily visualized in three dimensions such that eight nodes are located at eight corners of a simple cube. In general, the number of nodes N must be of the form 2^m, where m is the dimensionality of the hypercube. Each node is connected to m different nodes. The maximum number of hops is limited to m, while the average number of hops is about $m/2$ for large N. Each node requires m receivers. The number of receivers can be reduced by using a variant, known as the *deBruijn network*, but it requires more than $m/2$ hops on average. Another example of a multihop WDM network is provided by the *shuffle network* or its bidirectional equivalent—the Banyan network.

Figure 8.5 shows an example of the single-hop WDM network based on the use of a *broadcast star*. This network, called the *Lambdanet* [28], is an example of the *broadcast-and-select* network. The new feature of the Lambdanet is that each node is equipped with one transmitter emitting at a unique wavelength and N receivers operating at the N wavelengths, where N is the number of nodes. The output of all transmitters is combined in a passive star and distributed to all receivers equally. Each node receives the entire traffic flowing across the network. A tunable optical filter can be used to select the desired channel. In the case of the Lambdanet, each node uses a bank of receivers in place of a tunable filter. This feature creates a nonblocking network whose capacity and connectivity can be reconfigured electronically depending on the application. The network is also transparent to the bit rate or the modulation format. Different users can transmit data at different bit rates with different modulation formats. The flexibility of the Lambdanet makes it suitable for many applications. The main drawback of the Lambdanet is that the number of users is limited by the number

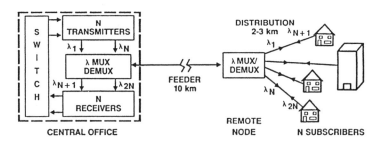

Figure 8.6: Passive photonic loop for local-loop applications. (After Ref. [31]; ©1988 IEE; reprinted with permission.)

of available wavelengths. Moreover, each node requires many receivers (equal to the number of nodes), resulting in a considerable investment in hardware costs.

A tunable receiver can reduce the cost and complexity of the Lambdanet. This is the approach adopted for the *Rainbow network* [29]. This network can support up to 32 nodes, each of which can transmit 1-Gb/s signals over 10–20 km. It makes use of a central passive star (see Fig. 8.5) together with the high-performance parallel interface for connecting multiple computers. A tunable optical filter is used to select the unique wavelength associated with each node. The main shortcoming of the Rainbow network is that tuning of receivers is a relatively slow process, making it difficult to use packet switching. An example of the WDM network that uses packet switching is provided by the *Starnet*. It can transmit data at bit rates of up to 1.25 Gb/s per node over a 10-km diameter while maintaining a signal-to-noise ratio (SNR) close to 24 dB [30].

WDM networks making use of a passive star coupler are often called *passive optical networks* (PONs) because they avoid active switching. PONs have the potential for bringing optical fibers to the home (or at least to the curb). In one scheme, called a *passive photonic loop* [31], multiple wavelengths are used for routing signals in the local loop. Figure 8.6 shows a block diagram of such a network. The central office contains N transmitters emitting at wavelengths $\lambda_1, \lambda_2, \ldots, \lambda_N$ and N receivers operating at wavelengths $\lambda_{N+1}, \ldots, \lambda_{2N}$ for a network of N subscribers. The signals to each subscriber are carried on separate wavelengths in each direction. A remote node multiplexes signals from the subscribers to send the combined signal to the central office. It also demultiplexes signals for individual subscribers. The remote node is passive and requires little maintenance if passive WDM components are used. A switch at the central office routes signals depending on their wavelengths.

The design of access networks for telecommunication applications was still evolving in 2001 [26]. The goal is to provide broadband access to each user and to deliver audio, video, and data channels on demand, while keeping the cost down. Indeed, many low-cost WDM components are being developed for this purpose. A technique known as *spectral slicing* uses the broad emission spectrum of an LED to provide multiple WDM channels inexpensively. A *waveguide-grating router* (WGR) can be used for wavelength routing. Spectral slicing and WGR devices are discussed in the next section devoted to WDM components.

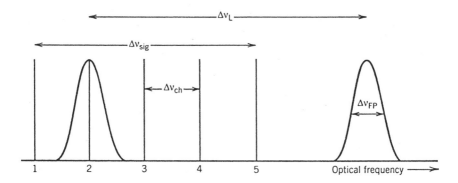

Figure 8.7: Channel selection through a tunable optical filter.

8.2 WDM Components

The implementation of WDM technology for fiber-optic communication systems requires several new optical components. Among them are multiplexers, which combine the output of several transmitters and launch it into an optical fiber (see Fig. 8.2); demultiplexers which split the received multichannel signal into individual channels destined to different receivers; star couplers which mix the output of several transmitters and broadcast the mixed signal to multiple receivers (see Fig. 8.5); tunable optical filters which filter out one channel at a specific wavelength that can be changed by tuning the passband of the optical filter; multiwavelength optical transmitters whose wavelength can be tuned over a few nanometers; add–drop multiplexers and WGRs which can distribute the WDM signal to different ports; and wavelength shifters which switch the channel wavelength. This section focuses on all such WDM components.

8.2.1 Tunable Optical Filters

It is instructive to consider optical filters first since they are often the building blocks of more complex WDM components. The role of a tunable optical filter in a WDM system is to select a desired channel at the receiver. Figure 8.7 shows the selection mechanism schematically. The filter bandwidth must be large enough to transmit the desired channel but, at the same time, small enough to block the neighboring channels.

All optical filters require a wavelength-selective mechanism and can be classified into two broad categories depending on whether optical interference or diffraction is the underlying physical mechanism. Each category can be further subdivided according to the scheme adopted. In this section we consider four kinds of optical filters; Fig. 8.8 shows an example of each kind. The desirable properties of a tunable optical filter include: (1) wide tuning range to maximize the number of channels that can be selected, (2) negligible crosstalk to avoid interference from adjacent channels, (3) fast tuning speed to minimize the access time, (4) small insertion loss, (5) polarization insensitivity, (6) stability against environmental changes (humidity, temperature, vibrations, etc.), and (7) last but not the least, low cost.

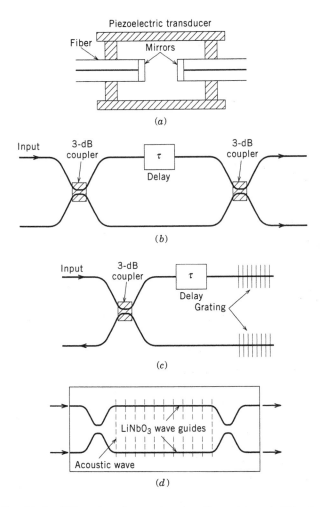

Figure 8.8: Four kinds of filters based on various interferometric and diffractive devices: (a) Fabry–Perot filter; (b) Mach–Zehnder filter; (c) grating-based Michelson filter; (d) acousto-optic filter. The shaded area represents a surface acoustic wave.

A Fabry–Perot (FP) interferometer—a cavity formed by using two mirrors—can act as a tunable optical filter if its length is controlled electronically by using a piezoelectric transducer [see Fig. 8.8(a)]. The transmittivity of a FP filter peaks at wavelengths that correspond to the longitudinal-mode frequencies given in Eq. (3.3.5). Hence, the frequency spacing between two successive transmission peaks, known as the *free spectral range*, is given by

$$\Delta v_L = c/(2n_g L), \tag{8.2.1}$$

where n_g is the group index of the intracavity material for a FP filter of length L.

If the filter is designed to pass a single channel (see Fig. 8.7), the combined bandwidth of the multichannel signal, $\Delta v_{\text{sig}} = N\Delta v_{\text{ch}} = NB/\eta_s$, must be less than Δv_L, where N is the number of channels, η_s is the spectral efficiency, and B is the bit rate. At the same time, the filter bandwidth Δv_{FP} (the width of the transmission peak in Fig. 8.7) should be large enough to pass the entire frequency contents of the selected channel. Typically, $\Delta v_{\text{FP}} \sim B$. The number of channels is thus limited by

$$N < \eta_s(\Delta v_L/\Delta v_{\text{FP}}) = \eta_s F, \qquad (8.2.2)$$

where $F = \Delta v_L/\Delta v_{\text{FP}}$ is the *finesse* of the FP filter. The concept of finesse is well known in the theory of FP interferometers [32]. If internal losses are neglected, the finesse is given by $F = \pi\sqrt{R}/(1-R)$ and is determined solely by the mirror reflectivity R, assumed to be the same for both mirrors [32].

Equation (8.2.2) provides a remarkably simple condition for the number of channels that a FP filter can resolve. As an example, if $\eta_s = \frac{1}{3}$, a FP filter with 99% reflecting mirrors can select up to 104 channels. Channel selection is made by changing the filter length L electronically. The length needs to be changed by only a fraction of the wavelength to tune the filter. The filter length L itself is determined from Eq. (8.2.1) together with the condition $\Delta v_L > \Delta v_{\text{sig}}$. As an example, for a 10-channel WDM signal with 0.8-nm channel spacing, $\Delta v_{\text{sig}} \approx 1$ THz. If $n_g = 1.5$ is used for the group index, L should be smaller than 100 μm. Such a short length together with the requirement of high mirror reflectivities underscores the complexity of the design of FP filters for WDM applications.

A practical all-fiber design of FP filters uses the air gap between two optical fibers (see Fig. 8.8). The two fiber ends forming the gap are coated to act as high-reflectivity mirrors [33]. The entire structure is enclosed in a piezoelectric chamber so that the gap length can be changed electronically for tuning and selecting a specific channel. The advantage of fiber FP filters is that they can be integrated within the system without incurring coupling losses. Such filters were used in commercial WDM fiber links starting in 1996. The number of channels is typically limited to below 100 ($F \approx 155$ for the 98% mirror reflectivity) but can be increased using two FP filters in tandem. Although tuning is relatively slow because of the mechanical nature of the tuning mechanism, it suffices for some applications.

Tunable FP filters can also be made using several other materials such as liquid crystals and semiconductor waveguides [34]–[39]. Liquid-crystal-based filters make use of the anisotropic nature of liquid crystals that makes it possible to change the refractive index electronically. A FP cavity is still formed by enclosing the liquid-crystal material within two high-reflectivity mirrors, but the tuning is done by changing the refractive index rather than the cavity length. Such FP filters can provide high finesse ($F \sim 300$) with a bandwidth of about 0.2 nm [34]. They can be tuned electrically over 50 nm, but switching time is typically ~ 1 ms or more when nematic liquid crystals are used. It can be reduced to below 10 μs by using smectic liquid crystals [35].

Thin dielectric films are commonly used for making narrow-band interference filters [36]. The basic idea is quite simple. A stack of suitably designed thin films acts as a high-reflectivity mirror. If two such mirrors are separated by a spacer dielectric layer, a FP cavity is formed that acts as an optical filter. The bandpass response can be tailored for a multicavity filter formed by using multiple thin-film mirrors separated by several spacer layers. Tuning can be realized in several different ways. In one approach, an InGaAsP/InP waveguide permits electronic tuning [37]. Silicon-based FP filters can be tuned using thermo-optic tuning [38]. Micromechanical tuning has also been used for InAlGaAs-based FP filters [39]. Such filters exhibited a tuning range of 40 nm with <0.35 nm bandwidth in the $1.55\text{-}\mu$m region.

A chain of Mach–Zehnder (MZ) interferometers can also be used for making a tunable optical filter. A MZ interferometer can be constructed simply by connecting the two output ports of a 3-dB coupler to the two input ports of another 3-dB coupler [see Fig. 8.8(b)]. The first coupler splits the input signal equally into two parts, which acquire different phase shifts (if the arm lengths are made different) before they interfere at the second coupler. Since the relative phase shift is wavelength dependent, the transmittivity $T(v)$ is also wavelength dependent. In fact, we can use Eq. (7.5.5) to find that $T(v) = |H(v)|^2 = \cos^2(\pi v \tau)$, where $v = \omega/2\pi$ is the frequency and τ is the relative delay in the two arms of the MZ interferometer [40]. A cascaded chain of such MZ interferometers with relative delays adjusted suitably acts as an optical filter that can be tuned by changing the arm lengths slightly. Mathematically, the transmittivity of a chain of M such interferometers is given by

$$T(v) = \prod_{m=1}^{M} \cos^2(\pi v \tau_m),\tag{8.2.3}$$

where τ_m is the relative delay in the mth member of the chain.

A commonly used method implements the relative delays τ_m such that each MZ stage blocks the alternate channels successively. This scheme requires $\tau_m = (2^m \Delta v_{\text{ch}})^{-1}$ for a channel spacing of Δv_{ch}. The resulting transmittivity of a 10-stage MZ chain has channel selectivity as good as that offered by a FP filter having a finesse of 1600. Moreover, such a filter is capable of selecting closely spaced channels. The MZ chain can be built by using fiber couplers or by using silica waveguides on a silicon substrate. The silica-on-silicon technology was exploited extensively during the 1990s to make many WDM components. Such devices are referred to as *planar lightwave circuits* because they use planar optical waveguides formed on a silicon substrate [41]–[45]. The underlying technology is sometimes called the *silicon optical-bench* technology [44]. Tuning in MZ filters is realized through a chromium heater deposited on one arm of each MZ interferometer (see Fig. 7.7). Since the tuning mechanism is thermal, it results in a slow response with a switching time of about 1 ms.

A separate class of tunable optical filters makes use of the wavelength selectivity provided by a Bragg grating. Fiber Bragg gratings provide a simple example of grating-based optical filters [46]; such filters have been discussed in Section 7.6. In its simplest form, a fiber grating acts as a reflection filter whose central wavelength can be controlled by changing the grating period, and whose bandwidth can be tailored by changing the grating strength or by chirping the grating period slightly. The reflective nature of fiber gratings is often a limitation in practice and requires the use of an *optical circulator*. A phase shift in the middle of the grating can convert a fiber grating into a narrowband transmission filter [47]. Many other schemes can be used to make transmission filters based on fiber gratings. In one approach, fiber gratings are used as mirrors of a FP filter, resulting in transmission filters whose free spectral range can vary over a wide range 0.1–10 nm [48]. In another design, a grating is inserted in each arm of a MZ interferometer to provide a transmission filter [46]. Other kinds of interferometers, such as the *Sagnac* and *Michelson interferometers*, can also be used to realize transmission filters. Figure 8.8(c) shows an example of the Michelson interferometer made by using a 3-dB fiber coupler and two fiber gratings acting as mirrors for the two arms of the Michelson interferometer [49]. Most of these schemes can also be implemented in the form of a planar lightwave circuit by forming silica waveguides on a silicon substrate.

Many other grating-based filters have been developed for WDM systems [50]–[54]. In one scheme, borrowed from the DFB-laser technology, the InGaAsP/InP material system is used to form planar waveguides functioning near 1.55 μm. The wavelength selectivity is provided by a built-in grating whose Bragg wavelength is tuned electrically through electrorefraction [50]. A phase-control section, similar to that used for multisegment DFB lasers, have also been used to tune distributed Bragg reflector (DBR) filters. Multiple gratings, each tunable independently, can also be used to make tunable filters [51]. Such filters can be tuned quickly (in a few nanoseconds) and can be designed to provide net gain since one or more amplifiers can be integrated with the filter. They can also be integrated with the receiver, as they use the same semiconductor material. These two properties of InGaAsP/InP filters make them quite attractive for WDM applications.

In another class of tunable filters, the grating is formed dynamically by using acoustic waves. Such filters, called *acousto-optic filters*, exhibit a wide tuning range ($>$ 100 nm) and are quite suitable for WDM applications [55]–[58]. The physical mechanism behind the operation of acousto-optic filters is the *photoelastic effect* through which an acoustic wave propagating through an acousto-optic material creates periodic changes in the refractive index (corresponding to the regions of local compression and rarefaction). In effect, the acoustic wave creates a periodic index grating that can diffract an optical beam. The wavelength selectivity stems from this acoustically induced grating. When a transverse electric (TE) wave with the propagation vector $\mathbf{k}$ is diffracted from this grating, its polarization can be changed from TE to transverse magnetic (TM) if the *phase-matching condition* $\mathbf{k}' = \mathbf{k} \pm \mathbf{K_a}$ is satisfied, where $\mathbf{k}'$ and $\mathbf{K_a}$ are the wave vectors associated with the TM and acoustic waves, respectively.

Acousto-optic tunable filters can be made by using bulk components as well as waveguides, and both kinds are available commercially. For WDM applications, the LiNbO$_3$ waveguide technology is often used since it can produce compact, polarization-independent, acousto-optic filters with a bandwidth of about 1 nm and a tuning range over 100 nm [56]. The basic design, shown schematically in Fig. 8.8(d), uses two polarization beam splitters, two LiNbO$_3$ waveguides, a surface-acoustic-wave transducer, all integrated on the same substrate. The incident WDM signal is split into its orthogonally polarized components by the first beam splitter. The channel whose wavelength λ satisfies the Bragg condition $\lambda = (\Delta n)\Lambda_a$ is directed to a different output port by the second beam splitter because of an acoustically induced change in its polarization direction; all other channels go to the other output port. The TE–TM index difference Δn is about 0.07 in LiNbO$_3$. Near $\lambda = 1.55$ μm, the acoustic wavelength Λ_a should be about 22 μm. This value corresponds to a frequency of about 170 MHz if we use the acoustic velocity of 3.75 km/s for LiNbO$_3$. Such a frequency can be easily applied. Moreover, its exact value can be changed electronically to change the wavelength that satisfies the Bragg condition. Tuning is relatively fast because of its electronic nature and can be accomplished in a switching time of less than 10 μs. Acousto-optic tunable filters are also suitable for wavelength routing and optical cross-connect applications in dense WDM systems.

Another category of tunable optical filters operates on the principle of amplification of a selected channel. Any amplifier with a gain bandwidth smaller than the channel spacing can be used as an optical filter. Tuning is realized by changing the wavelength at which the gain peak occurs. Stimulated Brillouin scattering (SBS), occurring naturally in silica fibers [59], can be used for selective amplification of one channel, but the gain bandwidth is quite small (< 100 MHz). The SBS phenomenon involves interaction between the optical and acoustic waves and is governed by a phase-matching condition similar to that found for acousto-optic filters. As discussed in Section 2.6, SBS occurs only in the backward direction and results in a frequency shift of about 10 GHz in the 1.55-μm region.

To use the SBS amplification as a tunable optical filter, a continuous-wave (CW) pump beam is launched at the receiver end of the optical fiber in a direction opposite to that of the multichannel signal, and the pump wavelength is tuned to select the channel. The pump beam transfers a part of its energy to a channel down-shifted from the pump frequency by exactly the Brillouin shift. A tunable pump laser is a prerequisite for this scheme. The bit rate of each channel is even then limited to 100 MHz or so. In a 1989 experiment in which a 128-channel WDM network was simulated by using two 8 × 8 star couplers [60], a 150-Mb/s channel could be selected with a channel spacing as small as 1.5 GHz.

Semiconductor optical amplifiers (SOAs) can also be used for channel selection provided that a DFB structure is used to narrow the gain bandwidth [61]. A built-in grating can easily provide a filter bandwidth below 1 nm. Tuning is achieved using a *phase-control section* in combination with a shift of Bragg wavelength through electrorefraction. In fact, such amplifiers are nothing but multisection semiconductor lasers (see Section 3.4.3) with antireflection coatings. In one experimental demonstration, two channels operating at 1 Gb/s and separated by 0.23 nm could be separated by selective amplification (> 10 dB) of one channel [62]. Four-wave mixing in an SOA can also be used to form a tunable filter whose center wavelength is determined by the pump laser [63].

8.2.2 Multiplexers and Demultiplexers

Multiplexers and demultiplexers are the essential components of a WDM system. Similar to the case of optical filters, demultiplexers require a wavelength-selective mechanism and can be classified into two broad categories. *Diffraction-based demultiplexers* use an angularly dispersive element, such as a diffraction grating, which disperses incident light spatially into various wavelength components. *Interference-based demultiplexers* make use of devices such as optical filters and directional couplers. In both cases, the same device can be used as a multiplexer or a demultiplexer, depending on the direction of propagation, because of the inherent reciprocity of optical waves in dielectric media.

Grating-based demultiplexers use the phenomenon of Bragg diffraction from an optical grating [64]–[67]. Figure 8.9 shows the design of two such demultiplexers. The

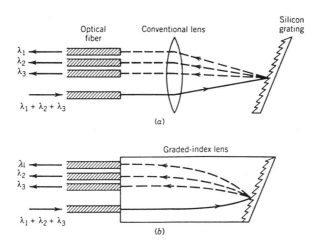

Figure 8.9: Grating-based demultiplexer making use of (a) a conventional lens and (b) a graded-index lens.

input WDM signal is focused onto a reflection grating, which separates various wavelength components spatially, and a lens focuses them onto individual fibers. Use of a graded-index lens simplifies alignment and provides a relatively compact device. The focusing lens can be eliminated altogether by using a concave grating. For a compact design, the concave grating can be integrated within a silicon slab waveguide [1]. In a different approach, multiple elliptical Bragg gratings are etched using the silicon technology [64]. The idea behind this approach is simple. If the input and output fibers are placed at the two foci of the elliptical grating, and the grating period Λ is adjusted to a specific wavelength λ_0 by using the Bragg condition $2\Lambda n_{\text{eff}} = \lambda_0$, where n_{eff} is the effective index of the waveguide mode, the grating would selectively reflect that wavelength and focus it onto the output fiber. Multiple gratings need to be etched, as each grating reflects only one wavelength. Because of the complexity of such a device, a single concave grating etched directly onto a silica waveguide is more practical. Such a grating can be designed to demultiplex up to 120 channels with a wavelength spacing of 0.3 nm [66].

A problem with grating demultiplexers is that their bandpass characteristics depend on the dimensions of the input and output fibers. In particular, the core size of output fibers must be large to ensure a flat passband and low insertion losses. For this reason, most early designs of multiplexers used multimode fibers. In a 1991 design, a microlens array was used to solve this problem and to demonstrate a 32-channel multiplexer for single-mode fiber applications [68]. The fiber array was produced by fixing single-mode fibers in V-shaped grooves etched into a silicon wafer. The microlens transforms the relatively small mode diameter of fibers ($\sim 10~\mu$m) into a much wider diameter (about 80 μm) just beyond the lens. This scheme provides a multiplexer that can work with channels spaced by only 1 nm in the wavelength region near 1.55 μm while accommodating a channel bandwidth of 0.7 nm.

Filter-based demultiplexers use the phenomenon of optical interference to select

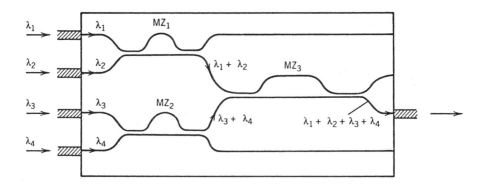

Figure 8.10: Layout of an integrated four-channel waveguide multiplexer based on Mach–Zehnder interferometers. (After Ref. [69]; ©1988 IEEE; reprinted with permission.)

the wavelength [1]. Demultiplexers based on the MZ filter have attracted the most attention. Similar to the case of a tunable optical filter, several MZ interferometers are combined to form a WDM demultiplexer [69]–[71]. A 128-channel multiplexer fabricated with the silica-waveguide technology was fabricated by 1989 [70]. Figure 8.10 illustrates the basic concept by showing the layout of a four-channel multiplexer. It consists of three MZ interferometers. One arm of each MZ interferometer is made longer than the other to provide a wavelength-dependent phase shift between the two arms. The path-length difference is chosen such that the total input power from two input ports at different wavelengths appears at only one output port. The whole structure can be fabricated on a silicone substrate using SiO_2 waveguides in the form of a planar lightwave circuit.

Fiber Bragg gratings can also be used for making all-fiber demultiplexers. In one approach, a $1 \times N$ fiber coupler is converted into a demultiplexer by forming a *phase-shifted grating* at the end of each output port, opening a narrowband transmission window (~ 0.1 nm) within the stop band [47]. The position of this window is varied by changing the amount of phase shift so that each arm of the $1 \times N$ fiber coupler transmits only one channel. The fiber-grating technology can be applied to form Bragg gratings directly on a planar silica waveguide. This approach has attracted attention since it permits integration of Bragg gratings within planar lightwave circuits. Such gratings were incorporated in an asymmetric MZ interferometer (unequal arm lengths) resulting in a compact multiplexer [72].

It is possible to construct multiplexers by using multiple directional couplers. The basic scheme is similar to that shown in Fig. 8.10 but simpler as MZ interferometers are not used. Furthermore, an all-fiber multiplexer made by using fiber couplers avoids coupling losses that occur whenever light is coupled into or out of an optical fiber. A *fused biconical taper* can also be used for making fiber couplers [73]. Multiplexers based on fiber couplers can be used only when channel spacing is relatively large ($>$ 10 nm) and are thus suitable mostly for coarse WDM applications.

From the standpoint of system design, integrated demultiplexers with low insertion losses are preferred. An interesting approach uses a *phased array* of optical waveguides

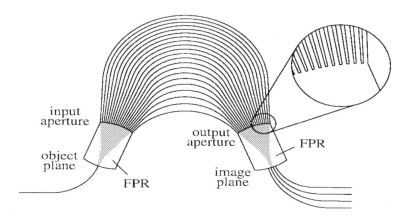

Figure 8.11: Schematic of a waveguide-grating demultiplexer consisting of an array of waveguides between two free-propagation regions (FPR). (After Ref. [74]; ©1996 IEEE; reprinted with permission.)

that acts as a grating. Such gratings are called *arrayed waveguide gratings* (AWGs) and have attracted considerable attention because they can be fabricated using the silicon, InP, or LiNbO$_3$ technology [74]–[81]. In the case of silica-on-silicon technology, they are useful for making planar lightwave circuits [79]. AWGs can be used for a variety of WDM applications and are discussed later in the context of WDM routers.

Figure 8.11 shows the design of a waveguide-grating demultiplexer, also known as a phased-array demultiplexer [74]. The incoming WDM signal is coupled into an array of planar waveguides after passing through a free-propagation region in the form of a lens. In each waveguide, the WDM signal experiences a different phase shift because of different lengths of waveguides. Moreover, the phase shifts are wavelength dependent because of the frequency dependence of the mode-propagation constant. As a result, different channels focus to different output waveguides when the light exiting from the array diffracts in another free-propagation region. The net result is that the WDM signal is demultiplexed into individual channels. Such demultiplexers were developed during the 1990s and became available commercially by 1999. They are able to resolve up to 256 channels with spacing as small as 0.2 nm. A combination of several suitably designed AWGs can increase the number of channels to more than 1000 while maintaining a 10-GHz resolution [82].

The performance of multiplexers is judged mainly by the amount of insertion loss for each channel. The performance criterion for demultiplexers is more stringent. First, the performance of a demultiplexer should be insensitive to the polarization of the incident WDM signal. Second, a demultiplexer should separate each channel without any leakage from the neighboring channels. In practice, some power leakage is likely to occur, especially in the case of dense WDM systems with small interchannel spacing. Such power leakage is referred to as crosstalk and should be quite small (< -20 dB) for a satisfactory system performance. The issue of interchannel crosstalk is discussed in Section 8.3.

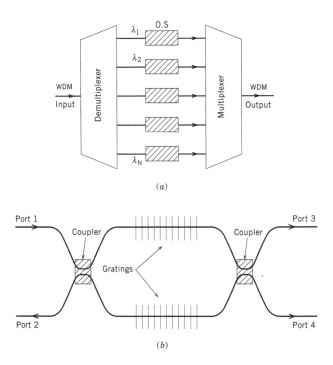

Figure 8.12: (a) A generic add–drop multiplexer based on optical switches (OS); (b) an add–drop filter made with a Mach–Zehnder interferometer and two identical fiber gratings.

8.2.3 Add–Drop Multiplexers

Add–drop multiplexers are needed for wide-area and metro-area networks in which one or more channels need to be dropped or added while preserving the integrity of other channels. Figure 8.12(a) shows a generic add–drop multiplexer schematically; it houses a bank of optical switches between a demultiplexer–multiplexer pair. The de-multiplexer separates all channels, optical switches drop, add, or pass individual chan-nels, and the multiplexer combines the entire signal back again. Any demultiplexer design discussed in the preceding subsection can be used to make add–drop multiplex-ers. It is even possible to amplify the WDM signal and equalize the channel powers at the add–drop multiplexer since each channel can be individually controlled [83]. The new component in such multiplexers is the optical switch, which can be made us-ing a variety of technologies including $LiNbO_3$ and InGaAsP waveguides. We discuss optical switches later in this section.

If a single channel needs to be demultiplexed, and no active control of individual channels is required, one can use a much simpler multiport device designed to send a single channel to one port while all other channels are transferred to another port. Such devices avoid the need for demultiplexing all channels and are called add–drop filters because they filter out a specific channel without affecting the WDM signal. If only a small portion of the channel power is filtered out, such a device acts as an "optical tap" as it leaves the contents of the WDM signal intact.

Several kinds of add–drop filters have been developed since the advent of WDM technology [84]–[94]. The simplest scheme uses a series of interconnected directional couplers, forming a MZ chain similar to that of a MZ filter discussed earlier. However, in contrast with the MZ filter of Section 8.2.1, the relative delay τ_m in Eq. (8.2.3) is made the same for each MZ interferometer. Such a device is sometimes referred to as a *resonant coupler* because it resonantly couples out a specific wavelength channel to one output port while the remainder of the channels appear at the other output port. Its performance can be optimized by controlling the coupling ratios of various directional couplers [86]. Although resonant couplers can be implemented in an all-fiber configuration using fiber couplers, the silica-on-silicon waveguide technology provides a compact alternative for designing such add–drop filters [87].

The wavelength selectivity of Bragg gratings can also be used to make add–drop filters. In one approach, referred to as the *grating-assisted* directional coupler, a Bragg grating is fabricated in the middle of a directional coupler [93]. Such devices can be made in a compact form using InGaAsP/InP or silica waveguides. However, an all-fiber device is often preferred for avoiding coupling losses. In a common approach, two identical Bragg gratings are formed on the two arms of a MZ interferometer made using two 3-dB fiber couplers. The operation of such an add–drop filter can be understood from Fig. 8.12(b). Assume that the WDM signal is incident on port 1 of the filter. The channel, whose wavelength λ_g falls within the stop band of the two identical Bragg gratings, is totally reflected and appears at port 2. The remaining channels are not affected by the gratings and appear at port 4. The same device can add a channel at the wavelength λ_g if the signal at that wavelength is injected from port 3. If the add and drop operations are performed simultaneously, it is important to make the gratings highly reflecting (close to 100%) to minimize the crosstalk. As early as 1995, such an all-fiber, add–drop filter exhibited the drop-off efficiency of more than 99%, while keeping the crosstalk level below 1% [88]. The crosstalk can be reduced below -50 dB by cascading several such devices [89].

Several other schemes use gratings to make add–drop filters. In one scheme, a waveguide with a built-in, phase-shifted grating is used to add or drop one channel from a WDM signal propagating in a neighboring waveguide [84]. In another, two identical AWGs are connected in series such that an optical amplifier connects each output port of one with the corresponding input port of the another [85]. The gain of amplifiers is adjusted such that only the channel to be dropped experiences amplification when passing through the device. Such a device is close to the generic add–drop multiplexer shown in Fig. 8.12(a) with the only difference that optical switches are replaced with optical amplifiers.

In another category of add–drop filters, optical circulators are used in combination with a fiber grating [92]. Such a device is simple in design and can be made by connecting each end of a fiber grating to a 3-port optical circulator. The channel reflected by the grating appears at the unused port of the input-end circulator. The same-wavelength channel can be added by injecting it from the output-end circulator. The device can also be made by using only one circulator provided it has more than three ports. Figure 8.13 shows two such schemes [94]. Scheme (a) uses a six-port circulator. The WDM signal entering from port 1 exits from port 2 and passes through a Bragg grating. The dropped channel appears at port 3 while the remaining channels re-enter the circulator at port 5

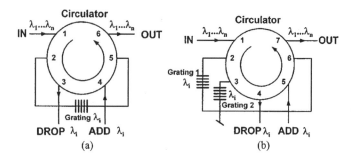

Figure 8.13: (a) Two designs of add–drop multiplexers using a single optical circulator in combination with fiber gratings. (After Ref. [94]; ©2001 IEEE; reprinted with permission.)

and leave the device from port 6. The channel to be added enters from port 4. Scheme (b) works in a similar way but uses two identical gratings to reduce the crosstalk level. Many other variants are possible.

8.2.4 Star Couplers

The role of a star coupler, as seen in Fig. 8.5, is to combine the optical signals entering from its multiple input ports and divide it equally among its output ports. In contrast with demultiplexers, star couplers do not contain wavelength-selective elements, as they do not attempt to separate individual channels. The number of input and output ports need not be the same. For example, in the case of video distribution, a relatively small number of video channels (say 100) may be sent to thousands of subscribers. The number of input and output ports is generally the same for the broadcast-and-select LANs in which each user wishes to receive all channels (see Fig. 8.5). Such a passive star coupler is referred to as an $N \times N$ broadcast star, where N is the number of input (or output) ports. A reflection star is sometimes used for LAN applications by reflecting the combined signal back to its input ports. Such a geometry saves considerable fiber when users are distributed over a large geographical area.

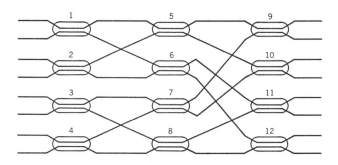

Figure 8.14: An 8×8 star coupler formed by using twelve 2×2 single-mode fiber couplers.

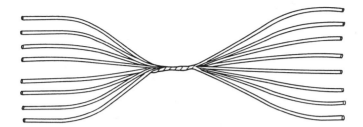

Figure 8.15: A star coupler formed by using the fused biconical tapering method.

Several kinds of star couplers have been developed for LAN applications [95]–[101]. An early approach made use of multiple 3-dB fiber couplers [96]. A 3-dB fiber coupler divides two input signals between its two output ports, the same functionality needed for a 2×2 star coupler. Higher-order $N \times N$ stars can be formed by combining several 2×2 couplers as long as N is a multiple of 2. Figure 8.14 shows an 8×8 star formed by interconnecting 12 fiber couplers. The complexity of such star couplers grows enormously with the number of ports.

Fused biconical-taper couplers can be used to make compact, monolithic, star couplers. Figure 8.15 shows schematically a star coupler formed using this technique. The idea is to fuse together a large number of fibers and elongate the fused portion to form a biconically tapered structure. In the tapered portion, signals from each fiber mix together and are shared almost equally among its output ports. Such a scheme works relatively well for multimode fibers [95] but is limited to only a few ports in the case of single-mode fibers. Fused 2×2 couplers were made as early as 1981 using single-mode fibers [73]; they can also be designed to operate over a wide wavelength range. Higher-order stars can be made using a combinatorial scheme similar to that shown in Fig. 8.12 [97].

A common approach for fabricating a compact broadcast star makes use of the silica-on-silicon technology in which two arrays of planar SiO_2 waveguides, separated by a central slab region, are formed on a silicon substrate. Such a star coupler was first demonstrated in 1989 in a 19×19 configuration [98]. The SiO_2 channel waveguides were 200 μm apart at the input end, but the final spacing near the central region was only 8 μm. The 3-cm-long star coupler had an efficiency of about 55%. A fiber amplifier can be integrated with the star coupler to amplify the output signals before broadcasting [99]. The silicon-on-insulator technology has been used for making star couplers. A 5×9 star made by using silicon rib waveguides exhibited low losses (1.3 dB) with relatively uniform coupling [100].

8.2.5 Wavelength Routers

An important WDM component is an $N \times N$ wavelength router, a device that combines the functionality of a star coupler with multiplexing and demultiplexing operations. Figure 8.16(a) shows the operation of such a wavelength router schematically for $N = 5$. The WDM signals entering from N input ports are demultiplexed into individual channels and directed toward the N output ports of the router in such a way

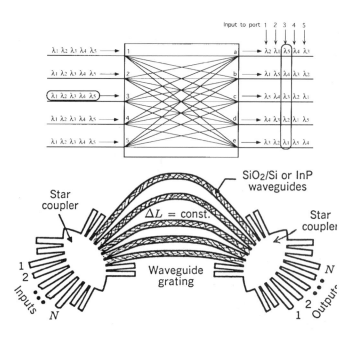

Figure 8.16: (a) Schematic illustration of a wavelength router and (b) its implementation using an AWG. (After Ref. [79]; ©1999 IEEE; reprinted with permission.)

that the WDM signal at each port is composed of channels entering at different input ports. This operation results in a cyclic form of demultiplexing. Such a device is an example of a passive router since its use does not involve any active element requiring electrical power. It is also called a *static router* since the routing topology is not dynamically reconfigurable. Despite its static nature, such a WDM device has many potential applications in WDM networks.

The most common design of a wavelength router uses a AWG demultiplexer shown in Fig. 8.11 modified to provide multiple input ports. Such a device, called the *waveguide-grating router* (WGR), is shown schematically in Fig. 8.16(b). It consists of two $N \times M$ star couplers such that M output ports of one star coupler are connected with M input ports of another star coupler through an array of M waveguides that acts as an AWG [74]. Such a device is a generalization of the MZ interferometer in the sense that a single input is divided coherently into M parts (rather than two), which acquire different phase shifts and interfere in the second free-propagation region to come out of N different ports depending on their wavelengths. The symmetric nature of the WGR permits to launch N WDM signals containing N different wavelengths simultaneously, and each WDM signal is demultiplexed to N output ports in a periodic fashion.

The physics behind the operation of a WGR requires a careful consideration of the phase changes as different wavelength signals diffract through the free-propagation region inside star couplers and propagate through the waveguide array [74]–[81]. The most important part of a WGR is the waveguide array designed such that the length difference ΔL between two neighboring waveguides remains constant from one wave-guide to next. The phase difference for a signal of wavelength λ, traveling from the pth input port to the qth output port through the mth waveguide (compared to the path connecting central ports), can be written as [13]

$$\phi_{pqm} = (2\pi m/\lambda)(n_1\delta_p + n_2\Delta L + n_1\delta_q'), \tag{8.2.4}$$

where n_1 and n_2 are the refractive indices in the regions occupied by the star couplers and waveguides, respectively. The lengths δ_p and δ_q' depend on the location of the input and output ports. When the condition

$$n_1(\delta_p + \delta_q') + n_2\Delta L = Q\lambda \tag{8.2.5}$$

is satisfied for some integer Q, the channel at the wavelength λ acquires phase shifts that are multiples of 2π while passing through different waveguides. As a result, all fields coming out of the M waveguides will interfere constructively at the qth port. Other wavelengths entering from the pth port will be directed to other output ports determined by the condition (8.2.5). Clearly, the device acts as a demultiplexer since a WDM signal entering from the pth port is distributed to different output ports depending on the channel wavelengths.

The routing function of a WGR results from the periodicity of the transmission spectrum. This property is also easily understood from Eq. (8.2.5). The phase condition for constructive interference can be satisfied for many integer values of Q. Thus, if Q is changed to $Q+1$, a different wavelength will satisfy Eq. (8.2.5) and will be directed toward the same port. The frequency difference between these two wavelengths is the free spectral range (FSR), analogous to that of FP filters. For a WGR, it is given by

$$\text{FSR} = \frac{c}{n_1(\delta_p + \delta_q') + n_2\Delta L}. \tag{8.2.6}$$

Strictly speaking, FSR is not the same for all ports, an undesirable feature from a practical standpoint. However, when δ_p and δ_q' are designed to be relatively small compared with ΔL, FSR becomes nearly constant for all ports. In that case, a WGR can be viewed as N demultiplexers working in parallel with the following property. If the WDM signal from the first input port is distributed to N output ports in the order $\lambda_1, \lambda_2, \ldots, \lambda_N$, the WDM signal from the second input port will be distributed as $\lambda_N, \lambda_1, \ldots, \lambda_{N-1}$, and the same cyclic pattern is followed for other input ports.

The optimization of a WGR requires precise control of many design parameters for reducing the crosstalk and maximizing the coupling efficiency. Despite the complexity of the design, WGRs are routinely fabricated in the form of a compact commercial device (each dimension ~ 1 cm) using either silica-on-silicon technology or InGaAsP/InP technology [74]–[81]. WGRs with 128 input and output ports were available by 1996 in the form of a planar lightwave circuit and were able to operate on WDM signals with a channel spacing as small as 0.2 nm while maintaining crosstalk below 16 dB. WGRs with 256 input and output ports have been fabricated using this technology [102]. WGRs can also be used for applications other than wavelength routing such as multichannel transmitters and receivers (discussed later in this section), tunable add–drop optical filters, and add–drop multiplexers.

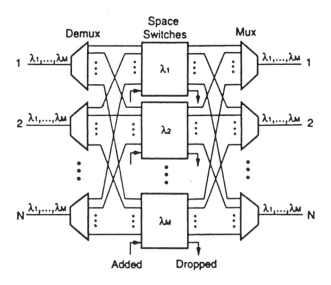

Figure 8.17: Schematic of an optical cross-connect based on optical switches.

8.2.6 Optical Cross-Connects

The development of wide-area WDM networks requires a dynamic wavelength routing scheme that can reconfigure the network while maintaining its nonblocking (transparent) nature. This functionality is provided by an *optical cross-connect* (OXC) which performs the same function as that provided by electronic digital switches in telephone networks. The use of dynamic routing also solves the problem of a limited number of available wavelengths through the wavelength-reuse technique. The design and fabrication of OXCs has remained a major topic of research since the advent of WDM systems [103]–[118].

Figure 8.17 shows the generic design of an OXC schematically. The device has N input ports, each port receiving a WDM signal consisting of M wavelengths. Demultiplexers split the signal into individual wavelengths and distribute each wavelength to the bank of M switching units, each unit receiving N input signals at the same wavelength. An extra input and output port is added to the switch to allow dropping or adding of a specific channel. Each switching unit contains N optical switches that can be configured to route the signals in any desirable fashion. The output of all switching units is sent to N multiplexers, which combine their M inputs to form the WDM signal. Such an OXC needs N multiplexers, N demultiplexers, and $M(N+1)^2$ optical switches. Switches used by an OXC are 2×2 space-division switches which switch an input signal to spatially separated output ports using a mechanical, thermo-optic, electro-optic, or all-optical technique. Many schemes have been developed for performing the switching operation. We discuss some of them next.

Mechanical switching is perhaps the simplest to understand. A simple mirror can act as a switch if the output direction can be changed by tilting the mirror. The use of "bulk" mirrors is impractical because of a large number of switches needed for mak-

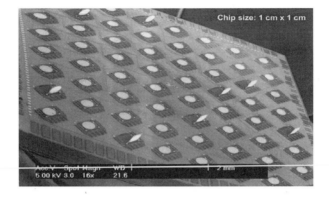

Figure 8.18: Scanning electron micrograph of an 8×8 MEMS optical switch based on free-rotating micromirrors. (After Ref. [112]; ©2000 IEEE; reprinted with permission.)

ing an OXC. For this reason, a micro-electro-mechanical system (MEMS) is used for switching [109]. Figure 8.18 shows an example of a 8×8 MEMS optical switch containing a two-dimensional array of free-rotating micromirrors [112]. Optical path lengths are far from being uniform in such a two-dimensional (2-D) geometry. This feature limits the switch size although multiple 2-D switches can be combined to increase the effective size. The three-dimensional (3-D) configuration in which the input and output fibers are located normal to the switching-fabric plane solves the size problem to a large extent. The switch size can be as large as 4096×4096 in the 3-D configuration. MEMS-based switches were becoming available commercially in 2002 and are likely to find applications in WDM networks. They are relatively slow to reconfigure (switching time > 10 ms) but that is not a major limitation in practice.

A MZ interferometer similar to that shown in Fig. 8.8(b) can also act as a 2×2 optical switch because the input signal can be directed toward different output ports by changing the delay in one of the arms by a small amount. The planar lightwave circuit technology uses the thermo-optic effect to change the refractive index of silica by heating. The temperature-induced change in the optical path length provides optical switching. As early as 1996, such optical switches were used to form a 8×16 OXC [103]. By 1998, such an OXC was packaged using switch boards of the standard $(33 \times 33 \text{ cm}^2)$ dimensions [107]. The extinction ratio can be increased by using two MZ interferometers in series, each with its own thermo-optic phase shifter, since the second unit blocks any light leaked through the first one [110]. Polymers are sometimes used in place of silica because of their large thermo-optic coefficient (more than 10 times larger compared with that of silica) for making OXCs [111]. Their use reduces both the fabrication cost and power consumption. The switching time is ~ 1 ms for all thermo-optic devices.

A directional coupler also acts as a 2×2 optical switch because it can direct an input signal toward different output ports in a controlled fashion. In LiNbO$_3$-based directional couplers, the refractive index can be changed by applying an external voltage through the electro-optic effect known as electrorefraction. The LiNbO$_3$ technology was used by 1988 to fabricate an 8×8 OXC [108]. Switching time of such cross-

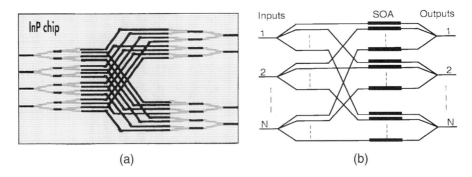

Figure 8.19: Examples of optical switches based on (a) Y-junction semiconductor waveguides and (b) SOAs with splitters. (After Ref. [105]; ©1996 IEEE; reprinted with permission.)

connects can be quite small (< 1 ns) as it is only limited by the speed with which electrical voltage can be changed. An OXC based on $LiNbO_3$ switches was used for the MONET project [21].

Semiconductor waveguides can also be used for making optical switches in the form of direction couplers, MZ interferometers, or Y junctions [105]. The InGaAsP/InP technology is most commonly used for such switches. Figure 8.19(a) shows a 4×4 switch based on the Y junctions; electrorefraction is used to switch the signal between the two arms of a Y junction. Since InGaAsP waveguides can provide amplification, SOAs can be used for compensating insertion losses. SOAs themselves can be used for making OXCs. The basic idea is shown schematically in Fig. 8.19(b) where SOAs act as a gate switch. Each input is divided into N branches using waveguide splitters, and each branch passes through an SOA, which either blocks light through absorption or transmits it while amplifying the signal simultaneously. Such OXCs have the advantage that all components can be integrated using the InGaAsP/InP technology while providing low insertion losses, or even a net gain, because of the use of SOAs. They can operate at high bit rates; operation at a bit rate of 2.5 Gb/s was demonstrated in 1996 within an installed fiber network [106].

Many other technologies can be used for making OXCs [115]. Examples include liquid crystals, bubbles, and electroholography. Liquid crystals in combination with polarizers either absorb or reflect the incident light depending on the electric voltage, and thus act as an optical switch. Although the liquid-crystal technology is well developed and is used routinely for computer-display applications, it has several disadvantages for making OXCs. It is relatively slow, is difficult to integrate with other optical components, and requires fixed input polarization. The last problem can be solved by splitting the input signal into orthogonally polarized components and switching each one separately, but only at the expense of increased complexity.

The bubble technology makes use of the phenomenon of *total internal reflection* for optical switching. A two-dimensional array of optical waveguides is formed in such a way that they intersect inside liquid-filled channels. When an air bubble is introduced at the intersection by vaporizing the liquid, light is reflected (i.e., switched) into another waveguide because of total internal reflection. This approach is appealing because of its low-cost potential (bubble-jet technology is used routinely for printers) but requires a careful design for reducing the crosstalk and insertion losses.

Electroholographic switches are similar to 2-D MEMS but employ a LiNbO$_3$ crystal for switching in place of a rotating mirror. Incident light can be switched at any point within the 2-D array of such crystals by applying an electric field and creating a Bragg grating at that location. Because of the wavelength selectivity of the Bragg grating, only a single wavelength can be switched by one device. This feature increases the complexity of such switching fabrics. Other issues are related to the polarization sensitivity of LiNbO$_3$-based devices.

Optical fibers themselves can be used for making OXCs if they are combined with fiber gratings and optical circulators [116]. The main drawback of any OXC is the large number of components and interconnections required that grows exponentially as the number of nodes and the number of wavelengths increase. Alternatively, the signal wavelength itself can be used for switching by making use of wavelength-division switches. Such a scheme makes use of static wavelength routers such as a WGR in combination with a new WDM component—the *wavelength converter*. We turn to this component next.

8.2.7 Wavelength Converters

A wavelength converter changes the input wavelength to a new wavelength without modifying the data content of the signal. Many schemes were developed during the 1990s for making wavelength converters [119]–[129]; four among them are shown schematically in Fig. 8.20.

A conceptually simple scheme uses an optoelectronic regenerator shown in Fig. 8.20(a). An optical receiver first converts the incident signal at the input wavelength λ_1 to an electrical bit pattern, which is then used by a transmitter to generate the optical signal at the desired wavelength λ_2. Such a scheme is relatively easy to implement as it uses standard components. Its other advantages include an insensitivity to input polarization and the possibility of net amplification. Among its disadvantages are limited transparency to bit rate and data format, speed limited by electronics, and a relatively high cost, all of which stem from the optoelectronic nature of wavelength conversion.

Several all-optical techniques for wavelength conversion make use of SOAs [119]–[122], amplifiers discussed in Section 6.2. The simplest scheme shown in Fig. 8.20(b) is based on cross-gain saturation occurring when a weak field is amplified inside the SOA together with a strong field, and the amplification of the weak field is affected by the strong field. To use this phenomenon, the pulsed signal whose wavelength λ_1 needs to be converted is launched into the SOA together with a low-power CW beam at the wavelength λ_2 at which the converted signal is desired. Amplifier gain is mostly saturated by the λ_1 beam. As a result, the CW beam is amplified by a large amount during 0 bits (no saturation) but by a much smaller amount during 1 bits. Clearly, the bit pattern of the incident signal will be transferred to the new wavelength with reverse polarity such that 1 and 0 bits are interchanged. This technique has been used in many experiments and can work at bit rates as high as 40 Gb/s. It can provide net gain to the wavelength-converted signal and can be made nearly polarization insensitive. Its main disadvantages are (i) relatively low on–off contrast, (ii) degradation due to spontaneous

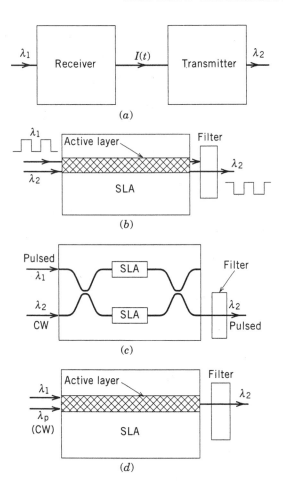

Figure 8.20: Four schemes for wavelength conversion: (a) optoelectronic regenerator; (b) gain saturation in a semiconductor laser amplifier (SLA); (c) phase modulation in a SLA placed in one arm of a Mach-Zehnder interferometer; (d) four-wave mixing inside a SLA.

emission, and (iii) phase distortion because of frequency chirping that invariably occurs in SOAs (see Section 3.5). The use of an absorbing medium in place of the SOA solves the polarity reversal problem. An electroabsorption modulator (see Section 3.6.4) has been used for wavelength conversion with success [127]. It works on the principle of cross-absorption saturation. The device blocks the CW signal at λ_2 because of high absorption except when the arrival of 1 bits at λ_1 saturates the absorption.

The contrast problem can be solved by using the MZ configuration of Fig. 8.20(c) in which an SOA is inserted in each arm of a MZ interferometer [119]. The pulsed signal at the wavelength λ_1 is split at the first coupler such that most power passes through one arm. At the same time, the CW signal at the wavelength λ_2 is split equally by this coupler and propagates simultaneously in the two arms. In the absence of the λ_1 beam, the CW beam exits from the cross port (upper port in the figure). However, when both beams are present simultaneously, all 1 bits are directed toward the bar port because of the refractive-index change induced by the λ_1 beam. The physical mechanism behind this behavior is the cross-phase modulation (XPM). Gain saturation induced by the λ_1 beam reduces the carrier density inside one SOA, which in turn increases the refractive index only in the arm through which the λ_1 beam passes. As a result, an additional π phase shift can be introduced on the CW beam because of cross-phase modulation, and the CW wave is directed toward the bar port during each 1 bit.

It should be evident from the preceding discussion that the output from the bar port of the MZ interferometer would consist of an exact replica of the incident signal with its wavelength converted to the new wavelength λ_2. An optical filter is placed in front of the bar port for blocking the original λ_1 signal. The MZ scheme is preferable over cross-gain saturation as it does not reverse the bit pattern and results in a higher on–off contrast simply because nothing exits from the bar port during 0 bits. In fact, the output from the cross port also has the same bit pattern but its polarity is reversed. Other types of interferometers (such as Michelson and Sagnac interferometers) can also be used with similar results. The MZ interferometer is often used in practice because it can be easily integrated by using SiO_2/Si or InGaAsP/InP waveguides, resulting in a compact device [125]. Such a device can operate at high bit rates (up to 80 Gb/s), offers a large contrast, and degrades the signal relatively little although spontaneous emission does affect the SNR. Its main disadvantage is a narrow dynamic range of the input power since the phase induced by the amplifier depends on it.

Another scheme employs the SOA as a nonlinear medium for four-wave mixing (FWM), the same nonlinear phenomenon that is a major source of interchannel crosstalk in WDM systems (see Section 8.3). The FWM technique has been discussed in Section 7.7 in the context of optical phase conjugation and dispersion compensation. As seen in Fig. 8.20(d), its use requires an intense CW pump beam that is launched into the SOA together with the signal whose wavelength needs to be converted [119]. If v_1 and v_2 are the frequencies of the input signal and the converted signal, the pump frequency v_p is chosen such that $v_p = (v_1 + v_2)/2$. At the amplifier output, a replica of the input signal appears at the carrier frequency v_2 because FWM requires the presence of both the pump and signal. One can understand the process physically as scattering of two pump photons of energy $2hv_p$ into two photons of energy hv_1 and hv_2. The nonlinearity responsible for the FWM has its origin in fast intraband relaxation processes occurring at a time scale of 0.1 ps [130]. As a result, frequency shifts as large as 10 THz, corresponding to wavelength conversion over a range of 80 nm, are possible. For the same reason, this technique can work at bit rates as high as 100 Gb/s and is transparent to both the bit rate and the data format. Because of the gain provided by the amplifier, conversion efficiency can be quite high, resulting even in a net gain. An added advantage of this technique is the reversal of the frequency chirp since its use inverts the signal spectrum (see Section 7.7). The performance can also be improved by using two SOAs in a tandem configuration.

The main disadvantage of any wavelength-conversion technique based on SOAs is that it requires a tunable laser source whose light should be coupled into the SOA, typically resulting in large coupling losses. An alternative is to integrate the functionality of a wavelength converter within a tunable semiconductor laser. Several such devices have been developed [119]. In the simplest scheme, the signal whose wavelength needs to be changed is injected into a tunable laser directly. The change in the laser threshold resulting from injection translates into modulation of the laser output, mimicking the bit pattern of the injected signal. Such a scheme requires relatively large input powers. Another scheme uses the low-power input signal to produce a frequency shift (typically, 10 GHz/mW) in the laser output for each 1 bit. The resulting frequency-modulated CW signal can be converted into amplitude modulation by using a MZ interferometer. Another scheme uses FWM inside the cavity of a tunable semiconductor laser, which also plays the role of the pump laser. A phase-shifted DFB laser provided wavelength conversion over a range of 30 nm with this technique [120]. A sampled grating within a distributed Bragg reflector has also been used for this purpose [123].

Another class of wavelength converters uses an optical fiber as the nonlinear medium. Both XPM and FWM can be employed for this purpose using the last two configurations shown in Fig. 8.20. In the FWM case, stimulated Raman scattering (SRS) affects the FWM if the frequency difference $|v_1 - v_2|$ falls within the Raman-gain bandwidth [124]. In the XPM case, the use of a Sagnac interferometer, also known as the nonlinear optical loop mirror [40], provides a wavelength converter capable of operating at bit rates up to 40 Gb/s for both the return-to-zero (RZ) and nonreturn-to-zero (NRZ) formats [126]. Such a device reflects all 0 bits but 1 bits are transmitted through the fiber loop because of the XPM-induced phase shift. In a 2001 experiment, wavelength conversion at the bit rate of 80 Gb/s was realized by using a 1-km-long optical fiber designed to have a large value of the nonlinear parameter γ [129]. A periodically poled LiNbO$_3$ waveguide has provided wavelength conversion at 160 Gb/s [128]. In principle, wavelength converters based on optical fibers can operate at bit rates as high as 1 Tb/s because of the fast nature of their nonlinear response.

8.2.8   WDM Transmitters and Receivers

Most WDM systems use a large number of DFB lasers whose frequencies are chosen to match the ITU frequency grid precisely. This approach becomes impractical when the number of channels becomes large. Two solutions are possible. In one approach, single-frequency lasers with a tuning range of 10 nm or more are employed (see Section 3.4.3). The use of such lasers reduces the inventory and maintenance problems. Alternatively, multiwavelength transmitters which generate light at 8 or more fixed wavelengths simultaneously can be used. Although such WDM transmitters attracted some attention in the 1980s, it was only during the 1990s that monolithically integrated WDM transmitters, operating near 1.55 μm with a channel spacing of 1 nm or less, were developed using the InP-based optoelectronic integrated-circuit (OEIC) technology [131]–[139].

Several different techniques have been pursued for designing WDM transmitters. In one approach, the output of several DFB or DBR semiconductor lasers, independently tunable through Bragg gratings, is combined by using passive waveguides [131]–[134]. A built-in amplifier boosts the power of the multiplexed signal to increase the transmitted power. In a 1993 device, the WDM transmitter not only integrated 16 DBR lasers with 0.8-nm wavelength spacing, but an electroabsorption modulator was also integrated with each laser [132]. In a 1996 device, 16 gain-coupled DFB lasers were integrated, and their wavelengths were controlled by changing the width of the ridge

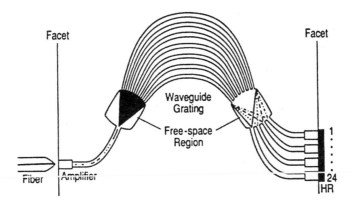

Figure 8.21: Schematic of a WDM laser made by integrating an AWG inside the laser cavity. (After Ref. [137]; ©1996 IEEE; reprinted with permission.)

waveguides and by tuning over a 1-nm range using a thin-film resistor [133]. In a different approach, sampled gratings with different periods are used to tune the wavelengths precisely of an array of DBR lasers [135]. The complexity of such devices makes it difficult to integrate more than 16 lasers on the same chip. The vertical-cavity surface-emitting laser (VCSEL) technology provides a unique approach to WDM transmitters since it can be used to produce a two-dimensional array of lasers covering a wide wavelength span at a relatively low cost [136]; it is well suited for LAN and data-transfer applications.

A waveguide grating integrated within the laser cavity can provide lasing at several wavelengths simultaneously. An AWG is often used for multiplexing the output of several optical amplifiers or DBR lasers [137]–[139]. In a 1996 demonstration of the basic idea, simultaneous operation at 18 wavelengths (spaced apart by 0.8 nm) was realized using an intracavity AWG [137]. Figure 8.21 shows the laser design schematically. Spontaneous emission of the amplifier located on the left side is demultiplexed into 18 spectral bands by the AWG through the technique of spectral slicing. The amplifier array on the right side selectively amplifies the set of 18 wavelengths, resulting in a laser emitting at all wavelengths simultaneously. A 16-wavelength transmitter with 50-GHz channel spacing was built in 1998 by this technique [138]. In a different approach, the AWG was not a part of the laser cavity but was used to multiplex the output of 10 DBR lasers, all produced on the same chip in an integrated fashion [139]. AWGs fabricated with the silica-on-silicon technology can also be used although they cannot be integrated on the InP substrate.

Fiber lasers can be designed to provide multiwavelength output and therefore act as a CW WDM source [140]. A ring-cavity fiber laser containing a frequency shifter (e.g., an acousto-optic device) and an optical filter with periodic transmission peaks (such as a FP filter, a sampled grating, or an AWG) can provide its output at a comb of frequencies set to coincide with the ITU grid. Up to 16 wavelengths have been obtained in practical lasers although the power is not generally uniform across them. A demultiplexer is still needed to separate the channels before data is imposed on them using individual modulators.

A unique approach to WDM sources exploits the technique of spectral slicing for realizing WDM transmitters and is capable of providing more than 1000 channels [141]–[145]. The output of a coherent, wide-bandwidth source is sliced spectrally using a mutipeak optical filter such as an AWG. In one implementation of this idea [141], picosecond pulses from a mode-locked fiber laser are first broadened spectrally to bandwidth as large as 200 nm through supercontinuum generation by exploiting the nonlinear effects in an optical fiber [59]. Spectral slicing of the output by an AWG then produces many WDM channels with a channel spacing of 1 nm or less. In a 2000 experiment, this technique produced 1000 channels with 12.5-GHz channel spacing [143]. In another experiment, 150 channels with 25-GHz channel spacing were realized within the C band covering the range 1530–1560 nm [145]. The SNR of each channel exceeded 28 dB, indicating that the source was suitable for dense WDM applications.

The generation of supercontinuum is not necessary if a mode-locked laser producing femtosecond pulses is employed. The spectral width of such pulses is quite large to begin with and can be enlarged to 50 nm or more by chirping them using 10–15 km of standard telecommunication fiber. Spectral slicing of the output by a demultiplexer can again provide many channels, each of which can be modulated independently. This technique also permits simultaneous modulation of all channels using a single modulator before the demultiplexer if the modulator is driven by a suitable electrical bit stream composed through TDM. A 32-channel WDM source was demonstrated in 1996 by using this method [142]. Since then, this technique has been used to provide sources with more than 1000 channels [144].

On the receiver end, multichannel WDM receivers have been developed because their use can simplify the system design and reduce the overall cost [146]. Monolithic receivers integrate a photodiode array with a demultiplexer on the same chip. Typically, A planar concave-grating demultiplexer or a WGR is integrated with the photodiode array. Even electronic amplifiers can be integrated within the same chip. The design of such monolithic receivers is similar to the transmitter shown in Fig. 8.21 except that no cavity is formed and the amplifier array is replaced with a photodiode array. Such a WDM receiver was first fabricated in 1995 by integrating an eight-channel WGR (with 0.8-nm channel spacing), eight p–i–n photodiodes, and eight preamplifiers using heterojunction-bipolar transistor technology [147].

8.3   System Performance Issues

The most important issue in the design of WDM lightwave systems is the *interchannel crosstalk*. The system performance degrades whenever crosstalk leads to transfer of power from one channel to another. Such a transfer can occur because of the nonlinear effects in optical fibers, a phenomenon referred to as *nonlinear crosstalk* as it depends on the nonlinear nature of the communication channel. However, some crosstalk occurs even in a perfectly linear channel because of the imperfect nature of various WDM components such as optical filters, demultiplexers, and switches. In this section we discuss both the linear and nonlinear crosstalk mechanisms and also consider other performance issues relevant for WDM systems.

8.3.1 Heterowavelength Linear Crosstalk

Linear crosstalk can be classified into two categories depending on its origin [148]–[163]. Optical filters and demultiplexers often let leak a fraction of the signal power from neighboring channels that interferes with the detection process. Such crosstalk is called *heterowavelength* or *out-of-band* crosstalk and is less of a problem because of its incoherent nature than the *homowavelength* or *in-band* crosstalk that occurs during routing of the WDM signal from multiple nodes. This subsection focuses on the heterowavelength crosstalk.

Consider the case in which a tunable optical filter is used to select a single channel among the N channels incident on it. If the optical filter is set to pass the mth channel, the optical power reaching the photodetector can be written as $P = P_m + \sum_{n \neq m}^{N} T_{mn} P_n$ where P_m is the power in the mth channel and T_{mn} is the filter transmittivity for channel n when channel m is selected. Crosstalk occurs if $T_{mn} \neq 0$ for $n \neq m$. It is called out-of-band crosstalk because it belongs to the channels lying outside the spectral band occupied by the channel detected. Its incoherent nature is also apparent from the fact that it depends only on the power of the neighboring channels.

To evaluate the impact of such crosstalk on system performance, one should consider the power penalty, defined as the additional power required at the receiver to counteract the effect of crosstalk. The photocurrent generated in response to the incident optical power is given by

$$I = R_m P_m + \sum_{n \neq m}^{N} R_n T_{mn} P_n \equiv I_{\text{ch}} + I_X, \qquad (8.3.1)$$

where $R_m = \eta_m q / h \nu_m$ is the photodetector responsivity for channel m at the optical frequency ν_m and η_m is the quantum efficiency. The second term I_X in Eq. (8.3.1) denotes the crosstalk contribution to the receiver current I. Its value depends on the bit pattern and becomes maximum when all interfering channels carry 1 bits simultaneously (the worst case).

A simple approach to calculating the crosstalk power penalty is based on the eye closure (see Section 4.3.3) occurring as a result of the crosstalk [148]. The eye closes most in the worst case for which I_X is maximum. In practice, I_{ch} is increased to maintain the system performance. If I_{ch} needs to be increased by a factor δ_X, the peak current corresponding to the top of the eye is $I_1 = \delta_X I_{\text{ch}} + I_X$. The decision threshold is set at $I_D = I_1/2$. The *eye opening* from I_D to the top level would be maintained at its original value $I_{\text{ch}}/2$ if

$$(\delta_X I_{\text{ch}} + I_X) - I_X - \tfrac{1}{2}(\delta_X I_{\text{ch}} + I_X) = \tfrac{1}{2} I_{\text{ch}}, \qquad (8.3.2)$$

or when $\delta_X = 1 + I_X / I_{\text{ch}}$. The quantity δ_X is just the power penalty for the mth channel. By using I_X and I_{ch} from Eq. (8.3.1), δ_X can be written (in dB) as

$$\delta_X = 10 \log_{10} \left(1 + \frac{\sum_{n \neq m}^{N} R_n T_{mn} P_n}{R_m P_m} \right), \qquad (8.3.3)$$

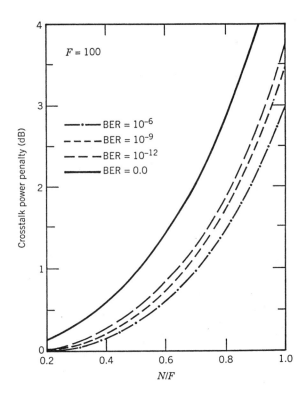

Figure 8.22: Crosstalk power penalty at four different values of the BER for a FP filter of finesse $F = 100$. (After Ref. [149]; ©1990 IEEE; reprinted with permission.)

where the powers correspond to their on-state values. If the peak power is assumed to be the same for all channels, the crosstalk penalty becomes power independent. Further, if the photodetector responsivity is nearly the same for all channels ($R_m \approx R_n$), δ_X is well approximated by

$$\delta_X \approx 10\log_{10}(1+X), \tag{8.3.4}$$

where $X = \sum_{n\neq m}^{N} T_{mn}$ is a measure of the out-of-band crosstalk; it represents the fraction of total power leaked into a specific channel from all other channels. The numerical value of X depends on the transmission characteristics of the specific optical filter. For a FP filter, X can be obtained in a closed form [149].

The preceding analysis of crosstalk penalty is based on the eye closure rather than the bit-error rate (BER). One can obtain an expression for the BER if I_X is treated as a random variable in Eq. (8.3.1). For a given value of I_X, the BER is obtained by using the analysis of Section 4.5.1. In particular, the BER is given by Eq. (4.5.6) with the on- and off-state currents given by $I_1 = I_{ch} + I_X$ and $I_0 = I_X$ if we assume that $I_{ch} = 0$ in the off-state. The decision threshold is set at $I_D = I_{ch}(1+X)/2$, which corresponds to the worst-case situation in which all neighboring channels are in the on state. The final BER is obtained by averaging over the distribution of the random variable I_X. The distribution of I_X has been calculated for a FP filter and is generally far from being Gaussian. The crosstalk power penalty δ_X can be calculated by finding the increase in I_{ch} needed to maintain a certain value of BER. Figure 8.22 shows the calculated penalty for several values of BER plotted as a function of N/F [149] with the choice $F = 100$. The solid curve corresponds to the error-free case (BER $= 0$). The power penalty can be kept below 0.2 dB to maintain a BER of 10^{-9} for values of N/F as large as 0.33. From Eq. (8.2.2) the channel spacing can be as little as three times the bit rate for such FP filters.

8.3.2 Homowavelength Linear Crosstalk

Homowavelength or in-band crosstalk results from WDM components used for rout-ing and switching along an optical network and has been of concern since the advent of WDM systems [150]–[163]. Its origin can be understood by considering a static wavelength router such as a WGR (see Fig. 8.16). For an $N \times N$ router, there exist N^2 combinations through which N-wavelength WDM signals can be split. Consider the output at one wavelength, say λ_m. Among the $N^2 - 1$ interfering signals that can ac-company the desired signal, $N - 1$ signals have the same carrier wavelength λ_m, while the remaining $N(N - 1)$ belong to different carrier wavelengths and are likely to be eliminated as they pass through other WDM components. The $N - 1$ crosstalk signals at the same wavelength (in-band crosstalk) originate from incomplete filtering through a WGR because of its partially overlapping transmission peaks [153]. The total optical field, including only the in-band crosstalk, can be written as

$$E_m(t) = \left(E_m + \sum_{n \neq m}^{N} E_n \right) \exp(-i\omega_m t), \tag{8.3.5}$$

where E_m is the desired signal and $\omega_m = 2\pi c/\lambda_m$. The coherent nature of the in-band crosstalk is obvious from Eq. (8.3.5).

To see the impact of in-band crosstalk on system performance, we should again consider the power penalty. The receiver current $I = R|E_m(t)|^2$ in this case contains interference or beat terms similar to the case of optical amplifiers (see Section 6.5). One can identify two types of beat terms; signal–crosstalk beating with terms like $E_m E_n$ and crosstalk–crosstalk beating with terms like $E_k E_n$, where $k \neq m$ and $n \neq m$. The latter terms are negligible in practice and can be ignored. The receiver current is then given approximately as

$$I(t) \approx RP_m(t) + 2R \sum_{n \neq m}^{N} \sqrt{P_m(t)P_n(t)} \cos[\phi_m(t) - \phi_n(t)], \tag{8.3.6}$$

where $P_n = |E_n|^2$ is the power and $\phi_n(t)$ is the phase. In practice, $P_n << P_m$ for $n \neq m$ because a WGR is built to reduce the crosstalk. Since phases are likely to fluctuate randomly, we can write Eq. (8.3.6) as $I(t) = R(P_m + \Delta P)$, treat the crosstalk as intensity noise, and use the approach of Section 4.6.2 for calculating the power penalty. In fact, the result is the same as in Eq. (4.6.11) and can be written as

$$\delta_X = -10 \log_{10}(1 - r_X^2 Q^2), \tag{8.3.7}$$

where

$$r_X^2 = \langle (\Delta P)^2 \rangle / P_m^2 = X(N-1), \qquad (8.3.8)$$

and $X = P_n/P_m$ is the crosstalk level defined as the fraction of power leaking through the WGR and is taken to be the same for all $N - 1$ sources of coherent in-band crosstalk by assuming equal powers. An average over the phases was performed by replacing $\cos^2 \theta = \frac{1}{2}$. In addition, r_X^2 was multiplied by another factor of $\frac{1}{2}$ to account for the fact that P_n is zero on average half of the times (during 0 bits). Experimental measurements of power penalty for a WGR agree with this simple model [153].

The impact of in-band crosstalk can be estimated from Fig. 4.19, where power penalty δ_X is plotted as a function of r_X. To keep the power penalty below 2 dB, $r_X <$ 0.07 is required, a condition that limits $X(N-1)$ to below -23 dB from Eq. (8.3.8). Thus, the crosstalk level X must be below -38 dB for $N = 16$ and below -43 dB for $N = 100$, rather stringent requirements.

The calculation of crosstalk penalty in the case of dynamic wavelength routing through optical cross-connects becomes quite complicated because of a large number of crosstalk elements that a signal can pass through in such WDM networks [155]. The worst-case analysis predicts a large power penalty (> 3 dB) when the number of crosstalk elements becomes more than 25 even if the crosstalk level of each component is only -40 dB. Clearly, the linear crosstalk is of primary concern in the design of WDM networks and should be controlled. A simple technique consists of modulating or scrambling the laser phase at the transmitter at a frequency much larger than the laser linewidth [164]. Both theory and experiments show that the acceptable crosstalk level exceeds 1% (-20 dB) with this technique [162].

8.3.3 Nonlinear Raman Crosstalk

Several nonlinear effects in optical fibers [59] can lead to interchannel and intrachannel crosstalk that affects the system performance considerably [165]–[171]. Section 2.6 discussed such nonlinear effects and their origin from a physical point of view. This subsection focuses on the Raman crosstalk.

As discussed in Section 2.6, stimulated Raman scattering (SRS) is generally not of concern for single-channel systems because of its relatively high threshold (about 500 mW near 1.55 μm). The situation is quite different for WDM systems in which the fiber acts as a Raman amplifier (see Section 6.3) such that the long-wavelength channels are amplified by the short-wavelength channels as long as the wavelength difference is within the bandwidth of the Raman gain. The Raman gain spectrum of silica fibers is so broad that amplification can occur for channels spaced as far apart as 100 nm. The shortest-wavelength channel is most depleted as it can pump many channels simultaneously. Such an energy transfer among channels can be detrimental for system performance as it depends on the bit pattern—amplification occurs only when 1 bits are present in both channels simultaneously. The Raman-induced crosstalk degrades the system performance and is of considerable concern for WDM systems [172]–[179].

Raman crosstalk can be avoided if channel powers are made so small that SRS-induced amplification is negligible over the entire fiber length. It is thus important to estimate the limiting value of the channel power. A simple model considers the depletion of the highest-frequency channel in the worst case in which 1 bits of all channels overlap completely simultaneously [165]. The amplification factor for each channel is $G_m = \exp(g_m L_{\text{eff}})$, where L_{eff} is the effective interaction length as defined in Eq. (2.6.2) and $g_m = g_R(\Omega_m)P_{\text{ch}}/A_{\text{eff}}$ is the Raman gain at $\Omega_m = \omega_1 - \omega_m$. For $g_m L_{\text{eff}} \ll 1$, the shortest-wavelength channel at ω_1 is depleted by a fraction $g_m L_{\text{eff}}$ due to Raman amplification of the mth channel. The total depletion for a M-channel WDM system is given by

$$D_R = \sum_{m=2}^{M} g_R(\Omega_m)P_{\text{ch}}L_{\text{eff}}/A_{\text{eff}}. \tag{8.3.9}$$

The summation in Eq. (8.3.9) can be carried out analytically if the Raman gain spectrum (see Fig. 2.18) is approximated by a triangular profile such that g_R increases linearly for frequencies up to 15 THz with a slope $S_R = dg_R/dv$ and then drops to zero. Using $g_R(\Omega_m) = mS_R\Delta v_{\text{ch}}$, the fractional power loss for the shortest-wavelength channel becomes [165]

$$D_R = \tfrac{1}{2}M(M-1)C_R P_{\text{ch}}L_{\text{eff}}, \tag{8.3.10}$$

where $C_R = S_R\Delta v_{\text{ch}}/(2A_{\text{eff}})$. In deriving this equation, channels were assumed to have a constant spacing Δv_{ch} and the Raman gain for each channel was reduced by a factor of 2 to account for random polarization states of different channels.

A more accurate analysis should consider not only depletion of each channel because of power transfer to longer-wavelength channels but also its own amplification by shorter-wavelength channels. If all other nonlinear effects are neglected along with GVD, evolution of the power P_n associated with the nth channel is governed by the following equation (see Section 6.3):

$$\frac{dP_n}{dz} + \alpha P_n = C_R P_n \sum_{m=1}^{M} (n-m)P_m, \tag{8.3.11}$$

where α is assumed to be the same for all channels. This set of M coupled nonlinear equations can be solved analytically. For a fiber of length L, the result is given by [172]

$$P_n(L) = P_n(0)e^{-\alpha L} \frac{P_t \exp[(n-1)C_R P_t L_{\text{eff}}]}{\sum_{m=1}^{M} P_m(0) \exp[(m-1)C_R P_t L_{\text{eff}}]}, \tag{8.3.12}$$

where $P_t = \sum_{m=1}^{M} P_m(0)$ is the total input power in all channels. This equation shows that channel powers follow an exponential distribution because of Raman-induced coupling among all channels.

The depletion factor D_R for the shorter-wavelength channel ($n = 1$) is obtained using $D_R = (P_{10} - P_1)/P_{10}$, where $P_{10} = P_1(0)\exp(-\alpha L)$ is the channel power expected in the absence of SRS. In the case of equal input powers in all channels, $P_t = MP_{\text{ch}}$ in Eq. (8.3.12), and D_R is given by

$$D_R = 1 - \exp\left[-\frac{1}{2}M(M-1)C_R P_{\text{ch}}L_{\text{eff}}\right] \frac{M\sinh(\tfrac{1}{2}MC_R P_{\text{ch}}L_{\text{eff}})}{\sinh(\tfrac{1}{2}M^2 C_R P_{\text{ch}}L_{\text{eff}})}. \tag{8.3.13}$$

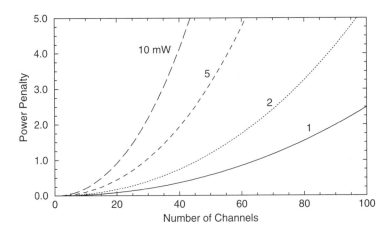

Figure 8.23: Raman-induced power penalty as a function of channel number for several values of P_{ch}. Channels are 100 GHz apart and launched with equal powers.

In the limit $M^2 C_R P_{\text{ch}} L_{\text{eff}} \ll 1$, this complicated expression reduces to the simple result in Eq. (8.3.10). In general, Eq. (8.3.10) overestimates the Raman crosstalk.

The Raman-induced power penalty is obtained using $\delta_R = -10 \log(1 - D_R)$ because the input channel power must be increased by a factor of $(1 - D_R)^{-1}$ to maintain the same system performance. Figure 8.23 shows how the power penalty increases with an increase in the channel power and the number of channels. The channel spacing is assumed to be 100 GHz. The slope of the Raman gain is estimated from the gain spectrum to be $S_R = 4.9 \times 10^{-18}$ m/(W-GHz) while $A_{\text{eff}} = 50 \ \mu\text{m}^2$ and $L_{\text{eff}} \approx 1/\alpha = 21.74$ km. As seen from Fig. 8.23, the power penalty becomes quite large for WDM systems with a large number of channels. If a value of at most 1 dB is considered acceptable, the limiting channel power P_{ch} exceeds 10 mW for 20 channels, but its value is reduced to below 1 mW when the number of WDM channels is larger than 70.

The foregoing analysis provides only a rough estimate of the Raman crosstalk as it neglects the fact that signals in each channel consist of a random sequence of 0 and 1 bits. A statistical analysis shows that the Raman crosstalk is lower by about a factor of 2 when signal modulation is taken into account [167]. The GVD effects that were neglected in the above analysis also reduce the Raman crosstalk since pulses in different channels travel at different speeds because of the group-velocity mismatch [173]. On the other hand, periodic amplification of the WDM signal can magnify the impact of SRS-induced degradation. The reason is that in-line amplifiers add noise which experiences less Raman loss than the signal itself, resulting in degradation of the SNR. The Raman crosstalk under realistic operating conditions was calculated in a 2001 study [179]. Numerical simulations showed that it can be reduced by inserting optical filters along the fiber link that block the low-frequency noise below the longest-wavelength channel [178]. Raman crosstalk can also be reduced using the technique of midway spectral inversion [174].

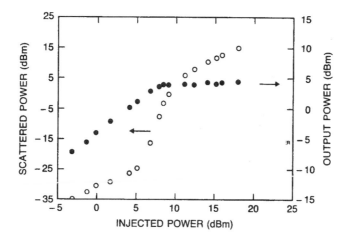

Figure 8.24: Output signal power (solid circles) and reflected SBS power (empty circles) as a function of power injected. (After Ref. [180]; ©1992 IEEE; reprinted with permission.)

8.3.4 Stimulated Brillouin Scattering

Stimulated Brillouin scattering (SBS) can also transfer energy from a high-frequency channel to a low-frequency one when the channel spacing equals the Brillouin shift. However, in contrast with the case of SRS, such an energy transfer is easily avoided with the proper design of multichannel communication systems. The reason is that the Brillouin-gain bandwidth is extremely narrow (~ 20 MHz) compared with the Raman-gain bandwidth (~ 5 THz). Thus, the channel spacing must match almost exactly the *Brillouin shift* (about 10 GHz in the 1.55-μm region) for SBS to occur; such an exact match is easily avoided. Furthermore, as discussed in Section 2.6, the two channels must be counterpropagating for Brillouin amplification to occur.

Although SBS does not induce interchannel crosstalk when all channels propagate in the forward direction, it nonetheless limits the channel powers. The reason is that a part of the channel power can be transferred to a backward-propagating Stokes wave generated from noise when the threshold condition $g_B P_{th} L_{eff}/A_{eff} \approx 21$ is satisfied (see Section 2.6). This condition is independent of the number and the presence of other channels. However, the threshold for each channel can be reached at low power levels. Figure 8.24 shows how the output power and power reflected backward through SBS vary in a 13-km-long dispersion-shifted fiber as the injected CW power is increased from 0.5 to 50 mW [180]. No more than 3 mW could be transmitted through the fiber in this experiment after the *Brillouin threshold*. For a fiber with $A_{eff} = 50$ μm^2 and $\alpha = 0.2$ dB/km, the threshold power is below 2 mW when the fiber length is long enough (> 20 km) that L_{eff} can be replaced by $1/\alpha$.

The preceding estimate applies to CW signals as it neglects the effects of signal modulation resulting in a random sequence of 0 and 1 bits. In general, the Brillouin threshold depends on the modulation format as well as on the ratio of the bit rate to the Brillouin-gain bandwidth [181]. It increases to above 5 mW for lightwave systems operating near 10 Gb/s. Some applications require launch powers in excess of 10 mW.

Several schemes have been proposed for raising the Brillouin threshold [182]–[187]. They rely on increasing either the Brillouin-gain bandwidth Δv_B or the spectral width of optical carrier. The former has a value of about 20 MHz for silica fibers, while the latter is typically < 10 MHz for DFB lasers used in WDM systems. The bandwidth of an optical carrier can be increased without affecting the system performance by modulating its phase at a frequency much lower than the bit rate. Typically, the modulation frequency Δv_m is in the range of 200–400 MHz. As the effective Brillouin gain is reduced by a factor of $(1 + \Delta v_m / \Delta v_B)$, the SBS threshold increases by the same factor. Since typically $\Delta v_B \sim 20$ MHz, the launched power can be increased by more than a factor of 10 by this technique.

If the Brillouin-gain bandwidth Δv_B of the fiber itself can be increased from its nominal value of 20 MHz to more than 200 MHz, the SBS threshold can be increased without requiring a phase modulator. One technique uses sinusoidal strain along the fiber length for this purpose. The applied strain changes the Brillouin shift v_B by a few percent in a periodic manner. The resulting Brillouin-gain spectrum is much broader than that occurring for a fixed value of v_B. The strain can be applied during cabling of the fiber. In one fiber cable, Δv_B was found to increase from 50 to 400 MHz [182]. The Brillouin shift v_B can also be changed by making the core radius nonuniform along the fiber length since the longitudinal acoustic frequency depends on the core radius. The same effect can be realized by changing the dopant concentration along the fiber length. This technique increased the SBS threshold of one fiber by 7 dB [183]. A side effect of varying the core radius or the dopant concentration is that the GVD parameter β_2 also changes along the fiber length. It is possible to vary both of them simultaneously in such a way that β_2 remains relatively uniform [185]. Phase modulation induced by a supervisory channel through the nonlinear phenomenon of cross-phase modulation (XPM) can also be used to suppress SBS [187]. XPM induced by neighboring channels can also help [184] but it is hard to control and is itself a source of crosstalk. In practice, a frequency modulator integrated within the transmitter provides the best method for suppressing SBS. Threshold levels >200 mW have been realized with this technique [186].

8.3.5 Cross-Phase Modulation

The SPM and XPM both affect the performance of WDM systems. The effects of SPM has been discussed in Sections 5.3 and 7.7 in the context of single-channel systems; they also apply to individual channels of a WDM system. The phenomenon of XPM is an important mechanism of nonlinear crosstalk in WDM lightwave systems and has been studied extensively in this context [188]–[199].

As discussed in Section 2.6, XPM originates from the intensity dependence of the refractive index, which produces an intensity-dependent phase shift as the signal propagates through the optical fiber. The phase shift for a specific channel depends not only on the power of that channel but also on the power of other channels [59]. The total phase shift for the jth channel is given by (see Section 2.6)

$$\phi_j^{NL} = \frac{\gamma}{\alpha} \left(P_j + 2 \sum_{m \neq j}^{N} P_m \right), \qquad (8.3.14)$$

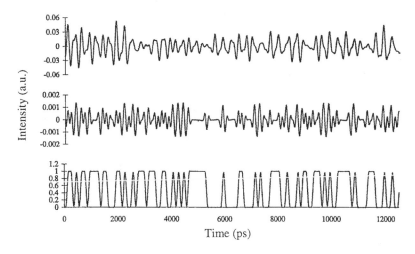

Figure 8.25: XPM-induced power fluctuations on a CW probe for a 130-km link (middle) and a 320-km link (top) with dispersion management. An NRZ bit stream in the pump channel is shown at the bottom. (After Ref. [191], ©1999 IEEE; reprinted with permission.)

where the first term is due to SPM and L_{eff} has been replaced with $1/\alpha$ assuming $\alpha L \gg 1$. The parameter γ is in the range 1–10 $W^{-1}km^{-1}$ depending on the type of fiber used, larger values occurring for dispersion-compensating fibers. The nonlinear phase shift depends on the bit pattern of various channels and can vary from zero to its maximum value $\phi_{max} = (\gamma/\alpha)(2N-1)P_j$ for N channels, if we assume equal channel powers.

Strictly speaking, the XPM-induced phase shift should not affect system performance if the GVD effects were negligible. However, any dispersion in fiber converts pattern-dependent phase shifts to power fluctuations, reducing the SNR at the receiver. This conversion can be understood by noting that time-dependent phase changes lead to frequency chirping that affects dispersion-induced broadening of the signal. Figure 8.25 shows XPM-induced fluctuations for a CW probe launched with a 10-Gb/s pump channel modulated using the NRZ format. The probe power fluctuates by as much as 6% after 320 km of dispersive fiber. The root-mean-square (RMS) value of fluctuations depends on the channel power and can be reduced by lowering it. As a rough estimate, if we use the condition $\phi_{max} < 1$, the channel power is restricted to

$$P_{ch} < \alpha/[\gamma(2N-1)]. \qquad (8.3.15)$$

For typical values of α and γ, P_{ch} should be below 10 mW even for five channels and reduces to below 1 mW for more than 50 channels.

The preceding analysis provides only a rough estimate as it ignores the fact that pulses belonging to different channels travel at different speeds and walk through each other at a rate that depends on their wavelength difference. Since XPM can occur only when pulses overlap in the time domain, its impact is reduced considerably by the walk-off effects. As a faster-moving pulse belonging to one channel collides with and passes through a specific pulse in another channel, the XPM-induced chirp shifts the pulse spectrum first toward red and then toward blue. In a lossless fiber, collisions of two pulses are perfectly symmetric, resulting in no net spectral shift at the end of the collision. In a loss-managed system with optical amplifiers placed periodically along the link, power variations make collisions between pulses of different channels asymmetric, resulting in a net frequency shift that depends on the channel spacing. Such frequency shifts lead to timing jitter (the speed of a channel depends on its frequency because of GVD) since their magnitude depends on the bit pattern as well as on the channel wavelengths. The combination of XPM-induced amplitude and timing jitter degrades the SNR at the receiver, especially for closely spaced channels, and leads to XPM-induced power penalty that depends on channel spacing and the type of fibers used for the WDM link. The power penalty increases for fibers with large GVD and for WDM systems designed with a small channel spacing and can exceed 5 dB even for 100-GHz spacing.

How can one control the XPM-induced crosstalk in WDM systems? Clearly, the use of low-GVD fibers will reduce this problem to some extent but is not practical because of the onset of FWM (see next subsection). In practice, dispersion management is employed in virtually all WDM systems such that the local dispersion is relatively large. Careful selection of the dispersion-map parameters may help from the XPM standpoint but may not be optimum from the SPM point of view [190]. A simple approach to XPM suppression consists of introducing relative time delays among the WDM channels after each map period such that the "1" bits in neighboring channels are unlikely to overlap most of the time [196]. The use of RZ format is quite helpful in this context because all 1 bits occupy only a fraction of the bit slot. In a 10-channel WDM experiment, time delays were introduced by using 10 fiber gratings spaced apart by varying distances chosen to enhance XPM suppression [198]. The BER floor observed after 500 km of transmission disappeared after the XPM suppressors (consisting of ten Bragg gratings) were inserted every 100 km. The residual power penalty at a BER of 10^{-10} was below 2 dB for all channels.

8.3.6 Four-Wave Mixing

As discussed in Section 2.6, the nonlinear phenomenon of FWM requires phase matching. It becomes a major source of nonlinear crosstalk whenever the channel spacing and fiber dispersion are small enough to satisfy the phase-matching condition approximately [59]. This is the case when a WDM system operates close to the zero-dispersion wavelength of dispersion-shifted fibers. For this reason, several techniques have been developed for reducing the impact of FWM in WDM systems [167].

The physical origin of FWM-induced crosstalk and the resulting system degradation can be understood by noting that FWM generates a new wave at the frequency $\omega_{ijk} = \omega_i + \omega_j - \omega_k$, whenever three waves at frequencies ω_i, ω_j, and ω_k copropagate inside the fiber. For an N-channel system, i, j, and k can vary from 1 to N, resulting in a large combination of new frequencies generated by FWM. In the case of equally spaced channels, the new frequencies coincide with the existing frequencies, leading to coherent in-band crosstalk. When channels are not equally spaced, most FWM components fall in between the channels and lead to incoherent out-of-band crosstalk. In both cases, the system performance is degraded because of a loss in the channel power, but the coherent crosstalk degrades system performance much more severely.

The FWM process in optical fibers is governed by a set of four coupled equations whose general solution requires a numerical approach [59]. If we neglect the phase shifts induced by SPM and XPM, assume that the three channels participating in the FWM process remain nearly undepleted, and include fiber losses, the amplitude A_F of the FWM component at the frequency ω_F is governed by

$$\frac{dA_F}{dz} = -\frac{\alpha}{2}A_F + d_F \gamma A_i A_j A_k^* \exp(-i\Delta kz), \qquad (8.3.16)$$

where $A_m(z) = A_m(0)\exp(-\alpha z/2)$ for $m = i, j, k$ and $d_F = 2 - \delta_{ij}$ is the degeneracy factor defined such that its value is 1 when $i = j$ but doubles when $i \neq j$. This equation can be easily integrated to obtain $A_F(z)$. The power transferred to the FWM component in a fiber of length L is given by [200]

$$P_F = |A_F(L)|^2 = \eta_F (d_F \gamma L)^2 P_i P_j P_k e^{-\alpha L}, \qquad (8.3.17)$$

where $P_m = |A_m(0)|^2$ is the launched power in the mth channel and η_F is a measure of the FWM efficiency defined as

$$\eta_F = \left| \frac{1 - \exp[-(\alpha + i\Delta k)L]}{(\alpha + i\Delta k)L} \right|^2. \qquad (8.3.18)$$

The FWM efficiency η_F depends on the channel spacing through the phase mismatch governed by

$$\Delta k = \beta_F + \beta_k - \beta_i - \beta_j \approx \beta_2(\omega_i - \omega_k)(\omega_j - \omega_k), \qquad (8.3.19)$$

where the propagation constants were expanded in a Taylor series around $\omega_c = (\omega_i + \omega_j)/2$ and β_2 is the GVD parameter at that frequency. If the GVD of the transmission fiber is relatively large, ($|\beta_2| > 5$ ps^2/km), η_F nearly vanishes for typical channel spacings of 50 GHz or more. In contrast, $\eta_F \approx 1$ close to the zero-dispersion wavelength of the fiber, resulting in considerable power in the FWM component, especially at high channel powers. In the case of equal channel powers, P_F increases as P_{ch}^3. This cubic dependence of the FWM component limits the channel powers to below 1 mW if FWM is nearly phase matched. Since the number of FWM components for an M-channel WDM system increases as $M^2(M - 1)/2$, the total power in all FWM components can be quite large.

A simple scheme for reducing the FWM-induced degradation consists of designing WDM systems with unequal channel spacings [167]. The main impact of FWM in this case is to reduce the channel power. This power depletion results in a power penalty that is relatively small compared with the case of equal channel spacings. Experimental measurements on WDM systems confirm the advantage of unequal channel spacings. In a 1999 experiment, this technique was used to transmit 22 channels, each operating at 10 Gb/s, over 320 km of dispersion-shifted fiber with 80-km amplifier spacing [201]. Channel spacings ranged from 125 to 275 GHz in the 1532- to 1562-nm wavelength region and were determined using a periodic allocation scheme [202]. The zero-dispersion wavelength of the fiber was close to 1548 nm, resulting in near phase matching of many FWM components. Nonetheless, the system performed quite well with less than 1.5-dB power penalty for all channels.

The use of a nonuniform channel spacing is not always practical because many WDM components, such as optical filters and waveguide-grating routers, require equal channel spacings. A practical solution is offered by the periodic dispersion-management technique discussed in Section 7.8. In this case, fibers with normal and anomalous GVD are combined to form a dispersion map such that GVD is high locally all along the fiber even though its average value is quite low. As a result, the FWM efficiency η_F is negligible throughout the fiber, resulting in little FWM-induced crosstalk. The use of dispersion management is common for suppressing FWM in WDM systems because of its practical simplicity. In fact, new kinds of fibers, called nonzero-dispersion-shifted fibers (NZDSFs), were designed and marketed after the advent of WDM systems. Typically, GVD is in the range of 4–8 ps/(km-nm) in such fibers to ensure that the FWM-induced crosstalk is minimized.

8.3.7 Other Design Issues

The design of WDM communication systems requires careful consideration of many transmitter and receiver characteristics. An important issue concerns the stability of the carrier frequency (or wavelength) associated with each channel. The frequency of light emitted from DFB or DBR semiconductor lasers can change considerably because of changes in the operating temperature (~ 10 GHz/$^\circ$C). Similar changes can also occur with the aging of lasers [203]. Such frequency changes are generally not of concern for single-channel systems. In the case of WDM lightwave systems it is important that the carrier frequencies of all channels remain stable, at least relatively, so that the channel spacing does not fluctuate with time.

A number of techniques have been used for frequency stabilization [204]–[209]. A common technique uses *electrical feedback* provided by a frequency discriminator using an atomic or molecular resonance to lock the laser frequency to the resonance frequency. For example, one can use ammonia, krypton, or acetylene for semiconductor lasers operating in the 1.55-μm region, as all three have resonances near that wavelength. Frequency stability to within 1 MHz can be achieved by this technique. Another technique makes use of the *optogalvanic effect* to lock the laser frequency to an atomic or molecular resonance. A phase-locked loop can also be used for frequency stabilization. In another scheme, a Michelson interferometer, calibrated by using a frequency-stabilized master DFB laser, provides a set of equally spaced reference frequencies [205]. A FP filter, an AWG, or any other filter with a comb-like periodic transmission spectrum can also be used for this purpose because it provides a reference set of equally spaced frequencies [206]. A fiber grating is useful for frequency stabilization but a separate grating is needed for each channel as its reflection spectrum is not periodic [207]. A frequency-dithered technique in combination with an AWG and an amplitude modulator can stabilize the channel frequency to within 0.3 GHz [209].

An important issue in the design of WDM networks is related to the loss of signal power that occurs because of insertion, distribution, and transmission losses. Optical amplifiers are used to compensate for such losses but not all channels are amplified by the same factor unless the gain spectrum is flat over the entire bandwidth of the WDM signal. Although gain-flattening techniques are commonly employed, channel powers can still deviate by 10 dB or more when the WDM signal passes through many optical amplifiers before being detected. It may then become necessary to control the power of individual channels (through selective attenuation) at each node within a WDM network to make the channel powers nearly uniform. The issue of power management in WDM networks is quite complex and requires attention to many details [210]–[212]. The buildup of amplifier noise can also become a limiting factor when the WDM signal passes through a large number of amplifiers.

Another major issue in the design of WDM systems concerns dispersion management. As discussed in Chapter 7, dispersion-management techniques are commonly used for WDM networks. However, in a reconfigurable network the exact path of a WDM channel can change in a dynamic fashion. Such networks will require compensation of residual dispersion at individual nodes. Network management is an active area of research and requires attention to many details [213].

8.4 Time-Division Multiplexing

As discussed in Section 1.2, TDM is commonly performed in the electrical domain to obtain digital hierarchies for telecommunication systems. In this sense, even single-channel lightwave systems carry multiple TDM channels. The electrical TDM becomes difficult to implement at bit rates above 10 Gb/s because of the limitations imposed by high-speed electronics. A solution is offered by the *optical* TDM (OTDM), a scheme that can increase the bit rate of a single optical carrier to values above 1 Tb/s. The OTDM technique was studied extensively during the 1990s [214]–[219]. Its commercial deployment requires new types of optical transmitters and receivers based on all-optical multiplexing and demultiplexing techniques. In this section we first discuss these new techniques and then focus on the design and performance issues related to OTDM lightwave systems.

8.4.1 Channel Multiplexing

In OTDM lightwave systems, several optical signals at a bit rate B share the same carrier frequency and are multiplexed optically to form a composite bit stream at the bit rate NB, where N is the number of channels. Several multiplexing techniques can be used for this purpose [219]. Figure 8.26 shows the design of an OTDM transmitter based on the delay-line technique. It requires a laser capable of generating a periodic pulse train at the repetition rate equal to the single-channel bit rate B. Moreover, the laser should produce pulses of width T_p such that $T_p < T_B = (NB)^{-1}$ to ensure that each pulse will fit within its allocated time slot T_B. The laser output is split equally into N branches, after amplification if necessary. A modulator in each branch blocks the pulses representing 0 bits and creates N independent bit streams at the bit rate B.

Multiplexing of N bit streams is achieved by a delay technique that can be implemented optically in a simple manner. In this scheme, the bit stream in the nth branch is delayed by an amount $(n-1)/(NB)$, where $n = 1, \ldots, N$. The output of all branches is

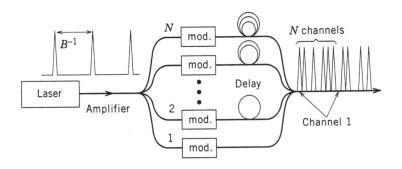

Figure 8.26: Design of an OTDM transmitter based on optical delay lines.

then combined to form a composite signal. It should be clear that the multiplexed bit stream produced using such a scheme has a bit slot corresponding to the bit rate NB. Furthermore, N consecutive bits in each interval of duration B^{-1} belong to N different channels, as required by the TDM scheme (see Section 1.2).

The entire OTDM multiplexer (except for modulators which require LiNbO$_3$ or semiconductor waveguides) can be built using single-mode fibers. Splitting and recombining of signals in N branches can be accomplished with $1 \times N$ fused fiber couplers. The optical delay lines can be implemented using fiber segments of controlled lengths. As an example, a 1-mm length difference introduces a delay of about 5 ps. Note that the delay lines can be relatively long (10 cm or more) because only the length difference has to be matched precisely. For a precision of 0.1 ps, typically required for a 40-Gb/s OTDM signal, the delay lengths should be controlled to within 20 μm. Such precision is hard to realize using optical fibers.

An alternative approach makes use of planar lightwave circuits fabricated using the silica-on-silicon technology [41]–[45]. Such devices can be made polarization insensitive while providing a precise control of the delay lengths. However, the entire multiplexer cannot be built in the form of a planar lightwave circuit as modulators cannot be integrated with this technology. A simple approach consists of inserting an InP chip containing an array of electroabsorption modulators in between the silica waveguides that are used for splitting, delaying and combining the multiple channels (see Fig. 8.26). The main problem with this approach is the spot-size mismatch as the optical signal passes from Si to InP waveguide (and vice versa). This problem can be solved by integrating spot-size converters with the modulators. Such an integrated OTDM multiplexer was used in a 160-Gb/s experiment in which 16 channels, each operating at 10 Gb/s were multiplexed [218].

An important difference between the OTDM and WDM techniques should be apparent from Fig. 8.26: The OTDM technique requires the use of the RZ format (see Section 1.2.3). In this respect, OTDM is similar to soliton systems (covered in Chapter 9), which must also use the RZ format. Historically, the NRZ format used before the advent of lightwave technology was retained even for optical communication systems. Starting in the late 1990s, the RZ format began to appear in dispersion-managed WDM systems in the form of CRZ format. The use of OTDM requires optical sources emitting a train of short optical pulses at a repetition rate as high as 40 GHz. Two types of lasers are commonly used for this purpose [219]. In one approach, gain switching or mode locking of a semiconductor laser provides 10–20 ps pulses at a high repetition rate, which can be compressed using a variety of techniques [40]. In another approach, a fiber laser is harmonically mode locked using an intracavity LiNbO$_3$ modulator [40]. Such lasers can provide pulse widths ~1 ps at a repetition rate of up to 40 GHz. More details on short-pulse transmitters are given in Section 9.2.4.

8.4.2 Channel Demultiplexing

Demultiplexing of individual channels from an OTDM signal requires electro-optic or all-optical techniques. Several schemes have been developed, each having its own merits and drawbacks [216]–[220]. Figure 8.27 shows three schemes discussed in this section. All demultiplexing techniques require a *clock signal*—a periodic pulse train at the single-channel bit rate. The clock signal is in the electric form for electro-optic demultiplexing but consists of an optical pulse train for all-optical demultiplexing.

The electro-optic technique uses several MZ-type LiNbO$_3$ modulators in series. Each modulator halves the bit rate by rejecting alternate bits in the incoming signal. Thus, an 8-channel OTDM system requires three modulators, driven by the same electrical clock signal (see Fig. 8.27), but with different voltages equal to $4V_0$, $2V_0$, and V_0, where V_0 is the voltage required for π phase shift in one arm of the MZ interferometer. Different channels can be selected by changing the phase of the clock signal. The main advantage of this technique is that it uses commercially available components. However, it has several disadvantages, the most important being that it is limited by the speed of modulators. The electro-optic technique also requires a large number of expensive components, some of which need high drive voltage.

Several all-optical techniques make use of a *nonlinear optical-loop mirror* (NOLM) constructed using a fiber loop whose ends are connected to the two output ports of a 3-dB fiber coupler as shown in Fig. 8.27(b). Such a device is also referred to as the Sagnac interferometer. The NOLM is called a mirror because it reflects its input entirely when the counterpropagating waves experience the same phase shift over one round trip. However, if the symmetry is broken by introducing a relative phase shift of π between them, the signal is fully transmitted by the NOLM. The demultiplexing operation of an NOLM is based on the XPM [59], the same nonlinear phenomenon that can lead to crosstalk in WDM systems.

Demultiplexing of an OTDM signal by an NOLM can be understood as follows. The clock signal consisting of a train of optical pulses at the single-channel bit rate is injected into the loop such that it propagates only in the clockwise direction. The OTDM signal enters the NOLM after being equally split into counterpropagating directions by the 3-dB coupler. The clock signal introduces a phase shift through XPM for pulses belonging to a specific channel within the OTDM signal. In the simplest case, optical fiber itself introduces XPM. The power of the optical signal and the loop length are made large enough to introduce a relative phase shift of π. As a result, a single channel is demultiplexed by the NOLM. In this sense, a NOLM is the TDM counterpart of the WDM add–drop multiplexers discussed in Section 8.2.3. All channels can be demultiplexed simultaneously by using several NOLMs in parallel [220]. Fiber nonlinearity is fast enough that such a device can respond at femtosecond time scales. De-

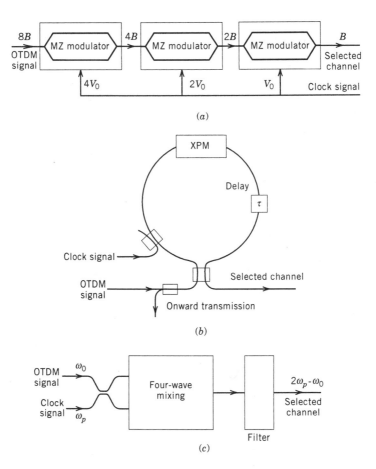

Figure 8.27: Demultiplexing schemes for OTDM signals based on (a) cascaded LiNbO$_3$ modulators, (b) XPM in a nonlinear optical-loop mirror, and (c) FWM in a nonlinear medium.

multiplexing of a 6.3-Gb/s channel from a 100-Gb/s OTDM signal was demonstrated in 1993. By 1998, the NOLM was used to demultiplex a 640-Gb/s OTDM signal [221].

The third scheme for demultiplexing in Fig. 8.27 makes use of FWM in a nonlinear medium and works in a way similar to the wavelength-conversion scheme discussed in Section 8.2.5. The OTDM signal is launched together with the clock signal (at a different wavelength) into a nonlinear medium. The clock signal plays the role of the pump for the FWM process. In time slots in which a clock pulse overlaps with the 1 bit of the channel that needs to be demultiplexed, FWM produces a pulse at the new wavelength. As a result, the pulse train at this new wavelength is an exact replica of the channel that needs to be demultiplexed. An optical filter is used to separate the demultiplexed channel from the OTDM and clock signals. A polarization-preserving fiber is often used as the nonlinear medium for FWM because of the ultrafast nature of its nonlinearity and its ability to preserve the state of polarization despite environmental fluctuations. As early as 1996, error-free demultiplexing of 10-Gb/s channels from

a 500-Gb/s OTDM signal was demonstrated by using clock pulses of about 1 ps duration [222]. This scheme can also amplify the demultiplexed channel (by up to 40 dB)
through parametric amplification inside the same fiber [223].

The main limitation of a fiber-based demultiplexer stems from the weak fiber nonlinearity. Typically, fiber length should be 5 km or more for the device to function at
practical power levels of the clock signal. This problem can be solved in two ways.
In one approach, the required fiber length is reduced by up to a factor of 10 by using
special fibers designed such that the nonlinear parameter γ is enhanced because of a
reduced spot size of the fiber mode [223]. Alternatively, a different nonlinear medium
can be used in place of the optical fiber. The nonlinear medium of choice in practice is
the SOA. Both the XPM and FWM schemes have been shown to work using SOAs. In
the case of a NOLM, an SOA is inserted within the fiber loop. The XPM-induced phase
shift occurs because of changes in the refractive index induced by the clock pulses as
they saturate the SOA gain (similar to the wavelength-conversion scheme discussed
earlier). As the phase shift occurs selectively only for the data bits belonging to a specific channel, that channel is demultiplexed. The refractive-index change induced by
the SOA is large enough that a relative phase shift of π can be induced at moderate
power levels by an SOA of $<$1-mm length.

The main limitation of an SOA results from its relatively slow temporal response
governed by the carrier lifetime (~ 1 ns). By injecting a CW signal with the clock
signal (at a different wavelength), the carrier lifetime can be reduced to below 100 ps.
Such demultiplexers can work at 10 Gb/s. Even faster response can be realized by
using a gating scheme. For example, by placing an SOA asymmetrically within the
NOLM such that the counterpropagating signals enter the SOA at different times, the
device can be made to respond at a time scale ~ 1 ps. Such a device is referred to
as the terahertz optical asymmetrical demultiplexer (TOAD). Its operation at bit rates
as high as 250 Gb/s was demonstrated by 1994 [224]. A MZ interferometer with two
SOAs in its two branches (see Fig. 8.20) can also demultiplex an OTDM signal at
high speeds and can be fabricated in the form an integrated compact chip using the
InGaAsP/InP technology [125]. The silica-on-silicon technology has also been used
to make a compact MZ demultiplexer in a symmetric configuration that was capable
of demultiplexing a 168-Gb/s signal [217]. If the SOAs are placed in an asymmetric
fashion, the device operates similar to a TOAD device. Figure 8.28(a) shows such a
MZ device fabricated with the InGaAsP/InP technology [225]. The offset between the
two SOAs plays a critical role in this device and is typically $<$1 mm.

The operating principle behind the MZ-TOAD device can be understood from
Fig. 8.28. The clock signal (control) enters from port 3 of the MZ interferometer and
is split into two branches. It enters the SOA1 first, saturates its gain, and opens the MZ
switch through XPM-induced phase shift. A few picoseconds later, the SOA2 is saturated by the clock signal. The resulting phase shift closes the MZ switch. The duration
of the switching window can be precisely controlled by the relative location of the two
SOAs as shown in Fig. 8.28(b). Such a device is not limited by the carrier lifetime and
can operate at high bit rates when designed properly.

Demultiplexing of an OTDM signal requires the recovery of a clock signal from the
OTDM signal itself. An all-optical scheme is needed because of the high bit rates associated with OTDM signals. An optical phase-locked loop based on the FWM process

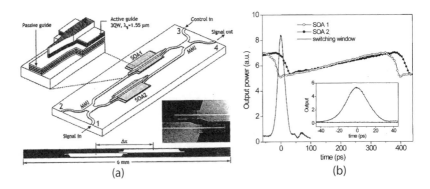

Figure 8.28: (a) Mach–Zehnder TOAD demultiplexer with two SOAs placed asymmetrically. Insets show the device structure. (b) Gain variations inside two SOAs and the resulting switching window. (After Ref. [225]; ©2001 IEEE; reprinted with permission.)

is commonly used for this purpose. Schemes based on a NOLM or an injection-locked laser can also be used [219]. Self-pulsing semiconductor lasers as well as mode-locked fiber lasers have been used for injection locking.

8.4.3 System Performance

The transmission distance L of OTDM signals is limited in practice by fiber dispersion because of the use of short optical pulses (~ 1 ps) dictated by relatively high bit rates. In fact, an OTDM signal carrying N channels at the bit rate B is equivalent to transmitting a single channel at the composite bit rate of NB, and the bit rate–distance product NBL is restricted by the dispersion limits found in Sections 2.4.3. As an example, it is evident from Fig. 2.13 that a 200-Gb/s system is limited to $L < 50$ km even when the system is designed to operate exactly at the zero-dispersion wavelength of the fiber. Thus, OTDM systems require not only dispersion-shifted fibers but also the use of dispersion-management techniques capable of reducing the impact of both the second- and third-order dispersive effects. Even then, PMD becomes a limiting factor for long fiber lengths and its compensation is often necessary. The intrachannel nonlinear effects also limit the performance of OTDM systems; the use of soliton-like pulses is often necessary for OTDM systems [217].

In spite of the difficulties inherent in propagation of single-carrier OTDM systems operating at bit rates exceeding 100 Gb/s, many laboratory experiments have realized high-speed transmission using the OTDM technique [219]. In a 1996 experiment, a 100-Gb/s OTDM signal consisting of 16 channels at 6.3 Gb/s was transmitted over 560 km by using optical amplifiers (80-km spacing) together with dispersion management. The laser source in this experiment was a mode-locked fiber laser producing 3.5-ps pulses at a repetition rate of 6.3 GHz (the bit rate of each multiplexed channel). A multiplexing scheme similar to that shown in Fig. 7.26 was used to generate the 100-Gb/s OTDM signal. The total bit rate was later extended to 400 Gb/s (forty 10-Gb/s channels) by using a supercontinuum pulse source producing 1-ps pulses [226]. Such short pulses are needed since the bit slot is only 2.5-ps wide at 400 Gb/s. It was necessary to compensate for the dispersion slope (third-order dispersion β_3) as 1-ps pulses were severely distorted and exhibited oscillatory tails extending to beyond 5 ps (typical characteristic of the third-order dispersion) in the absence of such compensation. Even then, the transmission distance was limited to 40 km. In a 2000 experiment, a 1.28-Tb/s ODTM signal could be transmitted over 70 km but it required compensation of second-, third-, and fourth-order dispersion simultaneously [227]. In a 2001 field trial, the bit rate of the OTDM system was limited to only 160 Gb/s but the signal was transmitted over 116 km using a standard two-fiber dispersion map [228].

A simple method for realizing high bit rates exceeding 1 Tb/s consists of combining the OTDM and WDM techniques. For example, a WDM signal consisting of M separate optical carriers such that each carrier carries N OTDM channels at the bit rate B has the total capacity $B_{tot} = MNB$. The dispersion limitations of such a system are set by the OTDM-signal bit rate of NB. In a 1999 experiment, this approach was used to realize a total capacity of 3 Tb/s by using $M = 19$, $N = 16$, and $B = 10$ Gb/s [219]. The channels were spaced 450 GHz apart (about 3.6 nm) to avoid overlap between neighboring WDM channels at the 160-Gb/s bit rate. The 70-nm WDM signal occupied both the C and L bands. The total capacity of such OTDM/WDM systems can exceed 10 Tb/s if the S band is also used although many factors such as various nonlinear effects in fibers and the practicality of dispersion compensation over a wide bandwidth are likely to limit the system performance.

OTDM has also been used for designing transparent optical networks capable of connecting multiple nodes for random bidirectional access [215]. Its use is especially practical for packet-based networks employing the ATM and TCP/IP protocols. Similar to the case of WDM networks, both single-hop and multihop architectures have been considered. Single-hop OTDM networks use passive star couplers to distribute the signal from one node to all other nodes. In contrast, multihop OTDM networks require signal processing at each node to route the traffic. A packet-switching technique is commonly used for such networks. Considerable effort was under way in 2001 for developing packet-switched OTDM networks [229]. Their implementation requires several new all-optical components for storage, compression, and decompression of individual packets [230].

8.5 Subcarrier Multiplexing

In some LAN and MAN applications the bit rate of each channel should be relatively low but the number of channels can become quite large. An example is provided by common-antenna (cable) television (CATV) networks that have used historically electrical communication techniques. The basic concept behind *subcarrier multiplexing* (SCM) is borrowed from microwave technology, which employs multiple microwave carriers for transmission of multiple channels (electrical FDM) over coaxial cables or free space. The total bandwidth is limited to well below 1 GHz when coaxial cables are used to transmit a multichannel microwave signal. However, if the microwave signal is transmitted optically by using optical fibers, the signal bandwidth can easily exceed 10 GHz for a single optical carrier. Such a scheme is referred to as SCM, since multiplexing is done by using microwave subcarriers rather than the optical carrier. It has

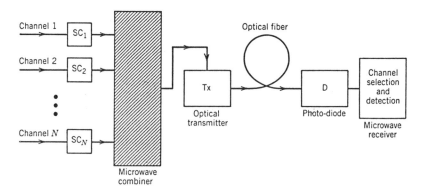

Figure 8.29: Schematic illustration of subcarrier multiplexing. Multiple microwave subcarriers (SC) are modulated, and the composite electrical signal is used to modulate an optical carrier at the transmitter (Tx).

been used commercially by the CATV industry since 1992 and can be combined with TDM or WDM. A combination of SCM and WDM can realize bandwidths in excess of 1 THz.

Figure 8.29 shows schematically a SCM lightwave system designed with a single optical carrier. The main advantage of SCM is the flexibility and the upgradability offered by it in the design of broadband networks. One can use analog or digital modulation, or a combination of the two, to transmit multiple voice, data, and video signals to a large number of users. Each user can be served by a single subcarrier, or the multichannel signal can be distributed to all users as done commonly by the CATV industry. The SCM technique has been studied extensively because of its wide-ranging practical applications [231]–[234]. In this section we describe both the analog and digital SCM systems with emphasis on their design and performance.

8.5.1 Analog SCM Systems

This book focuses mostly on digital modulation techniques as they are employed almost universally for lightwave systems. An exception occurs in the case of SCM systems designed for video distribution. Most CATV networks distribute television channels by using analog techniques based on frequency modulation (FM) or amplitude modulation with vestigial sideband (AM-VSB) formats [232]. As the wave form of an analog signal must be preserved during transmission, analog SCM systems require a high SNR at the receiver and impose strict linearity requirements on the optical source and the communication channel.

In analog SCM lightwave systems, each microwave subcarrier is modulated using an analog format, and the output of all subcarriers is summed using a microwave power combiner (see Fig. 8.29). The composite signal is used to modulate the intensity of a semiconductor laser directly by adding it to the bias current. The transmitted power can be written as

$$P(t) = P_b \left[1 + \sum_{j=1}^{N} m_j a_j \cos(2\pi f_j t + \phi_j) \right], \qquad (8.5.1)$$

where P_b is the output power at the bias level and m_j, a_j, f_j, and ϕ_j are, respectively, the modulation index, amplitude, frequency, and phase associated with the jth microwave subcarrier; a_j, f_j, or ϕ_j is modulated to impose the signal depending on whether AM, FM, or phase modulation (PM) is used.

The power at the receiver would also be in the form of Eq. (8.5.1) if the communication channel were perfectly linear. In practice, the analog signal is distorted during its transmission through the fiber link. The distortion is referred to as *intermodulation distortion* (IMD) and is similar in nature to the FWM distortion discussed in Section 8.3. Any nonlinearity in the response of the semiconductor laser used inside the optical transmitter or in the propagation characteristics of fibers generates new frequencies of the form $f_i + f_j$ and $f_i + f_j \pm f_k$, some of which lie within the transmission bandwidth and distort the analog signal. The new frequencies are referred to as the *intermodulation products* (IMPs). These are further subdivided as two-tone IMPs and triple-beat IMPs, depending on whether two frequencies coincide or all three frequencies are distinct. The triple-beat IMPs tend to be a major source of distortion because of their large number. An N-channel SCM system generates $N(N-1)(N-2)/2$ triple-beat terms compared with $N(N-1)$ two-tone terms. The second-order IMD must also be considered if subcarriers occupy a large bandwidth.

IMD has its origin in several distinct nonlinear mechanisms. The dynamic response of semiconductor lasers is governed by the rate equations (see Section 3.5), which are intrinsically nonlinear. The solution of these equations provides expressions for the second- and third-order IMPs originating from this intrinsic nonlinearity. Their contribution is largest whenever the IMP frequency falls near the relaxation-oscillation frequency. A second source of IMD is the nonlinearity of the power-current curve (see Fig. 3.20). The magnitude of resulting IMPs can be calculated by expanding the output power in a Taylor series around the bias power [232]. Several other mechanisms, such as fiber dispersion, frequency chirp, and mode-partition noise can cause IMD, and their impact on the SCM systems has been studied extensively [235].

The IMD-induced degradation of the system performance depends on the inter-channel interference created by IMPs. Depending on the channel spacing among microwave subcarriers, some of the IMPs fall within the bandwidth of a specific channel and affect the signal recovery. It is common to introduce composite second-order (CSO) and composite triple-beat (CTB) distortion by adding the power for all IMPs that fall within the passband of a specific channel [232]. The CSO and CTB distortion values are normalized to the carrier power of that channel and expressed in dBc units, where the "c" in dBc denotes normalization with respect to the carrier power. Typically, CSO and CTB distortion values should be below -60 dBc for negligible impact on the system performance; both of them increase rapidly with an increase in the modulation index.

System performance depends on the SNR associated with the demodulated signal. In the case of SCM systems, the *carrier-to-noise ratio* (CNR) is often used in place of SNR. The CNR is defined as the ratio of RMS carrier power to RMS noise power at the receiver and can be written as

$$\text{CNR} = \frac{(mR\bar{P})^2/2}{\sigma_s^2 + \sigma_T^2 + \sigma_I^2 + \sigma_{\text{IMD}}^2}, \tag{8.5.2}$$

where m is the modulation index, R is the detector responsivity, $\bar{P}$ is the average received optical power, and σ_s, σ_T, σ_I, and σ_{IMD} are the RMS values of the noise currents associated with the shot noise, thermal noise, intensity noise, and IMD, respectively. The expressions for σ_s^2 and σ_T^2 are given in Section 4.4.1. The RMS value σ_I of the intensity noise can be obtained from Eq. (4.6.6) in Section 4.6.2. If we assume that the relative intensity noise (RIN) of the laser is nearly uniform within the receiver bandwidth,

$$\sigma_I^2 = (\text{RIN})(R\bar{P})^2(2\Delta f). \tag{8.5.3}$$

The RMS value of σ_{IMD} depends on the CSO and CTB distortion values.

The CNR requirements of SCM systems depend on the modulation format. In the case of AM-VSB format, the CNR should typically exceed 50 dB for satisfactory performance. Such large values can be realized only by increasing the received optical power $\bar{P}$ to a relatively large value (> 0.1 mW). This requirement has two effects. First, the power budget of AM-analog SCM systems is extremely limited unless the transmitter power is increased above 10 mW. Second, the intensity-noise contribution to the receiver noise dominates the system performance as σ_I^2 increases quadratically with $\bar{P}$. In fact, the CNR becomes independent of the received optical power when σ_I dominates. From Eqs. (8.5.2) and (8.5.3) the limited value of CNR is given by

$$\text{CNR} \approx \frac{m^2}{4(\text{RIN})\Delta f}. \tag{8.5.4}$$

As an example, the RIN of the transmitter laser should be below -150 dB/Hz to realize a CNR of 50 dB if $m = 0.1$ and $\Delta f = 50$ MHz are used as the representative values. Larger values of RIN can be tolerated only by increasing the modulation index m or by decreasing the receiver bandwidth. Indeed, DFB lasers with low values of the RIN were developed during the 1990s for CATV applications. In general, the DFB laser is biased high above threshold to provide a bias power P_b in excess of 5 mW because its RIN decreases as P_b^{-3}. High values of the bias power also permit an increase in the modulation index m.

The intensity noise can become a problem even when the transmitter laser is selected with a low RIN value to provide a large CNR in accordance with Eq. (8.5.4). The reason is that the RIN can be enhanced during signal transmission inside optical fibers. One such mechanism is related to multiple reflections between two reflecting surfaces along the fiber link. As discussed in Section 5.4.5, the two reflective surfaces act as an FP interferometer which converts the laser-frequency noise into intensity noise. The reflection-induced RIN depends on both the laser linewidth and the spacing between reflecting surfaces. It can be avoided by using fiber components (splices and connectors) with negligible parasitic reflections (< -40 dB) and by using lasers with a narrow linewidth (< 1 MHz). Another mechanism for the RIN enhancement is provided by the dispersive fiber itself. Because of GVD, different frequency components travel at slightly different speeds. As a result, frequency fluctuations are converted into intensity fluctuations during signal transmission. The dispersion-induced RIN depends on laser linewidth and increases quadratically with fiber length. Fiber dispersion also enhances CSO and CTB distortion for long link lengths [232]. It becomes necessary to use dispersion-management techniques for such SCM systems. In a 1996 experiment, the use of a chirped fiber grating for dispersion compensation reduced the RIN by more than 30 dB for fiber spans of 30 and 60 km [236]. Of course, other compensation techniques such as optical phase conjugation can also be used [237].

The CNR requirement can be relaxed by changing the modulation format from AM to FM. The bandwidth of a FM subcarrier is considerably larger (30 MHz in place of 4 MHz). However, the required CNR at the receiver is much lower (about 16 dB in place of 50 dB) because of the so-called FM advantage that yields a studio-quality video signal (> 50-dB SNR) with only 16-dB CNR. As a result, the optical power needed at the receiver can be as small as 10 μW. The RIN is not much of a problem for such systems as long as the RIN value is below -135 dB/Hz. In fact, the receiver noise of FM systems is generally dominated by the thermal noise. Both the AM and FM techniques have been used successfully for analog SCM lightwave systems [232]. The number of channels for AM systems is often limited by the clipping noise occurring when the modulated signal drops below the laser threshold [238].

8.5.2 Digital SCM Systems

During the 1990s, the emphasis of SCM systems shifted from analog to digital modulation. The frequency-shift keying (FSK) format was used for modulating microwave subcarriers [231] as early as 1990 but its use requires coherent detection techniques (see Chapter 10). Moreover, a single digital video channel requires a bit rate of more 100 Mb/s or more in contrast with the analog channel that occupies a bandwidth of only about 6 MHz. For this reason, other modulation formats such as quadrature AM (called QAM), carrierless AM/PM, and quadrature PSK have been explored. A common technique uses a multilevel QAM format. If M represents the number of discrete levels used, the resulting nonbinary digital signal is called M-ary because each bit can have M possible amplitudes (typically $M = 64$). Such a signal can be recovered at the receiver without using coherent detection and requires a lower CNR compared with that needed for analog AM-VSB systems. The capacity of an SCM system can be increased considerably by employing hybrid techniques that mix analog and digital formats [233].

To produce the QAM format from a binary bit stream, two or more neighboring bits are combined together to form a multilevel signal at a reduced bit rate. For example, if 2 bits are combined in pairs, one obtains a bit stream at the half bit rate but each symbol represents four possible combinations 00, 01, 10, 11. To distinguish between 01 and 10, the signal phase should be modified. This forces one to consider both quadratures of the microwave subcarrier (hence the name QAM). More specifically, the jth combination is represented as

$$s_j(t) = c_j \cos(\omega_c t + \theta_j) \equiv a_j \cos(\omega_c t) + b_j \sin(\omega_c t), \quad (8.5.5)$$

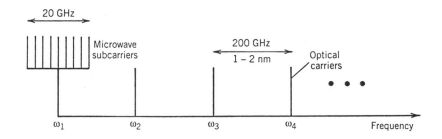

Figure 8.30: Frequency allocation in a multiwavelength SCM network.

where $j = 1$ to 4 and ω_c is the frequency of the subcarrier. As an example, θ_j can be $0, \pi/2, \pi$, and $3\pi/2$ for the four combinations. The main point is that the four combinations of the 2 bits are represented by four pairs of numbers of the form (a_j, b_j), where a_j and b_j have three possible values $(-1, 0, 1)$. The same idea can be extended to combine more than 2 bits. The combination of m bits yields the QAM format with $M = 2^m$ levels. It is common to refer such systems as M-QAM SCM systems.

The hybrid SCM systems that combine the analog AM-VSB format with the digital M-QAM format have attracted considerable attention because they can transmit a large number of video channels over the same fiber simultaneously [233]. The performance of such systems is affected by the clipping noise, multiple optical reflections, and the nonlinear mechanisms such as self-phase modulation (SPM) and SBS, all of which limit the total power and the number of channels that can be multiplexed. Nevertheless, hybrid SCM systems can transport up to 80 analog and 30 digital channels using a single optical transmitter. If only QAM format is employed, the number of digital channels is limited to about 80. In a 2000 experiment, 78 channels with the 64-QAM format were transmitted over 740 km [234]. Each channel had a bit rate of 30 Mb/s, resulting in a total capacity of 2.34 Gb/s. Such a SCM system can transport up to 500 compressed video channels. Further increase in the system capacity can be realized by combining the SCM and WDM techniques, a topic discussed next.

8.5.3 Multiwavelength SCM Systems

The combination of WDM and SCM provides the potential of designing broadband passive optical networks capable of providing integrated services (audio, video, data, etc.) to a large number of subscribers [239]–[243]. In this scheme, shown schematically in Fig. 8.30, multiple optical carriers are launched into the same optical fiber through the WDM technique. Each optical carrier carries multiple SCM channels using several microwave subcarriers. One can mix analog and digital signals using different subcarriers or different optical carriers. Such networks are extremely flexible and easy to upgrade as the demand grows. As early as 1990, 16 DFB lasers with a wavelength spacing of 2 nm in the 1.55-μm region were modulated with 100 analog video channels and six 622-Mb/s digital channels [240]. Video channels were multiplexed using the SCM technique such that one DFB laser carried 10 SCM channels over the bandwidth 300–700 MHz. The ultimate potential of such WDM systems was demonstrated in a

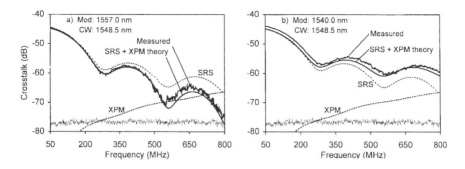

Figure 8.31: Predicted and measured crosstalk acquired over 25 km of fiber at 11-mW average power. The CW laser acts as a probe and its wavelength is (a) lower or (b) higher by 8.5 nm than the signal wavelength. (After Ref. [245]; ©1999 IEEE; reprinted with permission.)

2000 experiment in which a broadcast-and-select network was capable of delivering 10,000 channels, each operating at 20 Gb/s [242]. The network used 32 wavelengths (on the ITU grid) each of which could carry 310 microwave subcarriers by modulating at a composite bit rate of 20 Gb/s.

The limiting factor for multiwavelength SCM networks is interchannel crosstalk resulting from both the linear and nonlinear processes [244]–[246]. The nonlinear effects that produce interchannel crosstalk are SRS and XPM, both of which have been analyzed. Figure 8.31 shows the crosstalk measured in a two-channel experiment together with the theoretical prediction of the SRS- and XPM-induced crosstalk levels [245]. One channel is modulated and carries the actual signal while the other operates continuously (CW) but its power is low enough that it acts as a probe. The wavelength difference $\lambda_{\mathrm{mod}} - \lambda_{\mathrm{CW}}$ is ± 8.5 nm in the two cases shown in Fig. 8.31. The probe power varies with time because of SRS and XPM, and the crosstalk is defined as the ratio of radio-frequency (RF) powers in the two channels. The XPM-induced crosstalk increases and the Raman-induced crosstalk decreases with the modulation frequency but each has the same magnitude in the two cases shown in Fig. 8.31. The two crosstalks add up in phase only when $\lambda_{\mathrm{mod}} < \lambda_{\mathrm{CW}}$, resulting in a larger value of the total crosstalk in that case. The asymmetry seen in Fig. 8.31 is due to SRS and depends on whether the CW probe channel is being depleted or is being amplified by the other channel.

The linear crosstalk results from the phenomenon of optical beat interference. It occurs when two or more users transmit simultaneously on the same optical channel using different subcarrier frequencies. As the optical carrier frequencies are then slightly different, their beating produces a beat note in the photocurrent. If the beat-note frequency overlaps an active subcarrier channel, an interference signal would limit the detection process in a way similar to IMD. Statistical models have been used to estimate the probability of channel outage because of optical beat interference [244].

Multiwavelength SCM systems are quite useful for LAN and MAN applications [239]. They can provide multiple services (telephone, analog and digital TV channels, computer data, etc.) with only one optical transmitter and one optical receiver per user because different services can use different microwave subcarriers. This approach lowers the cost of terminal equipment in access networks. Different services can be offered without requiring synchronization, and microwave subcarriers can be processed using commercial electronic components. Each user is assigned a unique wavelength for transmitting multiple SCM messages but can receive multiple wavelengths. The main advantage of multiwavelength SCM is that the network can serve NM users, where N is the number of optical wavelengths and M is the number of microwave carriers by using only N distinct transmitter wavelengths. The optical wavelengths can be relatively far apart (coarse WDM) for reducing the cost of the terminal equipment. In another approach, the hybrid fiber/coaxial (HFC) technology is used to provide broadband integrated services to the subscriber. Digital video transport systems operating at 10 Gb/s by combining the WDM and SCM techniques are available commercially since 1996. The use of WDM and SCM for personal communication networks is quite appealing. The SCM technique is also being explored for network management and performance monitoring [247].

8.6   Code-Division Multiplexing

The multiplexing techniques discussed so far in this chapter can be classified as scheduled multiple-access techniques in which different users use the network according to a fixed assignment. Their major advantage is the simplicity of data routing among users. This simplicity, however, is achieved at the expense of an inefficient utilization of the channel bandwidth. This drawback can be overcome by using a random multiple-access technique that allows users to access any channel randomly at an arbitrary time. A multiplexing scheme well known in the domain of wireless communications makes use of the *spread-spectrum technique* [248]. It is referred to as *code-division multiplexing* (CDM) because each channel is coded in such a way that its spectrum spreads over a much wider region than occupied by the original signal.

Although spectrum spreading may appear counterintuitive from a spectral point of view, this is not the case because all users share the same spectrum, In fact, CDM is used extensively in microwave communications (e.g., cell phones) as it provides the most flexibility in a multiuser environment. The term *code-division multiple access* (CDMA) is often employed in place of CDM to emphasize the asynchronous and random nature of multiuser connections. Even though the use of CDMA for fiber-optic communications has been studied since 1986, it was only after 1995 that the technique of optical CDM was pursued seriously as an alternative to OTDM [249]–[271]. It can be easily combined with the WDM technique. Conceptually, the difference between the WDM, TDM, and CDM can be understood as follows. The WDM and TDM techniques partition the channel bandwidth or the time slots among users. In contrast, all users share the entire bandwidth and all time slots in a random fashion in the case of CDM.

8.6.1   Direct-Sequence Encoding

The new components needed for CDM systems are the encoders and decoders located at the transmitter and receiver ends, respectively. The encoder spreads the signal spectrum over a much wider region than the minimum bandwidth necessary for transmis-

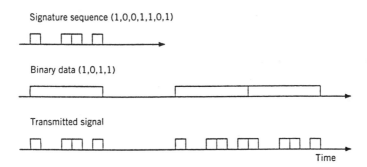

Figure 8.32: Coding of binary data in CDM systems using a signature sequence in the form of a 7-chip code.

sion. Spectral spreading is accomplished by means of a unique code that is independent of the signal itself. The decoder uses the same code for compressing the signal spectrum and recovering the data. The spectrum-spreading code is called a *signature sequence*. An advantage of the spread-spectrum method is that it is difficult to jam or intercept the signal because of its coded nature. The CDM technique is thus especially useful when security of the data is of concern.

Several methods can be used for data coding including direct-sequence encoding, time hopping, and frequency hopping. Figure 8.32 shows an example of the direct-sequence coding for optical CDM systems. Each bit of data is coded using a signature sequence consisting of a large number, say M, of shorter bits, called time "chips" borrowing the terminology used for wireless ($M = 7$ in the example shown). The effective bit rate (or the chip rate) increases by the factor of M because of coding. The signal spectrum is spread over a much wider region related to the bandwidth of individual chips. For example, the signal spectrum becomes broader by a factor of 64 if $M = 64$. Of course, the same spectral bandwidth is used by many users distinguished on the basis of different signature sequences assigned to them. The recovery of individual signals sharing the same bandwidth requires that the signature sequences come from a family of the orthogonal codes. The orthogonal nature of such codes ensures that each signal can be decoded accurately at the receiver end [263]. Transmitters are allowed to transmit messages at arbitrary times. The receiver recovers messages by decoding the received signal using the same signature sequence that was used at the transmitter. The decoding is accomplished using an optical correlation technique [254].

The encoders for direct-sequence coding typically use a delay-line scheme [249] that looks superficially similar to that shown in Fig. 8.26 for multiplexing several OTDM channels. The main difference is that a single modulator, placed after the laser, imposes the data on the pulse train. The resulting pulse train is split into several branches (equal to the number of code chips), and optical delay lines are used to encode the channel. At the receiver end, the decoder consists of the delay lines in the reverse order (matched-filter detection) such that it produces a peak in the correlation output whenever the user's code matches with a sequence of time chips in the received signal. Chip patterns of other users also produce a peak through cross-correlation but the amplitude of this peak is lower than the autocorrelation peak produced when the chip pattern matches precisely. An array of fiber Bragg gratings, designed with identical stop bands but different reflectivities, can also act as encoders and decoders [261]. Different gratings introduce different delays depending on their relative locations and produce a coded version of the signal. Such grating-based devices provide encoders and decoders in the form of a compact all-fiber device (except for the optical circulator needed to put the reflected coded signal back onto the transmission line).

The CDM pulse trains consisting of 0 and 1 chips suffer from two problems. First, only unipolar codes can be used simply because optical intensity or power cannot be negative. The number of such codes in a family of orthogonal codes is often not very large until the code length is increased to beyond 100 chips. Second, the cross-correlation function of the unipolar codes is relatively high, making the probability of an error also large. Both of these problems can be solved if the optical phase is used for coding in place of the amplitude. Such schemes are being pursued and are called coherent CDMA techniques [264]. An advantage of coherent CDM is that many families of bipolar orthogonal codes, developed for wireless systems and consisting of 1 and −1 chips, can be employed in the optical domain. When a CW laser source is used in combination with a phase modulator, another CW laser (called local oscillator) is required at the receiver for coherent detection (see Chapter 10). On the other hand, if ultrashort optical pulses are used as individual chips whose phase is shifted by π in chip slots corresponding to a −1 in the code, it is possible to decode the signal without using coherent detection techniques.

In a 2001 experiment, a coherent CDMA system was able to recover the 2.5 Gb/s signal transmitted using a 64-chip code [268]. A sampled (or superstructured) fiber grating was used for coding and decoding the data. Such a grating consists of an array of equally spaced smaller gratings so that a single pulse is split into multiple chips during reflection. Moreover, the phase of preselected chips can be changed by π so that each reflected pulse is converted into a phase-encoded train of chips. The decoder consists of a matched grating such that the reflected signal is converted into a single pulse through autocorrelation (constructive interference) for the signal bit while the cross-correlation or destructive interference produces no signal for signals belonging to other channels. The experiment used a NOLM (the same device used for demultiplexing of OTDM channels in Section 8.4) for improving the system performance. The NOLM passed the high-intensity autocorrelation peak but blocked the low-intensity cross-correlation peaks. The receiver was able to decode the 2.5-Gb/s bit stream from the 160-Gchip/s pulse train with less than 3-dB penalty at a BER of less than 10^{-9}. The use of time-gating detection helps to improve the performance in the presence of dispersive and crosstalk effects [269].

8.6.2 Spectral Encoding

Spectrum spreading can also be accomplished using the technique of frequency hopping in which the carrier frequency is shifted periodically according to a preassigned code [256]. The situation differs from WDM in the sense that a fixed frequency is not assigned to a given channel. Rather, all channels share the entire bandwidth by using different carrier frequencies at different times according to a code. A spectrally encoded signal can be represented in the form of a matrix shown schematically in

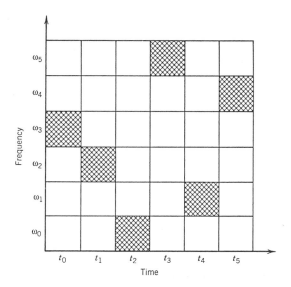

Figure 8.33: Frequency hopping in CDM lightwave systems. Filled square show frequencies for different time slots. A specific frequency-hop sequence (3, 2, 0, 5, 1, 4) is shown.

Fig. 8.33. The matrix rows correspond to assigned frequencies and the columns correspond to time slots. The matrix element m_{ij} equals 1 if and only if the frequency ω_i is transmitted in the interval t_j. Different users are assigned different frequency-hop patterns (or codes) to ensure that two users do not transmit at the same frequency during the same time slot. The code sequences that satisfy this property are said to be orthogonal codes. In the case of asynchronous transmission, complete orthogonality cannot be ensured. Such systems make use of pseudo-orthogonal codes with maximum auto-correlation and minimum cross-correlation to ensure the BER as low as possible. In general, the BER of such CDMA systems is relatively high (typically $> 10^{-6}$) but can be improved using a forward-error correction scheme.

Spectrally encoding of CDM lightwave systems requires a rapid change in the carrier frequency. It is difficult to make tunable semiconductor lasers whose wavelength can be changed over a wide range in a subnanosecond time scale. One possibility consists of hopping the frequency of a microwave subcarrier and then use the SCM technique for transmitting the CDM signal. This approach has the advantage that coding and decoding is done in the electrical domain, where the existing commercial microwave components can be used.

Several all-optical techniques have been developed for spectral encoding. They can be classified as coherent or incoherent depending on the the type of optical source used for the CDMA system. In the case of incoherent CDM, a broadband optical source such as an LED (or spontaneous emission from a fiber amplifier) is used in combination with a multipeak optical filter (such as an AWG) to create multiwavelength output [256]. Optical switches are then used to select different wavelengths for different chip slots. This technique can also be used to make CDM add–drop multiplexers [234]. An array of fiber gratings having different Bragg wavelengths can also be used for spectral

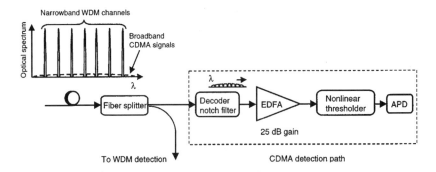

Figure 8.34: Receiver for a hybrid WDM–CDMA system sharing the same spectral bandwidth. A notch filter is used within the decoder to remove the WDM signal. (After Ref. [267]; ©2001 IEEE; reprinted with permission.)

encoding and decoding. A single chirped Moiré grating can replace the grating array because several gratings are written at the same location in such fiber gratings [46]. In a 2000 experiment, several Moiré gratings were used to demonstrate recovery of 622-Mb/s CDM channels [265].

In another approach called *coherence multiplexing* [258], a broadband optical source is used in combination with an unbalanced MZ interferometer that introduces a delay longer than the *coherence time* in one of its branches. Such CDM systems rely on coherence to discriminate among channels and are affected severely by the optical beat noise. In a demonstration of this technique, four 1-Gb/s channels were multiplexed. The optical source was an SOA operating below the laser threshold so that its output had bandwidth of 17 nm. A differential-detection technique was used to reduce the impact of optical beat noise. Indeed, bit-error rates below 10^{-9} could be achieved by using differential detection even when all four channels were operating simultaneously.

The coherent CDMA systems designed with spectral encoding make use of ultrashort optical pulses with a relatively broad spectrum [259]. The encoder splits the broad spectrum into many distinct wavelengths using a periodic optical filter (such as an AWG) and then assembles different frequency signals according to the code used for that channel. A matched-filter decoder at the receiver end performs the reverse operation so that a single ultrashort pulse is regenerated in a coherent fashion. This technique has a distinct advantage that the CDMA signal can be overlaid over a WDM signal such that both signals occupy the same wavelength range. Figure 8.34 shows schematically how such a hybrid scheme works [267]. The spectrum of the received signal consists of a broadband CDMA background and multiple sharp narrowband peaks that correspond to various WDM channels. The CDMA background does not affect the detection of WDM channels much because of its low amplitude. The CDMA receiver employs a notch filter to remove the WDM signal before decoding it. The hybrid WDM–CDMA scheme is spectrally efficient as it makes use of the unused extra bandwidth around each WDM channel. In a 2002 experiment, a spectral efficiency of 1.6 (b/s)/Hz and a capacity of 6.4 Tb/s were realized in the C band alone using the combination of CDMA and WDM techniques [271].

Problems

8.1 Dry fibers have acceptable losses over a spectral region extending from 1.3 to 1.6 μm. Estimate the capacity of a WDM system covering this entire region using 40-Gb/s channels spaced apart by 50 GHz.

8.2 The C and L spectral bands cover a wavelength range from 1.53 to 1.61 μm. How many channels can be transmitted through WDM when the channel spacing is 25 GHz? What is the effective bit rate–distance product when a WDM signal covering the two bands using 10-Gb/s channels is transmitted over 2000 km.

8.3 A 128 × 128 broadcast star is made by using 2 × 2 directional couplers, each having an insertion loss of 0.2 dB. Each channel transmits 1 mW of average power and requires 1 μW of average received power for operation at 1 Gb/s. What is the maximum transmission distance for each channel? Assume a cable loss of 0.25 dB/km and a loss of 3 dB from connectors and splices.

8.4 A Fabry–Perot filter of length L has equal reflectivities R for the two mirrors. Derive an expression for the transmission spectrum $T(v)$ considering multiple round trips inside the cavity containing air. Use it to show that the finesse is given by $F = \pi\sqrt{R}/(1-R)$.

8.5 A Fabry–Perot filter is used to select 100 channels spaced apart by 0.2 nm. What should be the length and the mirror reflectivities of the filter? Assume a refractive index of 1.5 and an operating wavelength of 1.55 μm.

8.6 The action of a fiber coupler is governed by the matrix equation $\mathbf{E}_{\text{out}} = \mathbf{T}\mathbf{E}_{\text{in}}$, where $\mathbf{T}$ is the 2 × 2 transfer matrix and $\mathbf{E}$ is a column vector whose two components represent the input (or output) fields at the two ports. Assuming that the total power is preserved, show that the transfer matrix $\mathbf{T}$ is given by

$$\mathbf{T} = \begin{pmatrix} \sqrt{1-f} & i\sqrt{f} \\ i\sqrt{f} & \sqrt{1-f} \end{pmatrix},$$

where f is the fraction of the power transferred to the cross port.

8.7 Explain how a Mach–Zehnder interferometer works. Prove that the transmission through a chain of M such interferometers is given by $T(v) = \prod_{m=1}^{M} \cos^2(\pi v \tau_m)$, where τ_m is the relative delay. Use the result of the preceding problem for the transfer matrix of a 3-dB fiber coupler.

8.8 Consider a fiber coupler with the transfer matrix given in Problem 8.6. Its two output ports are connected to each other to make a loop of length L. Find an expression for the transmittivity of the fiber loop. What happens when the coupler splits the input power equally? Provide a physical explanation.

8.9 The reflection coefficient of a fiber grating of length L is given by

$$r_g(\delta) = \frac{i\kappa \sin(qL)}{q\cos(qL) - i\delta\sin(qL)},$$

where $q^2 = \delta^2 - \kappa^2$, $\delta = (\omega - \omega_B)(\bar{n})/c$ is the detuning from the Bragg frequency ω_B, and κ is the coupling coefficient. Plot the reflectivity spectrum using

$\kappa = 8$ cm^{-1} and $\bar{n} = 1.45$, and a Bragg wavelength of 1.55 μm for $L = 3$, 5, and 8 mm. Estimate the grating bandwidth in GHz in the three cases.

8.10 You have been given ten 3-dB fiber couplers. Design a 4×4 demultiplexer with as few couplers as possible.

8.11 Explain how an array of planar waveguides can be used for demultiplexing WDM channels. Use diagrams as necessary.

8.12 Use a single fiber coupler and two fiber gratings to design an add–drop filter. Explain how such a device functions.

8.13 Use a waveguide-grating router to design an integrated WDM transmitter. How would the design change for a WDM receiver?

8.14 What is meant by the in-band linear crosstalk? Derive an expression for the power penalty induced by such crosstalk for a waveguide-grating router.

8.15 Explain how stimulated Raman scattering can cause crosstalk in multichannel lightwave systems. Derive Eq. (8.3.10) after approximating the Raman gain spectrum by a triangular profile.

8.16 Solve the set of M equations in Eq. (8.3.11) and show that the channel powers are given by Eq. (8.3.12).

8.17 Derive Eq. (8.3.14) by considering the nonlinear phase change induced by both self- and cross-phase modulation.

8.18 Solve Eq. (8.3.16) and show that the FWM efficiency is given by Eq. (8.3.18). Estimate its value for a 50-km fiber with $\alpha = 0.2$ dB/km and $\beta_2 = -1$ ps^2/km.

8.19 Derive an expression for the CNR of analog SCM lightwave systems by including thermal noise, shot noise, and intensity noise. Show that the CNR saturates to a constant value at high power levels.

8.20 Consider an analog SCM lightwave system operating at 1.55 μm. It uses a receiver of 90% quantum efficiency, 10 nA dark current, and thermal-noise RMS current of 0.1 mA over a 50-MHz bandwidth. The RIN of the transmitter laser is -150 dB/Hz. Calculate the average received power necessary to obtain 50-dB CNR for an AM–VSB system with a modulation index of 0.2.

References

[1] H. Ishio, J. Minowa, and K. Nosu, *J. Lightwave Technol.* **2**, 448 (1984).

[2] G. Winzer, *J. Lightwave Technol.* **2**, 369 (1984).

[3] N. A. Olsson, J. Hegarty, R. A. Logan, L. F. Johnson, K. L. Walker, L. G. Cohen, B. L. Kasper, and J. C. Campbell, *Electron. Lett.* **21**, 105 (1985).

[4] C. A. Brackett, *IEEE J. Sel. Areas Commun.* **8**, 948 (1990).

[5] P. E. Green, Jr., *Fiber-Optic Networks*, Prentice-Hall, Upper Saddle River, NJ, 1993.

[6] B. Mukherjee, *Optical Communication Networks*, McGraw-Hill, New York, 1997.

[7] A. Borella, G. Cancellieri, and F. Chiaraluce, *WDMA Optical Networks*, Artec House, Norwood, MA, 1998.

[8] R. A. Barry, *Optical Networks and Their Applications*, Optical Society of America, Washington, DC, 1998.

[9] T. E. Stern and K. Bala, *Multiwavelength Optical Networks*, Addison Wesley, Reading, MA, 1999.

[10] G. E. Keiser, *Optical Fiber Communications*, 3rd ed., McGraw-Hill, New York, 2000.

[11] K. M. Sivalingam and S. Subramaniam, Eds., *Optical WDM Networks: Principles and Practice*, Kluwer Academic, Norwell, MA, 2000.

[12] M. T. Fatehi and M. Wilson, *Optical Networking with WDM*, McGraw-Hill, New York, 2001.

[13] R. Ramaswami and K. Sivarajan, *Optical Networks: A Practical Perspective*, 2nd ed., Morgan Kaufmann Publishers, San Francisco, 2002.

[14] J. Zyskind and R. Berry, in *Optical Fiber Telecommunications IV*, Vol. B, I. P. Kaminow and T. Li, Eds., Academic Press, San Diego, CA, 2002.

[15] A. R. Chraplyvy, A. H. Gnauck, R. W. Tkach, R. M. Derosier, C. R. Giles, B. M. Nyman, G. A. Ferguson, J. W. Sulhoff, and J. L. Zyskind, *IEEE Photon. Technol. Lett.* **7**, 98 (1995).

[16] K. Fukuchi, T. Kasamatsu, M. Morie, R. Ohhira, T. Ito, K. Sekiya, D. Ogasahara, and T. Ono, Paper PD24, *Proc. Optical Fiber Commun. Conf.*, Optical Society of America, Washington, DC, 2001.

[17] N. S. Bergano and C. R. Davidson, *J. Lightwave Technol.* **14**, 1287 (1996).

[18] N. S. Bergano, in *Optical Fiber Telecommunications IV*, Vol. B, I. P. Kaminow and T. Li, Eds., Academic Press, San Diego, CA, 2002.

[19] I. S. Binetti, A. Bragheri, E. Iannone, and F. Bentivoglio, *J. Lightwave Technol.* **18**, 1677 (2000).

[20] A. A. M. Saleh and J. M. Simmons, *J. Lightwave Technol.* **17**, 2431 (1999).

[21] W. T. Anderson, J. Jackel, G. K. Chang, H. Dai, X. Wei. M. Goodman, C. Allyn, M. Alvarez, and O. Clarke, et. al., *J. Lightwave Technol.* **18**, 1988 (2000).

[22] P. E. Green, Jr., *IEEE J. Sel. Areas Commun.* **14**, 764 (1996).

[23] I. P. Kaminow, in *Optical Fiber Telecommunications III*, Vol. A, I. P. Kaminow and T. L. Koch, Eds., Academic Press, San Diego, CA, 1997, Chap. 15.

[24] T. Pfeiffer, J. Kissing, J. P. Elbers, B.Deppisch, M. Witte, H. Schmuck, and E. Voges, *J. Lightwave Technol.* **18**, 1928 (2000).

[25] N. Ghani, J. Y. Pan, and X. Cheng, in *Optical Fiber Telecommunications IV*, Vol. A, I. P. Kaminow and T. Li, Eds., Academic Press, San Diego, CA, 2002.

[26] E. Harstead and P. H. van Heyningen, in *Optical Fiber Telecommunications IV*, Vol. A, I. P. Kaminow and T. Li, Eds., Academic Press, San Diego, CA, 2002.

[27] P. W. Dowd, *IEEE Trans. Comput.* **41**, 1223 (1992).

[28] M. S. Goodman, H. Kobrinski, M. P. Vecchi, R. M. Bulley, and J. L. Gimlett, *IEEE J. Sel. Areas Commun.* **8**, 995 (1990).

[29] E. Hall, J. Kravitz, R. Ramaswami, M. Halvorson, S. Tenbrink, and R. Thomsen, *IEEE J. Sel. Areas Commun.* **14**, 814 (1996).

[30] D. Sadot and L. G. Kazovsky, *J. Lightwave Technol.* **15**, 1629 (1997).

[31] S. S. Wagner, H. Kobrinski, T. J. Robe, H. L. Lemberg, and L. S. Smoot, *Electron. Lett.* **24**, 344 (1988).

[32] M. Born and E. Wolf, *Principles of Optics*, 7th ed., Cambridge University Press, New York, 1999.

[33] J. Stone and L. W. Stulz, *Electron. Lett.* **23**, 781 (1987).

[34] K. Hirabayashi, H. Tsuda, and T. Kurokawa, *IEEE Photon. Technol. Lett.* **3**, 741 (1991); *J. Lightwave Technol.* **11**, 2033 (1993).

[35] A. Sneh and K. M. Johnson, *J. Lightwave Technol.* **14**, 1067 (1996).

[36] J. Ciosek, *Appl. Opt.* **39**, 135 (2000).

[37] H. K. Tsang, M. W. K. Mak, L. Y. Chan, J. B. D. Soole, C. Youtsey, and I. Adesida, *J. Lightwave Technol.* **17**, 1890 (1999).

[38] M. Iodice, G. Cocorullo, F. G. Della Corte, and I. Rendina, *Opt. Commun.* **183**, 415 (2000).

[39] J. Pfeiffer, J. Peerlings, R. Riemenschneider, R. Genovese, M. Aziz, E. Goutain, H. Kunzel, W. Gortz, G. Bohm, et al., *Mat. Sci. Semicond. Process.* **3**, 409 (2000).

[40] G. P. Agrawal, *Applications of Nonlinear Fiber Optics*, Academic Press, San Diego, CA, 2001.

[41] Y. Hibino, F. Hanawa, H. Nakagome, M. Ishii, and N. Takato, *J. Lightwave Technol.* **13**, 1728 (1995).

[42] S. Mino, K. Yoshino, Y. Yamada, T. Terui, M. Yasu, and K. Moriwaki, *J. Lightwave Technol.* **13**, 2320 (1995).

[43] M. Kawachi, *IEE Proc.* **143**, 257 (1996).

[44] Y. P. Li and C. H. Henry, *IEE Proc.* **143**, 263 (1996).

[45] K. Okamoto, *Opt. Quantum Electron.* **31**, 107 (1999).

[46] R. Kashyap, *Fiber Bragg Gratings*, Academic Press, San Diego, CA, 1999.

[47] G. P. Agrawal and S. Radic, *IEEE Photon. Technol. Lett.* **6**, 995 (1994).

[48] G. E. Town, K. Sugde, J. A. R. Williams, I. Bennion, and S. B. Poole, *IEEE Photon. Technol. Lett.* **7**, 78 (1995).

[49] F. Bilodeau, K. O. Hill, B. Malo, D. C. Johnson, and J. Albert, *IEEE Photon. Technol. Lett.* **6**, 80 (1994).

[50] T. Numai, S. Murata, and I. Mito, *Appl. Phys. Lett.* **53**, 83 (1988); **54**, 1859 (1989).

[51] J.-P. Weber, B. Stoltz, and M. Dasler, *Electron. Lett.* **31**, 220 (1995).

[52] K. N. Park, Y. T. Lee, M. H. Kim, K. S. Lee, and Y. H. Won, *IEEE Photon. Technol. Lett.* **10**, 555 (1998).

[53] B. Ortega, J. Capmany, and J. L. Cruz, *J. Lightwave Technol.* **17**, 1241 (1999).

[54] Y. Hibino, *IEICE Trans. Commun.* **E83B**, 2178 (2000).

[55] D. A. Smith, R. S. Chakravarthy, Z. Bao, J. E. Baran, J. L. Jackel, A. d'Alessandro, D. J. Fritz, S. H. Huang, X. Y. Zou, S.-M. Hwang, A. E. Willner, and K. D. Li, *J. Lightwave Technol.* **14**, 1005 (1996).

[56] J. L. Jackel, M. S. Goodman, J. E. Baran, W. J. Tomlinson, G.-K. Chang, M. Z. Iqbal, G. H. Song, K. Bala, C. A. Brackett, D. A. Smith, R. S. Chakravarthy, R. H. Hobbs, D. J. Fritz, R. W. Ade, and K. M. Kissa, *J. Lightwave Technol.* **14**, 1056 (1996).

[57] H. Herrmann, K. Schafer, and C. Schmidt, *IEEE Photon. Technol. Lett.* **10**, 120 (1998).

[58] T. E. Dimmick, G. Kakarantzas, T. A. Birks, and P. S. J. Russell, *IEEE Photon. Technol. Lett.* **12**, 1210 (2000).

[59] G. P. Agrawal, *Nonlinear Fiber Optics*, 3rd ed., Academic Press, San Diego, CA, 2001.

[60] R. W. Tkach, A. R. Chraplyvy, and R. M. Derosier, *IEEE Photon. Technol. Lett.* **1**, 111 (1989).

[61] K. Margari, H. Kawaguchi, K. Oe, Y. Nakano, and M. Fukuda, *Appl. Phys. Lett.* **51**, 1974 (1987); *IEEE J. Quantum Electron.* **24**, 2178 (1988).

[62] T. Numai, S. Murata, and I. Mito, *Appl. Phys. Lett.* **53**, 1168 (1988).

[63] S. Dubovitsky and W. H. Steier, *J. Lightwave Technol.* **14**, 1020 (1996).

[64] C. H. Henry, R. F. Kazarinov, Y. Shani, R. C. Kistler, V. Pol, and K. J. Orlowsky, *J. Lightwave Technol.* **8**, 748 (1990).

[65] K. A. McGreer, *IEEE Photon. Technol. Lett.* **8**, 553 (1996).

[66] S. J. Sun, K. A. McGreer, and J. N. Broughton, *IEEE Photon. Technol. Lett.* **10**, 90 (1998).

[67] F. N. Timofeev, E. G. Churin, P. Bayvel, V. Mikhailov, D. Rothnie, and J. E. Midwinter, *Opt. Quantum Electron.* **31**, 227 (1999).

[68] D. R. Wisely, *Electron. Lett.* **27**, 520 (1991).

[69] B. H. Verbeek, C. H. Henry, N. A. Olsson, K. J. Orlowsky, R. F. Kazarinov, and B. H. Johnson, *J. Lightwave Technol.* **6**, 1011 (1988).

[70] K. Oda, N. Tokato, T. Kominato, and H. Toba, *IEEE Photon. Technol. Lett.* **1**, 137 (1989).

[71] N. Takato, T. Kominato, A. Sugita, K. Jinguji, H. Toba, and M. Kawachi, *IEEE J. Sel. Areas Commun.* **8**, 1120 (1990).

[72] Y. Hibino, T. Kitagawa, K. O. Hill, F. Bilodeau, B. Malo, J. Albert, and D. C. Johnson, *IEEE Photon. Technol. Lett.* **8**, 84 (1996).

[73] B. S. Kawasaki, K. O. Hill, and R. G. Gaumont, *Opt. Lett.* **6**, 327 (1981).

[74] M. K. Smit and C. van Dam, *IEEE J. Sel. Topics Quantum Electron.* **2**, 251 (1996).

[75] R. Mestric, M. Renaud, M. Bachamann, B. Martin, and F. Goborit, *IEEE J. Sel. Topics Quantum Electron.* **2**, 257 (1996).

[76] H. Okayama, M. Kawahara, and T. Kamijoh, *J. Lightwave Technol.* **14**, 985 (1996).

[77] K. Okamoto, in *Photonic Networks*, G. Prati, Ed., Springer, New York, 1997.

[78] C. Dragone, *J. Lightwave Technol.* **16**, 1895 (1998).

[79] A. Kaneko, T. Goh, H. Yamada, T. Tanaka, and I. Ogawa, *IEEE J. Sel. Topics Quantum Electron.* **5**, 1227 (1999).

[80] P. Bernasconi, C. R. Doerr, C. Dragone, M. Cappuzzo, E. Laskowski, and A. Paunescu, *J. Lightwave Technol.* **18**, 985 (2000).

[81] K. Kato and Y. Tohmori, *IEEE J. Sel. Topics Quantum Electron.* **6**, 4 (2000).

[82] K. Takada, M. Abe, T. Shibata, and K. Okamoto, *IEEE Photon. Technol. Lett.* **13**, 577 (2001).

[83] F. Shehadeh, R. S. Vodhanel, M. Krain, C. Gibbons, R. E. Wagner, and M. Ali, *IEEE Photon. Technol. Lett.* **7**, 1075 (1995).

[84] H. A. Haus and Y. Lai, *J. Lightwave Technol.* **10**, 57 (1992).

[85] M. Zirngibl, C. H. Joyner, and B. Glance, *IEEE Photon. Technol. Lett.* **6**, 513 (1994).

[86] M. Kuznetsov, *J. Lightwave Technol.* **12**, 226 (1994).

[87] H. H. Yaffe, C. H. Henry, M. R. Serbin, and L. G. Cohen, *J. Lightwave Technol.* **12**, 1010 (1994).

[88] F. Bilodeau, D. C. Johnson, S. Thériault, B. Malo, J. Albert, and K. O. Hill, *IEEE Photon. Technol. Lett.* **7**, 388 (1995).

[89] T. Mizuochi, T. Kitayama, K. Shimizu, and K. Ito, *J. Lightwave Technol.* **16**, 265 (1998).

[90] T. Augustsson, *J. Lightwave Technol.* **16**, 1517 (1998).

[91] S. Rotolo, A. Tanzi, S. Brunazzi, D. DiMola, L. Cibinetto, M. Lenzi, G. L. Bona, B. J., Offrein, F. Horst, et al., *J. Lightwave Technol.* **18**, 569 (2000).

[92] Y. K. Chen, C. J. Hu, C. C. Lee, K. M. Feng, M. K. Lu, C. H. Chung, Y. K. Tu, and S. L. Tzeng, *IEEE Photon. Technol. Lett.* **12**, 1394 (2000).

[93] C. Riziotis and M. N. Zervas, *J. Lightwave Technol.* **19**, 92 (2001).

[94] A. V. Tran, W. D. Zhong, R. C. Tucker, and R. Lauder, *IEEE Photon. Technol. Lett.* **13**, 582 (2001).

[95] E. G. Rawson and M. D. Bailey, *Electron. Lett.* **15**, 432 (1979).

[96]   M. E. Marhic, *Opt. Lett.* **9**, 368 (1984).

[97]   D. B. Mortimore and J. W. Arkwright, *Appl. Opt.* **30**, 650 (1991).

[98]   C. Dragone, C. H. Henry, I. P. Kaminow, and R. C. Kistler, *IEEE Photon. Technol. Lett.* **1**, 241 (1989).

[99]   M. I. Irshid and M. Kavehrad, *IEEE Photon. Technol. Lett.* **4**, 48 (1992).

[100] P. D. Trinh, S. Yegnanaraynan, and B. Jalali, *IEEE Photon. Technol. Lett.* **8**, 794 (1996).

[101] J. M. H. Elmirghani and H. T. Mouftah, *IEEE Commun. Mag.* **38** (2), 58 (2000).

[102] Y. Hida, Y. Hibino, M. Itoh, A. Sugita, A. Himeno, and Y. Ohmori, *Electron. Lett.* **36**, 820 (2000).

[103] M. Koga, Y. Hamazumi, A. Watanabe, S. Okamoto, H. Obara, K. I. Sato, M. Okuno, and S. Suzuki, *J. Lightwave Technol.* **14**, 1106 (1996).

[104] A. Jourdan, F. Masetti, M. Garnot, G. Soulage, and M. Sotom, *J. Lightwave Technol.* **14**, 1198 (1996).

[105] M. Reanud, M. Bachmann, and M. Ermann, *IEEE J. Sel. Topics Quantum Electron.* **2**, 277 (1996).

[106] E. Almström, C. P. Larsen, L. Gillner, W. H. van Berlo, M. Gustavsson, and E. Berglind, *J. Lightwave Technol.* **14**, 996 (1996).

[107] M. Koga, A. Watanabe, T. Kawai, K. Sato, and Y. Ohmori, *IEEE J. Sel. Areas Commun.* **16**, 1260 (1998).

[108] T. Shiragaki, N. Henmi, T. Kato, M. Fujiwara, M. Misono, T. Shiozawa, and T. Suzuki, *IEEE J. Sel. Areas Commun.* **16**, 1179 (1998).

[109] M. C. Yu, N. F. de Rooij, and H. Fujita, Eds., *IEEE J. Sel. Topics Quantum Electron.* **5**, 2 (1999).

[110] S. Okamoto, M. Koga, H. Suzuki, and K. Kawai, *IEEE Commun. Mag.* **38** (3), 94 (2000).

[111] L. Eldada and L. W. Shaklette, *IEEE J. Sel. Topics Quantum Electron.* **6**, 54 (2000).

[112] L. Y. Lin, E. L. Goldstein, and R. W. Tkach, *J. Lightwave Technol.* **18**, 482 (2000).

[113] G. Wilfong, B. Mikkelsen, C. Doerr, and M. Zirngibl, *J. Lightwave Technol.* **18**, 1732 (2000).

[114] P. Helin, M. Mita, T. Bourouina, G. Reyne, and H. Fujita, *J. Lightwave Technol.* **18**, 1785 (2000).

[115] L. Eldada, *Opt. Eng.* **40**, 1165 (2001).

[116] E. Mutafungwa, *Opt. Fiber Technol.* **7**, 236 (2001).

[117] L. Wosinska, L. Thylen, and R. P. Holmstrom, *J. Lightwave Technol.* **19**, 1065 (2001).

[118] I. P. Kaminow and T. Li, Eds., *Optical Fiber Telecommunications IV*, Vol. A, Academic Press, San Diego, CA, 2002, Chaps. 8–12.

[119] G.-H. Duan, in *Semiconductor Lasers: Past, Present, and Future*, G. P. Agrawal, Ed., AIP Press, Woodbury, NY, 1995, Chap. 10.

[120] H. Kuwatsuka, H. Shoji, M. Matsuda, and H. Ishikawa, *Electron. Lett.* **31**, 2108 (1995).

[121] S. J. B. Yoo, *J. Lightwave Technol.* **14**, 955 (1996).

[122] C. Joergensen, S. L. Danielsen, K. E. Stubkjaer, M. Schilling, K. Daub, P. Doussiere, F. Pommerau, P. B. Hansen, H. N. Poulsen, et al., *IEEE J. Sel. Topics Quantum Electron.* **3**, 1168 (1997).

[123] H. Yasaka, H. Sanjoh, H. Ishii, Y. Yoshikuni, and K. Oe, *J. Lightwave Technol.* **15**, 334 (1997).

[124] A. Uchida, M. Takeoka, T. Nakata, and F. Kannari, *J. Lightwave Technol.* **16**, 92 (1998).

[125] J. Leuthold, P. A. Besse, E. Gamper, M. Dulk, S. Fischer, G. Guekos, and H. Melchior, *J. Lightwave Technol.* **17**, 1055 (1999).

[126] J. Yu, X. Zheng, C. Peucheret, A. T. Clausen, H. N. Poulsen, and P. Jeppesen, *J. Lightwave Technol.* **18**, 1001 (2000); *J. Lightwave Technol.* **18**, 1007 (2000).

[127] S. Hojfeldt, S. Bischoff, and J. Mork, *J. Lightwave Technol.* **18**, 1121 (2000).

[128] I. Brener, B. Mikkelsen, G. Raybon, R. Harel, K. Parameswaran, J. R. Kurz, and M. M. Fejer, *Electron. Lett.* **36**, 1788 (2000).

[129] J. Wu and P. Jeppesen, *IEEE Photon. Technol. Lett.* **13**, 833 (2001).

[130] G. P. Agrawal, *J. Opt. Soc. Am. B* **5**, 147 (1988).

[131] T. P. Lee, C. E. Zah, R. Bhat, W. C. Young, B. Pathak, F. Favire, P. S. D. Lin, N. C. Andreadakis, C. Caneau, A. W. Rahjel, M. Koza, J. K. Gamelin, L. Curtis, D. D. Mahoney, and A. Lepore, *J. Lightwave Technol.* **14**, 967 (1996).

[132] M. G. Young, U. Koren, B. I. Miller, M. A. Newkirk, M. Chien, M. Zirngibl, C. Dragone, B. Tell, H. M. Presby, and G. Raybon, *IEEE Photon. Technol. Lett.* **5**, 908 (1993).

[133] G. P. Li, T. Makino, A. Sarangan, and W. Huang, *IEEE Photon. Technol. Lett.* **8**, 22 (1996).

[134] T. L. Koch, in *Optical Fiber Telecommunications III*, Vol. B, I. P. Kaminow and T. L. Koch, Eds., Academic Press, San Diego, CA, 1997, Chap. 4.

[135] S. L. Lee, I. F. Jang, C. Y. Wang, C. T. Pien, and T. T. Shih, *IEEE J. Sel. Topics Quantum Electron.* **6**, 197 (2000).

[136] H. Li and K. Iga, *Vertical-Cavity Surface-Emitting Laser Devices*, Springer, New York, 2001.

[137] M. Zirngibl, C. H. Joyner, C. R. Doerr, L. W. Stulz, and H. M. Presby, *IEEE Photon. Technol. Lett.* **8**, 870 (1996).

[138] R. Monnard, A. K. Srivastava, C. R. Doerr, R. J. Essiambre, C. H. Joyner, L. W. Stulz, M. Zirngibl, Y. Sun, J. W. Sulhoff, J.L. Zyskind, and C. Wolf, *Electron. Lett.* **34**, 765 (1998).

[139] S. Menezo, A. Rigny, A. Talneau, F. Delorme, S. Grosmaire, H. Nakajima, E. Vergnol, F. Alexandre, and F. Gaborit, *IEEE J. Sel. Topics Quantum Electron.* **6**, 185 (2000).

[140] A. Bellemare, M. Karasek, M. Rochette, S. LaRochelle, and M. Tetu, *J. Lightwave Technol.* **18**, 825 (2000).

[141] T. Morioka, K. Uchiyama, S. Kawanishi, S. Suzuki, and M. Saruwatari, *Electron. Lett.* **31**, 1064 (1995).

[142] M. C. Nuss, W. H. Knox, and U. Koren, *Electron. Lett.* **32**, 1311 (1996).

[143] H. Takara, T. Ohara, K. Mori, K. Sato, E. Yamada, Y. Inoue, T. Shibata, M. Abe, T. Morioka, and K. I. Sato, *Electron. Lett.* **36**, 2089 (2000).

[144] L. Boivin and B. C. Collings, *Opt. Fiber Technol.* **7**, 1 (2001).

[145] E. Yamada, H. Takara, T. Ohara, K. Sato, T. Morioka, K. Jinguji, M. Itoh, and M. Ishii, *Electron. Lett.* **37**, 304 (2001).

[146] F. Tong, *IEEE Commun. Mag.* **36** (12), 42 (1998).

[147] S. Chandrasekhar, M. Zirngibl, A. G. Dentai, C. H. Joyner, F. Storz, C. A. Burrus, and L. M. Lunardi, *IEEE Photon. Technol. Lett.* **7**, 1342 (1995).

[148] P. A. Rosher and A. R. Hunwicks, *IEEE J. Sel. Areas Commun.* **8**, 1108 (1990).

[149] P. A. Humblet and W. M. Hamdy, *IEEE J. Sel. Areas Commun.* **8**, 1095 (1990).

[150] E. L. Goldstein and L. Eskildsen, *IEEE Photon. Technol. Lett.* **7**, 93 (1995).

[151] Y. D. Jin, Q. Jiang, and M. Kavehrad, *IEEE Photon. Technol. Lett.* **7**, 1210 (1995).

[152] D. J. Blumenthal, P. Granestrand, and L. Thylen, *IEEE Photon. Technol. Lett.* **8**, 284 (1996).

[153] H. Takahashi, K. Oda, and H. Toba, *J. Lightwave Technol.* **14**, 1097 (1996).

[154] C.-S. Li and F. Tong, *J. Lightwave Technol.* **14**, 1120 (1996).

[155] J. Zhou, R. Cadeddu, E. Casaccia, C. Cavazzoni, and M. J. O'Mahony, *J. Lightwave Technol.* **14**, 1423 (1996).

[156] K.-P. Ho and J. M. Kahn, *J. Lightwave Technol.* **14**, 1127 (1996).

[157] L. A. Buckman, L. P. Chen, and K. Y. Lau, *IEEE Photon. Technol. Lett.* **9**, 250 (1997).

[158] G. Murtaza and J. M. Senior, *Opt. Fiber Technol.* **3**, 471 (1997).

[159] M. Gustavsson, L. Gillner, and C. P. Larsen, *J. Lightwave Technol.* **15**, 2006 (1997).

[160] L. Gillner, C. P. Larsen, and M. Gustavsson, *J. Lightwave Technol.* **17**, 58 (1999).

[161] I. T. Monroy, E. Tangdiongga, and H. de Waardt, *J. Lightwave Technol.* **17**, 989 (1999).

[162] I. T. Monroy, E. Tangdiongga, R. Jonker, and H. de Waardt, *J. Lightwave Technol.* **18**, 637 (2000).

[163] K. Durnani and M. J. Holmes, *J. Lightwave Technol.* **18**, 1871 (2000).

[164] P. R. Pepeljugoski and K. Y. Lau, *J. Lightwave Technol.* **10**, 257 (1992).

[165] A. R. Chraplyvy, *J. Lightwave Technol.* **8**, 1548 (1990).

[166] N. Shibata, K. Nosu, K. Iwashita, and Y. Azuma, *IEEE J. Sel. Areas Commun.* **8**, 1068 (1990).

[167] F. Forghieri, R. W. Tkach, and A. R. Chraplyvy, in *Optical Fiber Telecommunications III*, Vol. A, I. P. Kaminow and T. L. Koch, Eds., Academic Press, San Diego, CA, 1997, Chap. 8.

[168] J. Kani, M. Jinno, T. Sakamoto, S. Aisawa, M. Fukui, K. Hattori, and K. Oguchi, *J. Lightwave Technol.* **17**, 2249 (1999).

[169] A. Mecozzi, C. B. Clausen, and M. Shtaif, *IEEE Photon. Technol. Lett.* **12**, 393 (2000).

[170] K. Tanaka, I. Morita, N. Edagawa, and M. Suzuki, *Electron. Lett.* **36**, 1217 (2000).

[171] P. Bayvel and R. I. Killey, in *Optical Fiber Telecommunications IV*, Vol. B, I. P. Kaminow and T. Li, Eds., Academic Press, San Diego, CA, 2002.

[172] D. N. Christodoulides and R. B. Jander, *IEEE Photon. Technol. Lett.* **8**, 1722 (1996).

[173] J. Wang, X. Sun, and M. Zhang, *IEEE Photon. Technol. Lett.* **10**, 540 (1998).

[174] M. E. Marhic, F. S. Yang, and L. G. Kazovsky, *J. Opt. Soc. Am. B* **15**, 957 (1998).

[175] A. G. Grandpierre, D. N. Christodoulides, and J. Toulouse, *IEEE Photon. Technol. Lett.* **11**, 1271 (1999).

[176] K. P. Ho, *J. Lightwave Technol.* **19**, 159 (2001).

[177] S. Norimatsu and T. Yamamoto, *J. Lightwave Technol.* **18**, 915 (2000).

[178] C. M. McIntosh, A. G. Grandpierre, D. N. Christodoulides, J. Toulouse, and J. M. P. Delvaux, *IEEE Photon. Technol. Lett.* **13**, 302 (2001).

[179] A. G. Grandpierre, D. N. Christodoulides, W. E. Schiesser, C. M. McIntosh, and J. Toulouse, *Opt. Commun.* **194**, 319 (2001).

[180] X. P. Mao, R. W. Tkach, A. R. Chraplyvy, R. M. Jopson, and R. M. Derosier, *IEEE Photon. Technol. Lett.* **4**, 66 (1992).

[181] D. A. Fishman and J. A. Nagel, *J. Lightwave Technol.* **11**, 1721 (1993).

[182] N. Yoshizawa and T. Imai, *J. Lightwave Technol.* **11**, 1518 (1993).

[183] K. Shiraki, M. Ohashi, and M. Tateda, *J. Lightwave Technol.* **14**, 50 (1996).

[184] Y. Horiuchi, S. Yamamoto, and S. Akiba, *Electron. Lett.* **34**, 390 (1998).

[185] K. Tsujikawa, K. Nakajima, Y. Miyajima, and M. Ohashi. *IEEE Photon. Technol. Lett.* **10**, 1139 (1998).

[186] L. E. Adams, G. Nykolak, T. Tanbun-Ek, A. J. Stentz, A M. Sergent, P. F. Sciortino, and L. Eskildsen, *Fiber Integ. Opt.* **17**, 311 (1998).

[187] S. S. Lee, H. J. Lee, W. Seo, and S. G. Lee, *IEEE Photon. Technol. Lett.* **13**, 741 (2001).

[188] T. Chiang, N. Kagi, M. E. Marhic, and L. G. Kazovsky, *J. Lightwave Technol.* **14**, 249 (1996).

[189] G. Bellotti, M. Varani, C. Francia, and A. Bononi, *IEEE Photon. Technol. Lett.* **10**, 1745 (1998).

[190] A. V. T. Cartaxo, *J. Lightwave Technol.* **17**, 178 (1999).

[191] R. Hui, K. R. Demarest, and C. T. Allen, *J. Lightwave Technol.* **17**, 1018 (1999).

[192] M. Eiselt, M. Shtaif, and L. D. Garettt, *IEEE Photon. Technol. Lett.* **11**, 748 (1999).

[193] L. E. Nelson, R. M. Jopson, A. H. Gnauck, and A. R. Chraplyvy, *IEEE Photon. Technol. Lett.* **11**, 907 (1999).

[194] M. Shtaif, M. Eiselt, and L. D. Garettt, *IEEE Photon. Technol. Lett.* **13**, 88 (2000).

[195] J. J. Yu and P. Jeppesen, *Opt. Commun.* **184**, 367 (2000).

[196] G. Bellotti and S. Bigo, *IEEE Photon. Technol. Lett.* **12**, 726 (2000).

[197] R. I. Killey, H. J. Thiele, V. Mikhailov, and P. Bayvel, *IEEE Photon. Technol. Lett.* **12**, 804 (2000).

[198] G. Bellotti, S. Bigo, P. Y. Cortes, S. Gauchard, and S. LaRochelle, *IEEE Photon. Technol. Lett.* **12**, 1403 (2000).

[199] S. Betti and M. Giaconi, *IEEE Photon. Technol. Lett.* **13**, 43 (2001); *IEEE Photon. Technol. Lett.* **13**, 305 (2001).

[200] N. Shibata, R. P. Braun, and R. G. Waarts, *IEEE J. Quantum Electron.* **23**, 1205 (1987).

[201] H. Suzuki, S. Ohteru, and N. Takachio, *IEEE Photon. Technol. Lett.* **11**, 1677 (1999).

[202] J. S. Lee, D. H. Lee, and C. S. Park, *IEEE Photon. Technol. Lett.* **10**, 825 (1998).

[203] H. Mawatari, M. Fukuda, F. Kano, Y. Tohmori, Y. Yoshikuni, and H. Toba, *J. Lightwave Technol.* **17**, 918 (1999).

[204] T. Ikegami, S. Sudo, and Y. Sakai, *Frequency Stabilization of Semiconductor Laser Diodes*, Artec House, Boston, 1995.

[205] M. Guy, B. Villeneuve, C. Latrasse, and M. Têtu, *J. Lightwave Technol.* **14**, 1136 (1996).

[206] H. J. Lee, G. Y. Lyu, S. Y. Park, and J. H. Lee, *IEEE Photon. Technol. Lett.* **10**, 276 (1998).

[207] Y. Park, S. T. Lee, and C. J. Chae, *IEEE Photon. Technol. Lett.* **10**, 1446 (1998).

[208] T. Ono and Y. Yano, *IEEE J. Quantum Electron.* **34**, 2080 (1998).

[209] Y. Horiuchi, S. Yamamoto, and M. Suzuki, *IEICE Trans. Comm.* **E84B**, 1145 (2001).

[210] J. Zhou and M. J. O'Mahony, *IEE Proc.* **143**, 178 (1996).

[211] Y. S. Fei, X. P. Zheng, H. Y. Zhang, Y. L. Guo, and B. K. Zhou, *IEEE Photon. Technol. Lett.* **11**, 1189 (1999).

[212] K. J. Zhang, D. L. Hart, K. I. Kang, and B. C. Moore, *Opt. Eng.* **40**, 1199 (2001).

[213] B. J. Wilson, N. G. Stoffel, J. L. Pastor, M. J. Post, K. H. Liu, T. Li, K. A. Walsh, J. Y. Wei, and Y. Tsai, *J. Lightwave Technol.* **18**, 2038 (2000).

[214] A. D. Ellis, D. M. Patrick, D. Flannery, R. J. Manning, D. A. O. Davies, and D. M. Spirit, *J. Lightwave Technol.* **13**, 761 (1995).

[215] V. W. S. Chan, K. L. Hall, E Modiano, and K. A. Rauschenbach, *J. Lightwave Technol.* **16**, 2146 (1998).

[216] S. Kawanishi, *IEEE J. Quantum Electron.* **34**, 2064 (1998).

[217] M. Nakazawa, H. Kubota, K. Suzuki, E. Yamada, and A. Sahara, *IEEE J. Sel. Topics Quantum Electron.* **6**, 363 (2000).

[218] T. G. Ulmer, M. C. Gross, K. M. Patel, , J.T. Simmons, P. W. Juodawlkis, B. R. Wasbburn, W. S. Astar, A. J. SpringThorpe, R. P. Kenan, C. M. Verber, and S. E. Ralph, *J. Lightwave Technol.* **18**, 1964 (2000).

[219] M. Saruwatari, *IEEE J. Sel. Topics Quantum Electron.* **6**, 1363 (2000).

[220] E. Bødtker and J. E. Bowers, *J. Lightwave Technol.* **13**, 1809 (1995).

[221] T. Yamamoto, E. Yoshida, and M. Nakazawa, *Electron. Lett.* **34**, 1013 (1998).

[222] T. Morioka, H. Takara, S. Kawanishi, T. Kitoh, and M. Saruwatari, *Electron. Lett.* **32**, 833 (1996).

[223] J. Hansryd and P. A. Andrekson, *IEEE Photon. Technol. Lett.* **13**, 732 (2001).

[224] J. Glesk, J. P. Sokoloff, and P. R. Prucnal, *Electron. Lett.* **30**, 339 (1994).

[225] P. V. Studenkov, M. R. Gokhale, J. Wei, W. Lin, I. Glesk, P. R. Prucnal, and S. R. Forrest, *IEEE Photon. Technol. Lett.* **13**, 600 (2001).

[226] S. Kawanishi, H. Takara, T. Morioka, O. Kamatani, K. Takaguchi, T. Kitoh, and M. Saruwatari, *Electron. Lett.* **32**, 916 (1996).

[227] M. Nakazawa, T. Yamamoto, and K. R. Tamura, *Electron. Lett.* **36**, 2027 (2000).

[228] U. Feiste, R. Ludwig, C. Schubert, J. Berger, C. Schmidt, H. G. Weber, B. Schmauss, A. Munk, B. Buchold, et al., *Electron. Lett.* **37**, 443 (2001).

[229] K. L. Deng, R. J. Runser, P. Toliver, I. Glesk, and P. R. Prucnal, *J. Lightwave Technol.* **18**, 1892 (2000).

[230] S. Aleksic, V. Krajinovic, and K. Bengi, *Photonic Network Commun.* **3**, 75 (2001).

[231] R. Gross and R. Olshansky, *J. Lightwave Technol.* **8**, 406 (1990).

[232] M. R. Phillips and T. E. Darcie, in *Optical Fiber Telecommunications IIIA*, I. P. Kaminow and T. L. Koch, Eds., Academic Press, San Diego, CA, 1997, Chap. 14.

[233] S. Ovadia, H. Dai, and C. Lin, *J. Lightwave Technol.* **16**, 2135 (1998).

[234] M. C. Wu, J. K. Wong, K. T. Tsai, Y. L. Chen, and W. I. Way, *IEEE Photon. Technol. Lett.* **12**, 1255 (2000).

[235] A. J. Rainal, *J. Lightwave Technol.* **14**, 474 (1996).

[236] J. Marti, A. Montero, J. Capmany, J. M. Fuster, and D. Pastor, *Electron. Lett.* **32**, 1605 (1996).

[237] H. Sotobayashi and K. Kitayama, *J. Lightwave Technol.* **17**, 2488 (1999).

[238] N. J. Frigo, M. R. Phillips, and G. E. Bodeep, *J. Lightwave Technol.* **11**, 138 (1993).

[239] W. I. Way, *Broadband Hybrid Fiber Coax Acess System Tecnologies*, Academic Press, San Diego, CA, 1998.

[240] W. I. Way, S. S. Wagner, M. M. Choy, C. Lin, R. C. Menendez, A. Tohme, A. Yi-Yan, A. C. Von Lehman, R. E. Spicer, et al., *IEEE Photon. Technol. Lett.* **2**, 665 (1990).

[241] C. C. Lee and S. Chi, *IEEE Photon. Technol. Lett.* **11**, 1066 (1999).

[242] M. Ogawara, M. Tsukada, J. Nishikido, A. Hiramatsu, M. Yamaguchi, and T. Matsunaga, *IEEE Photon. Technol. Lett.* **12**, 350 (2000).

[243] R. Hui, B. Zhu, R. Huang, C. Allen, K. Demarest, and D. Richards, *IEEE Photon. Technol. Lett.* **13**, 896 (2001).

[244] S. L. Woodward, X. Lu, T. E. Darcie, and G. E. Bodeep, *IEEE Photon. Technol. Lett.* **8**, 694 (1996).

[245] M. R. Phillips and D. M. Ott, *J. Lightwave Technol.* **17**, 1782 (1999).

[246] F. S. Yang, M. E. Marhic, and L. G. Kazovsky, *J. Lightwave Technol.* **18**, 512 (2000).

[247] G. Rossi, T. E. Dimmick, and D. J. Blumenthal, *J. Lightwave Technol.* **18**, 1639 (2000).

[248] R. E. Ziemer, R. L. Peterson, and D. E. Borth, *An Introduction to Spread Spectrum Communications*, Prentice Hall, Upper Saddle River, NJ, 1995.

[249] P. R. Prucnal, M. Santoro, and F. Tan, *J. Lightwave Technol.* **4**, 307 (1986).

[250] G. J. Foschinni and G. Vannucci, *J. Lightwave Technol.* **6**, 370 (1988).

[251] J. A. Salehi, *IEEE Trans. Commun.* **37**, 824 (1989).

[252] J. A. Salehi, A. M. Weiner, and J. P. Heritage, *J. Lightwave Technol.* **8**, 478 (1990).

[253] W. C. Kwong, P. A. Perrier, and P. R. Prucnal, *IEEE Trans. Commun.* **39**, 1625 (1991).

[254] R. M. Gagliardi, A. J. Mendez, M. R. Dale, and E. Park, *J. Lightwave Technol.* **11**, 20 (1993).

[255] M. E. Marhic, *J. Lightwave Technol.* **11**, 854 (1993).

[256] M. Kavehrad and D. Zaccarina, *J. Lightwave Technol.* **13**, 534 (1995).

[257] N. Karafolas and D. Uttamchandani, *Opt. Fiber Technol.* **2**, 149 (1996).

[258] D. D. Sampson, G. J. Pendock, and R. A. Griffin, *Fiber Integ. Opt.* **16**, 129 (1997).

[259] H. P. Sardesai, C. C. Desai, and A. M. Weiner, *J. Lightwave Technol.* **16**, 1953 (1998).

[260] W. Huang and I. Andonovic, *IEEE Trans. Commun.* **47**, 261 (1999).

[261] G. E. Town, K. Chan, and G. Yoffe, *IEEE J. Sel. Topics Quantum Electron.* **5**, 1325 (1999).

[262] N. Wada and K. I. Kitayama, *J. Lightwave Technol.* **17**, 1758 (1999).

[263] S. Kim, K. Yu, and N. Park, *J. Lightwave Technol.* **18**, 502 (2000).

[264] W. Huang, M. H. M. Nizam, I. Andonovic, and M. Tur, *J. Lightwave Technol.* **18**, 765 (2000).

[265] L. R. Chen and P. W. Smith, *IEEE Photon. Technol. Lett.* **12**, 1281 (2000); L. R. Chen, *IEEE Photon. Technol. Lett.* **13**, 1233 (2001).

[266] V. K. Jain and G. De Marchis, *Fiber Integ. Opt.* **20**, 1 (2001).

[267] S. Shen and A. M. Weiner, *IEEE Photon. Technol. Lett.* **13**, 82 (2001).

[268] P. C. Teh, P. Petropoulos, M. Ibsen, and D. J. Richardson, *J. Lightwave Technol.* **19**, 1352 (2001).

[269] P. Petropoulos, N. Wada, P. C. Teh, M. Ibsen, W. Chujo, K. I. Kitayama, and D. J. Richardson, *IEEE Photon. Technol. Lett.* **13**, 1239 (2001).

[270] Z. Wei, H. Ghafouri-Shiraz, and H. M. H Shalaby, *IEEE Photon. Technol. Lett.* **13**, 887 (2001); *ibid.* **13**, 890 (2001).

[271] H. Sotobayashi, W. Chujo, and K. I. Kitayama, *IEEE Photon. Technol. Lett.* **14**, 555 (2002).

Chapter 9

Soliton Systems

The word *soliton* was coined in 1965 to describe the particle-like properties of pulses propagating in a nonlinear medium [1]. The pulse envelope for solitons not only propagates undistorted but also survives collisions just as particles do. The existence of solitons in optical fibers and their use for optical communications were suggested in 1973 [2], and by 1980 solitons had been observed experimentally [3]. The potential of solitons for long-haul communication was first demonstrated in 1988 in an experiment in which fiber losses were compensated using the technique of Raman amplification [4]. Since then, a rapid progress during the 1990s has converted optical solitons into a practical candidate for modern lightwave systems [5]–[9]. In this chapter we focus on soliton communication systems with emphasis on the physics and design of such systems. The basic concepts behind fiber solitons are introduced in Section 9.1, where we also discuss the properties of such solitons. Section 9.2 shows how fiber solitons can be used for optical communications and how the design of such lightwave systems differs from that of conventional systems. The loss-managed and dispersion-managed solitons are considered in Sections 9.3 and 9.4, respectively. The effects of amplifier noise on such solitons are discussed in Section 9.5 with emphasis on the timing-jitter issue. Section 9.6 focuses on the design of high-capacity single-channel systems. The use of solitons for WDM lightwave systems is discussed in Section 9.7.

9.1 Fiber Solitons

The existence of solitons in optical fibers is the result of a balance between the group-velocity dispersion (GVD) and self-phase modulation (SPM), both of which, as discussed in Sections 2.4 and 5.3, limit the performance of fiber-optic communication systems when acting independently on optical pulses propagating inside fibers. One can develop an intuitive understanding of how such a balance is possible by following the analysis of Section 2.4. As shown there, the GVD broadens optical pulses during their propagation inside an optical fiber except when the pulse is initially chirped in the right way (see Fig. 2.12). More specifically, a chirped pulse can be compressed during the early stage of propagation whenever the GVD parameter β_2 and the chirp parameter

C happen to have opposite signs so that $\beta_2 C$ is negative. The nonlinear phenomenon of SPM imposes a chirp on the optical pulse such that $C > 0$. Since $\beta_2 < 0$ in the 1.55-μm wavelength region, the condition $\beta_2 C < 0$ is readily satisfied. Moreover, as the SPM-induced chirp is power dependent, it is not difficult to imagine that under certain conditions, SPM and GVD may cooperate in such a way that the SPM-induced chirp is just right to cancel the GVD-induced broadening of the pulse. The optical pulse would then propagate undistorted in the form of a soliton.

9.1.1 Nonlinear Schrödinger Equation

The mathematical description of solitons employs the nonlinear Schrödinger (NLS) equation, introduced in Section 5.3 [Eq. (5.3.1)] and satisfied by the pulse envelope $A(z,t)$ in the presence of GVD and SPM. This equation can be written as [10]

$$\frac{\partial A}{\partial z} + \frac{i\beta_2}{2}\frac{\partial^2 A}{\partial t^2} - \frac{\beta_3}{6}\frac{\partial^3 A}{\partial t^3} = i\gamma|A|^2 A - \frac{\alpha}{2}A, \qquad (9.1.1)$$

where fiber losses are included through the α parameter while β_2 and β_3 account for the second- and third-order dispersion (TOD) effects. The nonlinear parameter $\gamma = 2\pi n_2/(\lambda A_{\text{eff}})$ is defined in terms of the nonlinear-index coefficient n_2, the optical wavelength λ, and the effective core area A_{eff} introduced in Section 2.6.

To discuss the soliton solutions of Eq. (9.1.1) as simply as possible, we first set $\alpha = 0$ and $\beta_3 = 0$ (these parameters are included in later sections). It is useful to write this equation in a normalized form by introducing

$$\tau = \frac{t}{T_0}, \qquad \xi = \frac{z}{L_D}, \qquad U = \frac{A}{\sqrt{P_0}}, \qquad (9.1.2)$$

where T_0 is a measure of the pulse width, P_0 is the peak power of the pulse, and $L_D = T_0^2/|\beta_2|$ is the dispersion length. Equation (9.1.1) then takes the form

$$i\frac{\partial U}{\partial \xi} - \frac{s}{2}\frac{\partial^2 U}{\partial \tau^2} + N^2|U|^2 U = 0, \qquad (9.1.3)$$

where $s = \text{sgn}(\beta_2) = +1$ or -1, depending on whether β_2 is positive (normal GVD) or negative (anomalous GVD). The parameter N is defined as

$$N^2 = \gamma P_0 L_D = \gamma P_0 T_0^2/|\beta_2|. \qquad (9.1.4)$$

It represents a dimensionless combination of the pulse and fiber parameters. The physical significance of N will become clear later.

The NLS equation is well known in the soliton literature because it belongs to a special class of nonlinear partial differential equations that can be solved exactly with a mathematical technique known as the *inverse scattering method* [11]–[13]. Although the NLS equation supports solitons for both normal and anomalous GVD, pulse-like solitons are found only in the case of anomalous dispersion [14]. In the case of normal dispersion ($\beta_2 > 0$), the solutions exhibit a dip in a constant-intensity background. Such solutions, referred to as dark solitons, are discussed in Section 9.1.3. This chapter focuses mostly on pulse-like solitons, also called *bright* solitons.

9.1.2 Bright Solitons

Consider the case of anomalous GVD by setting $s = -1$ in Eq. (9.1.3). It is common to introduce $u = NU$ as a renormalized amplitude and write the NLS equation in its canonical form with no free parameters as

$$i\frac{\partial u}{\partial \xi} + \frac{1}{2}\frac{\partial^2 u}{\partial \tau^2} + |u|^2 u = 0. \tag{9.1.5}$$

This equation has been solved by the inverse scattering method [14]. Details of this method are available in several books devoted to solitons [11]–[13]. The main result can be summarized as follows. When an input pulse having an initial amplitude

$$u(0, \tau) = N\,\text{sech}(\tau) \tag{9.1.6}$$

is launched into the fiber, its shape remains unchanged during propagation when $N = 1$ but follows a periodic pattern for integer values of $N > 1$ such that the input shape is recovered at $\xi = m\pi/2$, where m is an integer.

An optical pulse whose parameters satisfy the condition $N = 1$ is called the *fundamental soliton*. Pulses corresponding to other integer values of N are called *higher-order solitons*. The parameter N represents the order of the soliton. By noting that $\xi = z/L_D$, the soliton period z_0, defined as the distance over which higher-order solitons recover their original shape, is given by

$$z_0 = \frac{\pi}{2}L_D = \frac{\pi}{2}\frac{T_0^2}{|\beta_2|}. \tag{9.1.7}$$

The *soliton period* z_0 and *soliton order* N play an important role in the theory of optical solitons. Figure 9.1 shows the pulse evolution for the first-order ($N = 1$) and third-order ($N = 3$) solitons over one soliton period by plotting the pulse intensity $|u(\xi, \tau)|^2$ (top row) and the frequency chirp (bottom row) defined as the time derivative of the soliton phase. Only a fundamental soliton maintains its shape and remains chirp-free during propagation inside optical fibers.

The solution corresponding to the fundamental soliton can be obtained by solving Eq. (9.1.5) directly, without recourse to the inverse scattering method. The approach consists of assuming that a solution of the form

$$u(\xi, \tau) = V(\tau)\exp[i\phi(\xi)] \tag{9.1.8}$$

exists, where V must be independent of ξ for Eq. (9.1.8) to represent a fundamental soliton that maintains its shape during propagation. The phase ϕ can depend on ξ but is assumed to be time independent. When Eq. (9.1.8) is substituted in Eq. (9.1.5) and the real and imaginary parts are separated, we obtain two real equations for V and ϕ. These equations show that ϕ should be of the form $\phi(\xi) = K\xi$, where K is a constant. The function $V(\tau)$ is then found to satisfy the nonlinear differential equation

$$\frac{d^2 V}{d\tau^2} = 2V(K - V^2). \tag{9.1.9}$$

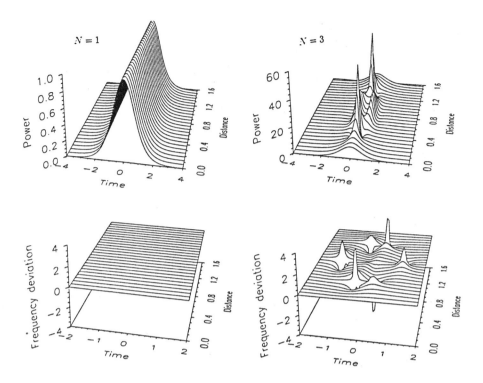

Figure 9.1: Evolution of the first-order (left column) and third-order (right column) solitons over one soliton period. Top and bottom rows show the pulse shape and chirp profile, respectively.

This equation can be solved by multiplying it by $2\,(dV/d\tau)$ and integrating over τ. The result is given as

$$(dV/d\tau)^2 = 2KV^2 - V^4 + C, \tag{9.1.10}$$

where C is a constant of integration. Using the boundary condition that both V and $dV/d\tau$ should vanish at $|\tau| = \infty$ for pulses, C is found to be 0. The constant K is determined using the other boundary condition that $V = 1$ and $dV/d\tau = 0$ at the soliton peak, assumed to occur at $\tau = 0$. Its use provides $K = \frac{1}{2}$, and hence $\phi = \xi/2$. Equation (9.1.10) is easily integrated to obtain $V(\tau) = \mathrm{sech}(\tau)$. We have thus found the well-known "sech" solution [11]–[13]

$$u(\xi, \tau) = \mathrm{sech}(\tau)\exp(i\xi/2) \tag{9.1.11}$$

for the fundamental soliton by integrating the NLS equation directly. It shows that the input pulse acquires a phase shift $\xi/2$ as it propagates inside the fiber, but its amplitude remains unchanged. It is this property of a fundamental soliton that makes it an ideal candidate for optical communications. In essence, the effects of fiber dispersion are exactly compensated by the fiber nonlinearity when the input pulse has a "sech" shape and its width and peak power are related by Eq. (9.1.4) in such a way that $N = 1$.

An important property of optical solitons is that they are remarkably stable against perturbations. Thus, even though the fundamental soliton requires a specific shape and

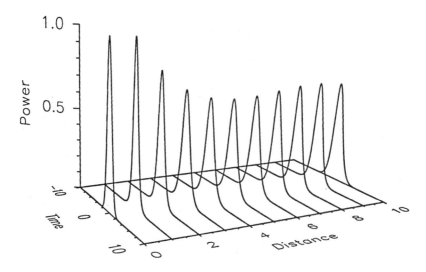

Figure 9.2: Evolution of a Gaussian pulse with $N = 1$ over the range $\xi = 0$–10. The pulse evolves toward the fundamental soliton by changing its shape, width, and peak power.

a certain peak power corresponding to $N = 1$ in Eq. (9.1.4), it can be created even when the pulse shape and the peak power deviate from the ideal conditions. Figure 9.2 shows the numerically simulated evolution of a Gaussian input pulse for which $N = 1$ but $u(0, \tau) = \exp(-\tau^2/2)$. As seen there, the pulse adjusts its shape and width in an attempt to become a fundamental soliton and attains a "sech" profile for $\xi \gg 1$. A similar behavior is observed when N deviates from 1. It turns out that the Nth-order soliton can be formed when the input value of N is in the range $N - \frac{1}{2}$ to $N + \frac{1}{2}$ [15]. In particular, the fundamental soliton can be excited for values of N in the range 0.5 to 1.5. Figure 9.3 shows the pulse evolution for $N = 1.2$ over the range $\xi = 0$–10 by solving the NLS equation numerically with the initial condition $u(0, \tau) = 1.2\,\text{sech}(\tau)$. The pulse width and the peak power oscillate initially but eventually become constant after the input pulse has adjusted itself to satisfy the condition $N = 1$ in Eq. (9.1.4).

It may seem mysterious that an optical fiber can force any input pulse to evolve toward a soliton. A simple way to understand this behavior is to think of optical solitons as the temporal modes of a nonlinear waveguide. Higher intensities in the pulse center create a temporal waveguide by increasing the refractive index only in the central part of the pulse. Such a waveguide supports temporal modes just as the core-cladding index difference led to spatial modes in Section 2.2. When an input pulse does not match a temporal mode precisely but is close to it, most of the pulse energy can still be coupled into that temporal mode. The rest of the energy spreads in the form of *dispersive waves*. It will be seen later that such dispersive waves affect the system performance and should be minimized by matching the input conditions as close to the ideal requirements as possible. When solitons adapt to perturbations adiabatically, perturbation theory developed specifically for solitons can be used to study how the soliton amplitude, width, frequency, speed, and phase evolve along the fiber.

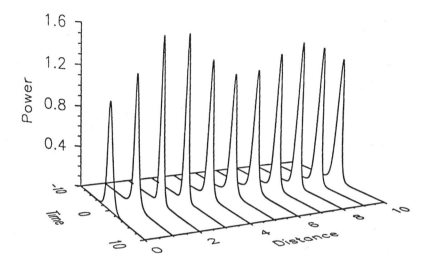

Figure 9.3: Pulse evolution for a "sech" pulse with $N = 1.2$ over the range $\xi = 0$–10. The pulse evolves toward the fundamental soliton ($N = 1$) by adjusting its width and peak power.

9.1.3 Dark Solitons

The NLS equation can be solved with the inverse scattering method even in the case of normal dispersion [16]. The intensity profile of the resulting solutions exhibits a dip in a uniform background, and it is the dip that remains unchanged during propagation inside the fiber [17]. For this reason, such solutions of the NLS equation are called *dark solitons*. Even though dark solitons were discovered in the 1970s, it was only after 1985 that they were studied thoroughly [18]–[28].

The NLS equation describing dark solitons is obtained from Eq. (9.1.5) by changing the sign of the second term. The resulting equation can again be solved by postulating a solution in the form of Eq. (9.1.8) and following the procedure outlined there. The general solution can be written as [28]

$$u_d(\xi, \tau) = (\eta \tanh \zeta - i\kappa) \exp(iu_0^2 \xi), \qquad (9.1.12)$$

where

$$\zeta = \eta(\tau - \kappa\xi), \quad \eta = u_0 \cos\phi, \quad \kappa = u_0 \sin\phi. \qquad (9.1.13)$$

Here, u_0 is the amplitude of the continuous-wave (CW) background and ϕ is an internal phase angle in the range 0 to $\pi/2$.

An important difference between the bright and dark solitons is that the speed of a dark soliton depends on its amplitude η through ϕ. For $\phi = 0$, Eq. (9.1.12) reduces to

$$u_d(\xi, \tau) = u_0 \tanh(u_0 \tau) \exp(iu_0^2 \xi). \qquad (9.1.14)$$

The peak power of the soliton drops to zero at the center of the dip only in the $\phi = 0$ case. Such a soliton is called the *black* soliton. When $\phi \neq 0$, the intensity does not drop to zero at the dip center; such solitons are referred to as the *gray* soliton. Another

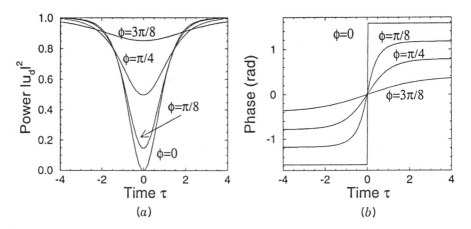

Figure 9.4: (a) Intensity and (b) phase profiles of dark solitons for several values of the internal phase ϕ. The intensity drops to zero at the center for black solitons.

interesting feature of dark solitons is related to their phase. In contrast with bright solitons which have a constant phase, the phase of a dark soliton changes across its width. Figure 9.4 shows the intensity and phase profiles for several values of ϕ. For a black soliton ($\phi = 0$), a phase shift of π occurs exactly at the center of the dip. For other values of ϕ, the phase changes by an amount $\pi - 2\phi$ in a more gradual fashion.

Dark solitons were observed during the 1980s in several experiments using broad optical pulses with a narrow dip at the pulse center. It is important to incorporate a π phase shift at the pulse center. Numerical simulations show that the central dip can propagate as a dark soliton despite the nonuniform background as long as the background intensity is uniform in the vicinity of the dip [18]. Higher-order dark solitons do not follow a periodic evolution pattern similar to that shown in Fig. 9.1 for the third-order bright soliton. The numerical results show that when $N > 1$, the input pulse forms a fundamental dark soliton by narrowing its width while ejecting several dark-soliton pairs in the process. In a 1993 experiment [19], 5.3-ps dark solitons, formed on a 36-ps wide pulse from a 850-nm Ti:sapphire laser, were propagated over 1 km of fiber. The same technique was later extended to transmit dark-soliton pulse trains over 2 km of fiber at a repetition rate of up to 60 GHz. These results show that dark solitons can be generated and maintained over considerable fiber lengths.

Several practical techniques were introduced during the 1990s for generating dark solitons. In one method, a Mach–Zehnder modulator driven by nearly rectangular electrical pulses, modulates the CW output of a semiconductor laser [20]. In an extension of this method, electric modulation is performed in one of the arms of a Mach–Zehnder interferometer. A simple all-optical technique consists of propagating two optical pulses, with a relative time delay between them, in the normal-GVD region of the fiber [21]. The two pulses broaden, become chirped, and acquire a nearly rectangular shape as they propagate inside the fiber. As these chirped pulses merge into each other, they interfere. The result at the fiber output is a train of isolated dark solitons. In another all-optical technique, nonlinear conversion of a beat signal in a dispersion-decreasing fiber was used to generate a train of dark solitons [22]. A 100-GHz train of 1.6-ps dark solitons was generated with this technique and propagated over 2.2 km of (two soliton periods) of a dispersion-shifted fiber. Optical switching using a fiber-loop mirror, in which a phase modulator is placed asymmetrically, can also produce dark solitons [23]. In another variation, a fiber with comb-like dispersion profile was used to generate dark soliton pulses with a width of 3.8 ps at the 48-GHz repetition rate [24].

An interesting scheme uses electronic circuitry to generate a coded train of dark solitons directly from the nonreturn-to-zero (NRZ) data in electric form [25]. First, the NRZ data and its clock at the bit rate are passed through an AND gate. The resulting signal is then sent to a flip-flop circuit in which all rising slopes flip the signal. The resulting electrical signal drives a Mach–Zehnder LiNbO$_3$ modulator and converts the CW output from a semiconductor laser into a coded train of dark solitons. This technique was used for data transmission, and a 10-Gb/s signal was transmitted over 1200 km by using dark solitons. Another relatively simple method uses spectral filtering of a mode-locked pulse train through a fiber grating [26]. This scheme has also been used to generate a 6.1-GHz train and propagate it over a 7-km-long fiber [27]. Numerical simulations show that dark solitons are more stable in the presence of noise and spread more slowly in the presence of fiber losses compared with bright solitons. Although these properties point to potential application of dark solitons for optical communications, only bright solitons were being pursued in 2002 for commercial applications.

9.2 Soliton-Based Communications

Solitons are attractive for optical communications because they are able to maintain their width even in the presence of fiber dispersion. However, their use requires substantial changes in system design compared with conventional nonsoliton systems. In this section we focus on several such issues.

9.2.1 Information Transmission with Solitons

As discussed in Section 1.2.3, two distinct modulation formats can be used to generate a digital bit stream. The NRZ format is commonly used because the signal bandwidth is about 50% smaller for it compared with that of the RZ format. However, the NRZ format cannot be used when solitons are used as information bits. The reason is easily understood by noting that the pulse width must be a small fraction of the bit slot to ensure that the neighboring solitons are well separated. Mathematically, the soliton solution in Eq. (9.1.11) is valid only when it occupies the entire time window ($-\infty < \tau < \infty$). It remains approximately valid for a train of solitons only when individual solitons are well isolated. This requirement can be used to relate the soliton width T_0 to the bit rate B as

$$B = \frac{1}{T_B} = \frac{1}{2q_0 T_0}, \qquad (9.2.1)$$

where T_B is the duration of the bit slot and $2q_0 = T_B/T_0$ is the separation between neighboring solitons in normalized units. Figure 9.5 shows a soliton bit stream in the

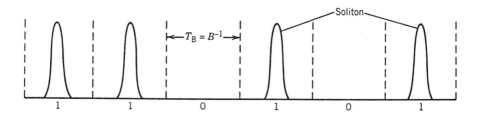

Figure 9.5: Soliton bit stream in RZ format. Each soliton occupies a small fraction of the bit slot so that neighboring soliton are spaced far apart.

RZ format. Typically, spacing between the solitons exceeds four times their full width at half maximum (FWHM).

The input pulse characteristics needed to excite the fundamental soliton can be obtained by setting $\xi = 0$ in Eq. (9.1.11). In physical units, the power across the pulse varies as

$$P(t) = |A(0,t)|^2 = P_0 \operatorname{sech}^2(t/T_0).\tag{9.2.2}$$

The required peak power P_0 is obtained from Eq. (9.1.4) by setting $N = 1$ and is related to the width T_0 and the fiber parameters as

$$P_0 = |\beta_2|/(\gamma T_0^2).\tag{9.2.3}$$

The width parameter T_0 is related to the FWHM of the soliton as

$$T_s = 2T_0 \ln(1 + \sqrt{2}) \simeq 1.763 T_0.\tag{9.2.4}$$

The pulse energy for the fundamental soliton is obtained using

$$E_s = \int_{-\infty}^{\infty} P(t)\,dt = 2P_0 T_0.\tag{9.2.5}$$

Assuming that 1 and 0 bits are equally likely to occur, the average power of the RZ signal becomes $\bar{P}_s = E_s(B/2) = P_0/2q_0$. As a simple example, $T_0 = 10$ ps for a 10-Gb/s soliton system if we choose $q_0 = 5$. The pulse FWHM is about 17.6 ps for $T_0 = 10$ ps. The peak power of the input pulse is 5 mW using $\beta_2 = -1$ ps^2/km and $\gamma = 2$ W^{-1}/km as typical values for dispersion-shifted fibers. This value of peak power corresponds to a pulse energy of 0.1 pJ and an average power level of only 0.5 mW.

9.2.2 Soliton Interaction

An important design parameter of soliton lightwave systems is the pulse width T_s. As discussed earlier, each soliton pulse occupies only a fraction of the bit slot. For practical reasons, one would like to pack solitons as tightly as possible. However, the presence of pulses in the neighboring bits perturbs the soliton simply because the combined optical field is not a solution of the NLS equation. This phenomenon, referred to as *soliton interaction*, has been studied extensively [29]–[33].

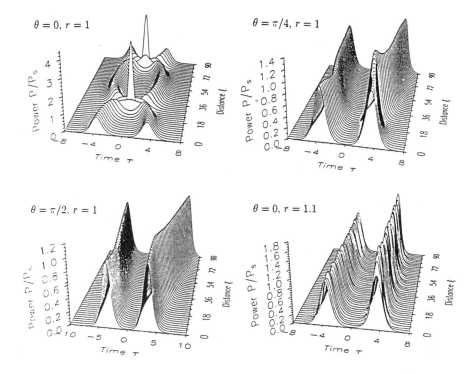

Figure 9.6: Evolution of a soliton pair over 90 dispersion lengths showing the effects of soliton interaction for four different choices of amplitude ratio r and relative phase θ. Initial spacing $q_0 = 3.5$ in all four cases.

One can understand the implications of soliton interaction by solving the NLS equation numerically with the input amplitude consisting of a soliton pair so that

$$u(0, \tau) = \text{sech}(\tau - q_0) + r\,\text{sech}[r(\tau + q_0)]\exp(i\theta), \tag{9.2.6}$$

where r is the relative amplitude of the two solitons, θ is the relative phase, and $2q_0$ is the initial (normalized) separation. Figure 9.6 shows the evolution of a soliton pair with $q_0 = 3.5$ for several values of the parameters r and θ. Clearly, soliton interaction depends strongly both on the relative phase θ and the amplitude ratio r.

Consider first the case of equal-amplitude solitons ($r = 1$). The two solitons attract each other in the in-phase case ($\theta = 0$) such that they collide periodically along the fiber length. However, for $\theta = \pi/4$, the solitons separate from each other after an initial attraction stage. For $\theta = \pi/2$, the solitons repel each other even more strongly, and their spacing increases with distance. From the standpoint of system design, such behavior is not acceptable. It would lead to jitter in the arrival time of solitons because the relative phase of neighboring solitons is not likely to remain well controlled. One way to avoid soliton interaction is to increase q_0 as the strength of interaction depends on soliton spacing. For sufficiently large q_0, deviations in the soliton position are expected to be small enough that the soliton remains at its initial position within the bit slot over the entire transmission distance.

The dependence of soliton separation on q_0 can be studied analytically by using the inverse scattering method [29]. A perturbative approach can be used for $q_0 \gg 1$. In the specific case of $r = 1$ and $\theta = 0$, the soliton separation $2q_s$ at any distance ξ is given by [30]

$$2\exp[2(q_s - q_0)] = 1 + \cos[4\xi \exp(-q_0)]. \qquad (9.2.7)$$

This relation shows that the spacing $q_s(\xi)$ between two neighboring solitons oscillates periodically with the period

$$\xi_p = (\pi/2)\exp(q_0). \qquad (9.2.8)$$

A more accurate expression, valid for arbitrary values of q_0, is given by [32]

$$\xi_p = \frac{\pi \sinh(2q_0) \cosh(q_0)}{2q_0 + \sinh(2q_0)}. \qquad (9.2.9)$$

Equation (9.2.8) is quite accurate for $q_0 > 3$. Its predictions are in agreement with the numerical results shown in Fig. 9.6 where $q_0 = 3.5$. It can be used for system design as follows. If $\xi_p L_D$ is much greater than the total transmission distance L_T, soliton interaction can be neglected since soliton spacing would deviate little from its initial value. For $q_0 = 6$, $\xi_p \approx 634$. Using $L_D = 100$ km for the dispersion length, $L_T \ll \xi_p L_D$ can be realized even for $L_T = 10{,}000$ km. If we use $L_D = T_0^2/|\beta_2|$ and $T_0 = (2Bq_0)^{-1}$ from Eq. (9.2.1), the condition $L_T \ll \xi_p L_D$ can be written in the form of a simple design criterion

$$B^2 L_T \ll \frac{\pi \exp(q_0)}{8q_0^2 |\beta_2|}. \qquad (9.2.10)$$

For the purpose of illustration, let us choose $\beta_2 = -1$ ps^2/km. Equation (9.2.10) then implies that $B^2 L_T \ll 4.4$ (Tb/s)^2-km if we use $q_0 = 6$ to minimize soliton interactions. The pulse width at a given bit rate B is determined from Eq. (9.2.1). For example, $T_s = 14.7$ ps at $B = 10$ Gb/s when $q_0 = 6$.

A relatively large soliton spacing, necessary to avoid soliton interaction, limits the bit rate of soliton communication systems. The spacing can be reduced by up to a factor of 2 by using unequal amplitudes for the neighboring solitons. As seen in Fig. 9.6, the separation for two in-phase solitons does not change by more than 10% for an initial soliton spacing as small as $q_0 = 3.5$ if their initial amplitudes differ by 10% ($r = 1.1$). Note that the peak powers or the energies of the two solitons deviate by only 1%. As discussed earlier, such small changes in the peak power are not detrimental for maintaining solitons. Thus, this scheme is feasible in practice and can be useful for increasing the system capacity. The design of such systems would, however, require attention to many details. Soliton interaction can also be modified by other factors, such as the initial frequency chirp imposed on input pulses.

9.2.3   Frequency Chirp

To propagate as a fundamental soliton inside the optical fiber, the input pulse should not only have a "sech" profile, but also be chirp-free. Many sources of short optical pulses have a frequency chirp imposed on them. The initial chirp can be detrimental to

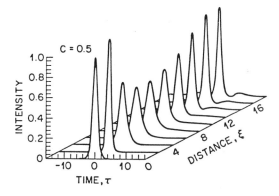

Figure 9.7: Evolution of a chirped optical pulse for the case $N = 1$ and $C = 0.5$. For $C = 0$ the pulse shape does not change, since the pulse propagates as a fundamental soliton.

soliton propagation simply because it disturbs the exact balance between the GVD and SPM [34]–[37].

The effect of an initial frequency chirp can be studied by solving Eq. (9.1.5) numerically with the input amplitude

$$u(0, \tau) = \text{sech}(\tau) \exp(-iC\tau^2/2), \tag{9.2.11}$$

where C is the chirp parameter introduced in Section 2.4.2. The quadratic form of phase variations corresponds to a linear frequency chirp such that the optical frequency increases with time (up-chirp) for positive values of C. Figure 9.7 shows the pulse evolution in the case $N = 1$ and $C = 0.5$. The pulse shape changes considerably even for $C = 0.5$. The pulse is initially compressed mainly because of the positive chirp; initial compression occurs even in the absence of nonlinear effects (see Section 2.4.2). The pulse then broadens but is eventually compressed a second time with the tails gradually separating from the main peak. The main peak evolves into a soliton over a propagation distance $\xi > 15$. A similar behavior occurs for negative values of C, although the initial compression does not occur in that case. The formation of a soliton is expected for small values of $|C|$ because solitons are stable under weak perturbations. But the input pulse does not evolve toward a soliton when $|C|$ exceeds a critical valve C_{crit}. The soliton seen in Fig. 9.7 does not form if C is increased from 0.5 to 2.

The critical value C_{crit} of the chirp parameter can be obtained by using the inverse scattering method [34]–[36]. It depends on N and is found to be $C_{\text{crit}} = 1.64$ for $N = 1$. It also depends on the form of the phase factor in Eq. (9.2.11). From the standpoint of system design, the initial chirp should be minimized as much as possible. This is necessary because even if the chirp is not detrimental for $|C| < C_{\text{crit}}$, a part of the pulse energy is shed as dispersive waves during the process of soliton formation [34]. For instance, only 83% of the input energy is converted into a soliton for the case $C = 0.5$ shown in Fig. 9.7, and this fraction reduces to 62% when $C = 0.8$.

9.2.4 Soliton Transmitters

Soliton communication systems require an optical source capable of producing chirp-free picosecond pulses at a high repetition rate with a shape as close to the "sech" shape as possible. The source should operate in the wavelength region near 1.55 μm, where fiber losses are minimum and where erbium-doped fiber amplifiers (EDFAs) can be used for compensating them. Semiconductor lasers, commonly used for nonsoliton lightwave systems, remain the lasers of choice even for soliton systems.

Early experiments on soliton transmission used the technique of gain switching for generating optical pulses of 20–30 ps duration by biasing the laser below threshold and pumping it high above threshold periodically [38]–[40]. The repetition rate was determined by the frequency of current modulation. A problem with the gain-switching technique is that each pulse becomes chirped because of the refractive-index changes governed by the linewidth enhancement factor (see Section 3.5.3). However, the pulse can be made nearly chirp-free by passing it through an optical fiber with normal GVD ($\beta_2 > 0$) such that it is compressed. The compression mechanism can be understood from the analysis of Section 2.4.2 by noting that gain switching produces pulses with a frequency chirp such that the chirp parameter C is negative. In a 1989 implementation of this technique [39], 14-ps optical pulses were obtained at a 3-GHz repetition rate by passing the gain-switched pulse through a 3.7-km-long fiber with $\beta_2 = 23$ ps^2/km near 1.55 μm. An EDFA amplified each pulse to the power level required for launching fundamental solitons. In another experiment, gain-switched pulses were simultaneously amplified and compressed inside an EDFA after first passing them through a narrowband optical filter [40]. It was possible to generate 17-ps-wide, nearly chirp-free, optical pulses at repetition rates in the range 6–24 GHz.

Mode-locked semiconductor lasers are also suitable for soliton communications and are often preferred because the pulse train emitted from such lasers is nearly chirp-free. The technique of active mode locking is generally used by modulating the laser current at a frequency equal to the frequency difference between the two neighboring longitudinal modes. However, most semiconductor lasers use a relatively short cavity length (< 0.5 mm typically), resulting in a modulation frequency of more than 50 GHz. An external-cavity configuration is often used to increase the cavity length and reduce the modulation frequency. In a practical approach, a chirped fiber grating is spliced to the pigtail attached to the optical transmitter to form the external cavity. Figure 9.8 shows the design of such a source of short optical pulses. The use of a chirped fiber grating provides wavelength stability to within 0.1 nm. The grating also offers a self-tuning mechanism that allows mode locking of the laser over a wide range of modulation frequencies [41]. A thermoelectric heater can be used to tune the operating wavelength over a range of 6–8 nm by changing the Bragg wavelength associated with the grating. Such a source produces soliton-like pulses of widths 12–18 ps at a repetition rate as large as 40 GHz and can be used at a bit rate of 40 Gb/s [42].

The main drawback of external-cavity semiconductor lasers stems from their hybrid nature. A monolithic source of picosecond pulses is preferred in practice. Several approaches have been used to produce such a source. Monolithic semiconductor lasers with a cavity length of about 4 mm can be actively mode-locked to produce a 10-GHz pulse train. Passive mode locking of a monolithic distributed Bragg reflector (DBR)

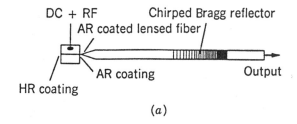

(a)

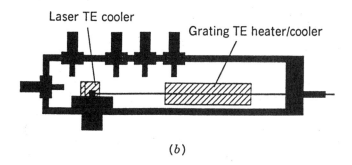

(b)

Figure 9.8: Schematic of (a) the device and (b) the package for a hybrid soliton pulse source. (After Ref. [41]; ©1995 IEEE; reprinted with permission.)

laser has produced 3.5-ps pulses at a repetition rate of 40 GHz [43]. An electroabsorption modulator, integrated with the semiconductor laser, offers another alternative. Such transmitters are commonly used for nonsoliton lightwave systems (see Section 3.6). They can also be used to produce a pulse train by using the nonlinear nature of the absorption response of the modulator. Chirp-free pulses of 10- to 20-ps duration at a repetition rate of 20 GHz were produced in 1993 with this technique [44]. By 1996, the repetition rate of modulator-integrated lasers could be increased to 50 GHz [45]. The *quantum-confinement Stark effect* in a multiquantum-well modulator can also be used to produce a pulse train suitable for soliton transmission [46].

Mode-locked fiber lasers provide an alternative to semiconductor sources although such lasers still need a semiconductor laser for pumping [47]. An EDFA is placed within the Fabry–Perot (FP) or ring cavity to make fiber lasers. Both active and passive mode-locking techniques have been used for producing short optical pulses. Active mode locking requires modulation at a high-order harmonic of the longitudinal-mode spacing because of relatively long cavity lengths (> 1 m) that are typically used for fiber lasers. Such harmonically mode-locked fiber lasers use an intracavity $LiNbO_3$ modulator and have been employed in soliton transmission experiments [48]. A semiconductor optical amplifier can also be used for active mode locking, producing pulses shorter than 10 ps at a repetition rate as high as 20 GHz [49]. Passively mode-locked fiber lasers either use a multiquantum-well device that acts as a fast saturable absorber or employ fiber nonlinearity to generate phase shifts that produce an effective saturable absorber.

In a different approach, nonlinear pulse shaping in a dispersion-decreasing fiber is used to produce a train of ultrashort pulses. The basic idea consists of injecting a CW beam, with weak sinusoidal modulation imposed on it, into such a fiber. The combination of GVD, SPM, and decreasing dispersion converts the sinusoidally modulated signal into a train of ultrashort solitons [50]. The repetition rate of pulses is governed by the frequency of initial sinusoidal modulation, often produced by beating two optical signals. Two distributed feedback (DFB) semiconductor lasers or a two-mode fiber laser can be used for this purpose. By 1993, this technique led to the development of an integrated fiber source capable of producing a soliton pulse train at high repetition rates by using a *comb-like* dispersion profile, created by splicing pieces of low- and high-dispersion fibers [50]. A *dual-frequency* fiber laser was used to generate the beat signal and to produce a 2.2-ps soliton train at the 59-GHz repetition rate. In another experiment, a 40-GHz soliton train of 3-ps pulses was generated using a single DFB laser whose output was modulated with a Mach–Zehnder modulator before launching it into a dispersion-tailored fiber with a comb-like GVD profile [51].

A simple method of pulse-train generation modulates the phase of the CW output obtained from a DFB semiconductor laser, followed by an optical bandpass filter [52]. Phase modulation generates frequency modulation (FM) sidebands on both sides of the carrier frequency, and the optical filter selects the sidebands on one side of the carrier. Such a device generates a stable pulse train of widths ~ 20 ps at a repetition rate that is controlled by the phase modulator. It can also be used as a dual-wavelength source by filtering sidebands on both sides of the carrier frequency, with a typical channel spacing of about 0.8 nm at the 1.55-μm wavelength. Another simple technique uses a single Mach–Zehnder modulator, driven by an electrical data stream in the NRZ format, to convert the CW output of a DFB laser into an optical bit stream in the RZ format [53]. Although optical pulses launched from such transmitters typically do not have the "sech" shape of a soliton, they can be used for soliton systems because of the soliton-formation capability of the fiber discussed earlier.

9.3 Loss-Managed Solitons

As discussed in Section 9.1, solitons use the nonlinear phenomenon of SPM to maintain their width even in the presence of fiber dispersion. However, this property holds only if fiber losses were negligible. It is not difficult to see that a decrease in soliton energy because of fiber losses would produce soliton broadening simply because a reduced peak power weakens the SPM effect necessary to counteract the GVD. Optical amplifiers can be used for compensating fiber losses. This section focuses on the management of losses through amplification of solitons.

9.3.1 Loss-Induced Soliton Broadening

Fiber losses are included through the last term in Eq. (9.1.1). In normalized units, the NLS equation becomes [see Eq. (9.1.5)]

$$i\frac{\partial u}{\partial \xi} + \frac{1}{2}\frac{\partial^2 u}{\partial \tau^2} + |u|^2 u = -\frac{i}{2}\Gamma u, \qquad (9.3.1)$$

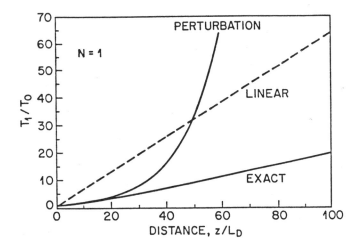

Figure 9.9: Broadening of fundamental solitons in lossy fibers ($\Gamma = 0.07$). The curve marked "exact" shows numerical results. Dashed curve shows the behavior expected in the absence of nonlinear effects. (After Ref. [55]; ©1985 Elsevier; reprinted with permission.)

where $\Gamma = \alpha L_D$ represents fiber losses over one dispersion length. When $\Gamma \ll 1$, the last term can be treated as a small perturbation [54]. The use of variational or perturbation methods results in the following approximate solution of Eq. (9.3.1):

$$u(\xi, \tau) \approx e^{-\Gamma\xi} \mathrm{sech}(\tau e^{-\Gamma\xi}) \exp[i(1 - e^{-2\Gamma\xi})/4\Gamma]. \qquad (9.3.2)$$

The solution (9.3.2) shows that the soliton width increases exponentially because of fiber losses as

$$T_1(\xi) = T_0 \exp(\Gamma\xi) = T_0 \exp(\alpha z). \qquad (9.3.3)$$

Such an exponential increase in the soliton width cannot be expected to continue for arbitrarily long distances. Numerical solutions of Eq. (9.3.1) indeed show a slower increase for $\xi \gg 1$ [55]. Figure 9.9 shows the broadening factor T_1/T_0 as a function of ξ when a fundamental soliton is launched into a fiber with $\Gamma = 0.07$. The perturbative result is also shown for comparison; it is reasonably accurate up to $\Gamma\xi = 1$. The dashed line in Fig. 9.9 shows the broadening expected in the absence of nonlinear effects. The important point to note is that soliton broadening is much less compared with the linear case. Thus, the nonlinear effects can be beneficial even when solitons cannot be maintained perfectly because of fiber losses. In a 1986 study, an increase in the repeater spacing by more than a factor of 2 was predicted using higher-order solitons [56].

In modern long-haul lightwave systems, pulses are transmitted over long fiber lengths without using electronic repeaters. To overcome the effect of fiber losses, solitons should be amplified periodically using either lumped or distributed amplification [57]–[60]. Figure 9.10 shows the two schemes schematically. The next two subsections focus on the design issues related to loss-managed solitons based on these two amplification schemes.

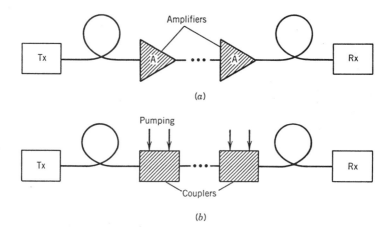

Figure 9.10: (a) Lumped and (b) distributed amplification schemes for compensation of fiber losses in soliton communication systems.

9.3.2 Lumped Amplification

The lumped amplification scheme shown in Fig. 9.10 is the same as that used for non-soliton systems. In both cases, optical amplifiers are placed periodically along the fiber link such that fiber losses between two amplifiers are exactly compensated by the amplifier gain. An important design parameter is the spacing L_A between amplifiers—it should be as large as possible to minimize the overall cost. For nonsoliton systems, L_A is typically 80–100 km. For soliton systems, L_A is restricted to much smaller values because of the soliton nature of signal propagation [57].

The physical reason behind smaller values of L_A is that optical amplifiers boost soliton energy to the input level over a length of few meters without allowing for gradual recovery of the fundamental soliton. The amplified soliton adjusts its width dynamically in the fiber section following the amplifier. However, it also sheds a part of its energy as dispersive waves during this adjustment phase. The dispersive part can accumulate to significant levels over a large number of amplification stages and must be avoided. One way to reduce the dispersive part is to reduce the amplifier spacing L_A such that the soliton is not perturbed much over this short length. Numerical simulations show [57] that this is the case when L_A is a small fraction of the dispersion length ($L_A \ll L_D$). The dispersion length L_D depends on both the pulse width T_0 and the GVD parameter β_2 and can vary from 10 to 1000 km depending on their values.

Periodic amplification of solitons can be treated mathematically by adding a gain term to Eq. (9.3.1) and writing it as [61]

$$i\frac{\partial u}{\partial \xi} + \frac{1}{2}\frac{\partial^2 u}{\partial \tau^2} + |u|^2 u = -\frac{i}{2}\Gamma u + \frac{i}{2}g(\xi)L_D u, \qquad (9.3.4)$$

where $g(\xi) = \sum_{m=1}^{N_A} g_m \delta(\xi - \xi_m)$, N_A is the total number of amplifiers, and g_m is the gain of the lumped amplifier located at ξ_m. If we assume that amplifiers are spaced uniformly, $\xi_m = m\xi_A$, where $\xi_A = L_A/L_D$ is the normalized amplifier spacing.

Because of rapid variations in the soliton energy introduced by periodic gain–loss changes, it is useful to make the transformation

$$u(\xi, \tau) = \sqrt{p(\xi)}\, v(\xi, \tau), \tag{9.3.5}$$

where $p(\xi)$ is a rapidly varying and $v(\xi, \tau)$ is a slowly varying function of ξ. Substituting Eq. (9.3.5) in Eq. (9.3.4), $v(\xi, \tau)$ is found to satisfy

$$i\frac{\partial v}{\partial \xi} + \frac{1}{2}\frac{\partial^2 v}{\partial \tau^2} + p(\xi)|v|^2 v = 0, \tag{9.3.6}$$

where $p(\xi)$ is obtained by solving the ordinary differential equation

$$\frac{dp}{d\xi} = [g(\xi)L_D - \Gamma]p. \tag{9.3.7}$$

The preceding equations can be solved analytically by noting that the amplifier gain is just large enough that $p(\xi)$ is a periodic function; it decreases exponentially in each period as $p(\xi) = \exp(-\Gamma\xi)$ but jumps to its initial value $p(0) = 1$ at the end of each period. Physically, $p(\xi)$ governs variations in the peak power (or the energy) of a soliton between two amplifiers. For a fiber with losses of 0.2 dB/km, $p(\xi)$ varies by a factor of 100 when $L_A = 100$ km.

In general, changes in soliton energy are accompanied by changes in the soliton width. Large rapid variations in $p(\xi)$ can destroy a soliton if its width changes rapidly through emission of dispersive waves. The concept of the path-averaged or guiding-center soliton makes use of the fact that solitons evolve little over a distance that is short compared with the dispersion length (or soliton period). Thus, when $\xi_A \ll 1$, the soliton width remains virtually unchanged even though its peak power $p(\xi)$ varies considerably in each section between two neighboring amplifiers. In effect, we can replace $p(\xi)$ by its average value $\bar{p}$ in Eq. (9.3.6) when $\xi_A \ll 1$. Introducing $u = \sqrt{\bar{p}}v$ as a new variable, this equation reduces to the standard NLS equation obtained for a lossless fiber.

From a practical viewpoint, a fundamental soliton can be excited if the input peak power P_s (or energy) of the path-averaged soliton is chosen to be larger by a factor $1/\bar{p}$. Introducing the amplifier gain as $G = \exp(\Gamma\xi_A)$ and using $\bar{p} = \xi_A^{-1}\int_0^{\xi_A} e^{-\Gamma\xi}\,d\xi$, the energy enhancement factor for loss-managed (LM) solitons is given by

$$f_{\mathrm{LM}} = \frac{P_s}{P_0} = \frac{1}{\bar{p}} = \frac{\Gamma\xi_A}{1 - \exp(-\Gamma\xi_A)} = \frac{G\ln G}{G - 1}, \tag{9.3.8}$$

where P_0 is the peak power in lossless fibers. Thus, soliton evolution in lossy fibers with periodic lumped amplification is identical to that in lossless fibers provided (i) amplifiers are spaced such that $L_A \ll L_D$ and (ii) the launched peak power is larger by a factor f_{LM}. As an example, $G = 10$ and $f_{\mathrm{LM}} \approx 2.56$ for 50-km amplifier spacing and fiber losses of 0.2 dB/km.

Figure 9.11 shows the evolution of a loss-managed soliton over a distance of 10 Mm assuming that solitons are amplified every 50 km. When the input pulse width corresponds to a dispersion length of 200 km, the soliton is preserved quite well even after

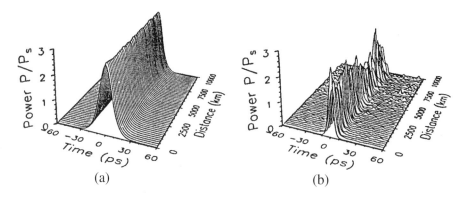

Figure 9.11: Evolution of loss-managed solitons over 10,000 km for (a) $L_D = 200$ km and (b) 25 km with $L_A = 50$ km, $\alpha = 0.22$ dB/km, and $\beta_2 = -0.5$ ps^2/km.

10 Mm because the condition $\xi_A \ll 1$ is reasonably well satisfied. However, if the dispersion length is reduced to 25 km ($\xi_A = 2$), the soliton is unable to sustain itself because of excessive emission of dispersive waves. The condition $\xi_A \ll 1$ or $L_A \ll L_D$, required to operate within the average-soliton regime, can be related to the width T_0 by using $L_D = T_0^2/|\beta_2|$. The resulting condition is

$$T_0 \gg \sqrt{|\beta_2|L_A}.\tag{9.3.9}$$

Since the bit rate B is related to T_0 through Eq. (9.2.1), the condition (9.3.9) can be written in the form of the following design criterion:

$$B^2 L_A \ll (4q_0^2|\beta_2|)^{-1}.\tag{9.3.10}$$

Choosing typical values $\beta_2 = -0.5$ ps^2/km, $L_A = 50$ km, and $q_0 = 5$, we obtain $T_0 \gg 5$ ps and $B \ll 20$ GHz. Clearly, the use of path-averaged solitons imposes a severe limitation on both the bit rate and the amplifier spacing for soliton communication systems.

9.3.3 Distributed Amplification

The condition $L_A \ll L_D$, imposed on loss-managed solitons when lumped amplifiers are used, becomes increasingly difficult to satisfy in practice as bit rates exceed 10 Gb/s. This condition can be relaxed considerably when distributed amplification is used. The distributed-amplification scheme is inherently superior to lumped amplification since its use provides a nearly lossless fiber by compensating losses locally at every point along the fiber link. In fact, this scheme was used as early as 1985 using the distributed gain provided by Raman amplification when the fiber carrying the signal was pumped at a wavelength of about 1.46 μm using a color-center laser [59]. Alternatively, the transmission fiber can be doped lightly with erbium ions and pumped periodically to provide distributed gain. Several experiments have demonstrated that solitons can be propagated in such active fibers over relatively long distances [62]–[66].

The advantage of distributed amplification can be seen from Eq. (9.3.7), which can be written in physical units as

$$\frac{dp}{dz} = [g(z) - \alpha]p. \qquad (9.3.11)$$

If $g(z)$ is constant and equal to α for all z, the peak power or energy of a soliton remains constant along the fiber link. This is the ideal situation in which the fiber is effectively lossless. In practice, distributed gain is realized by injecting pump power periodically into the fiber link. Since pump power does not remain constant because of fiber losses and pump depletion (e.g., absorption by dopants), $g(z)$ cannot be kept constant along the fiber. However, even though fiber losses cannot be compensated everywhere locally, they can be compensated fully over a distance L_A provided that

$$\int_0^{L_A} g(z)\,dz = \alpha L_A. \qquad (9.3.12)$$

A distributed-amplification scheme is designed to satisfy Eq. (9.3.12). The distance L_A is referred to as the *pump-station spacing*.

The important question is how much soliton energy varies during each gain–loss cycle. The extent of peak-power variations depends on L_A and on the pumping scheme adopted. Backward pumping is commonly used for distributed Raman amplification because such a configuration provides high gain where the signal is relatively weak. The gain coefficient $g(z)$ can be obtained following the discussion in Section 6.3. If we ignore pump depletion, the gain coefficient in Eq. (9.3.11) is given by $g(z) = g_0 \exp[-\alpha_p(L_A - z)]$, where α_p accounts for fiber losses at the pump wavelength. The resulting equation can be integrated analytically to obtain

$$p(z) = \exp\left\{ \alpha L_A \left[\frac{\exp(\alpha_p z) - 1}{\exp(\alpha_p L_A) - 1} \right] - \alpha z \right\}, \qquad (9.3.13)$$

where g_0 was chosen to ensure that $p(L_A) = 1$. Figure 9.12 shows how $p(z)$ varies along the fiber for $L_A = 50$ km using $\alpha = 0.2$ dB/km and $\alpha_p = 0.25$ dB/km. The case of lumped amplification is also shown for comparison. Whereas soliton energy varies by a factor of 10 in the lumped case, it varies by less than a factor of 2 in the case of distributed amplification.

The range of energy variations can be reduced further using a bidirectional pumping scheme. The gain coefficient $g(z)$ in this case can be approximated (neglecting pump depletion) as

$$g(z) = g_1 \exp(-\alpha_p z) + g_2 \exp[-\alpha_p(L_A - z)]. \qquad (9.3.14)$$

The constants g_1 and g_2 are related to the pump powers injected at both ends. Assuming equal pump powers and integrating Eq. (9.3.11), the soliton energy is found to vary as

$$p(z) = \exp\left[\alpha L_A \left(\frac{\sinh[\alpha_p(z - L_A/2)] + \sinh(\alpha_p L_A/2)}{2\sinh(\alpha_p L_A/2)} \right) - \alpha z \right]. \qquad (9.3.15)$$

This case is shown in Fig. 9.12 by a dashed line. Clearly, a bidirectional pumping scheme is the best as it reduces energy variations to below 15%. The range over which

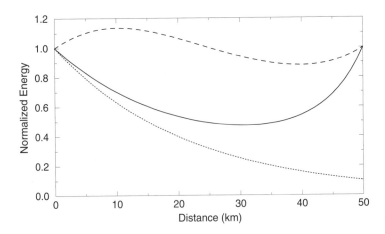

Figure 9.12: Variations in soliton energy for backward (solid line) and bidirectional (dashed line) pumping schemes with $L_A = 50$ km. The lumped-amplifier case is shown by the dotted line.

$p(z)$ varies increases with L_A. Nevertheless, it remains much smaller than that occurring in the lumped-amplification case. As an example, soliton energy varies by a factor of 100 or more when $L_A = 100$ km if lumped amplification is used but by less than a factor of 2 when the bidirectional pumping scheme is used for distributed amplification.

The effect of energy excursion on solitons depends on the ratio $\xi_A = L_A/L_D$. When $\xi_A < 1$, little soliton reshaping occurs. For $\xi_A \gg 1$, solitons evolve adiabatically with some emission of dispersive waves (the quasi-adiabatic regime). For intermediate values of ξ_A, a more complicated behavior occurs. In particular, dispersive waves and solitons are resonantly amplified when $\xi_A \simeq 4\pi$. Such a resonance can lead to unstable and chaotic behavior [60]. For this reason, distributed amplification is used with $\xi_A < 4\pi$ in practice [62]–[66].

Modeling of soliton communication systems making use of distributed amplification requires the addition of a gain term to the NLS equation, as in Eq. (9.3.4). In the case of soliton systems operating at bit rates $B > 20$ Gb/s such that $T_0 < 5$ ps, it is also necessary to include the effects of third-order dispersion (TOD) and a new nonlinear phenomenon known as the *soliton self-frequency shift* (SSFS). This effect was discovered in 1986 [67] and can be understood in terms of intrapulse Raman scattering [68]. The Raman effect leads to a continuous downshift of the soliton carrier frequency when the pulse spectrum becomes so broad that the high-frequency components of a pulse can transfer energy to the low-frequency components of the same pulse through Raman amplification. The Raman-induced frequency shift is negligible for $T_0 > 10$ ps but becomes of considerable importance for short solitons ($T_0 < 5$ ps). With the inclusion of SSFS and TOD, Eq. (9.3.4) takes the form [10]

$$i\frac{\partial u}{\partial \xi} + \frac{1}{2}\frac{\partial^2 u}{\partial \tau^2} + |u|^2 u = \frac{iL_D}{2}[g(\xi) - \alpha]u + i\delta_3 \frac{\partial^3 u}{\partial \tau^3} + \tau_R u \frac{\partial |u|^2}{\partial \tau}, \qquad (9.3.16)$$

where the TOD parameter δ_3 and the Raman parameter τ_R are defined as

$$\delta = \beta_3/(6|\beta_2|T_0), \qquad \tau_R = T_R/T_0. \qquad (9.3.17)$$

The quantity T_R is related to the slope of the Raman gain spectrum and has a value of about 3 fs for silica fibers [10].

Numerical simulations based on Eq. (9.3.16) show that the distributed-amplification scheme benefits considerably high-capacity soliton communication systems [69]. For example, when $L_D = 50$ km but amplifiers are placed 100 km apart, fundamental solitons with $T_0 = 5$ ps are destroyed after 500 km in the case of lumped amplifiers but can propagate over a distance of more than 5000 km when distributed amplification is used. For soliton widths below 5 ps, the Raman-induced spectral shift leads to considerable changes in the evolution of solitons as it modifies the gain and dispersion experienced by solitons. Fortunately, the finite gain bandwidth of amplifiers reduces the amount of spectral shift and stabilizes the soliton carrier frequency close to the gain peak [63]. Under certain conditions, the spectral shift can become so large that it cannot be compensated, and the soliton moves out of the gain window, loosing all its energy.

9.3.4 Experimental Progress

Early experiments on loss-managed solitons concentrated on the Raman-amplification scheme. An experiment in 1985 demonstrated that fiber losses can be compensated over 10 km by the Raman gain while maintaining the soliton width [59]. Two color-center lasers were used in this experiment. One laser produced 10-ps pulses at 1.56 μm, which were launched as fundamental solitons. The other laser operated continuously at 1.46 μm and acted as a pump for amplifying 1.56-μm solitons. In the absence of the Raman gain, the soliton broadened by about 50% because of loss-induced broadening. This amount of broadening was in agreement with Eq. (9.3.3), which predicts $T_1/T_0 = 1.51$ for $z = 10$ km and $\alpha = 0.18$ dB/km, the values used in the experiment. When the pump power was about 125 mW, the 1.8-dB Raman gain compensated the fiber losses and the output pulse was nearly identical with the input pulse.

A 1988 experiment transmitted solitons over 4000 km using the Raman-amplification scheme [4]. This experiment used a 42-km fiber loop whose loss was exactly compensated by injecting the CW pump light from a 1.46-μm color-center laser. The solitons were allowed to circulate many times along the fiber loop and their width was monitored after each round trip. The 55-ps solitons could be circulated along the loop up to 96 times without a significant increase in their pulse width, indicating soliton recovery over 4000 km. The distance could be increased to 6000 km with further optimization. This experiment was the first to demonstrate that solitons could be transmitted over transoceanic distances in principle. The main drawback was that Raman amplification required pump lasers emitting more than 500 mW of CW power near 1.46 μm. It was not possible to obtain such high powers from semiconductor lasers in 1988, and the color-center lasers used in the experiment were too bulky to be useful for practical lightwave systems.

The situation changed with the advent of EDFAs around 1989 when several experiments used them for loss-managed soliton systems [38]–[40]. These experiments can

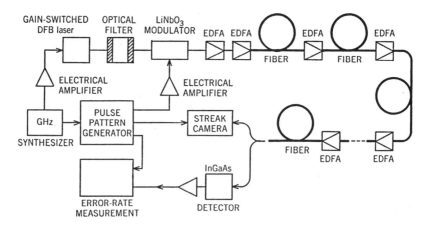

Figure 9.13: Setup used for soliton transmission in a 1990 experiment. Two EDFAs after the LiNbO$_3$ modulator boost pulse peak power to the level of fundamental solitons. (After Ref. [70]; ©1990 IEEE; reprinted with permission.)

be divided into two categories, depending on whether a linear fiber link or a recirculating fiber loop is used for the experiment. The experiments using fiber link are more realistic as they mimic the actual field conditions. Several 1990 experiments demonstrated soliton transmission over fiber lengths ∼100 km at bit rates of up to 5 Gb/s [70]–[72]. Figure 9.13 shows one such experimental setup in which a gain-switched laser is used for generating input pulses. The pulse train is filtered to reduce the frequency chirp and passed through a LiNbO$_3$ modulator to impose the RZ format on it. The resulting coded bit stream of solitons is transmitted through several fiber sections, and losses of each section are compensated by using an EDFA. The amplifier spacing is chosen to satisfy the criterion $L_A \ll L_D$ and is typically in the range 25–40 km. In a 1991 experiment, solitons were transmitted over 1000 km at 10 Gb/s [73]. The 45-ps-wide solitons permitted an amplifier spacing of 50 km in the average-soliton regime.

Since 1991, most soliton transmission experiments have used a recirculating fiber-loop configuration because of cost considerations. Figure 9.14 shows such an experimental setup schematically. A bit stream of solitons is launched into the loop and forced to circulate many times using optical switches. The quality of the signal is monitored after each round trip to ensure that the solitons maintain their width during transmission. In a 1991 experiment, 2.5-Gb/s solitons were transmitted over 12,000 km by using a 75-km fiber loop containing three EDFAs, spaced apart by 25 km [74]. In this experiment, the bit rate–distance product of $BL = 30$ (Tb/s)-km was limited mainly by the timing jitter induced by EDFAs. The use of amplifiers degrades the signal-to-noise ratio (SNR) and shifts the position of solitons in a random fashion. These issues are discussed in Section 9.5.

Because of the problems associated with the lumped amplifiers, several schemes were studied for reducing the timing jitter and improving the performance of soliton systems. Even the technique of Raman amplification was revived in 1999 and has

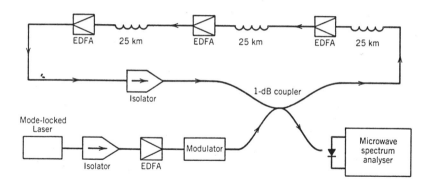

Figure 9.14: Recirculating-loop configuration used in a 1991 experiment for transmitting solitons over 12,000 km. (After Ref. [74]; ©1991 IEE; reprinted with permission.)

become quite common for both the soliton and nonsoliton systems. Its revival was possible because of the technological advances in the fields of semiconductor and fiber lasers, both of which can provide power levels in excess of 500 mW. The use of dispersion management also helps in reducing the timing jitter. We turn to dispersion-managed solitons next.

9.4 Dispersion-Managed Solitons

As discussed in Chapter 7, dispersion management is employed commonly for modern wavelength-division multiplexed (WDM) systems. It turns out that soliton systems benefit considerably if the GVD parameter β_2 varies along the link length. This section is devoted to such dispersion-managed solitons. We first consider dispersion-decreasing fibers and then focus on dispersion maps that consist of multiple sections of constant-dispersion fibers.

9.4.1 Dispersion-Decreasing Fibers

An interesting scheme proposed in 1987 relaxes completely the restriction $L_A \ll L_D$ imposed normally on loss-managed solitons, by decreasing the GVD along the fiber length [75]. Such fibers are called *dispersion-decreasing* fibers (DDFs) and are designed such that the decreasing GVD counteracts the reduced SPM experienced by solitons weakened from fiber losses.

Since dispersion management is used in combination with loss management, soliton evolution in a DDF is governed by Eq. (9.3.6) except that the second-derivative term has a new parameter d that is a function of ξ because of GVD variations along the fiber length. The modified NLS equation takes the form

$$i\frac{\partial v}{\partial \xi} + \frac{1}{2}d(\xi)\frac{\partial^2 v}{\partial \tau^2} + p(\xi)|v|^2 v = 0, \qquad (9.4.1)$$

where $v = u/\sqrt{p}$, $d(\xi) = \beta_2(\xi)/\beta_2(0)$, and $p(\xi)$ takes into account peak-power variations introduced by loss management. The distance ξ is normalized to the dispersion length, $L_D = T_0^2/|\beta_2(0)|$, defined using the GVD value at the fiber input.

Because of the ξ dependence of the second and third terms, Eq. (9.4.1) is not a standard NLS equation. However, it can be reduced to one if we introduce a new propagation variable as

$$\xi' = \int_0^\xi d(\xi)\,d\xi. \tag{9.4.2}$$

This transformation renormalizes the distance scale to the local value of GVD. In terms of ξ', Eq. (9.4.1) becomes

$$i\frac{\partial v}{\partial \xi'} + \frac{1}{2}\frac{\partial^2 v}{\partial \tau^2} + \frac{p(\xi)}{d(\xi)}|v|^2 v = 0. \tag{9.4.3}$$

If the GVD profile is chosen such that $d(\xi) = p(\xi) \equiv \exp(-\Gamma\xi)$, Eq. (9.4.3) reduces the standard NLS equation obtained in the absence of fiber losses. As a result, fiber losses have no effect on a soliton in spite of its reduced energy when DDFs are used. Lumped amplifiers can be placed at any distance and are not limited by the condition $L_A \ll L_D$.

The preceding analysis shows that fundamental solitons can be maintained in a lossy fiber provided its GVD decreases exponentially as

$$|\beta_2(z)| = |\beta_2(0)|\exp(-\alpha z). \tag{9.4.4}$$

This result can be understood qualitatively by noting that the soliton peak power P_0 decreases exponentially in a lossy fiber in exactly the same fashion. It is easy to deduce from Eq. (9.1.4) that the requirement $N = 1$ can be maintained, in spite of power losses, if both $|\beta_2|$ and γ decrease exponentially at the same rate. The fundamental soliton then maintains its shape and width even in a lossy fiber.

Fibers with a nearly exponential GVD profile have been fabricated [76]. A practical technique for making such DDFs consists of reducing the core diameter along the fiber length in a controlled manner during the fiber-drawing process. Variations in the fiber diameter change the waveguide contribution to β_2 and reduce its magnitude. Typically, GVD can be varied by a factor of 10 over a length of 20 to 40 km. The accuracy realized by the use of this technique is estimated to be better than 0.1 ps^2/km [77]. Propagation of solitons in DDFs has been demonstrated in several experiments [77]–[79]. In a 40-km DDF, solitons preserved their width and shape in spite of energy losses of more than 8 dB [78]. In a recirculating loop made using DDFs, a 6.5-ps soliton train at 10 Gb/s could be transmitted over 300 km [79].

Fibers with continuously varying GVD are not readily available. As an alternative, the exponential GVD profile of a DDF can be approximated with a staircase profile by splicing together several constant-dispersion fibers with different β_2 values. This approach was studied during the 1990s, and it was found that most of the benefits of DDFs can be realized using as few as four fiber segments [80]–[84]. How should one select the length and the GVD of each fiber used for emulating a DDF? The answer is not obvious, and several methods have been proposed. In one approach, power deviations are minimized in each section [80]. In another approach, fibers of different GVD values D_m and different lengths L_m are chosen such that the product $D_m L_m$ is the same for each section. In a third approach, D_m and L_m are selected to minimize shading of dispersive waves [81].

9.4.2 Periodic Dispersion Maps

A disadvantage of the DDF is that the average dispersion along the link is often relatively large. Generally speaking, operation of a soliton in the region of low average GVD improves system performance. Dispersion maps consisting of alternating-GVD fibers are attractive because their use lowers the average GVD of the entire link while keeping the GVD of each section large enough that the four-wave mixing (FWM) and TOD effects remain negligible.

The use of dispersion management forces each soliton to propagate in the normal-dispersion regime of a fiber during each map period. At first sight, such a scheme should not even work because the normal-GVD fibers do not support bright solitons and lead to considerable broadening and chirping of the pulse. So, why should solitons survive in a dispersion-managed fiber link? An intense theoretical effort devoted to this issue since 1996 has yielded an answer with a few surprises [85]–[102]. Physically speaking, if the map period is a fraction of the nonlinear length, the nonlinear effects are relatively small, and the pulse evolves in a linear fashion over one map period. On a longer length scale, solitons can still form if the SPM effects are balanced by the average dispersion. As a result, solitons can survive in an average sense, even though not only the peak power but also the width and shape of such solitons oscillate periodically. This section describes the properties of dispersion-managed (DM) solitons and the advantages offered by them.

Consider a simple dispersion map consisting of two fibers with positive and negative values of the GVD parameter β_2. Soliton evolution is still governed by Eq. (9.4.1) used earlier for DDFs. However, we cannot use ξ and τ as dimensionless parameters because the pulse width and GVD both vary along the fiber. It is better to use the physical units and write Eq. (9.4.1) as

$$i\frac{\partial B}{\partial z} - \frac{\beta_2(z)}{2}\frac{\partial^2 B}{\partial t^2} + \gamma p(z)|B|^2 B = 0, \tag{9.4.5}$$

where $B = A/\sqrt{p}$ and $p(z)$ is the solution of Eq. (9.3.11). The GVD parameter takes values β_{2a} and β_{2n} in the anomalous and normal sections of lengths l_a and l_n, respectively. The map period $L_{\text{map}} = l_a + l_n$ can be different from the amplifier spacing L_A. As is evident, the properties of DM solitons will depend on several map parameters even when only two types of fibers are used in each map period.

Equation (9.4.5) can be solved numerically using the split-step Fourier method. Numerical simulations show that a nearly periodic solution can often be found by adjusting input pulse parameters (width, chirp, and peak power) even though these parameters vary considerably in each map period. The shape of such DM solitons is typically closer to a Gaussian profile rather than the "sech" shape associated with standard solitons [86]–[88].

Numerical solutions, although essential, do not lead to much physical insight. Several techniques have been used to solve the NLS equation (9.4.5) approximately. A common approach makes use of the variational method [89]–[91]. Another approach expands $B(z,t)$ in terms of a complete set of the Hermite–Gauss functions that are solutions of the linear problem [92]. A third approach solves an integral equation, derived in the spectral domain using perturbation theory [94]–[96].

To simplify the following discussion, we focus on the variational method used earlier in Section 7.8.2. In fact, the Lagrangian density obtained there can be used directly for DM solitons as well as Eq. (9.4.5) is identical to Eq. (7.8.4). Because the shape of the DM soliton is close to a Gaussian pulse in numerical simulations, the soliton is assumed to evolve as

$$B(z,t) = a \exp[-(1+iC)t^2/2T^2 + i\phi], \qquad (9.4.6)$$

where a is the amplitude, T is the width, C is the chirp, and ϕ is the phase of the soliton. All four parameters vary with z because of perturbations produced by periodic variations of $\beta_2(z)$ and $p(z)$.

Following Section 7.8.2, we can obtain four ordinary differential equations for the four soliton parameters. The amplitude equation can be eliminated because $a^2T = a_0^2 T_0 = E_0/\sqrt{\pi}$ is independent of z and is related to the input pulse energy E_0. The phase equation can also be dropped since T and C do not depend on ϕ. The DM soliton then corresponds to a periodic solution of the following two equations for the pulse width T and chirp C:

$$\frac{dT}{dz} = \beta_2(z)\frac{C}{T}, \qquad (9.4.7)$$

$$\frac{dC}{dz} = \frac{\gamma E_0 p(z)}{\sqrt{2\pi}T} + \frac{\beta_2}{T^2}(1+C^2). \qquad (9.4.8)$$

These equations should be solved with the periodic boundary conditions

$$T_0 \equiv T(0) = T(L_A), \qquad C_0 \equiv C(0) = C(L_A) \qquad (9.4.9)$$

to ensure that the soliton recovers its initial state after each amplifier. The periodic boundary conditions fix the values of the initial width T_0 and the chirp C_0 at $z = 0$ for which a soliton can propagate in a periodic fashion for a given value of the pulse energy E_0. A new feature of the DM solitons is that the input pulse width depends on the dispersion map and cannot be chosen arbitrarily. In fact, T_0 cannot be below a critical value that is set by the map itself.

Figure 9.15 shows how the pulse width T_0 and the chirp C_0 of allowed periodic solutions vary with input pulse energy for a specific dispersion map. The minimum value T_m of the pulse width occurring in the middle of the anomalous-GVD section of the map is also shown. The map is suitable for 40-Gb/s systems and consists of alternating fibers with GVD of -4 and 4 ps^2/km and lengths $l_a \approx l_n = 5$ km such that the average GVD is -0.01 ps^2/km. The solid lines show the case of ideal distributed amplification for which $p(z) = 1$ in Eq. (9.4.8). The lumped-amplification case is shown by the dashed lines in Fig. 9.15 assuming 80-km amplifier spacing and 0.25 dB/km losses in each fiber section.

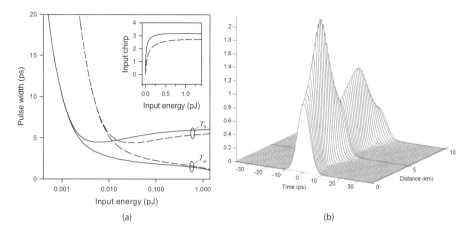

Figure 9.15: (a) Changes in T_0 (upper curve) and T_m (lower curve) with input pulse energy E_0 for $\alpha = 0$ (solid lines) and 0.25 dB/km (dashed lines). The inset shows the input chirp C_0 in the two cases. (b) Evolution of the DM soliton over one map period for $E_0 = 0.1$ pJ and $L_A = 80$ km.

Several conclusions can be drawn from Fig. 9.15. First, both T_0 and T_m decrease rapidly as pulse energy is increased. Second, T_0 attains its minimum value at a certain pulse energy E_c while T_m keeps decreasing slowly. Third, T_0 and T_m differ from each other considerably for $E_0 > E_c$. This behavior indicates that the pulse width changes considerably in each fiber section when this regime is approached. An example of pulse breathing is shown in Fig. 9.15(b) for $E_0 = 0.1$ pJ in the case of lumped amplification. The input chirp C_0 is relatively large ($C_0 \approx 1.8$) in this case. The most important feature of Fig. 9.15 is the existence of a minimum value of T_0 for a specific value of the pulse energy. The input chirp $C_0 = 1$ at that point. It is interesting to note that the minimum value of T_0 does not depend much on fiber losses and is about the same for the solid and dashed curves although the value of E_c is much larger in the lumped amplification case because of fiber losses.

As seen from Fig. 9.15, both the pulse width and the peak power of DM solitons vary considerably within each map period. Figure 9.16(a) shows the width and chirp variations over one map period for the DM soliton of Fig. 9.15(b). The pulse width varies by more than a factor of 2 and becomes minimum nearly in the middle of each fiber section where frequency chirp vanishes. The shortest pulse occurs in the middle of the anomalous-GVD section in the case of ideal distributed amplification in which fiber losses are compensated fully at every point along the fiber link. For comparison, Fig. 9.16(b) shows the width and chirp variations for a DM soliton whose input energy is close to E_c where the input pulse is shortest. Breathing of the pulse is reduced considerably together with the range of chirp variations. In both cases, the DM soliton is quite different from a standard fundamental soliton as it does not maintain its shape, width, or peak power. Nevertheless, its parameters repeat from period to period at any location within the map. For this reason, DM solitons can be used for optical communications in spite of oscillations in the pulse width. Moreover, such solitons perform better from a system standpoint.

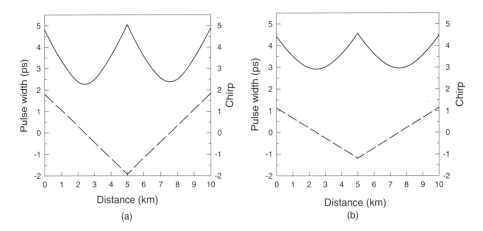

Figure 9.16: (a) Variations of pulse width and chirp over one map period for DM solitons with the input energy (a) $E_0 \gg E_c = 0.1$ pJ and (b) E_0 close to E_c.

9.4.3 Design Issues

Figures 9.15 and 9.16 show that Eqs. (9.4.7)–(9.4.9) permit periodic propagation of many different DM solitons in the same map by choosing different values of E_0, T_0, and C_0. How should one choose among these solutions when designing a soliton system? Pulse energies much smaller than E_c (corresponding to the minimum value of T_0) should be avoided because a low average power would then lead to rapid degradation of SNR as amplifier noise builds up with propagation. On the other hand, when $E_0 \gg E_c$, large variations in the pulse width in each fiber section would enhance the effects of soliton interaction if two neighboring solitons begin to overlap. Thus, the region near $E_0 = E_c$ is most suited for designing DM soliton systems. Numerical solutions of Eq. (9.4.5) confirm this conclusion.

The 40-Gb/s system design shown in Figs. 9.15 and 9.16 was possible only because the map period L_{map} was chosen to be much smaller than the amplifier spacing of 80 km, a configuration referred to as the *dense* dispersion management. When L_{map} is increased to 80 km using $l_a \approx l_b = 40$ km while keeping the same value of average dispersion, the minimum pulse width supported by the map increases by a factor of 3. The bit rate is then limited to about 20 Gb/s. In general, the required map period becomes shorter as the bit rate increases.

It is possible to find the values of T_0 and T_m by solving the variational equations (9.4.7)–(9.4.9) approximately. Equation (9.4.7) can be integrated to relate T and C as

$$T^2(z) = T_0^2 + 2 \int_0^z \beta_2(z) C(z) \, dz. \tag{9.4.10}$$

The chirp equation cannot be integrated but the numerical solutions show that $C(z)$ varies almost linearly in each fiber section. As seen in Fig. 9.16, $C(z)$ changes from C_0 to $-C_0$ in the first section and then back to C_0 in the second section. Noting that the ratio $(1 + C^2)/T^2$ is related to the spectral width that changes little over one map period when the nonlinear length is much larger than the local dispersion length, we average it over one map period and obtain the following relation between T_0 and C_0:

$$T_0 = T_{map}\sqrt{\frac{1+C_0^2}{|C_0|}}, \qquad T_{map} = \left(\frac{|\beta_{2n}\beta_{2a}l_n l_a|}{\beta_{2n}l_n - \beta_{2a}l_a}\right)^{1/2}, \qquad (9.4.11)$$

where T_{map} is a parameter with dimensions of time involving only the four map parameters. It provides a time scale associated with an arbitrary dispersion map in the sense that the stable periodic solutions supported by it have input pulse widths that are close to T_{map} (within a factor of 2 or so). The minimum value of T_0 occurs for $|C_0| = 1$ and is given by $T_0^{min} = \sqrt{2}T_{map}$.

Equation (9.4.11) can also be used to find the shortest pulse within the map. Recalling from Section 2.4 that the shortest pulse occurs at the point at which the propagating pulse becomes unchirped, $T_m = T_0/(1+C_0^2)^{1/2} = T_{map}/\sqrt{|C_0|}$. When the input pulse corresponds to its minimum value ($C_0 = 1$), T_m is exactly equal to T_{map}. The optimum value of the pulse stretching factor is equal to $\sqrt{2}$ under such conditions. These conclusions are in agreement with the numerical results shown in Fig. 9.16 for a specific map for which $T_{map} \approx 3.16$ ps. If dense dispersion management is not used for this map and L_{map} equals $L_A = 80$ km, this value of T_{map} increases to 9 ps. Since the FWHM of input pulses then exceeds 21 ps, such a map is unsuitable for 40-Gb/s soliton systems. In general, the required map period becomes shorter and shorter as the bit rate increases as is evident from the definition of T_{map} in Eq. (9.4.11).

It is useful to look for other combinations of the four map parameters that may play an important role in designing a DM soliton system. Two parameters that help for this purpose are defined as [89]

$$\bar{\beta}_2 = \frac{\beta_{2n}l_n + \beta_{2a}l_a}{l_n + l_a}, \qquad S_m = \frac{\beta_{2n}l_n - \beta_{2a}l_a}{T_{FWHM}^2}, \qquad (9.4.12)$$

where $T_{FWHM} \approx 1.665T_m$ is the FWHM at the location where pulse width is minimum in the anomalous-GVD section. Physically, $\bar{\beta}_2$ represents the average GVD of the entire link, while the map strength S_m is a measure of how much GVD varies between two fibers in each map period. The solutions of Eqs. (9.4.7)–(9.4.9) as a function of map strength S for different values of $\bar{\beta}_2$ reveal the surprising feature that DM solitons can exist even when the average GVD is normal provided the map strength exceeds a critical value S_{cr} [97]–[101]. Moreover, when $S_m > S_{cr}$ and $\bar{\beta}_2 > 0$, a periodic solution can exist for two different values of the input pulse energy. Numerical solutions of Eqs. (9.4.1) confirm these predictions but the critical value of the map strength is found to be only 3.9 instead of 4.8 obtained from the variational equations [89].

The existence of DM solitons in maps with normal average GVD is quite intriguing as one can envisage dispersion maps in which a soliton propagates in the normal-GVD regime most of the time. An example is provided by the dispersion map in which a short section of standard fiber ($\beta_{2a} \approx -20$ ps^2/km) is used with a long section of dispersion-shifted fiber ($\beta_{2n} \approx 1$ ps^2/km) such that $\bar{\beta}_2$ is close to zero but positive. The formation of DM solitons under such conditions can be understood by noting that when S_m exceeds 4, input energy of a pulse becomes large enough that its spectral width is considerably larger in the anomalous-GVD section compared with the normal-GVD section. Noting that the phase shift imposed on each spectral component varies as $\beta_2 \omega^2$ locally, one can define an effective value of the average GVD as [101]

$$\bar{\beta}_2^{\text{eff}} = \langle \beta_2 \Omega^2 \rangle / \langle \Omega^2 \rangle, \tag{9.4.13}$$

where Ω is the local value of the spectral width and the angle brackets indicate average over the map period. If $\bar{\beta}_2^{\text{eff}}$ is negative, the DM soliton can exist even if $\bar{\beta}_2$ is positive.

For map strengths below a critical value (about 3.9 numerically), the average GVD is anomalous for DM solitons. In that case, one is tempted to compare them with standard solitons forming in a uniform-GVD fiber link with $\beta_2 = \bar{\beta}_2$. For relatively small values of S_m, variations in the pulse width and chirp are small enough that one can ignore them. The main difference between the average-GVD and DM solitons then stems from the higher peak power required to sustain DM solitons. The energy enhancement factor for DM solitons is defined as [85]

$$f_{\text{DM}} = E_0^{\text{DM}} / E_0^{\text{av}} \tag{9.4.14}$$

and can exceed 10 depending on the system design. The larger energy of DM solitons benefits a soliton system in several ways. Among other things, it improves the SNR and decreases the timing jitter; these issues are discussed in Section 9.5.

Dispersion-management schemes were used for solitons as early as 1992 although they were referred to by names such as partial soliton communication and dispersion allocation [103]. In the simplest form of dispersion management, a relatively short segment of dispersion-compensating fiber (DCF) is added periodically to the transmission fiber, resulting in dispersion maps similar to those used for nonsoliton systems. It was found in a 1995 experiment that the use of DCFs reduced the timing jitter considerably [104]. In fact, in this 20-Gb/s experiment, the timing jitter became low enough when the average dispersion was reduced to a value near -0.025 ps^2/km that the 20-Gb/s signal could be transmitted over transoceanic distances.

Since 1996, a large number of experiments have shown the benefits of DM solitons for lightwave systems [105]–[114]. In one experiment, the use of a periodic dispersion map enabled transmission of a 20-Gb/s soliton bit stream over 5520 km of a fiber link containing amplifiers at 40-km intervals [105]. In another 20-Gb/s experiment [106], solitons could be transmitted over 9000 km without using any in-line optical filters since the periodic use of DCFs reduced timing jitter by more than a factor of 3. A 1997 experiment focused on transmission of DM solitons using dispersion maps such that solitons propagated most of the time in the normal-GVD regime [107]. This 10-Gb/s experiment transmitted signals over 28 Mm using a recirculating fiber loop consisting of 100 km of normal-GVD fiber and 8-km of anomalous-GVD fiber such that the average GVD was anomalous (about -0.1 ps^2/km). Periodic variations in the pulse width were also observed in such a fiber loop [108]. In a later experiment, the loop was modified to yield the average-GVD value of zero or slightly positive [109]. Stable transmission of 10-Gb/s solitons over 28 Mm was still observed. In all cases, experimental results were in excellent agreement with numerical simulations [110].

An important application of dispersion management consists for upgrading the existing terrestrial networks designed with standard fibers [111]–[114]. A 1997 experiment used fiber gratings for dispersion compensation and realized 10-Gb/s soliton transmission over 1000 km. Longer transmission distances were realized using a re-circulating fiber loop [112] consisting of 102 km of standard fiber with anomalous GVD ($\beta_2 \approx -21$ ps^2/km) and 17.3 km of DCF with normal GVD ($\beta_2 \approx 160$ ps^2/km). The map strength S was quite large in this experiment when 30-ps (FWHM) pulses were launched into the loop. By 1999, 10-Gb/s DM solitons could be transmitted over 16 Mm of standard fiber when soliton interactions were minimized by choosing the location of amplifiers appropriately [113].

9.5 Impact of Amplifier Noise

The use of in-line optical amplifiers affects the soliton evolution considerably. The reason is that amplifiers, needed to restore the soliton energy, also add noise originating from *amplified spontaneous emission* (ASE). As discussed in Section 6.5, the spectral density of ASE depends on the amplifier gain G itself and is given by Eq. (6.1.15). The ASE-induced noise degrades the SNR through amplitude fluctuations and introduces timing jitter through frequency fluctuations, both of which impact the performance of soliton systems. Timing jitter for solitons has been studied since 1986 and is referred to as the Gordon–Haus jitter [115]–[125]. The moment method is used in this section for studying the effects of amplifier noise.

9.5.1 Moment Method

The moment method has been introduced in Section 6.5.2 in the context of nonsoliton pulses. The same treatment can be extended for solitons [122]. In the case of Eq. (9.4.5), the three moments providing energy E, frequency shift Ω, and position q of the pulse are given by

$$E = \int_{-\infty}^{\infty} |B|^2 dt, \qquad q = \frac{1}{E} \int_{-\infty}^{\infty} t|B|^2 dt, \tag{9.5.1}$$

$$\Omega = \frac{i}{2E} \int_{-\infty}^{\infty} \left(B^* \frac{\partial B}{\partial t} - B \frac{\partial B^*}{\partial t} \right) dt. \tag{9.5.2}$$

The three quantities depend on z and vary along the fiber as the pulse shape governed by $|B(z,t)|^2$ evolves. Differentiating E, Ω, and q with respect to z and using Eq. (9.4.5), the three moments are found to evolve with z as [124]

$$\frac{dE}{dz} = \sum_n \delta E_n \delta(z - z_n), \tag{9.5.3}$$

$$\frac{d\Omega}{dz} = \sum_n \delta \Omega_n \delta(z - z_n), \tag{9.5.4}$$

$$\frac{dq}{dz} = \beta_2 \Omega + \sum_n \delta q_n \delta(z - z_n), \tag{9.5.5}$$

where δE_n, $\delta \Omega_n$, and δq_n are random changes induced by ASE at the nth amplifier located at z_n. The sum in these equations extends over the total number N_A of amplifiers.

Fiber losses do not appear in Eq. (9.5.3) because of the transformation $B = A/\sqrt{p}$ made in deriving Eq. (9.4.5); the actual pulse energy is given by pE, where $p(z)$ is obtained by solving Eq. (9.3.11).

The physical meaning of the moment equations is clear from Eqs. (9.5.3)–(9.5.5). Both E and Ω remain constant while propagating inside optical fibers but change in a random fashion at each amplifier location. Equation (9.5.5) shows how frequency fluctuations induced by an amplifier become temporal fluctuations because of the GVD. Physically speaking, the group velocity of the pulse depends on frequency. A random change in the group velocity results in a shift of the soliton position by a random amount within the bit slot. As a result, frequency fluctuations are converted into timing jitter by the GVD. The last term in Eq. (9.5.5) shows that ASE also shifts the soliton position directly.

Fluctuations in the position and frequency of a soliton at any amplifier vanish on average but their variances are finite. Moreover, the two fluctuations are not independent as they are produced by the same physical mechanism (spontaneous emission). We thus need to consider how the optical field $B(z,t)$ is affected by ASE and then calculate the variances and correlation functions of E, Ω, and q. At each amplifier, the field $B(z,t)$ changes by $\delta B(z,t)$ because of ASE. The fluctuation $\delta B(z,t)$ vanishes on average; its second-order moment can be written as

$$\langle \delta B(z_a,t)\delta B(z_a,t')\rangle = S_{\text{sp}}\delta(t-t'),\qquad(9.5.6)$$

where z_a denotes the location of an amplifier and

$$S_{\text{sp}} = n_{\text{sp}}h\nu_0(G-1)\qquad(9.5.7)$$

is the spectral density of ASE noise assumed to be constant (white noise) by treating the ASE process as a Markovian stochastic process [9]. This is justified in view of the independent nature of each spontaneous-emission event. The angle brackets in Eq. (9.5.6) denote an ensemble average over all such events. In Eq. (9.5.7), G represents the amplifier gain, $h\nu_0$ is the photon energy, and the spontaneous emission factor n_{sp} is related to the noise figure F_n of the amplifier as $F_n = 2n_{\text{sp}}$.

The moments of δE_n, δq_n, and $\delta \Omega_n$ are obtained by replacing B in Eqs. (9.5.1) and (9.5.2) with $B + \delta B$ and linearizing in δB. For an arbitrary pulse shape, the second-order moments are given by [122]

$$\langle(\delta E)^2\rangle = 2S_{\text{sp}}\int_{-\infty}^{\infty}|B|^2\,dt,\qquad \langle(\delta q)^2\rangle = \frac{2S_{\text{sp}}}{E_0^2}\int_{-\infty}^{\infty}(t-q)^2|B|^2\,dt,\quad(9.5.8)$$

$$\langle(\delta\Omega)^2\rangle = \frac{2S_{\text{sp}}}{E_0^2}\int_{-\infty}^{\infty}\left|\frac{\partial B}{\partial t}\right|^2\,dt,\qquad \langle\delta E\,\delta q\rangle = \frac{2S_{\text{sp}}}{E_0}\int_{-\infty}^{\infty}(t-q)|B|^2\,dt,\quad(9.5.9)$$

$$\langle\delta E\,\delta\Omega\rangle = \frac{iS_{\text{sp}}}{E_0}\int_{-\infty}^{\infty}\left(V^*\frac{\partial V}{\partial t}-V\frac{\partial V^*}{\partial t}\right)dt,\qquad(9.5.10)$$

$$\langle\delta\Omega\,\delta q\rangle = \frac{iS_{\text{sp}}}{2E_0^2}\int_{-\infty}^{\infty}(t-q)\left(V^*\frac{\partial V}{\partial t}-V\frac{\partial V^*}{\partial t}\right)dt,\qquad(9.5.11)$$

where $V = B\exp(i\Omega t)$. The integrals in these equations can be calculated if $B(z_a,t)$ is known at the amplifier location. The variances and correlations of fluctuations are the same for all amplifiers because ASE in any two amplifiers is not correlated (the subscript n has been dropped for this reason).

The pulse shape depends on whether the GVD is constant along the entire link or is changing in a periodic manner through a dispersion map. In the case of DM solitons, the exact form of $B(z_a, t)$ can only be obtained by solving Eq. (9.4.5) numerically. The use of Gaussian approximation for the pulse shape simplifies the analysis without introducing too much error because the pulse shape deviates from Gaussian only in the pulse wings (which contribute little to the integrals because of their low intensity levels). However, Eq. (9.4.6) should be modified as

$$B(z,t) = a \exp[-(1+iC)(t-q)^2/2T^2 - i\Omega(t-q) + i\phi] \qquad (9.5.12)$$

to include the frequency shift Ω and the position shift q explicitly, both of which are zero in the absence of optical amplifiers. The six parameters (a, C, T, q, Ω, and ϕ) vary with z in a periodic fashion. Using Eq. (9.5.12) in Eqs. (9.5.8)–(9.5.11), the variances and correlations of fluctuations are found to be

$$\langle(\delta E)^2\rangle = 2S_{sp}E_0, \qquad \langle(\delta\Omega)^2\rangle = (S_{sp}/E_0)[(1+C_0^2)/T_0^2], \qquad (9.5.13)$$

$$\langle(\delta q)^2\rangle = (S_{sp}/E_0)T_0^2, \qquad \langle\delta E\delta q\rangle = 0, \qquad (9.5.14)$$

$$\langle\delta\Omega\delta q\rangle\rangle = (S_{sp}/E_0)C_0, \qquad \langle\delta E\delta\Omega\rangle = (2\pi^{-1/2})S_{sp}C_0/T_0. \qquad (9.5.15)$$

The input pulse parameters appear in these equations because the pulse recovers its original form at each amplifier for DM solitons.

In the case of constant-dispersion fibers or DDFs, the soliton remains unchirped and maintains a "sech" shape. In this case, Eq. (9.5.12) should be replaced with

$$B(z,t) = a \operatorname{sech}[(t-q)/T] - i\Omega(t-q) + i\phi]. \qquad (9.5.16)$$

Using Eq. (9.5.16) in Eqs. (9.5.8)–(9.5.11), the variances are given by

$$\langle(\delta E)^2\rangle = 2S_{sp}E_0, \qquad \langle(\delta\Omega)^2\rangle = \frac{2S_{sp}}{3E_0T_0^2}, \qquad \langle(\delta q)^2\rangle = \frac{\pi^2 S_{sp}}{6E_0}T_0^2, \qquad (9.5.17)$$

but all three cross-correlations are zero. The presence of chirp on the DM solitons is responsible for producing cross-correlations.

9.5.2 Energy and Frequency Fluctuations

Energy fluctuations induced by optical amplifiers degrade the optical SNR. To find the SNR, we integrate Eq. (9.5.3) between two neighboring amplifiers and obtain the recurrence relation

$$E(z_n) = E(z_{n-1}) + \delta E_n, \qquad (9.5.18)$$

where $E(z_n)$ denotes energy at the output of the nth amplifier. It is easy to solve this recurrence relation for a cascaded chain of N_A amplifiers to obtain

$$E_f = E_0 + \sum_{n=1}^{N_A} \delta E_n, \qquad (9.5.19)$$

where E_f is the output energy and E_0 is the input energy of the pulse. The energy variance is calculated using Eq. (9.5.13) with $\langle \delta E_n \rangle = S_{\text{sp}}$ and is given by

$$\sigma_E^2 \equiv \langle E_f^2 \rangle - \langle E_f \rangle^2 = 2 N_A S_{\text{sp}} E_0. \tag{9.5.20}$$

The optical SNR is obtained in a standard manner and is given by

$$\text{SNR} = E_0/\sigma_E = (E_0/2 N_A S_{\text{sp}})^{1/2}. \tag{9.5.21}$$

Two conclusions can be drawn from this equation. First, the SNR decreases as the number of in-line amplifiers increases because of the accumulation of ASE along the link. Second, even though Eq. (9.5.21) applies for both the standard and DM solitons, the SNR is improved for DM solitons because of their higher energies. In fact, the improvement factor is given by $f_{\text{DM}}^{1/2}$, where f_{DM} is the energy enhancement factor associated with the DM solitons. As an example, the SNR is 14 dB after 100 amplifiers spaced 80-km apart for DM solitons with 0.1-pJ energy using $n_{\text{sp}} = 1.5$ and $\alpha = 0.2$ dB/km.

Frequency fluctuations induced by optical amplifiers are found by integrating Eq. (9.5.4) over one amplifier spacing, resulting in the recurrence relation

$$\Omega(z_n) = \Omega(z_{n-1}) + \delta \Omega_n, \tag{9.5.22}$$

where $\Omega(z_n)$ denotes frequency shift at the output of the nth amplifier. As before, the total frequency shift Ω_f for a cascaded chain of N_A amplifiers is given by $\Omega_f = \sum_{n=1}^{N_A} \delta \Omega_n$, where the initial frequency shift at $z = 0$ is taken to be zero because the soliton frequency equals the carrier frequency at the input end.

The variance of frequency fluctuations can be calculated using

$$\sigma_\Omega^2 \equiv \langle \Omega_f^2 \rangle - \langle \Omega_f \rangle^2 = \sum_{n=1}^{N_A} \sum_{m=1}^{N_A} \langle \delta \Omega_n \delta \Omega_m \rangle, \tag{9.5.23}$$

where we used $\langle \Omega_f \rangle = 0$. The average in this equation can be performed by noting that frequency fluctuations at two different amplifiers are not correlated. Using $\langle \delta \Omega_n \delta \Omega_m \rangle = \langle (\delta \Omega)^2 \rangle \delta_{nm}$ with Eq. (9.5.13) and performing the double sum in Eq. (9.5.23), the frequency variance for DM solitons is given by

$$\sigma_\Omega^2 = N_A (S_{\text{sp}}/E_0)[(1 + C_0^2)/T_0^2] = N_A S_{\text{sp}}/(E_0 T_m^2), \tag{9.5.24}$$

where T_m is the minimum pulse width within the dispersion map at the location where the pulse is transform-limited (no chirp). The variance increases linearly with the number of amplifiers. It also depends on the width T_0 and C_0 of the input pulse. However, the chirp parameter can be eliminated if σ_Ω^2 is written in terms of the minimum pulse width. In practice, T_m is also the width of the pulse at the optical transmitter before it is prechirped.

In the case of standard solitons, we should use Eq. (9.5.17) while performing the average in Eq. (9.5.23). The variance of frequency fluctuations in this case becomes

$$\sigma_\Omega^2 = N_A S_{\text{sp}}/(3 E_0 T_0^2). \tag{9.5.25}$$

Notice that T_0 is also equal to T_m for standard solitons which remain unchirped during propagation and maintain their width all along the fiber. At first site, it appears that DM solitons have a variance larger by a factor of 3 compared with the standard solitons. However, this is not the case if we recall that the input pulse energy E_0 is enhanced for DM solitons by a factor typically exceeding 3. As a result, the variance of frequency fluctuations is expected to be smaller for DM solitons.

Frequency fluctuations do not affect a soliton system directly unless a coherent detection scheme with frequency or phase modulation is employed (see Chapter 10). Nevertheless, they play a significant indirect role by inducing timing jitter such that the pulse in each 1 bit shifts from the center of its assigned bit slot in a random fashion. We turn to this issue next.

9.5.3 Timing Jitter

If optical amplifiers compensate for fiber losses, one may ask what limits the total transmission distance of a soliton link. The answer is provided by the timing jitter induced by optical amplifiers [115]–[125]. The origin of timing jitter can be understood by noting that a change in the soliton frequency by Ω affects the group velocity or the speed at which the pulse propagates through the fiber. If Ω fluctuates because of amplifier noise, soliton transit time through the fiber link also becomes random.

To calculate the variance of pulse-position fluctuations, we integrate Eq. (9.5.5) over the fiber section between two amplifiers and obtain the recurrence relation

$$q(z_n) = q(z_{n-1}) + \Omega(z_{n-1}) \int_{z_{n-1}}^{z_n} \beta_2(z)\, dz + \delta q_n, \qquad (9.5.26)$$

where $q(z_n)$ denotes the position at the output of the nth amplifier. This equation shows that the pulse position changes between any two amplifiers for two reasons. First, the cumulative frequency shift $\Omega(z_{n-1})$ produces a temporal shift if the GVD is not zero because of changes in the group velocity. Second, the nth amplifier shifts the position randomly by δq_n. It is easy to solve this recurrence relation for a cascaded chain of N_A amplifiers to obtain the final position in the form

$$q_f = \sum_{n=1}^{N_A} \delta q_n + \bar{\beta}_2 L_A \sum_{n=1}^{N_A} \sum_{i=1}^{n-1} \delta \Omega_i, \qquad (9.5.27)$$

where $\bar{\beta}_2$ is the average value of the GVD and the double sum stems from the cumulative frequency shift appearing in Eq. (9.5.26).

Timing jitter is calculated from this equation using $\sigma_t^2 = \langle q_f^2 \rangle - \langle q_f \rangle^2$ together with $\langle q_f \rangle = 0$. As before, the average can be performed by noting that fluctuations at two different amplifiers are not correlated. However, the timing jitter depends not only on the variances of position and frequency fluctuations but also on the cross-correlation function $\langle \delta q \, \delta \Omega \rangle$ at the same amplifier. The result can be written as

$$\sigma_t^2 = \sum_{n=1}^{N_A} \langle (\delta q)^2 \rangle + \bar{\beta}_2 L_A \sum_{n=1}^{N_A} (n-1) \langle \delta q \, \delta \Omega \rangle + (\bar{\beta}_2 L_A)^2 \sum_{n=1}^{N_A} (n-1)^2 \langle (\delta \Omega)^2 \rangle. \quad (9.5.28)$$

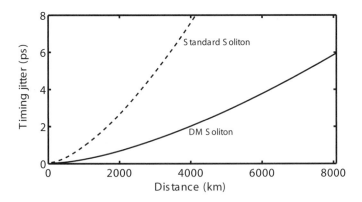

Figure 9.17: (a) ASE-induced timing jitter as a function of length for a 40-Gb/s system designed with DM (solid curve) and standard (dashed line) solitons.

Using Eqs. (9.5.13)–(9.5.15) in this equation and performing the sums, the timing jitter of DM solitons is given by [124]

$$\sigma_t^2 = \frac{S_{\rm sp}T_m^2}{E_0}[N_A(1+C_0^2) + N_A(N_A-1)C_0d + \tfrac{1}{6}N_A(N_A-1)(2N_A-1)d^2], \quad (9.5.29)$$

where the normalized parameter d is related to the accumulated dispersion over one amplifier spacing as

$$d = \frac{1}{T_m^2}\int_0^{L_A}\beta_2(z)\,dz = \frac{\bar{\beta}_2 L_A}{T_m^2}. \quad (9.5.30)$$

The three terms in Eq. (9.5.29) have the following origins. The first term inside the square brackets results from direct position fluctuations of a soliton within each amplifier. The second term is related to the cross-correlation between frequency and position fluctuations because both are produced by the same ASE noise. The third term is solely due to frequency fluctuations. For a long chain of cascaded amplifiers ($N_A \gg 1$), the jitter is dominated by the last term in Eq. (9.5.30) because of its N_A^3 dependence and is approximately given by

$$\frac{\sigma_t^2}{T_m^2} \approx \frac{S_{\rm sp}}{3E_0}N_A^3 d^2 = \frac{S_{\rm sp}L_T^3}{3E_0 L_D^2 L_A}, \quad (9.5.31)$$

where Eq. (9.5.30) was used together with $L_D = T_m^2/|\bar{\beta}_2|$ and $N_A = L_T/L_A$ for a lightwave system with the total transmission distance L_T.

Because of the cubic dependence of σ_t^2 on the system length L_T, the timing jitter can become an appreciable fraction of the bit slot for long-haul systems, especially at bit rates exceeding 10 Gb/s for which the bit slot is shorter than 100 ps. Such jitter would lead to large power penalties if left uncontrolled. As discussed in Section 6.5.2, jitter should be less than 10% of the bit slot in practice. Figure 9.17 shows how timing jitter increases with L_T for a 40-Gb/s DM soliton system designed using a dispersion map consisting of 10.5 km of anomalous-GVD fiber and 9.7 km of normal-GVD fiber

[$D = \pm 4$ ps/(km-nm)]. Optical amplifiers with $n_{sp} = 1.3$ (noise figure 4.1 dB) are placed every 80.8 km (4 map periods) along the fiber link for compensating 0.2-dB/km losses. Variational equations were used to find the input pulse parameters for which the soliton recovers periodically after each map period: $T_0 = 6.87$ ps, $C_0 = 0.56$, and $E_0 = 0.402$ pJ. The nonlinear parameter γ was 1.7 W^{-1}/km.

An important question is whether the use of dispersion management is helpful or harmful from the standpoint of timing jitter. The timing jitter for standard solitons can be found in a closed form by using Eq. (9.5.17) in Eq. (9.5.28) and is given by

$$\sigma_t^2 = \frac{S_{sp}T_0^2}{3E_s}[N_A + \tfrac{1}{6}N_A(N_A - 1)(2N_A - 1)d^2], \tag{9.5.32}$$

where we have used E_s for the input soliton energy to emphasize that it is different from the DM soliton energy E_0 used in Eq. (9.5.29). For a fair comparison of the DM and standard solitons, we consider an identical soliton system except that the dispersion map is replaced by a single fiber whose GVD is constant and equal to the average value $\bar{\beta}_2$. The soliton energy E_s can be found by using Eq. (9.2.3) in Eq. (9.2.5) and is given by

$$E_s = 2f_{LM}|\bar{\beta}_2|/(\gamma T_0), \tag{9.5.33}$$

where the factor f_{LM} is the enhancement factor resulting from loss management ($f_{LM} \approx 3.8$ for a 16-dB gain). The dashed line in Fig. 9.17 shows the timing jitter using Eqs. (9.5.32) and (9.5.33). A comparison of the two curves shows that the jitter is considerably smaller for DM solitons. The physical reason behind the jitter reduction is related to the enhanced energy of the DM solitons. In fact, the energy ratio E_0/E_s equals the energy enhancement factor f_{DM} introduced earlier in Eq. (9.4.14). From a practical standpoint, reduced jitter of DM solitons permits much longer transmission distances as evident from Fig. 9.17. Note that Eq. (9.5.32) also applies for DDFs because the GVD variations along the fiber can be included through the parameter d as defined in Eq. (9.5.30).

For long-haul soliton systems, the number of amplifiers is large enough that the N_A^3 term dominates in Eq. (9.5.32), and the timing jitter for standard solitons is approximately given by [116]

$$\frac{\sigma_t^2}{T_0^2} = \frac{S_{sp}L_T^3}{9E_sL_D^2L_A}. \tag{9.5.34}$$

Comparing Eqs. (9.5.31) and (9.5.34), one can conclude that the timing jitter is reduced by a factor of $(f_{DM}/3)^{1/2}$ when DM solitons are used. The factor of 3 has its origin in the nearly Gaussian shape of the DM solitons.

To find a simple design rule, we can use Eq. (9.5.34) with the condition $\sigma_t < b_j/B$, where b_j is the fraction of the bit slot by which a soliton can move without affecting the system performance adversely. Using $B = (2q_0T_0)^{-1}$ and E_s from Eq. (9.5.33), the bit rate–distance product BL_T for standard solitons is found to be limited by

$$BL_T < \left(\frac{9b_j^2 f_{LM}L_A}{S_{sp}q_0\gamma\bar{\beta}_2}\right)^{1/3}. \tag{9.5.35}$$

For DM solitons the energy enhancement factor f_{LM} is replaced by $f_{LM}f_{DM}/3$. The tolerable value of b_j depends on the acceptable BER and on details of the receiver design; typically, $b_j < 0.1$. To see how amplifier noise limits the total transmission distance, consider a standard soliton system operating close to the zero-dispersion wavelength with parameter values $q_0 = 5$, $\alpha = 0.2$ dB/km, $\gamma = 2$ W^{-1}/km, $\bar{\beta}_2 = -1$ ps/(km-nm), $n_{sp} = 1.5$, $L_A = 80$ km, and $b_j = 0.1$. Using $G = 16$ dB, we find $f_{LM} = 3.78$ and $S_{sp} = 6.46 \times 10^{-6}$ pJ. With these values, BL_T must be below 132 (Tb/s)-km. For a 40-Gb/s system, the transmission distance is limited to below 3300 km. This value can be increased to above 10,000 km for DM solitons.

9.5.4 Control of Timing Jitter

As the timing jitter ultimately limits the performance of soliton systems, it is essential to find a solution to the timing-jitter problem before the use of solitons can become practical. Several techniques were developed during the 1990s for controlling the timing jitter [126]–[146]. This section is devoted to a brief discussion of them.

The use of optical filters for controlling the timing jitter of solitons was proposed as early as 1991 [126]–[128]. This approach makes use of the fact that the ASE occurs over the entire amplifier bandwidth but the soliton spectrum occupies only a small fraction of it. The bandwidth of optical filters is chosen such that the soliton bit stream passes through the filter but most of the ASE is blocked. If an optical filter is placed after each amplifier, it improves the SNR because of the reduced ASE and also reduces the timing jitter simultaneously. This was indeed found to be the case in a 1991 experiment [127] but the reduction in timing jitter was less than 50%.

The filter technique can be improved dramatically by allowing the center frequency of the successive optical filters to slide slowly along the link. Such *sliding-frequency* filters avoid the accumulation of ASE within the filter bandwidth and, at the same time, reduce the growth of timing jitter [129]. The physical mechanism behind the operation of such filters can be understood as follows. As the filter passband shifts, solitons shift their spectrum as well to minimize filter-induced losses. In contrast, the spectrum of ASE cannot change. The net result is that the ASE noise accumulated over a few amplifiers is filtered out later when the soliton spectrum has shifted by more than its own bandwidth.

The moment method can be extended to include the effects of optical filters by noting that each filter modifies the soliton field such that

$$B_f(z_f, t) = \frac{1}{2\pi} \int_{-\infty}^{\infty} H_f(\omega - \omega_f)\tilde{B}(z_f, \omega)e^{-i\omega t}\, d\omega, \tag{9.5.36}$$

where $\tilde{B}(z_f, \omega)$ is the pulse spectrum and H_f is the transfer function of the optical filter located at ξ_f. The filter passband is shifted by ω_f from the soliton carrier frequency. If we approximate the filter spectrum by a parabola over the soliton spectrum and use $H_f(\omega - \omega_f) = 1 - b(\omega - \omega_f)^2$, it is easy to see that the filter introduces an additional loss for the soliton that should be compensated by increasing the gain of optical amplifiers. The analysis of timing jitter shows that sliding-frequency filters reduce jitter considerably for both the standard and DM solitons [142].

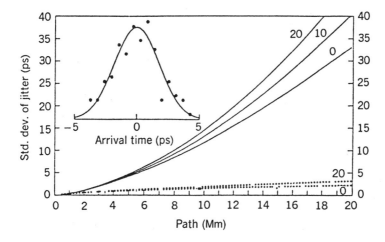

Figure 9.18: Timing jitter with (dotted curves) and without (solid curves) sliding-frequency filters at several bit rates as a function of distance. The inset shows a Gaussian fit to the numerically simulated jitter at 10,000 km for a 10-Gb/s system. (After Ref. [129]; ©1992 OSA; reprinted with permission.)

Figure 9.18 shows the predicted reduction in the timing jitter for standard solitons. The bit-rate dependence is due to the acoustic jitter (discussed later); the $B = 0$ curves show the contribution of the Gordon–Haus jitter alone. Optical filters help in reducing both types of timing jitter and permit transmission of 10-Gb/s solitons over more than 20 Mm. In the absence of filters, timing jitter becomes so large that a 10-Gb/s soliton system cannot be operated beyond 8000 km. The inset in Fig. 9.18 shows a Gaussian fit to the timing jitter of 10-Gb/s solitons at a distance of 10 Mm calculated by solving the NLS equation numerically after including the effects of both the ASE and sliding-frequency filters [129]. The timing-jitter distribution is approximately Gaussian with a standard deviation of about 1.76 ps. In the absence of filters, the jitter exceeds 10 ps under the same conditions.

Optical filters benefit a soliton system in many other ways. Their use reduces interaction between neighboring solitons [130]. The physical mechanism behind the reduced interaction is related to the change in the soliton phase at each filter. A rapid variation of the relative phase between neighboring solitons, occurring as a result of filtering, averages out the soliton interaction by alternating the nature of the interaction force from attractive to repulsive. Optical filters also help in reducing the accumulation of dispersive waves [131]. The reason is easy to understand. As the soliton spectrum shifts with the filters, dispersive waves produced at earlier stages are blocked by filters together with the ASE.

Solitons can also be controlled in the time domain using the technique of *synchronous* amplitude modulation, implemented in practice using a LiNbO$_3$ modulator [132]. The technique works by introducing additional losses for those solitons that have shifted from their original position (center of the bit slot). The modulator forces solitons to move toward its transmission peak where the loss is minimum. Mathematically, the action of modulator is to change the soliton amplitude as

$$B(z_m, t) \rightarrow T_m(t - t_m)B(z_m, t), \qquad (9.5.37)$$

where $T_m(\tau)$ is the transmission coefficient of the modulator located at $\xi = \xi_m$. The moment method or perturbation theory can be used to show that timing jitter is reduced considerably by modulators.

The synchronous modulation technique can also be implemented by using a phase modulator [133]. One can understand the effect of periodic phase modulation by recalling that a frequency shift, $\delta\omega = -d\phi(t)/dt$, is associated with any phase variation $\phi(t)$. Since a change in soliton frequency is equivalent to a change in the group velocity, phase modulation induces a temporal displacement. Synchronous phase modulation is implemented in such a way that the soliton experiences a frequency shift only if it moves away from the center of the bit slot, which confines it to its original position despite the timing jitter induced by ASE and other sources. Intensity and phase modulations can be combined together to further improve the system performance [134].

Synchronous modulation can be combined with optical filters to control solitons simultaneously in both the time and frequency domains. In fact, this combination permits arbitrarily long transmission distances [135]. The use of intensity modulators also permits a relatively large amplifier spacing by reducing the impact of dispersive waves. This property of modulators was exploited in 1995 to transmit a 20-Gb/s soliton train over 150,000 km with an amplifier spacing of 105 km [136]. Synchronous modulators also help in reducing the soliton interaction and in clamping the level of amplifier noise. The main drawback of modulators is that they require a clock signal that is synchronized with the original bit stream.

A relatively simple approach uses postcompensation of accumulated dispersion for reducing the timing jitter [137]. The basic idea can be understood from Eq. (9.5.29) or Eq. (9.5.32) obtained for the timing jitter of DM and standard solitons, respectively. The cubic term that dominates the jitter at long distances depends on the accumulated dispersion through the parameter d defined in Eq. (9.5.30). If a fiber is added at the end of the fiber link such that it reduces the accumulated GVD, it should help in reducing the jitter. It is easy to include the contribution of the postcompensation fiber to the timing jitter using the moment method. In the case of DM solitons, the jitter variance at the end of a postcompensation fiber of length L_c and GVD β_{2c} is given by [124]

$$\sigma_c^2 = \sigma_t^2 + (S_{sp}T_m^2/E_0)[2N_A C_0 d_c + N_A(N_A - 1)dd_c + N_A d_c^2], \qquad (9.5.38)$$

where σ_t^2 is given by Eq. (9.5.29) and $d_c = \beta_{2c}L_c/T_M^2$. If we define $y = -d_c/(N_A d)$ as the fraction by which the accumulated dispersion $N_A d$ is compensated and retain only the dominant cubic terms in Eq. (9.5.38), this equation can be written as

$$\sigma_c^2 = N_A^3 d^2 T_m^2 \frac{S_{sp}}{E_0} \left(\frac{1}{3} - y + y^2 \right). \qquad (9.5.39)$$

The minimum value occurs for $y = 0.5$ for which σ_c^2 is reduced by a factor of 4. Thus, timing jitter of solitons can be reduced by a factor of 2 by postcompensating the accumulated dispersion by 50%. The same conclusion holds for standard solitons [137].

Several other techniques can be used for controlling the timing jitter. One approach consists of inserting a fast saturable absorber periodically along the fiber link. Such a device absorbs low-intensity light such as ASE and dispersive waves but leaves the solitons intact by becoming transparent at high intensities. To be effective, it should respond at a time scale shorter than the soliton width. It is difficult to find an absorber that can respond at such short time scales. A nonlinear optical-loop mirror (see Section 8.4) can act as a fast saturable absorber and reduces the timing jitter of solitons while also stabilizing their amplitude [138]. Re-timing of a soliton train can also be accomplished by taking advantage of cross-phase modulation [139]. The technique overlaps the soliton data stream and another pulse train composed of only 1 bits (an optical clock) inside a fiber where cross-phase modulation (XPM) induces a nonlinear phase shift on each soliton in the signal bit stream. Such a phase modulation translates into a net frequency shift only when the soliton does not lie in the middle of the bit slot. Similar to the case of synchronous phase modulation, the direction of the frequency shift is such that the soliton is confined to the center of the bit slot. Other nonlinear effects such as stimulated Raman scattering [140] and four-wave mixing (FWM) can also be exploited for controlling the soliton parameters [141]. The technique of distributed amplification also helps in reducing the timing jitter. As an example, if solitons are amplified using distributed Raman amplification, timing jitter can be reduced by about a factor of 2 [125].

9.6 High-Speed Soliton Systems

As seen in Section 8.4, the optical time-division multiplexing (OTDM) technique can be used to increase the single-channel bit rate to beyond 10 Gb/s. As early as 1993, the bit rate of a soliton-based system was extended to 80 Gb/s with this method [147]. The major limitation of such systems stems from nonlinear interaction between two neighboring solitons. This problem is often solved by using a variant of polarization multiplexing in which neighboring bit slots carry orthogonally polarized pulses. The 80-Gb/s signal could be transmitted over 80 km with this technique. The same technique was later used to extend the bit rate to 160 Gb/s [148]. At such high bit rates, the bit slot is so small that the soliton width is typically reduced to below 5 ps, and several higher-order nonlinear and dispersive effects should be considered. This section is devoted to such issues.

9.6.1 System Design Issues

Both fiber losses and dispersion need to be managed properly in soliton systems designed to operate at high bit rates. The main design issues are related to the choice of a dispersion map and the relationship between the map period L_{map} and amplifier spacing L_A. As discussed in Section 9.4.2, the parameter T_{map} given in Eq. (9.4.11) sets the scale for the shortest pulse that can be propagated in a periodic fashion. Thus, the important design parameters are the local GVD and the length of various fiber sections used to form the dispersion map.

In the case of terrestrial systems operating over standard fibers and dispersion-managed using DCFs, $|\beta_2|$ exceeds 20 ps^2/km in both sections of the dispersion map. Consider, as an example, the typical situation in which the map consists of 60–70 km of standard fiber whose dispersion is compensated with 10–15 km of DCFs. The parameter T_{map} exceeds 25 ps for such maps. At the same time, the map strength is large enough that the pulse width oscillates over a wide range. These features make it difficult to realize a bit rate of even 40 Gb/s although such a map permitted by 1997 transmission of four 20-Gb/s channels over 2000 km [149]. Numerical simulations show the possibility of transmitting 40-Gb/s DM solitons over 2000 km of standard fiber if the average GVD of the dispersion map is kept relatively low [150]. In a 1999 experiment, 40-Gb/s DM solitons were indeed transmitted over standard fibers but the distance was limited to only 1160 km [151]. Interaction between solitons was the most limiting factor in this experiment. By 2000, the use of highly nonlinear fibers together with synchronous in-line modulation permitted transmission of 40-Gb/s solitons over transoceanic distance [152].

To design high-speed soliton systems operating at 40 Gb/s or more, the T_{map} parameter should be reduced to below 10 ps. If only a single-channel is transmitted, one can use fibers with values of local GVD below 1 ps^2/km. However, in the case of WDM systems the GVD parameter should be relatively large ($|\beta_2| > 4$ ps^2/km) in both the normal and anomalous-GVD fiber sections for suppressing the nonlinear effects such as cross-phase modulation (XPM) and four-wave mixing (FWM). It follows from Eq. (9.4.11) that the map period should become smaller and smaller as the bit rate increases. For example, when $|\beta_2| = 5$ ps^2/km in the two sections, the map period should be less than 15 km for realizing $T_{\text{map}} < 6$ ps. Such a map is suitable at bit rates of 40 Gb/s with only 25-ps bit slot. As an example, Fig. 9.19 shows the evolution of a DM soliton with the map used earlier for Figs. 9.15 and 9.16; the pulse parameters correspond to those of Fig. 9.16(b). The system design employs 8 map periods per 80-km amplifier spacing. Pulses at each amplifier maintain their shape even after 10 Mm even though pulse width and chirp oscillate within each map period as seen in Fig. 9.16(b).

The map used for Fig. 9.19 is suitable for 40-Gb/s systems but would not work at a bit rate of 100 Gb/s. A 100-Gb/s system has a bit period of only 10 ps and would require a dispersion map with $T_{\text{map}} < 4$ ps. Such systems require the use of *dense* or *short-period* dispersion management, a scheme in which the map period is a small fraction of the amplifier spacing [153]–[158]. Numerical simulation show the possibility of transmission at bit rates as high as 320 Gb/s [157] if the map period is reduced to below 1 km. Such fiber links may be hard to construct although a short-period DM cable with the map period of 9 km has been made [159]. In this cable, the dispersion map consisted of two 4.5-km fiber sections with GVD values of 17 and -15 ps/(km-nm), resulting in an average dispersion of 1 ps/(km-nm). It was used to demonstrate 640-Gb/s WDM transmission (32 channels at 20 Gb/s) over 280 km. The parameter T_{map} is about 6 ps for this dispersion map and can be reduced further by decreasing the local and the average GVD. In principle, such a fiber cable should be able to operate at 40 Gb/s per channel.

If a lightwave system is designed to support a single channel only, the dispersion map can be made using low-GVD fibers. For example, when $\beta_2 = \pm 0.5$ ps^2/km and section lengths are nearly equal to 40 km, $T_{\text{map}} \approx 3.2$ ps. Such a soliton system can op-

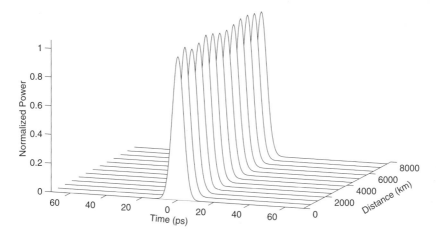

Figure 9.19: Evolution of a DM soliton over 8000 km with dense dispersion management (8 map periods per amplifier spacing). The map and pulse parameters correspond to those of Fig. 9.16(b).

erate at 40 Gb/s if the average GVD is kept low. In a 1998 experiment, 40-Gb/s solitons were indeed transmitted over 8600 km using a 140-km-long fiber loop with an average dispersion of only -0.03 ps^2/km [160]. Interaction between solitons was the most limiting factor in this experiment. As discussed later in this section, it can be reduced by alternating the polarization of neighboring bits. Indeed, the use of this technique permitted by 1999 the transmission of 40-Gb/s solitons over more than 10 Mm without employing any jitter-control technique [161]. The use of synchronous modulation allowed transmission of even 80-Gb/s solitons over 10 Mm [162]. Much higher capacities can be realized using the combination of WDM and OTDM techniques [163]. In a 2000 experiment, a single OTDM channel at a bit rate of 1.28 Tb/s was transmitted over 70 km using 380-fs optical pulses [164]. The transmission distance of such systems is limited by fiber dispersion; it was necessary to compensate for dispersion up to fourth order.

9.6.2 Soliton Interaction

The interaction among neighboring pulses becomes a critical issue as the bit rate increases. The reason is that the bit slot becomes so small (only 10 ps at 100 Gb/s) that one is often forced to pack the solitons closely. The interaction between two DM solitons can be studied numerically or by using a variational technique [165]–[170]. The qualitative features are similar to those discussed in Section 9.2.2 for the standard solitons. In general, the parameter q_0, related to the bit rate as $B = (2q_0 T_0)^{-1}$, should exceed 4 for the system to work properly.

A new feature of the interaction among DM solitons is that the collision length depends on details of the dispersion map. As a result, soliton systems can be optimized by choosing the pulse and the map parameters appropriately. Numerical simulations show that the system performance is optimum when the map strength is chosen to

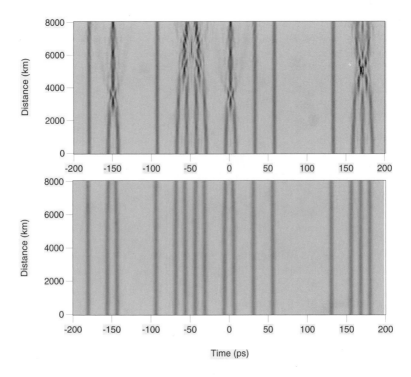

Figure 9.20: Propagation of a 32-bit soliton stream over 8000 km when the input pulse parameters correspond to those used in Fig. 9.16. Pulse energy equals 0.1 pJ in the top panel but has its optimum for the bottom panel.

be around 1.6 [165]. This value corresponds to a pulse energy in the vicinity of the minimum seen in Fig. 9.15. As an example, Fig. 9.20 shows propagation of a pulse train consisting of a 32-bit pattern over 8000 km for $E_0 = 0.1$ pJ (top panel) and its optimum value (bottom panel). The periodic evolution of the chirp and pulse width in these two cases is shown in Fig. 9.16. When the pulse energy is larger than the optimum value (higher map strength), solitons begin to collide after 3000 km. In contrast, solitons can propagate more than 8000 km before colliding when the pulse parameters are suitably optimized.

A new multiplexing technique, called intrachannel polarization multiplexing, can be used to reduce interaction among solitons. This technique is different from the conventional polarization-division multiplexing in which two neighboring channels at different wavelengths are made orthogonally polarized. In the case of intrachannel polarization multiplexing, the bits of a single-wavelength channel are interleaved in such a way that any two neighboring bits are orthogonally polarized. The technique was used for solitons as early as 1992 and has been studied extensively since then [171]–[179].

Figure 9.21 shows the basic idea behind polarization multiplexing schematically. At first glance, such a scheme should not work unless polarization-maintaining fibers

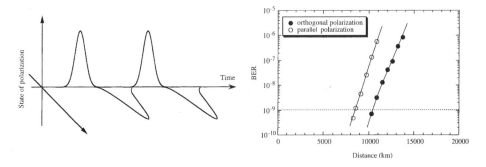

Figure 9.21: (a) Polarization multiplexing scheme and (b) the improvement realized with its use for a 40-Gb/s soliton system. (After Ref. [161]; ©1999 IEEE; reprinted with permission.)

are used since the polarization state of light changes randomly because of birefringence fluctuations resulting in polarization-mode dispersion (PMD). It turns out that even though the polarization states of the bit train does change in an unpredictable manner, the orthogonal nature of any two neighboring bits is nearly preserved. Because of this orthogonality, the interaction among solitons is much weaker compared with the copolarized-solitons case. Figure 9.21 shows how the reduced interaction lowers the bit-error rate (BER) and increases the transmission distance of a 40-Gb/s soliton system.

The use of polarization multiplexing helps to increase the bit rate as solitons can be packed more tightly because of reduced interaction among them [172]. Its implementation is not difficult in practice when the OTDM technique is used. It requires the generation of two bit streams using orthogonally polarized optical carriers and then interleave them using an optical delay line (see Section 8.4). A polarizing beam splitter can be used in combination with a polarization controller to demultiplex the two orthogonally polarized channels at the receiver end.

An important factor limiting the performance of polarization-multiplexed soliton systems is the PMD induced by random changes in the fiber birefringence [177]. In fact, PMD seriously limits the use of this technique for nonsoliton systems through pulse depolarization (different parts of the pulse have different polarizations). The situation is different for solitons which are known to be much more robust to the PMD effects [180]. The natural tendency of a soliton to preserve its integrity under various perturbations also holds for perturbations affecting its state of polarization. Unlike linear pulses, the state of polarization remains constant across the entire soliton (no depolarization across the pulse), and the effect of PMD is to induce a small change in the state of polarization of the entire soliton (a manifestation of its particle-like nature). Such resistance of solitons to PMD, however, breaks down for large amounts of PMD. The breakdown occurs for $D_p > 0.3D^{1/2}$ [181], where D_p is the PMD parameter introduced in Section 2.3.5 and expressed in $ps/\sqrt{km}$ and D is the dispersion parameter in units of ps/(nm-km). Since typically $D_p < 0.1\ ps/\sqrt{km}$ for high-quality optical fibers, D must exceed 0.06 ps/(nm-km). In the case of DM solitons, D corresponds to the average GVD of the link.

An extension of the polarization-multiplexing technique, called polarization-multi-level coding, has also been suggested [175]. In this technique, the information coded in each bit is contained in the angle that the soliton state of polarization makes with one of the principal birefringence axes. This technique is also limited by random variations of fiber birefringence and by randomization of the polarization angle by the amplifier noise and has not yet been implemented because of its complexity.

9.6.3   Impact of Higher-Order Effects

Higher-order effects such as TOD and intrapulse Raman scattering become quite important at high bit rates and must be accounted for a correct description of DM solitons [182]–[185]. These effects can be included by adding two new terms to the standard NLS equation as was done in Eq. (9.3.16) in the context of distributed amplification of standard solitons. In the case of DM solitons, the Raman and TOD terms are added to Eq. (9.4.5) to obtain the following generalized NLS equation:

$$
i\frac{\partial B}{\partial z} - \frac{\beta_2(z)}{2}\frac{\partial^2 B}{\partial t^2} + \gamma p(z)|B|^2 B = \frac{i\beta_3}{6}\frac{\partial^3 B}{\partial t^3} + T_R \gamma p(z) B \frac{\partial |B|^2}{\partial t}, \tag{9.6.1}
$$

where T_R is the Raman parameter with a typical value of 3 fs [10]. The effective nonlinear parameter $\bar{\gamma} \equiv \gamma p$ is z dependent because of variations in the soliton energy along the fiber link. The Raman term leads to the Raman-induced frequency shift known as the SSFS. This shift is negligible for 10-Gb/s systems but becomes increasingly important as the bit rate increases to 40 Gb/s and beyond. The TOD term also becomes important at high bit rates, especially when the average GVD of the fiber link is close to zero.

To understand the impact of TOD and SSFS on solitons, we use the moment method of Section 9.5 and assume that the last two terms in Eq. (9.6.1) are small enough that the pulse shape remains approximately Gaussian. Using Eq. (9.6.1) in Eqs. (9.5.1) and (9.5.2), the frequency shift Ω, and soliton position q are found to evolve with z as

$$
\frac{d\Omega}{dz} = -\frac{\gamma p T_R}{E} \int_{-\infty}^{\infty} \left(\frac{\partial |B|^2}{\partial t}\right)^2 dt + \sum_n \delta\Omega_n \delta(z - z_n), \tag{9.6.2}
$$

$$
\frac{dq}{dz} = \beta_2 \Omega + \frac{\beta_3}{6E} \int_{-\infty}^{\infty} \left(\frac{\partial |B|}{\partial t}\right)^2 dt + \sum_n \delta q_n \delta(z - z_n), \tag{9.6.3}
$$

where the last term accounts for fluctuations induced by the amplifier noise. These equations show that the Raman term in Eq. (9.6.1) leads to a frequency shift while the TOD term produces a shift in the soliton position.

Consider the case of standard loss-managed solitons. If we ignore the noise terms and integrate Eqs. (9.6.2) and (9.6.3) after using $B(z,t)$ from Eq. (9.5.16), the Raman-induced frequency shift Ω and the soliton position q are found to evolve along the fiber length as

$$
\Omega(z) = -\frac{4\gamma T_R E_s}{15T_0^3} \int_0^z p(z)\,dz \equiv -\frac{8f_{\mathrm{LM}}|\beta_2|}{15T_0^4} \int_0^z p(z)\,dz, \tag{9.6.4}
$$

$$
q(z) = \beta_2 \int_0^z \Omega(z)\,dz + \frac{\beta_3 z}{18T_0^2} + \beta_3 \int_0^z \Omega^2\,dz, \tag{9.6.5}
$$

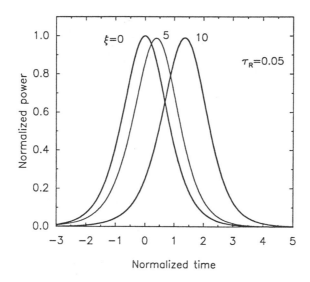

Figure 9.22: Pulse envelopes at distances $\xi = z/L_D = 0$, 5, and 10 showing the delay of a soliton because of the Raman-induced-frequency shift.

where Eq. (9.5.33) was used for the input soliton energy. The frequency shift grows linearly with z in the case of perfect distributed amplification ($p = 1$) but follows energy variations in the case of lumped amplification. In both cases, its magnitude scales with pulse width as T_0^{-4}. It is this feature that makes the SSFS increasingly more important as the bit rate increases. The soliton position q is affected by both the Raman-induced frequency shift and the TOD parameter β_3. This shift is deterministic in nature, in contrast with the position shift induced by amplifier-induced frequency fluctuations. The dominant contribution to q in Eq. (9.6.5) comes from the β_2 term.

One can understand the origin of changes in the soliton position by noting that the Raman-induced frequency shift in the carrier frequency toward longer wavelengths slows down a pulse propagating in the anomalous-GVD regime of the fiber. Figure 9.22 shows such a slowing down of a standard soliton ($N = 1$) by solving Eq. (9.6.1) numerically with $\beta_3 = 0$ and $T_R/T_0 = 0.05$. By the time the soliton has propagated over 10 dispersion lengths, it has been delayed by a significant fraction of its own width. A delay in the arrival time of a soliton is not of much concern when all bits are delayed by the same amount. However, the pulse energy and width fluctuate from bit to bit because of fluctuations induced by the amplifier noise. Such fluctuations are converted into timing jitter by the SSFS. The Raman-induced timing jitter is discussed later in this section.

How is the Raman-induced frequency shift affected by dispersion management? The width of DM solitons is not constant but oscillates in a periodic manner. Since the SSFS depends on the pulse width, it is clear that it will also vary significantly within each map period. The highest frequency shift occurs in the central region of each section where the pulse is nearly unchirped and the width is shortest. The total frequency shift will be less for DM solitons compared with the standard solitons for which the pulse remains unchirped throughout the link. Moreover, the SSFS will depend on the map strength and will be lower for stronger maps [182].

The effects of TOD on DM solitons is apparent from Eq. (9.6.5). The second term in this equation shows that the TOD produces a shift in the soliton position even in the absence of the Raman term ($T_R = 0$). The shift increases linearly with distance as $(\beta_3/18T_0^2)z$ and is negligible until T_0 becomes much shorter than 10 ps. As an example, the temporal shift is 5 fs/km for $T_0 = 1$ ps and $\beta_3 = 0.09$ ps^3/km and becomes comparable to the pulse width after 200 km of propagation. The shift increases in the presence of SSFS as apparent from the last term in Eq. (9.6.5). This shift is relatively small as it requires the presence of both the SSFS and TOD and can be neglected in most cases of practical interest. The TOD also distorts the DM soliton and generates dispersive waves [184]. Under certain conditions, a DM soliton can propagate over long distances after some energy has been shed in the form of dispersive waves [183].

Numerical simulations based on Eq. (9.6.1) show that 80-Gb/s solitons can propagate stably over 9000 km in the presence of higher-order effects if (i) TOD is compensated within the map, (ii) optical filters are used to reduce soliton interaction, timing jitter, and the Raman-induced frequency shift, and (iii) the map period L_{map} is reduced to a fraction of amplifier spacing [153]. The bit rate can even be increased to 160 Gb/s by controlling the GVD slope and PMD, but the distance is limited to about 2000 km [155]. These results show that soliton systems operating at 160 Gb/s are possible if their performance is not limited by timing jitter. We turn to this issue next.

9.6.4   Timing Jitter

Timing jitter discussed earlier in Section 9.5.4 increases considerably at higher bit rates because of the use of shorter optical pulses. Both the Raman effect and TOD lead to additional jitter that increases rapidly with the bit rate [186]–[189]. Moreover, several other mechanisms begin to contribute to the timing jitter. In this section we consider these additional sources of timing jitter.

Raman Jitter

The Raman jitter is a new source of timing jitter that dominates at high bit rates requiring short optical pulses ($T_0 < 5$ ps). Its origin can be understood as follows [187]. The Raman-induced frequency shift depends on the pulse energy as seen in Eq. (9.6.4). This frequency shift by itself does not introduce jitter because of its deterministic nature. However, fluctuations in the pulse energy introduced by amplifier noise can be converted into fluctuations in the soliton frequency through the Raman effect, which are in turn translated into position fluctuations by the GVD. The Raman jitter occurs for both the standard and DM solitons. In the case of standard solitons, the use of DDFs is often necessary but the analysis is simplified because the solitons are unchirped and maintain their width during propagation [187].

In the case of DM solitons, the pulse energy, width, and chirp oscillate in a periodic manner. To keep the discussion simple, the effects of TOD are ignored although they can be easily included. Using Eqs. (9.6.2) and (9.6.3) and following the method used in Section 9.5.4, the frequency shift and position of the soliton at the end of the nth amplifier can be written as

$$\Omega(z_n) = \Omega(z_{n-1}) + b_R E(z_{n-1}) + \delta\Omega_n, \tag{9.6.6}$$

$$q(z_n) = q(z_{n-1}) + b_2\Omega(z_{n-1}) + b_{2R}E(z_{n-1}) + \delta q_n, \tag{9.6.7}$$

where the parameters b_2, b_R, and b_{2R} are defined as

$$b_2 = \int_0^{L_A} \beta_2(z)\,dz, \quad b_R = -\frac{T_R}{2\sqrt{\pi}} \int_0^{L_A} \frac{dz\,\gamma(z)}{T^3(z)} \left[\int_0^z p(z_1)\,dz_1 \right], \tag{9.6.8}$$

$$b_{2R} = -\frac{T_R}{2\sqrt{\pi}} \int_0^{L_A} dz\,\beta_2(z)\gamma(z) \int_0^z \frac{dz_1}{T^3(z_1)} \left[\int_0^{z_1} p(z_2)\,dz_2 \right], \tag{9.6.9}$$

where $p(z)$ takes into account variations in the pulse energy. In the case of lumped amplification, $p(z) = \exp(-\alpha z)$.

The physical meaning of Eqs. (9.6.6) and (9.6.7) is quite clear. Between any two amplifiers, the Raman-induced frequency shift within the fiber should be added to the ASE-induced frequency shift. The former, however, depends on the pulse energy and would change randomly in response to energy fluctuations. The position of the pulse changes because of the cumulative frequency shifts induced by the Raman effect and amplifier noise. The two recurrence relations in Eqs. (9.6.6) and (9.6.7) should be solved to find the final pulse position at the end of the last amplifier. The timing jitter depends not only on the variances of δE $\delta\Omega$, and δq but also on their cross-correlations given in Eqs. (9.5.13)–(9.5.15). It can be written as

$$\sigma_t^2 = \sigma_{GH}^2 + R_1\langle(\delta E)^2\rangle + R_2\langle\delta E\,\delta\Omega\rangle \tag{9.6.10}$$

where σ_{GH} is the Gordon–Haus jitter obtained earlier in Eq. (9.5.29).

The Raman contribution to the timing jitter adds two new terms whose magnitude is governed by

$$R_1 = b_2^2 b_R^2 N_A(N_A - 1)(N_A^3 - 10N_A^2 + 29N_A - 9)/120 + b_{2R}N_A(N_A - 1)$$
$$\times [b_2 b_R(19N_A^2 - 65N_A + 48)/96 + b_{2R}(2N_A - 1)/6], \tag{9.6.11}$$

$$R_2 = b_2 N_A(N_A - 1)[b_2 b_R(N_A - 2)(3N_A - 1)/12 + b_{2R}(2N_A - 1)/3]. \tag{9.6.12}$$

The R_1 term depends on the variance of energy fluctuations and scales as N_A^5 for $N_A \gg 1$. The R_2 term has its origin in the cross-correlation between energy and frequency fluctuations and scales as N_A^4 for large N_A. The R_1 term will generally dominate at long distances for high bit rates requiring pulses shorter than 5 ps.

Figure 9.23 shows the growth of timing jitter with distance for a 160-Gb/s DM soliton system. The dispersion map consists of alternating 1-km fiber sections with GVD of -3.18 and 3.08 ps^2/km, resulting in an average dispersion of -0.05 ps^2/km and $T_{map} \approx 1.25$ ps. Fiber losses of 0.2 dB/km are compensated using lumped amplifiers every 50 km. The nonlinear parameters have values $\gamma = 2.25$ W^{-1}/km and $T_R = 3$ fs. The input pulse parameters are found by solving the variational equations of Section 9.4.2 and have values $T_0 = 1.25$ ps, $C_0 = 1$, and $E_0 = 0.12$ pJ. The Raman and Gordon–Haus contributions to the jitter are also shown separately to indicate when the Raman effects begin to dominate. Because of the N_A^5 dependence of the Raman contribution,

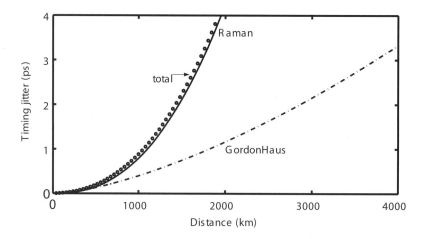

Figure 9.23: Timing jitter as a function of distance for a 160-Gb/s DM system designed with $T_0 = 1.25$ ps. The Raman and Gordon–Haus contributions to timing jitter are also shown.

most of the timing jitter results from the SSFS after 1000 km. Noting that the tolerable value of the timing jitter is less than 1 ps at 160 Gb/s, it is clear that the total transmission distance of such systems would be limited to below 1000 km in the absence of any jitter-control mechanism. Longer distances are possible at a lower bit rate of 80 Gb/s. The main conclusion is that the Raman jitter must be included whenever the width (FWHM) of solitons is shorter than 5 ps.

Acoustic Jitter

A timing-jitter mechanism that limits the total transmission distance at high bit rates has its origin in the generation of acoustic waves inside optical fibers [190]. Confinement of the optical mode within the fiber core creates an electric-field gradient in the radial direction of the fiber. This gradient creates an acoustic wave through *electrostriction*, a phenomenon that produces density variations in response to changes in the electric field. As the refractive index of any material depends on its density, the group velocity also changes with the generation of acoustic waves.

The acoustic-wave-induced index changes produced by a single pulse can last for more than 100 ns—the damping time associated with acoustic phonons. They can be measured through the XPM-induced frequency shift $\Delta\nu$ imposed on a probe signal and given by $\Delta\nu = -(L_{\text{eff}}/\lambda)d(\delta n_a)/dt$, where L_{eff} is the effective length of the fiber [191]. As seen in Fig. 9.24(a), the measured frequency shift is in the form of multiple spikes, each lasting for about 2 ns, roughly the time required for the acoustic wave to traverse the fiber core. The 21-ns period between the spikes corresponds to the round-trip time taken by the acoustic wave to reflect from the fiber cladding. Noting that $L_{\text{eff}} \approx 20$ km for 0.2-dB/km fiber losses, acoustically induced changes are relatively small ($\sim 10^{-14}$) but they lead to measurable jitter even at a bit rate of 10 Gb/s [6].

The origin of acoustic jitter can be understood by noting that pulses follow one another with a time interval that is only 25 ps at $B = 40$ Gb/s and becomes shorter

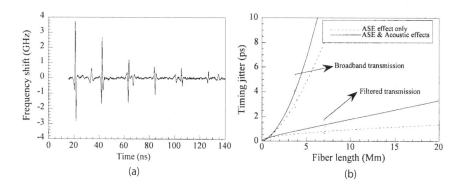

Figure 9.24: (a) Temporal changes in the XPM-induced frequency shift produced by acoustic waves in a 30-km-long fiber; (b) Increase in timing jitter produced by acoustic waves. (After Ref. [191]; ©2001 Academic; reprinted with permission.)

at higher bit rates. As a result, the acoustic wave generated by a single pulse affects hundreds of the following pulses. Physically, time-dependent changes in the refractive index induced by acoustic waves modulate the optical phase and manifest as a shift in the frequency (chirping) by a small amount ($\sim$1 GHz in Fig. 9.24). Similar to the case of SSFS and amplifier-induced noise, a frequency shift manifests as a shift in the pulse position because of GVD. The main difference is that acoustic jitter has a deterministic origin in contrast with the amplifier-induced jitter. In fact, if a pulse were to occupy each bit slot, all pulses would be shifted in time by the same amount through acoustic waves, resulting in a uniform shift of the pulse train but no timing jitter. However, any information-coded bit stream consists of a random string of 1 and 0 bits. As the change in the group velocity of each pulse depends on the presence or absence of pulses in the preceding hundreds of bit slots, different solitons arrive at slightly different times, resulting in timing jitter. The deterministic nature of acoustic jitter makes it possible to reduce its impact in practice by moving the detection window at the receiver through an automatic tracking circuit [192] or by using a coding scheme [193].

Acoustic jitter has been studied for both the standard and DM solitons [191]. In the case of standard solitons propagating inside a constant-GVD fiber of length L, the jitter scales as $\sigma_{\text{acou}} = C_a|D|L^2$, where the parameter C_a depends on the index change δn_a among other things. Figure 9.24(b) shows how the timing jitter is enhanced by the acoustic effect for a 10-Gb/s system designed with $D = 0.45$ ps/(km-nm), $T_0 = 10$ ps, $n_{\text{sp}} = 1.6$, and $L_A = 25$ km. Sliding-frequency filters reduce both the ASE-induced and acoustically induced timing jitters. The acoustic jitter then dominates because it increases linearly with the system length L while the ASE-induced jitter scales as $\sqrt{L}$.

PMD-Induced Jitter

PMD is another mechanism that can generate timing jitter through random fluctuations in the fiber birefringence [194]. As discussed in Section 2.3.5, the PMD effects are quantified through the PMD parameter D_p. In a practical lightwave system, all solitons are launched with the same state of polarization at the input end of a fiber link. However, as solitons are periodically amplified, their state of polarization becomes random because of the ASE added at every amplifier. Such polarization fluctuations lead to timing jitter in the arrival time of individual solitons through fiber birefringence because the two orthogonally polarized components travel with slightly different group velocities. This timing jitter, introduced by the combination of ASE and PMD, can be written as [194]

$$\sigma_{pol} = (\pi S_{sp} f_{LM} / 16 E_s L_A)^{1/2} D_p L. \qquad (9.6.13)$$

As a rough estimate, $\sigma_{pol} \approx 0.4$ ps for a standard-soliton system designed with $\alpha = 0.2$ dB/km, $L_A = 50$ km, $D_p = 0.1$ ps/$\sqrt{km}$, and $L = 10$ Mm. Such low values of σ_{pol} are unlikely to affect 10-Gb/s soliton systems for which the bit slot is 100 ps wide. However, for fibers with larger values of the PMD parameter ($D_p > 1$ ps/$\sqrt{km}$), the PMD-induced timing jitter becomes important enough that it should be considered together with other sources of the timing jitter. If we assume that each source of timing jitter is statistically independent, total timing jitter is obtained using

$$\sigma_{tot}^2 = \sigma_t^2 + \sigma_{acou}^2 + \sigma_{pol}^2. \qquad (9.6.14)$$

Both σ_{acou} and σ_{pol} increase linearly with transmission distance L. The ASE-induced jitter normally dominates because of its superlinear dependence on L but the situation changes when in-line filters are used to suppress it.

An important issue is related to *polarization-dependent* loss and gain. If a lightwave system contains multiple elements that amplify or attenuate the two polarization components of a pulse differently, the polarization state is easily altered. In the worst situation in which the polarization is oriented at $45°$ from the low-loss (or high-gain) direction, the state of polarization rotates by $45°$ and gets aligned with the low-loss direction only after 30–40 amplifiers for a gain–loss anisotropy of < 0.2 dB [195]. Even though the axes of polarization-dependent gain or loss are likely to be evenly distributed along any soliton link, such effects may still become an important source of timing jitter.

Interaction-Induced Jitter

In the preceding discussion of the timing jitter, individual pulses are assumed to be sufficiently far apart that their position is not affected by the phenomenon of soliton–soliton interaction. This is not always the case in practice. As seen in Section 9.2.2, even in the absence of amplifier noise, solitons may shift their position because of the attractive or repulsive forces experienced by them. As the interaction force between two solitons is strongly dependent on their separation and relative phase, both of which fluctuate because of amplifier noise, soliton–soliton interactions modify the ASE-induced timing jitter. By considering noise-induced fluctuations of the relative phase of neighboring solitons, timing jitter of interacting solitons is generally found to be enhanced by amplifier noise [196]. In contrast, when the input phase difference is close to π between neighboring solitons, phase randomization leads to reduction in the timing jitter.

An important consequence of soliton–soliton interaction is that the probability density of the timing jitter deviates considerably from the Gaussian statistics expected for the Gordon–Haus jitter in the absence of such interactions [197]. The non-Gaussian corrections can occur even when the interaction is relatively weak ($q_0 > 5$). They affect the bit-error rate of the system and must be included for an accurate estimate of the system performance. When solitons are packed so tightly that their interaction becomes quite important, the probability density function of the timing jitter develops a five-peak structure [198]. The use of numerical simulations is essential to study the impact of timing jitter on a bit stream composed of interacting solitons.

In the case of DM solitons, interaction-induced jitter becomes quite important, especially for strong maps for which the pulse width varies by a large factor within each map period [199]. It can be reduced by a proper design of the dispersion map. In fact, local dispersion of each fiber section and the average GVD of the entire link need to be optimized to reduce both the soliton–soliton interaction and the ASE-induced jitter simultaneously [161].

Control of Timing Jitter

The increase in the timing jitter brought by the Raman and TOD effects and a shorter bit slot at higher bit rates make the control of timing jitter essential before high-speed systems can become practical. As discussed in Section 9.5, both optical filters and synchronous modulators help in reducing the timing jitter. The technique of optical phase conjugation (OPC), discussed in Section 7.7 in the context of dispersion compensation, is also quite effective in improving the performance of soliton systems by reducing the soliton–soliton interaction and the Raman-induced timing jitter [200]–[204].

The implementation of OPC requires either parametric amplifiers [10] in place of EDFAs or insertion of a nonlinear optical device before each amplifier that changes the soliton amplitude from A to A^* while preserving all other features of the bit stream. Such a change is equivalent to inverting the soliton spectrum around the wavelength of the pump laser used for the FWM process. As discussed in Section 7.7, a few-kilometer-long fiber can be used for spectral inversion. The timing jitter changes considerably because of OPC. The moment method can be used to find the timing jitter in the presence of parametric amplifiers. The dependence of the Gordon–Haus contribution on the number of amplifiers changes from N_A^3 to N_A. The Raman-induced jitter is reduced even more dramatically—from N_A^5 to N_A [202]. These reductions result from the OPC-induced spectral inversion, which provides compensation for the effects of both the GVD and SSFS. However, OPC does not compensate for the effects of TOD.

The effects of TOD on the jitter are shown in Fig. 9.25 by plotting the timing jitter of 2-ps (FWHM) solitons propagating inside DDFs and amplified using parametric amplifiers. The Gordon–Haus timing jitter is shown by the thick solid line. Other curves correspond to different values of the TOD parameter. For $\beta_3 = 0.05$ ps^3/km, a typical value for dispersion-shifted fibers, the transmission distance is limited by TOD to below 1500 km. However, considerable improvement occurs when β_3 is reduced. Transmission over 7500 km is possible for $\beta_3 = 0$, and the distance can be increased further for slightly negative values of β_3. These results show that the compensation of both β_2 and β_3 is necessary at high bit rates (see Section 7.9).

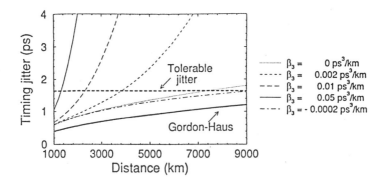

Figure 9.25: Effect of third-order dispersion on timing jitter in a periodically amplified DDF-based soliton system designed using parametric amplifiers.

9.7 WDM Soliton Systems

As discussed in Chapter 8, the capacity of a lightwave system can be increased considerably by using the WDM technique. A WDM soliton system transmits several soliton bit streams over the same fiber using different carrier frequencies. This section focuses on the issues relevant for designing WDM soliton systems [205].

9.7.1 Interchannel Collisions

A new feature that becomes important for WDM systems is the possibility of collisions among solitons belonging to different channels and traveling different group velocities. To understand the importance of such collisions as simply as possible, we first focus on standard solitons propagating in DDFs and use Eq. (9.4.3) as it includes the effects of both loss and dispersion variations. Dropping prime over ξ for notational convenience, this equation can be written as

$$i\frac{\partial v}{\partial \xi} + \frac{1}{2}\frac{\partial^2 v}{\partial \tau^2} + b(\xi)|v|^2 v = 0, \tag{9.7.1}$$

where $b(\xi) = p(\xi)/d(\xi)$. The functional form of $b(\xi)$ depends on the details of the loss- and dispersion-management schemes.

The effects of interchannel collisions on the performance of WDM systems can be best understood by considering the simplest case of two WDM channels separated by f_{ch}. In normalized units, solitons are separated in frequency by $\Omega_{\text{ch}} = 2\pi f_{\text{ch}} T_0$. Replacing v by $u_1 + u_2$ in Eq. (9.7.1) and neglecting the FWM terms, solitons in each channel evolve according to the following two coupled equations [206]:

$$i\frac{\partial u_1}{\partial \xi} + \frac{1}{2}\frac{\partial^2 u_1}{\partial \tau^2} + b(\xi)(|u_1|^2 + 2|u_2|^2)u_1 = 0, \tag{9.7.2}$$

$$i\frac{\partial u_2}{\partial \xi} + \frac{1}{2}\frac{\partial^2 u_2}{\partial \tau^2} + b(\xi)(|u_2|^2 + 2|u_1|^2)u_2 = 0. \tag{9.7.3}$$

Note that the two solitons are propagating at different speeds because of their different frequencies. As a result, the XPM term is important only when solitons belonging to different channels overlap during a collision.

It is useful to define the *collision length* L_{coll} as the distance over which two solitons interact (overlap) before separating. It is difficult to determine precisely the instant at which a collision begins or ends. A common convention uses $2T_s$ for the duration of the collision, assuming that a collision begins and ends when the two solitons overlap at their half-power points [206]. Since the relative speed of two solitons is $\Delta V = (|\beta_2|\Omega_{ch}/T_0)^{-1}$, the collision length is given by $L_{coll} = (\Delta V)(2T_s)$ or

$$L_{coll} = \frac{2T_s T_0}{|\beta_2|\Omega_{ch}} \approx \frac{0.28}{q_0|\beta_2|Bf_{ch}}, \tag{9.7.4}$$

where the relations $T_s = 1.763T_0$ and $B = (2q_0T_0)^{-1}$ were used. As an example, for $B = 10$ Gb/s, $q_0 = 5$, and $\beta_2 = -0.5$ ps^2/km, $L_{coll} \sim 100$ km for a channel spacing of 100 GHz but reduces to below 10 km when channels are separated by more than 1 THz.

Since XPM induces a time-dependent phase shift on each soliton, it leads to a shift in the soliton frequency during a collision. One can use a perturbation technique or the moment method to calculate this frequency shift. If we assume that the two solitons are identical before they collide, u_1 and u_2 are given by

$$u_m(\xi, \tau) = \text{sech}(\tau + \delta_m \xi)\exp[-i\delta_m\tau + i(1 - \delta_m^2)\xi/2 + i\phi_m], \tag{9.7.5}$$

where $\delta_m = \frac{1}{2}\Omega_{ch}$ for $m = 1$ and $-\frac{1}{2}\Omega_{ch}$ for $m = 2$. Using the moment method, the frequency shift for the first channel evolves with distance as

$$\frac{d\delta_1}{d\xi} = b(\xi) \int_{-\infty}^{\infty} \frac{\partial|u_1|^2}{\partial\tau}|u_2|^2 d\tau. \tag{9.7.6}$$

The equation for δ_2 is obtained by interchanging the subscripts 1 and 2. Noting from Eq. (9.7.5) that

$$\frac{\partial|u_m|^2}{\partial\tau} = \frac{1}{\delta_m}\frac{\partial|u_m|^2}{\partial\xi}, \tag{9.7.7}$$

and using $\delta_m = \pm\frac{1}{2}\Omega_{ch}$ in Eq. (9.7.6), the collision-induced frequency shift for the slower moving soliton is governed by [206]

$$\frac{d\delta_1}{d\xi} = \frac{b(\xi)}{\Omega_{ch}}\frac{d}{d\xi}\left[\int_{-\infty}^{\infty}\text{sech}^2\left(\tau - \frac{\Omega_{ch}\xi}{2}\right)\text{sech}^2\left(\tau + \frac{\Omega_{ch}\xi}{2}\right)d\tau\right]. \tag{9.7.8}$$

The change in δ_2 occurs by the same amount but in the opposite direction. The integral over τ can be performed analytically to obtain

$$\frac{d\delta_1}{dZ} = \frac{4b(Z)}{\Omega_{ch}}\frac{d}{dZ}\left(\frac{Z\cosh Z - \sinh Z}{\sinh^3 Z}\right), \tag{9.7.9}$$

where $Z = \Omega_{ch}\xi$. This equation provides changes in soliton frequency during inter-channel collisions under quite general conditions.

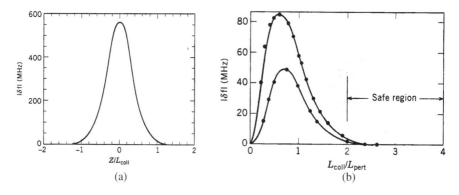

Figure 9.26: (a) Frequency shift during collision of two 50-ps solitons with 75-GHz channel spacing. (b) Residual frequency shift after a collision because of lumped amplifiers ($L_A = 20$ and 40 km for lower and upper curves, respectively). Numerical results are shown by solid dots. (After Ref. [206]; ©1991 IEEE; reprinted with permission.)

Consider first the ideal case of constant-dispersion lossless fibers so that $b = 1$ in Eq. (9.7.9). In that case, integration in Eq. (9.7.9) is trivial, and the frequency shift is given by

$$\Delta \delta_1(Z) = 4(Z \cosh Z - \sinh Z)/(\Omega_{ch} \sinh^3 Z). \qquad (9.7.10)$$

Figure 9.26(a) shows how the frequency of the slow-moving soliton changes during the collision of two 50-ps solitons when channel spacing is 75 GHz. The frequency shifts up over one collision length as two solitons approach each other, reaches a peak value of about 0.6 GHz at the point of maximum overlap, and then decreases back to zero as the two solitons separate. The maximum frequency shift depends on the channel spacing. It occurs at $Z = 0$ in Eq. (9.7.10) and is found to be $4/(3\Omega_{ch})$. In physical units, the maximum frequency shift becomes

$$\Delta f_{max} = (3\pi^2 T_0^2 f_{ch})^{-1}. \qquad (9.7.11)$$

Since the velocity of a soliton changes with its frequency, collisions speed up or slow down a soliton, depending on whether its frequency increases or decreases. At the end of the collision, each soliton recovers the frequency and speed it had before the collision, but its position within the bit slot changes. The temporal shift can be calculated by integrating Eq. (9.7.9). In physical units, it is given by

$$\Delta t = -T_0 \int_{-\infty}^{\infty} \Delta \delta_1(\xi)\, d\xi = \frac{4T_0}{\Omega_{ch}^2} = \frac{1}{\pi^2 T_0 f_{ch}^2}. \qquad (9.7.12)$$

If all bit slots were occupied, such collision-induced temporal shifts would be of no consequence since all solitons of a channel would be shifted by the same amount. However, 1 and 0 bits occur randomly in real bit streams. Since different solitons of a channel shift by different amounts, interchannel collisions induce some timing jitter even in lossless fibers.

9.7.2 Effect of Lumped Amplification

The situation is worse in loss-managed soliton systems in which fiber losses are compensated periodically through lumped amplifiers. The reason is that soliton collisions are affected adversely by variations in the pulse energy. Physically, large energy variations occurring over a collision length destroy the symmetric nature of the collision. Mathematically, the ξ dependence of $b(\xi)$ in Eq. (9.7.8) changes the frequency shift. As a result, solitons do not recover their original frequency and velocity after the collision is over. Equation (9.7.9) can be used to calculate the *residual* frequency shift for a given form of $b(\xi)$. Figure 9.26(b) shows the residual shift as a function of the ratio $L_{\text{coll}}/L_{\text{pert}}$, where L_{pert} is equal to the amplifier spacing L_A [206]. Numerical simulations (circles) agree with the prediction of Eq. (9.7.9). The residual frequency shift increases rapidly as L_{coll} approaches L_A and can become ~ 0.1 GHz. Such shifts are not acceptable in practice since they accumulate over multiple collisions and produce velocity changes large enough to move the soliton out of the bit slot.

When L_{coll} is large enough that a collision lasts over several amplifier spacings, effects of gain–loss variations begin to average out, and the residual frequency shift decreases. As seen in Fig. 9.26, it virtually vanishes for $L_{\text{coll}} > 2L_A$ (safe region). Since L_{coll} is inversely related to the channel spacing Ω_{ch}, this condition sets a limit on the maximum separation between the two outermost channels of a WDM system. The shortest collision length is obtained by replacing Ω_{ch} in Eq. (9.7.4) with $N_{\text{ch}}\Omega_{\text{ch}}$. Using $L_{\text{coll}} > 2L_A$, the number of WDM channels is limited to

$$N_{\text{ch}} < \frac{T_s L_D}{T_0 \Omega_{\text{ch}} L_A}. \tag{9.7.13}$$

One may think that the number of channels can be increased by reducing Ω_{ch}. However, its minimum value is limited to about $\Omega_{\text{ch}} = 5\Delta\omega_s$, where $\Delta\omega_s$ is the spectral width (FWHM) of solitons, because of interchannel crosstalk [207]. Using this condition in Eq. (9.7.13), the number of WDM channels is limited such that $N_{\text{ch}} < L_D/3L_A$. Using $L_D = T_0^2/|\beta_2|$ and $B = (2q_0 T_0)^{-1}$, this condition can be written as a simple design rule:

$$N_{\text{ch}} B^2 L_A < (12 q_0^2 |\beta_2|)^{-1}. \tag{9.7.14}$$

For the typical values $q_0 = 5$, $|\beta_2| = 0.8$ ps^2/km, and $L_A = 40$ km, the condition becomes $B\sqrt{N_{\text{ch}}} < 10$ Gb/s. The number of channels can be as large as 16 at a relatively low bit rate of 2.5 Gb/s but only a single channel is allowed at 10 Gb/s. Clearly, interchannel collisions limit the usefulness of the WDM technique severely.

9.7.3 Timing Jitter

In addition to the sources of timing jitter discussed in Section 9.6 for a single isolated channel, several other sources of jitter become important for WDM systems [208]–[213]. First, each interchannel collision generates a temporal shift [see Eq. (9.7.12)] of the same magnitude for both solitons but in opposite directions. Even though the temporal shift scales as Ω_{ch}^{-2} and decreases rapidly with increasing Ω_{ch}, the number of collisions increases linearly with Ω_{ch}. As a result, the total time shift scales as

Ω_{ch}^{-1}. Second, the number of collisions that two neighboring solitons in a given channel undergo is slightly different. This difference arises because adjacent solitons in a given channel interact with two different bit groups, shifted by one bit period. Since 1 and 0 bits occur in a random fashion, different solitons of the same channel are shifted by different amounts. This source of timing jitter is unique to WDM systems because of its dependence on the bit patterns of neighboring channels [211]. Third, collisions involving more than two solitons can occur and should be considered. In the limit of a large channel spacing (negligible overlap of soliton spectra), multisoliton interactions are well described by pairwise collisions [210].

Two other mechanisms of timing jitter should be considered for realistic WDM systems. As discussed earlier, energy variations due to gain–loss cycles make collisions asymmetric when L_{coll} becomes shorter than or comparable to the amplifier spacing L_A. Asymmetric collisions leave residual frequency shifts that affect a soliton all along the fiber link because of a change in its group velocity. This mechanism can be made ineffective by ensuring that L_{coll} exceeds $2L_A$. The second mechanism produces a residual frequency shift when solitons from different channels overlap at the input of the transmission link, resulting in an incomplete collision [208]. This situation occurs in all WDM solitons for some bits. For instance, two solitons overlapping completely at the input end of a fiber link will acquire a net frequency shift of $4/(3\Omega_{ch})$ since the first half of the collision is absent. Such residual frequency shifts are generated only over the first few amplification stages but pertain over the whole transmission length and become an important source of timing jitter [209].

Similar to the case of single-channel systems, sliding-frequency filters can reduce timing jitter in WDM systems [214]–[218]. Typically, Fabry–Perot filters are used since their periodic transmission windows allow filtering of all channels simultaneously. For best operation, the mirror reflectivities are kept low (below 25%) to reduce the finesse. Such low-contrast filters remove less energy from solitons but are as effective as filters with higher contrast. Their use allows channel spacing to be as little as five times the spectral width of the solitons [218]. The physical mechanism remains the same as for single-channel systems. More specifically, collision-induced frequency shifts are reduced because the filter forces the soliton frequency to move toward its transmission peak. The net result is that filters reduce the timing jitter considerably even for WDM systems [215]. Filtering can also relax the condition in Eq. (9.7.13), allowing L_{coll} to approach L_A, and thus helps to increase the number of channels in a WDM system [217].

The technique of synchronous modulation can also be applied to WDM systems for controlling timing jitter [219]. In a 1996 experiment involving four channels, each operating at 10 Gb/s, transmission over transoceanic distances was achieved by using modulators every 500 km [220]. When modulators were inserted every 250 km, three channels, each operating at 20 Gb/s, could be transmitted over transoceanic distances [221]. The main disadvantage of modulators is that demultiplexing of individual channels is necessary. Moreover, they require a clock signal that is synchronized to the bit stream. For this reason, the technique of synchronous modulation is rarely used in practice.

9.7.4 Dispersion Management

As discussed in Section 8.3.6, FWM is the most limiting factor for WDM systems when GVD is kept constant along the fiber link. The FWM problem virtually disappears when the dispersion-management technique is used. In fact, dispersion management is essential if a WDM soliton system is designed to transmit more than two or three channels. Starting in 1996, dispersion management was used for WDM soliton systems almost exclusively.

Dispersion-Decreasing Fibers

It is intuitively clear that DDFs with a continuously varying GVD profile should help a WDM system. We can use Eq. (9.7.1) for finding the optimum GVD profile. By tailoring the fiber dispersion as $p(\xi) = \exp(-\Gamma\xi)$, the same exponential profile encountered in Section 8.4.1, the parameter b becomes 1 all along the fiber link, resulting in an unperturbed NLS equation. As a result, soliton collisions become symmetric despite fiber losses, irrespective of the ratio L_{coll}/L_A. Consequently, no residual frequency shifts occur after a soliton collision for WDM systems making use of DDFs with an exponentially decreasing GVD.

Lumped amplifiers introduce a new mechanism of FWM in WDM systems. In their presence, soliton energy varies in a periodic fashion over each loss–amplification cycle. Such periodic variations in the peak power of solitons create a nonlinear-index grating that can nearly phase-match the FWM process [222]. The phase-matching condition can be written as

$$|\beta_2|(\Omega_{\text{ch}}/T_0)^2 = 2\pi m/L_A, \qquad (9.7.15)$$

where m is an integer and the amplifier spacing L_A is the period of the index grating. As a result of such phase matching, a few percent of soliton energy can be transferred to the FWM sidebands even when GVD is relatively large [222]. Moreover, FWM occurring during simultaneous collision of three solitons leads to permanent frequency shifts for the slowest- and fastest-moving solitons together with an energy exchange among all three channels [223].

FWM phase-matched by the nonlinear-index grating can also be avoided by using DDFs with an exponential GVD profile. The reason is related to the symmetric nature of soliton collisions in such systems. When collisions are symmetric, energy transferred to the FWM sidebands during the first half of a collision is returned back to the soliton during the second half of the same collision. Thus, the spectral sidebands generated through FWM do not grow with propagation of solitons. In practice, the staircase approximation for the exponential profile is used, employing multiple constant-dispersion fibers between two amplifiers.

Figure 9.27 shows the residual energy remaining in a FWM sideband as a function of amplifier length L_A when the exponential GVD profile is approximated using $m = 2$, 3, and 4 fiber sections chosen such that the product $D_m L_m$ is the same for all m [222]. Here D_m is the dispersion parameter in the mth section of length L_m. The case of constant-dispersion fibers is also shown for comparison. The average dispersion is 0.5 ps/(km-nm) in all cases. The double-peak nature of the curve in this case is due to the phase-matching condition in Eq. (9.7.15), which can be satisfied for different values

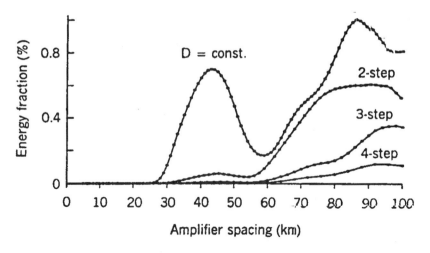

Figure 9.27: Fraction of soliton energy in an FWM sideband during a single collision when the exponential GVD profile is approximated by a staircase with two, three, and four steps. The case of constant-dispersion fibers is shown for comparison. (After Ref. [222]); ©1996 OSA; reprinted with permission.)

of the integer m because a peak occurs whenever $L_A = 2\pi m L_D / \Omega_{ch}^2$. Numerical simulations consider 20-ps solitons in two channels, spaced 75 GHz apart. Clearly, FWM can be nearly suppressed, with as few as three fiber sections, for an amplifier spacing below 60 km. An experiment in 1996 achieved transmission of seven 10-Gb/s channels over 9400 km using only four fiber segments in a recirculating fiber loop [224]. In a 1998 experiment, eight 20-Gb/s channels were transmitted over 10,000 km by using the same four-segment approach in combination with optical filters and modulators [225].

Periodic Dispersion Maps

Similar to the single-channel soliton systems discussed in Section 9.4.2, periodic dispersion maps consisting of two fiber segments with opposite GVD benefit the WDM soliton systems enormously. Issues such as interchannel collisions, timing jitter, and optimum dispersion maps were studied extensively during the 1990s [226]–[250]. The use of design optimization techniques has resulted in WDM soliton systems capable of operating at bit rates close to 1 Tb/s [251]–[262].

An important issue for WDM systems making use of DM solitons is how a dispersion map consisting of opposite-GVD fibers affects interchannel collisions and the timing jitter. It is easy to see that the nature of soliton collisions is changed drastically in such systems. Consider solitons in two different channels. A shorter-wavelength soliton travels faster in the anomalous-GVD section but slower in the normal-GVD section. Moreover, because of high local GVD, the speed difference is large. Also, the pulse width changes in each map period and can become quite large in some regions. The net result is that two colliding solitons move in a zigzag fashion and pass through each other many times before they separate from each other because of the

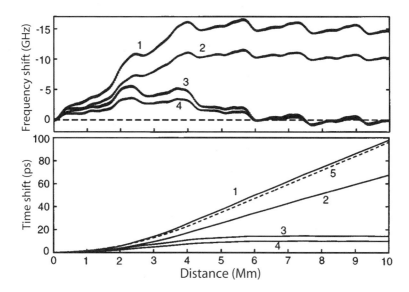

Figure 9.28: Collision-induced frequency and temporal shifts for a soliton surrounded by four channels on each side (75 GHz spacing). Curves 1 and 2 represent the case copolarized and orthogonally polarized solitons in neighboring channels, respectively. Curves 3 and 4 show the improvement realized with sliding-frequency filters. The dotted line shows the prediction of an analytical model. (After Ref. [238]); ©1999 OSA; reprinted with permission.)

much slower relative motion governed by the average value of GVD. Since the effective collision length is much larger than the map period (and the amplifier spacing), the condition $L_{coll} > 2L_A$ is satisfied even when soliton wavelengths differ by 20 nm or more. This feature makes it possible to design WDM soliton systems with a large number of high-bit-rate channels.

The residual frequency shift introduced during such a process depends on a large number of parameters including the map period, map strength, and amplifier spacing [236]–[240]. Physically speaking, residual frequency shifts occurring during complete collisions average out to zero. However, not all collisions are complete. For example, if solitons overlap initially, the incomplete nature of the collision will produce some residual frequency shift. The zigzag motion of solitons can also produce frequency shifts if the solitons approach each other near the junction of opposite-GVD fibers since they will reverse direction before crossing each other. Such partial collisions can result in large frequency shifts, which can shift solitons by a large amount within their bit slots. This behavior is unacceptable from a system standpoint.

A simple solution to this problem is provided by sliding-frequency filters [238]. Such filters reduce the frequency and temporal shifts to manageable levels in the same way they mitigate the effects of ASE-induced frequency shifts. Curve 1 in Figure 9.28 shows the frequency and temporal shifts (calculated numerically) for the middle channel surrounded by four channels on each side (channel spacing 75 GHz). The soliton shifts by 100 ps (one bit slot) over 10,000 km because its frequency shifts by more than 10 GHz. The use of orthogonally polarized solitons (curve 2) improves the situation somewhat but does not solve the problem. However, if sliding-frequency filters are employed, the temporal shift is reduced to below 15 ps for copolarized solitons (curve 3) and to below 10 ps for orthogonally polarized solitons (curve 4). In these numerical simulations, the map period and amplifier spacing are equal to 40 km. The dispersion map consists of 36 km of anomalous-GVD fiber and 4 km of DCF ($\beta_{2n} \approx 130$ ps^2/km) such that average value of dispersion is 0.1 ps/(km-nm).

Many issues need to be addressed in designing WDM systems. These include the SNR degradation because of the accumulation of ASE, timing jitter induced by ASE and other sources (acoustic waves, ASE, Raman-induced frequency shift, PMD, etc.), XPM-induced intrachannel interactions, and interchannel collisions. Most of them depend on the choice of the map period, local value of the GVD in each fiber section, and average dispersion of the entire link. The choice of loss-management scheme (lumped versus distributed amplification) also impacts the system performance. These issues are common to all WDM systems and can only be addressed by solving the underlying NLS equation numerically (see Appendix E).

Figure 9.29 shows the role played by local dispersion by comparing the maximum transmission distances for the lumped (EDFA) and distributed (Raman) amplification schemes [248]. The dispersion map consists of a long fiber section with GVD in the range 2–17 ps/(km-nm) and a short fiber section with the dispersion of -25 ps/(km-nm) whose length is chosen to yield an average GVD of 0.04 ps/(km-nm). The map period and amplifier spacing are equal to 50 km. Pulse parameters at a bit rate of 40 Gb/s correspond to the periodic propagation of DM solitons. The curves marked "interactions" provide the distance at which solitons have shifted by 30% of the 25-ps bit slot because of intrachannel pulse-to-pulse interactions. The curves marked "XPM" denote the limiting distance set by the XPM-induced interchannel interactions. The curves marked "PIM" show the improvement realized by polarization-interleaved multiplexing (PIM) of WDM channels. The two circles marked A and B denote the optimum values of local GVD in the cases of lumped and distributed amplification, respectively. Several points are noteworthy in Fig. 9.29. First, Raman amplification improves the transmission distance from the standpoint of intrachannel interactions but has a negative impact when interchannel collisions are considered. Second, polarization multiplexing helps for both lumped and distributed amplification. Third, the local value of GVD plays an important role and its optimum value is different for lumped and distributed amplification. The main conclusion is that numerical simulations are essential for optimizing any WDM system.

On the experimental side, 16 channels at 20 Gb/s were transmitted in 1997 over 1300 km of standard fiber with a map period of 100 km using a DCF that compensated partially both GVD and its slope [251]. In a 1998 experiment, 20 channels at 20 Gb/s were transmitted over 2000 km using dispersion-flattened fiber with a channel spacing of 0.8 nm [252]. A capacity of 640 Gb/s was realized in a 2000 experiment in which 16 channels at 40 Gb/s were transmitted over 1000 km [257]. In a later experiment, the system capacity was extended to 1 Tb/s by transmitting 25 channels at 40 Gb/s over 1500 km with 100-GHz channel spacing [261]. The 250-km recirculating fiber loop employed a dispersion map with the 50-km map period and a relatively low value of average dispersion for all channels. Dispersion slope (TOD) was nearly compensated and had a value of less than 0.005 ps/(km-nm^2). A BER of 10^{-9} could

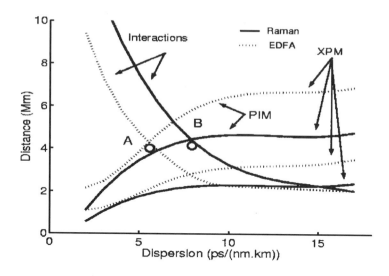

Figure 9.29: Maximum transmission distance as a function of local dispersion for a WDM soliton system with 40-Gb/s channels spaced apart by 100 GHz. The curves with labels "Interactions" and "XPM" show limitations due to intrachannel and interchannel pulse interactions, respectively. The PIM curves correspond to the case of polarization multiplexing. (After Ref. [248]); ©2001 OSA; reprinted with permission.)

be achieved for all channels because of dispersion-slope compensation realized using reverse-dispersion fibers. In a 2001 experiment, a system capacity of 2.56 Tb/s was realized (32 channels at 80 Gb/s) by interleaving two orthogonally polarized 40-Gb/s WDM pulse trains but the transmission distance was limited to 120 km [262]. The use of polarization multiplexing in combination with the carrier-suppressed RZ format permitted a spectral efficiency of 0.8 (b/s)/Hz in this experiment.

Many experiments have focused on soliton systems for transoceanic applications. The total bit rate is lower for such systems because of long distances over which solitons must travel. Transmission of eight channels at 10-Gb/s over transoceanic distances was realized as early as 1996 [205]. Eight 20-Gb/s channels were transmitted in a 1998 experiment but the distance was limited to 4000 km [254]. By 2000, the 160-Gb/s capacity was attained by transmitting eight 20-Gb/s channels over 10,000 km using optical filters and synchronous modulators inside a 250-km recirculating fiber loop [258]. It was necessary to use a polarization scrambler and a phase modulator at the input end. The 160-Gb/s capacity was also realized using two 80-Gb/s channels. In another experiment, up to 27 WDM channels were transmitted over 9000 km using a hybrid amplification scheme in which distributed Raman amplification (with backward pumping) compensated for nearly 70% of losses incurred over the 56-km map period [259]. In general, the use of distributed Raman amplification improves the system performance considerably as it reduces the XPM-induced interactions among solitons [248]. These experiments show that the use of DM solitons has the potential of realizing transoceanic lightwave systems capable of operating with a capacity of 1 Tb/s or more.

Problems

9.1 A 10-Gb/s soliton system is operating at 1.55 μm using fibers with $D = 2$ ps/(km-nm). The effective core area of the fiber is 50 μm^2. Calculate the peak power and the pulse energy required for fundamental solitons of 30-ps width (FWHM). Use $n_2 = 2.6 \times 10^{-20}$ m^2/W.

9.2 The soliton system of Problem 9.1 needs to be upgraded to 40 Gb/s. Calculate the pulse width, peak power, and the energy of solitons using $q_0 = 4$. What is the average launched power for this system?

9.3 Verify by direct substitution that the soliton solution given in Eq. (9.1.11) satisfies the NLS equation.

9.4 Solve the NLS equation using the split-step Fourier method (see Section 2.4 of Ref. [10] for details on this method). Reproduce Figs. 9.1–9.3 using your program. Any programming language, including software packages such as Mathematica and Matlab, can be used.

9.5 Verify numerically by propagating a fundamental soliton over 100 dispersion lengths that the shape of the soliton does not change on propagation. Repeat the simulation using a Gaussian input pulse shape with the same peak power and explain the results.

9.6 A 10-Gb/s soliton lightwave system is designed with $q_0 = 5$ to ensure well-separated solitons in the RZ bit stream. Calculate pulse width, peak power, pulse energy, and the average power of the RZ signal assuming $\beta_2 = -1$ ps^2/km and $\gamma = 2$ W^{-1}/km for the dispersion-shifted fiber.

9.7 A soliton communication system is designed to transmit data over 5000 km at $B = 10$ Gb/s. What should be the pulse width (FWHM) to ensure that the neighboring solitons do not interact during transmission? The dispersion parameter $D = 1$ ps/(km-nm) at the operating wavelength. Assume that soliton interaction is negligible when $B^2 L_T$ in Eq. (9.2.10) is 10% of its maximum allowed value.

9.8 Prove that the energy of standard solitons should be increased by the factor $G \ln G/(G-1)$ when the fiber loss α is compensated using optical amplifiers. Here $G = \exp(\alpha L_A)$ is the amplifier gain and L_A is the spacing between amplifiers assumed to be much smaller than the dispersion length.

9.9 A 10-Gb/s soliton communication system is designed with 50-km amplifier spacing. What should be the peak power of the input pulse to ensure that a fundamental soliton is maintained in an average sense in a fiber with 0.2 dB/km loss? Assume that $T_s = 20$ ps, $\beta_2 = -0.5$ ps^2/km and $\gamma = 2$ W^{-1}/km. What is the average launched power for such a system?

9.10 Calculate the maximum bit rate for a soliton lightwave system designed with $q_0 = 5$, $\beta_2 = -1$ ps^2/km, and $L_A = 50$ km. Assume that the condition (9.3.10) is satisfied when $B^2 L_A$ is at the 20% level. What is the soliton width at the maximum bit rate?

9.11 Derive Eq. (9.3.15) by integrating Eq. (9.3.11) in the case of bidirectional pumping. Plot $p(z)$ for $L_A = 20,\ 40,\ 60,$ and 80 km using $\alpha = 0.2$ dB/km and $\alpha_p = 0.25$ dB/km.

9.12 Use Eq. (9.3.15) to determine the pump-station spacing L_A for which the soliton energy deviates at most 20% from its input value.

9.13 Consider soliton evolution in a dispersion-decreasing fiber using the NLS equation and prove that soliton remains unperturbed when the fiber dispersion decreases exponentially as $\beta_2(z) = \beta_2(0)\exp(-\alpha z)$.

9.14 Starting from the NLS equation (9.4.5), derive the variational equations for the pulse width T and the chirp C using the Gaussian ansatz given in Eq. (9.4.6).

9.15 Solve Eqs. (9.4.7) and (9.4.8) numerically by imposing the periodicity condition given in Eq. (9.4.9). Plot T_0 and C_0 as a function of E_0 for a dispersion map made using 70 km of the standard fiber with $D = 17$ ps/(km-nm) and 10 km of dispersion-compensating fiber with $D = -115$ ps/(km-nm). Use $\gamma = 2$ W^{-1}/km and $\alpha = 0.2$ dB/km for the standard fiber and $\gamma = 6$ W^{-1}/km and $\alpha = 0.5$ dB/km for the other fiber.

9.16 Calculate the map strength S and the map parameter T_{map} for the map used in the preceding problem. Estimate the maximum bit rate that this map can support.

9.17 Verify using Eqs. (9.5.8)–(9.5.12) that the variances and correlations of amplifier-induced fluctuations are indeed given by Eqs. (9.5.13)–(9.5.15).

9.18 Prove that the variances of E, Ω, and q are given by Eq. (9.5.17) for the standard solitons using Eq. (9.5.16) in Eqs. (9.5.8)–(9.5.11).

9.19 Derive Eq. (9.5.29) for the timing jitter starting from the recurrence relation in Eq. (9.5.26). Show all the steps clearly.

9.20 Find the peak value of the collision-induced frequency and temporal shifts by integrating Eq. (9.7.8) with $b = 1$.

9.21 Explain how soliton collisions limit the number of channels in a WDM soliton system. Find how the maximum number of channels depends on the channel and amplifier spacings using the condition $L_{\text{coll}} > 2L_A$.

References

[1] N. Zabusky and M. D. Kruskal, *Phys. Rev. Lett.* **15**, 240 (1965).

[2] A. Hasegawa and F. Tappert, *Appl. Phys. Lett.* **23**, 142 (1973).

[3] L. F. Mollenauer, R. H. Stolen, and J. P. Gordon, *Phys. Rev. Lett.* **45**, 1095 (1980).

[4] L. F. Mollenauer and K. Smith, *Opt. Lett.* **13**, 675 (1988).

[5] A. Hasegawa and Y. Kodama, *Solitons in Optical Communications*, Clarendon Press, Oxford, 1995.

[6] L. F. Mollenauer, J. P. Gordon, and P. V. Mamychev, in *Optical Fiber Telecommunications IIIA*, I. P. Kaminow and T. L. Koch, Eds., Academic Press, San Diego, CA, 1997, Chap. 12.

[7] R. J. Essiambre and G. P. Agrawal, in *Progress in Optics*, E. Wolf, Ed., Vol. 37, Elsevier, Amsterdam, 1997, Chap. 4

[8] E. Iannone, F. Matera, A. Mecozzi, and M. Settembre, *Nonlinear Optical Communication Networks*, Wiley, New York, 1998, Chap. 5.

[9] G. P. Agrawal, *Applications of Nonlinear Fiber Optics*, Academic Press, San Diego, CA, 2001, Chap. 8.

[10] G. P. Agrawal, *Nonlinear Fiber Optics*, 3rd ed., Academic Press, San Diego, CA, 2001.

[11] M. J. Ablowitz and P. A. Clarkson, *Solitons, Nonlinear Evolution Equations, and Inverse Scattering*, Cambridge University Press, New York, 1991.

[12] G. L. Lamb, Jr., *Elements of Soliton Theory*, Dover, New York, 1994.

[13] T. Miwa, *Mathematics of Solitons*, Cambridge University Press, New York, 1999.

[14] V. E. Zakharov and A. B. Shabat, *Sov. Phys. JETP* **34**, 62 (1972).

[15] J. Satsuma and N. Yajima, *Prog. Theor. Phys.* **55**, 284 (1974).

[16] V. E. Zakharov and A. B. Shabat, *Sov. Phys. JETP* **37**, 823 (1973).

[17] A. Hasegawa and F. Tappert, *Appl. Phys. Lett.* **23**, 171 (1973).

[18] W. J. Tomlinson, R. J. Hawkins, A. M. Weiner, J. P. Heritage, and R. N. Thurston, *J. Opt. Soc. Am. B* **6**, 329 (1989).

[19] P. Emplit, M. Haelterman, and J. P. Hamaide, *Opt. Lett.* **18**, 1047 (1993).

[20] W. Zhao and E. Bourkoff, *J. Opt. Soc. Am. B* **9**, 1134 (1992).

[21] J. E. Rothenberg and H. K. Heinrich, *Opt. Lett.* **17**, 261 (1992).

[22] D. J. Richardson, R. P. Chamberlin, L. Dong, and D. N. Payne, *Electron. Lett.* **30**, 1326 (1994).

[23] O. G. Okhotnikov and F. M. Araujo, *Electron. Lett.* **31**, 2187 (1995).

[24] A. K. Atieh, P. Myslinski, J. Chrostowski, and P. Galko, *Opt. Commun.* **133**, 541 (1997).

[25] M. Nakazawa and K. Suzuki, *Electron. Lett.* **31**, 1084 (1995); *Electron. Lett.* **31**, 1076 (1995).

[26] P. Emplit, M. Haelterman, R. Kashyap, and M. DeLathouwer, *IEEE Photon. Technol. Lett.* **9**, 1122 (1997).

[27] R. Leners, P. Emplit, D. Foursa, M. Haelterman, and R. Kashyap, *J. Opt. Soc. Am. B* **14**, 2339 (1997).

[28] Y. S. Kivshar and B. Luther-Davies, *Phys. Rep.* **298**, 81 (1998).

[29] V. I. Karpman and V. V. Solovev, *Physica* **3D**, 487 (1981).

[30] J. P. Gordon, *Opt. Lett.* **8**, 596 (1983).

[31] F. M. Mitschke and L. F. Mollenauer, *Opt. Lett.* **12**, 355 (1987).

[32] C. Desem and P. L. Chu, *IEE Proc.* **134**, 145 (1987).

[33] Y. Kodama and K. Nozaki, *Opt. Lett.* **12**, 1038 (1987).

[34] C. Desem and P. L. Chu, *Opt. Lett.* **11**, 248 (1986).

[35] K. J. Blow and D. Wood, *Opt. Commun.* **58**, 349 (1986).

[36] A. I. Maimistov and Y. M. Sklyarov, *Sov. J. Quantum Electron.* **17**, 500 (1987).

[37] A. S. Gouveia-Neto, A. S. L. Gomes, and J. R. Taylor, *Opt. Commun.* **64**, 383 (1987).

[38] K. Iwatsuki, A. Takada, and M. Saruwatari, *Electron. Lett.* **24**, 1572 (1988).

[39] K. Iwatsuki, A. Takada, S. Nishi, and M. Saruwatari, *Electron. Lett.* **25**, 1003 (1989).

[40] M. Nakazawa, K. Suzuki, and Y. Kimura, *Opt. Lett.* **15**, 588 (1990).

[41] P. A. Morton, V. Mizrahi, G. T. Harvey, L. F. Mollenauer, T. Tanbun-Ek, R. A. Logan, H. M. Presby, T. Erdogan, A. M. Sergent, and K. W. Wecht, *IEEE Photon. Technol. Lett.* **7**, 111 (1995).

[42] V. Mikhailov, P. Bayvel, R. Wyatt, and I. Lealman, *Electron. Lett.* **37**, 909 (2001).

[43] S. Arahira, Y. Matsui, T. Kunii, S. Oshiba, and Y. Ogawa, *IEEE Photon. Technol. Lett.* **5**, 1362 (1993).

[44] M. Suzuki, H. Tanaka, N. Edagawa, Y. Matsushima, and H. Wakabayashi, *Fiber Integ. Opt.* **12**, 358 (1993).

[45] K. Sato, I. Kotaka, Y. Kondo, and M. Yamamoto, *Appl. Phys. Lett.* **69**, 2626 (1996).

[46] K. Wakita and I. Kotaka, *Microwave Opt. Tech. Lett.* **7**, 120 (1995).

[47] I. N. Duling, Ed., *Compact Sources of Ultrashort Pulses*, Cambridge University Press, New York, 1995.

[48] G. T. Harvey and L. F. Mollenauer, *Opt. Lett.* **18**, 107 (1993).

[49] D. M. Patrick, *Electron. Lett.* **30**, 43 (1994).

[50] S. V. Chernikov, J. R. Taylor, and R. Kashyap, *Electron. Lett.* **29**, 1788 (1993); *Opt. Lett.* **19**, 539 (1994).

[51] E. A. Swanson and S. R. Chinn, *IEEE Photon. Technol. Lett.* **7**, 114 (1995).

[52] P. V. Mamyshev, *Opt. Lett.* **19**, 2074 (1994).

[53] J. J. Veselka, S. K. Korotky, P. V. Mamyshev, A. H. Gnauck, G. Raybon, and N. M. Froberg, *IEEE Photon. Technol. Lett.* **8**, 950 (1996).

[54] A. Hasegawa and Y. Kodama, *Proc. IEEE* **69**, 1145 (1981); *Opt. Lett.* **7**, 285 (1982).

[55] K. J. Blow and N. J . Doran, *Opt. Commun.* **52**, 367 (1985).

[56] M. J. Potasek and G. P. Agrawal, *Electron. Lett.* **22**, 759 (1986).

[57] Y. Kodama and A. Hasegawa, *Opt. Lett.* **7**, 339 (1982); **8**, 342 (1983).

[58] A. Hasegawa, *Opt. Lett.* **8**, 650 (1983); *Appl. Opt.* **23**, 3302 (1984).

[59] L. F. Mollenauer, R. H. Stolen, and M. N. Islam, *Opt. Lett.* **10**, 229 (1985).

[60] L. F. Mollenauer, J. P. Gordon, and M. N. Islam, *IEEE J. Quantum Electron.* **22**, 157 (1986).

[61] A. Hasegawa and Y. Kodama, *Phys. Rev. Lett.* **66**, 161 (1991).

[62] D. M. Spirit, I. W. Marshall, P. D. Constantine, D. L. Williams, S. T. Davey, and B. J. Ainslie, *Electron. Lett.* **27**, 222 (1991).

[63] M. Nakazawa, H. Kubota, K. Kurakawa, and E. Yamada, *J. Opt. Soc. Am. B* **8**, 1811 (1991).

[64] K. Kurokawa and M. Nakazawa, *IEEE J. Quantum Electron.* **28**, 1922 (1992).

[65] K. Rottwitt, J. H. Povlsen, S. Gundersen, and A. Bjarklev, *Opt. Lett.* **18**, 867 (1993).

[66] C. Lester, K. Bertilsson, K. Rottwitt, P. A. Andrekson, M. A. Newhouse, and A. J. Antos, *Electron. Lett.* **31**, 219 (1995).

[67] F. M. Mitschke and L. F. Mollenauer, *Opt. Lett.* **11**, 659 (1986).

[68] J. P. Gordon, *Opt. Lett.* **11**, 662 (1986).

[69] Z. M. Liao and G. P. Agrawal, *IEEE Photon. Technol. Lett.* **11**, 818 (1999).

[70] M. Nakazawa, K. Suzuki, and Y. Kimura, *IEEE Photon. Technol. Lett.* **2**, 216 (1990).

[71] N. A. Olsson, P. A. Andrekson, P. C. Becker, J. R. Simpson, T. Tanbun-Ek, R. A. Logan, H. Presby, and K. Wecht, *IEEE Photon. Technol. Lett.* **2**, 358 (1990).

[72] K. Iwatsuki, S. Nishi, and K. Nakagawa, *IEEE Photon. Technol. Lett.* **2**, 355 (1990).

[73] E. Yamada, K. Suzuki, and M. Nakazawa, *Electron. Lett.* **27**, 1289 (1991).

[74] L. F. Mollenauer, B. M. Nyman, M. J. Neubelt, G. Raybon, and S. G. Evangelides, *Electron. Lett.* **27**, 178 (1991).

[75] K. Tajima, *Opt. Lett.* **12**, 54 (1987).

[76] V. A. Bogatyrjov, M. M. Bubnov, E. M. Dianov, and A. A. Sysoliatin, *Pure Appl. Opt.* **4**, 345 (1995).

[77] D. J. Richardson, R. P. Chamberlin, L. Dong, and D. N. Payne, *Electron. Lett.* **31**, 1681 (1995).

[78] A. J. Stentz, R. Boyd, and A. F. Evans, *Opt. Lett.* **20**, 1770 (1995).

[79] D. J. Richardson, L. Dong, R. P. Chamberlin, A. D. Ellis, T. Widdowson, and W. A. Pender, *Electron. Lett.* **32**, 373 (1996).

[80] W. Forysiak, F. M. Knox, and N. J. Doran, *Opt. Lett.* **19**, 174 (1994).

[81] T. Georges and B. Charbonnier, *Opt. Lett.* **21**, 1232 (1996); *IEEE Photon. Technol. Lett.* **9**, 127 (1997).

[82] S. Cardinal, E. Desurvire, J. P. Hamaide, and O. Audouin, *Electron. Lett.* **33**, 77 (1997).

[83] A. Hasegawa, Y. Kodama, and A. Murata, *Opt. Fiber Technol.* **3**, 197 (1997).

[84] S. Kumar, Y. Kodama, and A. Hasegawa, *Electron. Lett.* **33**, 459 (1997).

[85] N. J. Smith, F. M. Knox, N. J. Doran, K. J. Blow, and I. Bennion, *Electron. Lett.* **32**, 54 (1996).

[86] M. Nakazawa, H. Kubota, and K. Tamura, *IEEE Photon. Technol. Lett.* **8**, 452 (1996).

[87] M. Nakazawa, H. Kubota, A. Sahara, and K. Tamura, *IEEE Photon. Technol. Lett.* **8**, 1088 (1996).

[88] A. B. Grudinin and I. A. Goncharenko, *Electron. Lett.* **32**, 1602 (1996).

[89] A. Berntson, N. J. Doran, W. Forysiak, and J. H. B. Nijhof, *Opt. Lett.* **23**, 900 (1998).

[90] J. N. Kutz, P. Holmes, S. G. Evangelides, and J. P. Gordon, *J. Opt. Soc. Am. B* **15**, 87 (1998).

[91] S. K. Turitsyn, I. Gabitov, E. W. Laedke, V. K. Mezentsev, S. L. Musher, E. G. Shapiro, T. Schafer, and K. H. Spatschek, *Opt. Commun.* **151**, 117 (1998).

[92] T. I. Lakoba and D. J. Kaup, *Phys. Rev. E* **58**, 6728 (1998).

[93] S. K. Turitsyn and E. G. Shapiro, *J. Opt. Soc. Am. B* **16**, 1321 (1999).

[94] I. R. Gabitov, E. G. Shapiro, and S. K. Turitsyn, *Phys. Rev. E* **55**, 3624 (1997).

[95] M. J. Ablowitz and G. Bioindini, *Opt. Lett.* **23**, 1668 (1998).

[96] C. Paré and P. A. Belangér, *Opt. Lett.* **25**, 881 (2000).

[97] J. H. B. Nijhof, N. J. Doran, W. Forysiak, and F. M. Knox, *Electron. Lett.* **33**, 1726 (1997).

[98] V. S. Grigoryan and C. R. Menyuk, *Opt. Lett.* **23**, 609 (1998).

[99] J. N. Kutz and S. G. Evangelides, Jr., *Opt. Lett.* **23**, 685 (1998).

[100] Y. Chen and H. A. Haus, *Opt. Lett.* **23**, 1013 (1998).

[101] J. H. B. Nijhof, W. Forysiak, and N. J. Doran, *Opt. Lett.* **23**, 1674 (1998).

[102] S. K. Turitsyn, J. H. B. Nijhof, V. K. Mezentsev, and N. J. Doran, *Opt. Lett.* **24**, 1871 (1999).

[103] H. Kubota and M. Nakazawa, *Opt. Commun.* **87**, 15 (1992); M. Nakazawa and H. Kubota, *Electron. Lett.* **31**, 216 (1995).

[104] M. Suzuki, I. Morita, N. Edagawa, S. Yamamoto, H. Taga, and S. Akiba, *Electron. Lett.* **31**, 2027 (1995).

[105] A. Naka, T. Matsuda, and S. Saito, *Electron. Lett.* **32**, 1694 (1996).

[106] I. Morita, M. Suzuki, N. Edagawa, S. Yamamoto, H. Taga, and S. Akiba, *IEEE Photon. Technol. Lett.* **8**, 1573 (1996).

[107] J. M. Jacob, E. A. Golovchenko, A. N. Pilipetskii, G. M. Carter, and C. R. Menyuk, *IEEE Photon. Technol. Lett.* **9**, 130 (1997).

[108] G. M. Carter and J. M. Jacob, *IEEE Photon. Technol. Lett.* **10**, 546 (1998).

[109] V. S. Grigoryan, R. M. Mu, G. M. Carter, and C. R. Menyuk, *IEEE Photon. Technol. Lett.* **10**, 45 (2000).

[110] R. M. Mu, C. R. Menyuk, G. M. Carter, and J. M. Jacob, *IEEE J. Sel. Topics Quantum Electron.* **6**, 248 (2000).

[111] A. B. Grudinin, M. Durkin, M. Isben, R. I. Laming, A. Schiffini, P. Franco, E. Grandi, and M. Romagnoli, *Electron. Lett.* **33**, 1572 (1997).

[112] F. Favre, D. Le Guen, and T. Georges, *J. Lightwave Technol.* **17**, 1032 (1999).

[113] I. S. Penketh, P. Harper, S. B. Aleston, A. M. Niculae, I. Bennion, and N. J. Doran, *Opt. Lett.* **24**, 803 (1999).

[114] M. Zitelli, F. Favre, D. Le Guen, and S. Del Burgo, *IEEE Photon. Technol. Lett.* **9**, 904 (1999).

[115] J. P. Gordon and H. A. Haus, *Opt. Lett.* **11**, 665 (1986).

[116] D. Marcuse, *J. Lightwave Technol.* **10**, 273 (1992).

[117] N. J. Smith, W. Foryisak, and N. J. Doran, *Electron. Lett.* **32**, 2085 (1996).

[118] G. M. Carter, J. M. Jacob, C. R. Menyuk, E. A. Golovchenko, A. N. Pilipetskii, *Opt. Lett.* **22**, 513 (1997).

[119] S. Kumar and F. Lederer, *Opt. Lett.* **22**, 1870 (1997).

[120] J. N. Kutz and P. K. A. Wai, *IEEE Photon. Technol. Lett.* **10**, 702 (1998).

[121] T. Okamawari, A. Maruta, and Y. Kodama, *Opt. Lett.* **23**, 694 (1998); *Opt. Commun.* **149**, 261 (1998).

[122] V. S. Grigoryan, C. R. Menyuk, and R. M. Mu, *J. Lightwave Technol.* **17**, 1347 (1999).

[123] M. F. S. Ferreira and S. C. V. Latas, *J. Lightwave Technol.* **19**, 332 (2001).

[124] J. Santhanam, C. J. McKinstrie, T. I. Lakoba, and G. P. Agrawal, *Opt. Lett.* **26**, 1131 (2001).

[125] E. Poutrina and G. P. Agrawal, *IEEE Photon. Technol. Lett.* **14**, 39 (2002).

[126] A. Mecozzi, J. D. Moores, H. A. Haus, and Y. Lai, *Opt. Lett.* **16**, 1841 (1991).

[127] L. F. Mollenauer, M. J. Neubelt, M. Haner, E. Lichtman, S. G. Evangelides, and B. M. Nyman, *Electron. Lett.* **27**, 2055 (1991).

[128] Y. Kodama and A. Hasegawa, *Opt. Lett.* **17**, 31 (1992).

[129] L. F. Mollenauer, J. P. Gordon, and S. G. Evangelides, *Opt. Lett.* **17**, 1575 (1992).

[130] V. V. Afanasjev, *Opt. Lett.* **18**, 790 (1993).

[131] M. Romagnoli, S. Wabnitz, and M. Midrio, *Opt. Commun.* **104**, 293 (1994).

[132] M. Nakazawa, E. Yamada, H. Kubota, and K. Suzuki, *Electron. Lett.* **27**, 1270 (1991).

[133] N. J. Smith, W. J. Firth, K. J. Blow, and K. Smith, *Opt. Lett.* **19**, 16 (1994).

[134] S. Bigo, O. Audouin, and E. Desurvire, *Electron. Lett.* **31**, 2191 (1995).

[135] M. Nakazawa, K. Suzuki, E. Yamada, H. Kubota, Y. Kimura, and M. Takaya, *Electron. Lett.* **29**, 729 (1993).

[136] G. Aubin, E. Jeanny, T. Montalant, J. Moulu, F. Pirio, J.-B. Thomine, and F. Devaux, *Electron. Lett.* **31**, 1079 (1995).

[137] W. Forysiak, K. J. Blow. and N. J. Doran, *Electron. Lett.* **29**, 1225 (1993).

[138] M. Matsumoto, H. Ikeda, and A. Hasegawa, *Opt. Lett.* **19**, 183 (1994).

[139] T. Widdowson, D. J. Malyon, A. D. Ellis, K. Smith, and K. J. Blow, *Electron. Lett.* **30**, 990 (1994).

[140] S. Kumar and A. Hasegawa, *Opt. Lett.* **20**, 1856 (1995).

[141] V. S. Grigoryan, A. Hasegawa, and A. Maruta, *Opt. Lett.* **20**, 857 (1995).

[142] M. Matsumoto, *J. Opt. Soc. Am. B* **15**, 2831 (1998); *Opt. Lett.* **23**, 1901 (1998).

[143] S. K. Turitsyn and E. G. Shapiro, *J. Opt. Soc. Am. B* **16**, 1321 (1999).

[144] S. Waiyapot and M. Matsumoto, *IEEE Photon. Technol. Lett.* **11**, 1408 (1999).

[145] M. F. S. Ferreira and S. H. Sousa, *Electron. Lett.* **37**, 1184 (2001).

[146] M. Matsumoto, *Opt. Lett.* **23**, 1901 (2001).

[147] K. Iwatsuki, K. Suzuki, S. Nishi, and M. Saruwatari, *IEEE Photon. Technol. Lett.* **5**, 245 (1993).

[148] M. Nakazawa, K. Suzuki, E. Yoshida, E. Yamada, T. Kitoh, and M. Kawachi, *Electron. Lett.* **31**, 565 (1995).

[149] F. Favre, D. LeGuen, and F. Devaux, *Electron. Lett.* **33**, 1234 (1997).

[150] D. S. Govan, W. Forysiak, and N. J. Doran, *Opt. Lett.* **23**, 1523 (1998).

[151] P. Harper, S. B. Alleston, I. Bennion, and N. J. Doran, *Electron. Lett.* **35**, 2128 (1999).

[152] A. Sahara, T. Inui, T. Komukai, H. Kubota, and M. Nakazawa, *IEEE Photon. Technol. Lett.* **12**, 720 (2000).

[153] A. H. Liang, H. Toda, and A. Hasegawa, *Opt. Lett.* **24**, 799 (1999).

[154] S. K. Turitsyn, M. P. Fedoruk, and A. Gornakova, *Opt. Lett.* **24**, 869 (1999).

[155] T. Hirooka, T. Nakada, and A. Hasegawa, *IEEE Photon. Technol. Lett.* **12**, 633 (2000).

[156] L. J. Richardson, W. Forysiak, and N. J. Doran, *Opt. Lett.* **25**, 1010 (2000); *IEEE Photon. Technol. Lett.* **13**, 209 (2001).

[157] L. J. Richardson, W. Forysiak, and K. J. Blow, *Electron. Lett.* **36**, 2029 (2000).

[158] J. Martensson and A. Berntson, *IEEE Photon. Technol. Lett.* **13**, 666 (2001).

[159] H. S. Chung, S. K. Shin, D. W. Lee, D. W. Kim, and Y. C. Chung, *Electron. Lett.* **37**, 618 (2001).

[160] I. Morita. K. Tanaka, N. Edagawa, S. Yamamoto, and M. Suzuki, *Electron. Lett.* **34**, 1863 (1998).

[161] I. Morita. K. Tanaka, N. Edagawa, and M. Suzuki, *J. Lightwave Technol.* **17**, 2506 (1999).

[162] M. Nakazawa, K. Suzuki, and H. Kubota, *Electron. Lett.* **35**, 162 (1999).

[163] M. Nakazawa, H. Kubota, K. Suzuki, E. Yamada, and A. Sahara, *IEEE J. Sel. Topics Quantum Electron.* **6**, 363 (2000).

[164] M. Nakazawa, T. Yamamoto, and K. R. Tamura, *Electron. Lett.* **36**, 2027 (2000).

[165] T. Yu, E. A. Golovchenko, A. N. Pilipetskii, and C. R. Menyuk, *Opt. Lett.* **22**, 793 (1997).

[166] T. Georges, *J. Opt. Soc. Am. B* **15**, 1553 (1998).

[167] S. Kumar, M. Wald, F. Lederer, and A. Hasegawa, *Opt. Lett.* **23**, 1019 (1998).

[168] M. Romagnoli, L. Socci, M. Midrio, P. Franco, and T. Georges, *Opt. Lett.* **23**, 1182 (1998).

[169] T. Inoue, H. Sugahara, A. Maruta, and Y. Kodama, *IEEE Photon. Technol. Lett.* **12**, 299 (2000).

[170] Y. Takushima, T. Douke, and K. Kikuchi, *Electron. Lett.* **37**, 849 (2001).

[171] S. G. Evangelides, L. F. Mollenauer, J. P. Gordon, and N. S. Bergano, *J. Lightwave Technol.* **10**, 28 (1992).

[172] C. De Angelis, S. Wabnitz, and M. Haelterman, *Electron. Lett.* **29**, 1568 (1993).

[173] S. Wabnitz, *Opt. Lett.* **20**, 261 (1995).

[174] C. De Angelis and S. Wabnitz, *Opt. Commun.* **125**, 186 (1996).

[175] M. Midrio, P. Franco, M. Crivellari, M. Romagnoli, and F. Matera, *J. Opt. Soc. Am. B* **13**, 1526 (1996).

[176] C. F. Chen and S. Chi, *J. Opt.* **4**, 278 (1998).

[177] C. A. Eleftherianos, D. Syvridis, T. Sphicopoulos, and C. Caroubalos, *Opt. Commun.* **154**, 14 (1998).

[178] J. P. Silmon-Clyde and J. N. Elgin, *J. Opt. Soc. Am. B* **16**, 1348 (1999).

[179] F. Matera, M. Settembre, M. Tamburrini, F. Favre, D. Le Guen, T. Georges, M. Henry, G. Michaud, P. Franco, et al., *J. Lightwave Technol.* **17**, 2225 (1999).

[180] H. Sunnerud, J. Li, C. J. Xie, and P. A. Andrekson, *J. Lightwave Technol.* **19**, 1453 (2001).

[181] L. F. Mollenauer, K. Smith, J. P. Gordon, and C. R. Menyuk, *Opt. Lett.* **14**, 1219 (1989).

[182] T. Lakoba and D. T. Kaup, *Opt. Lett.* **24**, 808 (1999).

[183] Y. Takushima, X. Wang, and K. Kikuchi, *Electron. Lett.* **35**, 739 (1999).

[184] T. Lakoba and G. P. Agrawal, *J. Opt. Soc. Am. B* **16**, 1332 (1999).

[185] J. Pina, B. Abueva, and C. G. Goedde, *Opt. Commun.* **176**, 397 (2000).

[186] D. M. Baboiu, D. Mihalache, and N. C. Panoiu, *Opt. Lett.* **20**, 1865 (1995).

[187] R.-J. Essiambre and G. P. Agrawal, *J. Opt. Soc. Am. B* **14**, 314 (1997).

[188] M. Facao and M. Ferreira, *J. Nonlinear Math. Phys.* **8**, 112 (2001).

[189] J. F. Corney and P. D. Drummond, *J. Opt. Soc. Am. B* **18**, 153 (2001).

[190] E. M. Dianov, A. V. Luchnikov, A. N. Pilipetskii, and A. M. Prokhorov, *Sov. Lightwave Commun.* **1**, 235 (1991); *Appl. Phys.* **54**, 175 (1992).

[191] Y. Jaouen and L. du Mouza, *Opt. Fiber Technol.* **7**, 141 (2001).

[192] L. F. Mollenauer, *Opt. Lett.* **21**, 384 (1996).

[193] T. Adali, B. Wang, A. N. Pilipetskii, and C. R. Menyuk, *J. Lightwave Technol.* **16**, 986 (1998).

[194] L. F. Mollenauer and J. P. Gordon, *Opt. Lett.* **19**, 375 (1994).

[195] T. Widdowson, A. Lord, and D. J. Malyon, *Electron. Lett.* **30**, 879 (1994).

[196] T. Georges and F. Fabre, *Opt. Lett.* **16**, 1656 (1991).

[197] C. R. Menyuk, *Opt. Lett.* **20**, 285 (1995).

[198] A. N. Pinto, G. P. Agrawal, and J. F. da Rocha, *J. Lightwave Technol.* **16**, 515 (1998).

[199] J. Martensson, A. Berntson, M. Westlund, A. Danielsson, P. Johannisson, D. Anderson, and M. Lisak, *Opt. Lett.* **26**, 55 (2001).

[200] W. Forysiak and N. J. Doran, *J. Lightwave Technol.* **13**, 850 (1995).

[201] S. F. Wen and S. Chi, *Opt. Lett.* **20**, 976 (1995).

[202] R.-J. Essiambre and G. P. Agrawal, *J. Opt. Soc. Am. B* **14**, 323 (1997).

[203] T. Hirooka and A. Hasegawa, *Electron. Lett.* **34**, 2056 (1998).

[204] Y. S. Liu and Z. W. Zhang, *Int. J. Infrared Milli. Waves* **20**, 1673 (1999).

[205] L. F. Mollenauer and P. V. Mamyshev, *IEEE J. Quantum Electron.* **34**, 2089 (1998).

[206] L. F. Mollenauer, S. G. Evangelides, and J. P. Gordon, *J. Lightwave Technol.* **9**, 362 (1991).

[207] A. F. Benner, J. R. Sauer, and M. J. Ablowitz, *J. Opt. Soc. Am. B* **10**, 2331 (1993).

[208] Y. Kodama and A. Hasegawa, *Opt. Lett.* **16**, 208 (1991).

[209] T. Aakjer, J. H. Povlsen, and K. Rottwitt, *Opt. Lett.* **18**, 1908 (1993).

[210] S. Chakravarty, M. J. Ablowitz, J. R. Sauer, and R. B. Jenkins, *Opt. Lett.* **20**, 136 (1995).

[211] R. B. Jenkins, J. R. Sauer, S. Chakravarty, and M. J. Ablowitz, *Opt. Lett.* **20**, 1964 (1995).

[212] X. Y. Tang and M. K. Chin, *Opt. Commun.* **119**, 41 (1995).

[213] P. K. A. Wai, C. R. Menyuk, and B. Raghavan, *J. Lightwave Technol.* **14**, 1449 (1996).

[214] L. F. Mollenauer, E. Lichtman, G. T. Harvey, M. J. Neubelt, and B. M. Nyman, *Electron. Lett.* **28**, 792 (1992).

[215] A. Mecozzi and H. A. Haus, *Opt. Lett.* **17**, 988 (1992).

[216] R. Ohhira, M. Matsumoto, and A. Hasegawa, *Opt. Commun.* **111**, 39 (1994).

[217] M. Midrio, P. Franco, F. Matera, M. Romagnoli, and M. Settembre, *Opt. Commun.* **112**, 283 (1994).

[218] E. A. Golovchenko, A. N. Pilipetskii, and C. R. Menyuk, *Opt. Lett.* **21**, 195 (1996).

[219] E. Desurvire, O. Leclerc, and O. Audouin, *Opt. Lett.* **21**, 1026 (1996).

[220] M. Nakazawa, K. Suzuki, H. Kubota, Y. Kimura, E. Yamada, K. Tamura, T. Komukai, and T. Imai, *Electron. Lett.* **32**, 828 (1996).

[221] M. Nakazawa, K. Suzuki, H. Kubota, and E. Yamada, *Electron. Lett.* **32**, 1686 (1996).

[222] P. V. Mamyshev and L. F. Mollenauer, *Opt. Lett.* **21**, 396 (1996).

[223] S. G. Evangelides and J. P. Gordon, *J. Lightwave Technol.* **14**, 1639 (1996).

[224] L. F. Mollenauer, P. V. Mamyshev, and M. J. Neubelt, *Electron. Lett.* **32**, 471 (1996).

[225] M. Nakazawa, K. Suzuki, H. Kubota, A. Sahara, and E. Yamada, *Electron. Lett.* **34**, 103 (1998).

[226] X. Tang, R. Mu, and P. Ye, *Opt. Quantum Electron.* **26**, 969 (1994).

[227] E. Kolltveit, J. P. Hamaide, and O. Audouin, *Electron. Lett.* **32**, 1858 (1996).

[228] S. Wabnitz, *Opt. Lett.* **21**, 638 (1996).

[229] E. A. Golovchenko, A. N. Pilipetskii, and C. R. Menyuk, *Electron. Lett.* **33**, 735 (1997); *Opt. Lett.* **22**, 1156 (1997).

[230] Y. Kodama, A. V. Mikhailov, and S. Wabnitz, *Opt. Commun.* **143**, 53 (1997).

[231] J. F. L. Devaney, W. Forysiak, A. M. Niculae, and N. J. Doran, *Opt. Lett.* **22**, 1695 (1997).

[232] T. S. Yang, W. L. Kath, and S. K. Turitsyn, *Opt. Lett.* **23**, 597 (1998).

[233] M. J. Ablowitz, G. Bioindini, S. Chakravarty, and R. L. Horne, *Opt. Commun.* **150**, 305 (1998).

[234] A. Mecozzi, *J. Opt. Soc. Am. B* **15**, 152 (1998).

[235] S. Kumar, *Opt. Lett.* **23**, 1450 (1998).

[236] T. Hirooka and A. Hasegawa, *Opt. Lett.* **23**, 768 (1998).

[237] A. M. Niculae, W. Forysiak, A. J. Golág, J. H. B. Nijhof, and N. J. Doran, *Opt. Lett.* **23**, 1354 (1998).

[238] P. V. Mamyshev and L. F. Mollenauer, *Opt. Lett.* **24**, 1 (1999).

[239] H. Sugahara, A. Maruta, and Y. Kodama, *Opt. Lett.* **24**, 145 (1999).

[240] Y. Chen and H. A. Haus, *Opt. Lett.* **24**, 217 (1999).

[241] H. Sugahara, H. Kato, T. Inoue, A. Maruta, and Y. Kodama, *J. Lightwave Technol.* **17**, 1547 (1999).

[242] D. J. Kaup, B. A. Malomed, and J. Yang, *Opt. Lett.* **23**, 1600 (1998); *J. Opt. Soc. Am. B* **16**, 1628 (1999).

[243] A. Hasegawa and T. Hirooka, *Electron. Lett.* **36**, 68 (2000).

[244] S. K. Turitsyn, M. Fedoruk, T. S. Yang, and W. L. Kath, *IEEE J. Quantum Electron.* **36**, 290 (2000).

[245] S. Wabnitz and F. Neddam, *Opt. Commun.* **183**, 295 (2000).

[246] V. S. Grigoryan and A. Richter, *J. Lightwave Technol.* **18**, 1148 (2000).

[247] M. J. Ablowitz, G. Biondini, and E. S. Olson, *J. Opt. Soc. Am. B* **18**, 577 (2001).

[248] S. Wabnitz nad G. Le Meur, *Opt. Lett.* **26**, 777 (2001).

[249] M. Wald, B. A. Malomed, and F. Lederer, *Opt. Lett.* **26**, 965 (2001).

[250] H. Sugahara, *IEEE Photon. Technol. Lett.* **13**, 963 (2001).

[251] F. Favre, D. Le Guen, M. L. Moulinard, M. Henry, and T. Georges, *Electron. Lett.* **33**, 2135 (1997).

[252] K. Tanaka, I. Morita, M. Suzuki, N. Edagawa, and S. Yamamoto, *Electron. Lett.* **34**, 2257 (1998).

[253] L. F. Mollenauer, R. Bonney, J. P. Gordon, and P. V. Mamyshev, *Opt. Lett.* **24**, 285 (1999).

[254] I. Morita, M. Suzuki, N. Edagawa, K. Tanaka, and S. Yamamoto, *J. Lightwave Technol.* **17**, 80 (1999).

[255] F. Neddam, P. Le Lourec, B. Biotteau, P. Poignant, E. Pincemin, J. P. Hamaide, O. Audouin, E. Desurvire, T. Georges, et al., *Electron. Lett.* **35**, 1093 (1999).

[256] M. L. Dennis, W. I. Kaechele, L. Goldberg, T. F. Carruthers, and I. N. Duling, *IEEE Photon. Technol. Lett.* **11**, 1680 (1999).

[257] K. Suzuki, H. Kubota, A. Sahara, and M. Nakazawa, *Electron. Lett.* **36**, 443 (2000).

[258] M. Nakazawa, H. Kubota, K. Suzuki, E. Yamada, and A. Sahara, *IEEE J. Sel. Topics Quantum Electron.* **6**, 363 (2000).

[259] L. F. Mollenauer, P. V. Mamyshev, J. Gripp, M. J. Neubelt, N. Mamysheva, L. Grüner-Nielsen, and T. Veng, *Opt. Lett.* **25**, 704 (2000).

[260] P. Leclerc, B. Dany, D. Rouvillain, P. Brindel, E. Desurvire, C. Duchet, A. Shen, F. Devaux, E. Coquelin, et al., *Electron. Lett.* **36**, 1574 (2000).

[261] M. Nakazawa, *IEEE J. Sel. Topics Quantum Electron.* **6**, 1332 (2000).

[262] W. S. Lee, Y. Zhu, B. Shaw, D. Watley, C. Scahill, J. Homan, C. R. S. Fludger, M. Jones, and A. Hadjifotiou, *Electron. Lett.* **37**, 964 (2001).

Chapter 10

Coherent Lightwave Systems

The lightwave systems discussed so far are based on a simple digital modulation scheme in which an electrical bit stream modulates the intensity of an optical carrier inside the optical transmitter and the optical signal transmitted through the fiber link is incident directly on an optical receiver, which converts it to the original digital signal in the electrical domain. Such a scheme is referred to as *intensity modulation with direct detection* (IM/DD). Many alternative schemes, well known in the context of radio and microwave communication systems [1]–[6], transmit information by modulating the frequency or the phase of the optical carrier and detect the transmitted signal by using homodyne or heterodyne detection techniques. Since phase coherence of the optical carrier plays an important role in the implementation of such schemes, such optical communication systems are called coherent lightwave systems. Coherent transmission techniques were studied during the 1980s extensively [7]–[16]. Commercial deployment of coherent systems, however, has been delayed with the advent of optical amplifiers although the research and development phase has continued worldwide.

The motivation behind using the coherent communication techniques is two-fold. First, the receiver sensitivity can be improved by up to 20 dB compared with that of IM/DD systems. Second, the use of coherent detection may allow a more efficient use of fiber bandwidth by increasing the spectral efficiency of WDM systems. In this chapter we focus on the design of coherent lightwave systems. The basic concepts behind coherent detection are discussed in Section 10.1. In Section 10.2 we present new modulation formats possible with the use of coherent detection. Section 10.3 is devoted to synchronous and asynchronous demodulation schemes used by coherent receivers. The bit-error rate (BER) for various modulation and demodulation schemes is considered in Section 10.4. Section 10.5 focuses on the degradation of receiver sensitivity through mechanisms such as phase noise, intensity noise, polarization mismatch, fiber dispersion, and reflection feedback. The performance aspects of coherent lightwave systems are reviewed in Section 10.6 where we also discuss the status of such systems at the end of 2001.

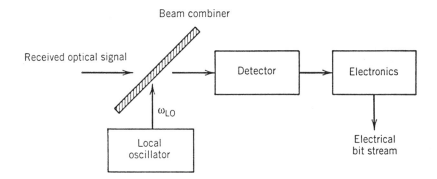

Figure 10.1: Schematic illustration of a coherent detection scheme.

10.1 Basic Concepts

10.1.1 Local Oscillator

The basic idea behind coherent detection consists of combining the optical signal coherently with a continuous-wave (CW) optical field before it falls on the photodetector (see Fig. 10.1). The CW field is generated locally at the receiver using a narrow-linewidth laser, called the *local oscillator* (LO), a term borrowed from the radio and microwave literature. To see how the mixing of the received optical signal with the LO output can improve the receiver performance, let us write the optical signal using complex notation as

$$E_s = A_s \exp[-i(\omega_0 t + \phi_s)], \tag{10.1.1}$$

where ω_0 is the carrier frequency, A_s is the amplitude, and ϕ_s is the phase. The optical field associated with the local oscillator is given by a similar expression,

$$E_{LO} = A_{LO} \exp[-i(\omega_{LO} t + \phi_{LO})], \tag{10.1.2}$$

where A_{LO}, ω_{LO}, and ϕ_{LO} represent the amplitude, frequency, and phase of the local oscillator, respectively. The scalar notation is used for both E_s and E_{LO} after assuming that the two fields are identically polarized (polarization-mismatch issues are discussed later in Section 10.5.3). Since a photodetector responds to the optical intensity, the optical power incident at the photodetector is given by $P = K|E_s + E_{LO}|^2$, where K is a constant of proportionality. Using Eqs. (10.1.1) and (10.1.2),

$$P(t) = P_s + P_{LO} + 2\sqrt{P_s P_{LO}} \cos(\omega_{IF} t + \phi_s - \phi_{LO}), \tag{10.1.3}$$

where

$$P_s = KA_s^2, \quad P_{LO} = KA_{LO}^2, \quad \omega_{IF} = \omega_0 - \omega_{LO}. \tag{10.1.4}$$

The frequency $\nu_{IF} \equiv \omega_{IF}/2\pi$ is known as the *intermediate frequency* (IF). When $\omega_0 \neq \omega_{LO}$, the optical signal is demodulated in two stages; its carrier frequency is first converted to an intermediate frequency ν_{IF} (typically 0.1–5 GHz) before the signal is demodulated to the baseband. It is not always necessary to use an intermediate frequency.

In fact, there are two different coherent detection techniques to choose from, depending on whether or not ω_{IF} equals zero. They are known as *homodyne* and *heterodyne* detection techniques.

10.1.2 Homodyne Detection

In this coherent-detection technique, the local-oscillator frequency ω_{LO} is selected to coincide with the signal-carrier frequency ω_0 so that $\omega_{IF} = 0$. From Eq. (10.1.3), the photocurrent ($I = RP$, where R is the detector responsivity) is given by

$$I(t) = R(P_s + P_{LO}) + 2R\sqrt{P_sP_{LO}}\cos(\phi_s - \phi_{LO}). \qquad (10.1.5)$$

Typically, $P_{LO} \gg P_s$, and $P_s + P_{LO} \approx P_{LO}$. The last term in Eq. (10.1.5) contains the information transmitted and is used by the decision circuit. Consider the case in which the local-oscillator phase is locked to the signal phase so that $\phi_s = \phi_{LO}$. The homodyne signal is then given by

$$I_p(t) = 2R\sqrt{P_sP_{LO}}. \qquad (10.1.6)$$

The main advantage of homodyne detection is evident from Eq. (10.1.6) if we note that the signal current in the direct-detection case is given by $I_{dd}(t) = RP_s(t)$. Denoting the average optical power by $\bar{P}_s$, the average electrical power is increased by a factor of $4P_{LO}/\bar{P}_s$ with the use of homodyne detection. Since P_{LO} can be made much larger than $\bar{P}_s$, the power enhancement can exceed 20 dB. Although shot noise is also enhanced, it is shown later in this section that homodyne detection improves the signal-to-noise ratio (SNR) by a large factor.

Another advantage of coherent detection is evident from Eq. (10.1.5). Because the last term in this equation contains the signal phase explicitly, it is possible to transmit information by modulating the phase or frequency of the optical carrier. Direct detection does not allow phase or frequency modulation, as all information about the signal phase is lost. The new modulation formats for coherent systems are discussed in Section 10.2.

A disadvantage of homodyne detection also results from its phase sensitivity. Since the last term in Eq. (10.1.5) contains the local-oscillator phase ϕ_{LO} explicitly, clearly ϕ_{LO} should be controlled. Ideally, ϕ_s and ϕ_{LO} should stay constant except for the intentional modulation of ϕ_s. In practice, both ϕ_s and ϕ_{LO} fluctuate with time in a random manner. However, their difference $\phi_s - \phi_{LO}$ can be forced to remain nearly constant through an optical phase-locked loop. The implementation of such a loop is not simple and makes the design of optical homodyne receivers quite complicated. In addition, matching of the transmitter and local-oscillator frequencies puts stringent requirements on the two optical sources. These problems can be overcome by the use of heterodyne detection, discussed next.

10.1.3 Heterodyne Detection

In the case of heterodyne detection the local-oscillator frequency ω_{LO} is chosen to differ form the signal-carrier frequency ω_0 such that the intermediate frequency ω_{IF} is in the microwave region ($v_{IF} \sim 1$ GHz). Using Eq. (10.1.3) together with $I = RP$, the photocurrent is now given by

$$I(t) = R(P_s + P_{LO}) + 2R\sqrt{P_s P_{LO}}\cos(\omega_{IF}t + \phi_s - \phi_{LO}). \qquad (10.1.7)$$

Since $P_{LO} \gg P_s$ in practice, the direct-current (dc) term is nearly constant and can be removed easily using bandpass filters. The heterodyne signal is then given by the alternating-current (ac) term in Eq. (10.1.7) or by

$$I_{ac}(t) = 2R\sqrt{P_s P_{LO}}\cos(\omega_{IF}t + \phi_s - \phi_{LO}). \qquad (10.1.8)$$

Similar to the case of homodyne detection, information can be transmitted through amplitude, phase, or frequency modulation of the optical carrier. More importantly, the local oscillator still amplifies the received signal by a large factor, thereby improving the SNR. However, the SNR improvement is lower by a factor of 2 (or by 3 dB) compared with the homodyne case. This reduction is referred to as the heterodyne-detection penalty. The origin of the 3-dB penalty can be seen by considering the signal power (proportional to the square of the current). Because of the ac nature of I_{ac}, the average signal power is reduced by a factor of 2 when I_{ac}^2 is averaged over a full cycle at the intermediate frequency (recall that the average of $\cos^2\theta$ over θ is $\frac{1}{2}$).

The advantage gained at the expense of the 3-dB penalty is that the receiver design is considerably simplified because an optical phase-locked loop is no longer needed. Fluctuations in both ϕ_s and ϕ_{LO} still need to be controlled using narrow-linewidth semiconductor lasers for both optical sources. However, as discussed in Section 10.5.1, the linewidth requirements are quite moderate when an asynchronous demodulation scheme is used. This feature makes the heterodyne-detection scheme quite suitable for practical implementation in coherent lightwave systems.

10.1.4 Signal-to-Noise Ratio

The advantage of coherent detection for lightwave systems can be made more quantitative by considering the SNR of the receiver current. For this purpose, it is necessary to extend the analysis of Section 4.4 to the case of heterodyne detection. The receiver current fluctuates because of shot noise and thermal noise. The variance σ^2 of current fluctuations is obtained by adding the two contributions so that

$$\sigma^2 = \sigma_s^2 + \sigma_T^2, \qquad (10.1.9)$$

where

$$\sigma_s^2 = 2q(I + I_d)\Delta f, \quad \sigma_T^2 = (4k_B T/R_L)F_n\Delta f. \qquad (10.1.10)$$

The notation used here is the same as in Section 4.4. The main difference from the analysis of Section 4.4 occurs in the shot-noise contribution. The current I in Eq. (10.1.10) is the total photocurrent generated at the detector and is given by Eq. (10.1.5) or Eq. (10.1.7), depending on whether homodyne or heterodyne detection is employed. In practice, $P_{LO} \gg P_s$, and I in Eq. (10.1.10) can be replaced by the dominant term RP_{LO} for both cases.

The SNR is obtained by dividing the average signal power by the average noise power. In the heterodyne case, it is given by

$$\text{SNR} = \frac{\langle I_{ac}^2 \rangle}{\sigma^2} = \frac{2R^2 \bar{P}_s P_{LO}}{2q(RP_{LO} + I_d)\Delta f + \sigma_T^2}. \tag{10.1.11}$$

In the homodyne case, the SNR is larger by a factor of 2 if we assume that $\phi_s = \phi_{LO}$ in Eq. (10.1.5). The main advantage of coherent detection can be seen from Eq. (10.1.11). Since the local-oscillator power P_{LO} can be controlled at the receiver, it can be made large enough that the receiver noise is dominated by shot noise. More specifically, $\sigma_s^2 \gg \sigma_T^2$ when

$$P_{LO} \gg \sigma_T^2/(2qR\Delta f). \tag{10.1.12}$$

Under the same conditions, the dark-current contribution to the shot noise is negligible ($I_d \ll RP_{LO}$). The SNR is then given by

$$\text{SNR} \approx \frac{R\bar{P}_s}{q\Delta f} = \frac{\eta \bar{P}_s}{h\nu\Delta f}, \tag{10.1.13}$$

where $R = \eta q/h\nu$ was used from Eq. (4.1.3). The use of coherent detection allows one to achieve the shot-noise limit even for p–i–n receivers whose performance is generally limited by thermal noise. Moreover, in contrast with the case of avalanche photodiode (APD) receivers, this limit is realized without adding any excess shot noise.

It is useful to express the SNR in terms of the number of photons, N_p, received within a single bit. At the bit rate B, the signal power $\bar{P}_s$ is related to N_p as $\bar{P}_s = N_p h\nu B$. Typically, $\Delta f \approx B/2$. By using these values of $\bar{P}_s$ and Δf in Eq. (10.1.13), the SNR is given by a simple expression

$$\text{SNR} = 2\eta N_p. \tag{10.1.14}$$

In the case of homodyne detection, SNR is larger by a factor of 2 and is given by SNR = $4\eta N_p$. Section 10.4 discusses the dependence of the BER on SNR and shows how receiver sensitivity is improved by the use of coherent detection.

10.2   Modulation Formats

As discussed in Section 10.1, an important advantage of using the coherent detection techniques is that both the amplitude and the phase of the received optical signal can be detected and measured. This feature opens up the possibility of sending information by modulating either the amplitude, or the phase, or the frequency of an optical carrier. In the case of digital communication systems, the three possibilities give rise to three modulation formats known as amplitude-shift keying (ASK), phase-shift keying (PSK), and frequency-shift keying (FSK) [1]–[6]. Figure 10.2 shows schematically the three modulation formats for a specific bit pattern. In the following subsections we consider each format separately and discuss its implementation in practical lightwave systems.

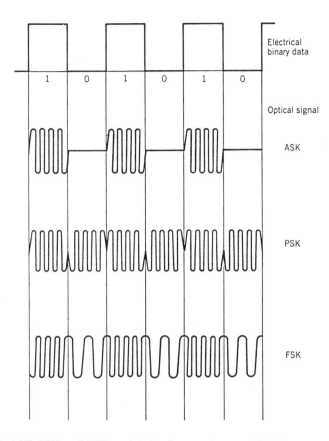

Figure 10.2: ASK, PSK, and FSK modulation formats for a specific bit pattern shown on the top.

10.2.1 ASK Format

The electric field associated with an optical signal can be written as [by taking the real part of Eq. (10.1.1)]

$$E_s(t) = A_s(t)\cos[\omega_0 t + \phi_s(t)]. \qquad (10.2.1)$$

In the case of ASK format, the amplitude A_s is modulated while keeping ω_0 and ϕ_s constant. For binary digital modulation, A_s takes one of the two fixed values during each bit period, depending on whether 1 or 0 bit is being transmitted. In most practical situations, A_s is set to zero during transmission of 0 bits. The ASK format is then called *on–off keying* (OOK) and is identical with the modulation scheme commonly used for noncoherent (IM/DD) digital lightwave systems.

The implementation of ASK for coherent systems differs from the case of the direct-detection systems in one important aspect. Whereas the optical bit stream for direct-detection systems can be generated by modulating a light-emitting diode (LED) or a semiconductor laser directly, external modulation is necessary for coherent communication systems. The reason behind this necessity is related to phase changes that invariably occur when the amplitude A_s (or the power) is changed by modulating the current applied to a semiconductor laser (see Section 3.5.3). For IM/DD systems, such unintentional phase changes are not seen by the detector (as the detector responds only to the optical power) and are not of major concern except for the chirp-induced power penalty discussed in Section 5.4.4. The situation is entirely different in the case of coherent systems, where the detector response depends on the phase of the received signal. The implementation of ASK format for coherent systems requires the phase ϕ_s to remain nearly constant. This is achieved by operating the semiconductor laser continuously at a constant current and modulating its output by using an external modulator (see Section 3.6.4). Since all external modulators have some insertion losses, a power penalty incurs whenever an external modulator is used; it can be reduced to below 1 dB for monolithically integrated modulators.

As discussed in Section 3.64, a commonly used external modulator makes use of LiNbO$_3$ waveguides in a Mach–Zehnder (MZ) configuration [17]. The performance of external modulators is quantified through the on–off ratio (also called extinction ratio) and the modulation bandwidth. LiNbO$_3$ modulators provide an on–off ratio in excess of 20 and can be modulated at speeds up to 75 GHz [18]. The driving voltage is typically 5 V but can be reduced to near 3 V with a suitable design. Other materials can also be used to make external modulators. For example, a polymeric electro-optic MZ modulator required only 1.8 V for shifting the phase of a 1.55-μm signal by π in one of the arms of the MZ interferometer [19].

Electroabsorption modulators, made using semiconductors, are often preferred because they do not require the use of an interferometer and can be integrated monolithically with the laser (see Section 3.6.4). Optical transmitters with an integrated electroabsorption modulator capable of modulating at 10 Gb/s were available commercially by 1999 and are used routinely for IM/DD lightwave systems [20]. By 2001, such integrated modulators exhibited a bandwidth of more than 50 GHz and had the potential of operating at bit rates of up to 100 Gb/s [21]. They are likely to be employed for coherent systems as well.

10.2.2 PSK Format

In the case of PSK format, the optical bit stream is generated by modulating the phase ϕ_s in Eq. (10.2.1) while the amplitude A_s and the frequency ω_0 of the optical carrier are kept constant. For binary PSK, the phase ϕ_s takes two values, commonly chosen to be 0 and π. Figure 10.2 shows the binary PSK format schematically for a specific bit pattern. An interesting aspect of the PSK format is that the optical intensity remains constant during all bits and the signal appears to have a CW form. Coherent detection is a necessity for PSK as all information would be lost if the optical signal were detected directly without mixing it with the output of a local oscillator.

The implementation of PSK requires an external modulator capable of changing the optical phase in response to an applied voltage. The physical mechanism used by such modulators is called electrorefraction. Any electro-optic crystal with proper orientation can be used for phase modulation. A LiNbO$_3$ crystal is commonly used in practice. The design of LiNbO$_3$-based phase modulators is much simpler than that of an amplitude modulator as a Mach–Zehnder interferometer is no longer needed, and a single waveguide can be used. The phase shift $\delta\phi$ occurring while the CW signal passes through the waveguide is related to the index change δn by the simple relation

$$\delta\phi = (2\pi/\lambda)(\delta n)l_m, \qquad (10.2.2)$$

where l_m is the length over which index change is induced by the applied voltage. The index change δn is proportional to the applied voltage, which is chosen such that $\delta\phi = \pi$. Thus, a phase shift of π can be imposed on the optical carrier by applying the required voltage for the duration of each "1" bit.

Semiconductors can also be used to make phase modulators, especially if a multi-quantum-well (MQW) structure is used. The electrorefraction effect originating from the *quantum-confinement Stark effect* is enhanced for a quantum-well design. Such MQW phase modulators have been developed [22]–[27] and are able to operate at a bit rate of up to 40 Gb/s in the wavelength range 1.3–1.6 μm. Already in 1992, MQW devices had a modulation bandwidth of 20 GHz and required only 3.85 V for introducing a π phase shift when operated near 1.55 μm [22]. The operating voltage was reduced to 2.8 V in a phase modulator based on the electroabsorption effect in a MQW waveguide [23]. A spot-size converter is sometimes integrated with the phase modulator to reduce coupling losses [24]. The best performance is achieved when a semiconductor phase modulator is monolithically integrated within the transmitter [25]. Such transmitters are quite useful for coherent lightwave systems.

The use of PSK format requires that the phase of the optical carrier remain stable so that phase information can be extracted at the receiver without ambiguity. This requirement puts a stringent condition on the tolerable linewidths of the transmitter laser and the local oscillator. As discussed later in Section 10.5.1, the linewidth requirement can be somewhat relaxed by using a variant of the PSK format, known as *differential phase-shift keying* (DPSK). In the case of DPSK, information is coded by using the phase difference between two neighboring bits. For instance, if ϕ_k represents the phase of the kth bit, the phase difference $\Delta\phi = \phi_k - \phi_{k-1}$ is changed by π or 0, depending on whether kth bit is a 1 or 0 bit. The advantage of DPSK is that the transmittal signal can be demodulated successfully as long as the carrier phase remains relatively stable over a duration of two bits.

10.2.3 FSK Format

In the case of FSK modulation, information is coded on the optical carrier by shifting the carrier frequency ω_0 itself [see Eq. (10.2.1)]. For a binary digital signal, ω_0 takes two values, $\omega_0 + \Delta\omega$ and $\omega_0 - \Delta\omega$, depending on whether a 1 or 0 bit is being transmitted. The shift $\Delta f = \Delta\omega/2\pi$ is called the *frequency deviation*. The quantity $2\Delta f$ is sometimes called *tone spacing*, as it represents the frequency spacing between 1 and 0 bits. The optical field for FSK format can be written as

$$E_s(t) = A_s \cos[(\omega_0 \pm \Delta\omega)t + \phi_s], \qquad (10.2.3)$$

where $+$ and $-$ signs correspond to 1 and 0 bits. By noting that the argument of cosine can be written as $\omega_0 t + (\phi_s \pm \Delta\omega t)$, the FSK format can also be viewed as a kind of

PSK modulation such that the carrier phase increases or decreases linearly over the bit duration.

The choice of the frequency deviation Δf depends on the available bandwidth. The total bandwidth of a FSK signal is given approximately by $2\Delta f + 2B$, where B is the bit rate [1]. When $\Delta f \gg B$, the bandwidth approaches $2\Delta f$ and is nearly independent of the bit rate. This case is often referred to as *wide-deviation* or wideband FSK. In the opposite case of $\Delta f \ll B$, called *narrow-deviation* or narrowband FSK, the bandwidth approaches $2B$. The ratio $\beta_{FM} = \Delta f/B$, called the frequency modulation (FM) index, serves to distinguish the two cases, depending on whether $\beta_{FM} \gg 1$ or $\beta_{FM} \ll 1$.

The implementation of FSK requires modulators capable of shifting the frequency of the incident optical signal. Electro-optic materials such as LiNbO$_3$ normally produce a phase shift proportional to the applied voltage. They can be used for FSK by applying a triangular voltage pulse (sawtooth-like), since a linear phase change corresponds to a frequency shift. An alternative technique makes use of Bragg scattering from acoustic waves. Such modulators are called acousto-optic modulators. Their use is somewhat cumbersome in the bulk form. However, they can be fabricated in compact form using surface acoustic waves on a slab waveguide. The device structure is similar to that of an acousto-optic filter used for wavelength-division multiplexing (WDM) applications (see Section 8.3.1). The maximum frequency shift is typically limited to below 1 GHz for such modulators.

The simplest method for producing an FSK signal makes use of the direct-modulation capability of semiconductor lasers. As discussed in Section 3.5.2, a change in the operating current of a semiconductor laser leads to changes in both the amplitude and frequency of emitted light. In the case of ASK, the frequency shift or the chirp of the emitted optical pulse is undesirable. But the same frequency shift can be used to advantage for the purpose of FSK. Typical values of frequency shifts are ~ 1 GHz/mA. Therefore, only a small change in the operating current (~ 1 mA) is required for producing the FSK signal. Such current changes are small enough that the amplitude does not change much from from bit to bit.

For the purpose of FSK, the FM response of a distributed feedback (DFB) laser should be flat over a bandwidth equal to the bit rate. As seen in Fig. 10.3, most DFB lasers exhibit a dip in their FM response at a frequency near 1 MHz [28]. The reason is that two different physical phenomena contribute to the frequency shift when the device current is changed. Changes in the refractive index, responsible for the frequency shift, can occur either because of a temperature shift or because of a change in the carrier density. The thermal effects contribute only up to modulation frequencies of about 1 MHz because of their slow response. The FM response decreases in the frequency range 0.1–10 MHz because the thermal contribution and the carrier-density contribution occur with opposite phases.

Several techniques can be used to make the FM response more uniform. An equalization circuit improves uniformity but also reduces the modulation efficiency. Another technique makes use of transmission codes which reduce the low-frequency components of the data where distortion is highest. Multisection DFB lasers have been developed to realize a uniform FM response [29]–[35]. Figure 10.3 shows the FM response of a two-section DFB laser. It is not only uniform up to about 1 GHz, but its modulation efficiency is also high. Even better performance is realized by using three-section

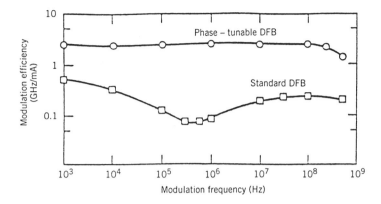

Figure 10.3: FM response of a typical DFB semiconductor laser exhibiting a dip in the frequency range 0.1–10 MHz. (After Ref. [12]; ©1988 IEEE; reprinted with permission.)

DBR lasers described in Section 3.4.3 in the context of tunable lasers. Flat FM response from 100 kHz to 15 GHz was demonstrated [29] in 1990 in such lasers. By 1995, the use of gain-coupled, phase-shifted, DFB lasers extended the range of uniform FM response from 10 kHz to 20 GHz [33]. When FSK is performed through direct modulation, the carrier phase varies continuously from bit to bit. This case is often referred to as continuous-phase FSK (CPFSK). When the tone spacing $2\Delta f$ is chosen to be $B/2$ ($\beta_{FM} = \frac{1}{2}$), CPFSK is also called minimum-shift keying (MSK).

10.3 Demodulation Schemes

As discussed in Section 10.1, either homodyne or heterodyne detection can be used to convert the received optical signal into an electrical form. In the case of homodyne detection, the optical signal is demodulated directly to the baseband. Although simple in concept, homodyne detection is difficult to implement in practice, as it requires a local oscillator whose frequency matches the carrier frequency exactly and whose phase is locked to the incoming signal. Such a demodulation scheme is called synchronous and is essential for homodyne detection. Although optical phase-locked loops have been developed for this purpose, their use is complicated in practice. Heterodyne detection simplifies the receiver design, as neither optical phase locking nor frequency matching of the local oscillator is required. However, the electrical signal oscillates rapidly at microwave frequencies and must be demodulated from the IF band to the baseband using techniques similar to those developed for microwave communication systems [1]–[6]. Demodulation can be carried out either synchronously or asynchronously. Asynchronous demodulation is also called incoherent in the radio communication literature. In the optical communication literature, the term *coherent detection* is used in a wider sense. A lightwave system is called coherent as long as it uses a local oscillator irrespective of the demodulation technique used to convert the IF signal to baseband frequencies. This section focuses on the synchronous and asynchronous demodulation schemes for heterodyne systems.

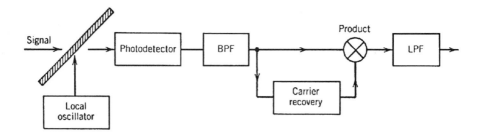

Figure 10.4: Block diagram of a synchronous heterodyne receiver.

10.3.1 Heterodyne Synchronous Demodulation

Figure 10.4 shows a synchronous heterodyne receiver schematically. The current generated at the photodiode is passed through a bandpass filter (BPF) centered at the intermediate frequency ω_{IF}. The filtered current in the absence of noise can be written as [see Eq. (10.1.8)]

$$I_f(t) = I_p \cos(\omega_{IF}t - \phi), \tag{10.3.1}$$

where $I_p = 2R\sqrt{P_sP_{LO}}$ and $\phi = \phi_{LO} - \phi_s$ is the phase difference between the local oscillator and the signal. The noise is also filtered by the BPF. Using the in-phase and out-of-phase quadrature components of the filtered Gaussian noise [1], the receiver noise is included through

$$I_f(t) = (I_p \cos\phi + i_c)\cos(\omega_{IF}t) + (I_p \sin\phi + i_s)\sin(\omega_{IF}t), \tag{10.3.2}$$

where i_c and i_s are Gaussian random variables of zero mean with variance σ^2 given by Eq. (10.1.9). For synchronous demodulation, $I_f(t)$ is multiplied by $\cos(\omega_{IF}t)$ and filtered by a low-pass filter. The resulting baseband signal is

$$I_d = \langle I_f \cos(\omega_{IF}t)\rangle = \tfrac{1}{2}(I_p \cos\phi + i_c), \tag{10.3.3}$$

where angle brackets denote low-pass filtering used for rejecting the ac components oscillating at $2\omega_{IF}$. Equation (10.3.3) shows that only the in-phase noise component affects the performance of synchronous heterodyne receivers.

Synchronous demodulation requires recovery of the microwave carrier at the intermediate frequency ω_{IF}. Several electronic schemes can be used for this purpose, all requiring a kind of electrical phase-locked loop [36]. Two commonly used loops are the *squaring loop* and the *Costas loop*. A squaring loop uses a square-law device to obtain a signal of the form $\cos^2(\omega_{IF}t)$ that has a frequency component at $2\omega_{IF}$. This component can be used to generate a microwave signal at ω_{IF}.

10.3.2 Heterodyne Asynchronous Demodulation

Figure 10.5 shows an asynchronous heterodyne receiver schematically. It does not require recovery of the microwave carrier at the intermediate frequency, resulting in a much simpler receiver design. The filtered signal $I_f(t)$ is converted to the baseband by

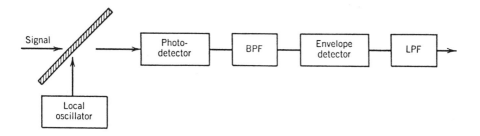

Figure 10.5: Block diagram of an asynchronous heterodyne receiver.

using an *envelope detector*, followed by a low-pass filter. The signal received by the decision circuit is just $I_d = |I_f|$, where I_f is given by Eq. (10.3.2). It can be written as

$$I_d = |I_f| = [(I_p \cos\phi + i_c)^2 + (I_p \sin\phi + i_s)^2]^{1/2}. \tag{10.3.4}$$

The main difference is that both the in-phase and out-of-phase quadrature components of the receiver noise affect the signal. The SNR is thus degraded compared with the case of synchronous demodulation. As discussed in Section 10.4, sensitivity degradation resulting from the reduced SNR is quite small (about 0.5 dB). As the phase-stability requirements are quite modest in the case of asynchronous demodulation, this scheme is commonly used for coherent lightwave systems.

The asynchronous heterodyne receiver shown in Fig. 10.5 requires modifications when the FSK and PSK modulation formats are used. Figure 10.6 shows two demodulation schemes. The FSK *dual-filter* receiver uses two separate branches to process the 1 and 0 bits whose carrier frequencies, and hence the intermediate frequencies, are different. The scheme can be used whenever the tone spacing is much larger than the bit rates, so that the spectra of 1 and 0 bits have negligible overlap (wide-deviation FSK). The two BPFs have their center frequencies separated exactly by the tone spacing so that each BPF passes either 1 or 0 bits only. The FSK dual-filter receiver can be thought of as two ASK single-filter receivers in parallel whose outputs are combined before reaching the decision circuit. A single-filter receiver of Fig. 10.5 can be used for FSK demodulation if its bandwidth is chosen to be wide enough to pass the entire bit stream. The signal is then processed by a frequency discriminator to identify 1 and 0 bits. This scheme works well only for narrow-deviation FSK, for which tone spacing is less than or comparable to the bit rate ($\beta_{FM} \leq 1$).

Asynchronous demodulation cannot be used in the case the PSK format because the phase of the transmitter laser and the local oscillator are not locked and can drift with time. However, the use of DPSK format permits asynchronous demodulation by using the delay scheme shown in Fig. 10.6(b). The idea is to multiply the received bit stream by a replica of it that has been delayed by one bit period. The resulting signal has a component of the form $\cos(\phi_k - \phi_{k-1})$, where ϕ_k is the phase of the kth bit, which can be used to recover the bit pattern since information is encoded in the phase difference $\phi_k - \phi_{k-1}$. Such a scheme requires phase stability only over a few bits and can be implemented by using DFB semiconductor lasers. The delay-demodulation

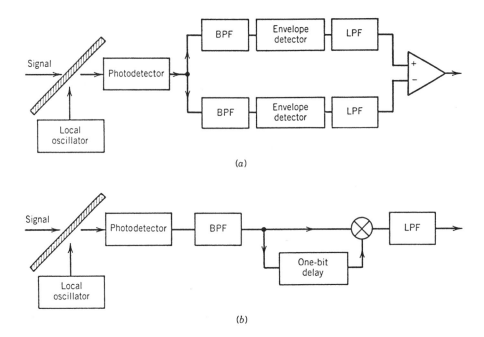

Figure 10.6: (a) Dual-filter FSK and (b) DPSK asynchronous heterodyne receivers.

scheme can also be used for CPFSK. The amount of delay in that case depends on the tone spacing and is chosen such that the phase is shifted by π for the delayed signal.

10.4   Bit-Error Rate

The preceding three sections have provided enough background material for calculating the bit-error rate (BER) of coherent lightwave systems. However, the BER, and hence the receiver sensitivity, depend on the modulation format as well as on the demodulation scheme used by the coherent receiver. The section considers each case separately.

10.4.1   Synchronous ASK Receivers

Consider first the case of heterodyne detection. The signal used by the decision circuit is given by Eq. (10.3.3). The phase ϕ generally varies randomly because of phase fluctuations associated with the transmitter laser and the local oscillator. As discussed in Section 10.5, the effect of phase fluctuations can be made negligible by using semiconductor lasers whose linewidth is a small fraction of the bit rate. Assuming this to be the case and setting $\phi = 0$ in Eq. (10.3.2), the decision signal is given by

$$I_d = \tfrac{1}{2}(I_p + i_c), \qquad\qquad (10.4.1)$$

where $I_p \equiv 2R(P_sP_{LO})^{1/2}$ takes values I_1 or I_0 depending on whether a 1 or 0 bit is being detected.

Consider the case $I_0 = 0$ in which no power is transmitted during the 0 bits. Except for the factor of $\frac{1}{2}$ in Eq. (10.4.1), the situation is analogous to the case of direct detection discussed in Section 4.5. The factor of $\frac{1}{2}$ does not affect the BER since both the signal and the noise are reduced by the same factor, leaving the SNR unchanged. In fact, one can use the same result [Eq. (4.5.9)],

$$\text{BER} = \frac{1}{2}\text{erfc}\left(\frac{Q}{\sqrt{2}}\right), \qquad (10.4.2)$$

where Q is given by Eq. (4.5.10) and can be written as

$$Q = \frac{I_1 - I_0}{\sigma_1 + \sigma_0} \approx \frac{I_1}{2\sigma_1} = \frac{1}{2}(\text{SNR})^{1/2}. \qquad (10.4.3)$$

In relating Q to SNR, we used $I_0 = 0$ and set $\sigma_0 \approx \sigma_1$. The latter approximation is justified for most coherent receivers whose noise is dominated by the shot noise induced by local-oscillator power and remains the same irrespective of the received signal power. Indeed, as shown in Section 10.1.4, the SNR of such receivers can be related to the number of photons received during each 1 bit by the simple relation $\text{SNR} = 2\eta N_p$ [see Eq. (10.1.14)]. Equations (10.4.2) and (10.4.3) then provide the following expression for the BER:

$$\text{BER} = \frac{1}{2}\text{erfc}(\sqrt{\eta N_p/4}). \qquad \text{[ASK heterodyne]} \qquad (10.4.4)$$

One can use the same method to calculate the BER in the case of ASK homodyne receivers. Equations (10.4.2) and (10.4.3) still remain applicable. However, the SNR is improved by 3 dB for the homodyne case, so that $\text{SNR} = 4\eta N_p$ and

$$\text{BER} = \frac{1}{2}\text{erfc}(\sqrt{\eta N_p/2}). \qquad \text{[ASK homodyne]} \qquad (10.4.5)$$

Equations (10.4.4) and (10.4.5) can be used to calculate the receiver sensitivity at a specific BER. Similar to the direct-detection case discussed in Section 4.4, we can define the receiver sensitivity $\bar{P}_{rec}$ as the average received power required for realizing a BER of 10^{-9} or less. From Eqs. (10.4.2) and (10.4.3), $\text{BER} = 10^{-9}$ when $Q \approx 6$ or when $\text{SNR} = 144$ (21.6 dB). For the ASK heterodyne case we can use Eq. (10.1.14) to relate SNR to $\bar{P}_{rec}$ if we note that $\bar{P}_{rec} = \bar{P}_s/2$ simply because signal power is zero during the 0 bits. The result is

$$\bar{P}_{rec} = 2Q^2 h\nu\Delta f/\eta = 72h\nu\Delta f/\eta. \qquad (10.4.6)$$

For the ASK homodyne case, $\bar{P}_{rec}$ is smaller by a factor of 2 because of the 3-dB homodyne-detection advantage discussed in Section 10.1.3. As an example, for a 1.55-μm ASK heterodyne receiver with $\eta = 0.8$ and $\Delta f = 1$ GHz, the receiver sensitivity is about 12 nW and reduces to 6 nW if homodyne detection is used.

The receiver sensitivity is often quoted in terms of the number of photons N_p using Eqs. (10.4.4) and (10.4.5) as such a choice makes it independent of the receiver bandwidth and the operating wavelength. Furthermore, η is also set to 1 so that the sensitivity corresponds to an ideal photodetector. It is easy to verify that for realizing a BER of $= 10^{-9}$, N_p should be 72 and 36 in the heterodyne and homodyne cases, respectively. It is important to remember that N_p corresponds to the number of photons within a single 1 bit. The average number of photons per bit, $\bar{N}_p$, is reduced by a factor of 2 if we assume that 0 and 1 bits are equally likely to occur in a long bit sequence.

10.4.2 Synchronous PSK Receivers

Consider first the case of heterodyne detection. The signal at the decision circuit is given by Eq. (10.3.3) or by

$$I_d = \tfrac{1}{2}(I_p \cos\phi + i_c). \tag{10.4.7}$$

The main difference from the ASK case is that I_p is constant, but the phase ϕ takes values 0 or π depending on whether a 1 or 0 is transmitted. In both cases, I_d is a Gaussian random variable but its average value is either $I_p/2$ or $-I_p/2$, depending on the received bit. The situation is analogous to the ASK case with the difference that $I_0 = -I_1$ in place of being zero. In fact, one can use Eq. (10.4.2) for the BER, but Q is now given by

$$Q = \frac{I_1 - I_0}{\sigma_1 + \sigma_0} \approx \frac{2I_1}{2\sigma_1} = (\text{SNR})^{1/2}, \tag{10.4.8}$$

where $I_0 = -I_1$ and $\sigma_0 = \sigma_1$ was used. By using $\text{SNR} = 2\eta N_p$ from Eq. (10.1.14), the BER is given by

$$\text{BER} = \tfrac{1}{2}\,\text{erfc}(\sqrt{\eta N_p}). \quad \text{[PSK heterodyne]} \tag{10.4.9}$$

As before, the SNR is improved by 3 dB, or by a factor of 2, in the case of PSK homodyne detection, so that

$$\text{BER} = \tfrac{1}{2}\,\text{erfc}(\sqrt{2\eta N_p}). \quad \text{[PSK homodyne]} \tag{10.4.10}$$

The receiver sensitivity at a BER of 10^{-9} can be obtained by using $Q = 6$ and Eq. (10.1.14) for SNR. For the purpose of comparison, it is useful to express the receiver sensitivity in terms of the number of photons N_p. It is easy to verify that $N_p = 18$ and 9 for the cases of heterodyne and homodyne PSK detection, respectively. The average number of photons/bit, $\bar{N}_p$, equals N_p for the PSK format because the same power is transmitted during 1 and 0 bits. A PSK homodyne receiver is the most sensitive receiver, requiring only 9 photons/bit. It should be emphasized that this conclusion is based on the Gaussian approximation for the receiver noise [37].

It is interesting to compare the sensitivity of coherent receivers with that of a direct-detection receiver. Table 10.1 shows such a comparison. As discussed in Section 4.5.3, an ideal direct-detection receiver requires 10 photons/bit to operate at a BER of $\leq 10^{-9}$. This value is only slightly inferior to the best case of a PSK homodyne receiver and considerably superior to that of heterodyne schemes. However, it is never achieved in practice because of thermal noise, dark current, and many other factors, which degrade the sensitivity to the extent that $\bar{N}_p > 1000$ is usually required. In the case of coherent receivers, $\bar{N}_p$ below 100 can be realized simply because shot noise can be made dominant by increasing the local-oscillator power. The performance of coherent receivers is discussed in Section 10.6.

Table 10.1 Sensitivity of synchronous receivers

Modulation Format	Bit-Error Rate	N_p	$\bar{N}_p$
ASK heterodyne	$\frac{1}{2}\text{erfc}(\sqrt{\eta N_p/4})$	72	36
ASK homodyne	$\frac{1}{2}\text{erfc}(\sqrt{\eta N_p/2})$	36	18
PSK heterodyne	$\frac{1}{2}\text{erfc}(\sqrt{\eta N_p})$	18	18
PSK homodyne	$\frac{1}{2}\text{erfc}(\sqrt{2\eta N_p})$	9	9
FSK heterodyne	$\frac{1}{2}\text{erfc}(\sqrt{\eta N_p/2})$	36	36
Direct detection	$\frac{1}{2}\exp(-\eta N_p)$	20	10

10.4.3 Synchronous FSK Receivers

Synchronous FSK receivers generally use a dual-filter scheme similar to that shown in Fig. 10.6(a) for the asynchronous case. Each filter passes only 1 or 0 bits. The scheme is equivalent to two complementary ASK heterodyne receivers operating in parallel. This feature can be used to calculate the BER of dual-filter synchronous FSK receivers. Indeed, one can use Eqs. (10.4.2) and (10.4.3) for the FSK case also. However, the SNR is improved by a factor of 2 compared with the ASK case. The improvement is due to the fact that whereas no power is received, on average, half the time for ASK receivers, the same amount of power is received all the time for FSK receivers. Hence the signal power is enhanced by a factor of 2, whereas the noise power remains the same if we assume the same receiver bandwidth in the two cases. By using SNR $= 4\eta N_p$ in Eq. (10.4.3), the BER is given by

$$\text{BER} = \tfrac{1}{2}\text{erfc}(\sqrt{\eta N_p/2}). \qquad \text{[FSK heterodyne]} \qquad (10.4.11)$$

The receiver sensitivity is obtained from Eq. (10.4.6) by replacing the factor of 72 by 36. In terms of the number of photons, the sensitivity is given by $N_p = 36$. The average number of photons/bit, $\bar{N}_p$, also equals 36, since each bit carries the same energy. A comparison of ASK and FSK heterodyne schemes in Table 10.1 shows that $\bar{N}_p = 36$ for both schemes. Therefore even though the ASK heterodyne receiver requires 72 photons within the 1 bit, the receiver sensitivity (average received power) is the same for both the ASK and FSK schemes. Figure 10.7 plots the BER as a function of N_p for the ASK, PSK, and FSK formats by using Eqs. (10.4.4), (10.4.9), and (10.4.11). The dotted curve shows the BER for the case of synchronous PSK homodyne receiver discussed in Section 10.4.2. The dashed curves correspond to the case of asynchronous receivers discussed in the following subsections.

10.4.4 Asynchronous ASK Receivers

The BER calculation for asynchronous receivers is slightly more complicated than for synchronous receivers because the noise statistics does not remain Gaussian when an

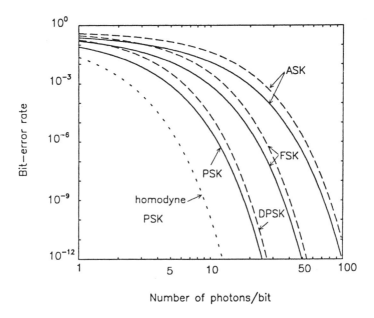

Figure 10.7: Bit-error-rate curves for various modulation formats. The solid and dashed lines correspond to the cases of synchronous and asynchronous demodulation, respectively.

envelope detector is used (see Fig. 10.5). The reason can be understood from Eq. (10.3.4), which shows the signal received by the decision circuit. In the case of an ideal ASK heterodyne receiver without phase fluctuations, ϕ can be set to zero so that (subscript d is dropped for simplicity of notation)

$$I = [(I_p + i_c)^2 + i_s^2]^{1/2}. \qquad (10.4.12)$$

Even though both $I_p + i_c$ and i_s are Gaussian random variables, the probability density function (PDF) of I is not Gaussian. It can be calculated by using a standard technique [38] and is found to be given by [39]

$$p(I, I_p) = \frac{I}{\sigma^2} \exp\left(-\frac{I^2 + I_p^2}{2\sigma^2}\right) I_0\left(\frac{I_p I}{\sigma^2}\right), \qquad (10.4.13)$$

where I_0 represents the modified Bessel function of the first kind. Both i_c and i_s are assumed to have a Gaussian PDF with zero mean and the same standard deviation σ, where σ is the RMS noise current. The PDF given by Eq. (10.4.13) is known as the *Rice distribution* [39]. Note that I varies in the range 0 to ∞, since the output of an envelope detector can have only positive values. When $I_p = 0$, the Rice distribution reduces to the *Rayleigh distribution*, well known in statistical optics [38].

The BER calculation follows the analysis of Section 4.5.1 with the only difference that the Rice distribution needs to be used in place of the Gaussian distribution. The BER is given by Eq. (4.5.2) or by

$$\text{BER} = \tfrac{1}{2}[P(0/1) + P(1/0)], \qquad (10.4.14)$$

where

$$P(0/1) = \int_0^{I_D} p(I, I_1) \, dI, \qquad P(1/0) = \int_{I_D}^{\infty} P(I, I_0) \, dI. \tag{10.4.15}$$

The notation is the same as that of Section 4.5.1. In particular, I_D is the decision level and I_1 and I_0 are values of I_p for 1 and 0 bits. The noise is the same for all bits ($\sigma_0 = \sigma_1 = \sigma$) because it is dominated by the local oscillator power. The integrals in Eq. (10.4.15) can be expressed in terms of Marcum's Q function defined as [40]

$$Q(\alpha, \beta) = \int_{\beta}^{\infty} x \exp\left(-\frac{x^2 + \alpha^2}{2}\right) I_0(\alpha x) \, dx. \tag{10.4.16}$$

The result for the BER is

$$\mathrm{BER} = \frac{1}{2}\left[1 - Q\left(\frac{I_1}{\sigma}, \frac{I_D}{\sigma}\right) + Q\left(\frac{I_0}{\sigma}, \frac{I_D}{\sigma}\right)\right]. \tag{10.4.17}$$

The decision level I_D is chosen such that the BER is minimum for given values of I_1, I_0, and σ. It is difficult to obtain an analytic expression of I_D under general conditions. However, under typical operating conditions, $I_0 \approx 0$, $I_1/\sigma \gg 1$, and I_D is well approximated by $I_1/2$. The BER then becomes

$$\mathrm{BER} \approx \tfrac{1}{2}\exp(-I_1^2/8\sigma^2) = \tfrac{1}{2}\exp(-\mathrm{SNR}/8). \tag{10.4.18}$$

When the receiver noise σ is dominated by the shot noise, the SNR is given by Eq. (10.1.14). Using $\mathrm{SNR} = 2\eta N_p$, we obtain the final result,

$$\mathrm{BER} = \tfrac{1}{2}\exp(-\eta N_p/4), \tag{10.4.19}$$

which should be compared with Eq. (10.4.4) obtained for the case of synchronous ASK heterodyne receivers. Equation (10.4.19) is plotted in Fig. 10.7 with a dashed line. It shows that the BER is larger in the asynchronous case for the same value of ηN_p. However, the difference is so small that the receiver sensitivity at a BER of 10^{-9} is degraded by only about 0.5 dB. If we assume that $\eta = 1$, Eq. (10.4.19) shows that BER $= 10^{-9}$ for $N_p = 80$ ($N_p = 72$ for the synchronous case). Asynchronous receivers hence provide performance comparable to that of synchronous receivers and are often used in practice because of their simpler design.

10.4.5 Asynchronous FSK Receivers

Although a single-filter heterodyne receiver can be used for FSK, it has the disadvantage that one-half of the received power is rejected, resulting in an obvious 3-dB penalty. For this reason, a dual-filter FSK receiver [see Fig. 10.6(a)] is commonly employed in which 1 and 0 bits pass through separate filters. The output of two envelope detectors are subtracted, and the resulting signal is used by the decision circuit. Since the average current takes values I_p and $-I_p$ for 1 and 0 bits, the decision threshold is set in the middle ($I_D = 0$). Let I and I' be the currents generated in the upper and lower branches of the dual filter receiver, where both of them include noise currents through Eq. (10.4.12). Consider the case in which 1 bits are received in the upper branch. The current I is then given by Eq. (10.4.12) and follows a Rice distribution with $I_p = I_1$ in Eq. (10.4.13). On the other hand, I' consists only of noise and its distribution is obtained by setting $I_p = 0$ in Eq. (10.4.13). An error is made when $I' > I$, as the signal is then below the decision level, resulting in

$$P(0/1) = \int_0^\infty p(I, I_1) \left[\int_I^\infty p(I', 0) \, dI' \right] dI, \tag{10.4.20}$$

where the inner integral provides the error probability for a fixed value of I and the outer integral sums it over all possible values of I. The probability $P(1/0)$ can be obtained similarly. In fact, $P(1/0) = P(0/1)$ because of the symmetric nature of a dual-filter receiver.

The integral in Eq. (10.4.20) can be evaluated analytically. By using Eq. (10.4.13) in the inner integral with $I_p = 0$, it is easy to verify that

$$\int_I^\infty p(I', 0) \, dI' = \exp\left(-\frac{I^2}{2\sigma^2}\right). \tag{10.4.21}$$

By using Eqs. (10.4.14), (10.4.20), and (10.4.21) with $P(1/0) = P(0/1)$, the BER is given by

$$\mathrm{BER} = \int_0^\infty \frac{I}{\sigma^2} \exp\left(-\frac{I^2 + I_1^2}{2\sigma^2}\right) I_0\left(\frac{I_1 I}{\sigma^2}\right) \exp\left(-\frac{I^2}{2\sigma^2}\right) dI, \tag{10.4.22}$$

where $p(I, I_p)$ was substituted from Eq. (10.4.13). By introducing the variable $x = \sqrt{2} I$, Eq. (10.4.22) can be written as

$$\mathrm{BER} = \frac{1}{2} \exp\left(-\frac{I^2}{4\sigma^2}\right) \int_0^\infty \frac{x}{\sigma^2} \exp\left(-\frac{x^2 + I_1^2/2}{2\sigma^2}\right) I_0\left(\frac{I_1 x}{\sigma^2 \sqrt{2}}\right) dx. \tag{10.4.23}$$

The integrand in Eq. (10.4.23) is just $p(x, I_1/\sqrt{2})$ and the integral must be 1. The BER is thus simply given by

$$\mathrm{BER} = \tfrac{1}{2} \exp(-I_1^2/4\sigma^2) = \tfrac{1}{2} \exp(-\mathrm{SNR}/4). \tag{10.4.24}$$

By using $\mathrm{SNR} = 2\eta N_p$ from Eq. (10.1.14), we obtain the final result

$$\mathrm{BER} = \tfrac{1}{2} \exp(-\eta N_p/2), \tag{10.4.25}$$

which should be compared with Eq. (10.4.11) obtained for the case of synchronous FSK heterodyne receivers. Figure 10.7 compares the BER in the two cases. Just as in the ASK case, the BER is larger for asynchronous demodulation. However, the difference is small, and the receiver sensitivity is degraded by only about 0.5 dB compared with the synchronous case. If we assume that $\eta = 1$, $N_p = 40$ at a BER of 10^{-9} ($N_p = 36$ in the synchronous case). $\bar{N}_p$ also equals 40, since the same number of photons are received during 1 and 0 bits. Similar to the synchronous case, $\bar{N}_p$ is the same for both the ASK and FSK formats.

Table 10.2 Sensitivity of asynchronous receivers

Modulation Format	Bit-Error Rate	N_p	$\bar{N}_p$
ASK heterodyne	$\frac{1}{2}\exp(-\eta N_p/4)$	80	40
FSK heterodyne	$\frac{1}{2}\exp(-\eta N_p/2)$	40	40
DPSK heterodyne	$\frac{1}{2}\exp(-\eta N_p)$	20	20
Direct detection	$\frac{1}{2}\exp(-\eta N_p)$	20	10

10.4.6 Asynchronous DPSK Receivers

As mentioned in Section 10.2.2, asynchronous demodulation cannot be used for PSK signals. A variant of PSK, known as DPSK, can be demodulated by using an asynchronous DPSK receiver [see Fig. 10.6(b)]. The filtered current is divided into two parts, and one part is delayed by exactly one bit period. The product of two currents contains information about the phase difference between the two neighboring bits and is used by the decision current to determine the bit pattern.

The BER calculation is more complicated for the DPSK case because the signal is formed by the product of two currents. The final result is, however, quite simple and is given by [11]

$$\text{BER} = \tfrac{1}{2}\exp(-\eta N_p). \tag{10.4.26}$$

It can be obtained from the FSK result, Eq. (10.4.24), by using a simple argument which shows that the demodulated DPSK signal corresponds to the FSK case if we replace I_1 by $2I_1$ and σ^2 by $2\sigma^2$ [13]. Figure 10.7 shows the BER by a dashed line (the curve marked DPSK). For $\eta = 1$, a BER of 10^{-9} is obtained for $N_p = 20$. Thus, a DPSK receiver is more sensitive by 3 dB compared with both ASK and FSK receivers. Table 10.2 lists the BER and the receiver sensitivity for the three modulation schemes used with asynchronous demodulation. The quantum limit of a direct-detection receiver is also listed for comparison. The sensitivity of an asynchronous DPSK receiver is only 3 dB away from this quantum limit.

10.5 Sensitivity Degradation

The sensitivity analysis of the preceding section assumes ideal operating conditions for a coherent lightwave system with perfect components. Many physical mechanisms degrade the receiver sensitivity in practical coherent systems; among them are phase noise, intensity noise, polarization mismatch, and fiber dispersion. In this section we discuss the sensitivity-degradation mechanisms and the techniques used to improve the performance with a proper receiver design.

10.5.1 Phase Noise

An important source of sensitivity degradation in coherent lightwave systems is the phase noise associated with the transmitter laser and the local oscillator. The reason can be understood from Eqs. (10.1.5) and (10.1.7), which show the current generated at the photodetector for homodyne and heterodyne receivers, respectively. In both cases, phase fluctuations lead to current fluctuations and degrade the SNR. Both the signal phase ϕ_s and the local-oscillator phase ϕ_{LO} should remain relatively stable to avoid the sensitivity degradation. A measure of the duration over which the laser phase remains relatively stable is provided by the *coherence time*. As the coherence time is inversely related to the laser linewidth Δv, it is common to use the linewidth-to-bit rate ratio, $\Delta v/B$, to characterize the effects of phase noise on the performance of coherent lightwave systems. Since both ϕ_s and ϕ_{LO} fluctuate independently, Δv is actually the sum of the linewidths Δv_T and Δv_{LO} associated with the transmitter and the local oscillator, respectively. The quantity $\Delta v = \Delta v_T + \Delta v_{LO}$ is often called the IF linewidth.

Considerable attention has been paid to calculate the BER in the presence of phase noise and to estimate the dependence of the power penalty on the ratio $\Delta v/B$ [41]–[55]. The tolerable value of $\Delta v/B$ for which the power penalty remains below 1 dB depends on the modulation format as well as on the demodulation technique. In general, the linewidth requirements are most stringent for homodyne receivers. Although the tolerable linewidth depends to some extent on the design of phase-locked loop, typically $\Delta v/B$ should be $< 5 \times 10^{-4}$ to realize a power penalty of less than 1 dB [43]. The requirement becomes $\Delta v/B < 1 \times 10^{-4}$ if the penalty is to be kept below 0.5 dB [44].

The linewidth requirements are relaxed considerably for heterodyne receivers, especially in the case of asynchronous demodulation with the ASK or FSK modulation format. For synchronous heterodyne receivers $\Delta v/B < 5 \times 10^{-3}$ is required [46]. In contrast, $\Delta v/B$ can exceed 0.1 for asynchronous ASK and FSK receivers [49]–[52]. The reason is related to the fact that such receivers use an envelope detector (see Fig. 10.5) that throws away the phase information. The effect of phase fluctuations is mainly to broaden the signal bandwidth. The signal can be recovered by increasing the bandwidth of the bandpass filter (BPF). In principle, any linewidth can be tolerated if the BPF bandwidth is suitably increased. However, a penalty must be paid since receiver noise increases with an increase in the BPF bandwidth. Figure 10.8 shows how the receiver sensitivity (expressed in average number of photons/bit, $\bar{N}_p$) degrades with $\Delta v/B$ for the ASK and FSK formats. The BER calculation is rather cumbersome and requires numerical simulations [51]. Approximate methods have been developed to provide the analytic results accurate to within 1 dB [52].

The DPSK format requires narrower linewidths compared with the ASK and FSK formats when asynchronous demodulation based on the delay scheme [see Fig. 10.6(b)] is used. The reason is that information is contained in the phase difference between the two neighboring bits, and the phase should remain stable at least over the duration of two bits. Theoretical estimates show that generally $\Delta v/B$ should be less than 1% to operate with a < 1 dB power penalty [43]. For a 1-Gb/s bit rate, the required linewidth is ~ 1 MHz but becomes < 1 MHz at lower bit rates.

The design of coherent lightwave systems requires semiconductor lasers that oper-

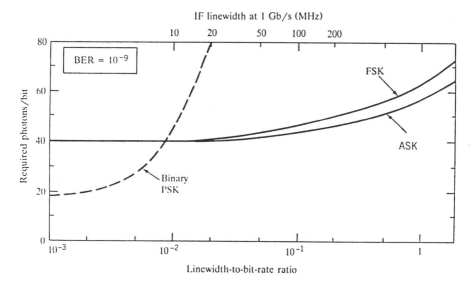

Figure 10.8: Receiver sensitivity $\bar{N}_p$ versus $\Delta v/B$ for asynchronous ASK and FSK heterodyne receivers. The dashed line shows the sensitivity degradation for a synchronous PSK heterodyne receiver. (After Ref. [49]; ©1988 IEEE; reprinted with permission.)

ate in a single longitudinal mode with a narrow linewidth and whose wavelength can be tuned (at least over a few nanometers) to match the carrier frequency ω_0 and the local-oscillator frequency ω_{LO} either exactly (homodyne detection) or to the required intermediate frequency. Multisection DFB lasers have been developed to meet these requirements (see Section 3.4.3). Narrow linewidth can also be obtained using a MQW design for the active region of a single-section DFB laser. Values as small as 0.1 MHz have been realized using strained MQW lasers [56].

An alternative approach solves the phase-noise problem by designing special receivers known as *phase-diversity receivers* [57]–[61]. Such receivers use two or more photodetectors whose outputs are combined to produce a signal that is independent of the phase difference $\phi_{IF} = \phi_s - \phi_{LO}$. The technique works quite well for ASK, FSK, and DPSK formats. Figure 10.9 shows schematically a multiport phase-diversity receiver. An optical component known as an *optical hybrid* combines the signal and local-oscillator inputs and provides its output through several ports with appropriate phase shifts introduced into different branches. The output from each port is processed electronically and combined to provide a current that is independent of ϕ_{IF}. In the case of a two-port homodyne receiver, the two output branches have a relative phase shift of 90°, so that the currents in the two branches vary as $I_p \cos\phi_{IF}$ and $I_p \sin\phi_{IF}$. When the two currents are squared and added, the signal becomes independent of ϕ_{IF}. In the case of three-port receivers, the three branches have relative phase shifts of 0, 120°, and 240°. Again, when the currents are added and squared, the signal becomes independent of ϕ_{IF}.

The preceding concept can be extended to design receivers with four or more branches. However, the receiver design becomes increasingly complex as more branches

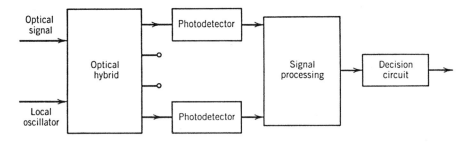

Figure 10.9: Schematic of a multiport phase-diversity receiver.

are added. Moreover, high-power local oscillators are needed to supply enough power to each branch. For these reasons, most phase-diversity receivers use two or three ports. Several system experiments have shown that the linewidth can approach the bit rate without introducing a significant power penalty even for homodyne receivers [58]–[61]. Numerical simulations of phase-diversity receivers show that the noise is far from being Gaussian [62]. In general, the BER is affected not only by the laser linewidth but also by other factors, such as the the BPF bandwidth.

10.5.2 Intensity Noise

The effect of intensity noise on the performance of direct-detection receivers was discussed in Section 4.6.2 and found to be negligible in most cases of practical interest. This is not the case for coherent receivers [63]–[67]. To understand why intensity noise plays such an important role in coherent receivers, we follow the analysis of Section 4.6.2 and write the current variance as

$$\sigma^2 = \sigma_s^2 + \sigma_T^2 + \sigma_I^2, \tag{10.5.1}$$

where $\sigma_I = RP_{LO}r_I$ and r_I is related to the *relative intensity noise* (RIN) of the local oscillator as defined in Eq. (4.6.7). If the RIN spectrum is flat up to the receiver bandwidth Δf, r_I^2 can be approximated by $2(\text{RIN})\Delta f$. The SNR is obtained by using Eq. (10.5.1) in Eq. (10.1.11) and is given by

$$\text{SNR} = \frac{2R^2 \bar{P}_s P_{LO}}{2q(RP_{LO} + I_d)\Delta f + \sigma_T^2 + 2R^2 P_{LO}^2 (\text{RIN})\Delta f}. \tag{10.5.2}$$

The local-oscillator power P_{LO} should be large enough to satisfy Eq. (10.1.12) if the receiver were to operate in the shot-noise limit. However, an increase in P_{LO} increases the contribution of intensity noise quadratically as seen from Eq. (10.5.2). If the intensity-noise contribution becomes comparable to shot noise, the SNR would decrease unless the signal power $\bar{P}_s$ is increased to offset the increase in receiver noise. This increase in $\bar{P}_s$ is just the power penalty δ_I resulting from the local-oscillator intensity noise. If we neglect I_d and σ_T^2 in Eq. (10.5.2) for a receiver designed to operate in the shot-noise limit, the power penalty (in dB) is given by the simple expression

$$\delta_I = 10\log_{10}[1 + (\eta/hv)P_{LO}(\text{RIN})]. \tag{10.5.3}$$

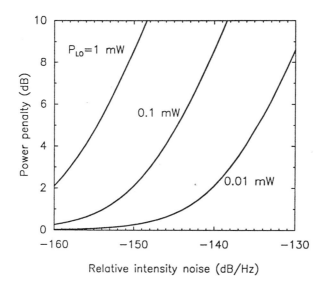

Figure 10.10: Power penalty versus RIN for several values of the local-oscillator power.

Figure 10.10 shows δ_I as a function of RIN for several values of P_{LO} using $\eta = 0.8$ and $h\nu = 0.8$ eV for 1.55-μm coherent receivers. The power penalty exceeds 2 dB when $P_{LO} = 1$ mW even for a local oscillator with a RIN of -160 dB/Hz, a value difficult to realize for DFB semiconductor lasers. For a local oscillator with a RIN of -150 dB/Hz, P_{LO} should be less than 0.1 mW to keep the power penalty below 2 dB. The power penalty can be made negligible at a RIN of -150 dB/Hz if only 10 μW of local-oscillator power is used. However, Eq. (10.1.13) is unlikely to be satisfied for such small values of P_{LO}, and receiver performance would be limited by thermal noise. Sensitivity degradation from local-oscillator intensity noise was observed in 1987 in a two-port ASK homodyne receiver [63]. The power penalty is reduced for three-port receivers but intensity noise remains a limiting factor for $P_{LO} > 0.1$ mW [61]. It should be stressed that the derivation of Eq. (10.5.3) is based on the assumption that the receiver noise is Gaussian. A numerical approach is necessary for a more accurate analysis of the intensity noise [65]–[67].

A solution to the intensity-noise problem is offered by the *balanced coherent receiver* [68] made with two photodetectors [69]–[71]. Figure 10.11 shows the receiver design schematically. A 3-dB fiber coupler mixes the optical signal with the local oscillator and splits the combined optical signal into two equal parts with a 90° relative phase shift. The operation of a balanced receiver can be understood by considering the photocurrents I_+ and I_- generated in each branch. Using the transfer matrix of a 3-dB coupler, the currents I_+ and I_- are given by

$$I_+ = \tfrac{1}{2}R(P_s + P_{LO}) + R\sqrt{P_s P_{LO}}\cos(\omega_{IF}t + \phi_{IF}), \qquad (10.5.4)$$

$$I_- = \tfrac{1}{2}R(P_s + P_{LO}) - R\sqrt{P_s P_{LO}}\cos(\omega_{IF}t + \phi_{IF}), \qquad (10.5.5)$$

where $\phi_{IF} = \phi_s - \phi_{LO} + \pi/2$.

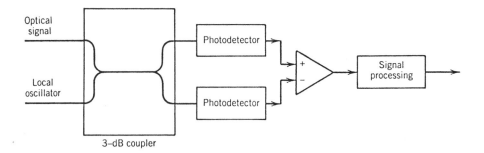

Figure 10.11: Schematic of a two-port balanced coherent receiver.

The subtraction of the two currents provides the heterodyne signal. The dc term is eliminated completely during the subtraction process when the two branches are balanced in such a way that each branch receives equal signal and local-oscillator powers. More importantly, the intensity noise associated with the dc term is also eliminated during the subtraction process. The reason is related to the fact that the same local oscillator provides power to each branch. As a result, intensity fluctuations in the two branches are perfectly correlated and cancel out during subtraction of the photocurrents I_+ and I_-. It should be noted that intensity fluctuations associated with the ac term are not canceled even in a balanced receiver. However, their impact is less severe on the system performance because of the square-root dependence of the ac term on the local-oscillator power.

Balanced receivers are commonly used while designing a coherent lightwave system because of the two advantages offered by them. First, the intensity-noise problem is nearly eliminated. Second, all of the signal and local-oscillator power is used effectively. A single-port receiver such as that shown in Fig. 10.1 rejects half of the signal power P_s (and half of P_{LO}) during the mixing process. This power loss is equivalent to a 3-dB power penalty. Balanced receivers use all of the signal power and avoid this power penalty. At the same time, all of the local-oscillator power is used by the balanced receiver, making it easier to operate in the shot-noise limit.

10.5.3   Polarization Mismatch

The polarization state of the received optical signal plays no role in direct-detection receivers simply because the photocurrent generated in such receivers depends only on the number of incident photons. This is not the case for coherent receivers, whose operation requires matching the state of polarization of the local oscillator to that of the signal received. The polarization-matching requirement can be understood from the analysis of Section 10.1, where the use of scalar fields E_s and E_{LO} implicitly assumed the same polarization state for the two optical fields. If $\hat{e}_s$ and $\hat{e}_{LO}$ represent the unit vectors along the direction of polarization of E_s and E_{LO}, respectively, the interference term in Eq. (10.1.3) contains an additional factor $\cos\theta$, where θ is the angle between $\hat{e}_s$ and $\hat{e}_{LO}$. Since the interference term is used by the decision circuit to reconstruct the transmitted bit stream, any change in θ from its ideal value of $\theta = 0$ reduces the signal

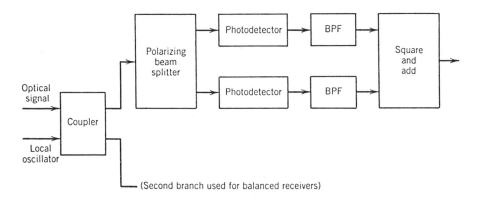

Figure 10.12: Schematic of a polarization-diversity coherent receiver.

and affects the receiver performance. In particular, if the polarization states of E_s and E_{LO} are orthogonal to each other ($\theta = 90°$), the signal disappears (complete fading). Any change in θ affects the BER through changes in the receiver current and SNR.

The polarization state $\hat{e}_{LO}$ of the local oscillator is determined by the laser and remains fixed. This is also the case for the transmitted signal before it is launched into the fiber. However, at the fiber output, the polarization state $\hat{e}_s$ of the signal received differs from that of the signal transmitted because of fiber birefringence, as discussed in Section 2.2.3 in the context of single-mode fibers. Such a change would not be a problem if $\hat{e}_s$ remained constant with time because one could match it with $\hat{e}_{LO}$ by simple optical techniques. The source of the problem lies in the polarization-mode dispersion (PMD) or the fact that $\hat{e}_s$ changes randomly in most fibers because of birefringence fluctuations related to environmental changes (nonuniform stress, temperature variations, etc.). Such changes occur on a time scale ranging from seconds to microseconds. They lead to random changes in the BER and render coherent receivers unusable unless some scheme is devised to make the BER independent of polarization fluctuations. Although polarization fluctuations do not occur in polarization-maintaining fibers, such fibers are not used in practice because they are difficult to work with and have higher losses than those of conventional fibers. Thus, a different solution to the polarization-mismatch problem is required.

Several schemes have been developed for solving the polarization-mismatch problem [72]–[77]. In one scheme [72], the polarization state of the optical signal received is tracked electronically and a feedback-control technique is used to match $\hat{e}_{LO}$ with $\hat{e}_s$. In another, polarization scrambling or spreading is used to force $\hat{e}_s$ to change randomly during a bit period [73]–[76]. Rapid changes of $\hat{e}_s$ are less of a problem than slow changes because, on average, the same power is received during each bit. A third scheme makes use of optical phase conjugation to solve the polarization problem [77]. The phase-conjugated signal can be generated inside a dispersion-shifted fiber through four-wave mixing (see Section 7.7). The pump laser used for four-wave mixing can also play the role of the local oscillator. The resulting photocurrent has a frequency component at twice the pump-signal detuning that can be used for recovering the bit stream.

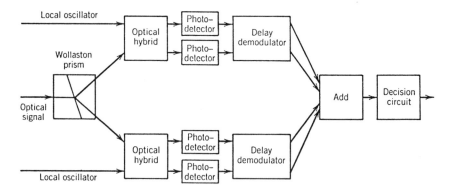

Figure 10.13: Four-port coherent DPSK receiver employing both phase and polarization diversity. (After Ref. [80]; ©1987 IEE; reprinted with permission.)

The most commonly used approach solves the polarization problem by using a two-port receiver, similar to that shown in Fig. 10.11, with the difference that the two branches process orthogonal polarization components. Such receivers are called *polarization-diversity receivers* [78]–[82] as their operation is independent of the polarization state of the signal received. The polarization-control problem has been studied extensively because of its importance for coherent lightwave systems [83]–[90].

Figure 10.12 shows the block diagram of a polarization-diversity receiver. A polarization beam splitter is used to separate the orthogonally polarized components which are processed by separate branches of the two-port receiver. When the photocurrents generated in the two branches are squared and added, the electrical signal becomes polarization independent. The power penalty incurred in following this technique depends on the modulation and demodulation techniques used by the receiver. In the case of synchronous demodulation, the power penalty can be as large as 3 dB [85]. However, the penalty is only 0.4–0.6 dB for optimized asynchronous receivers [78].

The technique of polarization diversity can be combined with phase diversity to realize a receiver that is independent of both phase and polarization fluctuations of the signal received [91]. Figure 10.13 shows such a four-port receiver having four branches, each with its own photodetector. The performance of such receivers would be limited by the intensity noise of the local oscillator, as discussed in Section 10.5.2. The next step consists of designing a balanced phase- and polarization-diversity receiver by using eight branches with their own photodetectors. Such a receiver has been demonstrated using a compact bulk optical hybrid [92]. In practical coherent systems, a balanced, polarization-diversity receiver is used in combination with narrow-linewidth lasers to simplify the receiver design, yet avoid the limitations imposed by intensity noise and polarization fluctuations.

10.5.4 Fiber Dispersion

Section 5.4 discussed how fiber dispersion limits the bit-rate–distance product (*BL*) of direct-detection (IM/DD) systems. Fiber dispersion also affects the performance of

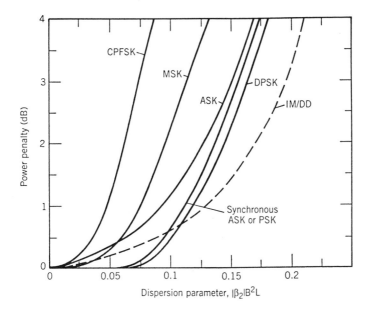

Figure 10.14: Dispersion-induced power penalty as a function of the dimensionless parameter $|\beta_2|B^2L$ for several modulation formats. The dashed line shows power penalty for a direct-detection system. (After Ref. [95]; ©1988 IEEE; reprinted with permission.)

coherent systems although its impact is less severe than for IM/DD systems [93]–[97]. The reason is easily understood by noting that coherent systems, by necessity, use a semiconductor laser operating in a single longitudinal mode with a narrow linewidth. Frequency chirping is avoided by using external modulators. Moreover, it is possible to compensate for fiber dispersion (see Section 7.2) through electronic equalization techniques in the IF domain [98].

The effect of fiber dispersion on the transmitted signal can be calculated by using the analysis of Section 2.4. In particular, Eq. (2.4.15) can be used to calculate the optical field at the fiber output for any modulation technique. The power penalty has been calculated for various modulation formats [95] through numerical simulations of the eye degradation occurring when a pseudo-random bit sequence is propagated through a single-mode fiber and demodulated by using a synchronous or asynchronous receiver. Figure 10.14 shows the power penalty as a function of the dimensionless parameter combination $|\beta_2|B^2L$ for several kinds of modulation formats. The dashed line shows, for comparison, the case of an IM/DD system. In all cases, the low-pass filter (before the decision circuit) is taken to be a second-order *Butterworth filter* [99], with the 3-dB bandwidth equal to 65% of the bit rate.

As seen in Fig. 10.14, fiber dispersion affects the performance of a coherent lightwave system qualitatively in the same way for all modulation formats, although quantitative differences do occur. The power penalty increases most rapidly for CPFSK and MSK formats, for which tone spacing is smaller than the bit rate. In all cases system performance depends on the product B^2L rather than BL. One can estimate the limiting value of B^2L by noting that the power penalty can be reduced to below 1 dB in most cases if the system is designed such that $|\beta_2|B^2L < 0.1$. For standard fibers with $\beta_2 = -20$ ps^2/km near 1.55 μm, B^2L is limited to 5000 (Gb/s)^2-km, and L should be <50 km at $B = 10$ Gb/s. Clearly, dispersion becomes a major limiting factor for systems designed with standard fibers when transmission distance is increased using optical amplifiers. Dispersion management would solve this problem. Electronic equalization can be used for compensating dispersion in coherent systems [100]. The basic idea is to pass the intermediate-frequency signal through a filter whose transfer function is the inverse of the transfer function associated with the fiber (see Section 7.2). It is also possible to compensate fiber dispersion through optical techniques such as dispersion management [101]. PMD then becomes a limiting factor for long-haul coherent systems [102]–[104].

10.5.5 Other Limiting Factors

Several other factors can degrade the performance of coherent lightwave systems and should be considered during system design. Reflection feedback is one such limiting factor. The effect of reflection feedback on IM/DD systems has been discussed in Section 5.4.5. Essentially the same discussion applies to coherent lightwave systems. Any feedback into the laser transmitter or the local oscillator must be avoided as it can lead to linewidth broadening or multimode operation of the semiconductor laser, both of which cannot be tolerated for coherent systems. The use of optical isolators within the transmitter may be necessary for controlling the effects of optical feedback.

Multiple reflections between two reflecting surfaces along the fiber cable can convert phase noise into intensity noise and affect system performance as discussed in Section 5.4.5. For coherent systems such conversion can occur even inside the receiver, where short fiber segments are used to connect the local oscillator to other receiver components, such as an optical hybrid (see Fig. 10.10). Calculations for phase-diversity receivers show that the reflectivity of splices and connectors should be below -35 dB under typical operating conditions [105]. Such reflection effects become less important for balanced receivers, where the impact of intensity noise on receiver performance is considerably reduced. Conversion of phase noise into intensity noise can occur even without parasitic reflections. However, the power penalty can be reduced to below 0.5 dB by ensuring that the ratio $\Delta v/B$ is below 20% in phase diversity ASK receivers [106].

Nonlinear effects in optical fibers discussed in Section 2.6 also limit the coherent system, depending on the optical power launched into the fiber [107]. Stimulated Raman scattering is not likely to be a limiting factor for single-channel coherent systems but becomes important for multichannel coherent systems (see Section 7.3.3). On the other hand, stimulated Brillouin scattering (SBS) has a low threshold and can affect even single-channel coherent systems. The SBS threshold depends on both the modulation format and the bit rate, and its effects on coherent systems have been studied extensively [108]–[110]. Nonlinear refraction converts intensity fluctuations into phase fluctuation through self- (SPM) and cross-phase modulation (XPM) [107]. The effects of SPM become important for long-haul systems using cascaded optical amplifiers [111]. Even XPM effects become significant in coherent FSK systems [112].

Four-wave mixing also becomes a limiting factor for WDM coherent systems [113] and need to be controlled employing high-dispersion locally but keeping the average dispersion low through dispersion management.

10.6 System Performance

A large number of transmission experiments were performed during the 1980s to demonstrate the potential of coherent lightwave systems. Their main objective was to show that coherent receivers are more sensitive than IM/DD receivers. This section focuses on the system performance issues while reviewing the state of the art of coherent lightwave systems.

10.6.1 Asynchronous Heterodyne Systems

Asynchronous heterodyne systems have attracted the most attention in practice simply because the linewidth requirements for the transmitter laser and the local oscillator are so relaxed that standard DFB lasers can be used. Experiments have been performed with the ASK, FSK, and DPSK modulation formats [114]–[116]. An ASK experiment in 1990 showed a baseline sensitivity (without the fiber) of 175 photons/bit at 4 Gb/s [116]. This value is only 10.4 dB away from the quantum limit of 40 photons/bit obtained in Section 10.4.4. The sensitivity degraded by only 1 dB when the signal was transmitted through 160 km of standard fiber with $D \approx 17$ ps/(nm-km). The system performance was similar when the FSK format was used in place of ASK. The frequency separation (tone spacing) was equal to the bit rate in this experiment.

The same experiment was repeated with the DPSK format using a LiNbO$_3$ phase modulator [116]. The baseline receiver sensitivity at 4 Gb/s was 209 photons/bit and degraded by 1.8 dB when the signal was transmitted over 160 km of standard fiber. Even better performance is possible for DPSK systems operating at lower bit rates. A record sensitivity of only 45 photons/bit was realized in 1986 at 400 Mb/s [114]. This value is only 3.5 dB away from the quantum limit of 20 photons/bit. For comparison, the receiver sensitivity of IM/DD receivers is such that $\bar{N}_p$ typically exceeds 1000 photons/bit even when APDs are used.

DPSK receivers have continued to attract attention because of their high sensitivity and relative ease of implementation [117]–[125]. The DPSK signal at the transmitter can be generated through direct modulation of a DFB laser [117]. Demodulation of the DPSK signal can be done optically using a Mach-Zehnder interferometer with a one-bit delay in one arm, followed by two photodetectors at each output port of the interferometer. Such receivers are called direct-detection DPSK receivers because they do not use a local oscillator and exhibit performance comparable to their heterodyne counterparts [118]. In a 3-Gb/s experiment making use of this scheme, only 62 photons/bit were needed by an optically demodulated DPSK receiver designed with an optical preamplifier [119]. In another variant, the transmitter sends a PSK signal but the receiver is designed to detect the phase difference such that a local oscillator is not needed [120]. Considerable work has been done to quantify the performance of various

DPSK and FSK schemes through numerical models that include the effects of phase noise and the preamplifier noise [121]–[125].

Asynchronous heterodyne schemes have also been used for long-haul coherent systems using in-line optical amplifiers for increasing the transmission distance. A 1991 experiment realized a transmission distance of 2223 km at 2.5 Gb/s by using 25 erbium-doped fiber amplifiers at approximately 80-km intervals [126]. The performance of long-haul coherent systems is affected by the amplifier noise as well as by the non-linear effects in optical fibers. Their design requires optimization of many operating parameters, such as amplifier spacing, launch power, laser linewidth, IF bandwidth, and decision threshold [127]–[129]. In the case of WDM systems, the use of DPSK can reduce the XPM-induced interaction among channels and improve the system performance [130].

10.6.2 Synchronous Heterodyne Systems

As discussed in Section 10.4, synchronous heterodyne receivers are more sensitive than asynchronous receivers. They are also more difficult to implement as the microwave carrier must be recovered from the received data for synchronous demodulation. Since the sensitivity advantage is minimal (less than 0.5 dB) for ASK and FSK formats (compare Tables 10.1 and 10.2), most of the laboratory experiments have focused on the PSK format [131]–[135] for which the receiver sensitivity is only 18 photons/bit. A problem with the PSK format is that the carrier is suppressed when the phase shift between 1 and 0 bits is exactly 180° because the transmitted power is then entirely contained in the modulation sidebands. This feature poses a problem for carrier recovery. A solution is offered by the pilot-carrier scheme in which the phase shift is reduced below 180° (typically 150–160°) so that a few percent of the power remains in the carrier and can be used for synchronous demodulation at the receiver.

Phase noise is a serious problem for synchronous heterodyne receivers. As discussed in Section 10.5.1, the ratio $\Delta v/B$ must be less than 5×10^{-3}, where $\Delta v = \Delta v_T + \Delta v_{LO}$ is the IF linewidth. For bit rates below 1 Gb/s, the laser linewidth should be less than 2 MHz. External-cavity semiconductor lasers are often used in the synchronous experiments, as they can provide linewidths below 0.1 MHz. Several experiments have been performed using diode-pumped Nd:YAG lasers [131]–[133], which operate at a fixed wavelength near 1.32 μm but provide linewidths as small as 1 kHz. In one experiment, the bit rate was 4 Gb/s, but the receiver sensitivity of 631 photons/bit was 15.4 dB away from the quantum limit of 18 photons/bit, mainly because of the residual thermal noise and the intensity noise as a balanced configuration was not used [133]. The receiver sensitivity could be improved to 235 photons/bit at a lower bit rate of 2 Gb/s. This sensitivity is still not as good as that obtained for asynchronous heterodyne receivers. The performance of multichannel heterodyne systems has also been analyzed [134].

10.6.3 Homodyne Systems

As seen in Table 10.1, homodyne systems with the PSK format offer the best receiver sensitivity as they require, in principle, only 9 photons/bit. Implementation

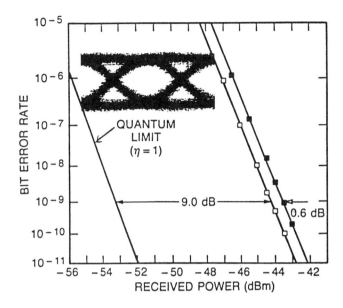

Figure 10.15: BER curves for a 4-Gb/s PSK homodyne transmission experiment with 5 m (empty squares) and 167 km (filled squares) of fiber. The quantum limit is shown for comparison. The inset shows the eye diagram after 167 km of fiber at −44 dBm received power. (After Ref. [145]; ©1990 IEEE; reprinted with permission.)

of such systems requires an optical phase-locked loop [136]–[141]. Many transmission experiments have shown the potential of PSK homodyne systems using He–Ne lasers, Nd:YAG lasers, and semiconductor lasers [142]–[150]. The receiver sensitivity achieved in these experiments depends on the bit rate. At a relatively low bit rate of 140 Mb/s, receiver sensitivities of 26 photons/bit at 1.52 μm [143] and 25 photons/bit at 1.32 μm [144] have been obtained using He–Ne and Nd:YAG lasers, respectively. In a 1992 experiment, a sensitivity of 20 photons/bit at 565 Mb/s was realized using synchronization bits for phase locking [147]. These values, although about 4 dB away from the quantum limit of 9 photons/bit, illustrate the potential of homodyne systems. In terms of the bit energy, 20 photons at 1.52 μm correspond to an energy of only 3 attojoules!

The sensitivity of PSK homodyne receivers decreases as the bit rate increases. A sensitivity of 46 photons/bit was found in a 1-Gb/s experiment that used external-cavity semiconductor lasers operating near 1.5 μm and transmitted the signal over 209 km of a standard fiber [142]. Dispersion penalty was negligible (about 0.1 dB) in this experiment, as expected from Fig. 10.15. In another experiment, the bit rate was extended to 4 Gb/s [145]. The baseline receiver sensitivity (without the fiber) was 72 photons/bit. When the signal was transmitted over 167 km of standard single-mode fiber, the receiver sensitivity degraded by only 0.6 dB (83 photons/bit), indicating that dispersion was not a problem even at 4 Gb/s. Figure 10.15 shows the BER curves obtained in this experiment with and without the fiber together with the eye diagram (inset) obtained after 167 km of fiber. Another experiment increased the bit rate to 10 Gb/s by using a 1.55-μm external-cavity DFB laser whose output was phase-modulated through a LiNbO$_3$ external modulator [146]. The receiver sensitivity at 10 Gb/s was 297 photons/bit. The signal was transmitted through 151 km of dispersion-shifted fiber without any dispersion-induced power penalty.

Long-haul homodyne systems use optical amplifiers for compensating fiber losses together with a dispersion-compensation scheme. In a 1993 experiment, a 6-Gb/s PSK signal was transmitted over 270 km using multiple in-line amplifiers [149]. A microstrip line was used as a delay equalizer (see Section 7.2) for compensating the fiber dispersion. Its use was feasible because of the implementation of the single-sideband technique. In a later experiment, the bit rate was extended to 10 Gb/s by using the vestigial-sideband technique [150]. The 1.55-μm PSK signal could be transmitted over 126 km of standard telecommunication fiber with dispersion compensation provided by a 10-cm microstrip line. The design of long-haul homodyne systems with in-line amplifiers requires consideration of many factors such as phase noise, shot noise, imperfect phase recovery, and amplifier noise. Numerical simulations are often used to optimize the system performance [151]–[154].

10.6.4 Current Status

Any new technology must be tested through field trials before it can be commercialized. Several field trials for coherent lightwave systems were carried out in the early 1990s [155]–[161]. In all cases, an asynchronous heterodyne receiver was used because of its simplicity and not-so-stringent linewidth requirements. The modulation format of choice was the CPFSK format. This choice avoids the use of an external modulator, thereby simplifying the transmitter design. Furthermore, the laboratory experiments have shown that high-sensitivity receivers can be designed at bit rates as high as 10 Gb/s. A balanced polarization-diversity heterodyne receiver is used to demodulate the transmitted signal.

Field trials have included testing of both land- and sea-based telecommunication systems. In the case of one submarine trial [159], the system was operated at 560 Mb/s with the CPFSK format over 90 km of fiber cable. In another submarine trial [160], the system was operated at 2.5 Gb/s with the CPFSK format over fiber lengths of up to 431 km by using regenerators. Both trials showed that the use of polarization-diversity receivers is essential for practical coherent systems. In addition, the receiver incorporated electronic circuitry for automatic gain and frequency controls.

In spite of the successful field trials, coherent lightwave systems had not reached the commercial stage in 2002. As mentioned earlier, the main reason is related to the success of the WDM technology with the advent of the erbium-doped fiber amplifiers. A second reason can be attributed to the complexity of coherent transmitters and receivers. The integration of these components on a single chip should address the reliability concerns. Considerable development effort was directed in the 1990s toward designing optoelectronic integrated circuits (OEICs) for coherent lightwave systems [162]–[170]. By 1994, a balanced, polarization-diversity heterodyne receiver containing four photodiodes and made by using the InP/InGaAsP material system, exhibited a bandwidth in excess of 10 GHz [164]. A tunable local oscillator can also be integrated on the same chip. Such a tunable polarization diversity heterodyne OEIC receiver was used in a 140-Mb/s system experiment, intended mainly for video distribution [165]. A balanced heterodyne OEIC receiver with a 9-GHz bandwidth was fabricated in 1996 by integrating a local oscillator and two photodiodes with a 3-dB coupler [168]. The coherent techniques may turn out to be more suitable for multichannel access networks for which bit rates per channel are relatively low but number of channels can be quite large [161]. An integrated transceiver suitable for bidirectional access networks has been developed for such applications [170].

Problems

10.1 Prove the 3-dB advantage of homodyne detection by showing that the average electrical power generated by a coherent receiver is twice as large for homodyne detection as for heterodyne detection under identical operating conditions.

10.2 Derive an expression for the SNR of a homodyne receiver by taking into account both the shot noise and the thermal noise.

10.3 Consider a 1.55-μm heterodyne receiver with a p–i–n photodiode of 90% quantum efficiency connected to a 50-Ω load resistance. How much local-oscillator power is needed to operate in the shot-noise limit? Assume that shot-noise limit is achieved when the thermal-noise contribution at room temperature to the noise power is below 1%.

10.4 Prove that the SNR of an ideal PSK homodyne receiver (perfect phase locking and 100% quantum efficiency) approaches $4\bar{N}_p$, where $\bar{N}_p$ is the average number of photons/bit. Assume that the receiver bandwidth equals half the bit rate and that the receiver operates in the shot-noise limit.

10.5 Show how an electro-optic material such as LiNbO$_3$ can be used for generating optical bit streams with ASK, PSK, and FSK modulation formats. Use diagrams as necessary.

10.6 A 1.55-μm DFB laser is used for the FSK modulation at 100 Mb/s with a tone spacing of 300 MHz. The modulation efficiency is 500 MHz/mA and the differential quantum efficiency equals 50% at the bias level of 3 mW. Estimate the power change associated with FSK by assuming that the two facets emit equal powers.

10.7 Derive an expression for the BER of a synchronous heterodyne ASK receiver by assuming that the in-phase noise component i_c has a probability density function

$$p(i_c) = \frac{1}{\sigma\sqrt{2}}\exp\left(-\frac{\sqrt{2}}{\sigma}|i_c|\right).$$

Determine the SNR required to achieve a BER of 10^{-9}.

10.8 Calculate the sensitivity (in dBm units) of a homodyne ASK receiver operating at 1.55 μm in the shot-noise limit by using the SNR expression obtained in Problem 10.2. Assume that $\eta = 0.8$ and $\Delta f = 1$ GHz. What is the receiver sensitivity when the PSK format is used in place of ASK?

10.9 Derive the Rice distribution [Eq. (10.4.13)] for the signal current I given Eq. (10.4.12) for an asynchronous heterodyne ASK receiver. Assume that both quadrature components of noise obey Gaussian statistics with standard deviation σ.

10.10 Show that the BER of an asynchronous heterodyne ASK receiver [Eq. (10.4.17)] can be approximated as

$$\text{BER} = \tfrac{1}{2}\exp[-I_1^2/(8\sigma^2)]$$

when $I_1/\sigma \gg 1$ and $I_0 = 0$. Assume that $I_D = I_1/2$.

10.11 Asynchronous heterodyne FSK receivers are commonly used for coherent lightwave systems. What is the SNR required by such receivers to operate at a BER of 10^{-9}? Calculate the receiver sensitivity (in dBm units) at 2 Gb/s in the shot-noise limit by assuming 1.2-GHz receiver bandwidth, 80% quantum efficiency, and a 1.55-μm operating wavelength.

10.12 Derive an expression for the SNR in terms of the intensity noise parameter r_I by including intensity noise through Eq. (10.5.1). Prove that the optimum value of P_{LO} at which the SNR is maximum is given by $P_{\text{LO}} = \sigma_T/(Rr_I)$ when the dark-current contribution to the shot noise is neglected.

10.13 Derive an expression for the power penalty as a function of r_I by using the SNR obtained in Problem 10.12.

10.14 Consider an optical carrier whose amplitude and frequency are constant but whose phase is modulated sinusoidally as $\phi(t) = \phi_0\sin(\omega_m t)$. Show that the amplitude becomes modulated during propagation inside the fiber because of fiber dispersion.

10.15 Discuss the effect of laser linewidth on coherent communication systems. Why is the homodyne PSK receiver most sensitive to phase fluctuations? How is this sensitivity reduced for asynchronous heterodyne receivers?

References

[1] M. Schwartz, *Information Transmission, Modulation, and Noise*, 4th ed., McGraw-Hill, New York, 1990.

[2] R. E. Ziemer, *Principles of Communications; Systems, Modulation and Noise*, Wiley, New York, 1994.

[3] L. W. Couch II, *Digital and Analog Communication Systems*, 5th ed., Prentice Hall, Upper Saddle River, NJ, 1995.

[4] M. S. Roden, *Analog and Digital Communication Systems*, Prentice Hall, Upper Saddle River, NJ, 1995.

[5] B. P. Lathi, *Modern Digital and Analog Communication Systems*, Oxford University Press, New York, 1995.

[6] W. R. Bennett, *Communication Systems and Techniques*, IEEE Press, Piscataway, NJ, 1995.

[7] J. Salz, *AT&T Tech. J.* **64**, 2153 (1985); *IEEE Commun. Mag.* **24** (6), 38 (1986).

[8] E. Basch and T. Brown, in *Optical Fiber Transmission*, E. E. Basch, Ed., SAMS, Indianapolis, IN, 1986, Chap. 16.

[9] T. Okoshi, *J. Lightwave Technol.* **5**, 44 (1987).

[10] T. Kimura, *J. Lightwave Technol.* **5**, 414 (1987)

[11] T. Okoshi and K. Kikuchi, *Coherent Optical Fiber Communications*, Kluwer Academic, Boston, 1988.

[12] R. A. Linke and A. H. Gnauck, *J. Lightwave Technol.* **6**, 1750 (1988).

[13] J. R. Barry and E. A. Lee, *Proc. IEEE* **78**, 1369 (1990).

[14] P. S. Henry and S. D. Persoinick, Eds., *Coherent Lightwave Communications*, IEEE Press, Piscataway, NJ, 1990.

[15] S. Betti, G. de Marchis, and E. Iannone, *Coherent Optical Communication Systems*, Wiley, New York, 1995.

[16] S. Ryu, *Coherent Lightwave Communication Systems*, Artec House, Boston, 1995.

[17] F. Heismann, S. K. Korotky, and J. J. Veselka, in *Optical Fiber Telecommunications III*, Vol. B, I. P. Kaminow and T. L. Loch, Eds., Academic Press, San Diego, CA, 1997, Chap. 8.

[18] K. Noguchi, O. Mitomi, and H. Miyazawa, *J. Lightwave Technol.* **16**, 615 (1998).

[19] H. Zhang, M. C. Oh, A. Szep, W. H. Steier, L. R. Dalton, H. Erlig, Y. Chang, D. H. Chang, and H. R. Fetterman, *Appl. Phys. Lett.* **78**, 3136 (2001).

[20] Y. Kim, S. K. Kim, J. Lee, Y. Kim, J. Kang, W. Choi, and J. Jeong, *Opt. Fiber Technol.* **7**, 84 (2001).

[21] Y. Akage, K. Kawano, S. Oku, R. Iga, H. Okamoto, Y. Miyamoto, and H. Takeuchi, *Electron. Lett.* **37**, 299 (2001).

[22] K. Wakita, I. Kotaka, and H. Asai, *IEEE Photon. Technol. Lett.* **4**, 29 (1992).

[23] S. Yoshida, Y. Tada, I. Kotaka, and K. Wakita, *Electron. Lett.* **30**, 1795 (1994).

[24] N. Yoshimoto, K. Kawano, Y. Hasumi, H. Takeuchi, S. Kondo, and Y. Noguchi, *IEEE Photon. Technol. Lett.* **6**, 208 (1994).

[25] K. Wakita and I. Kotaka, *Microwave Opt. Tech. Lett.* **7**, 120 (1994).

[26] E. M. Goldys and T. L. Tansley, *Microelectron. J.* **25**, 697 (1994).

[27] A. Segev, A. Saar, J. Oiknine-Schlesinger, and E. Ehrenfreund, *Superlattices Microstruct.* **19**, 47 (1996).

[28] S. Kobayashi, Y. Yamamoto, M. Ito, and T. Kimura, *IEEE J. Quantum Electron.* **18**, 582 (1982).

[29] S. Ogita, Y. Kotaki, M. Matsuda, Y. Kuwahara, H. Onaka, H. Miyata, and H. Ishikawa, *IEEE Photon. Technol. Lett.* **2**, 165 (1990).

[30] M. Kitamura, H. Yamazaki, H. Yamada, S. Takano, K. Kosuge, Y. Sugiyama, M. Yamaguchi, and I. Mito, *IEEE J. Quantum Electron.* **29**, 1728 (1993).

[31] M. Okai, M. Suzuki, and T. Taniwatari, *Electron. Lett.* **30**, 1135 (1994).

[32] B. Tromborg, H. E. Lassen, and H. Olesen, *IEEE J. Quantum Electron.* **30**, 939 (1994).

[33] M. Okai, M. Suzuki, and M. Aoki, *IEEE J. Sel. Topics Quantum Electron.* **1**, 461 (1995).

[34] J.-I. Shim, H. Olesen, H. Yamazaki, M. Yamaguchi, and M. Kitamura, *IEEE J. Sel. Topics Quantum Electron.* **1**, 516 (1995).

[35] M. Ferreira, *IEEE J. Quantum Electron.* **32**, 851 (1996).

[36] F. M. Gardner, *Phaselock Techniques*, Wiley, New York, 1985.

[37] M. I. Irshid and S. Y. Helo, *J. Opt. Commun.* **15**, 133 (1994).

[38] J. W. Goodman, *Statistical Optics*, Wiley, New York, 1985.

[39] S. O. Rice, *Bell Syst. Tech. J.* **23**, 282 (1944); **24**, 96 (1945).

[40] J. I. Marcum, *IRE Trans. Inform. Theory* **6**, 259 (1960).

[41] K. Kikuchi, T. Okoshi, M. Nagamatsu, and H. Henmi, *J. Lightwave Technol.* **2**, 1024 (1984).

[42] G. Nicholson, *Electron. Lett.* **20**, 1005 (1984).

[43] L. G. Kazovsky, *J. Lightwave Technol.* **3**, 1238 (1985); *J. Opt. Commun.* **7**, 66 (1986); *J. Lightwave Technol.* **4**, 415 (1986).

[44] B. Glance, *J. Lightwave Technol.* **4**, 228 (1986).

[45] I. Garrett and G. Jacobsen, *Electron. Lett.* **21**, 280 (1985); *J. Lightwave Technol.* **4**, 323 (1986); **5**, 551 (1987).

[46] T. G. Hodgkinson, *J. Lightwave Technol.* **5**, 573 (1987).

[47] G. Jacobsen and I. Garrett, *IEE Proc.* **134**, Pt. J, 303 (1987); *J. Lightwave Technol.* **5**, 478 (1987).

[48] L. G. Kazovsky, P. Meissner, and E. Patzak, *J. Lightwave Technol.* **5**, 770 (1987).

[49] G. J. Foschini, L. J. Greenstein, and G. Vannuchi, *IEEE Trans. Commun.* **36**, 306 (1988).

[50] L. J. Greenstein, G. Vannuchi, and G. J. Foschini, *IEEE Trans. Commun.* **37**, 405 (1989).

[51] I. Garrett, D. J. Bond, J. B. Waite, D. S. L. Lettis, and G. Jacobsen, *J. Lightwave Technol.* **8**, 329 (1990).

[52] L. G. Kazovsky and O. K. Tonguz, *J. Lightwave Technol.* **8**, 338 (1990).

[53] R. Corvaja and G. L. Pierobon, *J. Lightwave Technol.* **12**, 519 (1994).

[54] R. Corvaja, G. L. Pierobon, and L. Tomba, *J. Lightwave Technol.* **12**, 1665 (1994).

[55] H. Ghafouri-Shiraz, Y. H. Heng, and T. Aruga, *Microwave Opt. Tech. Lett.* **11**, 14 (1996).

[56] F. Kano, T. Yamanaka, N. Yamamoto, H. Mawatan, Y. Tohmori, and Y. Yoshikuni, *IEEE J. Quantum Electron.* **30**, 533 (1994).

[57] T. G. Hodgkinson, R. A. Harmon, and D.W. Smith, *Electron. Lett.* **21**, 867 (1985).

[58] A. W. Davis and S. Wright, *Electron. Lett.* **22**, 9 (1986).

[59] A. W. Davis, M. J. Pettitt, J. P. King, and S. Wright, *J. Lightwave Technol.* **5**, 561 (1987).

[60] L. G. Kazovsky, R. Welter, A. F. Elrefaie, and W. Sessa, *J. Lightwave Technol.* **6**, 1527 (1988).

[61] L. G. Kazovsky, *J. Lightwave Technol.* **7**, 279 (1989).

[62] C.-L. Ho and H.-N. Wang, *J. Lightwave Technol.* **13**, 971 (1995).

[63] L. G. Kazovsky, A. F. Elrefaie, R. Welter, P. Crepso, J. Gimlett, and R. W. Smith, *Electron. Lett.* **23**, 871 (1987).

[64] A. F. Elrefaie, D. A. Atlas, L. G. Kazovsky, and R. E. Wagner, *Electron. Lett.* **24**, 158 (1988).

[65] R. Gross, P. Meissner, and E. Patzak, *J. Lightwave Technol.* **6**, 521 (1988).

[66] W. H. C. de Krom, *J. Lightwave Technol.* **9**, 641 (1991).

[67] Y.-H. Lee, C.-C. Kuo, and H.-W. Tsao, *Microwave Opt. Tech. Lett.* **5**, 168 (1992).

[68] H. Van de Stadt, *Astron. Astrophys.* **36**, 341 (1974).

[69] G. L. Abbas, V. W. Chan, and T. K. Yee, *J. Lightwave Technol.* **3**, 1110 (1985).

[70] B. L. Kasper, C. A. Burrus, J. R. Talman, and K. L. Hall, *Electron. Lett.* **22**, 413 (1986).

[71] S. B. Alexander, *J. Lightwave Technol.* **5**, 523 (1987).

[72] T. Okoshi, *J. Lightwave Technol.* **3**, 1232 (1985).

[73] T. G. Hodgkinson, R. A. Harmon, and D. W. Smith, *Electron. Lett.* **23**, 513 (1987).

[74] M. W. Maeda and D. A. Smith, *Electron. Lett.* **27**, 10 (1991).

[75] P. Poggiolini and S. Benedetto, *IEEE Trans. Commun.* **42**, 2105 (1994).

[76] S. Benedetto and P. Poggiolini, *IEEE Trans. Commun.* **42**, 2915 (1994).

[77] G. P. Agrawal, *Quantum Semiclass. Opt.* **8**, 383 (1996).

[78] B. Glance, *J. Lightwave Technol.* **5**, 274 (1987).

[79] D. Kreit and R. C. Youngquist, *Electron. Lett.* **23**, 168 (1987).

[80] T. Okoshi and Y. C. Cheng, *Electron. Lett.* **23**, 377 (1987).

[81] A. D. Kersey, A. M. Yurek, A. Dandridge, and J. F. Weller, *Electron. Lett.* **23**, 924 (1987).

[82] S. Ryu, S. Yamamoto, and K. Mochizuki, *Electron. Lett.* **23**, 1382 (1987).

[83] M. Kavehrad and B. Glance, *J. Lightwave Technol.* **6**, 1386 (1988).

[84] I. M. I. Habbab and L. J. Cimini, *J. Lightwave Technol.* **6**, 1537 (1988).

[85] B. Enning, R. S. Vodhanel, E. Dietrich, E. Patzak, P. Meissner, and G. Wenke, *J. Lightwave Technol.* **7**, 459 (1989).

[86] N. G. Walker and G. R. Walker, *Electron. Lett.* **3**, 290 (1987); *J. Lightwave Technol.* **8**, 438 (1990).

[87] R. Noé, H. Heidrich, and D. Hoffman, *J. Lightwave Technol.* **6**, 1199 (1988).

[88] T. Pikaar, K. Van Bochove, A. Van Rooyen, H. Frankena, and F. Groen, *J. Lightwave Technol.* **7**, 1982 (1989).

[89] H. W. Tsao, J. Wu, S. C. Yang, and Y. H. Lee, *J. Lightwave Technol.* **8**, 385 (1990).

[90] T. Imai, *J. Lightwave Technol.* **9**, 650 (1991).

[91] Y. H. Cheng, T. Okoshi, and O. Ishida, *J. Lightwave Technol.* **7**, 368 (1989).

[92] R. Langenhorst, W. Pieper, M. Eiselt, D. Rhode, and H. G. Weber, *IEEE Photon. Technol. Lett.* **3**, 80 (1991).

[93] A. R. Chraplyvy, R. W. Tkaeh, L. L. Buhl, and R. C. Alferness, *Electron. Lett.* **22**, 409 (1986).

[94] K. Tajima, *J. Lightwave Technol.* **6**, 322 (1988).

[95] A. A. Elrefaie, R. E. Wagner, D. A. Atlas, and D. G. Daut, *J. Lightwave Technol.* **6**, 704 (1988).

[96] K. Nosu and K. Iwashita, *J. Lightwave Technol.* **6**, 686 (1988).

[97] B. Pal, R. Gangopadhyay, and G. Prati, *J. Lightwave Technol.* **18**, 530 (2000).

[98] R. G. Priest and T. G. Giallorenzi, *Opt. Lett.* **12**, 622 (1987).

[99] G. S. Mosehytz and P. Horn, *Active Filter Design Handbook*, Wiley, New York, 1981.

[100] N. Takachio and K. Iwashita, *Electron. Lett.* **24**, 108 (1988).

[101] A. H. Gnauck, L. J. Cimini, Jr., J. Stone, and L. W. Stulz, *IEEE Photon. Technol. Lett.* **2**, 585 (1990).

[102] M. S. Kao and J. S. Wu, *J. Lightwave Technol.* **11**, 303 (1993).

[103] T. Ono, S. Yamazaki, H. Shimizu, and K. Emura, *J. Lightwave Technol.* **12**, 891 (1994).

[104] C. De Angelis, A. Galtarossa, C. Campanile, and F. Matera, *J. Opt. Commun.* **16**, 173 (1995).

[105] L. Kazovsky, *Electron. Lett.* **24**, 522 (1988).

[106] F. N. Farokhrooz and J. P. Raina, *Int. J. Optoelectron.* **10**, 115 (1995).

[107] G. P. Agrawal, *Nonlinear Fiber Optics*, 3rd ed., Academic Press, San Diego, CA, 2001.

[108] Y. Aoki, K. Tajima, and I. Mito, *J. Lightwave Technol.* **6**, 710 (1988).

[109] T. Sugie, *J. Lightwave Technol.* **9**, 1145 (1991); *Opt. Quantum Electron.* **27**, 643 (1995).

[110] N. Okhawa and Y. Hayashi, *J. Lightwave Technol.* **13**, 914 (1995).

[111] N. Takachio, S. Norimatsu, K. Iwashita, and K. Yonenaga, *J. Lightwave Technol.* **12**, 247 (1994).

[112] K. Kuboki and Y. Uchida, *IEICE Trans. Commun.* **E78B**, 654 (1995).

[113] D. G. Schadt and T. D. Stephens, *J. Lightwave Technol.* **9**, 1187 (1991).

[114] R. A. Linke, B. L. Kasper, N. A. Olsson, and R. C. Alferness, *Electron. Lett.* **22**, 30 (1986).

[115] N. A. Olsson, M. G. Oberg, L. A. Koszi, and G. Przyblek, *Electron. Lett.* **24**, 36 (1988).

[116] A. H. Gnauck, K. C. Reichmann, J. M. Kahn, S. K. Korotky, J. J. Veselka, and T. L. Koch, *IEEE Photon. Technol. Lett.* **2**, 908 (1990).

[117] R. Noé, E. Meissner, B. Borchert, and H. Rodler, *IEEE Photon. Technol. Lett.* **4**, 1151 (1992).

[118] J. J. O. Pires and J. R. F. de Rocha, *J. Lightwave Technol.* **10**, 1722 (1992).

[119] E. A. Swanson, J. C. Livas, and R. S. Bondurant, *IEEE Photon. Technol. Lett.* **6**, 263 (1994).

[120] I. Bar-David, *IEE Proc.* **141**, Pt. J, 38 (1994).

[121] G. Jacobsen, *Electron. Lett.* **28**, 254 (1992); *IEEE Photon. Technol. Lett.* **5**, 105 (1993).

[122] A. Murat, P. A. Humblet, and J. S. Young, *J. Lightwave Technol.* **11**, 290 (1993).

[123] C. P. Kaiser, P. J. Smith, and M. Shafi, *J. Lightwave Technol.* **13**, 525 (1995).

[124] P. J. Smith, M. Shafi, and C. P. Kaiser, *IEEE J. Sel. Areas Commun.* **13**, 557 (1995).

[125] S. R. Chinn, D. M. Bronson, and J. C. Livas, *J. Lightwave Technol.* **14**, 370 (1996).

[126] S. Saito, T. Imai, and T. Ito, *J. Lightwave Technol.* **9**, 161 (1991).

[127] E. Iannone, F. S. Locati, F. Matera, M. Romagnoli, and M. Settembre, *Electron. Lett.* **28**, 645 (1992); *J. Lightwave Technol.* **11**, 1478 (1993).

[128] J. Farre, G. Jacobsen, E. Bodtker, and K. Stubkjaer, *J. Opt. Commun.* **14**, 65 (1993).

[129] F. Matera, M. Settembre, B. Daino, and G. De Marchis, *Opt. Commun.* **119**, 289 (1995).

[130] J. K. Rhee, D. Chowdhury, K. S. Cheng, and U. Gliese, *IEEE Photon. Technol. Lett.* **12**, 1627 (2000).

[131] A. Schoepflin, *Electron. Lett.* **26** 255 (1990).

[132] L. G. Kazovsky and D. A. Atlas, *IEEE Photon. Technol. Lett.* **2**, 431 (1990).

[133] L. G. Kazovsky, D. A. Atlas, and R. W. Smith, *IEEE Photon. Technol. Lett.* **2**, 589 (1990); D. A. Atlas and L. G. Kazovsky, *Electron. Lett.* **26**, 1032 (1990).

[134] I. B. Djordjevic and M. C. Stefanovic, *J. Lightwave Technol.* **17**, 2470 (1999).

[135] T. Chikama, S. Watanabe, T. Naito, H. Onaka, T. Kiyonaga, Y. Onada, H. Miyata, M. Suyuma, M. Seimo, and H. Kuwahara, *J. Lightwave Technol.* **8**, 309 (1990).

[136] D. J. Maylon, D. W. Smith, and R. Wyatt, *Electron. Lett.* **22**, 421 (1986).

[137] L. G. Kazovsky, *J. Lightwave Technol.* **4**, 182 (1986); L. G. Kazovsky and D. A. Atlas, *IEEE Photon. Technol. Lett.* **1**, 395 (1989).

[138] J. M. Kahn, B. L. Kasper, and K. J. Pollock, *Electron. Lett.* **25**, 626 (1989).

[139] S. Norimatsu, K. Iwashita, and K. Sato, *IEEE Photon. Technol. Lett.* **2**, 374 (1990).

[140] C.-H. Shin and M. Ohtsu, *IEEE Photon. Technol. Lett.* **2**, 297 (1990); *IEEE J. Quantum Electron.* **29**, 374 (1991).

[141] S. Norimatsu, *J. Lightwave Technol.* **13**, 2183 (1995).

[142] J. M. Kahn, *IEEE Photon. Technol. Lett.* **1**, 340 (1989).

[143] A. Schoepflin, S. Kugelmeier, G. Gottwald, D. Feicio, and G. Fischer, *Electron. Lett.* **26**, 395 (1990).

[144] D. A. Atlas and L. G. Kazovsky, *IEEE Photon. Technol. Lett.* **2**, 367 (1990).

[145] J. M. Kahn, A. H. Gnauck, J. J. Veselka, S. K. Korotky, and B. L. Kasper, *IEEE Photon. Technol. Lett.* **2**, 285 (1990).

[146] S. Norimatsu, K. Iwashita, and K. Noguchi, *Electron. Lett.* **26**, 648 (1990).

[147] B. Wandernorth, *Electron. Lett.* **27**, 1693 (1991); **28**, 387 (1992).

[148] S. L. Miller, *IEE Proc.* **139**, Pt. J, 215 (1992).

[149] K. Yonenaga and N. Takachio, *IEEE Photon. Technol. Lett.* **5**, 949 (1993).

[150] K. Yonenaga and S. Norimatsu, *IEEE Photon. Technol. Lett.* **7**, 929 (1995).

[151] C.-C. Huang and L. K. Wang, *J. Lightwave Technol.* **13**, 1963 (1995).

[152] S. C. Huang and L. K. Wang, *J. Lightwave Technol.* **14**, 661 (1996).

[153] F. N. Farokhrooz, H. A. Tafti, and J. P. Raina, *Opt. Quantum Electron.* **19**, 777 (1997).

[154] B. N. Biswas and A. Bhattacharya, *IETE J. Res.* **45** 311 (1999).

[155] S. Ryu, S. Yamamoto, Y. Namahira, K. Mochizuki, and M. Wakabayashi, *Electron. Lett.* **24**, 399 (1988).

[156] M. C. Brain, M. J. Creaner, R. Steele, N. G. Walker, G. R. Walker, J. Mellis, S. Al-Chalabi, J. Davidson, M. Rutherford, and I. C. Sturgess, *J. Lightwave Technol.* **8**, 423 (1990).

[157] T. W. Cline, J. M. P. Delavaux, N. K. Dutta, P. V. Eijk, C. Y. Kuo, B. Owen, Y. K. Park, T. C. Pleiss, R. S. Riggs, R. E. Tench, Y. Twu, L. D. Tzeng, and E. J. Wagner, *IEEE Photon. Technol. Lett.* **2**, 425 (1990).

[158] T. Imai, Y. Hayashi, N. Ohkawa, T. Sugie, Y. Ichihashi, and T. Ito, *Electron. Lett.* **26**, 1407 (1990).

[159] S. Ryu, S. Yamamoto, Y. Namihira, K. Mochizuki, and H. Wakabayashi, *J. Lightwave Technol.* **9**, 675 (1991).

[160] T. Imai, N. Ohkawa, Y. Hayashi, and Y. Ichihashi, *J. Lightwave Technol.* **9**, 761 (1991).

[161] E. J. Bachus, T. Almeida, P. Demeester, G. Depovere, A. Ebberg, M. R. Ferreira, G.-D. Khoe, O. Koning, R. Marsden, J. Rawsthorne, and N. Wauters, *J. Lightwave Technol.* **14**, 1309 (1996).

[162] T. L. Koch and U. Koren, *IEEE J. Quantum Electron.* **27**, 641 (1991).

[163] D. Trommer, A. Umbach, W. Passenberg, and G. Unterbörsch, *IEEE Photon. Technol. Lett.* **5**, 1038 (1993).

[164] F. Ghirardi, J. Brandon, F. Huet, M. Carré, A. Bruno, and A. Carenco, *IEEE Photon. Technol. Lett.* **6**, 814 (1994).

[165] U. Hilbk, T. Hermes, P. Meissner, C. Jacumeit, R. Stentel, and G. Unterbörsch, *IEEE Photon. Technol. Lett.* **7**, 129 (1995).

[166] F. Ghirardi, A. Bruno, B. Mersali, J. Brandon, L. Giraudet, A. Scavennec, and A. Carenco, *J. Lightwave Technol.* **13**, 1536 (1995).

[167] E. C. M. Pennings, D. Schouten, G.-D. Khoe, R. A. van Gils, and G. F. G. Depovere, *J. Lightwave Technol.* **13**, 1985 (1995).

[168] M. Hamacher, D. Trommer, K. Li, H. Schroeter-Janssen, W. Rehbein, and H. Heidrich, *IEEE Photon. Technol. Lett.* **8**, 75 (1996).

[169] R. Kaiser, D. Trommer, H. Heidrich, F. Fidorra, and M. Hamacher, *Opt. Quantum Electron.* **28**, 565 (1996).

[170] L. M. Wang, F. S. Choa, J. H. Chen, Y. J. Chai, M. H. Shih, *J. Lightwave Technol.* **17**, 1724 (1999).

Appendix A

System of Units

The international system of units (known as the SI, short for *Système International*) is used in this book. The three fundamental units in the SI are meter (m), second (s), and kilogram (kg). A prefix can be added to each of them to change its magnitude by a multiple of 10. Mass units are rarely required in this book. Most common measures of distance used are km (10^3 m) and Mm (10^6 m). On the other hand, common time measures are ns (10^{-9} s), ps (10^{-12} s), and fs (10^{-15} s). Other common units in this book are Watt (W) for optical power and W/m^2 for optical intensity. They can be related to the fundamental units through energy because optical power represents the rate of energy flow (1 W = 1 J/s). The energy can be expressed in several other ways using $E = h\nu = k_B T = mc^2$, where h is the Planck constant, k_B is the Boltzmann constant, and c is the speed of light. The frequency ν is expressed in hertz (1 Hz = 1 s^{-1}). Of course, because of the large frequencies associated with the optical waves, most frequencies in this book are expressed in GHz or THz.

In the design of optical communication systems the optical power can vary over several orders of magnitude as the signal travels from the transmitter to the receiver. Such large variations are handled most conveniently using decibel units, abbreviated dB, commonly used by engineers in many different fields. Any ratio R can be converted into decibels by using the general definition

$$R \text{ (in dB)} = 10 \log_{10} R. \tag{A.1}$$

The logarithmic nature of the decibel allows a large ratio to be expressed as a much smaller number. For example, 10^9 and 10^{-9} correspond to 90 dB and −90 dB, respectively. Since $R = 1$ corresponds to 0 dB, ratios smaller than 1 are negative in the decibel system. Furthermore, negative ratios cannot be written using decibel units.

The most common use of the decibel scale occurs for power ratios. For instance, the signal-to-noise ratio (SNR) of an optical or electrical signal is given by

$$\text{SNR} = 10 \log_{10}(P_S/P_N), \tag{A.2}$$

where P_S and P_N are the signal and noise powers, respectively. The fiber loss can also be expressed in decibel units by noting that the loss corresponds to a decrease in the optical power during transmission and thus can be expressed as a power ratio. For example, if a 1-mW signal reduces to 1 μW after transmission over 100 km of fiber, the 30-dB loss over the entire fiber span translates into a loss of 0.3 dB/km. The same technique can be used to define the insertion loss of any component. For instance, a 1-dB loss of a fiber connector implies that the optical power is reduced by 1 dB (by about 20%) when the signal passes through the connector. The bandwidth of an optical filter is defined at the 3-dB point, corresponding to 50% reduction in the signal power. The modulation bandwidth of ight-emitting diodes (LEDs) in Section 3.2 and of semiconductor lasers in Section 3.5 is also defined at the 3-dB point, at which the modulated powers drops by 50%.

Since the losses of all components in a fiber-optic communication systems are expressed in dB, it is useful to express the transmitted and received powers also by using a decibel scale. This is achieved by using a derived unit, denoted as dBm and defined as

$$\text{power (in dBm)} = 10 \log_{10}\left(\frac{\text{power}}{1 \text{ mW}}\right), \tag{A.3}$$

where the reference level of 1 mW is chosen simply because typical values of the transmitted power are in that range (the letter m in dBm is a reminder of the 1-mW reference level). In this decibel scale for the absolute power, 1 mW corresponds to 0 dBm, whereas powers below 1 mW are expressed as negative numbers. For example, a 10-μW power corresponds to -20 dBm. The advantage of decibel units becomes clear when the power budget of lightwave systems is considered in Chapter 5. Because of the logarithmic nature of the decibel scale, the power budget can be made simply by subtracting various losses from the transmitter power expressed in dBm units.

Appendix B

Acronyms

Each scientific field has its own jargon, and the field of optical communications is not an exception. Although an attempt was made to avoid extensive use of acronyms, many still appear throughout the book. Each acronym is defined the first time it appears in a chapter so that the reader does not have to search the entire text to find its meaning. As a further help, we list all acronyms here, in alphabetical order.

ac	alternating current
AM	amplitude modulation
AON	all-optical network
APD	avalanche photodiode
ASE	amplified spontaneous emission
ASK	amplitude-shift keying
ATM	asynchronous transfer mode
AWG	arrayed-waveguide grating
BER	bit-error rate
BH	buried heterostructure
BPF	bandpass filter
CATV	common-antenna (cable) television
CDM	code-division multiplexing
CDMA	code-division multiple access
CNR	carrier-to-noise ratio
CPFSK	continuous-phase frequency-shift keying
CRZ	chirped return-to-zero
CSMA	carrier-sense multiple access
CSO	composite second-order
CVD	chemical vapor deposition
CW	continuous wave
CTB	composite triple beat
DBR	distributed Bragg reflector
dc	direct current
DCF	dispersion-compensating fiber

DDF	dispersion-decreasing fiber
DFB	distributed feedback
DGD	differential group delay
DIP	dual in-line package
DM	dispersion-managed
DPSK	differential phase-shift keying
EDFA	erbium-doped fiber amplifier
FDDI	fiber distributed data interface
FDM	frequency-division multiplexing
FET	field-effect transistor
FM	frequency modulation
FP	Fabry–Perot
FSK	frequency-shift keying
FWHM	full-width at half-maximum
FWM	four-wave mixing
GVD	group-velocity dispersion
HBT	heterojunction-bipolar transistor
HDTV	high-definition television
HEMT	high-electron-mobility transistor
HFC	hybrid fiber-coaxial
HIPPI	high-performance parallel interface
IC	integrated circuit
IF	intermediate frequency
IMD	intermodulation distortion
IM/DD	intensity modulation with direct detection
IMP	intermodulation product
IP	Internet protocol
ISDN	integrated services digital network
ISI	intersymbol interference
ITU	International Telecommunication Union
LAN	local-area network
LEAF	large effective-area fiber
LED	light-emitting diode
LO	local oscillator
LPE	liquid-phase epitaxy
LPF	low-pass filter
MAN	metropolitan-area network
MBE	molecular-beam epitaxy
MCVD	modified chemical vapor deposition
MEMS	micro-electro-mechanical system
MOCVD	metal-organic chemical vapor deposition
MONET	multiwavelength optical network
MPEG	motion-picture entertainment group
MPN	mode-partition noise
MQW	multiquantum well
MSK	minimum-shift keying

MSM	metal–semiconductor–metal
MSR	mode-suppression ratio
MTTF	mean time to failure
MZ	Mach–Zehnder
NA	numerical aperture
NEP	noise-equivalent power
NLS	nonlinear Schrödinger
NOLM	nonlinear optical-loop mirror
NRZ	nonreturn to zero
NSE	nonlinear Schrödinger equation
NSDSF	nonzero-dispersion-shifted fiber
OC	optical carrier
OEIC	opto-electronic integrated circuit
OOK	on–off keying
OPC	optical phase conjugation
OTDM	optical time-division multiplexing
OVD	outside-vapor deposition
OXC	optical cross-connect
PCM	pulse-code modulation
PDF	probability density function
PDM	polarization-division multiplexing
$P\text{--}I$	power–current
PIC	photonic integrated circuit
PM	phase modulation
PMD	polarization-mode dispersion
PON	passive optical network
PSK	phase-shift keying
PSP	principal state of polarization
QAM	quadrature amplitude modulation
RDF	reverse-dispersion fiber
RF	radio frequency
RIN	relative intensity noise
RMS	root-mean-square
RZ	return to zero
SAGCM	separate absorption, grading, charge, and multiplication
SAGM	separate absorption, grading, and multiplication
SAM	separate absorption and multiplication
SBS	stimulated Brillouin scattering
SCM	subcarrier multiplexing
SDH	synchronous digital hierarchy
SI	*Système International*
SLM	single longitudinal mode
SNR	signal-to-noise ratio
SOA	semiconductor optical amplifier
SONET	synchronized optical network
SPM	self-phase modulation

SRS	stimulated Raman scattering
SSFS	soliton self-frequency shift
STM	synchronous transport module
STS	synchronous transport signal
TCP	transmission control protocol
TDM	time-division multiplexing
TE	transverse electric
TM	transverse magnetic
TOAD	terahertz optical asymmetric demultiplexer
TOD	third-order dispersion
TW	traveling wave
VAD	vapor-axial deposition
VCSEL	vertical-cavity surface-emitting laser
VPE	vapor-phase epitaxy
VSB	vestigial sideband
WAN	wide-area network
WDM	wavelength-division multiplexing
WDMA	wavelength-division multiple access
WGR	waveguide-grating router
XPM	cross-phase modulation
YAG	yttrium aluminium garnet
YIG	yttrium iron garnet
ZDWL	zero-dispersion wavelength

Appendix C

General Formula for Pulse Broadening

The discussion of pulse broadening in Section 2.4 assumes the Gaussian-shape pulses and includes dispersive effects only up to the third order. In this appendix, a general formula is derived that can be used for pulses of arbitrary shape. Moreover, it makes no assumption about the dispersive properties of the fiber and can be used to include dispersion to any order. The basic idea behind the derivation consists of the observation that the pulse spectrum does not change in a linear dispersive medium irrespective of what happens to the pulse shape. It is thus better to calculate the changes in the pulse width in the spectral domain.

For pulses of arbitrary shapes, a measure of the pulse width is provided by the quantity $\sigma^2 = \langle t^2 \rangle - \langle t \rangle^2$, where the first and second moments are calculated using the pulse shape as indicated in Eq. (2.4.21). These moments can also be defined in terms of the pulse spectrum as

$$\langle t \rangle = \int_{-\infty}^{\infty} t |A(z,t)|^2 \, dt \equiv \frac{-i}{2\pi} \int_{-\infty}^{\infty} \tilde{A}^*(z,\omega) \tilde{A}_\omega(z,\omega) \, d\omega, \qquad (C.1)$$

$$\langle t^2 \rangle = \int_{-\infty}^{\infty} t^2 |A(z,t)|^2 \, dt \equiv \frac{1}{2\pi} \int_{-\infty}^{\infty} |\tilde{A}_\omega(z,\omega)|^2 \, d\omega, \qquad (C.2)$$

where $\tilde{A}(z,\omega)$ is the Fourier transform of $A(z,t)$ and the subscript ω denotes partial derivative with respect to ω. For simplicity of discussion, we normalize A and $\tilde{A}$ such that

$$\int_{-\infty}^{\infty} |A(z,t)|^2 \, dt = \frac{1}{2\pi} \int_{-\infty}^{\infty} |\tilde{A}(z,\omega)|^2 \, d\omega = 1. \qquad (C.3)$$

As discussed in Section 2.4, when nonlinear effects are negligible, different spectral components propagate inside the fiber according to the simple relation

$$\tilde{A}(z,\omega) = \tilde{A}(0,\omega) \exp(i\beta z) = [S(\omega)e^{i\theta}] \exp(i\beta z), \qquad (C.4)$$

where $S(\omega)$ represents the spectrum of the input pulse and $\theta(\omega)$ accounts for the effects of input chirp. As seen in Eq. (2.4.13), the spectrum of chirped pulses acquires a frequency-dependent phase. The propagation constant β depends on frequency because of dispersion. It can also depend on z when dispersion management is used or when fiber parameters such as the core diameter are not uniform along the fiber.

If we substitute Eq. (C.4) in Eqs. (C.1) and (C.2), perform the derivatives as indicated, and calculate $\sigma^2 = \langle t^2 \rangle - \langle t \rangle^2$, we obtain

$$\sigma^2 = \sigma_0^2 + [\langle \tau^2 \rangle - \langle \tau \rangle^2] + 2[\langle \tau \theta_\omega \rangle - \langle \tau \rangle \langle \theta_\omega \rangle], \tag{C.5}$$

where the angle brackets now denote average over the input pulse spectrum such that

$$\langle f \rangle = \frac{1}{2\pi} \int_{-\infty}^{\infty} f(\omega) |S(\omega)|^2 \, d\omega. \tag{C.6}$$

In Eq. (C.5), σ_0 is the root-mean-square (RMS) width of input pulses, $\theta_\omega = d\theta/d\omega$, and τ is the group delay defined as

$$\tau(\omega) = \int_0^L \frac{\partial \beta(z, \omega)}{\partial \omega} \, dz \tag{C.7}$$

for a fiber of length L. Equation (C.5) can be used for pulses of arbitrary shape, width, and chirp. It makes no assumption about the form of $\beta(z, \omega)$ and thus can be used for dispersion-managed fiber links containing fibers with arbitrary dispersion characteristics.

As a simple application of Eq. (C.5), one can use it to derive Eq. (2.4.22). Assuming uniform dispersion and expanding $\beta(z, \omega)$ to third-order in ω, the group delay is given by

$$\tau(\omega) = (\beta_1 + \beta_2 \omega + \tfrac{1}{2}\beta_3 \omega^2) L. \tag{C.8}$$

For a chirped Gaussian pulse, Eq. (2.4.13) provides the following expressions for S and θ:

$$S(\omega) = \sqrt{\frac{4\pi T_0^2}{1 + C^2}} \exp\left[-\frac{\omega^2 T_0^2}{2(1 + C^2)} \right], \quad \theta(\omega) = \frac{C\omega^2 T_0^2}{2(1 + C^2)} - \tan^{-1} C. \tag{C.9}$$

The averages in Eq. (C.5) can be performed analytically using Eqs. (C.8) and (C.9) and result in Eq. (2.4.22).

As another application of Eq. (C.5), consider the derivation of Eq. (2.4.23) that includes the effects of a wide source spectrum. For such a pulse, the input field can be written as $A(0,t) = A_0(t) f(t)$, where $f(t)$ represents the pulse shape and $A_0(t)$ is fluctuating because of the partially coherent nature of the source. The spectrum $S(\omega)$ now becomes a convolution of the pulse spectrum and the source spectrum such that

$$S(\omega) = \frac{1}{2\pi} \int_{-\infty}^{\infty} S_p(\omega - \omega_1) F(\omega_1) \, d\omega_1, \tag{C.10}$$

where S_p is the pulse spectrum and $F(\omega_s)$ is the fluctuating field spectral component at the source with the correlation function of the form

$$\langle F^*(\omega_1) F(\omega_2) \rangle_s = G(\omega_1) \delta(\omega_1 - \omega_2). \tag{C.11}$$

The quantity $G(\omega)$ represents the source spectrum. The subscript s in Eq. (C.11) is a reminder that the angle brackets now denote an ensemble average over the field fluctuations.

The moments $\langle t \rangle$ and $\langle t^2 \rangle$ are now replaced by $\langle\langle t \rangle\rangle_s$ and $\langle\langle t^2 \rangle\rangle_s$ where the outer angle brackets stand for the ensemble average over field fluctuations. Both of them can be calculated in the special case in which the source spectrum is assumed to be Gaussian, i.e.,

$$G(\omega) = \frac{1}{\sigma_\omega \sqrt{2\pi}} \exp\left(-\frac{\omega^2}{2\sigma_\omega^2}\right), \tag{C.12}$$

where σ_ω is the RMS spectral width of the source. For example,

$$\langle\langle t \rangle\rangle_s = \int_{-\infty}^{\infty} \tau(\omega) \langle |S(\omega)|^2 \rangle_s \, d\omega - i \int_{-\infty}^{\infty} \langle S^*(\omega) S_\omega(\omega) \rangle_s \, d\omega$$

$$= L \iint_{-\infty}^{\infty} (\beta_1 + \beta_2\omega + \tfrac{1}{2}\beta_3\omega^2) |S_p(\omega - \omega_1)|^2 G(\omega_1) \, d\omega_1 \, d\omega \tag{C.13}$$

Since both the pulse spectrum and the source spectrum are assumed to be Gaussian, the integral over ω_1 can be performed first, resulting in another Gaussian spectrum. The integral over ω is then straightforward in Eq. (C.13) and yields

$$\langle\langle t \rangle\rangle_s = L\left[\beta_1 + \frac{\beta_3}{8\sigma_0^2}(1 + C^2 + V_\omega^2)\right], \tag{C.14}$$

where $V_\omega = 2\sigma_\omega\sigma_0$. Repeating the same procedure for $\langle\langle t^2 \rangle\rangle_s$, we recover Eq. (2.4.13) for the ratio σ/σ_0.

Appendix D

Ultimate System Capacity

With the advent of wavelength-division multiplexing (WDM) technology, lightwave systems with a capacity of more than 1 Tb/s have become available commercially. Moreover, system capacities in excess of 10 Tb/s have been demonstrated in several laboratory experiments. Every communication channel has a finite bandwidth, and optical fibers are no exception to this general rule. One may thus ask: What is the ultimate capacity of a fiber-optic communication system? This appendix focuses on this question.

The performance of any communication system is ultimately limited by the noise of the received signal. This limitation can be stated more formally by using the concept of *channel capacity* introduced within the framework of information theory [1]. It turns out that a maximum possible bit rate exists for error-free transmission of a binary digital signal in the presence of Gaussian noise. This rate is called the channel capacity. More specifically, the maximum capacity of a noisy communication channel is given by

$$C_s = \Delta f_{\text{ch}} \log_2(1 + S/N), \tag{D.1}$$

where Δf_{ch} is the channel bandwidth, S is the average signal power, and N is the average noise power. Equation (D.1) is valid for a linear channel with additive noise. It shows that the system capacity (i.e., the bit rate) can exceed the bandwidth of the transmission channel if the noise level is low enough to maintain a high signal-to-noise ratio (SNR). In fact, it is common to define the spectral efficiency of a communication channel as $\eta_s = C_s/(\Delta f_{\text{ch}})$ that is a measure of bits transmitted per second per unit bandwidth and is measured in units of (b/s)/Hz. For a SNR of >30 dB, η_s exceeds 10 (b/s)/Hz.

Equation (D.1) does not always apply to fiber-optic communication systems because of the nonlinear effects occurring inside optical fibers. It can nonetheless be used to provide an upper limit on the system capacity. The bandwidth Δf_{ch} of modern lightwave systems is limited by the bandwidth of optical amplifiers and is below 10 Tb/s (80 nm) even when both the C and L bands are used simultaneously. With the advent of new kinds of fibers and amplification techniques, one may expect that eventually Δf_{ch} will approach 50 THz by using the entire low-loss region extending from 1.25 to1.65 μm. The SNR should exceed 100 in practice to realize a bit-error rate below 10^{-9}. Using these values in Eq. (D.1), the maximum system capacity is close to

350 Tb/s assuming that the optical fiber acts as a linear channel. The highest capacity realized for WDM systems in a 2001 experiment was 10.9 Tb/s [2]. The most limiting factor is the spectral efficiency; it is typically limited to below 0.8 (b/s)/Hz because of the use of the binary amplitude-shift keying (ASK) format (on–off keying). The use of phase-shift and frequency-shift keying (PSK and FSK) formats in combination with advanced coding techniques is likely to improve the spectral efficiency.

The impact of the nonlinear effects and the amplifier noise on the ultimate capacity of lightwave systems has been studied in several recent publications [3]–[8]. Because the amplifier noise builds up in long-haul systems, the maximum channel capacity depends on the transmission distance and is generally larger for shorter distances. It also depends on the number of channels transmitted. The most important conclusion is that the maximum spectral efficiency is limited to 3–4 (b/s)/Hz because of the nonlinear effects. As a result, the system capacity may remain below 150 Tb/s even under the best operating conditions.

References

[1] C. E. Shannon, *Proc. IRE* **37**, 10 (1949).

[2] K. Fukuchi, T. Kasamatsu, M. Morie, R. Ohhira, T. Ito, K. Sekiya, D. Ogasahara, and T. Ono, Paper PD24, *Proc. Optical Fiber Commun. Conf.*, Optical Society of America, Washington, DC, 2001.

[3] P. P. Mitra and J. B. Stark, *Nature* **411**, 1027 (2001).

[4] A. Mecozzi, *IEEE Photon. Technol. Lett.* **13**, 1029 (2001).

[5] J. Tang, *J. Lightwave Technol.* **19**, 1104 (2001).

[6] J. Tang, *J. Lightwave Technol.* **19**, 1110 (2001).

[7] J. B. Stark, P. P. Mitra, and A. Sengupta, *Opt. Fiber Technol.* **7**, 275 (2001).

[8] E. E. Narimanov and P. P. Mitra, *J. Lightwave Technol.* **20**, 530 (2002).

Appendix E

Software Package

The back cover of the book contains a software package for designing fiber-optic communication system on a compact disk (CD) provided by the Optiwave Corporation (Website: www.optiwave.com). This is an especially prepared version of the software package called "OptiSystem Lite" and marketed commercially by Optiwave under the name OptiSystem. The CD contains a set of problems for each chapter that are appropriate for the readers of this book. The reader is encouraged to try these numerical exercises as they will help in understanding the important issues involved in the design of lightwave systems.

The CD should work on any PC running the Microsoft Windows software. The first step is to install the software package. The installation procedure should be straightforward for most users. Simply insert the CD in the CD-ROM drive, and follow the instructions. If the installer does not start automatically for some reason, one may have to click on the "setup" program in the root directory of the CD. After the installation, the user simply has to click on the icon named "OptiSys_Design 1.0" to start the program.

The philosophy behind the computer-aid design of lightwave systems has been discussed in Section 5.5. Similar to the setup seen in Fig. 5.15, the main window of the program is used to layout the lightwave system using various components from the component library. Once the layout is complete, the optical bit stream is propagated through the fiber link by solving the nonlinear Schrödinger (NLS) equation as discussed in Section 5.5. It is possible to record the temporal and spectral features of the bit stream at any location along the fiber link by inserting the appropriate data-visualization components.

The OptiSystem Lite software can be used for solving many problems assigned at the end of each chapter. Consider, for example, the simple problem of the propagation of optical pulses inside optical fibers as discussed in Section 2.4. Figure E.1 shows the layout for solving this problem. The input bit pattern should have the return-to-zero (RZ) format and be of the form "000010000" so that a single isolated pulse is propagated. The shape of this pulse can be specified directly or calculated in the case of direct modulation by solving the rate equations associated with semiconductor lasers. In the case of external modulation, a Mach–Zehnder modulator module should be used. The

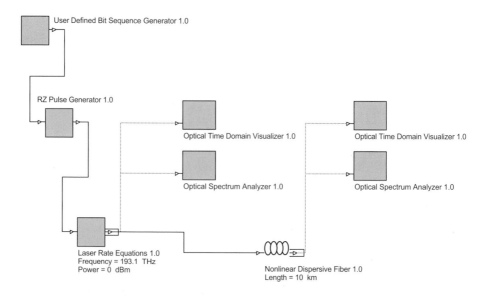

Figure E.1: An example of the layout for studying the propagation of optical pulses inside optical fibers. All the components are available in the standard component library supplied with the OptiSystem Lite software.

output of the laser or the modulator is connected to an "optical time-domain visualizer" and an "optical spectrum analyzer" so that the input pulse shape and spectrum can be observed graphically. The optical bit stream is transmitted through a fixed length of optical fiber whose output is again connected to the two visualizers to record the pulse shape and its spectrum. If the nonlinear effects are turned off, the spectrum should not change but the pulse would exhibit considerable broadening. For Gaussian pulse, the results would then agree with the theory of Section 2.4.

There are a large number of prepared samples that the user can use to understand how the program functions. On most computers, they will be stored in the directory "C:/Program Files/Optiwave/OptiSystem Lite 1.0/Book." Folders in this directory are named "Samples" and "Technical Descriptions." There are subfolders within the "Samples" folder for individual chapters which contain problems that can be solved by using the software. Most users of this book will benefit by solving these problems and analyzing the graphical output. A portable-data format (PDF) file is also included for each chapter within the folder "Technical Descriptions"; it can be consulted to find more details about each problem.

Index

absorption coefficient, 134
absorption rate, 80
accelerated aging, 124
acoustic frequency, 370
acoustic jitter, *see* timing jitter
acoustic waves, 59, 343, 454, 486
activation energy, 125
amplification factor, 227, 234, 238, 245, 270
amplified spontaneous emission, 252, 256,
      264, 435
amplifier noise, *see* noise
amplifier spacing, 265, 420, 421, 426
amplifiers
    applications of, 231
    bandwidth of, 227
    C-band, 259
    cascaded, 264–272
    doped-fiber, *see* fiber amplifiers
    gain of, 227
    in-line, 195, 241, 264–272, 280, 435
    L-band, 259, 272
    noise in, 230
    parametric, 249, 305, 457
    power, 231, 263
    properties of, 226–231
    Raman, 243–250, 259
    S-band, 259
    saturation characteristics of, 229
    semiconductor, *see* semiconductor op-
        tical amplifiers
amplitude-phase coupling, 110, 117
amplitude-shift keying, *see* modulation for-
       mat
anticorrelation, 116, 205
antireflection coating, 92, 103, 233, 344
APD, 142–148
    physical mechanism behind, 142
    bandwidth of, 144
    design of, 143
    enhanced shot noise in, 159

    excess noise factor for, 159
    gain of, 144
    optimum gain for, 161, 166
    reach-through, 145
    responsivity of, 144
    SAGCM, 146
    SAGM, 146
    SAM, 145
    staircase, 146
    superlattice, 147
apodization technique, 294
ASCII code, 9
ATM protocol, 334, 336, 381
attenuation coefficient, 55
Auger recombination, 83, 84, 109
autocorrelation function, 115, 116, 156, 157,
      389, 391
avalanche breakdown, 144
avalanche photodiode, *see* APD

bandgap discontinuity, 82
bandwidth
    amplifier, 227, 228, 257
    APD, 144
    Brillouin-gain, 370
    fiber, 53, 194
    filter, 151, 271, 341, 344
    front-end, 149
    gain, 227
    grating, 296
    LED, 91
    modulator, 485
    noise, 156
    photodetector, 136
    photodiode, 139
    Raman-amplifier, 243
    Raman-gain, 63
    RC circuit, 194
    receiver, 384
    semiconductor laser, 112